FREAM'S AGRICULTURE

FREAM'S
AGRICULTURE

A textbook prepared under the authority
of the Royal Agricultural Society of England

Edited by

C. R. W. Spedding
MSc PhD DSC FIBiol

Professor of Agricultural Systems
University of Reading

Foreword by
HRH The Duke of Edinburgh, KG, KT, OM

JOHN MURRAY

© Royal Agricultural Society of England 1983

First Edition (*Fream's Elements of Agriculture*) 1892
Sixteenth Edition (*Fream's Agriculture*) 1983

Printed in Great Britain
by Butler & Tanner Ltd, Frome and London

Fream, William
    Fream's agriculture.—16th ed
    1. Agriculture—Great Britain
    I. Title   II. Spedding, C.R.W.
    III. Royal Agricultural Society of England
    630'.941   S455
    ISBN 0-7195-4034-8

CONTENTS

Contents

ACKNOWLEDGEMENTS

The Editor wishes to acknowledge the help received from many colleagues during the preparation of this edition, but especially that of Miss Angela Hoxey, who, in addition to many other contributions, also prepared the index. It is a pleasure, too, to acknowledge the help of Mrs Valerie Craig in typing succesive drafts of the book.

We are also indebted to the following for permission to use the plate numbers shown in brackets:
RHM Research Ltd (1); the Grassland Research Institute and CAB (2, 3, 4, & 5); the *Farmers Weekly* (6, 8, 9, 11, 12, 13, 15, 18, 19, 26, 45, & 46); John Tarlton (10); Bob Sollars, via the *Farmers Weekly* (7); Commercial Camera Craft, via the *Farmers Weekly* (14); Douglas Low (16); the RASE (17, 20, 21 & 44); Aerofilms Ltd (22, 23); John Young and MAFF (24); Rosemaund Experimental Husbandry Farm (25); John Matthews, NIAE (27-36, 38-43); Smallford Planters Ltd (37).

We also wish to acknowledge the following for allowing us to use the Figures given in brackets: the Carnegie Institution of Washington (Fig. 12.6); Academic Press (part of Fig. 12.11 and Fig. 12.13); W. H. Freeman (part of Fig. 12.11); the British Grassland Society and O. R. Jewiss (Fig. 12.9); Hutchinson & Co (Fig. 13.1); Granada Publishing Ltd (Figs. 16.6, 16.7, 16.8, 16.9, 39.1-39.5); McGraw-Hill and J. K. Loosli (Figs. 13.2 & 13.5); Macmillan Publishers Ltd (Figs. 19.6 & 19.9); PUDOC & D. E. Van der Zaag (Fig. 19.10); Edward Arnold and M. J. Samways (Fig. 19.13); Cambridge University Press and R. J. Jones & R. L. Sandland (Fig. 21.7).

Finally, it is a pleasure to record the helpful collaboration of the publishers, John Murray.

FOREWORD

When Fream's *Elements of Agriculture* was first published in 1892 the technical and scientific aspects of cultivation and stock management were fairly primitive by today's standards, but the purpose and the structure of the book are just as relevant 90 years later.

This latest edition brings the story up to date and takes account of the increasingly international nature of agricultural organisation; although whether it satisfactorily explains the complexities of the Common Agricultural Policy will depend on the opinion and the patience of the reader.

The very rapid development of agricultural practices in the last 20 years is self-evident but it does raise the question whether this increasing intensification is not beginning to have consequences beyond the purely agricultural ambit. The treatment of stock, the consequences for wildlife, the effect of a combination of chemicals on the human metabolism and a number of similar issues, virtually unheard of even 50 years ago, are now becoming matters of public debate.

I am sure that this 16th edition of Fream's produced under the able guidance of Professor Colin Spedding, will help people both within the agricultural and food-processing industries as well as those in other walks of life to a better understanding of the contemporary methods of renewable resource production. I also hope that it will provide a comprehensible background for the sensible discussion of the wider and sometimes controversial issues generated by the introduction and the rapid spread of these methods.

Buckingham Palace   HRH The Duke of Edinburgh, KG, KT, OM
1983

PREFACE

This new edition of Fream maintains the objectives of earlier books in that it seeks to provide within a single volume a comprehensive picture of current agricultural practice.

In the treatment of the subject, however, it differs from previous editions – not only does it illustrate the complexity of present-day agricultural technology, but it aims to give the reader an appreciation of scientific and economic principles and of their application in farming systems. It also sets the United Kingdom farming scene in European and World perspective and relates it to the food industry and the national economy.

The Society hopes that this edition will prove of continuing value to those involved in the agricultural industry as well as to members of a discerning public to ensure a better understanding of the science and practice of agriculture.

The Editorial Panel placed the preparation of the book in the capable hands of Professor C. R. W. Spedding who undertook an enormous task with great enthusiasm. The Society is particularly grateful to the Editor, to the individual contributors, whose names are listed elsewhere in the book, and to the other members of the Editorial Panel: N. Coward, L. G. C. Gilling, J. D. F. Green, J. D. M. Hearth, G. H. Jackson, N. F. McCann, Professor M. McG. Cooper, J. R. G. Murray, Professor D. W. Wright and S. B. Young.

Special thanks are due to Mr J. D. F. Green, Chairman of the Society's Education and General Purposes Committee, which authorised the work, for his sustained interest; and to Mr G. H. Jackson, the Society's Agricultural Director, for his support throughout the project.

Royal Agricultural
Society of England
35 Belgrave Square
London SW1

A. J. Davies, BSc(Hons) Agric, FIBiol
Chairman
Editorial Panel

ILLUSTRATIONS

CONTRIBUTORS

Fream Essay by Mr Gwyn E. Jones

1 Professor C. R. W. Spedding
2 Professor M. McG. Cooper
3 Professor M. McG. Cooper
4 Professor M. McG. Cooper
5 Professor M. McG. Cooper
6 Mr R. B. Tranter
7 Mr J. B. Tyldesley
8 Mr Alan Harrison
9 Dr Ruth Gasson
10 Dr D. J. White
11 Dr G. W. Cooke
12 Professor C. R. W. Spedding & Miss A. M. Hoxey
13 Professor C. R. W. Spedding & Miss A. M. Hoxey
14 Professor P. N. Wilson
15 Dr H. C. Gough & Dr W. J. Brinley Morgan
16 Mr J. Matthews
17 Mr J. D. Young
18 Mr J. A. Burns
19 Mr P. M. Harris
20 Professor J. F. D. Greenhalgh
21 Dr J. Hodgson
22 Mr A. K. Giles
23 Mr D. J. Corbett
24 Dr P. V. Biscoe & Mr R. G. Dawson
25 Dr J. M. Walsingham
26 Mr R. S. Tayler
27 Dr R. K. Scott & Mr E. J. Allen
28 Mr B. Bastiman
29 Mr R. M. Deakins
30 Dr P. J. Salter
31 Mr W. R. Buckley
32 Mr G. F. Sheard
33 Dr A. R. Rees
34 Mr S. P. Carruthers
35 Mr J. M. Stansfield

WILLIAM FREAM (1854-1906)

The second half of the nineteenth century witnessed the growth in Great Britain of an influential movement which established a comprehensive system of scientific and technical education, organised mainly by the Government's Department of Science and Art. At its elementary level this became generally available in the schools of science created in almost every town and city throughout the country, attended largely by working men and women. To provide advanced courses of study (including the training of science teachers), colleges of science were developed in London and Dublin. During the closing two decades of the Victorian era, a small group of agriculturists adapted the general movement to create the basis of the modern system of agricultural education. They also applied the methods and findings of general scientific developments to agriculture, while they also widely disseminated the new knowledge through their writings. They formed the bridge between the age of a broadly based natural history and of farming dependent largely on tradition with any improved practices being empirically derived, and that of modern agricultural science specialisms and an efficiency-oriented farming. Foremost among these few agriculturists was William Fream.

During the 1890s, Fream had become well known as the country's most authoritative writer on agricultural subjects. At the beginning of the decade he had been appointed Associate Editor of the *Journal* of the Royal Agricultural Society of England (when it had been decided to publish it at quarterly intervals). Two years later, he was given complete responsibility for it as its Editor. In the meantime, he had also been commissioned by the Society in March 1891 to write a general illustrated textbook on agriculture, and given four months to complete the task. When *Elements of Agriculture* was published on 1 January 1892 it was an immediate success, and ran rapidly through several editions; before Fream died in 1906 over 35,000 copies had been sold. In addition, he also undertook the massive task of rewriting the nineteenth century's major work on practical farming, *The Complete Grazier*, published in 1893, and wrote numerous lengthy papers on a wide diversity of agricultural topics for his *Journal*. As a learned agriculturist, in contact with all scientific and practical developments in British agriculture, as well as abroad, and with a high reputation for his ability to write attractively, avoiding hasty or rash judgements, he was also offered the post of agricultural correspondent to *The Times*. From the beginning of 1894 to his death in 1906 he was widely known and respected as 'Dr Fream of *The Times*'.

His prodigious output during the 1890s was possible only as a consequence of his earlier experience as a student and as a teacher of agricultural students. Fream had been born in Gloucester, the second son of a small building contractor. His early education became assured when, as a result of his good voice, he was accepted as a chorister at the Cathedral. On leaving school he gained employment as a clerk in a local firm of corn merchants. This may have awoken his interest in agriculture; it certainly led him to try to improve himself by attending evening classes at the school of science in the city. As a result of his studies, in 1873 he gained one of the few coveted Royal Exhibitions which enabled him to study for three years at the Royal College of Science in Dublin, at the time the only place in the United Kingdom where agriculture and associated biological disciplines could be studied at an advanced level. It was while in Dublin that he also developed his life-long passion for field studies. On completing his course and gaining the Associateship of the College, he entered himself as an external student at London University for the B.Sc. degree with honours in chemistry, which he was awarded in 1877. Over the next two years he taught natural history at the Royal Agricultural College, Cirencester, followed by a year in London teaching botany at Guy's Hospital Medical School. During this brief period in London he also took the opportunity of pursuing further courses at the Royal School of Mines, under Professor T. H. Huxley, one of the late nineteenth century's most distinguished biological scientists and a staunch advocate of applied science and technical education.

It was during the following ten years that Fream's skills as a teacher and writer were to be fully developed. In 1880 he joined John Wrightson, who was then establishing a private, residential College of Agriculture at Downton, near Salisbury. This College, which continued until 1906, was soon to become one of the major centres of agricultural education in England. At its beginning, Wrightson collected together a small group of staff each of whom was to become a recognised authority in his field, all firm believers in the essential importance of integrating scientific studies with their application to practical farming. Fream, by now becoming a rather portly figure in his late twenties, with a fresh face, a trim beard and moustache, and balding slightly, was responsible for teaching the biological sciences (though, at times, he also taught surveying, and bookkeeping). With his students, he emphasised the practical relevance of what he taught in the classroom and laboratory through field studies on the farm attached to the College and in the surrounding area – a notable innovation in teaching methods for the time. His 'botany trots' over the Downs and along the Avon Valley became one of the most memorable parts of student life at the Downton College of Agriculture. In this stimulating atmosphere, Fream and his colleagues were all exceedingly active, writing copious articles for the press, as well as lecturing widely, undertaking research, and producing learned papers and books. In 1884 and 1888, Fream also toured Canada and wrote comprehensive reports for the Canadian government on the Dominion's agricultural resources and potentialities. On the occasion of the second visit, in recognition of this work, McGill University, Montreal, awarded him the honorary degree of Doctor of Laws.

When he left the College at Downton in 1890 to join the staff of the Royal Agricultural Society of England, Dr Fream was aware that he was also likely to be offered another appointment. This was the Steven Lectureship in agricultural science which had been established that year at Edinburgh University. He was formally appointed to this towards the end of the year, and for the remainder of his life each winter taught a course at Edinburgh of twenty lectures on entomology in relation to agriculture and forestry. His extremely active life in the 1890s, which thus mainly involved his work in London and Edinburgh, also included a six-month period in 1893 when, as an Assistant Commissioner, he wrote two reports for the 1893–7 Royal Commission on Agriculture. Nevertheless, though a bachelor, he continued to maintain his home in Downton. It was there that he died suddenly in late May 1906. For the previous five years his work had been confined mainly to *The Times*, apart from his lecturing commitments in Edinburgh, having resigned from the editorship of the *Journal* of the Royal Agricultural Society at the end of 1900.

His death at the early age of fifty-two came as a great shock to his wide circle of agricultural and scientific friends and acquaintances in Britain and abroad. Over the following two years a fund was collected sufficient to endow a memorial prize awarded annually to the best candidate in the National Diploma in Agriculture examination. This would undoubtedly have pleased Fream. Although in latter years he had gained his living largely as a journalist, above all else he had been an educator, a teacher and, at a time when agricultural advisory work was in its infancy, a popular and reputable disseminator of agricultural knowledge and information. His main memorial, however, remains the foundation which he laid with *Elements of Agriculture*. Though altered and enlarged through successive new editions, and now entirely rewritten, it retains the comprehensive character which William Fream established almost a century ago.

ABBREVIATIONS

ADAS	Agricultural Development and Advisory Service
ai	active ingredient (of a pesticide)
AI	artificial insemination
BVA	British Veterinary Association
BYDV	Barley Yellow Dwarf Virus
CAP	Common Agricultural Policy
CAS	Centre for Agricultural Strategy
CCC	Chlormequat
CDA	controlled droplet application
CEC	Commission of the European Communities
CF	crude fibre
CIMMYT	Centro Internacional de Mejoramiento de Maiz y Trigo
CP	crude protein
DAFS	Department of Agriculture and Fisheries for Scotland
DANI	Department of Agriculture, Northern Ireland
dBA	sound level in A-weighted decibels
DoE	Department of the Environment
DM	dry matter
EDC	Economic Development Committee
EE	ether extractive
EEC	European Economic Community
FCE	food conversion efficiency
GDP	Gross Domestic Product
g	gram
ha	hectare
ISO	International Standards Organisation
IU	international units
JCO	Joint Consultative Organisation
kg	kilogram
km	kilometre
kt	kilo-tonnes
l	litre
m	metre
m.	million
MAFF	Ministry of Agriculture, Fisheries and Food

Abbreviations

MLC	Meat and Livestock Commission
MMB	Milk Marketing Board
m.t	million tonnes
NEDO	National Economic Development Office
NFE	nitrogen-free extractives
NFU	National Farmers' Union
NIAB	National Institute of Agricultural Botany
NPN	non-protein nitrogen
OC	organo-chlorine
OP	organo-phosphate
PMB	Potato Marketing Board
pto	power-take-off (from a tractor)
SMD	Standard Man Day 1 SMD = 8 labour hours
SMD	soil moisture deficit
SWD	soil water deficit
t	tonne
UK	United Kingdom
UKASTA	United Kingdom Agricultural Supply Trade Association
ULV	ultra low volume
WMO	World Meteorological Organisation

GLOSSARY

Additives: Substances added to feedingstuffs.

Aerobic digestion: Decomposition of organic matter by micro-organisms in the presence of oxygen.

Aflatoxin: Toxin produced by the fungus *Aspergillus flavus*.

Anaerobic digestion: The breaking down of organic matter in oxygen-free conditions.

Anthelmintics: Drugs used to remove parasitic worms (helminths) from their hosts.

Anthesis: The time of flower opening.

Apical meristem: Growing point at tip of root and stem in vascular plants (see also *Meristem*)

Ark: A moveable, often triangular, shelter for pigs or poultry.

Artificial insemination (AI): The collection of sperm and its use in impregnating females.

Bagasse: Fibrous residue remaining after crushing sugar cane and removing the juice.

Bagasse pith: Central part of the sugar cane stem after crushing.

Barn-drying: Hay-making involving artificial drying in a building.

Battery cages: Cages, usually metal, in which poultry are housed.

Biological control: The control of one organism by deliberate use of another.

Bloat: Disorder of ruminants, involving persistent froth and consequent gas pressure in the rumen.

Brassicas: Plants of the group *Brassica* (e.g. turnip).

Break crops: Crops grown between periods of continuous cultivation of a main crop.

Breeding season: Period when animals are sexually active.

Broiler: Young chicken (usually 9-12 weeks of age) grown for meat.

Broken-mouthed ewes: Ewes having some teeth missing.

Calving interval: The interval between one calving and the next.

Carcass classification: Grouping of carcasses according to weight, fatness and conformation.

Carrying capacity: The number of animals that an area of land can support (feed).

Cash flow: Movement of funds through the business.

Catchcrop: Crop utilising land between two longer-term crops.

Chemotherapy: The treatment of disease by substances that have a specific antagonistic effect on the organism causing the disease.

Chlormequat (CCC): A plant growth regulator.

'Clean' pasture: Pasture free from animal parasites.

Coccidiosis: An intestinal disease of livestock caused by microscopic protozoa.

Complete budget (or *budget complete*): Budget in relation to a *whole* farm business.

Conacre: The letting of land in Northern Ireland (and Eire) for an eleven-month period.

Condition scoring: Description of animal condition in terms of lean and fat. Score ranges 0–5 cows, 1–5 ewes and 1–9 sows; the lowest figure indicates extreme thinness, the highest excessive fatness.

Conservation: Protection and preservation, in relation to (a) soil, (b) herbage or (c) the environment.

Controlled atmosphere (*CA*): Regulation of levels of oxygen and carbon dioxide as well as temperature to improve storage of fruit and vegetables.

Coprophagy: Consumption of own faeces (normal in the rabbit) (see also *Refection*).

Corium: The true skin or derma under the epidermis.

Cows: Adult female cattle which have had one or more calves.

Cowtel: Commercial name for a cow cubicle building.

Creep: A gap (in a fence or a barrier) through which only young animals can pass.

Creep-grazing: Grazing system involving 'creeps'.

Critical temperature (*lower*): The environmental temperature below which the metabolic rate must rise if the animals' deep-body temperature is to be maintained.

Crude fibre (*CF*): A constituent of animal feedstuffs comprising mainly cellulose, lignin and related compounds.

Crude protein (*CP*): An approximate assessment of the protein content of animal feed; usually calculated as $6.25 \times \% \text{N}$.

Culling: Removal of animals from a breeding population, generally on account of some physical or performance deficiency.

'Dedicated' fuel cropping: Use of resources solely towards the production of a crop destined for use as a fuel or fuel feedstock.

Deep litter: A system of keeping housed poultry on litter (about 15 cm deep).

Demesne (*land*): Possession of land with unrestricted rights of use.

Desiccant: A chemical causing drying.

Development (*biological*): Sequential organisational changes in an organism of a qualitative kind, often associated with growth.

Digestibility: Proportion of feed digested by animals, expressed as a ratio, a percentage or a coefficient.

Digestible energy (*DE*): That part of the feed energy that is available to the animal after digestion.

Diluent: Diluting agent.

Direct drilling: The sowing by drill, of seeds direct into a field, without previous cultivation.

Dressing-out percentage (*DO%*): The weight of carcass per 100 units of live weight (*see also* Killing-out percentage).

Dry (*cows, sheep*): Not producing milk.

D-value: The percentage of digestible organic matter in the dry matter.

Dystokia (*Dystocia*): Difficulty in the process of giving birth.

Ear emergence: Main heading date (e.g. for a sward, the date at which 50% of the inflorescences have emerged).

Ecto-parasite: A parasite which lives on the outside of its host.

Enclosures: The action of surrounding land with a fence (converting pieces of commonland into private property).

Endo-parasite: A parasite which lives within the body of the host.

Ergonomics: The scientific study of the efficiency of man in his working environment.

Estanciero: The keeper of an *estancia*; a cattle farmer in Spanish America.

Ether extractives (EE): A fraction of animal feedstuffs, containing mainly fats and oils.

European size unit (ESU): = 1,000 European units of account.

Evapo-transpiration: Loss of water by evaporation and transpiration.

Ewe lambs: Female lambs.

Extensive systems: Systems which use a large amount of land per unit of stock or output (see *Intensive systems*).

Fallopian tube: The tube that extends from the uterus to the ovary and into which the developing ova fall and are fertilised.

Fallowing: Resting land from cropping.

Farrowing: The act of parturition in the sow.

Fat lamb: In condition satisfactory for slaughter.

Feather-pecking (or *-picking*): An unfortunate habit of hens, usually when close-confined, of pecking at other birds' feathers, often resulting in considerable damage.

Field capacity: The state of saturated soil when all the soil moisture that is able to freely drain away has done so.

Fineness count (wool): A scale used to assess the fineness of wool fibre.

Finishing or *Fattening:* The feeding of cattle or sheep at a rate of growth which increases the ratio of muscle to bone, and increases the proportion of fatty tissue in the carcass to a level at which the animal is considered to be fit for slaughter.

Fodder: Generally refers to dried feeds such as hay.

Forage/Forage crops: Leafy crops that are grazed.

Forbs: Broad-leaved herbaceous plants.

Fossil fuels: Biological materials which have been subjected to long-term geological effects e.g. oil, coal, natural gas and peat.

Free-range: A system of poultry keeping in which hens are allowed to range over a relatively large area.

Frost 'heave': Loosening of the surface soil caused by frost.

Gasification: The heating of organic material in the presence of limited quantities of oxygen to liberate carbon monoxide and hydrogen.

Genetic engineering: The science of modifying the genetic constitution of plants and animals directly.

Gilt: A young female pig not having produced a litter.

Gimmers: Ewes $1-1\frac{1}{2}$ years old.

Goitrogenic: Goitrogen, Goitrogenic Factor, is one which gives rise to goitre.

Grain, 1000 grain weight: The weight of 1000 grains of a random sample of a cereal or grass variety.

Gram-negative: A bacterium which does not retain the stain in the Gram's staining technique used in classifying bacteria.

Gram-positive: A bacterium which retains the stain in the Gram's staining technique used in classifying bacteria.

Grazing: alternate: Grazing by different animal species alternately.

'Greenhouse effect': Global warming due to build-up of atmospheric carbon dioxide.

Gross Margin of an enterprise is the value of its Output less its Variable Costs.

Growth: Increase in size.

Growth-promoters: Substances given to farm animals to promote growth.

Growth regulator: A natural or chemical substance that regulates the enlargement, division or activation of plant cells.

Haemolytic anaemia: Disease involving destruction of red blood corpuscles and the consequent escape from them of haemoglobin.

Hay-box: An insulated box. The insulation material is usually hay but may be straw, paper or other materials.

Heat capacity: Heat-storing capacity of a material is a function of the mass of the material and its specific heat. Specific heat refers to the quantity of heat necessary to produce unit change of temperature in unit mass of material.

Heifers: Female cattle which have not calved or have calved for the first time.

Homeotherms: 'Warm-blooded' animals, whose body temperature is maintained above that of usual surroundings.

Hormone: A secretion from special glands within the animal's body that effects various body functions.

Husk: Parasitic bronchitis, mostly in cattle, caused by lungworms.

Hybrid: The first generation offspring of a cross between two individuals differing in one or more genes.

Hypocalcaemia: Disease caused by low blood calcium level.

Hypomagnesaemia: Disease caused by low magnesium content in the blood.

Index: For fertiliser requirement i.e. Nitrogen Index is based on the requirement of the crop to be grown, making allowance for residues available from previous cropping and manuring, and not on soil analysis. The index for phosphorus and potassium is based on soil analysis giving quantities in the soil that are available to the crop.

Intensive systems: Systems in which cropping is frequent and yields are high per ha; or where stock numbers are high per unit area.

Intercropping or *Mixed cropping:* The growing of more than one species on the same piece of land at the same time.

Inter-muscular fat: Fat laid down between muscles.

Intra-muscular fat: Fat laid down within muscular tissues.

Kemp: A very coarse fibre in a fleece.

Killing-out percentage (*KO%*): The weight of carcass per 100 units of live weight (see also *Dressing-out percentage*).

Labour profile: Graph or histogram showing the calculated standard man-hours required over time.

Lambing percentage: The number of lambs born (alive, dead, tailed, weaned) per 100 ewes (mated, marked, or put to the ram).

Latifundia: Large estates.

Leaching: Removal of nutrient materials in solution from the soil (usually in gravitational water).

Leaf Area Index (LAI): The area of green leaf per unit area of ground.

Leaseback: An arrangement whereby an owner sells his farm and it is immediately leased back to him by the new owner.

Legumes: Plants of the family Leguminosae (e.g. peas).

LD 50 (Lethal dose of 50%): Amount of active ingredient that kills 50% of the population treated.

Ley: Land temporarily sown to grass.

Lux: Unit of illumination

    1 lux = 0·0001 Phot
         0·1 Milliphot
         0·09 Foot-candle.

Marbling: Appearance of muscle with intra-muscular fat, especially of cut surface.

Meristem: Localised region of active cell-division in plants (see also *Apical meristem*).

Meslin, Meslen, Mashlum, Maslin or *Dredge:* A mixture of oats and barley and sometimes wheat, sown to provide grain for feeding to livestock.

Metabolic size or *weight:* The size of an animal to which its metabolic rate is proportional. Mean standard metabolic size of mammals is often expressed as body weight $^{0·75}$.

Metabolisable energy (ME): The energy of feed less the energy of faeces, urine and methane.

Micro-chip (micro-electronics): A large number of electronic circuits in miniature form.

Micro-nutrients or *Trace elements:* Nutrients which are required in very small amounts.

Minifundia: Small estates.

Minimal disease (Pigs): Pigs initially produced by hysterectomy and reared free from certain infections.

Mixed grazing: More than one type of animal grazing the same area at the same time.

Molasses: Dark brown syrup, by-product of sugar production.

Monocropping or *Monoculture:* The growing of the same, single crop species continuously on an area of land.

Monogastric, Non-Ruminant, or *Simple-stomached animals:* Animals having only one stomach (e.g. pig, poultry, man).

Monogerm 'seed': Multigerm 'seed' (e.g. sugar beet) after treatment to reduce the number of true seeds, ideally to one (see also *Multigerm 'seed'*).

Mulch: Material used to cover the bare soil between growing crop plants.

Multigerm 'seed': 'Seed' of some plants (e.g. sugar beet) is a fruit containing several true seeds (hence 'multigerm') (see also *Monogerm 'seed'*).

Neonatal losses: Deaths at or shortly after birth.

Net worth or *owner's equity* is the *Balance Sheet* value of *Assets* available to the owner of the business after all other claims against these *Assets* have been met.

Nitrogen fixation: Conversion of atmospheric N to plant compounds by micro-organisms (in soil, root nodules).

Nitrogen-free extractive (NFE): Fractions of feedstuff mainly containing soluble carbohydrates.

Non-protein nitrogen (NPN): Most commonly in the form of urea, used to supply part of the dietary nitrogen requirements of the ruminant at a lower cost than protein.

Nurse cow: Cow used to suckle calves of others.

Nutrient-Film Technique (NFT): A system of growing crops in which a very shallow stream of water containing all the dissolved nutrients required for growth is recirculated past the exposed roots of crop plants in a water-tight gulley.

Organic farming: Without the use of manufactured chemicals.

Ototoxicity: Damage to the ear tissues.

Ozone layer: A layer of ozone found in the stratosphere, where it absorbs harmful solar ultraviolet radiation.

Pallet: A moveable platform for use with fork-lift truck for lifting and moving goods.

Parenteral: Administration of a substance other than via the digestive system, e.g. by injection.

Partial budget or *Budget partial:* Relating to a partial change in a farm plan or system.

Permanent pasture: An established plant community in which the dominant species are perennial grasses, there are few or no shrubs, and trees are absent.

pH: Chemical measure of acidity ($<pH7$) and alkalinity ($>pH7$).

Pheromone: A chemical substance produced by one individual which affects the behaviour or physiology of another.

Photosynthesis: The process by which carbohydrates are manufactured by the chloroplasts of plants from CO_2 and water by means of the energy of sunlight. Net photosynthesis is equal to gross photosynthesis less the sum of respiration during the day and night.

Plantation crops: Subtropical and tropical perennial crops grown on plantations or large estates.

Poaching: Damage to herbage and soil caused by excessive treading in wet weather.

Poikilotherms: 'Cold-blooded' animals, whose body temperature varies to a large extent depending on the environment.

Primary energy: Fossil fuel energy extracted from the earth, expressed as total enthalpy.

Pullet: Young hen in first laying season.

Pulses or *Grain legumes:* Leguminous plants or their seeds, chiefly those plants with large seeds used for food.

Pyrolysis: The destructive distillation of organic material in the absence of oxygen to yield a variety of energy-rich products.

Raceme: An inflorescence in which the pedicelled flowers are arranged on a rachis or axis.

Refection: The consumption of own faeces (see also *Coprophagy*).

Relative humidity (RH): Water vapour in air compared with the amount of water vapour held at the same temperature when saturated.

Re-seeding: Sowing seeds of grassland species to re-establish a ley.

Respiration: The taking in of oxygen and giving off of carbon dioxide, for the purpose of releasing energy.

Rhizome: An elongated underground stem, usually horizontal, capable of producing new shoots and roots at the node.

Rogue: (a) A variation from the type of a variety or standard, usually inferior; (b) To eliminate such inferior individuals.

Rotation: The growing of a repeated sequence of different crops.

Rotational grazing: The practice of imposing a regular sequence of grazing and rest upon a series of grazing areas.

Rumen: Storage compartment of complicated stomach of ruminants.

Ruminant: Animals possessing a complicated stomach of four parts: rumen, reticulum, omasum, abomasum e.g. cow, sheep, deer, etc.

Scouring: Diarrhoea.

Selection pressure: Measure of the effectiveness of natural selection altering the genetic composition of a population.

Senescence (leaf senescence): Process in which leaves age and die, usually involving chlorophyll degradation.

Sensitisation: Induced sensitivity e.g. to drugs but also to sunlight (photosensitivity) causing dermatitis or conjunctivitis.

Set-stocking: Grazing system in which stock remain in one field or paddock for a prolonged period.

Shearlings: Sheep between their first and second shearings.

Slatted floors: Perforated floors formed by strips of reinforced concrete, steel or wood over a pit.

Slotted floors: Perforated floors formed by sheets of metal or plastic with slots of different size and shape, over a pit.

Slurry: The mixed   ec   ina   excre   c   st   ogether with any washing

ne yl cy   und in rape and stubble

                                                        to confined stock (see also

                                                        lent condition.

                                                        The amount of rain or

                                                        n associated with hypo-

                                                        to form a lock.
                                                        of the average length of

                                                        seed production.

                                                        land at a point in time.
                                                        nd over a given period.

Stolon (runner): A creeping stem above the soil surface (roots usually form at the nodes).

Store cattle (or *sheep*)*:* Animals which have been grown slowly so that their skeletal development has not been impaired, but muscular tissue is slightly below the animal's potential and fatty tissue is undeveloped. Such animals are purchased to be fattened at pasture or in yards. The fattening process involves some growth of bone, more of muscle, and rapid deposition of fat.

Stubble: That part of a crop left above ground after harvesting.

Sub-cutaneous (fat)*:* Fat layer beneath the skin.

Suckler herd: Beef cattle where the dam suckles its own calf (single suckling), another calf as well (double suckling) or several (multiple suckling).

Support energy: All forms of energy used other than direct solar radiation (including the 'fossil' fuels).

Swayback: Disease of lambs, involving inco-ordination of limbs, usually associated with copper deficiency.

Tedding: Tossing swaths of new-mown grass in haymaking in order to expose more grass to the sun and air and quicken the drying process.

Temporary grassland: Arable land sown to a ley for a limited period of years.

Terminal sire: Bull or ram used to sire progeny destined for slaughter.

Theaves: Female sheep from first to second shearing.

Thermal capacity: The ability to retain heat.

Thermal radiation. The transfer of heat from one body to another by radiation.

'Tied' accommodation: Accommodation which is dependent on employment.

Tiller: An aerial shoot of a grass-plant, arising from a leaf axil, normally at the base of an older tiller.

Total digestible nutrients (*TDN*)*:* A summation of the digestible crude protein, crude fibre, oil and nitrogen-free extract in a feedstuff.

Tramlines: Accurately spaced pathways left in a growing crop to provide wheel guide marks for subsequent operations.

Transpiration: The evaporation of moisture through the leaves.

Tup: An uncastrated male sheep. Also called a ram. To tup means to serve a ewe.

Ultra-high temperature (*UHT*)*:* Method of pasteurising milk using high temperature ($270°$F) for not less than one second.

U-value (*Thermal transmittance*)*:*
     1 Btu/ft^{2}h$°$F
     $5·678$ J/m^{2}S$°$C $= 5·678$W/m$^{2}°$C
     Watts per sq m degree Celcius.

Vegetable: Any edible plant stems, roots, leaves and fruits.

Volumes per million (*vpm*)*:* i.e. parts per million (ppm) by volume.

Ware potatoes: The largest potatoes in a crop sold for human consumption, as distinct from chats and seed potatoes.

Weaning: Removal of young mammals from their source of milk.

Weaning (*early*)*:* Removal of young mammals from the milk source before the normal time.

Wethers: Castrated adult male sheep.

Windrowing: Arranging herbage into rows.

Zadoks Scale: Decimal code for the growth stages of cereals (see Appendix Table 24.1).

Zero-grazing: Where grass and other forage is cut and carried to the animal (see also *Soilage*).

I

INTRODUCTION

Fream's *Elements of Agriculture* was first published in 1892 and with this volume enters its 16th edition.

Not only has the book been completely rewritten for this edition, but the structure and treatment have also been changed.

Agriculture is a large and complex subject and Fream's original book was a remarkable attempt to describe all its main elements. Since then the amount of factual information on its constituent parts has increased enormously and it is no longer feasible to include all the detail currently available within a single volume. Partly because of this, there is now a need for a synthesis of the factual material into descriptions of production systems. This edition has therefore been written with a major emphasis on the main production systems within agriculture as a whole and the principles that lie behind them.

A new edition is also greatly needed in order to reflect the considerable changes that have taken place in the agricultural industry in recent years.

Agriculture has always produced a wide range of products, with food as the most important. But some 75% of the food entering the home is now processed in one way or another and much farm produce is raw material for the food industry. Furthermore, farm businesses have grown in size and large amounts of capital are invested in land, machinery, buildings and livestock. In many cases, production systems involve a high degree of control – over the biological processes on which they are based, over livestock and even over the environment in which the livestock live.

In primitive systems, the degree of control is slight and plants and animals to a large extent fend for themselves. In many ways, it is the search for ever-greater control that has led to the so-called intensive production systems, characteristic of developed countries. Furthermore, the achievement of this greater control has usually depended on additional inputs, especially of equipment and energy. The latter, often called 'support' energy because it supports the main energy supply from the sun, has been used increasingly in developed agricultural systems, mainly to increase control and, thereby, increase output per unit of the main resources.

EFFICIENCY IN AGRICULTURE

It could be argued that agriculture, except perhaps in the most primitive societies, is never concerned solely with output but always with output per unit input – in other words, with efficiency. Indeed, it is thought by some non-agriculturists that agriculture has been *too* concerned with efficiency, to the exclusion of other important values. Such views are understandable but slightly miss the point. The fact is that although agriculture has to be concerned with efficiency, the latter cannot be represented by just one ratio. There are usually several outputs in agriculture and there are always many different inputs.

Expressing outputs and inputs in monetary terms allows one to arrive at an economic efficiency ratio, which, at least for current costs and prices, summarises the overall position, but it does not, of course, include those elements that have no cost or no price. So pollution, noise and appearance may be left out of the outputs, and sunlight, oxygen and atmospheric nitrogen left out of the inputs (except as reflected in the cost of land). Changes in the state of the system itself, such as soil fertility, may also be neglected.

The calculation of efficiency thus depends on one's individual standpoint and each of us can take several different views of an agricultural system.

This is an important point to recognise in a book that attempts to provide a comprehensive account of agriculture. There are many different ways of looking at agriculture and these views may be neither right nor wrong, though they may be appropriate or inappropriate for a particular purpose. So, to choose an enterprise and a particular way of carrying it out, it is essential to take an economic view in order to choose a system that will be profitable. On the other hand, if an enterprise is unprofitable, looking at an economic efficiency ratio will not necessarily indicate what action should be taken. This requires more detailed analysis and, ultimately, a consideration of technical or biological efficiency ratios, to pinpoint what action is required.

Nonetheless, the starting point in an assessment of agriculture generally has to be a whole system and its overall performance: the individual components make no sense except in relation to the system of which they are a part and within which they function.

It is because of the way in which component parts of a system interact that anyone primarily concerned with one part must have at least some understanding of the whole.

ABOUT THIS BOOK

The main purpose of this book is to give an overall picture of

British agriculture but it will be apparent from what has already been said that there is not going to be just one view. Some people will be interested in looking at it from their own particular standpoint, say, that of a student, a farmer, or a consumer, or a fertiliser salesman, or a banker, or a research worker or a veterinarian. Others will want a more synoptic approach to the whole that is a combination of all these views.

Part I starts with world agriculture and then deals, in increasing detail, with Continental European and then British agriculture. The idea is to present the international backcloth against which British agriculture has to operate: the activities of other countries affect the market for British agricultural products and the costs of inputs; they may therefore also influence the enterprises to be found on British farms.

This is not the only reason for starting the book with a picture of world agriculture. It also widens our own horizons just to be aware that others do things differently, that outputs can be valued differently and that certain developments are both feasible and economic in other climatic and economic environments.

Part II describes the resources available to agriculture, where they come from, who uses them, how they are used and what they cost. It deals with land (including the soil), labour and capital, the energy used to power agricultural operations, the crop plants and animals used for production and the fertilisers and feedstuffs involved in their husbandry. Again, these resources have much in common across the world and, in some cases (such as soil erosion), it is much easier to see what can happen when they are misused by looking at the more spectacular examples that occur in other countries.

All these resources represent the major elements from which agricultural systems are constructed but they do not themselves tell us much about the ways in which they interact or the ways in which they can profitably be combined. This requires guidelines or rules of how best to construct systems and these, in turn, depend upon organising facts in such a way that they can be used for a pre-selected purpose. A collection of unrelated facts is not very useful and the conversion of facts into useable knowledge depends upon the formulation of specific principles.

Part III is concerned with just this – and it is a relatively new venture, to attempt to state the main principles of crop and animal production, the principles governing their combination in grazing systems and those governing farm business management and the marketing of agricultural products.

Not everyone needs to be concerned with principles but they are vital to anyone who wishes to understand how agricultural systems function or who is going to be concerned, in any way, with their operation.

A principle may be defined as 'a statement that can be used as a basis for action' or, at the least, for further thought. Agricultural principles should therefore help in making the choices and decisions that are necessary and should provide guidance for action, leading to the achievement of agricultural objectives.

Achieving an objective involves decision on the balance of inputs and outputs in the interests of sustained profit and requires guiding principles (of management or farm economics) to help in making these decisions. But it also requires additional knowledge, in an applicable form, about the relationships between feed intake, its quality and quantity, and milk yield, for example.

The knowledge needed relates to both big and small decisions, to whole production systems and to processes within them: and for all of these guiding principles are required.

It nearly always turns out, therefore, that principles are hierarchical, that is, they can be arranged in the kind of diagram shown in Fig. 1.1. This classifies them, with major ones each giving rise to several lesser ones as the diagram is expanded. Thus there are principles involved in the breeding of crops and animals, just as there are principles governing the feeding of animals and the fertilising of crops.

In short, wherever one can identify an action in agriculture that needs to be carried out or a decision that needs to be taken, it is likely that a principle will be needed to guide one in how to apply what is already known and is relevant.

The purpose of this part of the book is to attempt a statement of the most useful principles and chapters in this section deal with the specific principles of crop and animal production, of the management of grazing systems (because there are special problems where animals interact with their feed supply as happens under grazing conditions), of managing the farm as a business and of marketing the products.

Parts IV and V deal with the main crop and animal production systems, under headings that relate to the major products (or groups of products). The logic behind this is the need to know what production systems exist, or are likely to operate in the future, how they work and what their main components are, and how they relate to soils and climate. Furthermore, the significance of particular weeds, pests, parasites or diseases can only be appreciated in the context of the systems in which they occur. For this reason the book does not devote whole chapters to any one of these topics but deals with them where they are important. This also means that the *ways* in which they are important become clear. So an insect life-cycle, if it is needed, is given where it interacts with the production system in which it is important (although full details are sometimes consigned to an appendix, where this provides an easier

Fig. 1.1 Principles of Agriculture

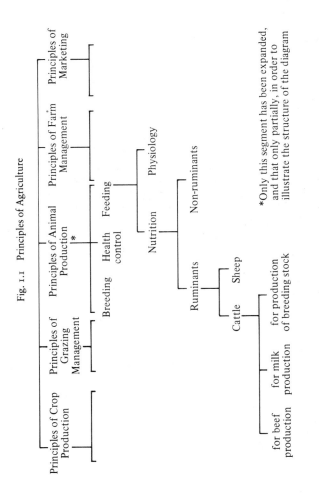

way to present information largely for reference). Thus if a reader is primarily interested in the life-cycle of an insect it is probably better to refer to a book on entomology but if the interest in an insect is in terms of its *role* in agriculture, then it will be best seen in relation to the production systems it affects.

Here, again, the intention is to convey descriptions of production systems and at least the purpose of each is clear: it is to produce a given product at a price that allows it to be sold and that provides the producer with an adequate return. It is in relation to this purpose that the parts of each system have to be judged. Even so, there are sometimes other aspects of a system that concern the community more than the producer. These may include alterations to the appearance of the countryside (by removal of hedges or the construction of tower silos, for example), pollution and even particular methods of keeping livestock. These features are also dealt with where they are most relevant.

Finally, Part VI has two aims: the first is to look into the future. Anyone involved in agriculture needs to make some assessment of where it is going, in relation to possible changes in demand for products, methods of farming, food processing and distribution in response to energy shortage, high energy costs, over-production in Europe, world hunger, pressures related to ecology and animal welfare, and the desire of individuals to be involved with the land.

The second, embodied in the last chapter, entitled 'Agriculture, Society and the Citizen' is intended to show why agriculture should be widely understood and not only by those people directly engaged in the agricultural industry. We are all consumers and have a basic interest in the future supply of food and other agricultural products. Furthermore, the community has a vital interest in any major land user because of the effects on our shared environment and should therefore be interested in agricultural policy and in influencing the debates and decisions relating to it.

But to take part effectively in such debates, the right and responsibility of the citizen, it is necessary to be adequately informed and to understand sufficiently how the agricultural industry works. It is for this reason that the last section ends with a discussion about agriculture and education, in the widest possible sense.

HOW TO USE THE BOOK

This book is certainly intended to be read from start to finish as an account of British agriculture, and is aimed at a wide readership. However, some readers will wish to study their own particular subjects. It could be that a manufacturer of herbicides will want to examine systematically where his products fit in, in which case he will have less interest in systems of animal production. Or a banker might wish to see where money is used or needed and for what purposes, or to assess whether a client's proposed yields are achievable.

Generally speaking, the readership will fall into three main groups. The first are those people who know very little about the subject but want an overall picture, coherent and comprehensive. The second group are those who know a part of the subject but want to see where it fits into the whole. Both of these groups will be best served by reading the book from start to finish. The third group are those who are already knowledgeable but want to refer to particular facts or systems, either to bring themselves up to date or to remind themselves of what they already know. This group may wish to use the book largely for reference and considerable effort has been put into making this both easy and satisfying.

The whole structure of the book is designed for easy reference, with cross-referencing wherever this is helpful. There is also a glossary of terms that not all readers will be familiar with. The text has been kept as free as possible from references to the literature: the latter have been listed separately, in two categories. Specific references are given wherever they are required as authorities for statements or sources of further detailed information. More general references and further reading are also given, so that a reader who, at any point, wishes to take a part of the subject further can do so.

PART I

THE STRUCTURE OF AGRICULTURE

2

WORLD AGRICULTURE: CROP PRODUCTION

Agriculture is a huge industry with a diversity arising out of three principal factors: climate, land quality and man, which interact one with the other. Climate, for instance, has a profound effect on soil conditions and also on the attitudes of people towards land. There is a world of difference between people living in the balmy environment of a tropical island, where acceptable food crops can be grown with little effort and fishing is a pleasant diversion as well as a source of protein, and Danish farmers working hard on land that barely a century before had been acid moorland.

Though soil and climate are the prime determinants of the range and quality of crops and livestock, man – in response to social and economic stimuli – has learned to modify basic environments to his advantage. Soil fertility can be improved by the use of fertilisers; depredations of pests and diseases can be lessened by crop rotation, hygiene and the use of chemicals; tender crops can be given protection from extremes of climate with glass or plastic; new varieties of crop plants have been evolved to extend their environmental tolerance; and irrigation schemes have brought abundance to deserts. Mankind has come a long way since the first static communities replaced nomadism and yet despite the extreme sophistication of developed countries there are still societies whose relationship with the land differs little from that at the dawn of civilisation.

In developing countries there is little scope for specialisation and nearly everyone is engaged directly in getting a living from land but only a small proportion of the populations of advanced societies are agriculturists. Contrast the situation of Malawi, with nearly 90% of her economically active people in agriculture, with that of the United States, with less than 3% of the total work-force in farming.

The USA provides an outstanding example of an advanced agriculture that has been able to increase output, despite a declining work-force. Sometimes it is said that movement from the land, a characteristic of all developed countries, is detrimental but this is not necessarily true. When other employment opportunities are available high labour productivity not only releases people from farming to provide other goods and services but also makes possible increased rewards for those who remain in the industry. Conversely, where there are few employment opportunities outside agriculture – which is the unfortunate situation of many developing countries – the emphasis must be on labour-intensive rather than capital-intensive agriculture to minimise the social evil of urban unemployment. Shack-dwelling communities on the outskirts of many South American and African cities are stark reminders of the dangers of urban drift with insufficient employment opportunities.

The USA has been fortunate in the progressive outlook of its farmers who have been able to exploit technical advances arising from research and development. Its pioneer settlers left behind the constraints of their mother countries, such as small fragmented holdings, leasehold arrangements favouring landlords, irregular fields and a peasant outlook, to create new styles of farming with a much improved structure; while expanding support industries have provided machinery as well as fertilisers and agro-chemicals. Improved communications and transport, with refrigeration for perishable commodities, have made possible rapid movements of produce and rationalisation of marketing.

There have been parallel developments in other countries of the New World, especially those mainly settled by immigrants from northern Europe. Some South American countries have developed rather differently with a preference on the part of many landowners for urban rather than rural living. Absentee ownership is detrimental to production and progress. As Confucius said a very long time ago the best fertiliser is a farmer's footprints! This situation will change, for of all the major regions South America has unquestionably the greatest potential for increasing output to meet the pressures of an expanding world population.

LAND RESOURCES

In 1978 there were 818 million people, some 46% of the world's economically active population, engaged in farming 4,565 million hectares to supply the needs of 4,258 million people (FAO, 1980). Farmland constituted 35% of the total land area of 13,074 m. ha with forest and woodland covering 4,056 m. ha accounting for a further 31%. Table 2.1 shows the regional distribution and the classification of farmland as well as the proportions of the total land area in agriculture.

Europe, with just over 30% of its land in temporary and permanent crops, is the most intensively farmed of the seven regions. Oceania with 59·9% has the highest proportion used for farming but this figure is distorted by including a large area of Australian permanent pastures which are often of very low productivity, for no less than 82% of Australia is classified as arid. Excluding polar

Table 2.1   Land distribution by regions, 1978 (m. ha)

Region	Total Land	Agric. Land Area	%	Arable Land Area	%	Perm. Crops Area	%	Perm. Pasture Area	%
Africa	2,965	975	32·9	154	5·1	15	0·5	806	27·4
N. America	2,136	627	29·9	265	12·6	7	0·3	355	17·3
S. America	1,754	550	32·3	83	5·0	22	1·3	445	26·0
Asia	2,677	1,072	40·0	426	15·8	23	0·9	623	23·3
Europe	473	229	48·8	127	27·0	15	3·2	87	16·6
Oceania	843	505	59·9	43	5·0	1	0·1	461	53·8
USSR	2,227	606	27·5	227	10·3	5	0·2	374	17·0

regions, this figure compares with 59% for Africa, 33% for Asia, 18% for North America, 16% for South America and 7% for Europe (Rado and Sinha, 1977).

Thanks to irrigation some of the land that was formerly desert is now amongst the most productive. Irrigation is a practice of great antiquity for it featured in the agriculture of the Nile and Euphrates valleys long before civilisation came to western Europe. Around 1400 B.C. Sumerian and Assyrian engineers were designing irrigation systems that still provide water for villagers in parts of Iran (Bronowski, 1973). Table 2.2 gives the regional distribution of irrigated land with estimates of recent increases in irrigation (FAO, 1980).

Irrigated land approximates to 14% of the area in temporary and permanent crops. Nearly two-thirds are accounted for by Asiatic countries, particularly China, India and Pakistan as well as smaller countries such as Thailand and Indonesia, where rice is the most important crop grown under irrigation. It is also important in some of the North African countries, especially in Egypt where virtually all cultivated land is irrigated. In neighbouring Sudan well over half the country's export income comes from cotton that is

Table 2.2   Regional distribution of irrigated land ('000 ha)

Region	1961–5	1978	Increase %
Africa	5,870	7,831	33
N. America	18,606	23,543	26
S. America	4,862	6,663	37
Asia	100,363	130,950	30
Europe	8,957	13,670	52
Oceania	1,198	1,656	39
USSR	9,618	16,600	73
World	149,474	200,913	34

principally grown in the Gezira which, prior to the First World War, was an arid waste with little agricultural value.

There has been a steady increase in the area under irrigation in recent years, especially in the USSR and, among European countries, in Spain and Romania. Between 1961 and 1975 Romania recorded an increase from 207,000 to 2,065,000 ha and over the same period the Spanish area increased from 2,089,000 to 2,943,000 ha. Together these two countries in 1978 had nearly two-fifths of Europe's irrigated land. Irrigation, often part of wider programmes including flood control works and the generation of electricity, is one of the most important measures that can be taken in the war against want. In Africa, for instance, there are huge areas with conveniently placed but as yet unharnessed sources of water and reasonably fertile soils and temperature and light regimes that favour plant growth. Apart from major projects, small irrigated areas can also be of great value to livestock production on semi-arid land because they provide reserves of fodder from high-yielding crops such as forage maize and alfalfa to safeguard livestock during droughts.

BALANCE OF CROP AND ANIMAL PRODUCTION

Canada, United States, Australia, New Zealand and most of the countries of western Europe enjoy combinations of foods of plant and animal origin but they are in the minority. Unfortunately, domestic livestock, even with the most sophisticated production systems, are not efficient converters of food; for instance, high-producing dairy cows and poultry, which are the highest ranking of our domesticated animals in this respect, are less than 25% efficient in terms of energy and protein recovery. Where populations exert heavy pressures on land, as they do in India, China and Indonesia, the emphasis is on crops for direct human consumption, especially cereals, legumes, root crops and oil seeds. Meat, an occasional luxury, is often a by-product of animals kept primarily for draught, while eggs and poultry meat are welcome produce from scavenging hens rather than organised poultry keeping.

A nation's access to food of animal origin is not wholly dependent on its land resources or its capacity to import; it is also influenced by the quality of farming in terms of both economic and technical efficiency. Both Denmark and The Netherlands have high population densities and yet they are net exporters of pig-meat, eggs and dairy produce. Generally, western Europeans have been concerned with animal husbandry for many generations and have developed a liking for their livestock as well as an understanding of their needs. When they migrated to new countries they took not only their skills but also the livestock of their home countries. In contrast, African people, with a few notable exceptions such as the Masai and Fulani people, are largely hunters and cultivators. Even in cattle-owning tribes numbers have been considered more important than quality because they determine a man's status, as well as his capacity to acquire wives.

Many countries have constraints against certain forms of animal production. Because of religious scruples, based originally on health considerations, neither Jews nor Muslims eat pigmeat while the sacred status of cows not only denies Hindus good red meat but also results in crop damage by wandering cattle that have outlived their usefulness as draught animals. India has 180 m. cattle and an annual output of only 70,000 tonnes of beef while Argentina has 60 m. cattle producing 3 m. tonnes of beef!

It is not only the priority of crops for direct human consumption that limits the development of animal production in many countries. Where there is no offsetting mineral wealth it is often necessary to grow crops for export in order to earn foreign exchange to pay for goods and services that cannot be provided domestically. Egypt, for instance, with 2·7 m. ha of arable land and over 40 m. people, finds it necessary to plant 0·5 m. ha of cotton every year which is well beyond her need for fibre. Similarly both India and Bangladesh divert land from food production to grow jute because of the value of this crop as an export earner.

Cereal production

Cereals are the most important source of carbohydrates for both human beings and monogastric farm animals and this is reflected in the world-wide emphasis placed on their production. According to FAO estimates there were 761 m. ha, just over 57% of the total area of arable land, planted in cereals in 1979. Yields averaged slightly more than 2 tonnes per ha to give a total of 1,553 m. tonnes or sufficient grain to provide, on a fair share basis, nearly 360 kg per head of human population.

There are eight cereals of varying national importance and their distribution on a regional basis is given in Table 2.3.

Despite the international importance of North American wheat this region accounted for less than one-sixth of the total area and about one-fifth of total production. Asia with 38% and USSR with 24% respectively account for more than three-fifths of the total wheat area. India with 22 m. ha and China with 40 m. ha together account for more than two-thirds of the Asian total. Asian yields with an average of 1,500 kg compare with those of the USSR but are appreciably below the average for North America with 2,100 kg and Europe with 3,400 kg. Though the USSR normally has at least twice the European area in wheat total yields for the two regions

Table 2.3   **Area and distribution of cereal crops, 1979 ('ooo ha)**

Region	Wheat	Rice	Maize	Sorghum	Millet	Barley	Oats	Rye
Africa	8,386	4,779	20,855	14,128	16,567	5,182	403	22
N. America	36,521	2,007	38,702	7,200	–	7,074	5,558	714
S. America	9,553	6,788	16,660	2,621	238	855	623	268
Asia	90,251	130,564	29,360	26,179	33,564	24,124	1,280	1,987
Europe	24,659	389	12,212	203	20	20,812	5,281	5,305
Oceania	11,670	130	83	470	33	2,694	1,348	26
USSR	57,682	610	2,667	80	2,784	37,005	12,239	6,476
World	238,722	145,267	120,539	50,881	53,206	97,746	26,732	14,798

are of about the same order. Yield differences result not only from climatic advantages but also from cultural practices, in particular in the use of fertilisers, especially nitrogen. The USSR is prone to have big fluctuations in its wheat output and Soviet demand is now an important factor determining world prices for wheat (Nove, 1977).

There have been substantial improvements in wheat yields in a number of countries with high population pressures as a result of the Mexican varieties developed by CIMMYT, the international centre for the improvement of maize and wheat, while corresponding improvements in rice production have resulted from the work of the International Rice Research Institute in the Philippines and these advances have been so spectacular that their adoption has been dubbed the Green Revolution. In the 1970s average wheat yields in Asia increased by nearly 40% and though increases in rice yields were lower nevertheless they were substantial.

Sixty per cent more land is used for the cultivation of wheat than of rice but there is a smaller total yield difference between the two because rice averages approximately 2,600 kg/ha as compared with wheat at 1,800 kg/ha. Under optimal conditions rice has only a marginal yield advantage but wheat often suffers from water deficiencies because it is a dry-land crop whereas rice is principally grown as paddy. Rice is at an advantage in the humid tropics whereas wheat is at its best in a temperate climate, though the highest rice yields are recorded in Japan, Korea and Spain, which are warm temperate countries.

Maize is the highest producer among cereals with an average yield now in excess of 3,000 kg/ha as a result of the development of high-producing varieties combined with better husbandry, in particular a greater use of nitrogenous fertilisers. Nowhere are these developments seen to better effect than in the USA where the average yield of 6,800 kg/ha is more than double the world figure. As a result, the USA, with just under one-quarter of the total maize area, produces more than half of world output. This is due not only to high technology but also to favourable climatic conditions. Because the crop normally has less than six months of photosynthetic activity it grows to best advantage under a long day régime, provided always there are high light intensities, adequate moisture and soil fertility, and freedom from frosts. These are conditions obtaining in the Corn Belt of the USA. Though maize is widely grown in the tropics, primarily as a food crop for man, it is at a disadvantage because of short days during the growing season. However, some tropical countries, with well-distributed rainfalls, can grow two maize crops in the year or one maize crop followed by grain sorghum where moisture is limiting for a second maize crop.

Rye and oats are declining in importance. The only countries with appreciable plantings of rye are the USSR and Poland which together account for two-thirds of the area devoted to this crop. Sorghum is grown as a stock food in the USA but in several African countries it is important in human diets. The crop matures more quickly than maize and this is an advantage where soil moisture deficiencies limit maize yields. Millet is primarily an Asian crop; China and India are the leading producers with 60% of the world crop, while barley is a crop of developed countries either for brewing or feeding livestock.

Roots and tubers

A wide range of roots and tubers is grown for direct human consumption, processing and stock feeding. They are usually labour-intensive crops and their area has tended to decline in developed countries. The three principal tubers used as human food are potatoes (18 m. ha), sweet potatoes (13 m. ha) and cassava (13 m. ha). Potatoes are a temperate region crop and perform to best advantage with long-day summers. Though the crop originated in South America it has great importance in Europe which accounts for nearly two-thirds of total plantings.

Sweet potatoes have a wide distribution in tropical and subtropical agriculture and China is the principal grower with over three-quarters of the world crop. It has long been an important item in the diet of Pacific island communities. Cassava, a tropical crop with a long growing season and drought tolerance, is well suited to the light and rainfall patterns of the humid tropics. It is important in Africa, especially in Zaire (1·8 m. ha) and Nigeria (1·2 m. ha). With high yields, despite minimal cultural inputs, it is a very suitable crop in a subsistence type of agriculture. It is also important in Indonesia with 1·4 m. ha and in Brazil with 2 m. ha.

Leguminous crops

There were 130 m. ha planted to grain legumes and groundnuts in 1979. Though this area is small relative to that for cereals, nevertheless these crops are of critical importance because they are important sources of protein in countries where animal proteins are in short supply. The principal crops are groundnuts, haricot beans, soyabeans, dried peas, broad beans, chick peas and lentils. In addition to their food value, legumes are important in rotations not only as break crops but also for residual nitrogen contributions that are particularly welcome in developing countries that cannot afford nitrogenous fertilisers.

Table 2.4 gives the regional distribution of these crops.

Table 2.4 Area and distribution of principal pulse crops, 1979 ('ooo ha)

Region	Haricot Beans	Broad Beans	Dry Peas	Chick Peas	Lentils	Soya	Ground nuts
Africa	2,157	758	472	339	116	287	6,196
N. America	2,718	86	124	143	58	29,211	754
S. America	4,924	276	170	48	98	9,452	762
Asia	14,072	5,460	5,327	9,667	1,469	16,448	11,166
Europe	1,559	420	296	167	95	448	15
Oceania	7	6	60	–	–	53	39
USSR	50	–	4,051	–	12	838	1
World	25,487	7,006	10,500	10,364	1,848	56,737	18,933

Soya is the most important of the grain legumes and is likely to increase largely because of the high biological value of its protein. The USA is the principal grower and accounts for about half the area and two-thirds of production. It originated in China, which is now in second place as a producer, followed by Brazil, but the lead in practically every aspect of soya production and processing has been taken by the USA, which has been conspicuously successful in developing varieties for specific climatic zones.

Dried haricot and broad beans are dietetically important in several Asian and Central American countries but the USSR favours the alternative of dried peas. Groundnuts, for the most part, are a tropical crop and their production is principally accounted for by African and Asian countries, in particular Nigeria, Senegal, Sudan, India and China. Both groundnuts and soya are important for their edible oils and the residues after oil extraction, especially that from soya which is rich in essential amino acids, are valuable in pig and poultry nutrition.

Plant oils

Some plant oils, like olive and sunflower, are widely used in cooking while others have industrial outlets, for instance, coconut oil in soaps and linseed oil in paints, while refined castor oil, in addition to its medicinal function, is a high-quality lubricant. Plant oils have become increasingly important in the manufacture of margarine and other animal fat substitutes because of the possibility that a high intake of animal fat such as butter may be linked with heart disease, a view that is hotly disputed by dairy interests.

There are five annual crops that are widely grown for oil extraction: sunflower and rape, each with an area of about 12 m. ha; sesame and linseed in the 6–7 m. ha range; and safflower with about 1·5 m. ha, but the most important oil seed crop on an area basis is cotton which is grown primarily for its fibre. Normally there are about 33 m. ha in cotton every year.

Mexico produces more than half the safflower, though India has the larger area, but yields there are only about one-quarter of the Mexican average. There is a similar situation with linseed, for here India, the leading grower with 2·4 m. ha, has an average yield of just over 200 kg/ha and yet a much lower total output than Canada with 0·9 m. ha and an average yield of about 1,000 kg/ha. Sesame, with a total output of 2 m. tonnes, has its greatest importance in India with about one-quarter of world production. China with 20% and Burma and Sudan, each with 10% of the world production, follow in order of importance.

The area in sunflower and rape increased by about 50% during the 1970s. Canada, by doubling its rape area, now accounts for about one-third of world production, but Canadian yields averaging about 1 tonne/ha do not compare with those of Western Europe, which typically are of the order of 2·5 tonnes/ha. Sunflowers now occupy about 12 m. ha and the USSR with 4·5 m. ha, yielding one-third of total output, is the largest producer and Soviet plant breeders have been conspicuously successful in breeding varieties with a high oil content. The crop is unimportant in Africa and Asia but the USA is now second to the USSR with over 2 m. ha in 1979 compared with just over 100,000 ha in 1969–71. It is now one of the most rapidly expanding crops in American agriculture.

Permanent oil crops

Principal products are olive, coconut, palm kernel and tung oils. Tung is of minor importance with an annual production of about 100,000 tonnes with China contributing about three-quarters of this amount. The production of copra, from which oil is extracted, is normally about 4·5 m. tonnes and the Philippines and Indonesia are together responsible for about three-fifths of this amount but there are small Pacific island communities, where copra is important because it is their principal export. It is relatively durable and this is important where transport is difficult.

The annual output of olive oil is about 1·7 m. tonnes and though many other warm temperate countries make small contributions the lion's share comes from Italy and Spain. Most of the balance comes from other countries of the Mediterranean region such as Syria, Tunisia, Morocco and Greece. Harvesting is a labour-demanding operation and for this reason olive oil production is at a disadvantage compared with other edible oils such as sunflower and soya where crops are harvested mechanically.

Palm kernels, like coconuts, are a tropical product and their production and the oil extracted from them are important exports for a number of developing countries, in particular Malaysia which produces about two-fifths of the world's total of about 4 m. tonnes. African countries, especially Nigeria, Ivory Coast and Zaire, account for nearly one-third of world production.

Fibre crops

The principal fibre crops are flax, hemp, jute, sisal and cotton. In 1979 world production of cotton lint totalled just over 14 m. tonnes from 33 m. ha. It is an important cash crop in many developing countries, particularly Egypt and Sudan in Africa, and China, India and Pakistan among Asian countries. The leading South American producer is Brazil usually with an output exceeding half a million tonnes, while Argentina and Colombia both produce more than 100,000 tonnes annually. The top producer, however, is the USA with 3·1 m. tonnes in 1979 followed by the USSR with 2·8 m. tonnes and China with 2·2 m. tonnes. Cotton is normally a very labour-intensive crop and it is surprising that it continues to be so important in the USA. The reasons lie in the remarkably successful joint efforts of agricultural engineers, biochemists and plant breeders which have enabled mechanical harvesting of the crop.

Annual production of flax fibre and tow is of the order of 600,000–700,000 tonnes with the USSR responsible for about half, and European countries, particularly France, Poland and Romania contributing about 200,000 tonnes. Jute is by far the most important of the coarse fibres, taking up about 2·6 m. ha with an annual output of 4 m. tonnes of which approximately 85% is of Asian origin. The principal growers are India and Bangladesh, each producing 1 m. tonnes or one-quarter of the world crop. Thailand, with 370,000 tonnes, is the only other significant producer.

The area in sisal fluctuates about 600,000 ha with a yield of approximately 450,000 tonnes of fibre. Brazil (280,000 ha) is the main producer but it is an important export crop in several African countries, notably Tanzania, Mozambique and Kenya. Hemp production, now about 220,000 tonnes, is declining. In 1969–71 there were 0·5 m. ha, but by 1979 the area had dropped to 360,000 ha. The USSR and India are the principal producers and together account for about half the total production.

Sugar cane and sugar beet

The world at large has a sweet tooth, for no less than 13 m. ha of cane are harvested each year for sugar extraction, mainly in tropical countries, while temperate countries with 8–9 m. ha of beet make a further contribution to an annual output of over 100 m. tonnes of raw sugar.

There are about a hundred countries growing sugar cane including island communities whose individual contributions are small but nevertheless vital to their economies – Mauritius, Barbados and Fiji are good examples. India, with 3·1 m. ha, has the largest area of harvested sugar cane and she is followed by Brazil with 2·5 m. ha. There are no very large growers in Africa though many African countries supply their own sugar needs, in particular Egypt with 100,000 ha and South Africa with 220,000 ha. Most Latin American countries have an appreciable area in cane, but, surprisingly, China, with nearly one-quarter of the world's population, has only 400,000 ha of cane and only 100,000 ha of beet to satisfy the sugar requirements of 945 m. people.

Beet production is dominated by Europe and the USSR which each harvest about 3·7 m. ha annually. Iran (200,000 ha) and Turkey (290,000 ha) are the only other beet sugar producers of any great consequence because generally tropical and subtropical countries with sugar industries favour sugar cane. The USA, with 450,000 ha, is not a large grower of sugar beet and this is attributable to a deliberate national policy of providing markets for neighbouring countries whose economies are greatly dependent on sugar. One unfortunate consequence of the UK joining the EEC, with its sugar surplus, is that there are now several developing countries, formerly part of the British Empire, that no longer have unrestricted access to their traditional sugar market.

Tobacco

Tobacco, like sugar, attracts a world-wide addiction, for there are well over a hundred countries contributing to an annual output of 5·4 m. tonnes of leaf which are harvested from 4·4 m. ha. Despite anti-smoking campaigns output is continuing to expand, which no doubt gratifies growers and governments depending heavily on tobacco revenues. There are two principal types of tobacco: the small-leaved Turkish tobacco with its very distinctive aroma and the large-leaved Virginian which is more widely grown. Asian countries account for more than half the world production and China is top of the league with an annual output of about 1 m. tonnes. The USA normally produces about 0·8 m. tonnes while Brazil, with 0·4 m. tonnes, is the leading South American grower. Tobacco is also important in Indonesia with an annual output of more than 450,000 tones. Zimbabwe is Africa's leading producer, with an annual output of 100,000 tonnes of very high-quality leaf.

Tea, coffee and cocoa

These are all crops of great importance in the economies of several developing countries. They are labour-intensive and high yields combined with quality require fairly specific environmental conditions and, in consequence, the main producers are unlikely to face severe competition. About 80% of tea production, amounting to 1·8 m. tonnes, comes from Asian countries. India (550,000 tonnes) is followed by China (300,000 tonnes) and Sri Lanka (200,000 tonnes). The only African countries with significant tea production are Kenya and Malawi but total African output is only 200,000 tonnes. The USSR produces about half this amount.

Cocoa is primarily a crop of Africa's humid tropics which account for two-thirds of a world output of 1·5 m. tonnes of beans. Ivory Coast, Ghana, Nigeria and Cameroon are the principal African contributors in descending order of importance. Outside Africa the main producer is Brazil with 300,000 tonnes. Attempts have been made to establish cocoa production in some of the smaller South American, Caribbean and Pacific countries but the only ones with an output in excess of 30,000 tonnes are Ecuador, Colombia, Dominican Republic, Mexico and Papua New Guinea.

Coffee production is more widespread than either tea or cocoa and there are about fifty countries making appreciable contributions to an output of nearly 5 m. tonnes of green beans. Brazil, which normally harvests about 1·2 m. tonnes, is the largest producer but there are about a dozen other countries contributing more than 100,000 tonnes. The leader is Colombia with 750,000 tonnes. Indonesia produces about half the Asian total of 500,000 tonnes while Ivory Coast and Ethiopia, the most important African producers, are responsible for about 18% of world production, which is showing a marked upward trend. Coffee – possibly due to the convenience of instant coffee as well as the proliferation of coffee bars that have an appeal to the young – is an increasingly popular drink in the Western World. It is difficult to believe that in the great 1930s depression Brazilians were using coffee beans for fuel!

Rubber

Despite advances in rubber substitutes natural rubber production continues to expand and now totals about 3·7 m. tonnes. Malaysia is the leading producer with 45% of world production. Other important sources are Indonesia and Thailand and together Asian countries produce about 90% of world output. Plantations have been established in several African countries but they produce only 6% of the world's rubber with Liberia and Nigeria responsible for most of this.

Tree nuts

Almonds and walnuts, each with a total output of 800,000 tonnes, are the most important of the tree nuts, and in each case the USA with nearly 300,000 tonnes of almonds and 180,000 tonnes of walnuts is the largest individual producer. European countries, however, provide over 40% of the almonds and 35% of the walnuts. Turkey is the principal producer of hazel nuts with nearly 60% of a world production of 450,000 tonnes. China produces about one-third of the chestnuts with approximately 170,000 tonnes. African and Asian countries dominate cashew nut production, which amounts to approximately 500,000 tonnes annually. India and Mozambique, each contributing about 30%, are the leading producers but Kenya and Tanzania are also important with annual outputs in excess of 50,000 tonnes. Pistachios, with an annual output of 112,000 tonnes in 1979, are quantitatively less important than the other five tree nuts but they are rapidly gaining in popularity and during the 1970s there was a threefold increase in their production.

Fruits

The world is blessed with many fruits adapted to a wide range of environmental conditions. Bananas, pineapples, avocados, papayas and mangoes, for instance, belong essentially to the humid tropics and subtropics, while citrus fruits are best suited to warm temperate conditions. Hard winters can benefit deciduous tree fruits because they reduce infections, while berry fruits are best suited to cool temperate climates provided there is adequate sunshine during the maturation period to develop full flavour as well as to limit the incidence of botrytis.

Bananas lead in volume of output, which in 1979 was nearly 40 m. tonnes. Not only are they important in the exchange balances of a number of small countries, such as some of the Caribbean communities, but also they and the closely related plantains, with an annual production of 20 m. tonnes, are important in local diets. Brazil, with an output of 6·5 m. tonnes is the leading producer of bananas, but India, Ecuador, Indonesia, Philippines and Thailand all have outputs in excess of 2 m. tonnes. About two-thirds of the plantains are produced by African countries, with Uganda the biggest single producer. Oranges, with an annual output of about 38 m. tonnes, are not far behind bananas in volume of production and there are no less than twenty-five countries with an output of more than 200,000 tonnes. Brazil, with just under 10 m. tonnes, is the leader of the group, followed by the USA with over 8 m. tonnes and Mexico with over 3 m. tonnes.

World production of apples is about 35 m. tonnes annually while pear production is about 7 m. tonnes. The USSR, with an apple output of about 7 m. tonnes, is the largest individual producer, with the USA in second place producing about half this amount. Peach and nectarine production total 7 m. tonnes as compared with just over 5 m. tonnes for plums and 1·6 m. tonnes for apricots.

Avocado production has been increasing steadily in recent years in response to a growing demand from the top end of the fruit trade and output is now about 1·3 m. tonnes. North America, in particular, Mexico and adjacent Caribbean countries, account for half of this total. Over three-quarters of the world's output of mangoes, some 14 m. tonnes, are produced in Asian countries, especially India with 9 m. tonnes. The annual output of nearly 8 m. tonnes of pineapple is more evenly spread with Asian countries producing just over half. China, with 900,000 tonnes, is the leading producer followed by the USA (mainly Hawaii) with 600,000 tonnes and Brazil 570,000 tonnes. It is an important crop in the East Indies especially Malaysia, Indonesia and Philippines.

Viticulture

Grapes, occupying 10 m. ha, have a high ranking among the permanent crops and Europe is by far the most important region with about three-fifths of this area. Most warm temperate countries settled by Europeans, in particular Argentina, the USA and South Africa, also have sizeable areas planted to grapes. There are also several predominantly Muslim countries, notably Turkey and Iran, with substantial plantings but in both instances raisins rather than wine are usually the end product, though Algeria, presumably because of a French influence, is a notable exception.

Wine, however, is certainly the principal product of European vineyards with an annual output of about 2·7 m. tonnes, which is about three-quarters of all production (3·6 m. tonnes). Turkey, with 250,000 tonnes, produces about 40% of the world's raisins with Greece and the USA each contributing about 15%. Grapes are an important fresh fruit and there is a steadily growing demand from developed countries which, thanks to the dovetailing of seasons between hemispheres and refrigeration with rapid transport, now enjoy a year-round supply of grapes as well as other seasonal fruits.

Vegetable production

World output of vegetables in 1979 was estimated to be 340 m. tonnes. This cannot be regarded as more than a guess because the huge variety of vegetables, their importance in subsistence cropping and the depredations of pests and diseases among other factors, makes estimation hazardous, but there can be no doubt that vegetable production, whether it is a field-scale operation, allotment or kitchen gardening, or the planting of small clearings following a bush fallow, is of vital importance in human diets.

The distribution of the various crops is profoundly affected by climate and food habits. Crucifers, which include cabbages, cauliflowers and turnips, are temperate crops and with the notable exception of cauliflowers when they are in head, are very frost hardy and this is important in countries with cold winters.

Cucurbitaceous vegetables include several widely grown frost-tender annuals that require summer warmth for full maturity. The commonest are cucumbers, gherkins, melons, pumpkins, squash and marrows. Most require some protection in cooler temperate regions and if out-of-season production is required some supplementary heating is usually essential. This also holds for tomatoes, now possibly the most popular of all vegetables. No less than 125 countries covering a wide climatic range of conditions, contribute to an annual output of nearly 50 m. tonnes. Chillies and peppers have a less popular appeal but nevertheless an annual production of more than 2 m. tonnes is very substantial. China accounts for about one-fifth of this amount while Spain is the principal European producer. The Spanish regard for peppers is reflected in their popularity in Latin America, especially in Mexico, but strangely neither Brazil nor Portugal appears to have much interest in them.

Onions have a geographical spread of production matching that of tomatoes and world production is now just short of 20 m. tonnes. The corresponding figure for garlic is 2·4 m. tonnes and China is the leading producer for both crops. Green peas and beans are mainly temperate vegetables and they have their greatest popularity in Europe and in countries such as the USA with predominantly European settlers. To some extent this is true of carrots but this vegetable is also favoured by the USSR and China, which together account for just over one-third of a world crop of 10 m. tonnes.

Many vegetables grown on a much smaller scale are collectively very important, for instance, yams, the two kinds of artichokes, aubergines, leeks, parsnips, celery, beets, spinach, calibrese, lettuce, edible fungi and a whole range of culinary herbs that give some kind of character to otherwise bland food. There is no lack of variety in the vegetables now available to man provided he makes the effort to grow them.

3

WORLD AGRICULTURE:
ANIMAL PRODUCTION

RUMINANT LIVESTOCK

Animal production in its various forms covers a wide spectrum of human effort. At one extreme there are nomads with their herds and flocks following traditional routes and sometimes ignoring international boundaries, in search of whatever nature may offer and often obtaining no more than a meagre living for animals and owners alike while, at the other end of the scale, there are highly sophisticated production systems with environmental controls and carefully formulated diets.

Common grazing as practised by settled African cattle-owning tribes is on a low rung of the production ladder. Often because of the social importance of cattle, with an emphasis on numbers rather than individual productivity, there is serious overgrazing, which is a hazard that may not only upset botanical balance but, more seriously, result in erosion. Enlightened veld management aims at limiting the intensity of stocking and periodic resting of blocks to allow recovery of desirable species but the difficult problem is one of convincing cattle owners that such a policy is in their long-term interest. There are still remains of common grazing in Europe, particularly in upland areas where usually there are grazing rights limiting the number of animals that have access to pastures. Sometimes these rights are held by farmers with adjacent enclosed land that provides winter accommodation as well as food reserves for periods of shortage. One of the great weaknesses of common grazing is the failure of participants to agree on improvements leading to better swards.

The extensive grazier with a full title to his land and a ring fence around it to exclude neighbours' stock is in a position to make such improvements as subdivisions, fertilising and reseeding and the creation of fodder reserves. More than this, he is able to take positive steps to improve the inherent quality of his stock by culling inferior animals and using selected sires, which is not always possible with common grazing. Some of the best examples of extensive grazing, where owners have full control of land and livestock, are the ranches, estancias and stations of the New World. They had their greatest importance in the nineteenth century when huge areas of virgin land, often savannah with useful natural pasture, were taken up in large blocks either for sheep or cattle production or, more occasionally, for the two species in combination. Progressively these very large enterprises have been scaled down as a result of economic and political pressures, except where the land is unsuitable for intensification because of natural conditions or poor communications which are still the principal barriers to more intensive farming on reasonably good land in northern Argentina and parts of Brazil, where there are still holdings of 50,000 ha or more.

New Zealand provides a good example of changes that have been made in pastoral farming over the past century. When Canterbury Province was first settled from about 1850 onwards, it was largely in native pasture which provided grazing for Merino sheep that had been imported from Australia where the breed had already been firmly established. Only people of some substance could afford these sheep in any great numbers with the result that the province was mainly held in large blocks by a comparatively small number of people. The Merino adapted well to prevailing conditions for it requires little shepherding and it has a valuable fleece. This was extremely important for, initially, wool was the only product of any consequence that could be dispatched to distant markets to arrive in an acceptable condition and leave a reasonable return. Mutton was then of minor importance because the small local market was incapable of absorbing the potential output of fat sheep. This style of farming prospered until about 1875 when there was a catastrophic fall in wool prices in the world-wide depression that followed the American Civil War.

The first reaction of farmers on good cultivated land, of which there was an abundance in the province, was a switch to cereal growing to exploit the fertility that had been augmented by pastoral farming. This provided some relief but eventual salvation came from another important development. In 1882 the first shipment of New Zealand frozen mutton was landed in London, an event that *The Times* described as 'a prodigious fact' which indeed it was for it initiated the export of perishable animal products on which the economy of New Zealand is now highly dependent. The emphasis in Canterbury and elsewhere then moved away from the single-purpose Merino to dual-purpose breeds such as Romney and Corriedale (a derivative from the Merino) to produce more acceptable carcasses.

The advent of refrigeration not only created another market outlet but also made possible intensification of farming. A situation where a privileged few had control of vast areas to deny a land-hungry majority the opportunity of working them to better effect could not continue in a country with universal suffrage and soon the land policy was directed towards closer settlement involving the breaking-up of large estates. This, with family farming, brought more changes, for instance closer subdivision of farms to give better control of pastures, the introduction of improved pasture species, the application of fertilisers, fodder conservation and supplementary cropping. It is not uncommon now to find farms with stocking rates in the order of fifteen breeding ewes to the hectare representing at least a tenfold increase in productivity over the last century. Canterbury farming has indeed come a long way since those halcyon days when station-owners, appropriately known as squatters, were able to sit in the shade for most of the year and watch their wool grow!

These changes were not limited to New Zealand but were widespread throughout neighbouring Australia, in the USA and in the temperate parts of South America. Large properties still remain but they very seldom consist of good land and the big Australian stations either lie beyond the wheat-growing belt towards the dead heart of the subcontinent or are in the as yet partially developed northern tropical region. Big cattle-breeding ranches in the USA still exist outside the fantasy world of film-making in the rangelands of the western states but, here again, they are on poorer quality land.

Many estancieros in Argentina complain that they, too, have been subject to pressures to reduce the size of their holdings to a point where their viability is impaired but they are still large by European standards except in the proximity of the cities such as Buenos Aires where there has been a mushroom growth of weekend farmlets. Much of the technological revolution that has transformed farming in developed countries seems to have by-passed Argentina and farmers there give the impression that they are still trying to exploit the declining natural fertility of the pampas and are not taking necessary steps to replace plant nutrients that have been lost by extractive cropping and the sale of livestock. The blame does not rest with the farmers alone for politicians have much to answer for in this connection.

Greater intensification in pastoral farming usually involves a change in the type of livestock that are carried on the farm. For instance breeding is biologically less efficient than fattening, which with a sensible buying and selling policy gives higher margins and a quicker turnover of capital. Breeding is recognised as being better suited to poorer pastures than those used for finishing but any form of beef and mutton production, be it the production of store stock or final fattening, is biologically an inefficient method of land use. For this reason such production on good quality land is often integrated with cash cropping in a system of alternate husbandry.

Unquestionably dairying is the most efficient form of pastoral farming, both economically and biologically. It is also the most labour intensive and for these reasons it is preferred to beef or sheep production on small farms, especially when there is a market for liquid milk which is usually the most lucrative outlet. Where there is a favourable relationship between the cost of concentrates and the price of milk there is a further advantage for dairying in that farmers are able to increase output by a combination of larger herds and higher yields without any further land requirement. The ultimate in this approach is the so-called 'cowtel' where large herds are kept in yards where they are trough-fed on concentrates and on fresh or conserved bulk feeds.

This style of dairying, which has its greatest vogue in dry areas such as California, where the costs of housing and handling manure are low and much of the fodder is grown under irrigation, contrasts with a system where there is a complete reliance on pasture, fodder crops and hay and silage. Nowhere is this better exemplified than in New Zealand's dairy industry which is primarily concerned with milk for the manufacture of butter, cheese and other milk products. As these are mainly destined for distant markets the emphasis is on low cost production which is achieved by good labour productivity, the use of cows of high genetic merit, good stock management and above all, thanks to a favourable climate, very productive pastures. It is not exceptional in the main dairying districts to have good grazing, without any great need for supplementary feeding, for nine months of the year and this without any application of nitrogenous fertilisers, which are not needed because of the active growth of soil-enriching white clover which is such an important component of New Zealand pastures. The aim in this type of dairying is less one of maximising yield per cow than of maximising output per unit area. This necessitates high stocking intensities and it is not exceptional to carry three cows and their share of herd replacements to a hectare of pasture and, in the process, to produce 9,000 litres of milk without any supplementary concentrates except perhaps a modicum of meal in the early stages of calf rearing.

NON-RUMINANT LIVESTOCK

This section is principally concerned with pigs and poultry but a mention must be made of rabbits kept intensively for meat production and white veal production where dairy-bred calves are reared on liquid diets and are slaughtered before they have developed ruminant function and with it a capacity to utilise fibrous feeds. This attribute is critical in appraising the role of ruminants as opposed to non-ruminants in agriculture. The former can use plant material of no direct value to man, often growing on land that cannot be brought under cultivation for the production of food crops – for instance, steep uplands with rocky outcrops – and in the process make a useful contribution to meat supplies. Non-ruminants, on the other hand, because of their digestive systems, compete with man who is also a monogastric animal. When they are kept under controlled environments aiming at maximum output, be it numbers of eggs or weight of meat, they require carefully formulated rations based on cereals and high-quality proteins and sometimes the question has to be asked whether the world at large, with its increasing population pressures, can continue to process cereals and grain legumes into eggs, milk and meat. Indeed, sometimes it seems we are in the shadows of Orwell's *Animal Farm* for there are already many people who are so far below the breadline that there is now an obscene situation where pigs can sometimes be much better nourished than human beings.

But pigs have not always been cosseted creatures for they have had a scavenger's role utilising edible wastes and oddments they have been able to find on their own account such as grubs, roots, acorns and beech-mast. There are still remnants of this in the cork oak country of the Iberian Peninsula where the best serrano ham is the product of pigs finished on acorns. The feeding of swill collected from large catering establishments, which reached an all-time high during the Second World War, continues to be important in Western Europe. There have been attempts to reduce concentrate use in pig fattening by satisfying appetite above a basic meal allowance with bulk food such as potatoes and fodder beet. This method, known as the Lehmann system, was developed in Germany during the First World War and, based on fodder beet, it had a limited vogue in Britain in the early 1950s. However it quickly went out of fashion as soon as barley and other coarse grains became freely available. The present position in Britain, as in all other developed countries, is that the pig industry is based largely on the feeding of concentrates.

It is, however, a highly competitive industry necessitating a determined effort to increase efficiency of feed conversion, because feed normally constitutes about four-fifths of the costs of production. There has been a five-pronged attack to this end: better management aimed at reduction of perinatal mortalities; early weaning to minimise double processing of food into milk and then into pig meat; improved prophylaxis and hygiene to safeguard thrift; a better understanding of nutritional requirements, especially in respect of quality of protein, and the value of additives such as copper, which has resulted in improved diets; improved housing particularly designed to avoid heat losses, and genetic improvements that have included carcass attributes as well as more efficient food conversion. It is not possible to give a precise assessment of the gains that pig producers have made since the mid-1950s but the management standards of the early 1980s differ greatly from those that were acceptable thirty years previously. A Danish or British pig farmer who, in 1950, was breeding his own fatteners and taking them to slaughter at 90 kg live weight for the Wiltshire bacon trade would have been satisfied with an annual crop of 18 weaners per sow, an average slaughter age of 180 days and a food conversion of 4·0 kg meal/kg of live-weight gain. His modern counterpart has his sights set on an annual crop of 22–3 weaners, 160-day slaughter and a conversion ratio better than 3·25:1.

Changes in poultry production have been even more dramatic. The traditional role of hens, still retained in some countries, is scratching a living around the farmyard and in the process producing a few dozen eggs which, by convention, were often a perquisite of the farmer's wife. Hens would receive household scraps and a scatter of tail corn and would have some form of accommodation, which, among other things, gave them protection from nocturnal predators. Often birds would be home-bred and sometimes the first intimation of additions to a flock would be the proud appearance of a clucking hen with her brood of chicks. As it was impossible to distinguish the sexes with any certainty until about six weeks of age cockerels in excess of breeding needs would be reared as table poultry. More progressive producers would sometimes buy settings of eggs from specialist breeders of purebred birds to put under their own broody hens.

Gradually from the beginning of this century poultry-keeping became a much more sophisticated undertaking. Specialist breeders, adopting such refinements as artificial incubation and brooding and sex determination at hatching, became the principal source of flock replacements purchased either as day-olds or point-of-lay pullets. Purebreds, notably White and Brown Leghorns, Rhode Island Reds and Light Sussex, were usually favoured by egg producers in developed countries but between the First and Second World Wars there was a swing to first-cross pullets in the belief that hybrid vigour improved both viability and performance. A favoured first cross in the UK was the Rhode Island Red on the Light Sussex where pullets and cockerels could be distinguished by colour differences at hatching.

Accommodation varied considerably. Some believed it advantageous to maintain flocks on free range with either fixed or movable houses for overnight accommodation. Another system, combining the benefits of an outdoor life on clean land with the enhancement of soil fertility through the immediate return of droppings, was containment of birds in folds that were moved daily. Other producers preferred to keep their birds completely under cover in open-fronted houses where they were safe from predators. Under these conditions the egg yolks did not have the rich colour of free-range eggs (through the ingestion of grass containing carotene) unless special precautions were taken, either by inclusion of dried grass in the poultry mash or daily provision of fresh green material.

Eventually enclosure became the preferred system of accommodating hens and this was due primarily to two main developments: rural electrification and an understanding of how light affects laying performance and with it the provision of lighting regimes to stimulate egg production. The keeping of hens on deep litter became popular in the early 1950s and usually this system worked reasonably well but it could fail with high humidities, giving rise to damp compacted litter. Many producers preferred the alternative of keeping birds on fine mesh netting over dropping pits which were cleared before another crop of pullets was introduced into a house.

Keeping hens in individual battery cages not only avoided feather pecking which can be a serious vice with high-intensity housing but also made possible the monitoring of individual production. Because of the high capital cost of cages in an increasingly competitive industry producers took the logical step of doubling up in cages and this in turn led to larger multiple cages so arranged as to have full mechanisation both of feeding and the removal of manure.

These developments, together with small profit margins per bird, have necessitated economies of scale, and poultry production is no longer, as it used to be, a modest sideline of mixed farming. Now in western countries it is usually a large-scale enterprise. There have been big changes, too, in the birds employed. No longer are they purchased as first crosses but are hybrids of complementary lines that have been evolved by testing and selection over many generations. Such breeding work cannot be undertaken with any great prospect of success without immense resources, and so the breeding and supply of chicks is no longer the livelihood of small independent breeders, working along conventional lines, but the function of large organisations operating on an international scale.

These hybrids are smaller birds than those they replaced and yet, due to selection, there has been no loss of egg size, but much has been gained in food economy. Because of their smaller adult size their food requirement from hatching to the point of lay is appreciably less and maintenance over the laying period is also reduced as compared with their predecessors. Overall, in terms of food use, modern layers are at least 70% more efficient than the laying poultry of the immediate pre-war period. However there is more to this improvement than the introduction of hybrids. There is now a better understanding of nutritional needs, which has resulted in improved feed formulations, and there is more effective control of disease with lower wastage and better environmental control, particularly in respect of exposure to light. The overall picture is that the modern poultry keeper, working with tens of thousands of birds compared with the hundreds at an earlier stage, is no longer satisfied with a hen-house average of less than twenty dozen eggs. Before the advent of the hybrids and environmental controls an average of twelve dozen eggs was considered to be reasonable and satisfactory.

There have been parallel changes in the production of table poultry, which at one time was a luxury, even in developed countries. Turkeys, in particular, were very special, reserved for such celebratory occasions as Thanksgiving or Christmas but today they constitute one of the cheapest forms of meat.

Farm-reared spring chickens, which were often by-products of rearing laying replacements, have declined in importance and mass-produced broilers now dominate the table poultry scene. Even in developing countries home-produced broilers make an acceptable contribution to the intake of animal protein because of huge advances in production techniques.

The mushroom growth of the industry in almost every country has depended less on participation by established poultry producers than on recruits and capital coming from outside farming. The very nature of the industry – small margins per bird necessitating economies of scale, the dependence of producers on the supply of improved hybrid chicks which is controlled by breeding companies, the large working capital requirements, the investment in special buildings, the precise food formulations and the need to have processing and marketing effectively organised – has introduced a novel situation for agriculture. The usual result has either been businessmen employing expert poultry managers or poultry-men working as links in a system of vertical integration controlled by business interests taking one profit from the supply of food and another from marketing and processing.

It is not surprising that the various forms of poultry production with large units concentrated on comparatively small areas and entirely dependent on purchased food and stock should acquire the description of 'factory farming' and in the process become a target for environmentalists, especially those with somewhat simplistic, back-to-nature attitudes. The other side of the fence is also important: the new technologies have reduced the cost of producing eggs and poultry meat and there are hundreds of millions now enjoying improved diets, that would have been denied them if poultry production had remained as it was before 1950.

TROPICAL ANIMAL PRODUCTION

Some of the greatest challenges facing world agriculture, especially in animal production, are to be met in the tropics where millions exist on nutritionally inadequate diets. The great forward strides in both crop and animal production over the past century have been made mainly in temperate countries, where, coincidentally, there has usually been the infrastructure necessary for a progressive agriculture, in the shape of research and extension, transport and marketing services and supply industries to provide the necessities of high-technology farming. Relatively few developing countries have either the export income or the local resources to finance developments that need to be made in their farming. Exceptions to this rule are countries with mineral wealth, especially from oil, but this can be a mixed blessing because national affluence sometimes leads to the neglect of agriculture. Events in Trinidad and in neighbouring Venezuela where there has been little or no progress in domestic agriculture since 1960 support this viewpoint.

Generally greater progress has been made in crop production than in animal production and this is understandable. Apart from the priority that has been given to the satisfaction of hunger rather than to dietary quality, the tropics are fortunate in having several high-yielding, well-adapted food crops such as rice, cassava, plantains and groundnuts, but the same cannot be said about domesticated animals. In the colonial era crops with an export potential received a lot of attention and governments were responsible not only for crop introductions but also for developing research programmes to ensure their successful establishment. Sugar cane, rubber, oil palms, tea, cocoa, coconuts, cotton and sisal are all crops that have been widely and successfully introduced. Though generally the emphasis was on export crops, invariably it was part of development strategy to upgrade the production of food crops. In contrast efforts to improve animal output have been puny except in respect of veterinary preventive medicine. Veterinarians, especially in British colonies, had control of animal production and naturally their preoccupation was with their main interest – the minimising of disease hazards – rather than the development of production systems. This, arguably, was the right approach initially because of the number of serious endemic diseases affecting livestock in the tropics.

Few cattle native to the tropics have dairy attributes remotely resembling those of temperate breeds such as the Friesian and Jersey but when these were introduced to tropical environments they quickly degenerated as a result of the combined impact of climate, diseases, poor nutrition and indifferent management. The position has been somewhat better with beef cattle, not only because they are less demanding in their management but also because some indigenous cattle with adaptation to tropical environments have a considerable beef potential, for instance the Boran of East Africa and some of the Zebu breeds originating in the Indian subcontinent. The latter have provided material for important cattle improvement work in the USA, in particular the development of the American Brahmin which is now a quite outstanding beef animal that has been widely exported to South and Central America and Africa and has been used in the creation of new breeds. Among these are the Charbray, a Brahmin–Charolais cross and the Santa Gertrudis which is a Beef Shorthorn–Brahmin combination. The Santa Gertrudis is the better known and the longer established and it has been exported to several tropical countries, notably in South America, Africa and Australia (which has made its own contribution to heat-tolerant cattle with its aptly named Droughtmaster).

Unfortunately there has been less progress in dairying where the most notable success has probably been the development of the Jamaica Hope based on the Jersey with a small contribution from the Sahiwal, long regarded as the most dairylike of Indian cattle. It retains many Jersey characteristics including high fat milk and an extreme dairy type of conformation. Under commercial conditions in Jamaica average annual yields compare favourably with those of Holsteins despite the latter's greater size and lower solids content of milk. This is due partly to the shorter calving interval in Jamaica Hope herds, possibly a reflection of superior adaptation to a tropical environment. Unfortunately, it is limited numerically and its specialised dairy function also reduces its suitability for many tropical countries where by-product beef is important in any dairy development programme.

There is a view that attempts to develop tropical dairy production on a European or American model are ill-conceived because they do not sufficiently take into account local needs and difficulties. Possibly it is more realistic to base production on bulk food rather than on concentrate feeding for, apart from the doubtful prospect of an economic response to feeding concentrates, sometimes the latter cannot be spared from human diets. There are already examples of this approach in some Central American countries where cows, mainly Zebu-Holstein crossbreds, are hand-milked in the mornings after being separated from their calves overnight but suckle their calves by day when they are at pasture. Average yields are low, and anything in advance of 1,000 l is considered satisfactory, but there is the important balancing consideration that calves get a reasonably good start in life, which is not often so with artificial rearing in the tropics where it is invariably expensive and plagued with ill-thrift.

Preston (1980), who is an advocate of this style of dairying, emphasises the fodder value of sugar cane with its extremely high energy yields under humid tropical conditions. There is also the alternative of a major reliance on pasture based on such species as pangola and African star-grass and using sugar cane, maize silage or fodder sorghum as a dry-season supplement. The great need is for a tropical pasture legume comparable to white clover in temperate grassland. Australian workers have made notable contributions in collecting and appraising tropical legumes but this particular challenge has not yet been fully met. When it is it will be an outstanding step forward in the utilisation of tropical pastures by ruminant livestock.

DAIRY PRODUCTION

Milk and its products have a very high ranking among foods of animal origin and in the most developed countries dairying is the most important single branch of animal production. In many countries where land resources are insufficient for specialised beef production the dairy industry is also the most important source of beef and veal. Milk is predominantly a product of dairy cows, among which Friesians are the most notable performers, but goats' and sheep milk also make a significant contribution, particularly in the Mediterranean region, while water-buffaloes are important in the Middle East and parts of Asia.

Estimates of milk production from the different sources are summarised in Table 3.1 (FAO, 1980). South Africa, with an

Table 3.1   Milk production according to region and species, 1979 ('ooo tonnes)

	Cows	Goats	Sheep	Buffalo
Africa	10·4	1·4	0·7	1·2
N. America	73·0	0·3	–	–
S. America	23·4	0·1	–	–
Asia	33·4	3·5	3·3	24·6
Europe	174·2	1·6	3·3	0·1
Oceania	12·2	–	–	–
USSR	92·8	0·4	0·1	–
World	419·4	7·3	7·4	25·9

annual output of 2·5 m. tonnes of cows' milk, is the only African country with a substantial dairy industry. North American output is dominated by the USA (56·1 m. tonnes) and Canada (7·1 m. tonnes). India, with an output of 10 m. tonnes, is the most important of the Asian countries and is followed by Japan with 6·5 m. tonnes. Australia and New Zealand, with outputs similar to Japan, account for almost all of Oceania's production. Brazil (10·6 m. tonnes) and Argentina (5·2 m. tonnes) are the principal South American producers. European production, which is increasing steadily, especially in EEC countries, now accounts for 40% of world production. The highest average yields are not recorded by the leading dairy countries but by Israel with 6,800 kg/cow and Japan 5,700 kg/cow. These figures compare with 5,000 kg in The Netherlands, 5,200 kg in the USA and a world average of about 2,000 kg.

India is the most important producer of buffalo milk with an annual output of 14·5 m. tonnes followed by Pakistan (7·6 m. tonnes), where buffalo milk production is almost three times greater than that of cows' milk. Egypt is the only African country producing buffalo milk and there it has about the same importance as cows' milk.

Milch goats are more widely distributed than milking buffaloes and no less than 75 countries record an output in excess of 1,000 tonnes of goat milk annually. The leading Asian producers are India (0·7 m. tonnes), Bangladesh (0·5 m. tonnes) and China (0·4 m. tonnes). There are some communities in the Mediterranean region where goats' milk is more important than cows' milk with much of it being consumed as a soft cheese.

The output of sheep milk at 7·3 m. tonnes approximates to that of goat milk and the leading producers are Asian and European countries, in particular Turkey, Iran, France, Greece and Italy. Again, it is consumed largely in the form of cheese. A high proportion of cows' milk is used for manufacture, especially of butter and cheese, but dried and condensed milk are also important and their export makes a valuable contribution to the diets of many developing countries.

The above figures are not exclusively based on cows' milk but include sheep and goat milk cheeses and butter and ghee made from buffalo milk. The most striking feature of Table 3.2 is Europe's dominant position in milk processing and within Europe the part played by members of the EEC, in particular France, West Germany and The Netherlands. India produces nearly half of the Asian

Table 3.2 **Milk products by regions, 1979** (m. tonnes)

	Cheese	Butter & Ghee	Evap. Milk	Dried Whole Milk	Dried Skim & Butter Milk
Africa	0·36	0·15	0·05	0·01	0·02
N. America	2·42	0·60	1·38	0·08	0·55
S. America	0·45	0·16	0·16	0·32	–
Asia	0·70	1·25	0·51	0·12	0·16
Europe	5·35	2·98	1·86	0·65	2·86
Oceania	0·23	0·36	0·09	0·15	0·28
USSR	1·50	1·41	0·53	0·22	0·29
World	11·01	6·92	4·57	1·55	4·17

output of butter and ghee and this reflects the importance of ghee in Indian diets. Though the output of dairy products from Oceania · is comparatively small nevertheless it is very important because this region, and New Zealand in particular, makes a major contribution to world trade in dairy products.

MEAT PRODUCTION

The principal sources of meat are cattle, pigs, sheep and poultry (Table 3.3), but there are other species, apart from various forms of game, which make significant contributions, in particular horses (0·6 m. tonnes), buffaloes (1·4 m. tonnes) and goats (1·6 m. tonnes) and these are included in the total shown in the last column of Table 3.3.

Table 3.3   **Meat production according to regions, 1979** (m. tonnes)

	Beef & Veal	Mutton & Lamb	Pigmeat	Poultry Meat	Total Meat
Africa	2·80	0·66	0·35	1·09	6·29
N. America	12·10	0·16	8·41	9·90	30·99
S. America	6·84	0·27	1·59	1·61	10·59
Asia	3·80	1·50	18·82	6·35	33·35
Europe	10·28	1·06	18·05	6·81	37·31
Oceania	2·53	1.00	0·27	0·28	4·12
USSR	7·00	0·86	5·30	2.00	15·50
World	45·36	5·50	52·79	28·04	138·14

Total meat supplies, at just over 138 m. tonnes, approximate to 32 kg/head on a world basis. It is a measure of the disparity between regions and countries that the USA and Canada, which together account for less than 6% of the world population, produce about one-fifth of all meat, virtually all for domestic consumption, and also import appreciable amounts of beef and lamb, mainly from Australasia. At the other end of the scale India with just over 15% of the world's population and, with virtually no offsetting imports, produces less than 1% of the world's output of meat.

Asia would be badly off for meat were it not for the big contribution made by pigs but this is mainly for the benefit of China, which, with an annual output of more than 15 m. tonnes, accounts for nearly 30% of world pigmeat production which increased by just over one-quarter between 1970 and 1979. This, however, does not match the 55% in poultry production over the same period when the production of sheep meat declined and there were only slight gains in beef and veal. This is a rather surprising situation, in the light of growing pressure on cereals for human consumption, and reflects in part the great forward strides that have been made recently in the efficiency of pig and poultry production.

EGG PRODUCTION

World production was estimated to be 26.6 m tonnes in 1979, an increase of approximately one-quarter over that at the beginning of the decade. Asia (8·3 m. tonnes), Europe (7·0 m. tonnes) and North America (5·2 m. tonnes) were the principal producers. The great popularity of eggs in human diets is reflected by the fact that F A O list 180 egg-producing countries with no less than 50 of these producing more than 30,000 tonnes annually. Eggs are unquestionably the most ubiquitous of all foods of animal origin.

WOOL PRODUCTION

The world's sheep population is in excess of 1,000 m. head but not all of these contribute to an annual production of 2·7 m. tonnes of greasy wool. A heavy fleece is an encumbrance in hot climates, especially where high humidities create an additional hazard in the shape of endoparasites. Consequently, most of the indigenous tropical sheep have short hairy coats while the principal wool breeds have been evolved in temperate countries: of these the Merino, which originated in Spain, has international recognition as the outstanding producer of high-quality wool. For many centuries Spain banned the export of Merinos in order to preserve a valuable monopoly but in the nineteenth century royal gifts were made to France, Saxony and Great Britain. Merinos were not suited to Britain's climate but fortunately British Merinos eventually found their way to South Africa, Australia and New Zealand. In the hands of Australian flockmasters they became a different sheep from their Spanish counterparts, producing much heavier fleeces of uniformly high-quality wool.

Details of the estimated output of greasy wool by regions in 1979 and changes that have occurred over the previous decade are given in Table 3.4.

The substantial decline in North American wool production is principally attributable to a fall in US sheep numbers. At the beginning of the decade there were over 20 m. head but by 1979 the total had fallen to 12 m., seemingly part of a continuing process because the national flock once exceeded 40 m. Australia is the leading producer with about one-eighth of the world's sheep producing more than one-quarter of the total wool, much of it from land that has no value for other forms of farming, on which it is common practice to run wether sheep because they can survive and

Table 3.4   **Output of greasy wool by regions, 1969–71 and 1979 ('000 tonnes)**

Region	1969–71 (Av.)	1979
Africa	210·8	195·4
N. America	94·2	55·3
S. America	330·0	333·5
Asia	277·0	320·8
Europe	255·0	272·4
Oceania	1,230·7	1,026·4
USSR	412·5	472·3
World	2,810·2	2,676·1

produce satisfactory fleeces under environmental conditions that are too harsh for breeding ewes.

Most Australian wools belong to the high Bradford fineness count categories required for fine fabrics but there is also stronger wool that is a by-product of fat lamb production, as it is in New Zealand where the Romney is the dominant breed, producing on average a fleece of about 5 kg which is appreciably heavier than the average clip of this breed in its native Kent. Not only have New Zealand flockmasters increased fleece weights but they have also reduced variability throughout the fleece. Wool continues to be that country's most important single export and the attention paid by farmers to wool attributes is a reflection of this importance.

WORLD FOOD SUPPLIES

Even in a world living under the threat of a nuclear holocaust, there is an even more serious danger presently affecting many communities – starvation. There are millions subsisting on inadequate diets and their numbers are being added to every year. In 1970 the estimated world population was 3,677 m. but by 1979 it had risen to 4,335 m., reflecting an annual increase of nearly 2%. By the end of this century it is estimated that the total population will be just under 6,000 m. if existing trends are maintained (Mauldin, 1977). Can the world's limited land resources cater for the needs of such a vast population so that malnutrition is no longer such a critical problem? The realistic answer is that it is probably no more than wishful thinking to suggest that this will ever be the case. Hunger has always been a spectre on this planet and people have suffered and will continue to suffer food shortages for a variety of reasons; for instance crop failures due to abnormal climatic conditions and natural catastrophes such as floods, hurricanes and earthquakes. The most that can reasonably be expected is an amelioration of the problem, partly through expansion of food production and partly through reduction in the rate of population increase by birth control, particularly in countries such as India and Bangladesh where nutritional inadequacy is endemic.

There is one comforting thought that past prophecies of impending doom, starting with Malthus nearly two centuries ago, have not been fully realised. In 1898 Sir William Crookes, when President of the British Association for the Advancement of Science, stated that there was almost no suitable land left to be reclaimed for food production. He predicted hunger on a global scale by 1930 because of population growth if it were not possible to improve supplies of nitrogenous fertilisers (Buringh, 1977). Fortunately, advances in chemical engineering have made this possible and the output of food has been increased enormously as a consequence. This is but one of the important technical advances that have been made during the twentieth century. There have been vast increases in the genetic potential of crop plants including possibly the most spectacular of all in this field, the development of hybrid maize. There have been notable improvements in crop protection and there is a better understanding of crop physiology and plant nutrition. Parallel developments have occurred in livestock production and it is probable that in the top ranks of dairy, pig and poultry production the efficiency of food utilisation has been more than doubled in the last fifty years.

Obviously in any biological situation further progress becomes more difficult with each step forward and there must come a time when a plateau is reached by those who practise the last word in farming technologies. However, the farming world in general is very far away from this situation, particularly in tropical environments where there is so much more to be done in optimising production systems. Buringh (1977) has reviewed the various estimates made of the world's food production potential and on the whole these are optimistic. Possibly one of the most important of the reports he considered is that prepared by the US President's Science Advisory Committee (US Report, 1967) which suggested that the total area of arable land could be more than doubled if the average level of technology used in the USA was applied generally.

This proviso is of great significance and underlines the importance not only of technical education and application in the task of producing more food, particularly in developing countries, but also the need for engineering and construction projects to provide the necessary infrastructure for development, for instance improved road, irrigation and drainage schemes, flood and erosion protection measures and the establishment of fertiliser and processing plants. It is in this sphere that the World Bank and its associated regional development banks can be of great service not only in providing necessary finance (usually on highly favourable terms) but also in ensuring that projects are viable and their progress is supervised adequately. The dynamic contribution made by Robert McNamara, while he was head of the World Bank, has contributed significantly in the campaign to help developing countries to help themselves which must be the continuing policy rather than one of providing food handouts by more fortunate nations, except in time of emergency.

4

EUROPEAN* AGRICULTURE

Except for the far north, which is not important agriculturally, Europe has a temperate climate and consequently its agriculture is more homogeneous than that of other continents with combinations of tropical and temperate conditions, as well as extremes of rainfall that Europe fortunately does not have. Factors apart from climate contribute to this uniformity, for instance the movement of people, mainly from east to west, that started in prehistoric times, and exchanges of knowledge between countries that has been facilitated by the literacy Europeans have enjoyed for many centuries. Wars and conquests with the resulting displacement and movement of people have also played a part; for instance British agriculture profited in the latter part of the seventeenth century following the Restoration when Royalist landowners, after exile in Europe, returned to their estates, with experience of new crops such as clover and turnips that eventually transformed the old three-course rotation of open-field farming.

Currently there is an increased exchange of plant and animal material as well as farming techniques between the countries of the world, especially in Europe. Many neighbouring countries have profited from the excellent work of French and Swedish cereal breeders while the USSR has been the primary source of the greatly improved varieties of sunflower now widely grown in Romania, Bulgaria and Spain. The USA has provided the knowledge as well as the hybrid stock for the intensive systems of egg and poultry-meat production now widely adopted in Europe. Canada has supplied improved strains of oil-seed rape while Canadian and American Holsteins are making an impact on Friesian herds that originated from stock initially imported from The Netherlands. Simmental and Charolais cattle now have a wide European distribution, Texel sheep have a growing importance outside their native Netherlands and improved British pigs have been exported to several other European countries. International organisations such

* The term Europe in this chapter covers all countries of this Continent and its associated islands with the exception of the Soviet Union, which is given separate treatment, and European Turkey. The latter area is included with the rest of Turkey, which is classified as an Asian country.

The Structure of Agriculture

as the European Association of Animal Production provide more than a forum for scientific discussions for they are also important in the dissemination of knowledge of production systems. These developments are receiving an impetus with the enlargement of the EEC and its unifying measures ranging from seed certification to grades and target prices for the principal agricultural commodities.

Nevertheless there are important regional differences not only in the range of crop and livestock production but also in the structure of agriculture, while between east and west climatic differences are confounded by political factors. Between north and south climate is the main determinant of agricultural practice but history and national characteristics also play important parts. In countries of the communist bloc collectivised farming has been imposed on what were once peasant communities either owning the land they worked or leasing it from owners of large estates. In many of these countries the changes have been dramatic. Bulgaria, for instance, before the advent of the present regime, had 6 m. ha of farm land divided into more than a million holdings, which required the labour of two-thirds of the total work-force. Now the proportion working in agriculture has been halved and there are only 1,200 large collectives.

Collectivisation may have released labour for other industries but it cannot be said that it has resulted in impressively high levels of output. Table 4.1 compares some yields of representative crops in communist countries with those of nine EEC countries.

Table 4.1 Some average crop yields for 1979 in communist and EEC countries (kg/ha)

Group	Wheat	Barley	Potatoes	Sugar Beet
Communist	3,360	2,840	16,300	32,000
EEC	4,900	4,200	29,000	43,500

Generally, average yields in the EEC are approximately 50% higher than those of the communist group and in no instance is the average yield of any single communist country greater than the EEC average while East Germany is the only one with anything like comparable results. The only available livestock performance data relate to milk yields and here again communist countries, with an average of 2,600 l per cow in 1979, compare unfavourably with the average EEC figure of 3,900 l.

These figures cannot be regarded as proof that collectivisation has failed for other factors such as quality of land and climate must be taken into account. Fertiliser usage in eastern Europe, including the Soviet Union is on a much lower level than it is in western

Europe and this unquestionably is an important contributory factor (Nove, 1977).

Though collectivisation has released labour for other industries, judged by the situation of western Europe, there is still a long way to go, for the seven communist countries have an average of 27% of their total work-force in agriculture with a range from 10% in East Germany to 48% in Romania. These figures compare with an 8% agricultural component in the EEC work-force with a range from 2% in the UK to 21.5% in Ireland, which was the least industrialised of the EEC countries until Greece's accession in 1981 with 38% of her work-force in agriculture.

Though one of the strongest arguments supporting collectives is the greater progress that can be made towards effective mechanisation with large compact units as opposed to the small and often fragmented holdings which they superseded, the data given in Table 4.2 do not show that farm mechanisation has proceeded as far in the communist bloc as it has in the EEC countries.

Table 4.2 Farm mechanisation in EEC and communist bloc countries, 1978

	EEC	Communist
Crops and permanent pasture (m.ha)	93·5	75·4
Total area in cereals (m.ha)	27·0	30·0
Agricultural workers	7·6	20·5
Area per worker (ha)	12·3	3·7
Tractors ('000)	4,960	1,399
Area per tractor (ha)	19	54
Harvesters and threshers ('000)	467	147
Areas of cereals per harvester or thresher (ha)	58	204

At first sight it might seem that communist countries with an average of one tractor to 54 ha are more efficient users of their tractor force than EEC countries with only 19 ha per tractor were it not for a substantially narrower labour ratio of 3·7 ha per worker, as opposed to 12·3 in EEC countries, that cannot be explained by more intensive farming. It is probable that EEC countries are over-lavish in their tractor provision for their farming does not match the standards set by the USA which probably has the highest effective level of farm mechanisation of the major agricultural countries. Here the ratio is one tractor to approximately 100 ha of crop and permanent pasture. Some of the disparity can be explained by the higher-powered tractors favoured by Americans who also enjoy the advantage of larger, more compact holdings. The US ratio of 107 ha of cereals per harvester, as opposed to 58 for the EEC, further reflects the better structure of American farming as compared with most western European countries which will long continue to suffer from constraints that have a foundation in their history.

FARM STRUCTURE

Communist regimes – with the notable exception of Poland, which retains a predominance of peasant farmers, often on very small holdings – resolved many problems due to small and excessively fragmented holdings and also, in some instances, to the tyranny exercised by estate owners who held their tenants in a system akin to thraldom. Whether they succeeded in getting the right solutions is debatable but certainly the methods by which they were achieved would not be acceptable to western democracies where the rights of individuals demand a less authoritarian approach to the rationalisation of farming structure. It is part and parcel of the Common Agricultural Policy of the EEC that those engaged in agriculture should have standards of living comparable with those of workers in other industries. The policy has been widely criticised because target prices have been set so high that some commodities, especially dairy produce, have been overproduced and subsequently disposed of at a discount outside the Community. Nevertheless through economic pressure and so-called 'golden handshakes' for the occupants of uneconomic holdings there has been a substantial migration from the land in EEC countries during the 1970s as figures in Table 4.3 show.

Table 4.3 Economically active farm populations in EEC countries, 1970 and 1979 ('000)

Country	1970	1979	Difference	Difference %
Belgium and Luxemburg	186	129	57	31
Denmark	258	180	78	30
France	2,876	2,089	787	27
West Germany	2,001	1,240	761	38
Eire	296	268	28	10
Italy	3,755	2,523	1,232	33
Netherlands	394	300	94	24
United Kingdom	728	542	186	26
Total	10,494	7,271	3,223	31

An average reduction of 3% per annum in the working population in EEC farms is a commendable rate of progress over a period that has also seen a substantial increase in the level of output. For instance, milk production rose by 17% and cereals by 26% over the decade. However, the accession of Greece, Spain and Portugal to the Community will bring in a large addition of small holdings.

The position in southern Spain, where latifundia predominates, is comparatively favourable but the north presents a very different picture; for instance, it is not exceptional in Galicia for a farmer to have a dozen or more separated plots of land totalling no more than 5–6 ha. Even if parcels were aggregrated, which is a course of action landowners are reluctant to take despite government attempts to encourage them, most holdings would be too small for economic milk production for which climatically the region is admirably suited. Northern Portugal is also affected by minifundia and Greece has fragmentation problems not unlike those of northern Spain. It is a particularly serious problem if irrigation schemes are envisaged for it is virtually impossible to devise rational systems of water distribution under such circumstances.

The UK is much better placed than its Continental partners in the EEC in respect of size and compactness of farms. This not only gives the UK greater opportunities for effecting economies of scale but also a considerable advantage in respect of the capital cost of land, as figures in Table 4.4 show.

Table 4.4   Relative prices of agricultural land in Europe (EUA per ha)* (Northfield 1979)

	1976	1977
England and Wales	1,654	1,975
Scotland	1,056	1,127
Germany	7,070	8,368
France	2,836	3,140
Netherlands	6,121	9,288
Belgium	8,832	9,969
Denmark	4,787	5,907

* *The value of units of account (EUA) varies slightly from time to time but in 1976 each unit of account was worth approximately £0·622 and in 1977 £0·654.*

The lower prices in Scotland, as compared with England and Wales, are attributable to the higher proportion of upland areas. The substantial rise in average prices, recorded in every country in 1977, reflects the general depreciation of currencies and the value of land as a hedge against inflation. This is something that peasant farmers in western Europe have appreciated for a long time and in recent years they have been joined in the competition for the limited amount of land coming on to the market by well-to-do urban dwellers seeking a home as well as an investment in a rural environment. This has been advanced as the principal reason for the 50% increase in land prices in The Netherlands experienced in 1977 (Northfield, 1979).

A high proportion of land coming on to the Continental market consists of small parcels whereas sales in Britain will usually relate to larger compact farms. Because the total value of a Continental land transaction is likely to be much less than it is for a land sale in Britain there will normally be many more interested in the purchase, including neighbouring farmers for whom an adjacent parcel of land consolidated into an established holding has a greater value than it has as a separate entity.

EFFECTS OF LATITUDE

Progressively, as one travels north in Europe, the feasible range of crops becomes more limited except where the aspect of the land or maritime effects create local conditions that are more favourable than those obtaining where latitude is the only consideration. The maritime effect is especially pronounced along Europe's western seaboard because of the Gulf Stream and the Canaries Current. Without it Brittany and Normandy would have winter conditions similar to those of Newfoundland. There are, as one would expect, greater temperature extremes over the course of the year as one moves away from the coast to experience the Continental impact on climate.

It is part of the long-term EEC plan that its more southerly members should be the main providers of frost-susceptible crops, rather than the northerly countries attempting to grow them with protective structures and artificial heating and lighting, and the logic of this has been underlined by the dramatic increases in the cost of fossil fuel. Nevertheless, long-established glasshouse industries in some northern countries, particularly in The Netherlands, are showing surprising resilience, probably because of very high levels of expertise. It seems that in the relatively affluent societies of northern Europe housewives are prepared to pay the substantial premiums that quality products from Dutch and Channel Island greenhouses command over imports from Italy, Spain and the Canary Islands.

There are, however, some crops that can be grown economically only in southern Europe; for instance, climatic conditions limit citrus production to Spain, Italy and Portugal and even in these countries citrus orchards require freedom from severe frosts so they are mainly located close to the sea. Though olives are more frost tolerant than citrus, nevertheless they are also a crop of mild temperature conditions and their production is dominated by Italy, Spain and Greece, which together are responsible for three-quarters of the world output of olive oil.

Deciduous stone fruits have much greater climatic tolerance than olives or citrus and they are able to withstand hard winter conditions provided there is freedom from frost at fruit set. Peaches, nectarines and apricots also require relatively warm summers for the development of full flavour, and Italy, Spain and southern France are the leading producers of these fruits. Plums, in a variety of forms, have a much wider range of production. Europe accounts for more than half the world production with Romania, Yugoslavia and West Germany as the major contributors to a total output of over 3 m. tonnes. Apples and pears are the most widely grown of the tree fruits and only Albania, Ireland, Finland and Norway grow less than 100,000 tonnes of pip fruits annually. France, with an average annual apple crop in the region of 3 m. tonnes, accounts for about one-quarter of Europe's production. Italy with just under 2 m. tonnes is second in importance for apples and with an annual pear crop in excess of 1 m. tonnes accounts for about one-third of Europe's output.

Grapes and wine have a very important place in Europe and vines occupy 6 m. ha, which is three-fifths of the world area in this crop. More than two-thirds of Europe's vineyards are in Spain, Italy and France in that descending order of importance, which is reversed in terms of wine production. Many Spanish vines are in low rainfall localities and often one vine will produce no more than 2–3 kg of grapes. In southern Spain, an area where the winters are mild and irrigation water is available, there is a growth of early-season table grape production, for with protection under plastic it is possible to harvest high-quality table grapes in late June–August to supply remunerative markets such as Germany and Sweden that would otherwise be dependent on expensively produced hothouse grapes at this time of the year.

Northern European countries are able to compete more favourably with southern in the production of soft fruits. This is due largely to their perishability and the need to market them quickly but some berry fruits, particularly raspberries, are better suited to cool temperate conditions than they are to mild winters followed by very hot dry summers. In part this is due to the lower incidence of vectors of virus diseases which reduce the productivity of vegetatively propagated crops. Nevertheless Italy produces over a quarter of Europe's strawberries with an increasing proportion of the crop being transported in refrigerated trucks to meet the early season demand in northern countries.

The movement of out-of-season produce from the south to the north, as well as that of other fruits and vegetables such as melons, peppers and egg fruit which require protection for their growth in northern climates, is likely to become more important as the barriers against international trade disappear and EEC nationals become reconciled to the objectives of CAP and abide by its rules. Highly significant in this respect is the growing involvement of Dutch interests in the horticulture of southern Europe especially that of Spain. With the growing cost and depletion of fossil energy the sensible policy must be one of exploiting solar energy and here countries of the Mediterranean basin have an enormous advantage. Logically, the agriculture of northern Europe should concentrate on such crops as cereals, roots and forages for which it is environmentally well suited and on various forms of animal production not only to utilise available food supplies but also to exploit the inherent stockmanship of northern people.

ARABLE CROPS

Cereals form the most important crop group and they account for 55% of Europe's 130 m. ha of arable land. Table 4.5 gives data

Table 4.5 Production of cereals in Europe, 1979

Cereal	Area (m. ha)	Yield (kg/ha)	Output (m. tonnes)
Wheat	24·66	3400	83·92
Rice	0·39	4970	1·93
Barley	20·81	3270	68·05
Maize	12·21	4610	56·29
Rye	5·31	2170	11·51
Oats	5·28	2630	13·88
Sorghum	0·20	3415	0·69
Total	68·86*	Av. 3495	236·27

* includes 34,000 t of millet

relating to their production. Wheat is the most important of the cereals and normally occupies about 25 m. ha extending from Greece to Finland. Barley occupies second place and is relatively more important in northern countries than in the south, where it is possible to grow reasonably good crops of maize which is third in order of importance. Maize is a crop well suited to countries with a continental climate and well over two-thirds of Europe's maize comes from Romania, the leading producer, Yugoslavia and Hungary. These three countries, on the other hand, account for only 5% of the barley. Here France, with an annual output of about 11 m. tonnes, leads the UK with 9·5 m. tonnes and West Germany with 8 m. tonnes.

Oats are a declining crop. Poland, with just over 1 m. ha producing 2·5 m. tonnes, has the largest oat area, but West Germany, as the result of much higher yields, is the largest producer with about 3 m. tonnes from 0·75 m. ha. Though barley is the Swedes' principal cereal, oats are also popular in that country, where plant breeders have been responsible for some of the best available varieties, and also in Finland, where it approaches barley in importance. Poland and East and West Germany dominate rye production and they account for about 80% of the European output. Rice is of minor importance and, as one might expect, is grown principally in countries bordering the Mediterranean. Italy is the largest producer but Spain has the highest-yielding crops with an average of over 6,000 kg/ha, which is not far behind South Korea and Japan, the world leaders. France produces about half of Europe's production of 700,000 tonnes of grain sorghum.

Except for maize, average cereal yields in northern European countries compare more than favourably with those of the south. The five countries that regularly average better than 500 kg/ha for wheat, namely The Netherlands, Belgium, Ireland, United Kingdom and Denmark, are all northerners and they also lead with barley and oat yields. Their superiority is not attributable to climate for it can be said fairly that with the exception of oats high yields are being obtained despite natural limitations and they reflect the very high husbandry standards achieved by northern farmers. Possibly this, also, is an effect of climate for as one moves north there is a greater environmental challenge which northern agriculturists have met with conspicuous success. This also includes one vitally important achievement, namely over many centuries of intensive use there has been little despoliation of land such as that experienced in many other parts of the world and it is to the credit of successive generations of farmers that fertility has not been exhausted with the passage of time. The Danes and the Scots, for instance, have created good arable land where once there was acid moorland; the Dutch have won rich land from the sea; while the English have transformed fens into some of the most productive land in Europe.

BREAK CROPS

Though cereals are the most important crops there is a need, even with such sophisticated aids as herbicides, fungicides and pesticides, to observe the rules of good husbandry, in particular the growing of crops with a function apart from their product value, namely providing a change from a long sequence of corn crops with its attendant risks such as loss of soil structure, increases in the level of pests and diseases and a high incidence of grass weeds. Broadly, break crops, as they are termed, can be placed in one of four main categories: first, temporary pastures of varying duration in a system of alternate husbandry; secondly, vegetable crops, especially legumes, grown on a field scale for drying, canning or freezing; thirdly, annual forages which include mangolds, fodder beets, turnips and the various kinds of rape and kale which may be

Table 4.6   European production of potatoes, sugar beet, oil-seed rape and sunflowers

| | Potatoes | | Sugar Beet | | Oil-seed Rape | | Sunflowers | |
	1969–71	1979	1969–71	1979	1969–71	1979	1969–71	1979
Area m. ha	7·2	5·8	3·0	3·7	1·0	1·2	1·4	2·1
Yield kg/ha	17,655	21,028	35,949	38,503	1,847	1,995	1,404	1,452
Total Production m. tonnes	126·6	121·8	107·2	141·0	1·9	2·4	1·9	3·0

either harvested for stock feed or utilised *in situ* by folded animals; and fourthly, a variety of cash crops such as potatoes, sugar beet, oil-seed rape, sunflowers, soyabeans, tobacco, cotton and linseed.

The most important in this last category in European agriculture are potatoes, sugar beet, oil-seed rape and sunflowers (see Table 4·6). With the exception of potatoes all have recently increased in importance but the area sown to potatoes has been offset largely by heavier crops which reflect the greater understanding growers now have of yield determinants as well as the availability of higher-yielding varieties. Not only has the area sown to sugar beet shown a substantial increase of the order of 20% over the decade but there has also been a steady increase in yields which has been especially marked in the countries which constituted the original membership of the E E C. This reflects a determined effort by the Community to be self-sufficient in sugar and this goal has been achieved, with the Community becoming an exporter of sugar in 1980. There is a very considerable range in yields with Belgium/Luxemburg consistently averaging better than 50 tonnes/ha and Romania trailing at 23 tonnes/ha.

The 1·2 m. ha of oil-seed rape makes Europe less dependent on tropical oils for the manufacture of margarine. The crop seems to have become less popular in several countries in recent years, notably in France where the area in 1979 was barely two-thirds of that ten years previously. On the other hand, there have been substantial increases in Sweden, Denmark and the U K, all countries where climate rules out alternative oil-seed crops such as soya or sunflower. The latter crop showed a 50% increase in both area and total yield. Romania, Yugoslavia, Hungary and Bulgaria (some distance behind) are the most important producers in the communist bloc while Spain is the leading western European grower with a spectacular increase from less than 200,000 ha at the beginning of the decade to 642,000 ha in 1979, but Spanish yields at about 800 kg/ha are disappointingly low as compared with communist countries with at least double this yield. This disparity is due primarily to the low rainfall in Andalusia, Spain's principal sunflower-growing region, where precipitation, mainly coming in the winter months, is usually of the order of 400-600 mm per annum. Southern France with a better rainfall pattern averages about 2000 kg/ha.

Europe, with 670,000 tonnes, produced less than 1% of the world's soya in 1979 but there is some evidence that this crop, so long dominated by the United States, is becoming more popular, especially in Bulgaria and Romania which are favourable climatically. Cotton is of little consequence in Europe with Greece accounting for rather more than half the total production of 500,000 tonnes of seed cotton. Tobacco is relatively more important with an annual output of 700,000 tonnes of leaf or about one-eighth of world production. The principal growers are Greece and Bulgaria, each with 125,000 tonnes, and Italy with 110,000 tonnes.

Vegetables are very important in European diets and the FAO estimate of 64 m. tonnes produced in 1979 is about one-sixth of estimated world production. There is a wide variety of vegetable crops with some rather characteristic national preferences. Spain for instance, produces nearly half of Europe's garlic and over one-fifth of the onions and green peppers. Italy is the principal grower of aubergines and globe artichokes (which scarcely figure in northern countries) and she accounts for about one-third of the tomatoes. Poland produces twice as many cabbages as any other country, while France and the UK account for about half Europe's output of green peas.

ANIMAL PRODUCTION

In most European countries animal products are important in the national diet. The UK is the only country with any large measure of dependence on animal products from outside Europe but the volume of these imports is being steadily reduced as production within the EEC expands. Soon the only animal products of any significance coming from outside Europe will be non-edible items like wool and hides, which are raw materials of industry.

Table 4.7 gives the estimated numbers of different kinds of

Table 4.7   Mammalian livestock in Europe ('000)

	1969-71	1979	Movement
Horses	7,714	5,620	− 2,094
Mules	1,143	672	− 471
Asses	1,775	1,391	− 384
Cattle	123,603	134,535	+ 10,932
Pigs	132,398	172,052	+ 39,754
Sheep	127,195	131,603	+ 4,408
Goats	12,378	11,579	− 799

mammalian livestock in Europe at the beginning and the end of the seventies.

There has been a substantial fall in draught animals and, with the exception of goats, an increase in livestock making direct contributions to food supplies. However, the decline in draught animals has been compensated by a 30% increase in Europe's tractor force over the same period. Countries of the communist bloc, in particular Poland (1·85 m.), Yugoslavia (0·7 m.) and Romania (0·6 m.), which together account for nearly three-fifths of Europe's horses, have the greatest dependence on draught animals. No figures are available for draught cattle but they continue to be important in parts of Spain and Italy and throughout the Balkans, where it is the common lot of the cow to be a milk producer as well as a source of energy and finally to end up as a carcass of rather tough beef. Spain, Greece and Italy between them account for three-quarters of the mules, which are now of minor importance. Donkeys have their greatest importance in southern Europe, particularly in Bulgaria, Spain, Greece and Portugal. Ireland is the only northern country where donkeys have any numerical importance.

Cattle have shown a steady increase of about 1% per annum over the decade. France, with nearly 24 m. head, accounts for nearly 17% of the European total with West Germany, the UK and Poland each contributing about 11%. Ireland, with 7 m. cattle and 3·2 m. people, has the greatest number per head of population, while Greece, at the other end of the scale, has only one cattle beast for ten people. Pigs are the most numerous of the larger farm animals and they also show the greatest relative increase over the period. West Germany, with 22·6 m. pigs, has nearly twice as many as neighbouring France, while Poland, with just over 21 m., almost matches the German total. Denmark and The Netherlands, both relatively small countries, each have over 9 m. pigs, which exceeds the normal UK population of about 8 m. Denmark has long been a major supplier of the UK bacon market and has, until recently, been a pace-setter in pig improvement. The Danish feat of converting its Landrace, which was once a short lard-type pig, into the long lean bacon pig it is today, with a greatly increased efficiency of food conversion, is a remarkable and successful example of purposeful livestock improvement. The combination of this breeding with suitable curing and efficient marketing of an acceptable product has created such goodwill for Danish bacon that it continues to sell at a premium over home-cured bacon, despite efforts to improve the latter product.

Greece, with 4·5 m. goats, has nearly 40% of Europe's total with Spain accounting for a further 20%. Other countries with significant numbers are France, Italy, Portugal, Albania, Bulgaria and Romania but goats have very little appeal in northern Europe. The

UK, for instance, has only 6,000 goats as compared with 2·3 m. in Spain.

There is a different situation with sheep for the UK, with 30 m. head, has practically the same number as the combined total for Romania and Spain which occupy second and third places respectively with 15·6 m. and 14·5 m. The only other countries with more than 10 m. sheep are France (11·5 m.) and Bulgaria (10·1 m.). Romania, in an effort to expand its sheep population, has been importing breeding stock from New Zealand with the aim of increasing mutton exports to oil-rich Middle East states.

United Kingdom sheep-men are principally interested in meat with wool as a by-product but for many Continental producers, notably France, Spain, Bulgaria, Greece and Italy, cheese made from sheep milk is important for small-scale farmers, with the French blue-veined Roquefort as the best known of these cheeses. Sheep used for such production are selected and managed primarily for their milking function and usually lambs are weaned at 4–5 weeks when they are able to thrive on solid food. In Spain and Italy these early-weaned lambs often go for immediate slaughter to cater for a luxury demand for baby lamb which is a delicious commodity but regrettably a sad waste of carcass potential.

On the Continent, unlike the UK, it is usual to herd sheep by day and fold them by night. Seldom are Continental sheep kept in fenced enclosures and this makes them labour-intensive requiring high prices for sheep products to ensure commercial viability. The great difficulty the EEC has experienced in securing sheep-meat arrangements acceptable both to France and the UK stems from this situation, for the UK with its differently structured sheep industry has such pronounced comparative advantage that it could very easily have undercut the luxury prices that French farmers must have to remain in business.

Transhumance, still practised in a number of European countries, is no longer on the scale it was last century when there were huge migrations of sheep from lowland winter bases to summer in the uplands. France, Spain and Italy still practise transhumance usually with Merino-type sheep which, with their herding instincts, are well adapted to the system. Romania is another country with a form of transhumance with flocks wintered on the Danubian plain moving in early summer to sub-alpine Carpathian grazings where, incidentally, wolves are still a problem, necessitating folding by night. There the primary function of dogs is not herding sheep but giving the flock protection from these marauders.

MEAT AND WOOL PRODUCTION

Table 4.8 summarises changes in meat production over the past

Table 4.8 **Meat and wool production in Europe** ('000 m. t)

	1969–71	1979
Beef and veal	8,799	10,277
Mutton and lamb	940	1,059
Goat-meat	83	87
Pig-meat	13,120	18,045
Horse-meat	211	180
Poultry-meat	4,175	6,812
Wool	225	272

decade. The most striking features are the increases of approximately 38% in pig-meat and over 60% in poultry meat. These are attributable to three principal factors: the development of genetically superior stock; better feeding and management with improved environmental controls that include containment of diseases; and the greater availability and relative cheapness of concentrated foods. Improvements in pig production have originated mainly within Europe and this is a sector where Europe has not needed American expertise.

The 16% increase for beef and veal is no more than can be expected because specialised beef production, based on single-suckled calves, is an extensive operation and Europe with its population pressures can only afford to use marginal land for such specialised beef production. The great bulk of Europe's beef must be a by-product of the dairy industry. This means more than a manufacturing quality of beef from animals that have outlived their usefulness for draught or milk production, for it also includes good-quality beef from animals, especially males surplus to herd-replacement needs. Continental farmers have long favoured dual-purpose animals and the Red Dane and Friesian, despite their dairy merit, have good fleshing qualities while other breeds, such as the Brown Swiss, Simmental and Normandy have even greater beefing propensities. There is a considerable likelihood that dual-purpose attributes will suffer with the increasing preoccupation of farmers with milk yields and the growing popularity of the American Holstein with its classic dairy conformation and promise of appreciably higher milk yields.

Increased beef production is not due to increased slaughterings but to heavier slaughter weights, especially in France and West Germany. At the beginning of the decade carcass weights averaged 190 kg but by 1979 they had risen to 219 kg reflecting a movement away from veal which, in contrast to UK tastes, has been a favoured meat on the Continent, where heavy carcasses were once synonymous with tough geriatric beef. There has been a growing opposition to the intensive production of white veal, especially with hormone implantation to promote faster growth rates, and at the same time, in western Europe at least, there has been a growing appreciation of beef from animals slaughtered at 15–24 months, which represents a more sensible exploitation of the growth potential of the calves available for meat production.

The UK, France and Spain, in descending order of importance, account for more than half of the sheep-meat. Here there is a wide variation in slaughter weights. In The Netherlands where the Texel, possibly the leanest of breeds used to sire a terminal generation, has a dominant influence, the average carcass weight is 25 kg as compared with 12 kg in Spain and only 9 kg in Italy. The low weights recorded both in Italy and Spain reflect not only the slaughter of early-weaned lambs from ewes used for milk production but also the European preferences, especially in Mediterranean countries, for very lean meat. Mutton fat is not appreciated by most Continentals and this must be realised by British lamb exporters if they are to create goodwill for their produce.

Horse-meat, which many British people regard with considerable repugnance, is declining in importance as the horse population falls. France and Italy are the principal producers and between them they account for about three-fifths of the total production. Horse-meat comes primarily from old draught animals but there are some horse herds in the Pyrenees that are maintained for meat production because of their reputedly superior adaptation, compared with cattle, to the mountain environment.

Wool production appears to be virtually stationary moving in response only to small changes in sheep numbers. By antipodean standards, where great emphasis is placed on both fleece and fibre traits, the European approach to wool is, to say the least, very casual. This is understandable when one remembers that sheep-meat enjoys protection from international competition but wool has to face unrestricted competition from such countries as Australia and New Zealand.

DAIRY PRODUCE

Dairying is a major growth sector in European agriculture and Table 4.9 gives details of quantitative changes in production during the seventies.

Over the period there has been a 4% increase in the number of dairy cows but total milk production went up by 17% in response to the increase in average yields which rose from 2,976 to 3,335 litres. There is still scope for even greater increases for there are several countries, principally in northern Europe, whose average yields exceed 4,500 litres. The most substantial increases are being made in the nine older members of the EEC. Thankfully, Greece

Table 4.9   Dairy production in Europe ('000 t)

	1969–71	1979	Movement
Fresh Cows' Milk	147,672	174,234	26,562 (18)
Cheese (all kinds)	3,762	5,348	1,586 (42)
Butter	2,388	2,984	596 (25)
Condensed Milk	1,809	1,855	46 (3)
Dried Whole Milk	426	649	223 (52)
Dried Skim & Butter Milk	1,728	2,862	1,134 (66)

Note: Figures in brackets are percentage increases

will contribute little to the Community's headaches in disposing of surplus dairy produce for, with an average milk yield of less than 1,400 litres from 0.5 m. cows, she competes with Albania for the wooden spoon of milk production. Sweden leads, with an average in excess of 5,000 litres, with neighbouring Norway and The Netherlands in close attendance and a rapidly improving UK not far behind. Unfortunately liquid milk consumption does not match the overall increase in milk production, so inevitably there have been substantial increases in butter and cheese production and also in dried milk, which has been particularly pronounced in the EEC, which has been forced to dispense of intervention stocks at a discount and in the process to upset the market prospects of countries that do not have any system of market subsidies. Europe now produces over 40% of the world output of butter and nearly half of the cheese with France and West Germany, in that order, making the largest contributions.

EGG PRODUCTION

Egg production has been increasing at an average rate of just under 2% per annum and the leading producers are now West Germany, the UK and France each with an output in excess of 800,000 tonnes. The most impressive increases are taking place in countries outside the EEC. During the 1970s Romania almost doubled her egg output while there have been increases of about 50% in Yugoslavia, Hungary and Bulgaria. Thanks to a rapid adoption of advanced methods of production with environmental controls, egg production no longer has the seasonal variations that were once so characteristic. Moreover the new technology has encouraged national self-sufficiency in both egg and poultry production since climatic factors no longer have their former relevance except in the range of locally produced feeds.

OUTLOOK FOR EUROPEAN AGRICULTURE

Europe, with lower population growth pressures on its land resources than that in Africa, Asia and the two Americas, is also in the fortunate position of being one of the best-fed regions in the world and this, except for produce of tropical origin, is based principally on domestic production. The U K, with its long tradition of international trading, is the only European country with any great dependence on temperate food imports from Third World countries but this is changing with full integration into the E E C and a determination on the part of France to make the highest possible contribution to self-sufficiency within the Community.

The main obstacles to this are the need to maintain and enlarge a reciprocal trading relationship with primary producing countries that are customers for Community goods and services and the capacity of the Community's farmers to supply food at competitive prices. Gradually the strength of the farming lobbies within the E E C is diminishing, with the considerable decline of those engaged in agriculture and the growing opposition to dear food which critics attribute to over-protection from international competition. With the growing belief that too high a proportion of the Community budget has been used to support its agricultural policy, at the expense of other sectors, it is inevitable that future pricing arrangements, particularly for items in apparent surplus, will not give full recoupment for rising costs. This means that there will be a much greater emphasis on efficiency; those who cannot reach required standards will fall by the wayside, except where it is believed that productivity is not the only criterion and that there is a need, on social grounds, to maintain a prosperous rural environment to be enjoyed by the nation at large. This viewpoint is likely to apply particularly in highly industrialised countries with small farms in marginal upland areas that have an important amenity value. The prospect, however, is that such sectors will have to be funded by individual countries and not by the Community chest.

The elimination of inefficient farming enterprises is unlikely to reduce the volume of production which, with technological advances, has been rising steadily in most European countries. It is more likely that the opposite situation will obtain for an obvious response to lower margins will be strenuous efforts to increase production and give a better spread of overheads. This is a characteristic of family farming in times of depression except where production is highly dependent on purchased inputs, as it is in so-called factory farming. It is unlikely that the dominant family-farm character of western European agriculture will change in the forseeable future. If anything it will be intensified.

Of all Community countries France, unquestionably, has the greatest potential for expanded production and also, on her recent record, the will to realise this potential. Industrially she is at a disadvantage because of her comparative lack of fossil fuel but no Community country has a larger endowment of good land with such a favourable climate and it is France's declared policy to exploit this great natural resource. Particularly, there is considerable scope for expanding dairying where average annual yields are of the order of 3,000 kg/cow, which is about two-thirds of the average achieved in Denmark, The Netherlands and the UK. France has a very considerable climatic advantage over her northern neighbours in that she is able to grow heavy crops of forage maize which, when conserved as silage, is such a valuable basis for winter-feeding, not only of milking-cows but also of fattening cattle. In addition, particularly in Normandy and Britanny and a wide belt of good land bordering the Bay of Biscay, there are excellent conditions for pastoral farming; if the competence of Dutch farmers in the management and utilisation of pasture could be matched by their French counterparts, France would dominate dairying in western Europe. Apart from replacing dairy imports from countries outside the Community, she could be a valuable source of milk protein for less fortunate countries in Asia and Africa that are suffering from the chronic inadequacy of their diets, especially those of children.

It must be reiterated, however, that the contribution European countries can give to developing countries must be much more than handouts of free or cheap food given in times of emergency. There is a need also to impart the wisdom and knowledge that has been accumulated over centuries and also the dedication, characteristic of European farmers, to the handling of land and livestock, so that countries where malnutrition and near-starvation are the common lot are better able to utilise their own resources. Europe is in a strong position to provide not only this help but also the capital essential for agricultural growth and development. Any problems that European agriculture may have, on either side of the Iron Curtain, are trivial compared with those of the poorer countries of Asia and Africa.

5

BRITISH AGRICULTURE

Stone Age people, nearly 6,000 years ago, were Britain's first farmers, for unlike their predecessors who depended entirely on hunting, fishing and the gathering of wild fruits, they also had domesticated livestock and cereals, which they planted in easily cultivated clearings in a countryside dominantly of either forest or fen. In their turn they were supplanted by invaders, in particular the Celts, who had an organised agriculture with enclosed fields of characteristic shape that sometimes can still be recognised. A succession of invaders followed with each making some contribution to the character of British agriculture. The Romans, for instance, introduced, along with other crops, vines to the south of England and their example has given recent encouragement to latter-day viticulturists.

The Romans also introduced cattle, primarily for draught purposes, because the Celtic cattle were not sufficiently robust for the heavy military transport required of them. Anglo-Saxons, in the wake of the departing Romans, also brought cattle, and Britain's all-red breeds, such as the Sussex and North Devon, are thought to have an Anglo-Saxon foundation. The polled factor, characteristic of the Galloway and Aberdeen Angus, came with Scandinavian cattle brought in by the Vikings. The livestock of Britain, like the human population, has an assorted gene pool derived from many sources.

The greatest impact made by the Anglo-Saxons on agriculture was unquestionably the introduction of the manorial system of land tenure, which was generally characteristic of Teutonic societies. In Anglo-Saxon times land was the possession of the lord of the manor but its use was shared with other members of the community. In its original form the arable land was divided into three large open fields which were worked on a simple three-course rotation of wheat, spring corn (usually barley) and fallow. During their cropping phase these open fields were further subdivided into furlong strips, about a chain wide, and these were cultivated either by the tenants or the lord of the manor as part of his demesne. Each tenant would have a number of separate strips and these were reallocated at random each year in an endeavour to secure a fair distribution of land. Initially, tenants worked on the demesne land instead of paying money rents and also had other obligations such as military service.

Each manor had its quota of common grazings and woodland, with the latter being a source of fuel and timber and also, in season, providing acorns and beech mast for herded pigs. As in the modern administration of commons the tenants had specified entitlements in respect of stock numbers. The usual practice was one of folding stock at night on the volunteer growth of the fallows, partly for security and partly to transfer fertility from the commons to the arable land which, otherwise, would have been impoverished by continuous cropping.

Under the general conditions of manorial farming crop yields and animal performance must have been miserably poor, judged by present-day criteria. Once grazing was exhausted with the onset of winter, the only keep for ruminants would be straw and possibly some hay saved from protected parts of the commons. In order to reduce the wintering burden all surplus animals were slaughtered in the autumn and salted down. With everybody's stock running together not only were there risks of spreading disease but also there were few opportunities for selective matings and, indeed, because a harsh environment is such a great leveller for livestock, as well as for man, it would have been very difficult in those days, as it is with primitive livestock ownership today, to distinguish genetically superior stock except perhaps in one respect, their power to survive on hard commons.

Though the manorial system remained for several centuries following the Norman Conquest it changed gradually in response to social and economic pressures (discussed by Trevelyan (1942) in *English Social History* and by Ernle (1972) in *English Farming, Past and Present*). As population pressures increased and the emphasis moved from subsistence to cash crop farming it became necessary to introduce new crops and new farming techniques but this was not easily achieved with common field farming where everyone subscribed to the same conservative programme. With a shift of people away from the land to other occupations, a movement that was greatly accelerated during the eighteenth century in what came to be known as the Industrial Revolution, there was increasing pressure to consolidate and enclose the open field strips and create the field system now so characteristic of Britain's farming scene.

Enclosure started as a gradual movement during the Middle Ages, particularly with the demesne land reserved for the lord of the manor, but with the expansion of wool production, which was to become the country's principal source of export income as well as a basis of local prosperity, enclosure was extended to grazing land, especially in Tudor times when there were complaints that

London merchants were active in securing title to farming land. Sometimes enclosure was achieved by local agreement but more often towards the end of the eighteenth century when the movement was at its zenith it was imposed and local resistance was overcome with private acts by parliaments that were more concerned with the interests of the rich than of those whose privileges were being eroded. The hardships at the lowest end of the farming ladder, though considerable, were infinitely less severe than those suffered by the crofters of Highland Scotland who were forcibly ejected from their smallholdings to create large sheep farms. Perhaps the best one can say about these sorry episodes is that countries of the New World profited greatly from this exodus from what at its best must have been a harsh style of life.

Though enclosures greatly strengthened the privileges of the landowning class and were at the time the cause of much social injustice, undoubtedly they brought great benefits to the nation's agriculture. In particular they gave Britain a farming structure that is the envy of Continental neighbours with a burden of small fragmented farms. In the normal course of estate management, in a process that continues, small uneconomic holdings have been consolidated to safeguard viability as farmers became increasingly dependent on capital-demanding mechanical aids. Generally, landlords have had the interests of their tenants at heart and they have taken a great pride in their land. Much of the great beauty of the British countryside is attributable to enlightened landlords and though much of the tree planting was designed to benefit the shooting, which continues to be an important consideration of land ownership, the overall effect is that the countryside gives pleasure to millions who have no direct links with the land. Their representatives in this highly urbanised society are now the watchful guardians of this heritage of beauty which, increasingly, is being accepted by landowners and farmers alike as an amenity to be shared with the nation at large.

THE GREAT IMPROVERS

The most important immediate consequence of the enclosure movement was the freedom that it gave tenants to innovate and, for landowners, opportunities for effecting improvements of which perhaps the most important was drainage. Francis, Earl of Bedford and his son William, who was the first Duke, made the drainage of the southern fenland one of their principal tasks during a combined life-span that extended over practically the whole of the seventeenth century. They had the advice of a Dutch engineer, Vermuyden, to complete a huge project that would have been demanding even with modern technology. By contemporary standards it was a truly colossal undertaking and it needed enormous courage to face and overcome the difficulties involved. Its eventual success stimulated further drainage projects during the following century when the northern fenland in Lincolnshire was reclaimed to create what is now some of the best arable land in Britain.

The eighteenth century was very fruitful in respect of advances in farming. Early on, Jethro Tull introduced drilling as opposed to broadcasting and this made possible horse-hoeing, which was another of his innovations, to reduce weed competition. He also advocated fine tilths in the mistaken belief that plant roots were able to ingest soil particles and though his theory was wrong the extra cultivations were beneficial. Viscount Townshend of Raynham in Norfolk, who became better known as Turnip Townshend, followed with his Norfolk four-course rotation where wheat was succeeded by roots (usually turnips) which, in turn, were succeeded by barley that was undersown with red clover which was later combined with ryegrass. This represented an enormous step forward for not only did turnips provide a break between the two white straw crops but also with good husbandry they constituted a cleaning crop at a time when labour was plentiful and there were no chemicals for controlling weeds.

In addition to providing a break the clover had the additional virtue of fixing atmospheric nitrogen to enrich the soil. The clover hay along with turnips provided something that had been lacking in the old three-course system, namely good fodder for wintering livestock. No longer was it necessary to slaughter surplus stock in the autumn. Instead it became an integral part of the farming policy in the arable districts of Britain to fatten cattle in fold yards not only for their direct profit but also to tread straw into farmyard manure which was usually applied to the land for the root break. Later the value of this muck was enhanced by the lavish feeding of linseed and other oil-seed residues. Farmers in the eighteenth and nineteenth centuries, before artificial fertilisers were freely available, appreciated fully the truth of the old adage 'A full fold maketh a full granary.'

The torch that Townshend lit was carried high by another Norfolk landowner, Coke of Holkham, later created Earl of Leicester, who not only encouraged the use of the four-course rotation in his part of the country but also introduced marling of light land to give it body. The joint efforts and example of Townshend and Coke transformed the once backward county of Norfolk and put it, along with East Lothian in Scotland, in the front rank of British agriculture. Coke's 'sheep-shearings' at Holkham became well known not only throughout Britain but also in Europe and were attended by hundreds of visitors who came to see and discuss his methods of farming and especially the use of folded sheep on crops such as turnips, rape, and vetches as agents in building up fertility on light land (Edgar, 1979). This latter practice was widely adopted on the gravels and light soils of southern England in the classical sheep/barley association of downland farming that continued to be profitable until cheap grain from North America and refrigerated meat from the southern hemisphere destroyed its viability.

The great Norfolk improvers had their disciples in practically every county. Successive Dukes of Bedford at Woburn continued the family tradition of estate improvement and Lord Egremont of Petworth and Earl Spencer of Althorp also had honoured places in the succession of great improvers. Most large estates had their home farms and enlightened landlords used these to demonstrate new techniques to their tenants. This was the time when many agricultural societies, of which the most notable is the Royal Agricultural Society of England, were started and these, in the first instance without the shows that later became their main feature, were sponsored and led by progressive landowners.

The dominant figure in livestock improvement was a Leicestershire farmer, Robert Bakewell of Dishley (1725–95), who indisputably is the father-figure of pedigree livestock breeding. He worked principally with Shire horses, Longhorn cattle and Leicester sheep and his breeding and selection methods anticipated by a century modern concepts of livestock improvement. His choice of the Longhorn was a little unfortunate because with its horn spread it was not suited to close confinement in yards which had become the standard method of wintering cattle in the arable districts. Sheep were his greatest success and he transformed the late-maturing, slab-sided local sheep into a very useful mutton breed, well suited to contemporary requirements but much too fat for present-day tastes. The Dishley Leicesters achieved international fame and were used for the amelioration of practically all the local races of long-wool sheep including the Lincoln, Romney and Border Leicester.

The Border Leicester was primarily the achievement of a progressive Border farmer, George Culley, and quite fortuitously the University of Newcastle-upon-Tyne became the fortunate possessor of thirty-two letters and other papers written by Bakewell to Culley which reveal the forward thinking of this quite remarkable man. Bakewell practised very close breeding, reputedly because he could not find unrelated mates of sufficient merit to use on his stock. He also put greater emphasis on progeny testing than on the phenotypic excellence of individuals and this is a lesson that British livestock breeders had to re-learn almost a century and a half later.

Bakewell's farm became a Mecca for animal breeders, not only from Britain but also from the Continent, who came to study his methods. Prominent among them were the Collings brothers who worked with their local Teeswater cattle to establish the Shorthorn, which was once the most important breed not only in Britain but also in several New World countries, especially Argentina. Other breeders, also working with local stock, followed in the wake of Bakewell and the Collings, notably Watson and McCombie in Scotland, who were pioneers of the Aberdeen Angus, and several breeders, notably members of the Tomkins family, established the Hereford as an ideal animal for pastoral beef production. There were others who worked with horses, pigs and sheep, and British breeds gained such a reputation that this country came to be regarded abroad as the livestock centre of the world. Breeding became a special interest of wealthy landowners and with the development of agricultural shows the exhibition of livestock became one of their most important attractions.

BRITISH AGRICULTURE AND THE INDUSTRIAL REVOLUTION

During the eighteenth century the population of Britain nearly doubled but such were the improvements in farming and the gains from land reclamation that the country was virtually self-sufficient for staple items such as cereals and meat. Farming continued to prosper in the first half of the nineteenth century, at least as far as landowners and larger farmers were concerned, but not for the labourers and smallholders. At first the landowning class, with its representation in Parliament, was able to safeguard its interests against the growing pressures from an increasingly industrialised society and commercial interests more concerned with international trade and the exchange of goods and services for cheap food and raw materials from abroad than with the interests of agriculturists who, for a long time, had enjoyed the benefits of protective corn laws and export bounties. Finally, in 1846, the Free Traders won the day with the repeal of the Corn Laws which had been one of the great political issues.

This did not immediately result in agricultural depression, for prices for the principal commodities held up remarkably well until the 1870s and the collapse of the boom conditions created by the Franco-Prussian War and the American Civil War. Peace in America brought a resurgence of settlement and agricultural development in the Middle West, which was brought closer to Europe with railways and shipping based on steam. A flood of cheap grain knocked the bottom out of cereal prices in Britain and much of the marginal corn-growing land such as the heavy clays of Essex and the Midlands and thin downland soils 'tumbled down' to grass. There was a swing from corn to livestock to capitalise on cheap grain but even here there was no great prosperity for the average farmer. The Danes and Dutch, on the whole, seemed to be superior to their British counterparts in converting imported grains and protein supplements into pig-meat, eggs and dairy produce, while another technological advance in the 1880s, refrigerated ocean transport, opened up fresh competition for livestock producers in the shape of milk products and meat, first of all from North America and later from Argentina and Australasia where there were large British investments. Accounts of British farming over the period from 1880 until the outbreak of the First World War (for instance Ernle's *English Farming, Past and Present* and Hall's *Pilgrimage of British Farming*) give the impression that farming had become a depressed occupation and that its interests were subservient to those of urban dwellers in a Britain that had hitched her star very firmly to a policy of exporting goods and services paid for by cheap food and raw materials from the New World.

FARMING BETWEEN THE WARS (1919-39)

Britain's neglect of her agriculture and the destruction of shipping by the German submarine offensive during the latter half of the First World War nearly brought disaster and as a result there was some belated encouragement for agriculture in the shape of better prices, but the minor boom did not last. In 1921 Britain returned to the pre-war situation where agriculture had very largely to fend for itself. In the immediate post-war period there was an increase in dairying and also some expansion of pig and poultry production and it seemed that British farmers were learning to make better use of the imported concentrates that were so freely available. But there was trouble in store not only for farming but for industry as a whole with the onset in 1929 of what was possibly the worst recession that the western world has ever experienced. In the early 1930s there were $3\frac{1}{2}$ m. unemployed and prices of produce slumped disastrously. The position of British farming was exacerbated by protection policies on the Continent and Britain's initial adherence to free trade, but eventually the government was forced to take action with some reversal of its traditional non-protectionist policy. Tariff protection was given for a range of horticultural products, quotas were fixed for imported bacon, hams and other meat and there were subsidies or price insurance schemes for wheat, cattle, sheep and manufacturing milk (Winifrith, 1962).

Important in this context were the Agricultural Marketing Acts of 1931 and 1933 which led to the setting up of marketing boards for hops, milk, potatoes, pigs and bacon. The main task of the Potato Marketing Board was the regulation of supplies by the simple device of varying riddle size and later by control of the area of main crop plantings. It was successful inasmuch as it brought some sort of order to the marketing of a very difficult commodity.

The marketing boards for bacon and pigs were less successful, partly because of difficulties arising from two distinct outlets for pig-meat, one for pork and the other for bacon. Contracts for the supply of pigs to bacon factories were introduced to ensure a regular supply for processing but producers were inclined to break these contracts if pork was more attractive.

Milk was the great success story among the marketing boards, of which there were five with one covering England and Wales, three in Scotland and one in Northern Ireland. Here again there are two market outlets: one for liquid milk which is more remunerative than the alternative of manufacturing milk. Between the wars there was a rapid growth of large milk-retailing concerns, especially in London, and naturally they aimed to get milk as cheaply as possible while farmers in their turn were anxious to supply them rather than processors who, faced with severe international competion, could offer only a fraction of the prices offered for liquid milk. The large retailers, taking advantage of improved transport, began to bypass traditional suppliers in close proximity to their market in favour of cheaper milk from manufacturing areas such as Somerset. The situation came to a head with the full impact of the slump at the beginning of the 1930s. One year an Essex farmer could be selling milk for fifteen pence a gallon (old money) to a London distributor but in the following year he had been supplanted by a Wiltshire farmer and sixpence a gallon (old money) was the price offered by the nearest cheese factory.

Theoretically Milk Boards became the owners of all milk at one point during its movement from farms to market. This allowed them to introduce price pooling and in each region producers received the net average price of all milk realisations. In other words milk going for manufacture had the same realisation value for farmers as that going for liquid consumption. Producer-retailers were required to pay a levy to the Boards so that their returns as milk producers, as opposed to distributors, approximated to the wholesale price of milk.

With the stability engendered by the Boards' activities milk production became the sheet anchor of British farming and at the outbreak of war in 1939 it was the industry's most important enterprise and of direct concern to more than half the farmers. Though production standards in 1939 were unimpressive, judged by present criteria, nevertheless the industry was poised to make a vital contribution not only to the prosperity of farming but, more important, to essential food supplies over the course of the war.

The marketing boards were not the only shafts to lighten the gloom of the 1930s. Contributions from agricultural scientists were opening up new vistas for farming. John Hammond and his associates at Cambridge were bringing artificial insemination to the point where this exciting development had a practical application. Sir George Stapledon and his team at Aberystwyth were producing improved cultivars of the more important grasses and clovers and advocating what he called 'ley farming' as a substitute for moribund 'dog and stick farming' on indifferent permanent pastures. At Jeallot's Hill in Berkshire Stephen Watson was effecting improvements in grass conservation while the veterinary profession was devising protection against such diseases as contagious abortion and pulpy kidney. Plant and animal physiologists were discovering the importance of minor elements such as cobalt, manganese and copper while agricultural engineers were starting to realise the full potential of tractors instead of regarding them simply as a substitute for horses.

There were pioneers among farmers, such as Arthur Hosier, who developed the mobile milking bail for dairying which became popular on downland pastures. Rex Paterson and his wife started dairying with one of these bails and before the 1930s were out he had, with great success, developed several of these units on downland farms that landlords were very happy to let for almost a nominal figure because there was so little demand for them. Other farmers, realising the potential of the downlands for mechanised corn production, had introduced the combine drill and the combine harvester to demonstrate something of the shape of things to come.

Then the government made its first notable acknowledgement of the importance of domestic agriculture with the growing conviction from 1936 onwards of the inevitability of war with Hitler's Germany. In 1937, it introduced a soil-fertility scheme involving subsidies for the application of lime and basic slag and behind-the-scenes contingency plans were made so that agriculture would immediately be in a position to make significant contributions to the nation's food supplies in the event of war. In 1937, two years before war actually started, there was an unmistakable feeling that British farming was once more on the move.

FARMING UNDER PRESSURE (1939-47)

The achievements of the industry during the war and immediately after were remarkable. These have been described in some detail by Murray (1955) and so their importance in the present context rests on the influence they have had on the shape and character of present-day agriculture. The first concern of the government was to safeguard energy intake and at the same time ensure that diets had sufficient protein of the right quality. This had to be done in spite of a sharp reduction in supplies of imported feeding stuffs on which much of the animal production had depended between the wars. Stapledon in one of his more cynical moods suggested then that British grassland was little more than exercise ground for stock fed on imported concentrates. This perhaps was over-harsh but unquestionably the quality of grassland management was deplorable except on traditional fattening pastures such as those in Leicestershire and Romney Marsh.

Government policy for war-time agriculture was a combination of carrot and stick, with the carrot taking the form of attractive prices for priority commodities such as milk, cereals, potatoes and sugar beet and the stick a disciplinary package of compulsory cropping orders, supervision and even dispossession. As a result 2.9 m. ha of permanent pasture, mostly of very poor quality, came under the plough and the energy output of British farming was virtually doubled in the space of four years. Both pigs and poultry suffered because they were in direct competition with humans for high-energy foods but the dairy industry made big advances despite the shortage of concentrates. Rather belatedly British dairy farmers were being forced into a realisation that pastures had a production function in feeding dairy cows.

Apart from invaluable contributions made to the nation's food perhaps the most important outcome of war-time agriculture was the changed attitude of farmers and their awareness of new technical possibilities. Because government policy had made available tractors and implements and had ensured supplies of fertilisers much of the dog and stick farming had given way to a system of alternate husbandry, widely advocated by Stapledon, where grass in the shape of short-term leys became a crop as well as a means of restoring heart to land after an arable phase. The move from horses to tractors, that had started between the wars, was greatly accelerated and the combine harvester was taking over from reaper and binder. No longer was the muck cart, with the occasional application of slag and lime, considered to be sufficient, for compound fertilisers were coming into their own and nowhere did they make a greater impact than on the southern downlands that at one time had depended on folded sheep as a source of fertility. On this and other light land, long regarded as being limited to barley, farmers were discovering that with a combination of ley farming and the use of artificial fertilisers they could grow crops of wheat that compared favourably with those on stronger traditional wheat land.

Though farming had acquired a spirit of adventure there was still an uncomfortable feeling, especially among older farmers, that history could be repeated and that with peace and the return to normal trade the government might let them down, as it did in 1921. This time it was to be different, with a realisation in high places that a strong agriculture was an important ingredient of national prosperity. The tangible manifestation of this change of heart was the Agriculture Act of 1947 which, among other things, established price guarantees for the more important commodities. Annual prices for review commodities and the levels of various production subsidies were fixed at the beginning of each year after joint discussions between the Ministry of Agriculture and farming representatives, which came to be known as the February Price Review.

This arrangement was similar to the procedure for price-fixing adopted during the war. In this and in the early years of guaranteed prices the farmers' delegations were led by a powerful and determined advocate in James Turner, later Lord Netherthorpe, who was one of the architects of the 1947 Act which, despite weaknesses, subsequently repaired in the Agriculture Act of 1957 with long-term assurances on prices, gave British farmers the confidence to invest and expand, as they did to such good effect over the next thirty years.

THE GOLDEN YEARS (1948-72)

This heading is no extravagance for there has been no period in the history of British agriculture with such a record of increased productivity achieved by the purposeful application of advances in knowledge. In the early years of the February Price Review it was claimed by James Turner, no doubt with pleasant memories of the cost-plus situation of the war years, that there was under-recoupment for rising costs and there were complaints that farmers were not getting a commensurate reward for their greater efficiency. The Ministry view essentially was that if the carrot was held too close to the figurative farming nose there would not be the same incentive to innovate and its objective was competitive farming. In any case it could argue that many advances originated not on farms but in state-financed establishments. Notable examples of this are provided by Proctor barley, bred at Cambridge by Dr G. D. H. Bell, which lifted the ceiling of yield for malting barley by at least 0.5 t/ha as compared with the varieties it replaced, and MCPB, the selective weed-killer for legumes, that was synthesised by Professor Louis Wain at Wye College. Essentially, the exciting strides made by British agriculture were the result of combined efforts by farmers, agricultural scientists and advisory services including those of commercial interests serving agriculture.

No branch of farming illustrates these technological changes to better effect than the dairy industry. In 1948 a new loose-housing unit for 60 cows, tended by two men, with relief help on days off, and milked in a two-level tandem parlour, warranted a full-page illustrated article in the farming press with conjecture in subsequent correspondence that this could be asking too much of man and beast. If a herd averaged 4,500 l with a stocking rate of a cow to 0·6 ha and a concentrate usage of 0·3 kg/l it also provided feature news. By the early 1970s sights had been raised dramatically with two men tending 120 cows running on 50 ha and averaging at least 5,500 l with appreciably less concentrate per litre.

Greater productivity was made possible by many important developments, for instance improved mechanical milking through a herring-bone parlour with a better understanding of the let-down process and the elimination of hand stripping, circulatory cleaning of milking equipment, cow cubicles and mechanical handling of slurry, self- or easy-feeding of silage of infinitely better quality than it was in the first tentative years of its adoption, and more effective control of diseases, especially mastitis. There was the change, expedited by artificial insemination, from the dual-purpose Short-horn, which once dominated the dairying scene, to the more specialised Friesian, which, fortunately, is also a useful source of beef.

Herd improvement was not confined to breed change because within the breeds, especially the Friesian, there were strenuous efforts to secure genetic gains and here the Milk Marketing Board for England and Wales played a leading role. During the war it was given responsibility for milk recording, until then a moribund service of the Ministry of Agriculture. Immediately the Board re-vitalised recording and it also took a major responsibility for another important service, artificial insemination. At the beginning when there was a predominance of small producers and an average milk-selling herd was only 15 cows the main attraction of artificial insemination was that it was cheaper than natural service except in large herds but it was soon realised that it was also a powerful tool for dairy herd improvement. With control of milk recording and a major involvement in artificial insemination the MMB was in a position to develop what is possibly the most effective system ever devised of dairy sire appraisal and exploitation.

Genetic improvement was not limited to dairy stock. An important feature of one of the early Price Reviews was finance for setting up pig progeny testing stations, organised on the earlier Danish model, but subsequently reorganised for boar performance testing stations, when it was realised that this was cheaper and much more effective than progeny testing for improving growth, food economy and muscle development. This, combined with pig-breeding companies supplying hybrids – also the products of performance testing – gave commercial pig producers stock that was at least of equal merit to that available to their Danish and Dutch counterparts.

There was less progress with beef cattle but a start was made in recording and central performance testing with the establishment in 1965 of the Beef Recording Association, later incorporated in the

Meat and Livestock Commission which also absorbed the Pig Industry Development Authority with its responsibility for pig improvement. The M L C also took responsibility for sheep recording but progress has not been comparable with that for other livestock, including poultry, which from 1965 have principally been products of multinational breeding companies.

One of the most interesting aspects of the animal-breeding scene over the period has been the growing acceptance by pedigree breeders of the advice of population geneticists with an understanding of the limitations under which breeders have to operate. Immediately after the war pedigree breeders, especially the well established, were inclined to write off geneticists rather scathingly as paper breeders with no eye for stock but twenty years later the position was different. Leading Friesian breeders had formed co-operative groups to use the selection techniques developed by M M B geneticists and beef breed societies were actively promoting performance testing. Perceptive breeders who had seen the eclipse of Shorthorns, both as dairy and beef animals, realised that the Charolais, in particular, constituted a threat even to well-established breeds such as the Aberdeen Angus and the Hereford unless positive steps were taken to improve growth rates.

This change of heart was in keeping with the progressive spirit of the times, reflected in so many farming activities and possibly to greater effect in the management and utilisation of grassland. When Professor M. McG. Cooper used the phrase 'farming at half cock' while addressing the British Association for the Advancement of Science in 1949 he was referring to the general failure of farmers to appreciate the value of well-managed pastures and their harking back to the pre-war situation of concentrates in plenty. Those were the days when attempts to make silage more often than not ended up with a very dubious, foul-smelling product and the bulk of pastures received little in the way of fertilisers. The criticism generated an angry reaction from over-complacent industry leaders but subsequently events have confirmed its validity because the net output of nutrients from grassland has substantially doubled since then.

Cereal yields, thanks largely to a succession of higher-yielding short-strawed varieties that respond safely to heavier nitrogen dressings than were once considered prudent, showed a substantial increase and by the beginning of the 1970s were practically twice those recorded thirty years previously. There was also a change in the balance of species with increases in barley at the expense of oats. This was due partly to the greater ease of harvesting barley by combine and a wider recognition of its value in ruminant nutrition. Perhaps the most serious impediment to production was that encountered by some farmers who ventured too far into continuous cereal growing and experienced reduced yields because of a deterioration of soil structure, and a greater incidence of fungal diseases and weeds, especially couch grass. Many had their fingers burnt and had to have second thoughts about this approach to the simplification of farming.

Nevertheless farming was becoming less complex with greater specialisation and a substantial reduction in the number of farming enterprises. In dairying, where average herd size was practically trebled, the number of milk-selling farms was about one-third of the 1947 tally. The reduction in commercial poultry producers was even more spectacular and the sideline flock, so long a feature of British farming, virtually disappeared. This trend has continued and by 1980 three-fifths of the laying birds were in flocks of 20,000 or more and the 2,200 holdings with broilers averaged 26,800 birds (*Annual Review of Agric.*, 1981). Guaranteed prices gave farmers the confidence to specialise while the change of emphasis in agricultural extension from husbandry to farm business analysis, which started in the mid-1950s, accelerated the swing from what was once described as mixed farming and muddled thinking to a concentration on enterprises best suited to farm and farmer alike.

Over the period, owner-occupation of farms increased at the expense of leaseholds, which were at a peak in 1914 when only 12% of the farms, including home farms of big estates, were owner-occupied. The swing started between the wars, stimulated by the imposition of what were considered to be swingeing death duties, and the trend increased markedly after the last war. By 1978, 63% of the holdings, accounting for 57% of the agricultural land, were farmed by owners (Northfield, 1979). This development not only reflected taxation pressures but also the favourable buying position of many tenants. Apart from prosperity arising from a buoyant agriculture, the 1947 Act gave tenants security of tenure and they were able to buy at a discount in comparison with the selling price of vacant possession farms. Though land values rose steadily over the period they were still comparatively reasonable when the UK joined the EEC in 1972 but after that there was a price explosion that more than matched the increase in commodity prices. Farming freehold land has now become a very capital-demanding operation.

FARMING AND THE EEC

Prior to the last war it was variously estimated that home agriculture was providing only about half of the nation's food supplies but the situation has changed radically with the revitalisation of the industry over the past forty years. The figures given in Table

Table 5.1 Domestic production of selected commodities as a percentage of total UK supplies (*Annual Review of Agric.*, 1981)

Commodity	1969-71 (av.)	1980	Main Source of Imports
Wheat	46	85	Countries outside EEC
Barley	90	113	EEC
All cereals	61	86	Countries outside EEC
Rape seed	12	62	Countries outside EEC
Potatoes	92	91	Countries outside EEC
Sugar	34	45	Countries outside EEC
Beef and veal	78	89	EEC
Mutton and lamb	40	66	Countries outside EEC
Pork	100	97	EEC
Bacon and ham	40	41	EEC
Poultry meat	99	99	EEC
Total meat	73	84	EEC
Fresh milk	100	100	
Butter	14	58	EEC & Countries outside
Cheese	47	72	EEC
Eggs	99	99	EEC
Wool	17	41	Countries outside EEC

5.1 not only reflect this change but also how the industry is reacting to Community membership.

A noteworthy change over the 1970s is the increase in cereal production, especially barley, as the result of two main factors: heavier crops and a higher area reflecting the attractive prices for cereals within the EEC, which showed a threefold increase over the period. Total cereals include considerable quantities of maize, mainly from the USA and hard wheat from Canada. Though the EEC has a wheat surplus most of it is soft wheat which is not suitable for the preferred type of British loaf.

The EEC now has surplus sugar but it accepts, albeit reluctantly, the UK's obligation to import sugar from Third World countries, mainly members of the Commonwealth, for whom sugar is an important earner of foreign exchange. There is a similar situation with lamb, one of New Zealand's principal exports, but here there is no EEC country capable of supplanting New Zealand as a source of reasonably priced lamb. The 66% self-sufficiency figure for lamb in 1980 as compared with 40% in 1969–71 does not reflect a material increase in home production. It is due to a 25% reduction in imports as New Zealand and Australia respond to the EEC import levies by exploiting new markets. Lamb is one of the commodities that will become scarcer and comparatively more expensive as a result of EEC membership.

Butter is in the same category. New Zealand, once a principal import source, is still permitted a diminishing share of the British market, in spite of objections from French and Irish interests, but over the period even with an increase of 170% in home production total butter supplies have fallen from 467,000 to 291,000 tonnes, a reflection of consumer resistance to rising prices. There has also been a 65% increase in home-produced cheese to compensate, in part, for the ending of imports of cheap Cheddar cheese from New Zealand, which again is part of the price consumers have had to pay for membership of the EEC – a complete negation of the cheap food policy this country maintained for more than a century. The bulk and perishability of whole milk, reinforced by health regulations, have up till 1981 kept out imports and so the UK has been able to preserve self-sufficiency, but there is a likelihood that this citadel may soon be breached. Whole-milk consumption has declined in recent years but meanwhile total milk production has increased from 12,328 m. l in 1969–71 to 15,334 m. l in 1980. This increase is largely due to a rise in average yields from 3,780 to 4,723 l per cow, which is evidence of the efficiency of the British dairy industry which compares favourably with that of any of its EEC partners.

Imported beef now comes mainly from Eire and in part this replaces the trade in store cattle which were once her principal export. Bacon and ham continue to come principally from Denmark while the few imported potatoes are extra-earlies mainly from the Canary Islands, Spain and Cyprus. The rise in the proportion of home-produced wool from 17 to 41% is due almost entirely to reduced imports of raw wool, a consequence of industrial recession. Oil-seed rape, like cereals, is a genuine growth commodity and this crop is now providing a very useful break in the sequence of white straw crops on many farms.

Generally, the figures in Table 5.1 do not suggest that output from British agriculture has been adversely affected by membership of the EEC. Whether this situation can continue will depend on two main factors: first, an ability to match her partners not only in efficiency of production but also in marketing, which has not always been her strongest point; secondly, that all members of the EEC will abide by the rules and eschew any measure that gives an unfair competitive advantage or any action, official or otherwise, that impedes trade in farm products. Britain must accept for her part that there are products where Mediterranean countries have an environmental advantage, especially those requiring artificial heating and lighting. Fossil fuel, with its all too rapid rate of exhaustion, is bound to become more expensive and the more Community agriculture can exploit solar energy the better it will be for future generations.

Despite initial shortcomings of the Common Agricultural Policy, which, by British criteria, has possibly tended to be over-protective towards small, relatively inefficient farms, and the nonsense of some aspects of intervention buying and selling, British farmers have good reason to be confident about their future. Technically they entered the E E C from a position of strength while membership and the foreseeable future of C A P are in the nature of an insurance against a return to the depressed years of the early 1930s when prices were ruinously low and production was being curtailed because of apparent gluts. But these were due to under-consumption, not over-production, in a world beset by social and economic injustices. Many of these remain but now, thankfully, not to any marked degree within the E E C whose next responsibility must be one of helping Third World countries to help themselves to the same fortunate situation.

PART II

THE RESOURCES OF
AGRICULTURE

6

LAND

The most significant characteristic of land as a resource is that it is fixed in supply. Of all man's economic activities, agriculture is the one most heavily dependent on land, for its availability, location and quality all affect the type and nature of farming practised.

The UK is one of the most densely populated and highly urbanised countries in the world, though some 80% of its land area is in agricultural use. Agriculture is thus an extensive user of land and contributes less than 3% to the GDP of the UK.

SOILS AND THEIR DISTRIBUTION

Soil is all important to agriculture. Not only is it the growing medium for grass and other crops, but its characteristics, together with location, relief and climate govern the range of activities that are possible.

Soil is made up of minerals, organic matter, air and water. The minerals include clay, silt and sand, the proportions of which determine the soil texture. Sandy soils have the largest particles and clayey soils the smallest – they are sometimes called 'light' and 'heavy' soils, respectively. In loamy soils there is a variation in the particle sizes present and they and silty soils are often described as of medium texture. The texture of a soil is important for it is closely correlated with other properties such as the ability to hold water, porosity and base content.

Soils are not static systems, for a series of physical, chemical and biological factors act upon them continually. The characteristics of a soil's parent material also affect formation, with base content and particle size being particularly important; the topography of the land has implications for the movement of soils and the amount of light received. Plants and animals affect soil formation by adding organic matter – and man himself often regulates their activities. Vegetation also acts to protect the surface of the soil and the influence of this and the various other soil-forming factors depend on the time over which they have operated.

The pattern of soils in the UK, when examined at the local level, is complex and soils frequently vary within both farms and fields. This wide variation reflects the range of parent materials, climate and local topography. However, in simple terms the soils can be divided into two zones – the highland zone in the north and west of the country and the lowland zone in the east, south and centre.

Figure 6.1 shows the distribution of British soils in more detail. (The soils of Northern Ireland developed, in the main, from glacial deposits and brown earths and brown podsolic soils are widespread.)

> Coppock (1971) classified British soils into four groups, with the upland zone (where climatic factors largely preclude cultivation) accounting for two. First, the podsols, peats and peaty podsols with their low fertility and impeded drainage, and second, the soils of the upland margins which are mainly acid-brown soils and have a low base status and low fertility. The third group (which is mainly in the lowland zone) contains the grey-brown podsolic soils which are similar to the second group, but have greater agricultural potential due to their more favourable climate, though at low levels they can suffer from poor drainage. The last group provides the best agricultural conditions and consists of the brown earths, calcareous soils and soils formed on alluvium and fen peat – most of these lowland soils are well drained and have a relatively high base status. Appendix Table 6.1 (pp. 769–72) provides a detailed soil classification.

The pattern of UK farming still reflects, in part, the distribution of soil types despite man's increasing ability to alter the natural properties of the soil. Thus the highland zone is still predominantly pastoral, with arable farming being practised in the lowland zone and especially in the east.

LAND CLASSIFICATION

The quality or usefulness of a piece of land for agriculture depends mainly on soil quality, location, climate, altitude and topography. In addition, the provision of fixed equipment such as buildings, drains and fences is also important. If the formulation of agricultural and land-use policy at both the national and farm level is to be meaningful, the quality and capability of the land should be assessed.

Two classifications of agricultural land exist in Britain. They differ in their approach and complexity due to their objectives and the time allowed for mapping. Both were intended to provide information for land-use planning.

The MAFF have an Agricultural Land Classification of England and Wales which is published in map form at the scale of 1:63,360. It classifies land into five grades according to its versatility as

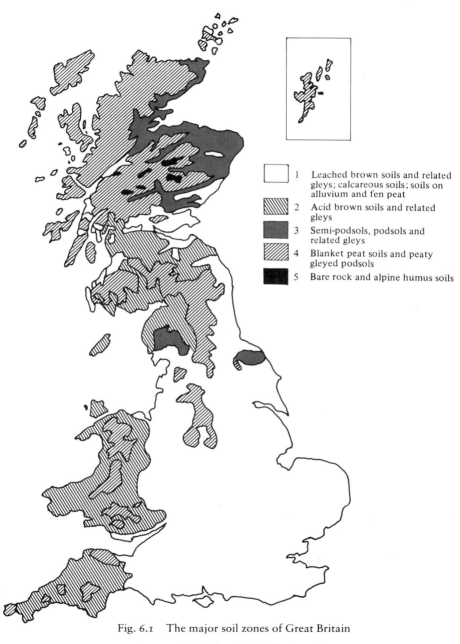

Fig. 6.1   The major soil zones of Great Britain

Source: Adapted from Curtis, Courtney &
Trudgill (1976) and Burnham (1970)

1   Leached brown soils and related
    gleys; calcareous soils; soils on
    alluvium and fen peat

2   Acid brown soils and related
    gleys

3   Semi-podsols, podsols and
    related gleys

4   Blanket peat soils and peaty
    gleyed podsols

5   Bare rock and alpine humus soils

　　　　　　　The Resources of Agriculture

determined by physical characteristics alone, such as climate, relief and soil. The advantage claimed for this classification was that such factors would not date. (Roughly equivalent classifications exist for Scotland and Northern Ireland.)

> Grades 1 and 2 are the best land in the country – a wide range of crops giving good yields at low cost can be grown on them. Grade 3 is valuable land suitable for a range of agricultural uses and is capable of growing good crops of cereals, roots and grass at the top of the category and moderate crops of oats, barley, roots and grass at the bottom of the category. Grade 4 has restricted potential – some can grow crops for winter stock maintenance and the poorest is only suitable for improved pasture if reseeded. Grade 5 land has very restricted potential limiting its use mainly to rough grazing. (Fuller details of this classification are provided in Appendix Table 6.2, pp. 772–3.)

Table 6.1 An estimate of UK agricultural land by grade, 1976

	England		Wales		Scotland		N. Ireland		UK	
Grade[1]	'000 ha	%	'000 ha	%	'000 ha	%	'000 ha	%	'000 ha	%
1	327	3·3	3	0·2	20	0·3	–	–	350	1·8
2	1652	16·7	39	2·3	155	2·4	36	3·3	1882	9·8
3	5343	54·0	295	17·5	880	13·6	457	42·0	6975	36·4
4	1553	15·7	745	44·2	660	10·2	531	49·0	3489	18·2
5	1019	10·3	604	35·8	4765	73·5	62	5·7	6450	33·7
Total[2]	9894	100·0	1686	100·0	6840	100·0	1086	100·0	19146	100·0

[1] MAFF Agricultural Land Classification grades for England and Wales
[2] Some totals are affected by rounding

Source: Agriculture EDC (1977)

Table 6.1 presents estimates of the distribution of land between MAFF grades for the UK. The most striking fact shown is that some 50% of the UK is in grades 4 and 5 and suffers from serious physical constraints such that only extensive livestock farming is possible. Within the UK there is a wide variation between the countries for, whilst only 25% of England is of this quality, over 80% of Scotland is. The great majority of grade 1 and 2 land is in England, particularly in the Eastern Region, and the East and West Midlands and the South-East and South-West Regions also have high proportions of such good land.

There have been many criticisms of the MAFF system. The most common one is that grade 3 is too wide, covering as it does some 50% of the agricultural area of England and Wales. Thus it was announced (MAFF, 1976) that steps would be taken to subdivide grade 3 into three subgroups. A further major limitation of the MAFF classification was that its minimum unit of evaluation is 81 ha, making it unsuitable for most planning decisions.

The Soil Survey of England and Wales also uses a classification scheme called the Land Use Capability Classification. This was developed 'to express the influence of soil, site and climate on farming' (Mackney, 1974). It divides land into seven classes depending on the severity of limitations affecting its use.

The classes range from class 1, which is land with very minor or no physical limitations to use, to class 7, which is land with very severe limitations that restrict use to rough grazing, forestry and recreation. (Fuller details of the system are provided in Appendix Table 6.2 and grades 1–4 of the MAFF systems are equivalent to classes 1–4 of the Soil Survey system.) The Soil Survey's seven classes are further subdivided (up to a maximum of two) into subclasses based on five physical factors which influence production or need correction. These factors are soil, wetness, climate, gradient and erosion. To date, maps showing this classification have been published at the scale of 1:25,000 for about 25% of England and Wales (and a similar classification exists for parts of Scotland), which is perhaps the main drawback of the system. However, an overall map showing the classification for the whole of England and Wales has been published at a scale of 1:1,000,000.

An advantage of the Soil Survey's classification is that it can be adapted fairly easily to suit different purposes or land uses. Figure 6.2 is an example of this for it shows the degree of suitability of different parts of England and Wales for grassland production and was developed out of the Land Use Capability Classification.

In summary, it can be seen that both the above classifications are only broad indications of the flexibility of the land for agriculture and its associated limiting factors for various types of farming. Both are suited for planning or policy making at the regional level, but they are useful for a preliminary general assessment only if one is involved in the planning of an individual site.

LAND USE

The land area of the UK, excluding inland water, is some 24 m. ha. British data on land-use matters are limited in both quality and quantity; those in the category often described as 'urban' are especially deficient. However, the estimate made by the Agriculture EDC (1977) provides the most recent official figures of the overall land-use position and is presented in Table 6.2.

Figure 6.3 shows clearly the division between cultivated land and rough grazing for the UK, with the second being predominant in the Highland and Islands of Scotland and parts of Wales. Figure

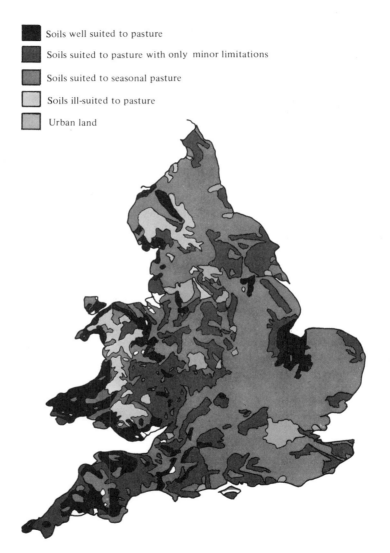

■ Soils well suited to pasture

Soils suited to pasture with only minor limitations

Soils suited to seasonal pasture

Soils ill-suited to pasture

Urban land

Fig. 6.2   Soil suitability for grassland in England and Wales

Source: Jollans (1981)

Table 6.2 An estimate of land uses[1] in the UK, 1976

Land uses	England '000 ha	%	Wales '000 ha	%	Scotland '000 ha	%	N. Ireland '000 ha	%	UK '000 ha	%
Crops and fallow	4021	31·0	107	5·2	595	7·7	79	5·9	4802	19·9
Temporary grass	1201	9·3	172	8·3	670	8·7	271	20·0	2314	9·6
Permanent grass	3240	25·0	775	37·5	450	5·8	485	36·0	4950	20·5
Rough grazing	1194	9·2	599	29·0	4765	61·8	210	15·6	6768	28·1
Other land[2]	238	1·8	33	1·6	–	–	41	3·0	312	1·3
Total agriculture	9894	76·3	1686	81·7	6480	84·0	1086	80·5	19146	79·4
Urban	1427	11·0	103	5·0	275	3·6	40	2·9	1845	7·7
Forestry and woodland	767	5·9	204	9·9	750	9·7	62	4·6	1783	7·4
Miscellaneous	885	6·8	71	3·4	210	2·7	160	12·0	1326	5·5
Total land[3]	12973	100·0	2064	100·0	7715	100·0	1348	100·0	24100	100·0

[1] Excluding inland water
[2] Includes 'woodland ancillary to farming', farm roads and buildings
[3] Not all figures add due to rounding

Source: Agriculture EDC (1977)

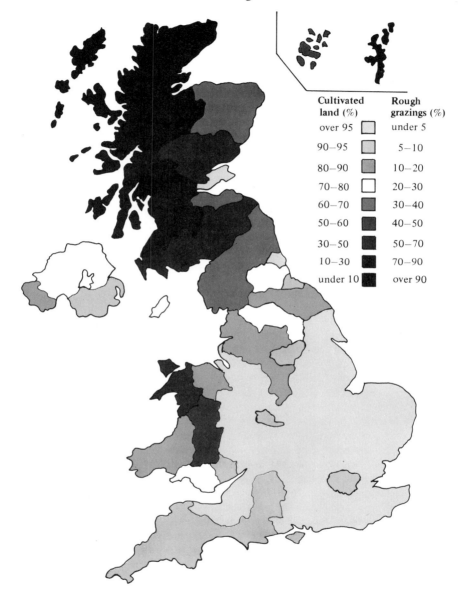

Fig. 6.3   The division of UK agricultural land in 1978 between cultivated land
and rough grazings (common rough grazings in England and Wales are not
included)
Source: Jollans (1981)

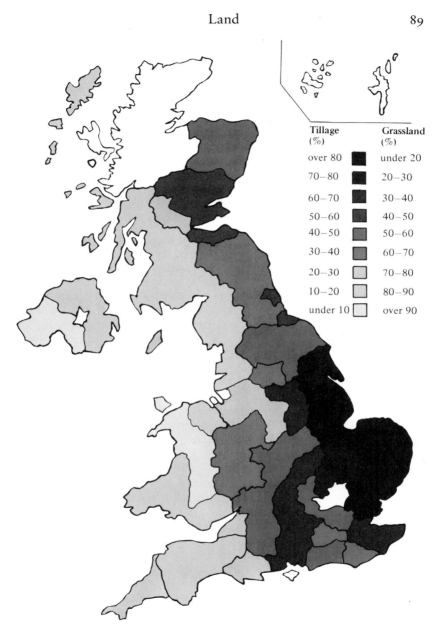

Fig. 6.4   The division of UK cultivated land between tillage and grassland in
1978
Source: Jollans (1981)

6.4 shows the division of cultivated land between tillage and grass-land.

The source for the agricultural part of Table 6.2 is the annual June agricultural census. Census forms are sent to occupiers of all agricultural holdings with an annual labour requirement of 40 or more smd. Forms are sent only to holdings of less than 4·05 ha if they have a significant agricultural output. Completion and return is a statutory requirement. During the last 15 years there have been several changes in the threshold of significance for inclusion in the census as well as in the definitions of the various categories of land use covered. Such changes, together with recent metrication, have inevitably impaired the accuracy and usefulness of the census returns.

There are some 2·1 m. ha of forest in the UK, about half of which are owned by the State and managed by the Forestry Commission or the Department of Agriculture, Northern Ireland (DANI). Almost all the State forest is 'productive' and 90% of it is coniferous. Some 400,000 ha in the UK are broadleaved woodlands and coppice and these are mainly in private ownership. Nearly 300,000 ha of forest land are unproductive and consist of scrub and felled woodland; most of this land is privately owned, often by farmers, and if better managed could make a useful contribution to incomes. Norway and Sitka spruce are the most common conifers although Scots, Corsican and Lodgepole pines are also fairly common. Oak, birch, beech and ash make up 70% of the area of broadleaved trees. (For further details about the UK forest industry see CAS, 1980.)

Much of the rural area of the UK is in multiple use with agriculture often predominating. Examples include defence lands, water-gathering grounds and common land. In 1977 there were some 100,000 ha of defence lands which for most of the year were in agricultural use – such as parts of Dartmoor and the Brecon Beacons. The Regional Water Authorities own over 130,000 ha of gathering grounds, though agricultural activities (mainly rough grazing) and some public access take place on this land. The Royal Commission on Common Lands (1958) concluded that there were some 609,000 ha of common land in England and Wales with about 80% of this area given over to grazing. Scotland has some 500,000 ha of common land, much of it associated with crofts. Much common land experiences *de facto* access by the general public and 67% of the common land in England is in the north.

Decisions affecting agriculture and forestry land use are largely free from planning controls. However, government financial assistance to agriculture is important in determining the pattern of agricultural land use and hence the type of farming practice. Much of government expenditure on agriculture is in the form of market regulation under the CAP and so at various times different crops are more favoured by farmers than others. The government also supports agriculture by a range of grants and allowances designed to help the industry achieve more efficient production. All these affect individual farmers' land-use decisions.

Over the period 1967-75 there was an overall fall in the area of agricultural land. More than half this fall was due to a loss of rough grazing land, though the area of arable land and of grassland also fell; in proportionate terms the fall in rough grazing was higher.

LAND TRANSFER FROM AGRICULTURE

The emotive subject of loss of agricultural land and its effect on national agricultural output has attracted much attention. Several bodies have published reports on the subject, and although their objectives and methods varied, they all concluded that care must be taken to ensure that the amounts of high-quality agricultural land transferred to other uses are minimised, if national agricultural output is not to suffer.

As with land-use data, it is disturbing that the amount and quality of data available on land transfer is not good. The only official source is calculated from, and depends on, the accuracy of the June census, although this was not primarily designed for the collection of such information. The most recent information for the UK is presented in Table 6.3: the accuracy of this information has never been fully assessed.

It can be seen that on average between 1967 and 1975 about 18,000 ha net have been transferred from agriculture to urban types of use each year (less since 1972), and about 25,000 ha net to forestry. When the residual transfers are included the average figure per year for net transfers of agricultural land in this period amounts to about 54,000 ha for the UK (about 0·2% of the total land area), with a range of some 46,000 ha between the largest and smallest annual totals. Comparable figures for later years are no longer being published, but it is understood that by 1980 the annual total had fallen by at least 10,000 ha.

Uncertainty surrounds the question of the quality of agricultural land transferred to urban development.

However, the Agriculture EDC (1977) concluded that it is likely that, in England and Wales as a whole, there has been no disproportionate loss of land in the top three MAFF grades despite the fact that such land is usually flat and well drained and therefore cheaper to build on. But in areas where there is relatively little better quality land (such as the North-West Region, Wales, Scotland and Northern Ireland) there is evidence to suggest that there does seem to have been a disproportionate loss of such land. This phenomenon, which seems likely to

Table 6.3　Analysis of net annual transfers to (+) and from (−) agricultural land in the UK, 1967–75 ('000 ha)

Year[4]	Urban, industrial & recreational development[1]	Government departments	Forestry & private woodlands	Land not previously recorded[2]	Other adjustments[3]	Total[6]
1967/8	−18·3	+0·2	−15·3		−8·0	−41·4
1968/9	−20·2	−0·4	−12·5		−5·0	−38·1
1969/70	−20·8	+1·5	−33·5		]−10·7[5]	]−57·6[5]
1970/1	−16·1	−0·4	−24·7			
1971/2	−19·9	+0·2	−47·1		−18·1	−84·6
1972/3	−22·1	0·0	−21·1	+0·3	−10·0	−53·2
1973/4	−14·7	−1·4	−22·0	0·0	−5·4	−43·5
1974/5	−15·1	+0·4	−24·2	0·0	−18·3	−57·2
8-year average 1967/8–74/5	−18·4	0·0	−25·0		−10·7	−54·2

[1] Includes mineral workings

[2] New category in 1973

[3] Includes corrections, re-classifications and unexplained differences which result from a recording system that does not permit a complete area reconciliation in a single year

[4] The figures are based on areas returned by farmers at June each year but, to preserve comparability with previous years, they are adjusted to discount changes in the coverage of the census since 1967

[5] An annual average for the two years is given as the separate figures for the individual years are not considered to be reliable

[6] Not all figures add due to rounding

Source: MAFF (1975 and 1978)

continue, is disturbing in view of the small amount of good-quality land in those areas. The majority of the land transferred to forestry throughout the U K has been of M A F F grades 4 and 5 or their equivalents.

In addition to the foregoing, there are three categories of loss of agricultural production from land, due to underutilisation, partial loss arising out of multiple use, such as water-gathering, and temporary loss to agriculture during, for example, open-cast mining. Such losses affect a considerable area.

LAND USE PLANNING AND AGRICULTURE

The Town and Country Planning Act, 1971, established a two-tier planning system in which the upper tier provides the strategic or structure plan and the lower tier the tactical or local plan.

Structure plans are made normally by the county planning authorities and local plans by the districts. Local plans must conform with the structure plan for their area. Both have to be drawn up following public consultation and structure plans have to be approved by the Secretary of State for the Environment. The D o E recognises that in almost every area the choice between agriculture and development will be regarded as a 'key' issue; in such cases planning authorities have to state their attitude towards agricultural land and to consult the M A F F on a continuing basis. Similarly, the M A F F has to be consulted by the district planning authorities during the preparation of local plans. (Similar legislation exists in Wales, Scotland and Northern Ireland, although each country has appropriate Secretaries of State.)

In general, permission has to be sought from the local planning authority for development.

With certain exceptions development is defined in Section 22 of the above Act as '... the carrying out of building, engineering, mining or other operations in, on, over and under land, or the making of any material change in the use of buildings or other land...'. Subsection 2 defines exemption to the above as 'use of land for agriculture or forestry and use of any building occupied together with land so used'. The General Development Order, 1977, defines 'permitted development' as 'the carrying out on agricultural land having an area of more than one acre and comprised in an agricultural unit of building or engineering operations requisite for the use of that land for purposes of agriculture (other than the placing on land of structures not designed for those purposes or the provision of dwellings) so long as: the ground area covered either by itself or by the addition of other buildings erected within the preceding two years and within 90 m of such buildings does not exceed 465 m²; or 3 m in height within 3 km of an aerodrome or 12 m in any other case; or is within 25 m of the metalled portion of a trunk or classified road.'

The above general permission can be removed by the relevant Secretary of State confirming an Article 4 Direction to that effect. Most such Directions have concerned the removal of agricultural buildings from the permitted category and they have been used most commonly in Areas of Outstanding Natural Beauty. The Town and Country Planning (Landscape Areas Special Development) Order, 1950, established regulations concerned with the restriction of agricultural and forestry buildings normally exempt from planning permission. The areas concerned were mainly in National Parks.

Tree felling is controlled by several pieces of legislation. Section 60 of the Town and Country Planning Act 1971 permits the local planning authority to serve a tree preservation order on a single or group of trees or an area of woodland to control felling or lopping. This cannot be made on dying trees or when a working plan exists as approved by the Forestry Commission. The Town and Country Amenities Act, 1974, Section 10, revises the penalties for breaking the above orders. The Forestry Act, 1967, prohibits all felling without Forestry Commission consent, exemptions being small trees or those within a planning permission.

Since 1945, governments have sought to safeguard good agricultural land from development. Accordingly, the MAFF has to be consulted on all planning applications for development on areas of over 4 ha of agricultural land.

If the MAFF objects to the application on the grounds that it is against the long-term agricultural interest and the local planning authority does not agree, either they or the MAFF can ask the DoE to 'call-in' the application for the Secretary of State to decide. If he does decide on 'calling-in' it results in a local planning inquiry. The MAFF can also reply to the local planning authority by stating the agricultural considerations that should be borne in mind or certain conditions (such as restoration after mineral extraction) that should, in their view, be imposed.

There has been much recent discussion about the need to rationalise the whole rural planning structure and, especially, about the need for planners to consider agriculture in more detail.

The most important other rural planning legislation is that which enables the designation of National Parks and Areas of Outstanding Natural Beauty. At present nearly 20% of England and Wales is so affected and farmers in such areas are faced with various constraints affecting, for instance, the design and siting of buildings and the types of cultivations and land improvements that can be carried out. However, compensation for such constraints is becoming more common. There are plans for similar designated areas for Scotland.

A range of public authorities have the power to acquire land for a variety of reasons, such as housing and highways. These compulsory purchase powers are infrequently used.

The countryside is no longer the sole preserve of those who produce food and timber. Demand for rural land for housing, industry and transport; for minerals and aggregates; for water; and for a wide range of leisure and recreation activities has been increasing and seems likely to continue.

All these activities, however, cause either temporary or permanent changes in the appearance or nature of the countryside and may easily result in conflict. Changes in agricultural and forestry practice, although often carried out for sound commercial reasons, also alter the rural landscape and affect wildlife. However, the Wildlife and Countryside Act, 1981, limits the powers of farmers and landowners in respect of some such practices.

Perhaps the most significant trend affecting rural land use has been the increase in participation in leisure and recreation, for many urban dwellers now use the countryside for formal and informal recreation activities on both daytrip and longer-stay basis. Such activity causes pressure on the rural environment – the very thing that people come to enjoy. The demand for minerals and aggregates in recent years has also caused land use problems. Much sand and gravel and coal for opencast extraction is found under good quality farm land. Such land is often the easiest and cheapest to build on. Thus, conflict often results where, as in National Parks and Areas of Outstanding Beauty, many hard rock aggregates and non-metalliferous minerals are found, extraction of which tends to scar the landscape.

Such a rapidly changing situation needs well-founded and efficient administrative machinery to resolve conflicts. However, in the UK there is a plethora of both national and local government agencies; professional and special interest groups; and pressure groups including the so-called 'amenity and conservation lobby'; all with different mandates, responsibilities and vested interests in the countryside. This often causes confusion, delays in land-use decisions and enlargement of conflicts. Two problem areas where this situation is probably at its worst – the urban fringe and the uplands – will be examined next.

The urban fringe has been defined as 'The land between continuous built-up areas of cities or large towns and the open country around'.

Agriculture is the major urban fringe land use. However, several factors often result in the land being farmed inefficiently and prevent the landscape from being maintained. Fragmentation of holdings by expanding urban development in some cases threatens the viability of the farm business. Trespass and vandalism are often cited as problems and so livestock farming is frequently abandoned. Much urban fringe farmland is underfarmed, the result of its being subject to planning permission for development or under a threat (whether real or imagined) of change of use so that farmers are unwilling to invest in improvements. This situation is sometimes self-imposed, with farmers deliberately farming 'to quit' in the hope that they will be able to sell land for development for a large capital gain. Where the above factors exist the quality of the landscape inevitably suffers from the poorly maintained buildings, pasture and field boundaries. In short, a range of land-use pressures interact to create uncertainty which results in an untidy and derelict landscape.

Planners, particularly in the London Green Belt and the North West, are now acutely aware of the problems of the urban fringe and are trying to protect the high-quality farmland that still exists.

The other areas of continuing conflict over rural land-use matters are the hills and uplands, particularly those areas designated as National Parks.

Livestock rearing and forestry are currently the most important commercial activities but recreation is becoming more important. Water supply, mineral extraction, nature conservation and defence purposes are also locally important and usually competing. Like the urban fringe, the hills and uplands suffer from acute physical, social and economic problems. These include low temperatures, a short growing season, steep slopes and poor soils, a declining and ageing population with few transport facilities as well as poor social, medical and educational services.

Policies to arrest depopulation and stabilise and improve incomes in these areas need to be discussed and implemented at a local level. Currently, the greatest proportion of government assistance for the hills and uplands goes directly to farming, but it is just one form of land use and seldom shows a good return in job creation terms for the investment of public funds.

Suggestions for the improvement of the problems of these areas – and indeed for all rural land-use problems – have tended to have one main point in common, the need for a national rural land-use strategy that has regional and local level components, implemented by multi-purpose development agencies with broad remits, such as the Highlands and Islands Development Board.

LANDOWNERSHIP AND OCCUPANCY

No current official information on agricultural landownership exists, though the 'New Domesday Books' of the 1870s provide comprehensive data for that period. Over time much disquiet has been expressed at this lack of knowledge, particularly since capital taxation has been in force to reduce perceived inequalities in wealth distribution: clearly without knowledge of this sort no serious attempt can be made to assess the success of the policies. Due to increasing concern, the *Committee of Inquiry into the Acquisition and Occupancy of Agricultural Land* was set up in 1977 to investigate such matters.

Table 6.4 summarises the findings of this committee, chaired by Lord Northfield. It will be seen that private individuals, companies and trusts own some 90% of the total.

Table 6.4. **Agricultural landownership in Great Britain, 1978**

Category of owner	Area owned	Proportion of total
	(m. ha)	(%)
Public and semi-public bodies and traditional institutions	1·5	8·5
Financial institutions	0·2	1·2
Private individuals, companies and trusts	16·0	90·3
All owners	17·7	100·0

Source: MAFF (1979)

Very little is known about the size stratification of areas in one ownership.

It is ironic that, although more is known about institutions than about other categories of owner, the former own only about 10% of the area. This is probably due, in part, to the complex forms of business organisation that have been created in the private sector in an attempt to minimise capital taxation. It appears that trusts of various forms are predominant in the case of let land and sole and joint traders are more common with smaller owner-occupied holdings. Companies are relatively rare.

A fear that the area of land owned by foreign nationals and institutions (especially financial) was increasing was one reason behind the setting up of the Northfield Committee. However, despite finding that in recent years there had been a rise in net purchases by these groups, the Committee concluded that even by 2020 the proportion owned by them would only have increased to about 15% (MAFF, 1979).

The number of agricultural holdings in the UK has fallen steadily this century. By 1980 the total was some 243,500, about 5% fewer than in 1975. This fall has been most pronounced in small-scale holdings, though such businesses are still important in Northern Ireland and Wales. Accompanying the fall in numbers has been an increase in the average size so that, if rough grazings are included, by 1980 the average full-time holding was 116 ha compared with 111 ha in 1975 (MAFF, 1981). Individual enterprise size has also risen and the size of holding seems likely to continue to grow.

Information on occupancy is also deficient although annual collection of such information is encompassed in the June Census. Table 6.5 shows the distribution of land between that owned and that rented in Great Britain since 1908. Northern Ireland is not included because, since the various Land Acts between 1870 and 1925, virtually all land has been owner-occupied. However, some 19% of land in Northern Ireland is let in conacre, a form of short-term letting.

Table 6.5 Agricultural land tenure in Great Britain[1], 1908–80

Year	Rented/mainly rented		Owned/mainly owned	
	% area	% holdings	% area	% holdings
1908	88	88	12	12
1922	82	86	18	14
1950	62	60	38	40
1960	51	46	49	54
1970	45	42	55	58
1976	44	38	56	62
1977	44	37	56	63
1978	43	37	57	63
1979	42	35	58	65
1980	42	34	58	66

[1] Tenure status determined on basis of: crops and grass only in 1908 and 1922 (England and Wales) but in 1908, 1922, 1950 and 1960 (Scotland); crops and grass and rough grazing in 1960 (England and Wales), and total area (including rough grazing and other land) in all other years throughout Great Britain.

Sources: MAFF (1979); Lund and Slater (1979); MAFF (1981).

Table 6.5 shows how the decline in the area and number of holdings rented has been fairly rapid, though in recent years it has slowed, being accompanied by a growing shortage of farms offered to let. The change towards owner-occupation has resulted from landlords either taking land in hand when tenancies expired (and in recent years the security of tenure legislation has encouraged this) or else selling to sitting tenants.

The position is complicated by an increase in mixed-tenure

farming and the growing number of holdings with father/son partnerships.

There are other regional tenurial peculiarities apart from con-acre. The 609,000 ha of common land in England and Wales is an example, land over which occupiers with common rights can graze livestock but who have few rights in respect of land improvement. Following a Royal Commission report, steps are being taken to clear up the disputes and clarify the somewhat confused situation that exists over common land. In the Highland districts of Scotland where crofting tenure is practised, there are some 500,000 ha of crofters' common grazing land, land which is used in conjunction with the long-leasehold land that goes with the croft house. The Crofting Reform (Scotland) Act, 1976, enabled crofters to become owner-occupiers and provided incentives to carry out improvements so that agricultural development of the crofting districts could be facilitated.

LAND PRICES AND RENTS

Agricultural land prices escalated between 1946-78; the average price increased some 23 times compared with a rise in the Consumer Price Index of some six times and in Net Farm Income of about nine times. This trend was particularly noticeable in the 1970s.

Whilst the area of land sold annually has fluctuated, it has tended to decrease over time so that only some 1½-2% of the agricultural area is now changing hands annually, resulting from 7,000-8,000 transactions. Probably due to the strong demand for land (reflecting its inelastic supply) more transactions are carried out by private treaty than by auction. MAFF (1979) estimated that some 60% of these transactions were cases of existing farmers expanding their businesses; there are also many cases where land changes hands without any financial transaction, as in the case of inheritance.

Since, in broad terms, 60% of the UK agricultural area is owner-occupied and 40% is tenanted, the land market can also be divided into these two groups and in recent years the area of tenanted land sold was about 20% of the total area of land sold. This disproportionate ratio is due probably to the recent succession of tenure legislation which has made potential purchasers some-what reluctant to purchase tenanted land and is reflected in the so-called 'vacant possession premium' of around 30%.

Land prices in the constituent parts of the UK are presented in Table 6.6. This shows a good deal of variation between countries, with, until recently, the highest price being paid in England.

The extent of the 'vacant possession premium' is also clear. The main factors affecting land prices are: land quality and location (farms in eastern England fetch higher prices than those in the

Table 6.6 Agricultural land prices[1] in the UK, 1972-8 (£/ha)

	England[2]			Wales[2]			N. Ireland[2,3]		Scotland	
	Vacant Possession	Tenanted	All	Vacant Possession	Tenanted	All	Vacant Possession	Vacant Possession	Tenanted	All
1972	1238	1050	1161	724	376	672	699	342	203	304
1973	1643	1354	1557	1013	465	952	831	692	386	613
1974	1330	961	1273	836	531	809	998	757	604	721
1975	1205	797	1125	926	642	814	1143	602	442	565
1976	1472	1019	1349	1019	587	910	1392	775	481	710
1977	1994	1563	1909	1327	628	1215	1847	1061	632	963
1978	2602	1687	2434	1788	818	1655	2618	1090	590	943

[1] From the statutory Inland Revenue series compiled and collated by Lund and Slater (1979). Series covers parcels of >5 ha in England, Wales and Scotland and >2 ha in N. Ireland

[2] There is a delay between the date on which sale is agreed and the date when it is included in the series – this delay is from 6–9 months in England and Wales and about 3 months in N. Ireland. The prices shown in the table relate to sales included in the series in the years ending the following September and March respectively

[3] Although the conacre system exists in N. Ireland (a form of short-term letting) there is virtually no land that in the legal sense is tenanted

north and west); presence of buildings and, in particular, quality of houses lead to higher prices; size of farm or parcel of land.

Between 1946–78 average farm rents rose some 10–12 times – not as sharp a rise as that shown by land prices and the rent series is smoother than that for land prices. It is thought that this is due to the regular statutory rent determination (every three years except in Scotland where it is five) and because the rental market is less affected by non-agricultural reasons for holding land.

Table 6.7 shows average rents paid in Great Britain and it can be seen that the highest rents were in England where in the last few years there was an annual increase of about 20%. As with land prices there is a wide variation in rent levels with, it is thought, the factors of land quality and location, farm size, provision of fixed equipment and type of landlord being responsible.

Table 6.7 Farm rents[1] in Great Britain[2], 1972–9 (£/ha)

	Average rent paid in calendar year		
	England[3]	Wales[3]	Scotland[4]
1972	15·17	9·02	10·28
1973	16·13	9·74	11·07
1974	17·49	10·18	11·92
1975	20·26	11·49	13·22
1976	24·66	13·68	15·62
1977	29·80	15·47	18·30
1978	35·60	17·71	(21·52)
1979	(42·10)[5]	(21·85)	(25·47)

[1] From MAFF and DAFS information compiled and collated by Lund and Slater (1979)
[2] Practically no land in N. Ireland is tenanted
[3] Derived as weighted averages
[4] Derived as weighted averages and relating to rent per ha of crops and grass on full-time farms
[5] Figures in brackets are provisional

7

CLIMATE

Introduction

The weather, always thought of by the farmer as the enemy of regular production and high yields, is in fact his best friend and there is ample scope for putting weather knowledge to profitable use, both in strategic planning on the farm, and in day-to-day tactics.

Climate is a summary of the weather: the central idea is of an underlying regularity, that the weather is patterned in space and time, and can therefore be apprehended by something more concise than a recital of all its vagaries. To appreciate the patterns, adequate observations are required. All observations of a set need to be *comparable, accurate, regular* and *sustained*. Historically, it took a long time for all those needs to be appreciated. Even today and in the developed countries, they are not always fully met. This important topic will be taken up in more detail later in this chapter. Agricultural problems will not wait for meteorology to attain a state of perfection and completeness, so often we have to do the best we can with imperfect data, always watching for the implied pitfalls.

Summarising begins by calculating averages (of temperature, say, or rainfall) at one station, which may extend over hours, days or months. If an average is formed over a long period of years of all the January temperatures at 1200 hours at Kew, the result will be a good estimate of the most likely midday January temperatures at Kew in the near future. Some care must be taken over the choice of the 'long period of years' because climate is continually changing slightly. The period must be long enough for extreme years not to have undue influence, but not so long as to include obsolete climate. Thirty years is traditional, but it may be that at present a recent 10 or 15 years is as good if not better.

Working out averages is a first step in describing climate, but what crop or animal ever experienced 'average' conditions during its lifetime? It is the fluctuations of weather (the moist spring, the dry summer, the stormy autumn) which make for notably good or bad crops, and for epidemics of pest and disease or the absence of troubles usually considered endemic. The second step is to describe these fluctuations. One way of doing this is to state the extremes. In the example above, the mean midday January temperature at Kew might be supplemented by the highest and lowest such temperatures encountered, say in the period 1941-70. But extremes change a lot when a different run, or a longer run, of years is taken. The temperatures exceeded in 25% and 75% of years – the lower and upper quartiles – would be more representative of the variability encountered over most of the period. In particular circumstances the use of standard (or root-mean-square) deviation, or the display of a sample decade of annual means, may be appropriate. The important point is that the inherent variability of weather should be recognised and described numerically in an appropriate way.

In the statistical handling of weather data, a vital question is: to what area of country is a particular weather measurement applicable? Micrometeorology deals with the fine detail of weather. Macrometeorology deals with the large weather systems, familiar from charts seen in newspapers and on television. Mesometeorology deals with systems such as tornadoes and individual thunderstorms. The farmer, concerned mainly with individual fields and individual crops, even perhaps parts of a field prone to frost or wind damage, has interests on the micrometeorological scale. In the practical side of agricultural meteorology, it is only rarely that time and money are available for detailed micrometeorological studies. The meteorologist working for agricultural advisory services is therefore obliged largely to use weather observations made at standard observing stations of existing networks. He is aware that these readings are not identical with those above the field he is studying, much less within the crop. Nevertheless if he works carefully he can use time and space scales that ensure that the micro- and macro-conditions move nearly in parallel, and so obtain practically useful crop-weather relations. At the same time he will be following carefully the work of more academic micrometeorologists, and incorporating in his work the physical and biological insights which they obtain.

WEATHER IN THE LIFE CYCLE OF A CROP

Weather affects every aspect of the growth of a crop but it is as well to remember that *previous* seasons' weather can affect current crops, through the seed and through the soil. The amount of seed available, its germination and vigour, its burden of pest and disease, all depend on weather during the lifetime of the seed crop, and subsequent storage conditions. Rainfall or the lack of it before a crop is sown can affect particularly a short-term crop for which little rainfall is expected during its life, and which therefore depends for its moisture supply on reserves in the soil. Weather conditions at harvest have long-lasting effects. In temperate latitudes the worst of these occur in a wet autumn when heavy machinery is used on clay soils, particularly for the harvesting of root crops. Severe damage to soil structure may occur, with poor root action in following years. In the worst cases, it may be impossible to sow the following crop.

Soil cultivations and sowing

Though the cycle of work and growth on the land is continuous, it will be convenient to start the survey in the autumn immediately after harvest. The farmer is faced then with the residues of the previous crop, stubble, perhaps straw or tops of root crops, and almost certainly weeds, pests and disease. He has to clear as much as possible (some would say as much as necessary) of this away, and create soil conditions favourable for the growth of the next crop. In arable areas cereal straw is now often burnt. This removes the residues quickly and at low cost, and often with a useful reduction of pest and disease. An additional cost which need not be borne is the destruction of hedges and trees, nor should the public roads be shrouded in smoke. It makes sense to obtain a weather forecast and to avoid occasions when the wind is variable in direction or likely to change suddenly during the operation.

On a few of the very best soils it is possible to put machinery on the land the day after rain, perhaps even while it is still raining, and do no damage to the soil. Such soils tend to be devoted to intensive horticulture, with several crops going in and coming out during the year, and those who farm them are a fortunate minority. On most soils autumn work must be completed before soils return to field capacity. A soil is said to be at field capacity when the loss of moisture by evaporation through the summer has been fully made good, and any extra rainfall runs into the drainage system. It does not refer to a waterlogged condition. A farmer should be aware of the average date of return to field capacity in his area, and its variability from year to year. In England and Wales this information is readily available (Smith, 1976). Some typical dates are given in Table 7.1. Return is later in the south and east of Britain, where summer evaporation is greater and autumn rainfall less, than it is in the north and west. This is one reason why farming in the south-east is largely arable, even on heavy land, while most of the north and west is in grass, except on the best soils.

In the period before return to field capacity, workability of the land depends largely on recent rainfall in relation to soil type, since these are the factors affecting traction of tackle and the possibility

Table 7.1 Typical dates of return of soil to field capacity

Area	Median date	Range including 50% of years
Cornwall	5 Oct	12 Sep-7 Nov
East Anglia	1 Dec	15 Nov-22 Dec
Vale of York	13 Nov	3 Oct-12 Dec
Yorkshire Pennines	27 Aug	22 Jul-20 Sep
Fylde of Lancashire	13 Oct	16 Sep-19 Nov

of smearing the soil. Of all the operations on bare soil, drilling is the most critical, since smear must be avoided at all costs or seedling emergence will be poor. In spring or autumn, 1-2 days of fair weather are needed after rain before drilling can take place on light land, and 3-4 days on heavy land. Harrowing and ploughing are progressively less critical, particularly if mouldboard ploughing is to be left through the winter for frost to do much of the work of tilth formation. Estimates have been made on this sort of basis of the 'available workdays' in different places and years, and these have given reasonable agreement with farmers' statements of when they would have been prepared to work the land. The striking result that comes out of this line of work is the great variability in workdays available on different farms and in different years.

Germination and emergence

Soil temperature and moisture are the important factors. The main temperate region crops are adapted to their climate so that in the spring, when soil temperatures rise high enough for germination, moisture is usually adequate also. One advantage of autumn sowing, for those hardy crops for which it is possible, is that adequate moisture is always ultimately available, and usually comes in advance of severe cold. Tender crops such as maize and runner beans require soil temperatures around 10°C for satisfactory germination, and these temperatures may not occur in England until late April or early May. By this time there is often a problem with dryness. Irrigation before or after sowing may help.

Development

This is the core of the subject of crop-weather influences. A useful first step is to define the length of the growing season, since on the whole crops give more, particularly of vegetative parts, when this is long. For temperate crops, the threshold for growth is often taken as 6°C. Rye will grow at 2°C, hence its usefulness as an early grazing crop. Ryegrass has been observed to grow slowly at temperatures down to 0°C, and casual observations of lawns at the end of winter confirm this. For maize, an air temperature of 10°C is often taken as the threshold for growth, the same as the soil temperature needed for its germination. But in general, once an air temperature of 6°C is reached, many temperate crops grow steadily. Average lengths of growing season based on days with mean air temperature above 6°C are readily available (Smith, 1976), and we have information too on the year-to-year variation. Table 7.2 gives some examples. They indicate the great variation which can occur in the climatic potential of land for cropping within quite small geographical areas. The shorter the growing season, the more variable it is – an additional hazard for the upland farmer.

Table 7.2   Typical lengths of growing season

Area	Median length in days	Median start and finish
Cornwall	322	20 Feb–8 Jan
East Anglia	248	26 Mar–29 Nov
Vale of York	245	27 Mar–27 Nov
Yorkshire Pennines	209	16 Apr–11 Nov
Fylde of Lancashire	258	26 Mar–6 Dec

In half the years the length of growing season is within about ±15 days of the median. Year-to-year variation is greater at the spring than at the autumn end of the growing season.

Once temperatures are high enough for growth, solar radiation becomes a dominant factor.

Only a small fraction of the available radiant energy, usually less than 1%, is directly used in photosynthesis. Most of the rest is used to evaporate water transpired from the plant, and to a lesser extent direct from the ground. This energy is not wasted however, because nutrients are carried up the plant in the dilute transpiration stream.

Disease

Development of plant disease requires three things: the plant must be in a susceptible state; there must be a source of infection (inoculum); and the environment must be right for infection. Virus diseases are generally transmitted by sap-sucking insects, and do not concern us directly. Bacterial diseases and particularly the very numerous fungal diseases nearly all require a liquid water film for division of the bacteria or germination of the fungal spores. This water film may be a result of rainfall or of dew formation, and its persistence will depend on the temperature and humidity of the air and the degree of ventilation of the surface. Empirical rules have been developed defining conditions favourable to the spread of several diseases, potato blight and apple scab being notable examples, and this permits timely and economical application of fungicide. A more recent success is forecasting of *Septoria nodorum* on winter wheat in England and Wales. This disease is severely damaging only in some years, so that routine fungicide applications are not worth while. The spores are spread by rainsplash. In late May the wheat plant elongates rapidly, and this gives the fungus its chance to spread up the plant. Figure 7.1 shows the relation between

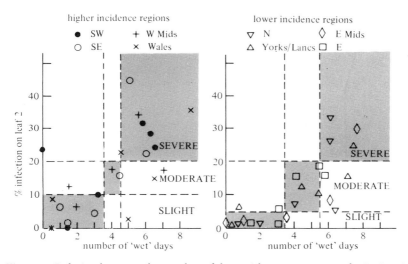

Fig. 7.1   Relation between the number of days with 1 mm or more of rain ('wet' days) in the second half of May and the percentage area of leaf two affected by Septoria. The disease assessment was made at about the milky ripe stage, 4–6 weeks before harvest. Shaded stippled areas show major categories of disease incidence

days with rain at this time and level of infection just before harvest, which is the basis of a successful forecast rule. It is perhaps an advance on previous work in this field in that it specifies the *number* and *timing* of the infection periods, and the *severity* of the resulting infection.

Chemical applications

Chemical applications for the control of weeds, pests and disease are all weather sensitive. The prime requirement is that as much as possible of the sprayed chemical should go where it is wanted, and as little as possible elsewhere. Very small drops are effectively suspended in the naturally turbulent wind, and so can drift for considerable distances. They can also be caught up in organised convection currents, for instance under cumulus clouds, and deposited at a distance when those clouds dissolve, say, half an hour later. In practice most spray equipment produces some small droplets, so spraying should be restricted to quite low wind speeds.

Generally there should be no rain at the time of spraying and for a period afterwards. Hormone weedkillers, in particular, are effectively absorbed only if temperatures are high enough for active growth. The meteorologist can turn this descriptive account of suitable weather for spraying into a fairly precise specification in terms of the observations made at a professional weather station, and past records can be scanned to find how often suitable conditions have occurred.

Agricultural climatology now demands information on the joint frequencies of several weather variables, such as rainfall, wind and temperature. This has been made possible by the development of computers and associated data banks of meteorological observations.

Flowering

Whereas vegetative development is controlled largely by solar radiation, progress towards flowering is influenced more often by temperature, and sometimes daylength. Water shortage, too, can accelerate flowering, though at the expense of crop yields.

A simple framework for temperature effects is to consider the situation where there is no progress (say towards flowering) until temperatures rise above a threshold, called the base temperature. Above the base, the rate is assumed to rise proportionately with rise of temperature. Each day's contribution is represented by its average temperature minus the base temperature, and these contributions are added from day to day to form what is known as an accumulated temperature or temperature sum. The procedure is straightforward when the air temperature is above the base all day. When the temperature is partly above and partly below, an adjustment is required to take account only of the part of the daily temperature wave which is above the base. This method often works quite well, even when the rate of progress towards flowering is not strictly linear with temperature. For maize in England, progress towards flowering is quite well represented by accumulated temperatures above $10°C$, the total depending on the variety. This has been helpful in delimiting the areas suitable for the crop in a region which is at the edge of its climatic range.

In temperate grasses, tillers which have received adequate winter cold will progress towards flowering at a rate controlled by temperature, but only after a critical daylength has been exceeded. The critical daylength depends on the cultivar. In high latitudes, there is a rapid change around the spring solstice from short days to long days, and therefore the sequence of initiation dates is telescoped compared with farther south (in the northern hemisphere). This goes a long way to explain the difficulty in selecting grasses to give a more extended conservation season, in high latitudes.

In perennial crops such as fruit trees, the flower buds are formed in the previous autumn and their number and quality depend on the previous year's weather. Weather at flowering may be critical for the whole year's crop, since frost may destroy the flowers entirely, while cold and wet can stop pollinating insects from working. Frequency of frosts is a very local phenomenon.

Ripening

As for flowering, warmth is the main requirement. For instance, the accumulated temperature scheme for maize mentioned in the previous section can be extended successfully to include the various stages of ripening. Since loss of moisture is an integral part of the ripening of cereals and pulses, dry conditions are also needed, but if these occur at too early a stage, yields are reduced. For fruit crops where attractive coloration is important to the consumer, sunshine as well as warmth is needed.

Harvest

The first important harvest in Britain is of grass for hay and silage. Dry conditions are required for cutting and wilting. In the case of hay, low humidity and adequate ventilation by wind help considerably the further progress towards readiness for baling. Studies in England of sequences of dry days in May, June and July have shown that late May and early June give the most runs of 1-5 dry days. This, plus an appreciation that digestibility of the forage decreases rapidly after heading, has led to a move away from traditional July haytime to earlier cutting. These studies have shown also that the number of short runs of 1-2 dry days (for silage) is several times more than runs of five days (for conventional hay) and that the number of short runs is also less variable from year to year, and so have made a contribution to the trend away from hay to silage making.

Direct combining of seed and grain crops is now almost universal, and the main weather requirement is a dry crop at the time of the operation. The availability of big machines has tempted farmers with suitable soil into planting large acreages of profitable cash crops such as potatoes and sugar beet. Ideally these should be out of the ground before autumn rains return the soil to field capacity. The modern heavy tackle can do the job in a fashion even in very wet conditions, but damage to the soil structure can be severe. Part of the problem lies in persistence of weather patterns, so that cool damp summers which make for late development (though not necessarily low yields) tend to be followed by wet autumns and difficult harvests. If in such an autumn a farmer is additionally committed to rapid cultivations and the sowing of a large area of winter cereals, and perhaps also rape and beans, the problems can be tremendous. To provide the answers that are needed on intensive and highly mechanised farms, the whole-crop and whole-farm approaches will increasingly be needed.

Storage

After harvest, the crop may still need protection from rain, frost and the effects of pest and disease, and nowadays permanent buildings are used mostly. The crop as harvested may need treatment to prolong storage. For instance potatoes need to be warm at first, so that harvesting wounds may heal and rots be kept at bay. Later, cool conditions are required to reduce weight loss and sprouting, but frost must be kept out. Blowing of outside air through the crop at times when the temperature is right is the most economical method of achieving the desired conditions. Knowledge of air temperatures at the place in question plays an important part in both the planning and execution of these operations. With grain, it is often necessary to remove moisture from the crop before storage, yet to leave it cool. Onions and apples are two minor but high-value crops, for which environmental control of storage is practised successfully.

WEATHER AND ANIMAL HUSBANDRY

Direct effects of exposure

Cattle and sheep, the species which are kept outside for much or all of the year in Britain, have, as mature animals, the ability to maintain a near-constant body temperature in the face of a wide range of external conditions. This is often specified as a range of ambient temperature (still air being assumed) but wind greatly increases heat loss, and it is in terms of heat loss that the thermal equilibrium of an animal is best specified. Wetting of the coats of animals is very effective in increasing heat loss. The coat functions like loft insulation, by trapping still air in small spaces, and if these spaces are filled with water, performance decreases dramatically.

Cattle and sheep have minimum rates of heat production, which have been found experimentally to be adequate to maintain the animals' thermal equilibrium at temperatures well below freezing-point. But these results apply to mature, healthy, well-fed animals with dry coats. In practice, stock will often benefit from shelter from wind and rain. Shelter from wind has been shown to reduce deaths at lambing, and is particularly valuable to the new-born animal. Once the coat is dry, plastic raincoats have been used to keep it so, and have proved their worth as a form of artificial shelter. Warm dry conditions are particularly valuable for young calves. Low or rapidly changing temperatures and draughts must be avoided, while maintaining adequate ventilation. Most pigs and poultry in Britain are now housed, since the reduced heat loss of animals can result in increased meat or egg production: but this must not be carried to extremes, for inadequate ventilation leads to disease problems.

Animals supplied with ample dry bedding have an instinctive ability to adjust their own microclimate, and the same applies to young stock kept with their dams. Intensive systems with little or no bedding and early weaning are therefore particularly at risk from any breakdown of the heating and ventilation equipment on which they depend.

Effects on grazing

Grazed crops are different, in that animals stand on them during harvest. In wet conditions hooves sink into the ground, bringing about a condition known as 'poaching' and destroying both crop and soil structure. The problem is serious because much grazing is on heavy soils in areas of high rainfall. Drainage helps, but some soils will poach at well below field capacity. In mild moist areas, where grass may grow both early and late in the year with soil very wet, it is soil moisture, and not growing temperatures, which determines the housing and turning out of cattle.

Effects on parasites

When parasites have part of their life-cycle external to the animal, the weather may affect timing and severity of infection. The parasite causing liver fluke disease of sheep spends the summer in a snail which flourishes in wet conditions, and consequently there is a well-established relation between wet summers and the influence of fluke. Nematodiriasis is a disease of lambs, caused by a nematode which is picked up from pastures in the spring. The timing of egg hatch is well related to soil temperature, with moisture as a secondary factor.

Forecasts of hatch dates of nematodes and likely severity of disease can play a useful part in control. They can guide farmers in the best dates for moving animals to clean pasture, and also serve as a warning of the need to undertake dosing with anthelmintics before infection becomes damaging.

Windborne disease

Certain animal diseases have been shown to be spread by the wind. The big success in this field has been the understanding of the spread of foot-and-mouth disease, epidemics of which have caused the slaughter of whole herds of cattle and great personal distress to the farmers involved.

> For a long time the source of infection was mysterious; infected food, the wheels of vehicles and boots of visitors to farms, and even the feet of starlings being variously blamed. Whatever the cause of a primary infection, secondary spread is nearly always by wind, the virus travelling by droplets coughed by infected animals. Survival of the virus depends on the droplet's water remaining unevaporated, so humid or rainy conditions are required. Rainfall also contributes to deposition of the droplets, which are then ingested by grazing animals. Overcast and night-time conditions are also favourable to survival of the virus, which may be damaged by sunlight. These criteria enable sectors down-wind of a primary outbreak to be defined, and veterinary inspections and precautions concentrated in the areas where they will be most effective. The usefulness of these arrangements was amply confirmed in 1981 when an outbreak originating in Brittany spread from the Channel Islands to the Isle of Wight and central southern England, but was soon brought under control.

WEATHER MODIFICATION

It is perhaps for the best that large-scale weather modification is out of reach. 'Good' and 'bad' weather are relative terms; one farmer may desire fine weather to dry his hay, while at the same time his neighbour hopes for rain to germinate his turnip seed. Manipulation of microclimate is another matter. The farmer can often do this to his advantage, but generally without upsetting his neighbours.

Drainage

Artificial drainage by clay or perforated plastic pipes has an important part to play in getting excess soil water away quickly. It cannot do more than bring the water content down to field capacity (by the definition of field capacity), but the process is speeded and waterlogging avoided. That is very important for effective root action, both on pasture and for arable crops. Excessive water replaces air in the pores of the soil, depriving roots of oxygen they need, and the soil may consolidate so that root extension is slowed down. In the spring, wet soils warm up slowly, because water in the soil pores has a much higher heat capacity than the air which it replaces. Growth is slow to start and again drainage will be beneficial.

Design of effective and economical drainage systems depends on a knowledge of the rainfall climate of area, as well as the soil characteristics. It is not economic to provide a drainage system to deal immediately with the most violent rainfall or the wettest of winters. A compromise has to be found, for instance a design may be found which is completely adequate nine years in ten, and does a fair job in the tenth year. Such an approach demands detailed knowledge of rainfall, its variation from year to year, and range of intensity over times ranging from a few hours to several days. This information is now available for England and Wales (Smith and Trafford, 1976).

Many agricultural enterprises now have large areas of impermeable surface (concrete standings, roofs of buildings and glasshouses). When heavy rain falls on these there is no buffering effect of soil absorption, and sophisticated drainage designs may be necessary.

Irrigation

The object is to provide the best conditions of soil moisture for the crop at each stage of its growth, or at least at those stages which are vital for successful cropping and high yields.

Irrigation planning is concerned with the *balance* of water gained by and lost from the soil. Gain is by rainfall plus any water applied. If the planning is effective loss by drainage will not occur; loss will be as water vapour from the soil and from the plants. Transpiration refers to water vapour loss through the plant surfaces. Evaporation refers to vapour loss from liquid water surfaces. Evapotranspiration is a term sometimes used for the total vapour loss from soil and crop surfaces. In agricultural work it is the water loss from crops which is of most importance, and the term transpiration will be used with the understanding that it may include a proportion of water loss direct from the soil.

Satisfactory estimation of water loss is not easy. The best approaches are those that use the work of the micrometeorologist on the detail of conditions in and above the crop, simplify them appropriately, and connect them to the macro-scale measurements made in the standard meteorological enclosure.

The starting point is a realisation that transpiration is limited and governed largely by the available energy for providing the necessary latent heat of vaporisation of water. Solar radiation is the energy source. From the incoming radiation is subtracted what is reflected at the ground, also outgoing thermal radiation. The result is called net

radiation, and this must now be split between latent heat of vaporisation of water, and heat going to warm the air and soil. Irrigation
planning deals in time scales of several days. Over such periods soil
heating and cooling largely averages out. In temperate climates, and
generally when irrigation is applied, the heat going to the air is a small
component of the energy balance and some short cuts may be taken in
its estimation. The remaining energy goes to transpiration.

The most widely used codification of these ideas is the so-called
Penman formula, which is in fact not a single algebraic formula but a
battery of formulae from which a selection may be made according to
the observations available. If net radiation measurements are available,
they may be used directly. If only incoming shortwave radiation is
measured, a reflection factor and an estimate of outgoing based on
temperature may be used to estimate net radiation, and, if shortwave
radiation is not measured, it in turn may be estimated from hours of
bright sunshine. Then the available energy is split between air heating
and transpiration by a crude representation of the profiles of wind,
temperature and humidity over the crop. In the course of this temperature and humidity at the leaf surface are connected in a way that
assumes that water is freely available to the crop. The result is *potential
transpiration*, which is the estimated transpiration from a crop not
under water stress. The widespread success of Penman's ideas is due to
their firm physical foundation in the surface energy balance, and the
aerodynamics of heat and vapour transfer over the crop. The minimum
observational requirements are standard measurements of hours of
bright sunshine, air temperature and humidity, and wind speed. The
calculations give daily values, but for reliability should be averaged
over periods of a week or more. Monthly and long-term averages of
potential transpiration can be formed in the usual way, and are now
readily available for England and Wales. Once such averages have been
formed, it is no longer necessary to perform the Penman calculations in
detail. For the summer months at least, it is found that deviations of
potential transpiration from average are simply related to the particular
month's sunshine hours, which then becomes the only weather element
which needs to be measured routinely.

With transpiration calculated and rainfall measured, a soil moisture balance can be struck.

The starting point is with soil at field capacity at the end of winter, and
the reduction in soil moisture below this level is known as soil moisture
deficit (SMD). Week by week calculated transpiration is accumulated
and any rainfall subtracted, to give the current deficit. The SMD
cannot become negative; any excess moisture is assumed to run away
down the drains. A maximum permissible SMD is decided depending
on the crop and the soil. Such maxima are fixed on the basis of field
experiments, and are typically 25 mm for grass and salad crops and

50 mm for deeper-rooted crops. When the maximum is approached, water is applied so as to bring the soil near to field capacity again, and the calculations continue.

These water-balance calculations have their first application in planning irrigation installations. By carrying out the calculations on past weather data, the range of irrigation need over a series of years can be found, and the capacities of reservoirs and application equipment planned rationally. As with drainage, it is usually not economic to provide for the most extreme years. Once the system is working, the calculations can be used with current weather data to decide the amount and timing of applications. Some refinements may need to be applied for periods when the soil is bare or only partially covered with foliage. This applies particularly to crops such as potatoes and sugar beet which are slow to cover the ground.

Irrigation planned in this way generally produces better crops than a policy of 'putting some on when it is needed'. By the time a crop is visibly in need of water, much potential yield can have been lost.

It is sometimes remarked that irrigation water does not do as much good to a crop as the same amount of natural rainfall. There may be two reasons for this. Rates of irrigation application are typically much greater than all but the heaviest natural rainfall, and the droplet size is larger. Some mechanical damage to foliage and soil structure may therefore result. Also, irrigated crops are often small moist areas surrounded by much larger dry ones. Warm dry air blowing from the unirrigated parts to irrigated fields can cause moisture to be extracted from the crop at much more than the potential rate, so that unexpected soil moisture deficits develop in the irrigated crop.

Mulches

Instead of applying water to combat unwelcome dryness, it may be more profitable to apply a mulch to the soil and retain what is already there. There is an excellent account of the effects of mulches in W M O Technical Note 136.

Frost protection

To avoid frost, it is useful to have some idea of how it develops. The ground (or crop) surface radiates heat to space all the time, but by day this is balanced or outweighed by incoming radiation. By night with clear skies, the surface temperature will fall unless the surface can gain heat from some source.

In windy conditions, this may come by turbulent transfer from a deep layer of air. In still conditions, heat is withdrawn from only a shallow layer of air near the ground, and an air frost may develop. Crop and soil conditions are important. Moist, compacted and bare soil readily conducts heat to the radiating surface, and frost may be retarded or averted. Dry, cultivated and crop- or weed-covered surfaces are full of air spaces, act as insulators, and may encourage frost. This is one reason why bare soil surfaces in orchards are now common. Straw mulches can be disastrous for a crop such as blackcurrants where one night of frost can destroy the year's crop, so it is now recommended that their application for water retention should be delayed until after flowering.

Once a cold air layer has formed near the ground, it may, being relatively dense, drain under gravity to lower levels. Certain hollows and valley bottoms regularly get persistent frosts in this way, and should be avoided for the planting of susceptible crops.

Where siting cannot eliminate the problem at source, some *ad hoc* measures are available. It is seldom economic to supply heat directly in the open. Smudge generators work largely through the smoke which envelops the crop and acts as a blanket to outgoing radiation. The most effective method is to cover the crop with a fine spray of water, which freezes with the release of latent heat. Sufficient water must be available to keep up the spraying for the duration of the frost, so that the plants are covered with an ice-water mixture at 0°C. Since plant-cell contents freeze at below 0°C, damage will be avoided provided the plants are strong enough mechanically to sustain the weight of ice. Further information is available in WMO Technical Note 51.

Shelter

Protection from the wind is usually provided primarily to avoid direct mechanical damage to crops or structures, and if properly designed it can be most effective. Secondary effects such as changes in temperature, humidity and transpiration rate depend on many factors and are more difficult to demonstrate. Solid fences and dense hedges have a large sheltering effect immediately downwind, but beyond the wind is strong and turbulent. More permeable barriers have a filtering effect which extends further downwind.

Livestock may benefit from shelter, as their heat loss is reduced, particularly lambs and calves in the first days of life. Mature stock sometimes take shelter when it would be better for them to be grazing.

Shelter can increase frost risk by stopping cold air drainage. Where this is likely, broken and staggered belts may be used. Shelter belts which divert draining cold air from susceptible crops can make a positive contribution.

Protected cropping

From the meteorological point of view, glasshouses, plastic tunnels and growth rooms are all climate modifiers. By enclosing the growing space with walls and roof, the rainfall is reduced to nil, and wind speed to a small and relatively controllable value. Water must then be supplied artificially. Incoming radiation is still the best basic guide to water use, but artificial leaves and other devices to estimate evaporation in the controlled environment work better than in the open. Radiation itself may be reduced by shading, increased by the use of lamps, or supplied entirely artificially as in growth rooms for young plants. Temperature control is usually vital. Glass is transparent to incoming shortwave solar radiation, but relatively opaque to outgoing thermal radiation – the so-called 'greenhouse effect'. Some temperature lift both by day and by night can therefore be achieved without artificial heating. The same applies by day to plastic used to cover tunnels, but most plastic is relatively transparent to thermal radiation and so tunnels may suffer from low night temperatures. Enclosing the air in structures to achieve warmth may cause high humidity, since water transpired by the plants cannot get away, and this can lead to disease problems.

In planning enclosed environments, a knowledge of the outside conditions to be expected is the essential starting point. Extreme temperature conditions give the maximum heating requirement; average conditions, the likely running cost. Wind climate should not be forgotten, both for its effect on heat loss and the damage which gales can cause to structures.

Buildings for crop storage and animal housing

In glasshouses admission of light is the main requirement, but in most other agricultural structures this is not so and the walls can be designed more to retain heat. Much attention is given in the literature to thermal conductances or U values. Thermal capacity (the ability to retain heat) is also important, particularly for young or stressed livestock, which respond badly to sudden changes of temperature. The correct approach is to form an energy balance for the building and its contents, not forgetting the metabolic heat of animals and stored crops.

WEATHER OBSERVATIONS AND OBSERVING NETWORKS

Although some weather stations are set up specifically for agricultural purposes, particularly on experimental stations, we are dependent largely on readings acquired for other purposes. At the top of the scale are professionally manned stations, often on airfields, reporting usually every hour, day and night. The readings taken will include wind speed and direction, air temperature and humidity, rainfall and sunshine. The observer himself will estimate and describe 'present weather' (rain, fog, snow, etc) and visibility.

In the UK there are about 80 such stations. Their distribution shows an understandable bias towards flat country suitable for airfields, and coastal stations which first encounter incoming weather systems. Next in order come some 400 climatological stations at which readings are taken once daily at 9 a.m. Some information on what has happened since the last reading is provided by instruments which record sunshine continuously, rainfall totals, 'run-of-wind', maximum and minimum temperatures. A proportion of these stations is set up for agricultural purposes, and these will record temperatures on the ground and in the soil, while others may not. The observers are usually part-timers, and the regularity and reliability of observations is less than at the professional stations. Finally, there are some 6,000 stations recording rainfall only, at daily (or if very isolated, longer) intervals. At all these stations there is some professional control over the vital matters of siting and exposure of the instruments. It is relatively easy to read a thermometer with great precision; much less so to ensure that it is at the temperature of the surrounding air, and that these immediate surroundings are representative of conditions in an area which, it is hoped, will take us most of the way to the next observing station. The approach has been through standard instruments exposed in standard housings over a standard surface (short grass), so far as possible on an open site. Such readings are at any rate comparable with each other and through time. Their interpretation and handling as a guide to what happens on a particular farm, over or within a crop or in a building, obviously requires skill and experience. Most national meteorological services have specialist agricultural meteorologists, and it is sensible to seek information and advice through them. In different countries they may be found at the national meteorological headquarters, at major offices of the agricultural advisory service, or at colleges of agriculture. An important part of their work is processing of primary meteorological data into agriculturally relevant forms such as soil moisture deficits, available workdays, disease infection periods, and so on.

Much progress has been made in automatic equipment, but it needs a lot of skilled attention and even so, reliability is not as high as with the trained human observer. Proposals for full automation of environmental measurements in association with agricultural experimentation should at present be viewed with caution. Moreover, such *ad hoc* measurements cannot be set against the climatological background which exists for the well-established routine measurements.

CLIMATIC CHANGE

Long ago the earth's climate was different from what it now is. Ice ages have occurred at intervals of the order of 100,000 years and involved a drop in mean temperature of *c.* 10°C. Less familiar, but important for agriculture, is the idea that the atmosphere is always fluctuating on a very wide range of time scales (Fig. 7.2).

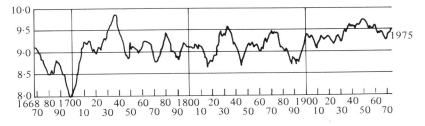

Fig. 7.2   Ten-year running means of Central England temperatures from 1650 to 1960 as compiled by Manley and now extended to 1975

To assess the likelihood of future changes in climate and their possible effects on agriculture, we must first know their causes. There is now general agreement that the ice ages are brought about by very slow changes in the pattern of receipt of solar energy at the earth's surface, caused by perturbations of its orbit by the outer planets. It is only 10,000 years since the last ice age, not long in geological terms, and one opinion is that the present temperate climate is only a brief intermission and that we might at any time be plunged back into deep cold. This seems unduly alarmist and it would be better to concentrate on the causes of the fluctuations of *c.* 1°C and periods from decades to centuries, which have been recorded in historical times.

At present there is not complete agreement on the causes of these, but likely candidates are large emissions of volcanic dust, and changes in cosmic ray activity affecting the ozone layer in the stratosphere and hence its circulation. These effects are essentially capricious. However, northern hemisphere temperatures rose about 0·7°C from 1880 to 1950, and fell about 0·4°C from 1950 to 1970, a decline which has probably now been arrested.

It would be sensible for agricultural planners to regard the conditions in the 1940s and 1950s as historically favourable, and be prepared for weather somewhat less favourable and somewhat more variable for the rest of the century. As always any deterioration of climate will bear hardest on marginal crops and marginal land.

8

CAPITAL

Prices of farmland in England and Wales rose threefold between the middle of the 1950s and 1960s and then trebled again over the next ten years so that by the middle to late 1970s they stood about ten times higher than they had in the early 1950s. Although they have been subject to wide and erratic fluctuations in recent years, nevertheless price movements were strongly upward taking the decade of the 1970s as a whole, in spite of falls of almost £1,000 per ha recorded in the second half of 1980 itself. At the market's peak that same year, farmland in England and Wales sold with vacant possession brought, on average, almost £4,500 per ha.

Such striking price changes have had an important bearing on the make-up of national farm balance sheets as can be seen from those set out overleaf for three years in the 1970s.

The fact that aggregate asset values rose sevenfold between 1970 and 1979 is mainly the result of changes in land prices. The direct and important consequence is that land and buildings have come to occupy dominant roles in total asset arrays while other inputs, especially livestock, have assumed correspondingly less important ones. So-called landlords' assets probably made up little more than half farming's total capital stock at the beginning of the 1950s whereas currently they make up over 80%. The result is that arrangements for holding land and for financing its transfer hold the key to the industry's financial and general business structure.

In spite of the massive increases in asset values, especially in more recent years (total assets probably rose only threefold over the 1950s and 1960s together) liabilities have not shown anything like the same rates of gain. Indeed, debts increased little more than threefold in the 1970s during which time assets rose sevenfold. Consequently, farming's net worth has itself grown dramatically from a little over £7,500 m. in 1970 to over £58,000 m. in 1979, and those figures are to be compared with a total net worth of probably less than £2,500 m. at the beginning of the 1950s. In real terms,

Table 8.1   National farm balance sheets: UK

	1970	1974	1979
Assets	%	%	%
Land and Buildings	66	75	82
Machinery	9	6	6
Livestock	15	11	7
Crops and Stores	6	4	3
Debtors and Cash	4	4	2
	100	100	100
Total Assets	£8,870 m.	£19,250 m.	£62,000 m.
	%	%	%
Liabilities			
Long term specialist institutions	14	15	} 13
Insurance Companies	3	2	
Banks	39	49	66
Private and Family	26	13	5
HP and trade	18	21	16
	100	100	100
Total Liabilities	£1,250 m. (14%)[1]	£1,865 m. (10%)	£3,800 m. (6%)

[1] Figures in brackets are liabilities as a % of assets
Sources: Harrison (1975); de Paula (1976); Bruce (1979)

total liabilities probably did not rise at all in the 1970s but fell to even less than they were in the 1950s and 1960s.

The banks are much the most important source of borrowed funds while borrowing from private sources and from within the farming family have become less important. However, the area is not well served with full and reliable statistics and some earlier estimates of informal borrowing from non-institutional sources were almost certainly overstated. Nevertheless, there is a good deal of evidence to show that the role of the family is still a central one in the financing of farms in spite of the fact that traditional family loyalties have tended to become less strong than in the past, so putting heavier burdens of debt and risk-bearing on individual farmers and their partners.

The government has seldom sought to support farming directly through credit institutions and borrowing arrangements to any significant degree although there were important Agricultural Credit Acts in 1923 and 1928. The first measures constituted very largely a farmer/banker rescue operation following the collapse of farming fortunes in the early 1920s (itself partly due in turn to the reduction of government price support to the industry). The second were instrumental in setting up the Agricultural Mortgage Corporation in 1929 following R. R. Enfield's study of the industry's finances. The Corporation is far and away the most important institution in the UK existing solely in order to finance purchases of farmland (and to a lesser extent capital improvements). Although it was slow to establish itself as a serious competitor to the banks in providing long-term lending and indeed had made only relatively modest progress by the 1950s, in recent years it has financed about a quarter of total sales each year. However, over a third of its loans more recently have been on a secondary, additional advances basis for which further security has not had to be lodged, thereby reflecting the radically different credit ranking of the person with land on which there is only a modest mortgage, and the would-be entrant to farming who, although probably well-fitted in every other way to farm, lacks capital of his own.

For many years Agricultural Mortgage Corporation lending was on a fixed-interest basis only, with repayments to be made on a constant, fully amortised basis over a period of up to 60 years with borrowing restricted to two-thirds of the Corporation's valuation of the property. Then repayment through an endowment life assurance policy was introduced and now it is possible to contract on a fixed or variable interest basis. In certain cases also, loans can be taken up on a straight five to ten year period.

However, trends in total liabilities and assets tell only part of the story of how farms are financed. Financial arrangements need to be examined for representative groups and, in the end, for individuals.

LAND TENURE AND FARM FINANCE

Just as the striking upward movement of land prices has made it the major determinant of farming's aggregate assets and liabilities so land tenure provides the key to individual farmers' wealth and indebtedness positions.

When rents average some £50 per ha and the same amount of farmland sells, as in 1981, for about 70 times that figure (and almost 90 times it a year earlier) then, area for area and farming system for farming system, the form of tenure which a farmer has to adopt in order to enter the industry must inevitably exert a major influence on the scale on which he can operate with any given amount of funds to which he has access. In the first decade or two of this century it would have been quite the exception to have come into farming by any means other than tenancy. Indeed, concern was expressed in the 1928 Enfield Report on Agricultural Credit that the farm tenancy system allowed people to begin farming on their own accounts on scales which were too great and therefore too risky for the funds which they owned personally. Now, it is owner-occupation that is the rule and tenancy the exception, and the fear is that access to personal funds is virtually the only test of fitness to farm so that many otherwise talented and worthy would-be entrants are being excluded as a consequence. The decline of agricultural landlordism has had many consequences, not the least of which has been to remove a large part of the supply of funds supporting and locked up in land. It has therefore, albeit slowly and over several decades, transformed the socio-financial structure of farming as landlords have sold (often to sitting tenants but more recently to institutions) or, alternatively, have taken land in hand to farm on their own accounts.

Even as recently as 1950 almost half of farms were entirely rented, although ten years later that number was estimated to have fallen to 37% and ten years later again to 31%. Currently it is probably under 25%. Over the same three decades entirely owner-occupied farms have increased from 36% to just under 50% while mixed-tenure farms have increased from around 15% to almost 30%. Nevertheless, even such striking figures as these almost certainly fail to reflect the full extent to which owner-occupier farming has developed today. In a recent official (Wyre Forest) study of land-ownership (the first of its kind for over a hundred years and even then on only a modest scale) it was found that almost 70% of land was owner-occupied and that no less than 80% was owned by the farmer or his family.

The extent to which official statistics fail to register fully the growth in owner-occupation can be gleaned only from independent and, unfortunately, relatively small studies but there can be no doubt that owner-farming is understated, and that such understatement stems from the growth of within-family tenancy arrangements designed to reduce capital taxation liabilities. Moreover, it has been found that, in the eastern counties, at any rate, it is particularly on larger farms that owner-occupation is much more common than official statistics show. The discrepancies between official statistics and the findings of independent studies of individual farm businesses arise from the fact that official statistics have sought to establish the *de jure* position regarding tenure while individual researchers have aimed to discover the *de facto* one. A further discrepancy arises over farm numbers in that holdings, which are the primary units of enumeration on which official statistics are based, overstate the numbers of small farms but understate the numbers of large ones. The problems are therefore related. In England in 1969 (Harrison, 1975) it was found that about two-thirds of farms below 20 ha were entirely owner-occupied against only about one-third of other size groups (slightly under for those 20 ha plus, slightly over for the even larger-sized ones). Purely rented farms featured less prominently and mixed-tenure farms more prominently as size of farm increased.

Such a dramatic revolution in land tenure, although bloodless, might have been expected to require the correspondingly dramatic creation of new, highly institutionalised financial sources and arrangements and to have resulted in increasing, and perhaps large-scale, indebtedness. But the national farm balance sheets (see Table 8.1) show that is not the case. In spite of the amount of borrowing that individuals have had to undertake in order to obtain land and the working capital to stock and crop it, it is the modest amount of total borrowing relative to farming's total assets base that is remarkable. There are a number of reasons why that should be so. First, the changeover has taken several decades to achieve so that the inevitably heavy indebtedness of more recent purchasers has come to be hidden by the enhanced net worth of earlier buyers. Second, although land prices have moved steadily and sometimes, as in the early seventies, rapidly upwards many purchases have been by sitting tenants at modest prices. Indeed, not only did many of the early purchasers in the 1920s and 1930s buy as sitting tenants, their purchases were made at what proved to be both low prices and rock-bottom interest rates. Third, financial burdens in farming have generally been shared by families, thus involving little outside borrowing. In more recent decades inheritance of actual properties has become the most common way into farming. Fourth, in spite of rising land prices and of farm incomes and prices generally, nevertheless, lending for farm purchase has not been indexed but has remained fixed in money terms so that inflation has quickly eroded the real burdens of indebtedness. Fifth, although farms have reduced in numbers and hence increased markedly in average size, those going out have been the smaller so that those remaining in the industry have not had to make the increases in scale apparently implied by the growth in average sizes.

The effects of having to finance landownership on the overall pattern of both assets and of liabilities are reflected in the figures for the three main tenure groups in the two most recent (1978/9 and 1979/80) studies of farm incomes in England carried out by agricultural economists in various regions on behalf of M A F F (see Table 8.2).

The farms making up the sample in the two years are not identical so that although there is a large measure of overlap, precise conclusions about changed assets and liabilities cannot be reached. Nevertheless a number of important points can be made on the basis of the evidence presented.

First, the total assets financed by the owner-farmer (and his creditors) far outstrip those financed by the tenant (and his creditors). Second, in both cases they amount to large sums in absolute

Table 8.2   Liabilities and assets per farm by the three major tenure groups in England

1978/9	
Tenanted farms, 91 ha average size	£71, 562 total assets
	£10, 341 total liabilities (14%)
Owner-occupied farms, 73 ha average size	£272,131 total assets
	£19,396 total liabilities (7%)
Mixed-tenure farms, 114 ha average size	£268,457 total assets
	£24,101 total liabilities (9%)
1979/80	
Tenanted farms, 104 ha average size	£89,868 total assets
	£15, 646 total liabilities (17%)
Owner-occupied farms, 80 ha average size	£337,021 total assets
	£30,134 total liabilities (9%)
Mixed-tenure farms, 111 ha average size	£294,886 total assets
	£27,644 total liabilities (9%)

Source: MAFF (1980); MAFF (1981)

terms whether measured against the assets employed in other non-farming family businesses employing only one or two workers, or the amounts of personal wealth which all but the most wealthy members of society own. Third, although owner-occupiers and farmers on mixed-tenure holdings are heavier borrowers in total than are tenants, nevertheless, tenants are about twice as heavily indebted as owners and part-owners when liabilities are related to assets employed.

The most important reason for this apparently heavier borrowing by tenants is mainly, if not entirely, due to the way land values have appreciated relative to historical debts, in turn largely mortgage-based in order to finance land purchases, thereby enhancing owners' net worths. However, it almost certainly has another significant reason, even if quantitatively less important, in that tenants are more inclined to borrow in order to invest more heavily in working capital simply because they do not have the owner's advantage of enjoying capital gains on land.

But the differences in financial arrangements between them do not end there. The relatively greater borrowing by tenants tends to be obtained from shorter-term and less specialist agricultural credit sources. Moreover, there are significant differences in capital employment also; in particular, tenants have tended, in the past at any rate, to invest relatively more in machinery while owner-occupiers have invested in larger and newer building arrays than are to be found on rented farms. In 1978/9 and 1979/80, for example, the two most recent years for which figures are available from *Farm Management Survey: Farm Liabilities and Assets* (itself part of the MAFF/Universities study of farm incomes), tenants invested

48·7% of newly disposable funds in machinery and equipment in the former year and 47·8% in the latter, while corresponding figures for owner-occupiers were 36·1% and 32·8%.

Although there are periodic surveys of machinery and equipment on farms as part of the routine collection of official statistics and these provide aggregate numbers of the more important items of machinery employed, long-run series for individual categories are often not available; moreover, unanswerable questions of comparability of type, working capacity and efficiency arise which make detailed conclusions impossible to draw. Nevertheless it is noteworthy that during the 1960s and 1970s total numbers of machines employed on farms have risen only modestly. Indeed, during the 1970s (when unfortunately the coverage of the figures was less good) tractors appear to be the only items of importance that increased, the remaining items registering marked falls in numbers.

The only recent study of farm buildings was for England and Wales (Hill & Kempson, 1977) and showed that, in 1973, no less than 36% of buildings (by floor area) dated from before 1915 while not far short of that amount dated from the introduction of the Farm Improvement Scheme (FIS) in 1957. Smaller, and therefore generally owner-occupied farms, tended to have more older buildings in spite of the fact that owner-occupiers collectively had made more use of government grants towards building costs (under the FIS) than had landlords and their tenants. Although the largest farms singled out for separate attention, those of over 200 ha, had only around one-ninth of the stock of pre-1915 buildings they accounted for more than one-fifth of the stock dating from the introduction of the FIS. Both these facts reflect the way in which that particular support scheme appealed to and was better able to assist the larger farms regardless of tenure, partly because it was aimed at not encouraging heavy investment in specialist (e.g., pig and poultry)-type buildings.

FARM TYPE AND CAPITAL EMPLOYMENT

It is not only size and tenure of farm which influence the amount of capital a farmer must find – it is type of farm as well. This is reflected in the figures of capital per farm for a sample of owner-occupied farms in 1973-4 (Table 8.3). Unfortunately, corresponding figures for more recent years are no longer calculated.

Size of farm as measured by the standardised workload which the farming system represents and by the area of land farmed provide roughly comparable scale indicators for livestock and cropping farms. At any rate to 600 SMDs area and workload rankings differ little, although, above that, the additional workload for every additional hectare is less on livestock farms than on cropping ones.

Table 8.3   Assets per farm by size and type of farm; owned farms, England, 1974

Type of Farm	Size in Standard Man Days	in Ha	Land & Buildings £	Other Fixed Assets (inc. Breeding Stock) £	Trading Livestock, Crops, Stores & Other £	Total Assets £
Dairy	275–599	24	16,464	21,671	1,826	40,161
	600–1,199	56	49,406	60,807	4,896	115,109
	1,200–4,199	104	107,197	129,376	10,572	247,145
Livestock	275–599	59	36,784	42,088	4,277	83,149
	600–1,199	125	62,698	74,716	8,325	145,739
	1,200–4,199	205	238,654	262,016	28,322	528,992
Cropping	275–599	57	84,834	87,834	10,069	182,737
	600–1,199	100	123,592	131,052	12,984	267,628
	1,200–4,199	176	253,156	267,395	26,619	547,170

Source: MAFF (1975)

However, dairy farms are much more labour demanding than both other systems. Standard man day grouping for standard man day grouping the area of the dairy farm is about half that of each of the other two types. More recent information on the amounts of tenant's capital employed is provided in the MAFF reports on farm incomes from which the figures for 1978/9 and 1979/80 are taken (Table 8.4).

The contrasts between the amounts of tenant's or working capital per hectare are remarkable. At the one extreme in 1979/80 a single hectare horticultural holding employed as much capital as a hill and upland farm 100 times as large in terms of area. Within other groups, working capital levels of £1,000/ha on specialist dairy farms compared with figures for pig and poultry farms that approached double that amount, but were somewhat more than for general cropping farms and more than double those for specialist cereal farms. Within individual type of farming groups the fall-off in level of capital employed per hectare with increasing area was by no means general and, even where it did occur, was not marked except in the case of horticultural holdings which, at the lower end of the scale, were employing over £20,000 of working capital per hectare but, at the upper, little more than a tenth of that rate.

FINANCING ENTRY INTO FARMING

As the total stock of farm capital has increased both through new capital formation and rising land values, and the percentage of

Table 8.4   Tenant's capital per farm by size and type of farm, 1978/9 and 1979/80

	1978/9		1979/80	
	ha	£	ha	£
Specialist Dairy Farms				
Small	18	15,280	19	17,623
Medium	51	53,837	51	61,704
Large	136	138,082	132	158,232
Hill & Uplands Cattle and Sheep Farms				
Small	113	28,960	126	33,454
Medium	381	73,321	359	79,195
Large	–	–	–	–
Specialist Cereal Farms				
Small	–	–	–	–
Medium	79	40,878	79	46,607
Large	205	106,049	203	129,544
General Cropping Farms				
Small	–	–	–	–
Medium	51	35,762	57	42,748
Large	166	112,564	164	125,622
Pig & Poultry Farms				
Small	–	–	–	–
Medium	32	39,919	26	46,231
Large	84	114,751	90	138,568
Horticultural Holdings				
Small	2	25,580	1	24,630
Medium	4	35,511	11	37,675
Large	59	68,926	39	90,873

Small farms are between 4 and 7·9 ESU; medium between 16 and 23·9; and large between 40 and 99·9 (ESU stands for European Size Units). They are based on standard gross margins at average 1972–4 values. One ESU equals 1,000 European Units of Account.

Source: MAFF (1980) and (1981)

owner-occupied land has risen, so the financial burden facing new entrants to farming has increased. Many have been compelled to incur levels of indebtedness of 50% and over of assets in order to gain entry. Their presence is obscured in the aggregate, apparently very reassuring, financial statements already examined, but independent surveys have found as few as 5% of farmers to be responsible for 60–70% of total borrowing. They constitute, therefore, a heavily indebted but small minority of farmers, most of whom have been farming relatively few years.

At such high levels of indebtedness farmers in that minority are at risk should incomes or land values fall. For example, a loss of £5,000 in any one year on a farm employing £100,000 of assets where the farmer had no debts would represent a 5% loss of equity to meet it. Had the farmer financed that same farm by borrowing

£50,000 and contributing the rest from his own funds then double that rate of loss of equity would have been required to meet it. Neither depressed incomes nor falling land values have occurred on a sustained basis as yet, so that up to now this at-risk minority has been a transient one, with its membership changing year by year as asset prices and incomes have risen. They probably represent a relatively recent phenomenon, at least since the periods of farming depression in the 1920s and 1930s when new entrants, and their creditors, were caught on falling markets. Clearly, however, they face massive risks should farmland and product markets reflect sustained fears of depression again. Moreover, if that did happen it could cause serious disruption for farms still employ the simplest business forms. Proprietorships and partnerships dominate the industry and the use of joint-stock financing is scarcely ever encountered. The consequence is that farmers' entire personal fortunes are at risk in the event of business fortunes deteriorating. In the general course of events business forms of this nature might be expected to result in relatively cautious investment and borrowing patterns. In the event of depression they result in proprietors and their families bearing the heaviest burden of risk whereas more sophisticated business forms would cause many of those costs (including sometimes the repudiation of debts and the laying-off of labour) to be shifted back on to the public in general.

Harrison (1975) found for a sample of farms in England in 1969 that 67% of farm businesses were proprietorships, 27% partnerships, just over 4% private companies, while the remaining $1\frac{1}{2}$% or so were either public companies for whom farming was merely an adjunct to other industrial or trading activities or they were prisons, remand schools or other similar institutions. The study provided little evidence that the company, and risk-reducing, form of business organisation was employed, or was going to be employed, to recruit capital or entrepreneurship which would not have been equally readily available via partnerships.

Outside the typical, within-family, farming partnerships, many novel person-to-person arrangements were encountered (Harrison, 1975) of partnerships which had been formed reflecting varying degrees of paternalism, capital contribution and management participation on the one hand, and the search for temperamentally compatible and trustworthy business colleagues on the other. But such examples served in the end only to emphasise the part played by families and by inheritance in achieving entry into farming. Nevertheless, there is a clear link between size of farm and business form as the figures in Table 8.5 show.

On 97 out of every 100 farms in that same sample all the principals, where there was more than one, were closely related by blood or marriage.

Table 8.5 Business form (%) and size of farm: England, 1969

Business form	<20 ha	20-40 ha	40-113 ha	Over 113 ha
Proprietorship	81	71	59	40
Partnership	15	27	36	42
Private company	3	–	4	17
Other	1	2	1	1
	100	100	100	100

Source: Harrison (1975)

Amongst other things, this close connection between – indeed virtual coincidence of – business and family structure has resulted in a large measure of business continuity from generation to generation. The study already quoted, for example, found that successors to farm businesses were identified early in their careers, and in good time so far as actual transfer was concerned even where that occurred late in a formal sense.

A further feature of farm business structure is that part-time farming is common. About 30% of farmers have been found to have other sources of earned income, 70% of those from second and by definition non-farming businesses. Those part-time farm businesses are more common amongst the smaller but they are to be found in all size groups and even more amongst the larger than the medium ones. The move to part-time farming tends to be an inter-generation phenomenon rather than a within-career one but is not to be regarded as the short-term, transient phenomenon which some observers declared it to be some years ago.

Just as farming is dominated by proprietorships and close family partnerships so also is new capital formation financed mainly from personal and family funds. It is not unusual for 80–90% of the costs of investment to be met from personal sources including the re-investment of profits, sale of assets and outside funding such as gifts and non-farm earnings. Arrangements for funding investment in 1979/80 are shown in Table 8.6.

Throughout the last three decades and more the upward movement of both land prices and incomes has reduced borrowers' risks although sometimes incomes have lagged behind land prices thereby adding to the burden of farming transfers and reducing the willingness of lenders to provide loans, for a time at any rate. More recently, however, the failure of incomes to maintain themselves in real terms has heralded an extremely risky situation for borrowers in the shorter term and for capital tax payers in the longer (Table 8.7).

The farms have been treated as rented ones and income represents the return to the farmer and his wife for their own labour,

Table 8.6 Sources of funds for investment by tenure and size of farm, 1979/80

	Tenanted farms			
ESU/farm	4–15·9	16–39·9	40–249·9	4–249·9
Average size (ESU)	10·9	25·1	70·4	29·2
Average size (ha)	67	94	192	104
	% of funds			
Sales of land	–	–	–	–
Sales of machinery	10·9	12·2	13·4	12·6
Depreciation provisions	17·6	33·0	33·7	32·4
Retained earnings	8·8	17·8	18·1	16·1
(of which sales of breeding stock)	(2.6)	(4·8)	(2·2)	(3·2)
Capital funds introduced	23·2	14·2	10·5	14·1
Grants on fixed assets	6·8	3·5	5·6	5·1
Increased loans	22·6	19·3	18·6	19·6
	Owner-occupied farms			
ESU/farm	4–15·9	16–39·9	40–249·9	4–349·9
Average size (ESU)	10·7	26·2	64·7	26·0
Average size (ha)	50	75	167	80
	% of funds			
Sales of land	3·8	6·2	5·6	5·5
Sales of machinery	6·7	9·4	7·6	8·0
Depreciation provisions	23·5	26·6	18·8	22·3
Retained earnings	−9·5	11·0	13·4	8·7
(of which sales of breeding stock)	(−3·3)	(2·2)	(1·2)	(1·5)
Capital funds introduced	48·3	13·9	9·4	17·5
Grants on fixed assets	2·3	6·2	4·3	4·6
Increased loans	24·9	26·7	40·9	33·3

(For definition of ESU see Table 8.4)

Source: MAFF (1981)

management and working capital. Any appreciation in the value of breeding livestock has been ignored. However, aggregate figures such as these, disturbing as they are, fail to reveal the extents to which incomes for different types and sizes of farms can vary from year to year (Table 8.8).

Further evidence of the way in which income can vary from year to year for different sizes and types of farms is provided in Table 8.9 setting out income, after deducting for the farmer's manual

Table 8.7 Movements in aggregate net farm income, 1968–78

| | Farming Net Income (£ million) | |
	At current prices	At 1975 prices
1968	448	927
1969	498	976
1970	473	870
1971	574	970
1972	590	926
1973	857	1,234
1974	745	924
1975	921	921
1976	1,136	977
1977	1,301	963
1978 (est)	1,256	854

Source: Annual Abstract of Statistics (1980)

Table 8.8 Indices of income/ha by type of farming in England and Wales, 1968/9–1977/8
(1975–6 = 100), current money terms

	Dairy	Livestock	Cropping	Mixed	Pigs & Poultry	Horticulture
1968–9	24	25	15	23	22	30
1969–70	25	25	21	23	22	34
1970–1	30	32	23	26	20	35
1971–2	54	53	29	41	31	36
1972–3	71	81	40	55	40	54
1973–4	63	80	76	71	59	62
1974–5	52	59	79	64	49	70
1975–6	100	100	100	100	100	100
1976–7	107	120	113	94	74	113
1977–8	130	137	73	82	68	77
Actual 1977–8						
£/ha	146	52	89	95	211	212

Source: MAFF (1979)

labour, as a percentage of tenant's capital for a wide range of types and sizes of farms in the 1970s.

In order to permit comparison between farms the above figures have been calculated on the assumption that all farms are rented and actual or notional rental charges made. This has the effect of obscuring the effects of owner-occupiers' actual finance costs and of capital gains on performance. It is a serious deficiency in the statistics which, among other things, makes the return on working capital appear high relative to the return on all factor inputs. Nevertheless, it is still quite clear that heavily indebted new entrants must be at risk during the early years of mortgage servicing

Table 8.9   Return on tenant's capital (%) by size and type of farm; selected years, England and Wales

	1973/4-1975/6		1975/6-1976/7		1976/7-1977/8	
	All	Best 50%	All	Best 50%	All	Best 50%
Mainly Dairy						
Under 60 ha	8	12	15	26	12	20
60–120 ha	14	21	22	30	18	26
Over 120 ha	17	25	19	29	19	28
Mainly Arable						
Under 100 ha	19	28	23	38	11	25
100–200 ha	29	44	18	28	11	22
Over 200 ha	30	48	36	55	19	31
Dairy & Arable						
Under 200 ha	24	36	31	46	21	29
Over 200 ha	24	36	26	36	23	30
Mainly Sheep & Cattle						
Under 100 ha	10	15	12	23	5	18
Over 100 ha	15	22	14	23	12	29
Sheep/Cattle & Arable						
Under 100 ha	17	35	22	39	12	24
Over 100 ha	20	30	15	24	13	23
Mainly Pigs/poultry	29	44	30	38	13	34
Mixed, Pigs/poultry	22	33	26	44	14	36
Intensive Arable						
Fruit	35	50	34	58	25	45
Vegetables	45	65	48	67	24	37

Source: Nix (1977, 1978, 1979)

whenever incomes move downwards, even if only temporarily. Moreover, this is so in spite of the fact that, with interest allowed as a tax-deductible expense and prices and incomes generally rising over recent decades, real interest costs have often been negative and seldom more than 2% or 3% when positive. The problem is that land prices have long been high relative to farm incomes. Thus, with land priced, as recently, at around £3,500/ha, farming income of £150/ha or even £200/ha before tax and before living costs would be quite inadequate to service even a two-thirds mortgage with nominal interest net of tax of 35% costing, say, 10% or 11% and requiring therefore an annual payment of £250/ha or more to meet it.

A not dissimilar conclusion would have been reached if that sort of calculation had been made at any time in recent decades. Thus, although in the 1950s average income would probably have sufficed to service long-term mortgages on small- to medium-sized farms and especially on smaller ones, it would almost certainly not have done so on the larger. By and large, the bigger the farm the less adequate would income have been. In general, the burden of new mortgages tended to increase relative to income up to the late 1960s then to fall before peaking again in 1974. However, although initial mortgage burdens have long looked daunting and increasingly so as time has progressed, nevertheless, once undertaken, the high-risk period of heavy indebtedness has tended to become shorter rather than longer as the 1960s and 1970s have progressed.

The explanation of this somewhat paradoxical situation lies in the fact that incomes in current terms rose five and sixfold between the mid-1960s and mid-1970s thereby reducing mortgage burdens to almost nominal proportions as they fell to be repaid in current and therefore greatly depreciated pounds. In this process the already well established and less indebted have fared best and farmers themselves have dominated the farmland market as they have made eight or nine of every ten purchases made. The result, as the global statistics show, is that the industry displays all the signs, in aggregate terms at any rate, of financial solidarity and strength.

LANDOWNERSHIP AND CAPITAL TAXATION

The position of the new entrant is still not reassuring, however, and the observation that the current generation's capital gains provide the next generation's entry finance problems sounds a particularly appropriate warning so long as proprietorships predominate because capital taxation will then tend to coincide with transfer of ownership and management and to be levied when the business is at its most vulnerable. This has been recognised by the government which, although failing to adjust tax thresholds in line with more general price rises has also granted special concessions to make inter-generation capital transfers easier. At first these concessions applied to owner-occupier farms only, then to small-scale family businesses generally. Landlords were not granted those privileges until a 20% valuation concession was granted to them in the Finance Act of 1981. This belated and less generous allowance than that extended to owner-farmers has to be seen in relation to the fact that rented farms regularly sell for only two-thirds of the price that farms with vacant possession fetch, although, as financial institutions made their full impact on the market in the early 1970s, that one-third premium was bid down to well under one-fifth.

Capital taxation arrangements in the UK stand in marked contrast to those in other member states of the EEC since here they are based on the donor's wealth and there on the recipient's. Moreover, tax concessions in the other member states are much more generous than here, especially for the typically smaller, family farm business being transferred between closely related members of that family.

Although the farm landlord in the UK still plays an important role in capital provision and in farmer selection his role as a supplier of new investment funds has become a greatly reduced one. Nevertheless, landlords and their tenants together supply, in about equal proportions, the 25% of funds required, over and above the 50% provided by owner-occupiers and the 25% by government grant, to finance new capital formation. Tenants tend to invest more heavily in short-term and specialist assets, while landlords tend to invest more in longer-term and more general-purpose ones. As farms have become bigger it has become increasingly common for a farmer to rent from one or more landlords, but single landlords renting to a number of tenants and concerned to manage those farms as a unit are a major part of the totality of the farm tenancy picture. Finally, institutional landlords are important; almost 10% of farmland in Britain is owned by public and semi-public institutions. Moreover, those landlords play an important role in helping to maintain flexibility in what would otherwise be an immobile, inheritance-motivated and owner-occupier dominated system of finance and tenure.

In all these respects landlordism in the UK is different from that in the other member states of the EEC where landlords are not only small-scale, simple rentiers exercising no management or capital formation financing role, they are often retired farmers and hardly ever include institutions. Indeed in some countries, such as Denmark, institutional landownership is illegal. The concept of an estate is one that is not readily understood in Continental Europe and is not encountered in practice. A number of attempts have been made in the various member states to provide some of the functions carried out by landlords here. For example, in France the formation of small ownership groups (GFA) is encouraged where the viability of the farm business is threatened by the size of the inter-generation transfer problem. The Dutch have tried to encourage leaseback arrangements through their Land Bank. In general, however, it has to be recognised that landlordism is feared and, with the exception of West Germany, not encouraged. Even there it is small-scale and mainly of an outgoing farmer type.

The major growth point of institutional landlordism is the financial institutions which, although owning less than 2% of land in total, have tended to buy 10% and more of land coming on to the market in recent years. By this means they have helped to create a sale and leaseback market thereby making it possible for young owner-occupiers to achieve financial stability when over-indebted relative to their scales of operations. The use of sale and leaseback has enabled some farmers to gain control of much larger areas of land than would otherwise have been possible. Whether or not a sale and leaseback transaction represents a good move from the farmer's point of view depends on how depressed the 'with tenant' price of land is compared to that with vacant possession and on what future movements of land prices are expected. That both landlord and farmer can regard a sale and leaseback as mutually beneficial will turn on their respective views of longer-term price movements and their respective costs of waiting (interest or capital costs).

As U K landlordism has declined those private landlords remaining have altered their roles in two important respects. In addition to their reduced investment in buildings and other fixed equipment, a function which tenants have tended to undertake in their stead, often securing a tacit agreement that rents would remain at low levels, thereby helping towards a lower, and capital taxation reducing valuation of the farm for the landlord, there has been a rapid movement out of Discretionary Trust ownership now penalised by Capital Transfer Tax. The other development in landlordism is that as the Agricultural Holdings Acts and the 1976 tenant's successor legislation have increasingly favoured tenants in possession so full tenancies have become harder to find. This generation's tenants have gained but at the expense of future generations of potential tenants.

Nevertheless, landlordism continues, albeit in a greatly modified form. As 'normal' farm tenancies have virtually disappeared so landlords have endeavoured to take land in hand either solely or through arrangements with their tenants, granting them in turn grass lettings or licences. Although a licence to occupy does not create a legal interest in land so that the licensee occupies the land only at the free will of the owner, without the protection of the Holdings Act (provided the agreement is for a period of less than twelve months and contains no right of renewal) nevertheless, tenancy by licence is becoming an increasingly popular arrangement. Moreover, in practice, security of tenure need not be reduced and might even be increased where the right to succession is not guaranteed under the 1976 Act.

Such arrangements can also form part of a move towards farming in partnership in which the holder of the licence is a company or partnership, usually including a member or members of the landowner's family. Often a licence is granted to the landowner and a farmer tenant (or his son) farming in partnership. In that way the tenant is likely to benefit from receipt of capital, the owner from improved tax status both on income and on capital account. Sometimes such arrangements are offered in order to induce a tenant to give up his tenancy or a son not to apply for succession.

While it needs to be borne in mind that such arrangements are only a small part of what is by now a minority tenure form, nevertheless they form an important element at the margin of change of an industry in which the total farmland market covers only about 1·5–2% of the total stock each year and inheritance is the mode both of successor and farmer selection and of capital transfer.

A parallel but even more minor development has been the introduction of city funds in order to form large-scale farming combines. The funds have been recruited by specialist land management companies from individual and group investors seeking a combination of profit sharing and capital gains from appreciating land values. However, the problems and policies of those conglomerates by no means typify the general characteristics of farms over 500 ha and they are totally exceptional by wider west European farming norms.

9

LABOUR

AGRICULTURAL EMPLOYMENT IN THE BRITISH ECONOMY

At the beginning of the nineteenth century, more than a third of the working people of Britain were employed on the land. Today the proportion is under 3%. In the past hundred years alone, the British working population has more than doubled while numbers engaged in agriculture have been halved. (In this chapter the term *agriculture* is intended to include horticulture and *farmer* to include grower.)

One reason why the agricultural industry has been able to maintain and even increase its output with a shrinking labour force is that many of its traditional tasks are now carried out beyond the farm gate. At an early stage of economic development, most of the activities involved in producing food are undertaken by members of farm households. As an economy develops, so the proportion of labour engaged in food production which is farm labour, goes down. In Britain at the present time, the farm labour force is declining but the sector providing agriculture with necessary inputs such as machinery, energy, fertilisers, chemicals and services is growing at about the same rate as the overall economy. The same is probably true of developments in processing and distribution of farm products. Overall, less than half the labour embodied in food production in Britain today is farm labour. Whilst this chapter is concerned only with those who work on farms, it must be remembered that they in turn depend on workers in many other industries.

Sources of information

In order to build up a picture of the agricultural labour force it is necessary to consult a number of sources. The longest-running statistical series is the decennial population census, dating back to the early nineteenth century. Most comprehensive information is provided by the agricultural census carried out in June each year by the Agricultural Departments of England and Wales, Scotland and Northern Ireland. The annual Wages and Employment Enquiries give supplementary data on wages and conditions of employment for hired workers. Besides periodic censuses and surveys such as these, isolated studies are sometimes undertaken to shed light on particular aspects of agricultural employment. For example, the Economic Development Committee for Agriculture carried out a sample survey of the farm labour force in England and Wales in 1970, with the aim of forecasting manpower trends in the industry more accurately (Agriculture EDC, 1972).

Labour statistics need careful interpretation. Definitions can vary from one source to another and between countries of the UK. Even within a given series, changes in coverage of definition can cause discontinuities and make it difficult to identify trends and to build up a picture of the farm labour force in the UK which is both accurate and consistent over a period of years.

Size and composition of the farm labour force

The farm labour force can be divided into farmers, partners and directors, their spouses working on the farm, other family workers, salaried managers, hired workers and seasonal or casual workers. Farmers, partners and directors and their spouses doing farm work, including managerial and office work, are regarded as part of the labour force but those who have only a financial interest in the business are not. Family workers are relatives of farmers, partners and directors or of their wives or husbands who work regularly on the holding but have no contract of employment. Salaried managers are those with contracts of employment as managers who are normally paid monthly and may have profit-sharing arrangements as well. Hired workers are all workers with contracts of employment, including relatives with contracts. Seasonal or casual workers include all hired and family workers who are occupied on the holding on census day but not regularly employed there. Table 9.1 gives numbers in each of these categories for the national farm in June 1979.

Table 9.1   Composition of the UK farm labour force, 1979

	'000s
Farmers, partners and directors	293
Spouses of farmers, partners and directors	71
Family workers	55
Salaried managers	8
Hired regular workers	196
Seasonal and casual workers	91
Total farm labour force	714

Source: MAFF Annual Review (1980).

Table 9.2 gives a further breakdown of the employed labour force, covering hired and family workers but not farm managers. It shows that just over a quarter of the employees are female and that women are more likely than men to be employed on a part-time or temporary basis.

Spouses of farmers have only recently been included in the agricultural census and although Table 9.1 records 71,000 wives or husbands of farmers doing farm work, evidence from other sources suggests the number of wives who help at least occasionally on the farm may be much higher. According to the Agriculture EDC survey, there were 81,200 working wives on full-time farms in

Table 9.2   Characteristics of the employed labour force on farms in the UK, June 1979

	Male '000s	Female '000s	Total '000s	% female
Regular whole-time	169	18	187	9·6
Regular part-time	32	32	64	50·0
Seasonal or casual	52	39	91	42·8
All employees	253	89	342	26·0

Source: MAFF Annual Review (1980)

England and Wales in 1970. More recent figures indicate that the occupiers of 59% of full-time holdings and 41% of part-time holdings in England and Wales are assisted by their spouses (Ministry of Agriculture, Fisheries and Food, 1976). On this basis, the number of spouses working on UK holdings could be as high as 128,000.

Hours of work

Table 9.1 does not give a complete picture of the labour supply for agriculture because some categories put in longer hours than others. Farmers, who were included in the Agriculture EDC survey, claimed to work an average 64-hour week, employees 52 hours and wives who helped on the farm, 17 hours. Among those enumerated in the 1979 census, 27% of farmers, partners and directors and 26% of regular employees worked only part time. Some of the seasonal and casual workers covered in the June census might work for only a few days in the year.

Using data from the Agriculture EDC survey, Sparrow (1972) estimated the total number of hours contributed by each category in the farm labour force. Table 9.3 indicates that regular workers

Table 9.3 Weekly labour input on full-time farms in England and Wales

	Average weekly labour input (million man hours)	% of weekly labour input
Farmers	10·1	36·5
Farmers' wives	1·4	5·1
Regular workers	13·4	48·2
Contractors and employees	0·9	3·3
Casual labour	1·9	6·9
Total hours worked	27·7	100·0

Source: Sparrow (1972)

supply nearly half the total hours, farmers a little over one-third, wives about 5% and others 10%.

Structure of the labour force

According to Table 9.1, 419,000 of the 714,000 persons engaged in farming in 1979, or 59%, were members of farm families. The way the statistics are collected and presented probably underestimates the family contribution to the labour force. If a more generous estimate were made of wives helping on the farm and if relatives with contracts of employment and those working casually were included with family workers, the farm family component of the labour force might be as high as two-thirds.

Although there is no precise definition, family farming is generally understood to mean the type of arrangement where members of one family provide all or most of the labour, capital and management and control the farm business. Family farming is contrasted with systems of large-scale farming where land, labour, capital, management and business control are supplied by separate individuals or organisations, none of whom need be related to one another.

Family farming is the dominant form in British agriculture today. More than three-quarters of all holdings employ no regular full-time workers but depend on farm family labour, supplemented in some cases by part-time, seasonal or casual help. Nearly half of all British farmers and their spouses are on holdings where they provide at least 80% of the labour. Over a quarter are on holdings where they constitute the entire labour force.

While family farming is widespread throughout the UK, it is most typical of the uplands and pastoral farming regions of the north and west where dairying and livestock farming predominate. In Wales, Northern Ireland and north-west Scotland, farm families provide more than three-quarters of the labour, farmers alone supplying more than half in Wales and Northern Ireland (see Appendix Table 9.1, p. 774). Large-scale farming is more highly developed in the arable farming regions of the east and south, where greater use is made of non-family labour. In east and south-east England and in south-east Scotland hired workers account for 60% of the total labour force; in Wales, Northern Ireland and north-west Scotland less than 25%. This means that hired labour is concentrated on the eastern side of Britain. East Anglia and south-east England alone account for about a quarter of the UK farm labour force but employ a third of the non-family workers.

Whilst this analytical distinction between large-scale and family farming can be useful, most farms combine characteristics of each type. The regional distinction is not clear cut, either. Family farming is common in the eastern and southern counties of England while many large-scale farming businesses can be found in upland areas of Britain.

The great majority of British farmers manage their businesses with family help alone. Only the small number of farmers who produce the bulk of the nation's agricultural output employ non-family workers. Table 9.4 shows that the 132,700 hired full-time workers in England and Wales in June 1979 were employed on a mere 44,000 holdings, leaving 78% of holdings with none. Even among labour-employing farms, large labour forces are uncommon. Only 1% of agricultural holdings employed 10 or more full-time workers in 1979, only 0·3% (500 holdings) having a permanent labour force of 20 or more. At the other end of the scale, 85% of labour-employing farms have fewer than five regular workers and 46% have only one. Looked at another way, 20,000 farm workers are the only employees on their respective farms and only 39,000 are employed in gangs of 10 or more.

The way the farm labour force is distributed has certain conse-

Table 9.4 Distribution of workers by size of labour force in England and Wales, 1979

No. full-time hired workers per holding	Workers		Holdings	
	'000s	%	'000s	%
None	–	–	153·5	77·7
1	20·3	15·3	20·3	10·3
2–4	44·6	33·6	17·2	8·7
5–9	28·9	21·8	4·6	2·3
10–19	18·1	13·6	1·4	0·7
20 and over	20·8	15·7	0·5	0·3
Total	132·7	100·0	197·5	100·0

Source: MAFF, results of June 1979 census

quences for labour relations in the industry. The majority of British farmers are never involved in managing workers, wage bargaining or settling labour disputes. Even on the majority of labour-employing farms, the number of employees is so small that labour relations tend to be of a close, personal nature, more reminiscent of a family enterprise than of a large industrial concern.

Trends in agricultural employment

One of the most striking features of the farm labour force has been its long-run decline, both in absolute numbers and relative to the rest of the economy. This is no new phenomenon; some historians trace the origins of 'the drift from the land' back to the Black Death. The run-down has been continuous since the early nineteenth century. Agriculture, horticulture and forestry absorbed over 20% of the UK working population in the mid-nineteenth century, only 3% in 1971. By 1979 the proportion was down to 2·6%.

In 1950 there were 918,000 men and women employed on British farms, not counting farmers and their wives. By 1979 the hired labour force was down to 352,000. The 1960s saw the most rapid rundown in hired labour, with a net reduction among full-time workers of 6% per annum. During the 1970s the rate of outflow was much slower.

Some categories have been depleted more rapidly than others. On the whole, men have been leaving farm employment faster than women since the Second World War and full-time workers faster than seasonal and casual workers or part-timers. Farms in the UK were employing virtually as many seasonal workers in June 1979 as in June 1960 whereas the number of full-time workers had fallen by more than 60%.

Farmers have not been leaving as rapidly as employees. The total number of farmers enumerated in the 1951 population census for Great Britain was practically the same as in 1851. Since that time their numbers have declined but not as sharply as for workers. During the 1960s the net reduction in the number of farmers was estimated to be about 2% per annum. Over the same period full-time workers were leaving the land at the rate of 6% per annum. One consequence is that farmers are becoming more important in the British farm labour force. In 1960 the ratio of workers to farmers was about 2:1 but today it is close to 1:1. This suggests that family farming is not only dominant but increasing in Britain at the expense of large-scale farming.

Reasons for 'the drift from the land'

'The drift from the land' has been the subject of a number of investigations by agricultural economists. All their findings have underlined the importance of low wages among reasons for workers leaving agriculture. Other reasons frequently quoted include poor promotion prospects, long and uncertain hours of work, weekend working, aspects of working conditions including dust, dirt and bad weather, poor housing and the tied cottage system. These drawbacks of farm employment have their counterpart in higher wages, better prospects, shorter hours, free weekends and pleasanter working conditions, actual or imagined, in other employment. Town living may attract farm workers' families if it holds out prospects of better housing and other amenities, shorter journeys to shops and schools and more job opportunities for workers' wives and children.

Attractions of non-farm jobs and town living and the drawbacks of working on a farm are often portrayed as the principal causes of 'the drift from the land'. In fact, dismissal and natural wastage have played a significant part, too. Demand for labour in agriculture has fallen sharply in the last thirty years as a result of farm enlargement, introduction of labour-saving buildings, machinery and methods, use of herbicides and pesticides and simplification of farming systems. Many farmers have made a conscious decision to reduce the labour force by dismissing workers or, more often, by not replacing them when they leave, for whatever reason. Only about half the workers leaving agriculture in the later 1960s were being replaced. This implied that pressures 'pushing' them out of farming were just as important as those 'pulling' them to other industries in accounting for 'the drift from the land'.

However attractive other jobs may seem relative to farming, changing employment depends on suitable vacancies occurring. The rate of outflow of farm labour is closely related to the level of unemployment in the economy at large. Workers left agriculture much faster during the booming 1960s than in the present era of high unemployment.

Housing can prove to be as much of a constraint on mobility as job opportunities. Competition from commuters, second-home owners and retired people has put ownership of a country cottage out of the reach of most farm workers whilst the supply of council housing and other rented accommodation in rural areas is usually inadequate. As a result, many workers who would like to leave agriculture, have been obliged to remain, living in tied accommodation. This may help to explain why, although the rate of labour turnover in farming may seem high, it is quite modest compared with construction or manufacturing industries.

Quality of labour

By the late 1960s the rate of decline in the farm labour force, especially among full-time male workers, was causing concern. Policy-makers were beginning to ask whether numbers of workers remaining on farms in the 1970s and 1980s would be sufficient to meet output targets such as those set in the 1965 National Plan. By the end of the 1970s these fears had largely subsided. The Economic Development Committee for Agriculture, asked to examine the feasibility of a 2·5% growth rate for agriculture in the 1980s, concluded that labour shortages were unlikely to be a constraint. Supplies of labour were expected to be no more and possibly less scarce than in the past as a result of increasing numbers in the workforce and continuing high levels of unemployment, while the demand for labour in agriculture was expected to continue its downward trend (Agriculture E D C, 1977).

Emphasis has now shifted from quantity to quality. The Manpower Working Group of the Agriculture E D C has recommended that every effort be made to increase the effectiveness of manpower in the industry. Workers remaining on farms in the 1980s must be capable of adapting to rapid technological change as well as assuming responsibility for increasing amounts of capital. (The total capital stock per person engaged in British agriculture, excluding the value of land, was put at just under £23,000 in 1979. This represented a 47% increase, at constant prices, on the amount in 1969 (Ministry of Agriculture, Fisheries and Food, 1980).)

Improvements in agricultural education, more widespread training to craft level, better training for management, especially man management, more attention to recruitment and placement and measures to eliminate unproductive use of family labour on small farms were among the recommendations put forward by the Manpower Working Group. Of these, the two most important indicators of manpower quality were thought to be the standard of management in the industry and the proportion of the workforce trained to craftsman level. Yet indicators like these need to be considered in the context of such factors as the age structure of the workforce, general levels of education and rates of turnover.

Age structure

Farmers are substantially older than the majority of the British working population, as might be expected in a population of business proprietors. Table 9.5 shows that compared with the national labour force, there are more farmers over 35 and fewer below. Farmers in the Agriculture E D C sample averaged 46 years, workers 37 years. Farm workers are a young group compared with the

Table 9.5 Age distribution of farmers, farm workers and the British working population in 1971

Age last birthday	Farmers, managers and market gardeners %	Agricultural workers %	Total economically active population %
15-19	2·3	12·7	8·7
20-24	5·0	11·1	12·3
25-34	15·8	17·6	18·9
35-44	21·2	19·2	19·7
45-54	23·6	17·8	20·9
55-64	22·3	16·2	16·3
65 and over	9·8	5·4	3·2
Total	100·0	100·0	100·0

Source: Derived from 1971 population census

British working population as a whole, with a particularly large proportion of under-20s. This reflects the fact that many of them enter the industry straight from school, only to leave again before they are 30.

Although fears have been expressed from time to time that the population of British farmers is an ageing one, the reverse appears to be the case. Harrison (1975) estimated that farmers in England in 1969 were on average younger and had begun farming earlier than their counterparts of thirty or forty years before. The hired labour force, too, seems to be getting younger. During the 1970s the proportion of under-35s rose while the proportion of over-55s dropped slightly (Britton, Burrell, Hill and Ray, 1980).

Education and qualifications

Only a minority of British farmers and farm workers left school with any formal qualifications. In the Agriculture EDC sample, only 18% of full-time farmers and 10% of workers held school-leaving certificates such as School Certificates or GCE. This reflects a widespread tendency for members of the agricultural labour force to have left school at the earliest opportunity to start work on a farm. The average school-leaving age was 15·0 years for farmers in the sample and 14·7 years for workers. Agriculture was the first regular job for 86% of the farmers and 80% of the workers.

Consequently, few members of the farm labour force have studied agricultural subjects at college or university. Only about 10% of farmers and 9% of workers interviewed in the Agriculture EDC survey had studied or were studying for agricultural qualifications. Farm managers as a group were better qualified, 26% holding school leaving certificates and 20% studying for or holding agricultural qualifications.

Although levels of formal education in the farm labour force are generally low, the situation is changing. Young farmers and workers entering the industry now are much more likely than the older generation to hold school-leaving certificates and to have attended agricultural college. Table 9.6 shows how rapidly the proportion of farmers with agricultural qualifications declines with age. With workers, the trend is even more striking.

Table 9.6 Distribution of agricultural qualifications according to age

Age now years	Proportion holding or studying for agricultural qualifications	
	Farmers %	Workers %
25 and under	35·1	24·2
26–34	24·1	8·6
35–44	9·8	3·3
45–54	5·5	1·1
55–64	2·8	0·2
65 and over	2·7	0·0
All age groups	10·3	9·5

Source: Agriculture ED C (1972)

Since the Agriculture ED C carried out its survey in 1970 there have been significant developments in agricultural education and training in Britain. Numbers of farmers and workers receiving further education in agricultural subjects have been growing steadily. In 1969/70 there were about 1,150 students attending Higher and Ordinary National Diploma courses in agriculture in the UK, a figure which had risen to almost 3,000 by 1975/6. In that year there were some 2,000 students taking National Certificates and Advanced National Certificate courses compared with about 1,650 in 1969/70. Numbers of students taking courses in agriculture, forestry and veterinary science in British universities rose from under 4,300 in 1969 to nearly 5,800 in 1978.

Agricultural training

Most farmers in Britain claim to have received some form of practical training but rarely has this training been organised in any formal sense. Of the farmers in the Agriculture ED C survey, 76% said they had been trained on the farm at home and only 14% had received formal training, as farm pupils or apprentices or by other recognised routes. Only about a third of the workers said they had been trained and again, most had gained their experience on the family farm.

Following the establishment of the Agricultural Training Board in 1966, a comprehensive pattern of formal training and associated further education has been developed in Great Britain. A similar service is available in Northern Ireland under the Agricultural Training Scheme. Anyone engaged in *bona fide* farm and horticultural production activities is eligible to attend these training courses. Among the main types of training provided are initial courses for newcomers to agriculture, craft skills training courses, training in staff management skills, training at the place of work and training for instructors. Table 9.7 gives details of some of the

Table 9.7   Training courses and trainees in Great Britain, 1979/80

Type of course	No. courses	No. training days	No. trainees
Initial courses	116	727	1,557
Craft skills	10,781	14,205	67,520
Staff management	168	394	1,864
Training at place of work	269	not known	3,278

Source: Agricultural Training Board Annual Report (1979/80)

training provided by the Board in the year ended March 1980. Demand for and provision of training have grown rapidly in recent years. In 1979/80, for instance, a total of 10,781 craft skills training courses was held, providing 67,520 trainee places. Comparable figures for 1968/9 were 830 courses and about 10,000 trainees. It is estimated that in 1975/6, about 15% of the regular agricultural workforce in Britain received short courses of training whereas in 1965, the year before the Board was established, the proportion was probably less than 1% (Agriculture E D C, 1977).

The Board runs an apprenticeship scheme for new entrants to agriculture – the New Entrant (Apprenticeship) Training Scheme. An apprenticeship lasts three years, during which time the young person receives training and supervised practice in a chosen branch of the industry, attends further education and is tested for proficiency in appropriate skills. Successful completion of an apprenticeship leads to craftsman status. In March 1980 there were 6,191 apprentices registered with the Board. A less formal route to craftsman status, the Craft Skills Training Scheme, has recently been introduced. There were 2,215 trainees in this scheme in March 1980 (Agricultural Training Board, 1980).

The Board set itself the target of training 32% of young workers entering the adult farm labour force to craftsman level by 1980/1. The turnout of qualified craftsmen was only 2·5% of the eligible 19-year olds in 1965/6, rising to 13·5% by 1975/6. In view of the importance it attached to labour quality, the Agriculture EDC Manpower Working Group proposed that the target should be raised to 60% for 1984/5.

Nowadays new entrants to the farm labour force can acquire craftsman status only by passing the necessary proficiency tests but those who were already working on farms in May 1972, when registration of craftsmen came into force, can be registered through an employer's declaration that they are of craftsman standard. Only a minority of those eligible are thought to have taken advantage of this offer, for no more than 36% of full-time male farm workers in England and Wales today are registered as craftsmen or higher grades. The proportion of workers who are of craftsmen standard could well be double this figure.

Fears are sometimes expressed that training workers will only increase their propensity to move, so that the employer who has borne much of the cost of training will not reap the benefit. Compared with other industries, the rate of labour turnover in British agriculture is quite modest. According to the Ministry of Agriculture's Wages and Employment Enquiry for 1979, 24% of all full-time hired male workers have been on the same holding or employed by the same farmer for twenty or more years and only 16% have had less than two years' service.

From the industry's point of view, circulation of skilled and experienced labour between farms ought not to detract from the chances of increasing manpower effectiveness. In view of the structure of the agricultural labour force with its preponderance of small units, an ambitious worker will be more likely to improve his prospects by moving to another farm than by waiting for a vacancy in his present employment. More serious for British agriculture is the loss of trained manpower to other industries. Farming has always tended to recruit more school-leavers than are needed to replenish the adult workforce. Many leave during their twenties, before they have had time to 'repay' the industry the full value of any training they have received. It is claimed that the more skilled and better-qualified young workers will be the first to leave, so that in effect agriculture is training a labour force for other industries. Available evidence points to the opposite conclusion, however. Those who have undertaken further education or training in agriculture have demonstrated a commitment to the industry which lessens the chances that they will leave. Jones and Peberdy (1979) monitored the progress of a batch of school-leavers who entered agriculture in 1974, half of whom joined the apprenticeship scheme and half did not. By 1978, 15% of the apprentices had left agriculture but among non-apprentices the wastage was 29%.

Earnings of agricultural workers

Farm work has traditionally been a low-paid occupation. At the beginning of the First World War, the average minimum wage for a man working on a farm was 90p for a 58-hour week or approximately 1·5p per hour. By the outbreak of the Second World War this had risen only to £1·75 per week. Agricultural wages improved considerably during the war, the minimum being set at £3·50 in 1946. Table 9.8 shows that this improvement has been maintained

Table 9.8 Average minimum weekly wage and hours for farm workers in England and Wales, 1914–80

	Average minimum weekly wage (£)	Hours in standard working week	Minimum hourly rate (p)
1914	0·90	58	1·5
1930	1·59	51	3·1
1939	1·75	50	3·5
1946	3·50	48	7·5
1950	5·00	47	10·8
1960	8·00	46	17·5
1970	13·15	43	30·8
1975	30·50	40	76·5
1980	58·00	40	145·0

Source: MAFF (1967) and Wages and Employment Enquiry (1979)

along with a gradual reduction in hours. By 1980 the minimum wage for an ordinary worker in England and Wales was £58 for a 40-hour week or £1·45 per hour. Scottish workers have traditionally earned a little more, those in Northern Ireland slightly less than their counterparts in England and Wales.

Despite a very substantial improvement in real as well as in money incomes, farm workers' earnings continue to lag behind those of other manual workers. The gap seemed to be closing during the 1940s and 1950s, widened during the 1960s and early 1970s and appears to be narrowing once more (see Appendix Table 9.2, p. 775). On average, farm workers today earn about three-quarters as much as manual workers in other British industries.

It is often claimed that low wages in agriculture are offset by the value of perquisites. Certain benefits such as housing, board and lodging, casual meals, milk and potatoes can be reckoned as part of the legally prescribed wage. Others, such as fuel, lighting and vegetables, are valued by agreement between farmer and worker.

Payments in kind are given quite nominal values for wage-fixing purposes. For instance, a farmer is entitled to deduct only £1·50 at present for a tied house and 3p for a pint of milk. These benefits therefore represent a hidden subsidy to workers, which goes some way to closing the wages gap between agriculture and other industries.

This argument needs to be treated with caution. In the first place, not all farm workers enjoy 'hidden subsidies'. While 51% of hired men live in tied accommodation, the remaining 49%, along with most female workers, are not subsidised in this way. It is the better-paid workers who tend to receive most payments in kind. Housing is the most substantial perquisite and currently 68% of farm foremen and 75% of dairy cowmen but only 42% of general farm workers and 12% of female farm workers live in tied cottages. Payments in kind seem to be offered as an inducement to attract the more skilled workers rather than being used as a means of improving the lot of the lowest paid.

Second, paying workers in kind may be on the decline. Since the Rent (Agriculture) Act of 1976 increased the security of those living in tied cottages, the proportion of hired men provided with accommodation has fallen from 55% to 51%. Farmers' wives and workers' wives may be less willing than in the past to provide subsidised board and lodging to farm workers. Trends in specialisation in farming systems, too, may reduce the possibility of workers receiving such perquisites as milk, eggs and potatoes.

Third, it must be remembered that workers in other industries, with whom agricultural earnings are compared, may also receive perquisites such as cheap housing or free travel. Fourthly, farm workers may *not* receive many of the benefits which workers in other industries take for granted. Superannuation schemes and private pension funds, subsidised canteens, free issue of protective clothing, standby payments for being on call at weekends are the exception rather than the rule in British agriculture today. On balance, some commentators believe that the extent to which payments in kind offset low pay in agriculture has been grossly exaggerated.

During the year ended June 1980, earnings of hired full-time men on farms in England and Wales averaged £78·95. The difference between this and the minimum wage of £58 is made up of overtime payments and premiums. The minimum wage relates to a 40-hour week but in the year in question, full-time male workers averaged 46·5 hours. The amount earned over and above the minimum entitlement for hours worked is the *premium*. For the year ended June 1980 the average premium for all full-time men was just under £10 per week.

The wages structure which was introduced in 1972 established statutory premiums for three grades of skill above the ordinary farm worker. Currently craftsmen are entitled to a basic minimum wage 15% above that of ordinary workers. Appointment Grade II workers (foremen, supervisors and unit managers) are entitled to a premium of 25% and Grade I farm managers to 35%.

The statutory wages structure regularised a system of payment which was already widespread in agriculture. Yet, typical of the industry, informal arrangements tend to persist alongside those which are sanctioned by law. Although only 36% of male workers are registered as craftsmen or higher grades, and therefore legally entitled to statutory premiums, the proportion of men earning premiums in 1979 was no less than 93%.

Wage bargaining and union activity

Since 1940, minimum wage rates and basic hours for farm workers have been determined nationally by Agricultural Wages Boards. A Board consists of 21 members with eight representing employers, eight the workers and five independent. Customarily five of the workers' representatives are nominated by the National Union of Agricultural & Allied Workers (N U A A W) and three by the Transport & General Workers' Union (T G W U) which has some farm workers among its members.* Wages Boards are typical of low pay industries. Besides protecting the interests of the small number of workers who remain on the minimum wage, their awards are decisive in improving the living standards of all farm workers because of the fairly narrow dispersion of earnings.

Although one of Britain's largest unions, the N U A A W has never been one of the most powerful. Membership figures are not readily available but in the early 1960s it was estimated that about one-third of the membership was not in agriculture but in allied occupations such as forestry, road maintenance, land drainage and gardening. Since then, recruitment of new members from the food-processing sector may have raised the non-agricultural membership to about 40%. On this basis, Newby calculates that the N U A A W contains approximately 40% of the full-time hired farm workers in England and Wales. Membership has always been more vigorous in areas with a greater concentration of workers, notably in the eastern counties. The three counties of Lincolnshire, Norfolk and Yorkshire alone account for nearly two-fifths of the current membership.

By the very structure and nature of agriculture, farm workers are not an easy group to organise. The wide geographical dispersion of members makes collection of subscriptions and notification of

_____
* The N U A A W is now the Agricultural Workers Section of the T G W U.

meetings difficult and, inevitably, branch activity tends to be weak. The NUAAW has the smallest average branch membership of any British trade union and only the TGWU has more branches. Merely keeping in touch with over 3,000 branches presents the union with problems of a kind and scale unusual for the trade union movement. Added to this, determination to keep subscription rates low means that union funds are always stretched.

The NUAAW is not a militant union. Even if funds were adequate to support a strike, it is highly unlikely that the union would contemplate calling one. It could not count on support of the 60% of agricultural workers who were not members and even among its members, it seems doubtful whether a strike call would be unanimously obeyed. The last time the union called a strike was in Norfolk in 1923, over the farmers' decision to reduce wages. On that occasion, only a quarter of the farm workers at most obeyed the call, and this was in the union's strongest area.

The fact that most farm workers are employed in ones and twos means that dealings with employers are mostly at a very personal level. This makes it hard for farmers and their employees to confront one another in a formal sense as representing their respective interest groups. Working relations in agriculture are generally felt to be stable and harmonious. Whether this should be a cause for congratulation or whether it reflects the farm worker's powerlessness to change the situation remains a vexed question.

10

SUPPORT ENERGY

Introduction

In the developed countries of the world, agriculture is an energy-intensive activity to an extent that the energy use per man employed is similar to that in heavy engineering (Leach, 1976). The so-called 'energy crisis' of 1973 was characterised by a temporary shortage of oil and a hitherto unknown escalation in price which served to warn that at present rates of consumption, world oil supplies have a finite life that is frighteningly short. This, perhaps, more than anything else, has stimulated the activity of energy analysis whereby attempts are made to determine the sources and flows of energy in a process right through to the finished product. Although common in the process industries, where the amounts of energy in the form of heat at the correct temperature need to be known for the correct design and operation of plant, the extension of such methods to determine the 'support energy' to a wide range of products is of recent origin and merits some further consideration by way of justification.

In any process which uses energy, the energy carries with it certain cost factors which depend on the form in which energy is used. Hence, conventional economic analysis will reflect the importance of energy in food production, through both the direct effects of fuel costs and their more indirect effects on the cost of fertilisers, machinery and other goods, but it will not tell us anything about the amounts of energy that are used or indicate more energy efficient means of production. Critics may argue that economic analysis provides all the information that is needed to determine the merits of alternative production systems. So it does, up to a point, but where alternative systems differ little in their financial economy, the one showing the greatest energy efficiency may be preferred if there is concern about resource depletion. Thus, energy analysis provides useful insights into the amounts and forms in which energy is used and, as such, it complements but does not replace economic analysis.

This chapter has two purposes: to introduce the reader to the methodology of energy analysis applied to agriculture and to sources of relevant data; and to present some facts and figures relating to support-energy use in UK agriculture. It is inevitable that a work such as this is compelled to draw heavily on the excellent book by Gerald Leach (1976) in pursuing these aims.

ENERGY ANALYSIS: METHODS AND DATA
Support energy

The support energy to produce a given crop is the sum of all the non-renewable energy inputs. Thus, fuels to power machines and to dry crops are accounted, as also is the energy sequestered in the manufacture and provision of fertilisers, machines, buildings and any other goods required. Solar energy is not included in the calculations because it is abundant and its capture by plants demands no expenditure of non-renewable energy.

Direct energy sources These comprise fossil fuels such as coal, petroleum and gas and also electricity. The provision of energy itself involves expenditure of energy so that each particular form has its own 'energy overhead'. Thus the direct energy will usually be a value over and above the energy content and this is called the primary direct energy. In the case of electricity, the fuel consumed at the power station and hence the primary direct energy is several times greater than the energy consumed at the point of delivery.

Indirect energy Energy is sequestered in the manufacture and provision of fertilisers, machines, buildings and many other items. To determine these reliably, it would be desirable to assess all the energy inputs to a particular item during its manufacture from raw materials to the finished product but this is not usually possible and alternative methods have been developed which are often rather approximate.

Energy content of labour It can be argued that the support energy required to maintain the life-style of the labour force should be considered as an energy input. Further, the definition could be extended to include their dependants. It is arguable where the line should be drawn but taking the practical point of view it is fortunate that in highly developed systems of mechanised agriculture, labour contributes little to energy costs and may safely be ignored. In subsistence agriculture it may be the dominant factor and its inclusion may be essential if calculations are to make sense.

Units of energy Many units are in common use to express the value of energy in its various forms. The basic units used here will be the joule and its multiples. Appropriate conversion factors to some other

Table 10.1   Energy conversion factors

	joule (J)	Wh	cal	Btu
1 J =	1	2.778×10^{-4}	0.2389	9.481×10^{-4}
1 Wh =	3600	1	859.8	3.412
1 cal =	4.187	1.163×10^{-3}	1	3.968×10^{-3}
1 Btu =	1055	0.293	252	1

1 therm = 10^5 Btu = 105.5 MJ

Table 10.2   Definitions of prefixes

Prefix	kilo	mega	giga	tera	peta	exa
Symbol	k	M	G	T	P	E
Factor by which unit is multiplied	10^3	10^6	10^9	10^{12}	10^{15}	10^{18}

Table 10.3   Approximate energy equivalents

1 million tonnes of oil (Mtoe) equals approximately
7.5 million barrels
425 million therms
1.7 million tonnes of coal (Mtce)
1200 million m³ of natural gas
44.8 PJ

commonly used units are given in Table 10.1. Table 10.2 defines the prefixes which are used to give multiples of the joule and of other quantities. For practical purposes, energies are often expressed in terms of tonnes of oil or coal equivalent and some approximate conversions to other units used are given in Table 10.3. These are only approximate because the conversions depend on the energy content of oil, coal and gas and purely notional values have been chosen to give the conversions shown.

Energy requirements for delivered fuels and power   The energy requirements for delivered fuels and power are not simply the enthalpies of the fuel burned or of the electricity sent out to the grid because energy is also consumed in extracting fuels from the ground, in processing, in transporting them and in building and running all the associated plant and equipment. 'Energy overhead' factors have been calculated (Leach, 1976) and these may be used as multipliers (Table 10.4) for converting the energy content of delivered fuels and electricity to their respective total energy requirements. The multipliers given are those derived by Leach for 1968-72, with the exception of electricity,

Table 10.4 Energy requirements for delivered fuels and power

	Energy content as used	Multiplier	Total energy input
Coal (to agriculture), 1979	30·1 GJ/t	1·04	31·3 GJ/t
Coke, 1979	28·1 GJ/t	1·18	33·1 GJ/t
Natural gas	38·5 MJ/m³	1·05	40·4 MJ/m³
Petrol/gasoline, 1979	34·9 MJ/l	1·134	39·6 MJ/l
	46·9 MJ/kg	1·134	53·2 MJ/kg
Diesel fuel (gas/oil), 1979	38·2 MJ/l	1·134	43·3 MJ/l
	45·5 MJ/kg	1·134	51·6 MJ/kg
Fuel oil, 1979	41·2 MJ/l	1·134	46·7 MJ/l
	42·8 MJ/kg	1·134	48·5 MJ/kg
Electricity, 1979	3·6 MJ/kWh	3·68	13·3 MJ/kWh

which has been updated, using Leach's earlier results for electricity and more recent data relating to the thermal efficiency of power stations and the efficiency of electricity transmission. These data and the enthalpies of the different fuels were obtained from the Digest of UK Energy Statistics.

Energy requirements for fertilisers and lime   Estimates of the energy required to manufacture and provide fertilisers and lime have been made by both Leach and ICI (Leach, 1976) and their values are in close agreement based on data for the period 1968–73. Table 10.5 gives more

Table 10.5 Energy requirements for fertiliser and lime

	Specific energy
	MJ/kg
Nitrogen (N)	73
Phosphates (P₂O₅)	14
Potash (K₂O)	8
Lime (ground limestone)	2

recent ICI data (Lewis & Tatchell, 1979) for nitrogen, phosphates and potash together with an earlier value for lime (Leach, 1976). These values are weighted averages derived from the compositions of the various compound fertilisers used. Some values for particular compounds are listed by Leach (1976) and these may be more convenient to use where the particular compound applied is known.

Nitrogen is the most important and by far the most energy-intensive element in fertilisers. Normally, it is applied as a 'straight' N fertiliser, usually as ammonium nitrate (34·5% N) but also as ammonium sulphate (21% N), urea (46·6% N) or liquid ammonia (82·4% N). The feedstock is natural gas and this is made into ammonia and nitric acid from which urea or ammonium nitrate may be made. The calculations include the energy costs of fuels, feedstocks, plant, transport and packaging.

Nearly all phosphate is applied to the soil in processed form as super-phosphate or in compound (N P K) formulations. The calculation includes costs of mining, concentration, processing, shipping and plant.

Potash is normally derived from potassium salts: the calculations take account of mining, transport, plant and processing.

Lime is used usually as ground chalk or limestone but also as 'burnt lime' (calcium oxide) or as 'slaked lime' (calcium hydroxide). The value given must be regarded as a rough approximation.

Goods and services   Estimates of the energy input to a good or service can be made in a number of ways. It is not possible within the compass of this chapter to explain these methods in any detail other than to give an indication of the use that may be made of them.

The crudest and quickest method is simply to multiply the monetary cost of a product by the prevailing figure for primary energy consumption per unit of Gross Domestic Product. In the UK this ratio for 1978 was 8865 PJ/142,000 M£ = 62 MJ/£. A large number of industries do in fact have MJ/£ performances close to the national average figure but the method can also be extremely inaccurate. This approach should be used very sparingly and only for minor items where no other information is available.

The method of 'process analysis' is to trace individually all outputs and inputs through the entire skein of processes, sales, purchases and transactions leading back from a finished product to fossil fuel energy sources. The method is extremely laborious and is subject to error because of cut-off of the 'trace-back' at a premature stage.

The most commonly used method is that based on input–output analysis. This depends on the use of national input–output tables which record monetary transactions between an array of industrial and commercial groupings as purchasers and the same array as sellers. Since the energy industries are also recorded, it is possible to compute both the direct purchases by any industry from these energy industries (e.g. agriculture from coal) and of all the second, third and higher order indirect purchases (e.g. agriculture from tractor manufacturing from iron and steel from coal).

This method also suffers from some serious drawbacks. The most important are the high degree of aggregation, so that individual products and inputs cannot be separated; the exclusion in many cases of capital purchases by industries; incomplete coverage of all relevant industries and services and the fact that the base information is obtained from the Census of Production Reports for 1968, since when, energy inputs per unit of production may have changed. Leach (1976) reports energy values in MJ/£ obtained by this method, the computations being

Table 10.6 **Energy inputs (MJ/£ cost) for some agricultural requirements**

	1968	1978
Agricultural machinery depreciation	200	53·2
Agricultural machinery repairs	200	53·2
Agricultural buildings	260	69·1
Lubricating oils and greases	550	146
Sundry goods	180	47·9

performed by the Open University Energy Research Group. To range outside the base year it is essential to allow for monetary inflation by dividing the energy value in MJ/£ by the inflation factor. Table 10.6 shows some useful energy values for the base year of 1968 and for 1978 using an inflation factor of 3.76 (Annual Abstract of Statistics) based on the wholesale price index for agricultural machinery. It so happens that this value also applies closely to buildings and is used throughout. Any other year can be calculated from the base data in a similar way. Such energy values should, of course, always be used only in conjunction with product costs or prices for the same year.

Some miscellaneous energy values which are of use in agricultural calculations are given in Table 10.7 and the source of each is indicated.

Table 10.7 **Miscellaneous energy values**

Item	Energy	Source
Herbicides (per kg of active ingredient)		
2, 4-D	85 MJ/kg	Green, 1978
Paraquat	460 MJ/kg	Green, 1978
Glyphosate	454 MJ/kg	Green, 1978
Formic acid	64 MJ/l	White, 1979
Propionic acid	53 MJ/l	White, 1979
Purchased compound feed (1970/1)	9·57 GJ/t	Leach, 1976
Feedstuff processing (off-farm) (1970/1)	3·39 GJ/t	Leach, 1976
Piped water (1968)	9·1 MJ/t	Leach, 1976
Grain drying (dried product)	520 MJ/t	Leach, 1976
Barn hay drying (DM of product)	2·6 GJ/t	White, 1979
High temperature green crop drying (DM of product)	19·3 GJ/t	White, 1979

Those for herbicides are purely representative and Green (1978) lists values for 22 different herbicides, fungicides and insecticides. The values for crop drying refer to 'typical' or average values for fuel use and seasonal moisture removal and include allowances for plant and storage. However, values vary widely with drier type and crop moisture removed and where specific conditions are known, the references should be consulted so that an appropriate calculation may be made.

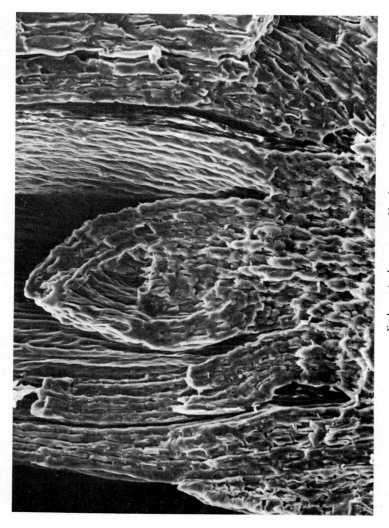

1 Embryonic shoot apex of barley

2 White clover

3 Red clover

4   Lucerne

5   Sainfoin

6   Cereal trials

7   Putting the wraps back on a strawberry crop

8   Friesian cows strip-grazing kale

9   A suckler herd on land reclaimed after opencast mining

10   Hampshire Down sheep

11   South Devon ewes with their lambs

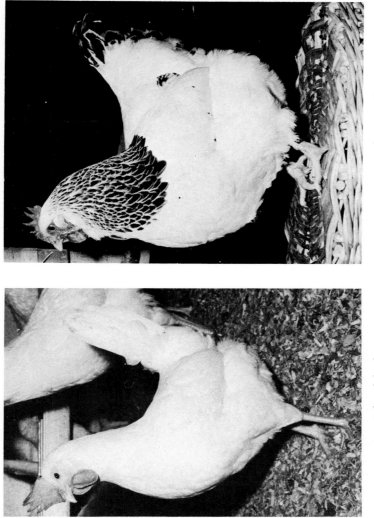

13 Light Sussex hen

12 Babcock bird

14   Landrace gilt

15   Iron age sow, or the closest thing to a wild pig, at Cotswold Farm Park

16 Fish farming

17 Clear felling in preparation for replanting. This site, a valley slope in south-west England, is ideal for Douglas Fir

18   Tree planting on a Cotswold farm

19   Land reclamation in Powys

20   Sheltered edge habitat in a well-wooded setting—a splendid combination for wildlife. The mature trees must be replaced if the landscape is to retain its character and conservation value

21   Mute swans feeding on a Warwickshire pond. Adequate light and freedom from pollution are essential to maintain water quality

22   A patchwork of small fields

23   Contrasting large fields

Energy used in field operations with tractors and tractor depreciation   On the basis of field tests, Rutherford (1974) has produced tractor work rates and fuel consumed for various field operations, using 55 kW tractors in average conditions, and the results are conveniently tabulated in Leach (1976) in terms of MJ/ha.

Tractors are used on the farm for many different tasks and Leach (1976) has devised a method of relating the energy due to depreciation, repairs, oil and grease to the actual fuel energy used. For tractors ranging from 37 to 67 kW, he found that the ratio of non-fuel items to fuel energy used was about 0·4:1. Thus, to calculate the energy equivalent of tractor depreciation etc, the fuel energy for the field operation

Table 10.8   Energy budget for silage: N applied 400 kg ha, crop yield 11·3 t DM/ha, harvested in 3 cuts taken over 50 ha

Fertilisers	MJ/ha	%
N 400 kg/ha × 73 MJ/kg	= 29,200	
P$_2$O$_5$ 112 kg/ha × 14 MJ/kg	= 1,568	
K$_2$O 168 kg/ha × 8 MJ/kg	= 1,344	
Fertiliser subtotal	32,112	72
Fuel used in field operations		
Chain harrowing 1 × 50 MJ/ha	= 50	
Fertiliser application 3 × 50 MJ/ha	= 150	
Mowing 3 × 210 MJ/ha	= 630	
Forage harvesting 3 × 880 MJ/ha	= 2,640	
Transport of silage 3 × 340 MJ/ha	= 1,020	
Ensiling 3 × 250 MJ/ha	= 750	
Fuel subtotal	5,240	12
Tractor depreciation, maintenance and repairs		
0·4 × fuel used = 0·4 × 5,240 MJ/ha	= 2,096	
Field machinery depreciation. Cost £7,000, depreciated over 5 years and used to cut 50 ha/yr (£7,000/(5 yr × 50 ha/yr)) × 53·2 MJ/£	= 1,490	
Machine depreciation subtotal	3,586	8
Storage silo (roofed bunker with concrete walls)		
For storage of 125 t DM, the known building cost is £7,500. Depreciated over 15 years with sheeting costs of £75/yr gives a cost per ha of ((£7,500/15 yr + 75)/125t) × 11·3 t/ha × 69·1 MJ/£	= 3,592	8
	44,530	100

$$\text{Total input energy} = \frac{44,530 \text{ MJ/ha}}{11\cdot3 \text{ t/ha}} = 3\cdot94 \text{ MJ/kg DM}$$

Metabolisable energy of conserved product assuming no losses = 9·30 MJ/kg DM

$$\text{Energy ratio} = \frac{\text{Metabolisable energy}}{\text{Input energy}} = \frac{9\cdot30 \text{ MJ/kg DM}}{3\cdot94 \text{ MJ/kg DM}} = 2\cdot36$$

is multiplied by 0·4. Values of both fuel and non-fuel energy are tabulated by Leach (1976).

Example energy budget   Table 10.8 gives an example of an energy budget for silage which uses methods and data contained in previous sections and the references given. More detail of the calculation and sources of information are given in White (1979). The present example has been altered to use Leach (1976) energy data for tractors and to update machine and building costs to 1978.

ENERGY USE IN FOOD PRODUCTION
Primary energy in agriculture

A calculation has been made of the primary energy consumed in UK agriculture in 1978 and this is shown in Table 10.9. Appropriate

Table 10.9   Primary energy consumed in UK agriculture, 1978

Item	PJ	PJ	%
Coal 7 M.therm × 105·5 MJ/therm × 1·04	= 0·8		
Coke 3 M.therm × 105·5 MJ/therm × 1·18	= 0·4		
Solid fuel total		1·2	0·3
Petroleum 585 M.therm × 105·5 MJ/therm × 1·134	=	70·0	17·5
Electricity 137 M.therm × 105·5 MJ/therm × 3·68 × 0·7	=	37·2	9·3
Nitrogen (N) 1155 Mt × 73 GJ/t	= 84·3		
Phosphates (P_2O_5) 410 Mt × 14 GJ/t	= 5·7		
Potash (K_2O) 412 Mt × 8 GJ/t	= 3·3		
Fertilisers total		93·3	23·4
Machinery (629 M.£ + 0·5 × 244 M.£) × 53·2 MJ/£	=	40·0	10·0
Feedstuff processing (off-farm) 15·4 Mt × 3·39 GJ/t	=	52·5	13·2
Chemicals		8·5	2·1
Buildings 332 M.£ × 69·1 MJ/£	=	22·9	5·7
Transport, distribution and services		16·3	4·1
Miscellaneous		4·3	1·1
Imported feedstuffs		53·2	13·3
		399·4	100·0

factors to convert fuels, fertilisers, machinery and buildings to energy are given in Tables 10.4–10.7.

For solid fuels, petroleum and electricity, the consumption by agriculture is given in the *Digest of United Kingdom Energy Statistics* in million therms. In the case of electricity, there is reason to believe that approximately 70% is used for agricultural purposes and 30% for domestic purposes on the farm. The electricity consumption has therefore been factored by 0·7.

The amounts of fertilisers used were obtained from *Fertiliser Statistics*, published by the Fertiliser Manufacturers Association. It will be noted that nitrogen is by far the most energy-intensive nutrient and accounts for 90% of the energy required in the manufacture of fertilisers.

For agricultural machinery, an approximate method of calculating the energy value (which checks quite well with 1968) is to take the sum of the gross capital formation (which represents the investment in machinery in the given year) and one-half of the amount spent on machinery repairs (on the assumption that spares and labour each constitute half of the sum) from the *Annual Review of Agriculture* and multiply by the deflated energy value for agricultural machinery depreciation and repairs. Buildings are treated in a similar manner through use of gross capital formation or investment.

For feedstuff processing off-farm the quantity is obtained from *Output and utilisation of farm produce in the UK* (HMSO). For the remaining items, the Leach (1976) values relating to 1968 are used.

Table 10.9 shows that the largest energy users in agriculture are fertilisers and petroleum, with imported feedstuffs, off-farm feedstuff processing, machinery and electricity next in line. The overall total to produce food at the farm gate is 400 PJ per annum and if this is divided by the total national energy consumption of 9,294 P J for 1978, agriculture's share of national consumption is 4·3%. Of course, not all of this energy is in fact expended in the U K because imported feedstuffs form a substantial item and there will also be imports of machinery and other goods. However, even if the largest single item, namely feedstuffs, is excluded, the energy consumed is still 3·7% of national consumption.

Primary energy in food production

It should be appreciated that further substantial inputs of energy are involved before food reaches the consumer, in processing, transport, packaging, distribution, retailing and household preparation. Leach (1976) has made estimates for 1968 of the primary energy consumed in the fishing industry, the food processing industries including any accompanying transport and packaging and in food distribution and these values are shown in Table 10.10. To these have been added an estimate of the energy involved in food storage and preparation in the home and in catering establishments. The table thus gives the primary energy consumed in the UK in making all foods, both indigenous and imported, available to the consumer. The total energy involved is nearly 19% of U K national energy consumption or approximately four times that involved in agricultural production to the farm gate. The U K food and drinks industries and food storage and preparation in the home and

Table 10.10   Primary energy involved in food production, UK, 1968

Item	PJ	%
Agriculture (less feedstuffs)	274	15·7
Feedstuff processing (off-farm)	51	2·9
Imported feedstuffs	53	3·0
Food imports	260	14·9
Fish imports	13	0·7
UK fishing	33	1·9
UK food and drinks industries (less feedstuff processing)	476	27·3
UK food shops	139	7·9
Food storage and preparation	450	25·7
	1,749	100·0

catering establishments both use more energy than agriculture itself. The combined imports of food, feedstuffs and fish also surpass agriculture's energy use. Although not a charge to the UK energy budget, we need to remind ourselves that these imports must be regarded as supplementary energy in much the same way as fuel, for if they were not available to us we would have to grow substitutes, assuming we had the land resources to do so, and that would involve an additional demand on our national energy supplies. Even so, if the imports are discounted, the direct demand on the UK energy resources is still of the order of 15·5%.

Food production in the UK

In 1978, 53% of our unprocessed food was produced in the UK, this figure being the monetary value of food moving into manufacture and distribution derived from home agriculture from all sources. Thus, we may say that an energy expenditure equivalent to 4·5% of national energy consumption produces a little more than half of our unprocessed food. In terms of indigenous type food supplies, that can be grown in the UK, this represents 65% of total consumption.

Energy in the production of particular commodities

An example of an energy budget is given in Table 10.8 and Leach (1976) presents a comprehensive collection of energy budgets for a range of crops, animal products and farming systems.

Table 10.11 shows estimates of agricultural use of support energy for a number of commodities. The second column gives the input of support energy on a per annum basis in relation to land area employed to raise the crop. The third column gives the output of metabolisable energy in the edible parts of the crop, per unit of land area, and provides a measure of the effectiveness with which land is used to produce energy from food. For animal products, the relationship to land is established through the energy required to grow all the feed the animals need. The fourth column gives the ratio of energy output to energy input and is a measure of the 'efficiency' of a food conversion process. The higher the value of E, the greater is the energy output for a given energy input. The fifth column gives the protein output per annum on a basis of land area (kg/ha y) and the sixth column the cost of the protein in energy terms (MJ/kg).

It is necessary to add the cautionary note that unique values of E for particular commodities are not to be expected since these estimates depend on particular agricultural systems and practices. However, an attempt has been made throughout to use average values, that is average fertiliser inputs and average crop yields. The present values are generally in accord with those given by Leach (1976) although there are some differences for animal products.

It is apparent from Table 10.11 that, in terms of both energy and protein, some commodities are produced more efficiently than

Table 10.11   Estimates of agricultural use of support energy

Commodity or product	Input of support energy GJ/ha yr	Output of ME GJ/ha yr	E = ME/ Support energy	Protein output kg/ha yr	Energy input to produce protein MJ/kg
Wheat	19·3	60·0	3·12	495	39
Barley	17·6	46·2	2·63	364	48
Potatoes	52·0	69·3	1·33	460	113
Sugar beet	25·2	82·5	3·28	–	–
Carrots	25·1	32·5	1·30	234	107
Brussels sprouts	32·4	10·9	0·34	296	109
Onions (bulb)	93·4	27·7	0·30	276	338
Tomatoes (glasshouse)	1300	62·0	0·05	945	1,360
Milk	32·5	18·5	0·57	201	162
Beef (from dairy herd)	10·4	3·2	0·31	40	257
Beef (from beef herd)	10·6	2·4	0·23	31	348
Pigs (pork and bacon)	18·0	11·4	0·63	76	238
Sheep (lamb and mutton)	10·1	2·5	0·25	22	465
Poultry (eggs)	22·5	6·0	0·26	113	200
Poultry (broilers)	29·4	4·3	0·15	145	203
Poultry (turkeys)	23·6	7·1	0·30	129	184

others; for example, the arable crops, cereals and potatoes, have E values ranging from 1·3 to 3·3, while animal products have lower E values generally in the range of 0·15 to 0·31 with pig products and milk being somewhat higher. More energy is required to produce protein from animal products, generally in the range of 162 to 465 MJ/kg, than from arable crops for which energy values range from 39 to 113 MJ/kg of protein. In relation to land area employed, both the energy output and protein output were significantly greater for arable crops than for animal products. None of these conclusions should cause any surprise since animals feed on plants and are bound to produce less energy as meat than that contained in the plants eaten. Some plants cannot be eaten by man directly and have to be processed through ruminants to produce an edible product.

The horticultural crops in Table 10.11 show wide contrasts but values generally lie between those for arable crops and animal products. The exception is glasshouse-grown tomatoes, where the energy supplied to maintain the required growing temperature was by far the dominant factor and resulted in an E value of 0·05.

While estimates based on both usable energy and on protein serve to illustrate our very considerable dependence on energy subsidies to capture the solar energy that goes into producing our food, results such as those in Table 10.11 must be treated with circumspection. Important though they are, energy and protein are not the only things that we get from food or even in some cases *the* most important. Food also supplies minerals and vitamins and, of course, there is the pleasure that it is derived from eating varied foods. Neither is it suggested that the results in Table 10.11 should be used to argue that we should cease to eat animal and horticultural products and live instead on a diet of cereals and potatoes. It is nevertheless of interest to note that alternative dietary strategies are possible, that would consume less energy, through an appropriate blend of plant and animal products.

Energy budget for UK agriculture

The presentation of energy budgets for particular commodities leads to the idea of presenting an overall budget for UK agriculture and this is given in Table 10.12. The inputs are those in Table 10.9. The outputs in Table 10.12 are the energy equivalents of the edible parts of the crops passing into human consumption. Thus, for crops such as cereals which are used to feed both humans and animals, only that portion used directly for humans is included. Intermediate outputs such as animal feedingstuffs, wastes and plant residues are not shown because they are used to feed animals or are returned to the land. Human food was chosen as the basis for

Table 10.12   Overall energy budget for UK agriculture up to the farm gate (PJ/yr)

Input of support energy		Output of energy available to man		
Solid fuels	1·2	Cereals	56·8	
Petroleum	70·0	Potatoes	13·0	
Electricity	37·2	Sugar beet	14·6	
Fertilisers (manufacture)	93·3	Arable crops		84·4
Machinery (manufacture)	40·0	Vegetables	2·4	
Feedstuff processing		Fruit	0·8	
(off-farm)	52·5	Horticulture		3·2
Chemicals (manufacture)	8·5	Milk	38·0	
Buildings (materials and		Beef	13·8	
construction)	22·9	Pigs (pork and bacon)	17·3	
Transport, distribution and		Sheep (lamb and mutton)	4·5	
services	16·3	Poultry (eggs)	4·5	
Miscellaneous	4·3	Poultry (broilers)	1·9	
Imported feedstuffs (includ-		Poultry (turkey)	0·6	
ing transport)	53·2	Livestock		80·6
Totals	399·4		168·2	168·2

$$E_a = \frac{\text{Energy available to man}}{\text{Agricultural support energy}} = \frac{168}{400} = 0·42$$

$$E_f = \frac{\text{Energy available to man}}{\text{Food system support energy}} = \frac{168}{1749} = 0·1$$

determining output because food production is the primary purpose of agriculture but it should not be overlooked that energy is also involved in the by-products that man uses directly such as hides and wool just as energy must be expended in the production, say, of man-made fibres. Thus, in that sense, it may be said that the usable energy is greater than that based on food alone.

Table 10.12 shows that the overall E ratio for agriculture is about 0·4. Leach (1976) and Blaxter (1975) have calculated lower output energies than those given in Table 10.12 and correspondingly lower E ratios of 0·35 and 0·34 respectively. Thus, to obtain the energy we consume in the 53% of our unprocessed food that is produced in the UK, we put in two-and-a-half to three times as much energy in the form of fossil fuels and imported feedstuffs. The outputs to man, in energy terms, of the plant and animal systems are comparable. Yet Leach (1976) shows that of all farmland in the UK, only 8·3% provides food directly to man, nearly all the remaining 92% being devoted to feeding livestock, which also consume around 7 m. tonnes of imported feedstuffs each year. Since this 8% provides half the energy output of UK farms, the vegetable system is energetically much more efficient in the use of land resources than is the animal system.

BENEFITS OF SUPPORT ENERGY USE AND THE FUTURE

At the turn of the century, very little fossil fuel was consumed in growing food and most energy inputs, in England and Wales, were provided by over 1 m. full-time workers and almost as many horses. Today the number of full-time workers is around 200,000. From 1930 onwards, the number of tractors in use increased rapidly, levelling off at around 1960, although the average power has continued to increase. The use of artificial fertilisers increased rapidly from 1940 onwards and the trend continues. Around 1900, typical crop and livestock yields per ha were one-half or less than those of today. Increases in crop yields can be attributed to the adoption of improved seed varieties coupled with enlightened use of fertilisers and crop protection chemicals.

In summary, it may be said that we have substituted energy for manpower through the increased use of machines (and thereby relieved much drudgery) and energy for land through increased fertilisation of crops. In so doing, we have released resources such as men and land for other purposes but at the expense of our resources of fossil fuel energy. This resource may itself become limiting one day and measures are under investigation to increase efficiency of energy use through conservation measures, alternative practices which use less energy, re-use of wastes within agriculture and alternative sources of energy (White 1979).

11

FERTILISERS AND MANURES

PLANT FOODS

Plants obtain carbon, hydrogen and oxygen from air and water. Their other essential nutrients are simple chemical ions which are taken up by the roots from soil. Six elements are needed by crops in quantities ranging from 10 kg to several hundred kilogrammes per hectare; these are calcium (Ca), magnesium (Mg), nitrogen (N), phosphorus (P), potassium (K) and sulphur (S); these are often called the major nutrients. Other elements, termed micronutrients or trace elements, are needed in quantities ranging from a few grammes to a kilogramme per hectare; they are boron (B), chlorine (Cl), cobalt (Co), copper (Cu), iron (Fe), manganese (Mn), molybdenum (Mo), and zinc (Zn). The need for the major nutrients was established in the last century; work in this century has shown that the trace elements are essential for plants (and for animals).

Nutrient reserves in soils

All soils contain a stock of these essential elements that are plant nutrients and in natural conditions the size of the reserves determines the fertility of the soil (which means its capacity to support plant growth). Where the annual growth is 'harvested' and removed the nutrients that have been taken up are lost to the ecosystem. Unless natural processes, the biological fixation of nitrogen, the release of other nutrients by weathering of minerals in soil, and return of nutrients in rainfall, make good the annual loss of nutrients in the 'harvest', the fertility of the site will decline and plant production will diminish.

The purpose of manures and fertilisers

Loss of nutrients from the cycle of growth and decay in the plant community was always a limiting factor for production from primitive agricultural systems. Thousands of years ago farmers learned that productivity was increased by applying wastes from crops, animals and humans to soil. These wastes are organic manures and we now realise that their function is to return plant nutrients to the soil. Further improvement came in Roman times when the value of marling (to neutralise soil acidity and supply calcium) and of growing legumes (to fix extra nitrogen) was established. There the improvement of soil fertility had to rest until chemical fertilisers were introduced in the first half of the last century. Fertilisers are simple salts either mined from natural deposits or manufactured by the chemical industry, which supply extra plant nutrients. By contrast the nutrients in organic manures have already been part of a natural cycle; using them to enrich one area must result in impoverishing the other areas from which the wastes came.

In short, the introduction of chemical fertilisers was a tremendous breakthrough. Soil productivity was no longer dependent on natural plant nutrient cycles. Fertilisers gave flexibility to farmers in planning their cropping systems, they now have the power to raise the productive potential of their land quickly without robbing other areas of their fertility. In the last century restrictive covenants in tenancy agreements were common; they prohibited the sale of some crops, wastes (such as straw), and manures, and prescribed the farming system which should be followed. Their purpose was to conserve nutrient stocks on the farm; fertilisers have made them unnecessary. By using fertilisers a farmer can grow crops in systems that suit him best and are most rewarding, while maintaining or improving the fertility of his soil.

THE FERTILISERS AVAILABLE TO FARMERS

Early history

By the 1830s materials that acted as fertilisers were being used by farmers in Britain. Bones were imported to supply phosphorus and sodium nitrate and guano (deposits of bird excreta) imported from South America supplied nitrogen. The fertiliser industry was established in Britain in the 1840s by John Bennett Lawes, the founder of Rothamsted Experimental Station. Bones were found to be satisfactory sources of phosphate on some soils and not on others. Lawes recognised that differences in soil acidity were responsible and he made a product (superphosphate) that was effective on all soils by dissolving the bones in sulphuric acid. He established a factory to make superphosphate, at first from bones, afterwards from indigenous and, later, imported mineral rock phosphates.

A century ago the fertiliser industry was well established. Nitrogen was supplied by ammonium sulphate (a by-product of making gas from coal) and by imported sodium nitrate and guano. Superphosphate supplied the phosphorus and was augmented late in the last century by basic slag (a by-product from steel-making). Potassium was at first supplied by plant ashes, but little was used. Deposits of potassium salts were, however, discovered in France and Germany and these were mined and used from the 1880s onwards.

The present position

A fertiliser may now be defined as a chemical substance that is a product of mining or manufacturing industry, which is added to soil to supply one or more of the essential major or micronutrients of plants. The present importance of fertilisers is shown by the fact that about one-eighth of the total expenses of UK farmers is for fertilisers and lime.

Nitrogen fertilisers

The development in Germany of the Haber-Bosch process for fixing nitrogen from air, and so synthesising ammonia, gave the fertiliser industry a new dimension. Provided the fossil fuels were available to operate the factory, the supply of N-fertilisers was no longer limited to a by-product from making gas from coal, and to the supply of sodium nitrate that could be mined. Industry has responded by making a whole range of new fertiliser materials from synthetic ammonia.

Ammonia (with 82% N) is a gas at ordinary temperatures and pressures but, when liquified and kept under pressure, can be transported and applied to soil. Liquid ammonia combines with soil but it must be injected at least 15 cm deep to avoid loss of the vapour. Much of this anhydrous ammonia is used directly as fertiliser in the USA, and in some other countries (e.g. Denmark); some has been used in Britain. Aqueous ammonia, made by dissolving the gas in water, is usually sold to contain 26% N; it is easier to handle having only a small vapour pressure, and need not be injected so deeply.

Ammonium sulphate is still made but much is exported to the tropics where it is favoured because it is not hygroscopic and is easy to handle in humid climates. Ammonium chloride, made by neutralising ammonia with hydrochloric acid, is not used in Britain but is used for fertilising rice in Asia.

Oxidising ammonia to nitric acid and combining this with more ammonia gives ammonium nitrate. When mixed with calcium carbonate this fertiliser became familiar in the 1930s as 'Nitro-Chalk'; the mixture is still sold but is now more concentrated with 26% N. 'Straight' ammonium nitrate is probably the most common nitrogen fertiliser now in Britain; it is sold in granulated or prilled forms and contains 33–34% N.

On a world scale the most striking recent development has been a great increase in the use of urea as fertiliser. It is made simply by combining ammonia with carbon dioxide under pressure. Containing 45–46% N, it is the most concentrated solid N-fertiliser. It hydrolyses quickly in soil to form ammonium carbonate, which is unstable and releases ammonia that may damage delicate roots, or may escape to the air from dressings applied to the surface. These disadvantages are overcome by placing the urea well below the surface of the soil, but not near seeds or the roots of young plants. Loss of ammonia is also avoided if broadcast urea is washed immediately into the soil by rain or irrigation. Some urea is made and used in Britain but it forms only a small proportion of the total N-fertiliser used. In recent years many new factories have been built in Asia to make nitrogen fertilisers, particularly for rice; nearly all of them produce urea. It seems likely that for the world as a whole urea will in future become the most important N-fertiliser and much research will be required to ensure that it is used efficiently.

Other solid nitrogen fertilisers are important in a few areas and for special crops. Sodium nitrate (16% N), potassium nitrate (13·5% N, 44% K_2O) and calcium nitrate (15·5% N) are very soluble and quick-acting and are used for top-dressing crops.

Slow-acting synthetic fertilisers have been developed as alternatives to the 'natural' organic fertilisers (hoof and horn meal, blood manure, and shoddy (wool waste)), which used to be important sources of nitrogen in horticulture. The best-known synthetic material is made by reacting urea and formaldehyde; a typical urea- formaldehyde fertiliser has 38% N, mostly insoluble in water and it is slow-acting. These kinds of fertilisers are expensive sources of nitrogen but are used in horticulture, for ornamental plants, and for lawns.

Phosphate fertilisers

Mineral rock phosphates are mined in several parts of the world; most of that used for fertilisers in Britain comes from North Africa; the deposits are calcium phosphates. A small proportion of the imported rock is used directly on soil; only certain deposits are suitable, the best-known comes from Gafsa in Tunisia and contains about 29% P_2O_5. For direct application the mineral phosphate must be finely ground so that practically all passes through the 100-mesh British Standard Sieve (apertures of 152 microns). Mineral phosphates should be used only on acid soils where they dissolve slowly and become available to plants; they are not dissolved by neutral and calcareous soils and should not be applied to such land.

The greater part of the imported mineral phosphate is dissolved in acid to make phosphates that are soluble in water and in ammonium citrate solutions and which are suitable for use on all kinds of soil. These are described below.

Superphosphate (with 16–22% P_2O_5) was the first manufactured phosphate; it is still made on a small scale by treating mineral phosphate with sulphuric acid, but it is becoming obsolete because of its low concentration. The active component of both ordinary and triple superphosphate is mono-calcium phosphate ($Ca(H_2PO_4)_2$) which is soluble in water. Triple (or concentrated) superphosphate (with 44–50% P_2O_5) is made by treating mineral phosphate with sufficient sulphuric acid to decompose the rock completely; the calcium sulphate formed is filtered from the strong solution of phosphoric acid that results. This acid is then used to attack a further quantity of rock phosphate to give 'triple super'. Both ordinary and triple superphosphate are valued by their solubility in water and in neutral ammonium citrate solution.

Ammonium phosphates are made by reacting ammonia with phosphoric acid; two products are made and nearly all is used to mix with other fertilisers to make 'compound' fertilisers. Mono-ammonium phosphate (MAP) usually contains about 11% N and 48% P_2O_5. Di-ammonium phosphate (DAP) often analyses at 18% N and 46% P_2O_5. These phosphates are soluble in water and are valued by this test and by their solubilities in neutral ammonium citrate solution.

Basic slag is a by-product of steel-making. The phosphorus comes from the iron ore used; it is not soluble in water and is slower in action than water-soluble phosphate. But in the best materials all the P becomes available to crops both in acid and in neutral soils. Slags are valued by their total content of P and by solubility in 2% citric acid solution; they must be finely ground. The best products have 80% or more of their total P soluble in citric acid solution and 80% should pass through the 100-mesh sieve. Basic slag also contains lime and has extra value for neutralising acid soils. Formerly much basic slag was used in Britain to improve grassland but recent changes in steel-making processes and in the ores used have greatly reduced the quantities of high-grade slags available. To overcome this shortage mixtures of lower-grade basic slags with mineral phosphates have been sold. A new product has recently been marketed in Britain; these fertilisers are made by calcining aluminium calcium phosphates mined in Senegal in West Africa; they are valued by solubility in alkaline ammonium citrate solution.

Experiments in other countries have shown that these phosphates are effective in neutral as well as in acid soils; there have not been enough field experiments in Britain to value them thoroughly for our conditions.

Potassium fertilisers

Deposits of potassium ('potash') salts are mined in several parts of the world. A mine in the Cleveland district of north-east England was opened a few years ago and it supplies part of our needs; most of the remainder of our potash comes from France, Germany and eastern Europe. The fertilisers sold are muriate of potash (which is essentially potassium chloride), containing 60-63% K_2O, and sulphate of potash (essentially potassium sulphate), containing 48-52% K_2O. The sulphate has to be made from the chloride and so is more expensive, but it is preferred for some horticultural crops and for potatoes as it can give better quality produce. Potassium nitrate (13% N and 44% K_2O) is also expensive but is recommended for some horticultural crops. Potash salts are crude minerals (often with 25-30% K_2O); they contain much sodium chloride which is useful to sugar beet and related crops. Kainite (with 15-20% K_2O) is a double salt that supplies magnesium and it is used on soils deficient in this element.

All potassium fertilisers are valued by the solubility of the potassium in water, the K they supply acts equally well. Differences in their values are caused by the other ions present, and by the value of any sodium and magnesium that they contain.

Other nutrients as fertilisers

When supplies of nutrients other than N, P and K are deficient in soils, they must be supplied to secure full crop yields and the salts used must be regarded as fertilisers.

Magnesium (Mg) is often deficient in light sandy soils in Britain and is supplied as the sulphate, oxide or carbonate of magnesium. In addition, about 100,000 tonnes of Mg are supplied annually in dolomitic limestone, much of which is applied to pasture to raise the concentration of Mg in herbage to benefit grazing livestock and prevent hypomagnesaemia (see Chapter 20).

Sulphur (S) is needed by crops in quantities about equal to the quantities of phosphorus needed. Much S is released when fossil fuels are burned and it is deposited on the soil in rain. Over much of England rainfall supplies 20-40 kg/ha of S annually. This is more than crops need and no examples of sulphur-deficiency have been found on farm crops in England. In regions remote from industry the atmosphere contributes much less sulphur and fertilisers have to be used to supply this nutrient; for example in Australia and New Zealand S-fertilisers often give large increases in yield. Parts of Ireland receive only small amounts of sulphur in rainfall and on light soils low in organic matter many responses to S-fertilisers have now been recorded.

Micronutrients (or trace elements) are boron, cobalt, copper, iron, manganese, molybdenum and zinc; when these are deficient in soils they must be applied as fertilisers. Often they are applied as sprays to foliage, but they may be mixed with solid fertilisers supplying N, P and K; if this is done the amounts added must be declared. The percentages of crop areas in England dressed with trace elements range from 25% of the sugar-beet area, to 5% of the cereal area, and to only 1% of the grassland (where the dressings are mainly of cobalt to benefit grazing animals).

At present in Britain the trace element fertilisers commonly used on agricultural crops are boron, copper and manganese. Iron is necessary for some horticultural and fruit crops on soils containing much calcium carbonate. Molybdenum is occasionally needed for brassica crops. Zinc deficiency is rare in Britain but is serious in other regions, particularly in Asia. All over the world the intensification and simplification of agricultural systems will increase the risk of trace element deficiencies. Larger yields make larger demands on reserves in the soils; deficiencies are most likely on light soils, particularly where all produce is sold, and no animal manures are returned. A dressing of farmyard manure once in four years will supply about as much micronutrients as crops remove.

Compound fertilisers

Fertilisers which supply only one nutrient are called 'straights'; those that supply two or more of the nutrients N, P and K, and perhaps other nutrients too, are termed 'compound' fertilisers in the UK; in the USA they are often called 'mixed' fertilisers. Until the 1940s most compounds were mixtures of the straights available, that is, ammonium sulphate, ordinary superphosphate and muriate of potash. All these were in powdered forms, the mixtures were hygroscopic, and often they set to lumps which had to be broken before application. Vigorous development work by the fertiliser industry has changed this. Both straights and compounds are now sold in granulated or prilled forms which store and handle well; in moisture-proof packages they retain their free-flowing characteristics. At the same time the development of more concentrated forms has resulted in the average concentration of compounds being double that of 40 years ago. These developments have cheapened transport and handling and have led to easier and more efficient application.

Liquid fertilisers

Anhydrous and aqueous ammonia, which have already been described, are liquid fertilisers; they must be injected into soil to prevent loss of ammonia, as must *low-pressure solutions* made from aqueous ammonia and urea. Other liquid fertilisers that are available are not under pressure and may be applied to soil in the ways that solid fertilisers are.

Liquids supplying nitrogen only are usually solutions of urea and ammonium nitrate. Solutions supplying two or three of the nutrients N, P and K are made from urea and/or ammonium nitrate, ammonium phosphate or ammonium polyphosphate, and potassium chloride. The disadvantage of liquid fertilisers is that they are less concentrated than solids and so greater weights have to be transported. This difficulty has been overcome by some suppliers by making suspensions of solids which are stabilised by adding special clays. Liquid fertilisers are as efficient as solids when both are correctly applied. The advantage of liquids is that when equipped with appropriate tanks and equipment for pumping and distribution, farmers find that labour can be saved in fertiliser application. In Britain, liquids are believed to supply about 5% of the fertiliser market, most of this trade being in solutions supplying nitrogen. In some other countries, notably the USA, liquids have a much larger share of the market.

DESCRIPTIONS AND VALUATIONS OF FERTILISERS

Official regulations

Since 1978 trade in fertilisers in Britain has been covered by the Fertilisers Regulations 1977 (Statutory Instruments No. 1489), which were introduced to include and implement the EEC Directive published in 1975. (The Regulations are well explained in a leaflet published by the MAFF (1978).) Solid fertilisers which comply with the specifications may be called EEC fertilisers. The Regulations also state what information must be given about fertilisers which are sold in Britain, but which are outside the EEC scheme; these are liming materials, liquid and semi-liquid fertilisers, fertilisers containing pesticide, or herbicide, or an organic-based nutrient (other than urea), or added trace elements.

The information that must be declared is more comprehensive than under the old (1973) Regulations and, for EEC fertilisers, is as follows:

For nitrogen: percentages of total-N, nitric-N, ammoniacal-N, ureic-N and cyanamide-N.

For phosphate: percentages of phosphorus pentoxide (P_2O_5), (with the equivalent as phosphorus (P) given in brackets), soluble in certain extractants. These extractants are mineral acids (giving total P), alkaline ammonium citrate solution, neutral ammonium citrate solution, water, 2% formic acid solution and 2% citric acid solution.

Potassium and magnesium soluble in water must be declared as oxides (K_2O and MgO) with the equivalents as elements (K and Mg) in brackets.

The Regulations for EEC fertilisers specify the analytical methods which must be used to value superphosphate (neutral ammonium citrate and water), basic slag (2% citric acid), rock phosphates (2% formic acid) and other sources of phosphate. They also specify minimum concentrations and solubilities. Any mixed (or compound) fertiliser must be described as an 'NPK', 'NP', 'NK', or 'PK' Fertiliser. (Officially the name 'Compound Fertiliser' is now restricted to materials containing organic matter and these are outside the EEC scheme.)

Non-EEC fertilisers may still be marketed in Britain; for nitrogen the total quantity must be stated and also the amount of N as urea where this is more than 10%, other forms of N need not be declared. These recent Regulations do not affect the actual quality of fertilisers and they make little difference to fertiliser practice which must still be based on advice leading to a correct choice of fertiliser and correct decisions on the amount to be applied, and on the timing and placing of the dressing. The advantage of the new Regulations is that farmers are now better informed about the nature and composition of the materials they buy.

Prices of fertilisers

Table 11.1 gives the analyses and prices of 'straight' fertilisers published in April 1981, together with the price of 1 kg of the

Table 11.1   Analyses and prices of straight fertilisers in April 1981

Fertiliser	Analysis	Price* £/tonne	Cost of 1 kg of nutrient (p)	
Nitrate of soda	16% N	80	50	} for N
'Nitro-Chalk'†	25% N	94	37.6	
'Nitram' and 'Nitra-Shell'	34.5% N	115	33.3	} for P_2O_5
Triple superphosphate	47% P_2O_5	150	31.9	
Muriate of potash	60% K_2O	95	15.8	} for K_2O
Sulphate of potash	50% K_2O	114	22.8	

* Nitrate of soda price is 'ex depot', other prices are 'delivered'
† 'Nitro-Chalk' is a mixture of ammonium nitrate with calcium carbonate; 'Nitram' and 'Nitra-Shell' are pure ammonium nitrate

nutrients calculated from these figures. N in nitrate of soda is seen to be more expensive than N in other fertilisers; K in sulphate of potash is more expensive than K in the muriate. Such calculations can also be used to value the nutrients in a multi-nutrient fertiliser in terms of the prices of 'straights'. For example, the price of Chilean potash nitrate (with 15% N and 14% K_2O) was published as £100/tonne. A tonne would contain 150 kg of N and 140 kg of K_2O; using the cheapest prices in Table 11.1 (for ammonium nitrate and muriate of potash) shows that an equal weight of N and K_2O could be bought for £72. (The chemical composition of the fertilisers however would be different since the N in Chilean potash nitrate would all be in nitrate form and the fertiliser contains no chloride (the muriate does); these are points which, for special purposes, may justify the higher priced material.) Fertilisers containing three nutrients may be valued in the same way.

LIME

Crop roots cannot grow satisfactorily in acid soils; liming materials have two functions. They neutralise soil acidity and they supply the major nutrient calcium. Surveys in the 1930s, when only about 0·5 million tons of liming materials were used each year in Britain, showed that much farmland was so acid that all crop growth was limited and some crops could not be grown at all. The government introduced a subsidy on lime under the Land Fertility Scheme in 1937. Johnston and Whinham (1980) have described the subsequent excellent progress with liming. Use of liming materials rose to 7 m. tonnes annually in 1960; it is now about 4 m. tonnes a year. Nearly all British arable land now has a satisfactory lime status but some permanent grass, particularly in the uplands, is still very acid and needs lime. The subsidy on lime stopped in 1976 and use has since diminished; there is a danger that, without this assistance, some farmers may not use sufficient lime.

Lime is supplied as calcium oxide (CaO), hydroxide (Ca(OH)$_2$), and carbonate ($CaCO_3$). Although 'burnt' and 'slaked' lime (oxide and hydroxide respectively) were much used in the past, almost all the lime used now is crushed and ground limestone from a variety of geological formations. Limes are valued by a laboratory test which measures their ability (in terms of 'neutralising value' expressed as CaO) to neutralise soil acidity. The cheapest effective source of lime should be used; in field experiments there have been no advantages from using the more expensive burnt or slaked limes as compared with the carbonate. Fine limes act more quickly than coarse materials. Current British Regulations require that the percentages of ground limestones and magnesian limestones passing through a standard sieve (with apertures of 152 microns) be declared.

ORGANIC MANURES

Until fertilisers were introduced organic manures originating in plant and animal wastes were the only means that farmers had of returning plant nutrients to the soil. That they still have a vital role can be seen by examining figures for the amounts of plant nutrients in the crops grown, in animal excreta, and in the fertilisers now used in the UK. A portion of the produce of arable land is fed to livestock; the plant nutrients in these feeds are augmented by those contained in imported feedingstuffs. While all of the grass grown is available to animals, a proportion will not be eaten; perhaps two-thirds of the nutrients in grass will pass through animals. Stock excrete much of the N and P, and nearly all the K which is in the food they eat. As Table 11·2 shows, excreta from stock in Britain

Table 11.2   Estimates for the UK of the plant nutrients in crops grown, in animal excreta and human sewage, and in the fertilisers used in 1979

	N	P	K
		'ooo t	
In crops			
arable	460	75	500
grassland	1100	170	1000
total	1560	245	1500
In animal excreta			
total	850	200	700
(applied to land in manures)	(375	75	240)
In human sewage			
total	150	65	60
in effluent (to rivers)	100	50	50
in sludge (some to land)	25	15	3
In fertilisers used in 1979	1186	182	345

contain about two-thirds as much N, about as much P and twice as much K as is in the fertilisers now used. Sewage from humans contains very much less N, P and K than is in wastes from animals.

Of the total animal excreta produced, rather more than half is dropped in patches on grassland by grazing cattle and sheep. These excreta are, of course, not uniformly distributed but, over a period of years the return of P and K may be expected to even out: however, there is a large and unavoidable loss of the N which volatilises as ammonia to the air from patches of both urine and faeces. The remainder of the excreta – from cattle when housed, and virtually all of that from pigs and poultry – is handled as organic manure. The traditional material was farmyard manure (F Y M), the result of allowing mixtures of excreta and bedding to ferment. The methods of keeping livestock have changed greatly in the last 25 years, however: bedding is used less commonly, stock are housed on slatted floors and their excreta are collected and handled as semi-liquid slurries. The practices and problems of handling and distributing these slurries are discussed elsewhere in this volume.

Some analyses of typical organic manures are given in Table 11·3. They must be regarded as average examples because actual

Table 11.3   Examples of average analyses of organic manures used in Britain

	Moisture	N	P†	K†
	% in fresh materials as received			
Poultry manures				
deep litter	32	1·7	0·9	1·1
battery	66	1·5	0·5	0·6
straw-droppings compost	65	1·1	0·9	0·8
slurry	75	1·4	0·5	0·5
Cattle manures				
farmyard manure	76	0·6	0·1	0·5
slurry	89	0·7	0·1	0·6
Pig manures				
Slurries { fed on whey	97	0·2	0·1	0·2
{ fed on dry meal	90	0·6	0·2	0·3
farmyard manure	75	0·6	0·3	0·3
Sewage sludge	55	1·0	0·3	0·2*
Municipal town refuse	35	0·5	0·2	0·3*

* In both sewage sludges and town refuses much of the potassium is contained in mineral particles present in contaminants; this fraction is useless to crops
† Analyses for P and K are given in element form; if the figures in this Table are used to calculate fertiliser dressings expressed in oxide form (P_2O_5 and K_2O), adjustments must be made

analyses vary greatly from sample to sample, but they do show the trends that exist. After allowing for differences in moisture content, poultry manures are seen to be relatively rich in phosphate, while manures made from straw, or those containing the excreta of grass-fed ruminants, are richer in potassium.

The value of nutrients in manures

Chemical analyses measure the total quantities of N, P and K in manures, but not their availabilities to crops, these can only be measured in field experiments. Most of the N in manures is combined with organic materials and is released only as they decay. In practice, about a third of the N in F Y M and similar manures is quickly released, but much is resistant and persists in the soil. About half of the P is quickly available; the remainder is combined with organic matter and acts slowly. Nearly all of the K is soluble in water and is immediately available to crops. For practical guidance it can be assumed that 25 t/ha of F Y M will supply to a first-year crop 40 kg of N, 20 kg of P (45 kg of P_2O_5) and 80 kg of K (95 kg of K_2O). In planning to use both fertilisers and manures, these figures can be used to adjust the amount of fertiliser given. It can be assumed that further but smaller quantities of N and P will be available to later crops.

Field tests have been made of the value of nutrients in other manures. About half of the N in poultry manure is available to a first-year crop. The efficiency of the N in slurries has been found to vary with the type and origin of the material and the time when it was applied. When cow slurry was applied to grassland in March it was most efficient; both earlier and later applications were less efficient as the following figures show:

Slurry applied in	% of N recovered in herbage
December	14
Jan./Feb.	26
March	49
April	39

Pig slurry has been tested on cereals. When applied in mid-winter only a quarter of the N was used by the crop; when applied immediately before sowing in spring up to three-quarters of the N was useful.

All farm manures and slurries should be regarded not as disposal problems, but as assets because they contain nutrients that would be expensive to buy. In the absence of guidance from analyses of their own slurries farmers may assume that fresh slurry contains about 0·7% N, 0·2% P_2O_5 and 0·7% K_2O. Therefore 1,000 l will supply 7 kg each of N and K_2O and 2 kg of P_2O_5. If applied in March the N should be equal in effect to half the weight of N applied as fertiliser, perhaps more. All the P and K should be useful in maintaining reserves in the soil.

Micronutrients are supplied by organic manures because they are made from the wastes of plants and animals to which the elements were essential. An average sample of F Y M contained these amounts:

Table 11.4

	Parts per million in dried manure	Approximate amount supplied by 45 t/ha of manure (kg/ha)
B	20	0·2
Co	6	0·07
Cu	60	0·7
Mn	400	4·4
Zn	120	1·3

Ordinary manufactured fertilisers do not supply worthwhile amounts of micronutrients; only basic slag contains sufficient to make a real contribution to crop nutrition. But a dressing of F Y M once in a four-year rotation can supply as much micronutrients as crops remove.

Sewage sludges

The effluent from sewage works which is discharged to rivers contains much of the N and P and nearly all the K that was in the sewage (Table 11·2). The semi-liquid sludge produced contains some N and P that is useful to crops; the K content of sludges is small and can be ignored. In field experiments sewage sludges and town refuses have been inferior to F Y M when similar weights were applied often because the urban wastes contained very little useful K. The value of sewage sludges and refuse products lies almost entirely in the plant nutrients they contain; the organic matter provided has no immediate value. In general, sewage sludges can be reckoned as useful sources of N and P but dusts and pulverised refuses are so poor in nutrients that they are of little use as manures.

Possible toxic effects from sewage sludges must be mentioned. Where sewage contains industrial effluents the sludges often contain enough heavy metals (lead, zinc, copper, chromium and nickel are commonly found) for large dressings to be toxic to crops, particularly if they are used repeatedly. This difficulty is now recognised and guidelines have been established which have been accepted by the Water Authorities who supply sludge and by farmers' advisers. The amounts of sludge which may safely be applied over a period of years are specified; if these guidelines are followed there should be no harm to crops.

ADVICE TO FARMERS ON THE USE OF FERTILISERS AND MANURES

Advice on fertilising is based on the results of field experiments where the effects of different rates and combinations of N, P and K fertilisers on crop yields have been measured. In addition, the effects of applying organic manures as well on the responses of crops to fertilisers have been investigated. When interpreting the results of such trials an adviser also takes account of soil type, climatic region, and local weather during the period of the experiments. The results of all work of this kind are used to provide general advice for all important crops, and indications of the modifications needed to allow for different soils, crop rotations and climates. A publication from the MAFF by Shaw (1979) gives the most recent advice for farmers in England and Wales.

General advice for a crop and a region is further refined by allowing for the reserves of nutrients in particular fields. Cropping and manuring history of the field and the type of soil determines how much nitrogen is likely to become available during the season. The reserves of phosphorus and potassium are determined by analysing a sample of soil. With this information well-based advice can be given on the best combination of fertilisers to apply, together with such organic manures as may be available.

The composition of plants also reflects their nutritional status and analyses of selected parts, such as leaves or stems, are used to guide fertiliser programmes for horticultural and glasshouse crops and for soft and tree fruit. Plant analyses are little used in advising on the use of major nutrients for British field crops, but they are useful for diagnosing micronutrient deficiencies. When plants are seriously deficient in a nutrient their leaves often show symptoms that are typical of the particular deficiency. Colour 'atlases' of these symptoms have been published and these are useful in diagnosing deficiencies, particularly of micronutrients.

CHANGES IN THE AMOUNTS OF FERTILISERS USED IN THE UK

Total amounts

Table 11·5 shows the amounts of fertilisers used in selected years during the last century. In the first 60 years fertiliser use grew only slowly because farming was unprofitable following the agricultural depression of the 1870s. Fertilisers were applied only to valuable crops; much of the land, and particularly grassland, received none. The urgent need for food in the Second World War caused the

Table 11.5   Fertilisers used in selected years during the past century in the UK

Year	N	P_2O_5	K_2O
		'000 t	
1874	35	91	3
1913	30	183	23
1939	61	173	76
1950	229	468	238
1958	320	392	354
1965	574	487	432
1973	947	482	416
1981	1335	440	450

government to encourage fertiliser use and to assist with prices so that the amounts applied increased to the limit set by the supplies that were available (all the P and K used had to be imported and these nutrients were rationed during the war). The prices of N and P were subsidised until 1974.

Between 1939 and 1950 the N used increased nearly four times while the P and K used increased roughly three times. The increases in phosphate were essential to establish arable cropping where neglected grassland was ploughed for the first time. In some areas K-fertilisers were essential to initiate cereal cropping. From 1950 to 1980 the total N used increased about five times. The K used increased up to about 1969; phosphate was already used heavily by 1950 and the total amounts then applied have not been materially exceeded.

Fertilisers used on selected crops in England and Wales

Surveys of Fertiliser Practice described by Church (1981) have provided the figures in Table 11.6. The amount of fertilisers used on arable crops, and particularly the N used, have increased in parallel with other innovations – notably, new varieties that responded better to N, and new chemicals to control weeds, pests and diseases of crops. The effects of these inputs to agriculture are not additive; they interact to produce the large yields that we now obtain. Results of research on the use of fertilisers have been quickly taken up by farmers with the result that our arable crops are now rationally and adequately fertilised.

Great changes have occurred in the use of fertilisers on grassland (Table 11.6). The use of P and K has become well established. But the most striking change has been in the use of nitrogen where the rate of growth reflects farmers' increasing ability to use profitably the extra grass grown by N-fertiliser. Accurate estimates of grassland yields are not available but we know that stock-carrying capacity of grassland has been increased greatly; we feed twice as

Table 11.6   Average amounts of fertilisers used on crops in England and Wales*

Year	Winter wheat	Spring barley	Potatoes	Sugar beet	Temporary grass	Permanent grass
			kg/ha of N			
1943/5	19	21	79	92	4	4
1957	51	35	124	134	26	11
1970	90	82	166	161	95	51
1980	145	86	185	137	167	93
			kg/ha of P_2O_5			
1943/5	30	36	92	88	11	11
1957	30	34	126	119	34	20
1970	41	40	181	117	44	28
1980	46	37	185	65	35	21
			kg/ha of K_2O			
1943/5	2	4	100	72	0	0
1957	33	45	200	203	21	9
1970	35	44	250	191	36	20
1980	39	40	259	144	37	20

* *Source:* from Church (1981)

many cattle as in pre-war days and 25% more sheep. There is, however, considerable scope for using more N-fertiliser on grassland, especially on the 5 m. ha of permanent grass; a quarter of this old grass received no nitrogen in 1980, and when N was applied the rates were less than for temporary grass.

THE FUTURE FOR FERTILISERS

Fertilisers are an essential resource for modern intensive agricultural systems; the levels of production now obtained in Britain and other developed countries will not be sustained, let alone increased, without their use. They will have a vital role in future in building even larger yields by the interactions of the nutrition of more productive varieties with better methods of controlling their pests and diseases. Organic wastes will also retain their historic and essential role of recirculating plant nutrients on the farm; used properly they will reduce the need for fertilisers.

Nitrogen

In spite of their vital role in food production, N-fertilisers have been much criticised because energy is used to make them and because they may 'leak' from the farming system into the environment. The problems of the energy resources needed to make nitrogen fertilisers are discussed in Chapter 10.

Further criticism of the use of N-fertilisers has been concerned with possible damage to the environment by the nitrate which leaks into natural waters, and by the nitrous oxide released by denitrification processes – which *might* upset the ozone layer in the stratosphere, which protects us from ultraviolet radiation. These issues have been discussed thoroughly by the Royal Commission on Environmental Pollution (1979).

The results of these pressures, and of increases in prices of fertilisers, is that research on sources of nitrogen for agriculture, and the efficiency of the nitrogen cycle, now has priority the world over. Research will improve the present level of efficiency, where arable crops commonly recover no more than 50% of the N in applied fertilisers, and often less. Nevertheless, losses of N by leaching of nitrate will not be completely prevented in arable-farming systems. When land is bare of crops in autumn and winter, any nitrate left over, together with that released by microbial decay of crop residues, is liable to be lost by leaching before another crop can be grown. Research on biological fixation of nitrogen has also been stimulated. This will produce more efficient strains of the *Rhizobium* bacteria which fix nitrogen from the air in the nodules on the roots of leguminous plants. There is no doubt that legumes will, in future, play a greater part in the nitrogen economy of our pastures. Other, longer-term research aims to transfer the nitrogen-fixing ability to other organisms with the purpose of endowing non-leguminous crops with the ability to obtain their supplies of N from the air.

Phosphorus and potassium

The world has good reserves of minerals to supply P and K, sufficient for several hundred years at current rates of use. Where phosphate has been used for long, as in Britain, reserves of P have accumulated in the soils; these are useful to crops and reduce the need for fresh fertiliser-P. Because our soils have been improved in this way, there is no need to use more P than at present, but there is a case for diverting some of the total now used to the poor soils of the uplands of Britain. Reserves of potassium have also been built up in some of our clay soils and recent experimentation indicates that the amount of K now used is sufficient to balance the losses from our farming systems. These assessments of the adequacy of our present manuring with P and K assume that the best use will be made of the amounts of these nutrients in crop and animal wastes.

Other nutrients

Deficiencies of other nutrients will tend to increase as larger yields are grown, particularly in systems where organic manures are not returned to the soil. We must keep a watching brief on the adequacy of the sulphur supplies from rainfall. A change to sulphur-free fuels, or the removal of sulphur dioxide from the products of combustion, could, in the long term, result in shortages of S for crops as has happened in other countries. The needs for adequate liming, and for magnesium on light soils, must not be forgotten. Trace element fertilisers will have to be used more often as larger yields make larger demands on reserves in soils.

Other systems

There is now much debate on the value of biological systems of farming where fertilisers are used in minimum quantities, or not at all. These discussions are valuable in drawing attention to the need to recycle all the nutrients in wastes, and to the low efficiency of nitrogen fertilisers in some farming systems. But in biological husbandry systems that avoid all use of fertilisers the extra nutrients needed to balance losses, or to increase total supplies to provide for larger yields, can only come from natural processes. This means that extra nitrogen must be fixed by biological mechanisms, and that extra supplies of other nutrients must come from the weathering of minerals in the soil. The rates of increase in nutrient supplies by these processes are too slow to cope with the world's future need for food. We are committed to fertilisers because they are the only means we have of quickly increasing the amounts of plant nutrients in the cycle of soil to crop to consumer.

12

CROP PLANTS

A great many plant species are used in UK agriculture, although the total used in world agriculture represents only about 6% of the plant species available. Within those used agriculturally, a very small number are of major importance.

Only about 2% of the UK agricultural land is used for horticulture: the main species are dealt with in Chapters 30, 31, 32 and 33 and will not be further detailed here.

About 8% of the total land area is used for forestry and, although this subject is not included here in any detail, the main species used are given in Table 12.1. Approximately 60% of the afforested area is currently administered by the Forestry Commission (1982).

Table 12.1   The main species used in UK forestry

Sitka spruce	*Picea sitchensis*
Norway spruce	*Picea excelsa*
Scots pine	*Pinus sylvestris*
Corsican pine	*Pinus laricio*
Lodgepole pine	*Pinus murrayana*
Larch	*Larix* spp.
Douglas fir	*Pseudotsuga douglasii*
Oak	*Quercus* spp.
Birch	*Betula* spp.
Beech	*Fagus* spp.
Ash	*Fraxinus* spp.

Source: CAS (1980)

With these exceptions, the major crop plants can be considered under six main headings: cereals, root crops, forage crops, grassland, fruit and vegetables. The last category includes both field vegetables (dealt with in Chapter 30) and horticultural vegetables (included in Chapter 32) so will be considered only briefly here.

CEREALS

The main cereals are wheat (*Triticum vulgare*), barley (*Hordeum distichum* and *H. polystichum*), oats (*Avena sativa*) and rye (*Secale cereale*) and the production systems based on them are described in Chapter 24. The main product is the seed, a one-seeded fruit, or caryopsis (see Fig. 12.1), but the straw may also be important, e.g. as a source of fuel. In the very young seedling stages cereals can be identified by examining the sprouted grains (Fig. 12.2). Later identification depends upon the appearance of the leaf (especially the auricles – see Fig. 12.3) or the ears (Fig. 12.4).

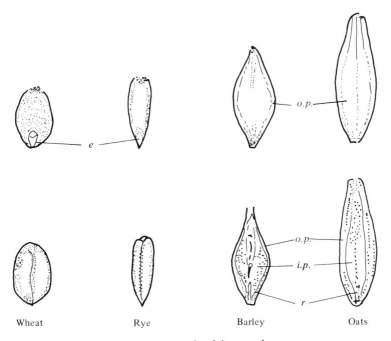

| Wheat | Rye | Barley | Oats |

Fig. 12.1   Seeds of the cereals
e, position of embryo. *o.p.*, outer pale. *i.p.*, inner pale. *r*, rachilla

ROOT CROPS

The true root crops (Table 12.2) are biennial and would use their root reserves in order to flower if they were left in the ground for a second year. The roots have characteristic shapes and occupy different positions in relation to the soil surface (Fig. 12.5).

The potato (*Solanum tuberosum*) is usually considered with the root crops but is, in fact, a tuber-bearing perennial, although it is grown as an annual: it is the only major crop in the UK to be propagated vegetatively. The potato plant has leafy aerial shoots

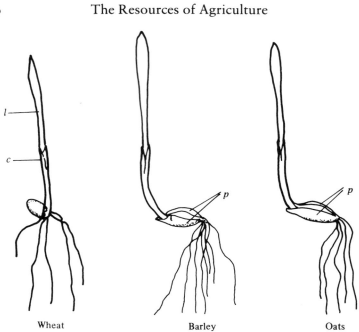

Fig. 12.2 Seedlings of the cereals
c, coleoptile or sheath; l, first leaf; p, pales

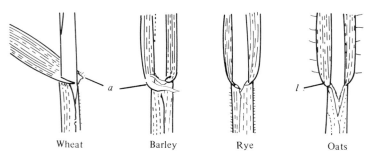

Fig. 12.3 Auricles and ligules of the cereals
a, auricle; l, ligule

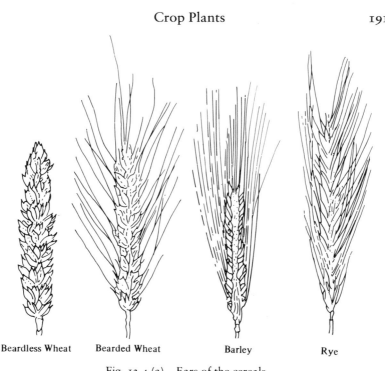

| Beardless Wheat | Bearded Wheat | Barley | Rye |

Fig. 12.4 (a)   Ears of the cereals

Table 12.2   **The true root crops**

Common name	Latin name	Colour
Sugar beet	*Beta vulgaris*	*White*
Fodder beet	*Beta vulgaris*	*Yellow, red, orange, white*
Mangel	*Beta vulgaris*	*Yellow, red, orange, white*
Turnip	*Brassica rapa* var. *rapa*	*White, yellow, green*
Swede	*Brassica napus* var. *napobrassica*	*Purple, bronze, green*
Carrot	*Daucus carota*	*Orange/red*

that die back at the end of a single season and a much-branched fibrous root system (Fig. 12.6). The tubers are borne on underground stolons that arise as lateral shoots from the most basal nodes.

FORAGE CROPS

The main forage crops are kale, rape, rye, turnips, swedes, fodder beet, mangels, maize and whole-crop cereal silage. (These are dealt

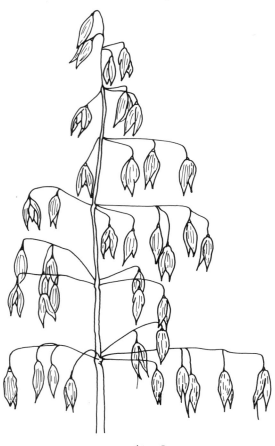

Fig. 12.4 (b)   Oat

with in Chapter 28 and the areas grown are given in Table 28.1. Their relative yields are shown in Table 28.2.)

By-products, such as sugar beet tops, are also used in a similar fashion to autumn-grazed forage crops and fodder radish and cabbage are also grazed in the autumn.

The structure of two of the major forage crops is illustrated in Fig. 12.7.

GRASSLAND

The major sown species in grassland are grasses (*Gramineae*) and legumes (*Leguminosae*) (see Chapter 29). A wide range of other

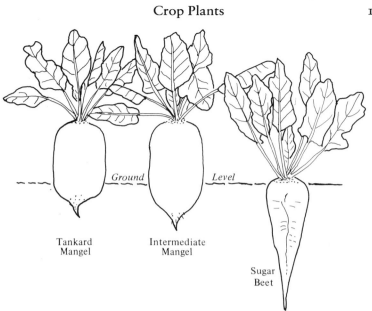

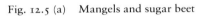

Ground   Level

Tankard
Mangel

Intermediate
Mangel

Sugar
Beet

Fig. 12.5 (a)   Mangels and sugar beet

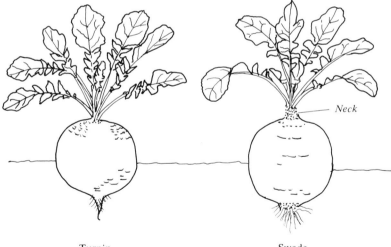

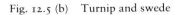

Neck

Turnip   Swede

Fig. 12.5 (b)   Turnip and swede

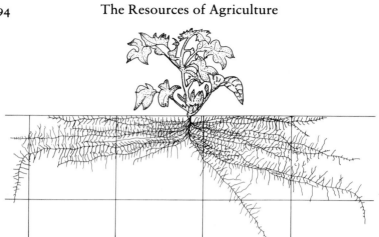

Fig. 12.6   One-half of root system of a potato plant 50 days old
Source: Weaver (1922)

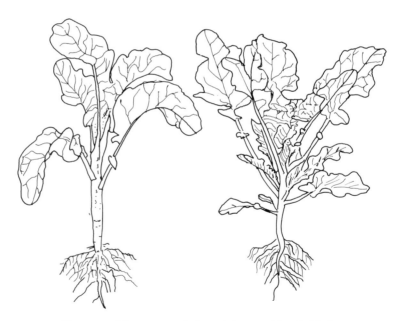

Fig. 12.7   Marrowstem kale and thousand-headed kale

Table 12.3 Some of the additional grass and legume species occurring in permanent pasture

Grasses	
Redtop bent	*Agrostis gigantea*
Creeping bent	*A. stolonifera*
Common bent	*A. tenuis*
Meadow foxtail	*Alopecurus pratensis*
Sweet Vernal-grass	*Anthoxanthum odoratum*
Tall oatgrass	*Arrhenatherum elatius*
Crested dogstail	*Cynosurus eristatus*
Red fescue & Chewings fescue	*Festuca rubra*
Yorkshire fog	*Holcus lanatus*
Smaller cat's tail	*Phleum bertolonii*
Smooth-stalked meadowgrass	*Poa pratensis*
Rough-stalked meadowgrass	*Poa trivialis*
Yellow oatgrass	*Trisetum flavescens*
Legumes	
Kidney vetch	*Anthyllis vulneraria* L.
Birdsfoot trefoil	*Lotus corniculatus* L.
Sweet clover	*Melilotus alba* Desr.

Source: Spedding & Diekmahns (1972)

species may be prominent in particular localities and a wider range of both grass and legume species occurs in permanent pasture (Table 12.3).

Grasses

The grasses (Fig. 12.8) usually have cylindrical jointed stems, with long, narrow leaves arranged in two opposite rows. The root system is fibrous and the fruits are of caryopsis type (see Hubbard, 1968).

The aerial parts of a grass consist of leaves, stem and inflorescence or ear. The leaf is made up of a sheath which clasps the stem, and a blade. At the junction of the sheath in many species there are two structures: the ligule, a colourless delicate membrane closely clasping the stem, and the auricles, which are projections of the leaf edges at the sheath juncture. The shape and size of the ligule, which is always present, and the extent to which the auricles are developed (absent in some species) are aids to the identification of grasses. The shape of the blade and its appearance in cross section are also useful characters for recognition of the species.

Inflorescences or ears are of two kinds. Most common is a branching ear, or panicle, with the individual spikelets borne on short or long branches, as in meadow fescue or the meadow grasses. In grasses such

Fig. 12.8   Perennial ryegrass (after Hubbard, 1968)

as perennial ryegrass and timothy the spikelets are fixed directly on the
stem rachis forming a spike. Each spikelet has an outer covering, the
glumes, which when separated from the seed become the chaff. They
enclose a short stalk at the base of the seed known as the rachilla, which
bears the florets, the size and shape of which also provide a means of
species identification.

Each floret has its own covering, the paleae or pales enclosing the
ovary and stamens. The seed consists of the ripened ovary (caryopsis),
enclosed by the pales, and sometimes, as in timothy, of the glumes as
well.

A most important feature of grasses is the tendency to branch or tiller (see Chapter 21), forming a dense sward. Growth takes place from a meristem very close to the ground and thus protected from all but the closest defoliation (see Fig. 12.9). Creeping grasses, however, have stolons or rhizomes and adventitious roots may develop from any of their nodes.

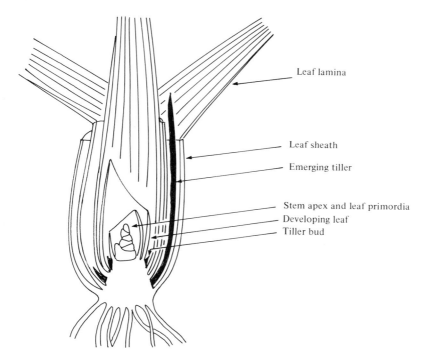

Leaf lamina

Leaf sheath

Emerging tiller

Stem apex and leaf primordia
Developing leaf
Tiller bud

Fig. 12.9   Grass growth from a meristem

Source: Jewiss (1981)

Legumes

The growth habit of the main legume species is described in Chapters 19 and 21. Structure varies between erect and creeping, leaves are arranged alternately on the stem and, apart from the first produced, are compound, usually forming three leaflets. The brightly coloured flowers are adapted to insect pollination and the fruit is a pod.

Legumes are able to fix atmospheric nitrogen by means of bacteria sited in root nodules (Fig. 12.10).

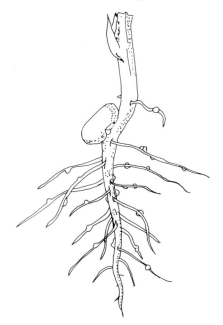

Fig. 12.10   Nodules on the roots of field bean

The nitrogen-fixing bacteria, of the genus *Rhizobium*, are host specific, so the correct strain must be present. When they are not present in the soil, seed may be inoculated. When nodules are ineffective, they are small, numerous and whitish; when effective, they tend to be few, large and reddish.

FRUIT AND HOPS

For convenience, these have been dealt with together in Chapter 31, in which the main species are listed. They are too numerous to discuss in detail but the main types of fruit are illustrated in Fig. 12.11.

VEGETABLES

The number of vegetable crops grown in the UK is considerable. (The most important field vegetables are listed in Table 30.1.)

The main brassicas are cabbage (see Fig. 12.12), Brussels sprouts, calabrese, cauliflower, broccoli and curly kale. Carrots are by far the most important root vegetable, onions are an important crop

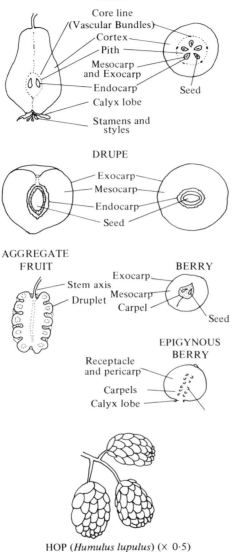

HOP (*Humulus lupulus*) (× 0·5)

Fig. 12.11   Main types of fruit and hops
Sources: Westwood (1978); Brouk (1975). From *Temperate-zone Pomology* by
Melvin N. Westwood. W. H. Freeman & Co. 1978

Fig. 12.12　Cabbage

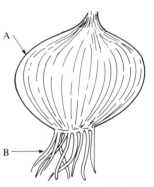

Entire tunicate onion bulb (× 0·5)
A. Coloured membranous scales
B. Adventitious roots

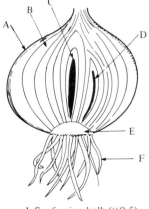

L.S. of onion bulb (× 0·5)
A. Membranous scales
B. Fleshy scales
C. Main bud
D. Lateral bud
E. Stem
F. Roots

Fig. 12.13　Onions (*Allium cepa*, etc)
Source: Brouk (1975)

(see Fig. 12.13) and peas and beans occupy a bigger area than any other category.

Peas and beans are legumes and, therefore, are able to fix their own nitrogen.

In addition to the foregoing crops and products, there are also a number of other plant species used for hedges, shelter belts, amenity and recreational purposes. Many plants are grown for gardens and cut flowers and potted plants are produced for indoor use. (Most of these are dealt with in Chapter 33.)

There are also some minor crops grown for purposes not so far mentioned, including flax, grown for the production of fibres.

Parts of the UK are well suited to the production of other fibre crops – notably the white-wooded bat willow (*Salix alba* 'Cerulea') for cricket bats, and the basket osier (*Salix triandra*) for the production of basket furniture and baskets of many types. These willows represent a form of agricultural production which has declined rapidly since the advent of man-made alternatives. They remain as small enterprises (about 14,000 mature cricket-bat willow trees are harvested each year), providing resources for craftsman-based industries including the production of brooms and brush-heads as well as cricket bats and baskets.

Flax was traditionally grown in Northern Ireland although there are few climatic reasons to prevent its production in other parts of the UK. Flax grown for fibre and linseed grown for oil seed are two different cultivated strains of the same species. The fibre flax plant grows to a height of 90–120 cm and has been bred selectively for maximum fibre content; it is harvested complete with roots by mechanical flax pullers. The flax grown for linseed is cut by combine harvesters when the first seed pods have become brittle; an attachment on the harvester cracks the seed bolls to release the seed. Flax seed is densely sown in April in a fine seed bed and the plants are ready to be harvested in about three months. Both the fibre and the seed crops are to some extent dual-purpose in that the seed from the fibre flax can be marketed as can the fibre of linseed.

In the face of intense competition from wool and imported cotton, and more recently from man-made fibres, the production of fibre flax has declined to negligible proportions in the UK. Almost half the world's flax fibre is grown in the USSR whilst the quarter of world production that is grown in Europe comes mainly from France and eastern Europe. However, apart from hemp, flax remains Britain's only commercially exploitable fibre crop.

13

FARM ANIMALS

Relatively few of the world's animal species have been domesticated and used in agriculture and only a few additional ones, such as the eland, are being actively investigated.

Most farm animals are warm-blooded (homeotherms) and only a very few are cold-blooded (poikilotherms).

The main species are listed in Table 13.1, with some of their characteristic attributes. More details will be found in the chapters dealing with the main animal production systems.

Although it is often possible to state the main reason why a particular animal is kept, it is rare for any animal to produce only one product. This is partly because most animals can actually produce more than one important product but also because, whatever the *main* product, most other parts of the animal have some use. This is illustrated for a beef animal in Table 13.2.

Thus if animals are kept mainly for meat, in most countries only about half the animal is consumed. Carcass weights are usually less than half the liveweight (see Table 13.3), this proportion being referred to as the killing-out percentage (KO %) or the dressing-out percentage (DO %). Other parts of the body are eaten and some parts of the carcass (e.g. the bones) may not be: all the parts not eaten, however, generally have a use and the economics of production may be greatly influenced by this.

Animals kept for milk may also grow hair or wool and both they and their progeny may eventually finish up as meat. Even animals kept for work may finally be eaten, although not all edible animal components are consumed by man.

It is therefore less easy to attribute products to animal species than is the case with plants but the origins of the main products are indicated in Table 13.4.

In the species of animals that have been subjected to most development, a great variety of breeds has arisen. Much of this has been the result of deliberate animal breeding but many breeds are primarily the result of natural selection in a particular environment. In the UK, a great many breeds of cattle, sheep, pigs, goats and horses were bred, carefully segregated, developed and used for quite specific purposes. This great variety has been preserved in

				Products	Normal Body Temp. °C	Mature L.wt ♀ (kg)	Birth wt. (kg)	Length of gestation or incubation (days)
Warm-blooded	Mammals	Ruminants	Cattle	Meat, Milk, Skin	38·9	350–700	25–54	273–286
			Sheep	Meat, Milk, Wool & Skin	40·0	32–76	2·5–6·4	140–149
			Goats	Milk, Meat, Skin	40·0	45–54	1·7–5·8	145–156
		Non-Ruminants	Pigs	Meat	39·8	130–220	0·8–2·0	112–118
			Horses	Traction, Transport, Recreation	38·1	136–1000	up to 102	325–341
			Rabbits	Meat, Skin	38·3	1–5·4	up to 0·058	31
	Birds		Chicken	Eggs, Meat	41·6	1·8–2·7		21
			Turkey	Meat		8–12		28
			Duck	Meat		4·0		28
			Geese	Eggs, Meat, Eggs		4·5–9·0		30–33
Cold-Blooded	Vertebrates		Trout	Fish meat	Ambient Temp.	0·2–3*		
			Carp	Fish meat	Ambient Temp.	up to 4*		
	Invertebrates		Bees	Honey, Wax, Pollination	Ambient Temp. Ambient Temp. Ambient Temp.			

*Harvested weight only

Table 13.2   **Useful by-products of the beef animal***

Product	Parts of the animal
Edible offal	Liver, heart, tongue, kidney, brains, cheek & head trimmings, tail, stomach
Leather	Hide
Insulin and Pharmaceuticals	Pancreas, pituitary, thyroid & thymus glands genitals, spinal cord, gall & gallstones
Edible fats, tallow and for processing and manufacture	Internal fat
Glue, gelatin, tallow, meal, pet-food, buttons, and handles	Bones
Pet-food, animal feed, fertiliser, glues and fire-suppressing foam	Blood
Cow heel jelly, neatsfoot oil etc	Feet
Meal	Horns
Surgical ligatures and pharmaceuticals	Small intestines

* Where the main product is taken to be the carcass

Table 13.3   **Live weights and carcass weights**

	Live weight at slaughter (kg)	K 0%	Carcass weight (kg)
Cattle	300–600	55–60	165–360
Sheep	39–43	45–50	17·5–21·5
Pigs { Bacon	90–140 }	70–75	63–105
Pigs { Pork	56 }	70–75	39–42
Rabbits	2–4	50–55	1–2·2
Hens	2	68–72	1·36–1·44

Table 13.4   **Main Products**

Product	Main sources
Milk	Cows, goats, sheep
Meat	Cattle, sheep, pigs, poultry
Wool	Sheep
Eggs	Chicken, ducks and geese
Skins	Cattle, sheep, goats
Honey	Bees

sheep, goats and horses, although many of the breeds are not numerically strong (see Table 13.5) but the situation has been quite different for other species. In the case of dairy cattle, farming has come to be dominated by a very few breeds: in the case of beef cattle, the number of breeds has, by contrast, increased by impor-

Table 13.5 **Breeds of sheep, goats and horses**

Species	Breeds	UK Pop. (1971)
Sheep	Scottish Blackface	2,338,000
	Welsh Mountain	1,974,000
	North Country Cheviot	505,000
	Swaledale	504,000
	Clun Forest	401,000
	Hardy Speckleface	327,000
	Beulah Speckled-face	206,000
	Black Welsh Mountain	2,000
	Bluefaced Leicester	2,000
	Border Leicester	9,596
	British Oldenburg	No figures available
	British Texel	"
	Cadzow Improver Ram	"
	Cambridge	"
	Cheviot-South Country	260,000
	Colbred	1,000
	Cotswold	No figures available
	Dales Bred	196,000
	Dartmoor-Greyface	33,000
	Improved Dartmoor	2,000
	Derbyshire Gritstone	20,000
	Devon Closewool	36,290
	Devon Longwool	134,000
	Dorset Down	8,600
	Dorset Horn	7,898
	Poll Dorset	8,501
	Exmoor Horn	35,000
	Finnish Landrace	500
	Hampshire Down	5,500
	Herdwick	58,000
	Hill Radnor	28,000
	Improved Welsh	41,000
	Jacob	2,000
	Kerry	209,000
	Leicester Longwool	300
	Lincoln Longwool	750
	Llanwenog	20,000
	Lleyn	7,000
	Lonk	12,000
	Oxford Down	600
	Romney	294,000
	Rough Fell	69,000
	Ryeland	700

The Resources of Agriculture

Species	Breeds	UK Pop (1971)
Sheep	Shetland	1,000
	Shropshire	No figures available
	South Devon	44,000
	Southdown	858
	South Wales Mountain	6,000
	Suffolk	36,000
	Teeswater	7,000
	Wensleydale Longwool	183
	White-faced Dartmoor	39,000
	Wiltshire Horn	1,000
	Masham	No figures available
	Scottish Greyface	"
	Scottish Half-bred	"
	Mule	"
	Welsh Halfbred	"

Species	Breeds	UK Pop. (1980 estimated)
Goat	Saanen	2,800
	British Saanen	19,707
	Toggenburg	1,200
	British Toggenburg	13,520
	British Alpine	5,360
	Anglo-Nubian	15,173
	British (Pedigree cross-breds)	26,240
	Golden Guernsey	1,547
	English Guernsey	80

Species	Breeds		UK Pop. (1981)*
Horse	Shire	10,000–15,000	
	Clydesdale	700+	
	Suffolk	350	
	British Percheron	325	
	Cleveland Bay	488	purebred
		940	part-bred
	Welsh Cob	1,799	foals registered 1980/1

* These figures are mostly estimates given by the breed societies

tation of other European breeds. In pigs and poultry, the trend has been towards an abandonment of most of the traditional breeds with reliance on hybrids (often nameless and bearing only a number), produced by specialist breeding companies from only a limited number of breeds. Rabbits are still represented by a variety of breeds, although hybrids are commercially available.

Cold-blooded animals are far less developed and deliberate breeding is still at a very early stage for fish and bees: it is practically non-existent for snails, crocodiles and earthworms, for example. Naturally, it depends upon their agricultural importance, a point that is well illustrated by the fact that, in Japan, there are more than 1,000 commercial varieties of silkworm (fed on 39 varieties of mulberry). None of the cold-blooded animals is of major economic importance in UK agriculture: those that may become so are dealt with in Chapter 40. One possible exception is the honey-bee (also dealt with in Chapter 40) because, although the production of honey is rarely a major enterprise, the importance of honey-bees in pollination has not been calculated and must be considerable.

The warm-blooded animals may most usefully be grouped as ruminants, non-ruminants, and birds.

RUMINANTS

The ruminants are all grazing animals, although they are not always kept in this way, and are structurally and physiologically adapted to the collection and digestion of fibrous vegetation.

Such adaptation is best illustrated by the structure of the jaw and the arrangement of the teeth (see Fig. 13.1) and the alimentary tract, with its characteristic rumen (Fig. 13.2). The latter is a greatly enlarged first part of the stomach, in which cellulose can be digested by bacteria and protozoa. Associated with rumen function is the equally characteristic cud-chewing (rumination) that ensures both fine comminution and mixing of the ingested herbage. This bacterial digestion of the feed has several important implications (see Chapters 14 and 20), including the fact that ruminants are able to make use of non-protein nitrogen in the feed.

The most important ruminants are cattle, sheep and goats.

Cattle

Traditionally, dairy and beef breeds of cattle were thought of as quite distinct but the Friesian is now the dominant dairy breed and is used as a dual-purpose animal, quite capable of producing beef.

Dairy breeds
The total number of dairy cows (excluding heifers) in 1979/80 was

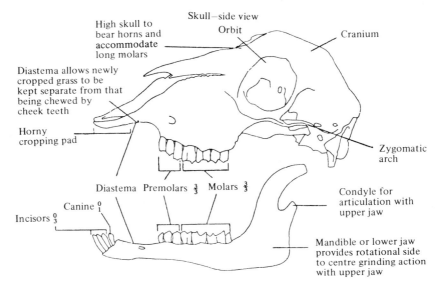

Skull—side view

High skull to bear horns and accommodate long molars

Orbit

Cranium

Diastema allows newly cropped grass to be kept separate from that being chewed by cheek teeth

Horny cropping pad

Zygomatic arch

Diastema   Premolars $\frac{3}{3}$ Molars $\frac{3}{3}$

Canine $\frac{0}{1}$

Incisors $\frac{0}{3}$

Condyle for articulation with upper jaw

Mandible or lower jaw provides rotational side to centre grinding action with upper jaw

Dental formula: I $\frac{0}{3}$, C $\frac{0}{1}$, Pm $\frac{3}{3}$, M $\frac{3}{3}$. Total number of teeth = 32

Incisors— On lower jaw only, crop off grass by biting against upper cropping pad.
Canine— On lower jaw only, assists in cropping grass.
Premolars ⌐ Consist of vertical folds of enamel with dentine and cement filling
Molars   ⌐ crevices—enamel, dentine and cement wear away at different rates to provide a rough grinding surface.

Fig. 13.1   The jaw of a ruminant (sheep)
Source: Robinson & Wiggins (1971)

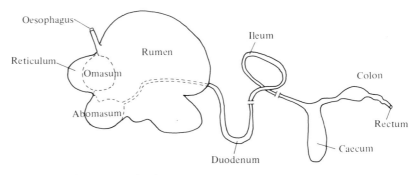

Oesophagus

Ileum

Reticulum

Rumen

Omasum

Colon

Abomasum

Rectum

Caecum

Duodenum

Fig. 13.2   The digestive tract of a ruminant (bovine)
Source: Maynard & Loosli (1965). From *Animal Nutrition* by L. A. Maynard & J. K. Loosli. Used with permission of McGraw-Hill Book Co. 1965

approximately 3.2 m., kept on some 62×10^3 holdings, with an average herd size of 52. Average gross yield of milk per cow was 4,740 l and the total milk supply just over 15,000 m.l. The typical lactation curve of a cow is shown in Fig. 13.3 and milk composition in Table 13.6.

The principal breeds are shown in Table 13.7. Friesians represent about 90% of the total in England, Wales and Northern Ireland but, in Scotland, some 36% are Ayrshires.

Beef breeds

The beef industry in the UK reflects a wide variety of resources and pressures. The large national dairy herd has developed to meet the need for fresh milk, and the use of calves from the dairy herd for beef production is biologically efficient. The efficiency of production of energy as milk and meat from the gross energy consumed by dairy cattle and their offspring is considerably higher (11%) than that produced by single-purpose suckler herds as beef only (4.5%) (Holmes, 1970). However, the value of a beef suckler herd as a means of utilising grassland or the by-products of arable cropping was reflected in the existence of some 81,800 herds of beef cows in the UK in 1980.

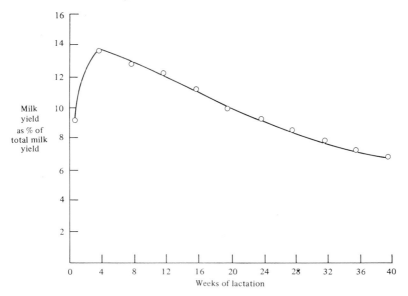

Fig. 13.3   Lactation curve of a cow (after Spedding, 1971)

Table 13.6   Milk composition* of cows, goats and sheep (after Jenness & Sloan, 1970)

Species	Total solids	Fat	Casein	Whey protein	Lactose	Ash
Cow	12·7	3·7	2·8	0·6	4·8	0·7
Goat	13·2	4·5	2·5	0·4	4·1	0·8
Sheep	19·3	7·4	4·6	0·9	4·8	1·0

* Grams per 100 g or grams per 100 ml: no attempt is made to distinguish these modes of expression; some figures are rounded off

Table 13.7   Principal breeds of dairy cow

	UK Pop. ('000) (1978/9)
British Friesian	2,939
Ayrshire	201
Guernsey	69
Jersey	58
Holstein	41
Dairy Shorthorn	14
Total cows	3,474˙

Source: *UK Dairy Facts and Figures* (1980)
Federation of UK MMBs

The national cattle population is made up of 3·9 m. cows and heifers in dairy herds and 1·7 m. in beef suckler herds (Fig. 13.4). These are the sole sources of young calves for beef systems, since only 8% of the total beef is derived from imported livestock, and these are store cattle from Eire. The growth and carcass characteristics of the cattle entering beef systems therefore reflect the breed structure of these two types of herds, and the breeding policy employed in them.

Calves from dairy herds   The records of breeds used to supply semen for artificial insemination (AI) give a good indication of the breeds of calves derived from the national dairy herd, since over 60% of dairy cow matings are by AI. Over 85% of dairy cows are of the British Friesian breed, with Ayrshire, Channel Islands, Dairy Shorthorn and crossbred cattle or other breeds comprising the remainder. The Friesian breed dominates inseminations, with 62% in 1979/80, to provide pure-bred replacement heifers; the beef breeds are used for most of the remaining inseminations (over 600,000) with 34%, to improve the value of the calves for beef

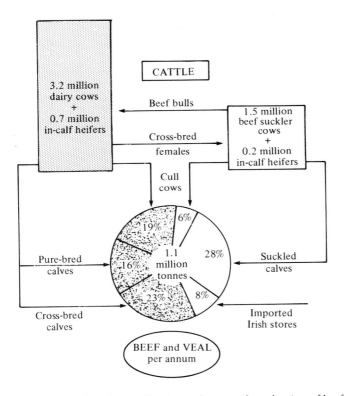

Fig. 13.4   Sources of cattle contributing to the annual production of beef and veal in the UK (based on Allen & Kilkenny, 1980; quantities for 1980, from MLC, 1981)

Table 13.8   **Proportion of cows and heifers inseminated by bulls of beef breeds, as a percentage of total beef inseminations (MMB 1980)**

Breed of bull	Dairy cows	Dairy heifers
Hereford	59	40
Aberdeen Angus	5	42
Charolais	23	2
Murray Grey	2	8
Limousin	6	4
Other breeds		

production. Within these beef inseminations, the Hereford and Aberdeen Angus breeds predominate (Table 13.8). The widespread use of the Hereford and Aberdeen Angus with heifers results partly from the reduced incidence of calving difficulties with calves sired by these two breeds, as compared with those sired by the Friesian or by some of the larger beef breeds such as the Charolais. The Charolais, however, transmits a high growth rate and good muscular development to the cross-bred calf, and is second to the Hereford in total number of beef inseminations. All of the beef breeds mentioned above, and to some extent the Simmental, have the further advantage of 'colour marking', i.e. their colouring, transmitted to their offspring, identifies the breed and thus assists trade in live cross-bred cattle with beef characteristics.

Much of the beef derived from calves born in dairy herds, therefore, is from cross-bred cattle (23%, the majority of which are Hereford × Friesian) and 16% is from pure-bred cattle, with the Friesian predominating. Some 19% of the total beef, however, is from cows culled from the dairy herd, and much of this is of manufacturing quality (see Fig. 13.4).

Calves from suckler herds   Commercial suckler herds contain a high proportion of cross-bred cows. Many of these are cross-bred female calves from dairy herds (i.e. mostly Hereford × Friesian and Angus × Friesian heifer calves) which are purchased to provide breeding replacements for culled cows. Other cross-breeds such as the Blue-Grey (Galloway × White Shorthorn) are widely used in the hills and uplands. There are also pure-bred commercial herds of breeds which are hardy in the cold and wet conditions at the higher altitudes and western latitudes of the British Isles. No more than about 100,000 of the 1·7 m. suckler cows and heifers are in pedigree herds which supply beef bulls for natural service in dairy and suckler herds or for the artificial insemination organisations.

The suckled calves which supply 28% of the total beef are cross-bred, and to a large extent are third-cross calves derived from a first-cross dam and a terminal sire. Culled cows from suckler herds produce only 6% of the total beef supply, but on average these tend to be older than the cows culled from dairy herds, and more of them are of manufacturing quality (see Fig. 13.4).

Breed and carcass characteristics
The attributes of breeds which must be taken into account in considering their suitability for specific systems are their growth rate and the relative rates of development of body tissues which affect the composition of the carcass at slaughter at a specified age.

Breeds with a low rate of live-weight gain as a result of their small skeletal size lay down fatty tissue at an early age, and are described as 'early maturing'. Breeds with a high rate of live-weight gain, linked with their large skeletal size and hence large muscular development, do not lay down fatty tissue until a later age and are relatively 'late maturing'. These characteristics are recorded in standardised performance tests for bulls, fed *ad lib*, as live weight at 400 days, height at shoulders and depth of fat over the back. The tests show that, on average, the Aberdeen Angus and Hereford breeds are smaller and earlier maturing than the Charolais (Table 13.9). Data for most of the beef breeds used in the UK are given in

Table 13.9 Liveweight, size and fatness of bulls of three breeds in performance tests, and effect of these breeds, when crossed with Friesian dams, on average growth rate of the cross-bred steer in comparison with Friesian steers (MLC, 1976, 1979)

Breed of bull in performance test or of sire crossed with Friesian cow	Performance test			Daily gain of cross-bred steer (% above or below Friesian steer)
	Liveweight (kg) (age 400 days)	Height at shoulders (cm) (age 1 year)	Depth of back-fat (mm) (age 1 year)	
Charolais	622	124	3·1	+ 10·4
Hereford	495	112	6·9	+ 3·1
Aberdeen Angus	470	110	6·5	− 14·2

Appendix Table 13.1 (see p. 776) with similar measurements recorded on farms, including birth weights. The on-farm records contain an environmental effect (e.g. lower weights, as a result of poorer nutrition on hill and upland farms, for the smaller breeds which are more widely used in these poorer environments) but a similar ranking can be seen for both sets of measurements, and the relative sizes of these breeds are shown.

These breed traits of growth-rate and fatness are transmitted to the cross-bred offspring when the beef breeds are mated with dairy breeds. The Charolais × Friesian steer, on average, grows 10.4% faster than the Friesian steer, and the Aberdeen Angus × Friesian 14.2% more slowly. The effect of cross-breeding on carcass composition in steers, fed in an 18-month system, is well illustrated by the data of R. W. Pomeroy, B. M. Scott and D. R. Williams, published by the Royal Smithfield Club (1964). During the fattening process from 12 to 18 months of age in all breeds there are considerable increases in live weight, carcass weight, muscular development (muscle:bone ratio) and fatness (% fatty tissue in carcass) (Table 13.10). The Friesian is slightly heavier than the Hereford at the same age but has a lower muscle:bone ratio and level of fatness. The Hereford × Friesian is intermediate, with the Hereford conferring improved beef characteristics on the cross-bred animal.

Table 13.10 Changes in weight and carcass composition during fattening from 12 to 18 months of age in steers of two different breeds and in the crossbred steer (Royal Smithfield Club, 1964)

	Age (months)	
	12	18
Liveweight (kg)		
Hereford	250	410
Hereford × Friesian	310	445
Friesian	310	455
Carcass weight (kg)		
Hereford	132	235
Hereford × Friesian	162	259
Friesian	154	260
Muscle: bone ratio in carcass		
Hereford	4·4	4·9
Hereford × Friesian	4·2	4·9
Friesian	3·8	4·4
% Fatty tissue in carcass		
Hereford	17·2	32·5
Hereford × Friesian	18·2	25·4
Friesian	13·9	25·4

The characteristics of meat development and fatness are assessed in the EEC beef carcass classification scheme introduced in 1981 under EEC Regulation 1208/81. This is operated in the UK by the Meat and Livestock Commission (MLC, 1981b). The method utilises a judgement by eye, by experienced operators, of the thickness of flesh (muscle plus fat) on the carcass (conformation classes) and the cover of subcutaneous fat over the carcass (fat classes). There are seven conformation classes based on the letters EUROP. These range from excellent (E), which applies to exceptional double-muscled carcasses, through U+ and U and through O and O− to P (the poorest). Fat classes range from 1 (least fat cover over the carcass – almost none) in seven steps through 4L (low) and 4H (high) to 5L and 5H (fattest). Both conformation and fatness affect the percentage of lean meat in the carcass, which has a major influence on its value in the meat trade. The purpose of the EUROP scheme is to allow dead-weight price reporting to be carried out on a common basis in the member states, and to facilitate the use of common standards for Intervention purchases. Carcass classification also provides an important link in marketing between producer, wholesaler and retailer.

There is some concern in the EEC that the rapidly increasing proportion of the North American Holstein in the Friesian populations of the EEC member states will have adverse effects upon carcass quality and on beef-production potential through reduced reproductive efficiency. This would lead to an increased need in future for cross-breeding with those breeds such as the Aberdeen Angus, Hereford, Limousin and Charolais (already widely used in dairy herds) which increase the beef qualities of the cross-breed.

Sheep

In June 1978 there were 30 m. ewes and lambs in the UK, and of this figure just over 14 m. were breeding ewes. Of these ewes 5·9 m. were in the hills, 2·2 m. on the uplands and 6·1 m. on the lowlands. Sheep contributed £333 m. to the nation's wealth, which was 4·7% of the agricultural economy, with 4·2% from the sale of sheep meat and 0·5% from wool. About 80,000 of the 272,000 agricultural holdings in the UK contained sheep, or about 1 in 3.

Sheep are rarely kept as the sole enterprise on the farm, except on some hill-farms, but are normally associated with other enterprises. Nearly two-thirds of the breeding ewes in England and Wales are kept on livestock farms, that is sheep with beef cattle, 10% on dairy farms and 8% on arable cropping farms. The remainder are kept on mixed farms or with pigs, poultry or horticulture.

Within the UK there are more sheep in western England (2·26 m.) than in the east (1·32 m.), but more again in the north (2·6 m.). Scotland and Wales both have about 3 m. sheep each, whilst Northern Ireland has half a million.

The main sheep breeds are given in Table 13.5 and the major crosses associated with particular production systems are detailed in Chapter 37. Most sheep are kept for meat and milk production is relatively unimportant in the UK. Wool, however, is an important by-product.

Wool

Of the estimated world wool production Britain's share is about 2%, a long way behind the big wool producers, Australia, the USSR, New Zealand and Argentina, but it is still a substantial quantity. Depending on the breed, the fleece is divided into wool and non-wool fibres, namely kemp and hair; the more primitive breeds of sheep have and retain a higher proportion of kemp and hair. Table 13.11 gives some details of the types of fleece found on British breeds. There is no breed with such a fine count as the Merino, a fineness that is aided by drought and undernutrition.

> The fineness count is based on the number of yarn lengths, each of 511·84 m that could be spun from 0·454 kg weight of wool prepared for spinning. The variety of wool characteristics is evident from the table. Amongst the hill breeds the Scottish Blackface has a long staple length

Table 13.11 Wool characteristics of British sheep breeds

Breed	Wool Classification	Mean fleece wt. (kg)	Mean Staple Length (cm)	Fineness Count
Scottish Blackface	Mountain and hill	2·5	27	28–40
Welsh Mountain	Mountain and hill	1·5	8	36–50
Swaledale	Mountain and hill	2·0	25	28–32
North Country Cheviot	Mountain and hill	2·0	9	50–56
Romney Marsh	Lustre Longwool	4·0	16	48–56
Devon Longwool	Lustre Longwool	6·0	30	32–36
Lincoln	Lustre Longwool	6·3	30	36–40
Teeswater	Lustre Longwool	6·0	30	46–48
Clun Forest	Shortwool and Down	2·5	9	56–58
Dorset Down	Shortwool and Down	2·5	6	56–58
Dorset Horn	Shortwool and Down	2·5	9	54–58
Southdown	Shortwool and Down	2·5	6	58–60
Suffolk	Shortwool and Down	3·0	8	54–58

which can be very coarse, whereas the Welsh Mountain has a finer, shorter fleece. The Longwools clip a heavy fleece (6·0–6·3 kg) with the Teeswater having the finest wool; of the shortwools the Southdown has marginally the finest fleece but it is also the lightest.

The wool produced in the world is divided into three main categories: (1) 'Merino' wool, with quality number 60s and over; (2) 'Crossbred' wool, 44s–58s; and (3) 'Carpet' wool, 44s and below.

Merino wool is made into best quality, light worsted and woollen garments, but there are no British breeds of sheep in this category. 'Crossbred' wool is used for good tweeds, lower quality worsted and knitting wool and finally there is the coarsest wool which is made into carpets and mattresses, although some may be used for tweeds.

Goats

The total goat herd of the country is not large (about 135,000) but the number of breeds is considerable (see Table 13.5).

Most of them are kept for milk production but there is a growing demand for goat meat amongst certain sections of the community.

NON-RUMINANTS

These include herbivorous animals such as horses and rabbits and the 'simple-stomached' animals, such as pigs. Of all these, the pig is by far the most important economically.

Pigs

The total number of pigs is about 7·7 m. and the number of holdings with breeding pigs is some 32 m., with an average size of breeding herd of 25. Pork production in the UK is normally about three times as great as that of bacon and ham. There are still a number of breeds of pigs (see Table 13.12) but the predominant breeds in

Table 13.12   Breeds of pigs

British Breeds	No. of litters recorded	Herd book registrations* 1981	
		Boars	Gilts
Large White	12,240	709	3,281
Middle White	58	15	55
Welsh	3,060	320	1,021
British Landrace	7,377	720	2,299
Large Black	75	25	86
British Saddleback	417	60	208
Gloucester Old Spot	309	36	246
Berkshire	68	22	65
British Lop**	–	–	–
Tamworth	117	34	86
Oxford Sandy & Black	1	–	1
Chester White	52	48	67
Foreign Breeds used			
Duroc	88	25	63
Hampshire	221	39	78
Lacombe**	–	–	–
Piétrain**	–	–	–
Poland**	–	–	–
China**	–	–	–

* National Pig Breeders' Association records for 1981. There will be many pedigree animals not recorded
** No records for these breeds from NPBA

the (predominantly) intensive production systems are hybrids of the Large White or Landrace breeds.

Pigs are omnivorous and their alimentary tracts (see Fig. 38.5) are not adapted in any special way to the digestion of fibrous feeds. They thus have one, relatively uncomplicated stomach and cannot digest cellulose or make use of non-protein nitrogen.

(Details relating to pig meat, carcass grading, the biology of pig reproduction and growth, and the nutrition of pigs of all ages, are given in Chapter 38.)

Horses

Although there are many breeds of horses in the UK (see Table 13.5 for working breeds) and more than o·5 m. in total, they are now of negligible importance in agriculture.

They may represent an important market for the sale of hay and dried grass but they are kept mainly for recreational purposes and a very few work on farms or for breweries.

The horse is a grazing animal with a different mechanism for digesting herbage, involving bacteria but in an enlarged colon (see Fig. 13.5).

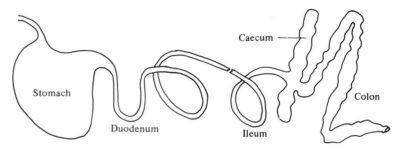

Fig. 13.5 The digestive tract of a horse
Source: Hayward & Loosli (1965). From *Animal Nutrition* by L. A. Maynard & J. K. Loosli. McGraw-Hill Book Co. 1965

Rabbits

Commerical rabbit production in the UK appears to be of growing importance and there is clearly scope for increased home production, since only about one-third of home consumption is currently home produced. However, at the same time, because of peculiarities in the carcass weights required, exports are also substantial (about half of those produced at home and equal to about 37% of the quantity imported).

The commonest breeds for large, intensive units are the Californian and the New Zealand White but many other breeds are kept (Table 13.13). The rabbit has a unique adaptation to a fibrous diet, partly in the form of an enlarged caecum (see Fig. 13.6) and partly by virtue of its habit of producing special 'night' faeces which are directly eaten as they are produced (refection) and thus subjected to digestion twice.

Table 13.13   **Breeds of rabbits with a value in meat production**

New Zealand White
Californian
Giant Blanc
Beveren
New Zealand Red
Dutch
Rex
Chinchilla Giganta
Havana
Flemish Giant

Source: MAFF (1973)

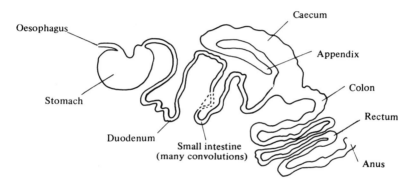

Fig. 13.6   The digestive tract of a rabbit
Source: Spedding, Walsingham & Hoxey (1981)

Birds

Poultry used in agriculture include hens, ducks, turkeys and geese, in that order of importance, with some minor enterprises based on pheasants or quail.

Hens

These are dealt with in some detail in Chapter 39 and their digestive system is illustrated in Fig. 39.1.

They are kept mainly for eggs and meat and tend to be in rather different systems for these two purposes.

Most of the eggs come from hybrid birds in intensive battery systems, although a number of different breeds still survive (Table 13.14). The number of holdings is still considerable, however (75,000 in 1977), and the average flock size is in consequence only about 650.

Table 13.14   **Breeds of chickens still found in UK**

White Leghorn
Rhode Island Red
Light Sussex
White Rock
New Hampshire
Orpington and Buff Orpington
Maran
Welsummer
Barnvelder
Australorp
Ancona

The number of broiler holdings, by contrast, is only about 2,000 and the average flock size nearer 25,000.

Ducks

Duck production is not negligible but, apart from very small flocks, tends to be concentrated in a few very large enterprises, involving only a few breeds (Table 13.15). As can be seen from the table, however, ducks can outyield hens for egg production per year, although their feed conversion efficiency is poorer.

Turkeys

Turkeys are kept almost entirely for meat production and the trade remains rather seasonal.

The main breeds used are British White, Norfolk Black, Beltsville White and Broad-Breasted Bronze.

Geese

Goose eggs are consumed, on a very small scale, but most geese are reared at grass for meat production. As with turkeys, lighter carcasses are now favoured and the main breeds are Embden, Toulouse and Chinese.

Table 13.15   **Breeds of duck and their attributes**

	Main product	FCE	Carcass wt. (kg)	Av. annual egg production
Aylesbury	Meat	—	2·27	100
Pekin	Meat	2·8	1·8	130
Khaki Campbell	Eggs	2·3–3·4	–	300
Hen (for comparison)	Eggs	2·6	–	240
	Meat	2·0	1·8–2·7	180

14

FEEDS FOR ANIMALS

Introduction

This chapter deals with the various feeds used in the diets of farm animals. It is important to appreciate the complementary nature of the various types of feed. Very few animals are fed on a single feed or even on a simple mixture. In the case of ruminants the diet normally consists of a combination of a forage and a concentrate. In the case of pigs and poultry kept intensively the diet will normally consist solely of a concentrate mixture, but the mixture itself may comprise many different feeds. If the mixture is produced on the farm it may be relatively simple, comprising about five or six feeds, but in a purchased compound the feeds used may number 20 or more.

All feed can be broken down into constituent chemical fractions as shown in Table 14.1. The various raw materials which are

Table 14.1   Components of Different Fractions in the Analysis of Feed (after McDonald et al., 1973)

Fraction	Components
Moisture	Water
Ash	Essential elements and minerals
	Non-essential elements and minerals
Crude Protein (CP)	Proteins, amino acids, amines, nitrates,
	nitrogenous glycosides, B-vitamins, nucleic acids
Ether Extractives (EE)	Fats, oils, waxes, organic acids, pigments, sterols,
	vitamins A, D, E, K
Crude Fibre (CF)	Cellulose, hemi-cellulose, lignin
Nitrogen-free extractives (NFE)	Cellulose, hemi-cellulose, lignin, sugars, starch,
	pectins, organic acids, resins, tannins, pigments,
	water-soluble vitamins

blended together to form the concentrated part of the diet cannot simply be classified as suppliers of 'energy' or 'protein', because very few feeds contain only a single nutrient. This point is made very clearly in Table 14.2 which indicates that more total protein is supplied to livestock from the so-called 'energy feeds' than from traditional 'protein feeds'. It is, therefore, convenient to group these separate raw materials according to their biological origin.

Table 14.2   Supply of crude protein (CP) to farm livestock in the UK (July 1973–June 1974) (kt)

	Home-produced from:		Imported	Total
	Home-grown	Imported	Imported	Total
High Energy Feeds				
Cereals				
Barley	611	–	55	666
Wheat	240	–	33	273
Oats	79	–	1	80
Maize and Sorghum	–	–	164	164
Other	20	–	–	20
Cereal Offals	93	111	24	228
Misc. Energy Feeds	91	4	22	117
Total	1,134	115	299	1,548
High Protein Feeds				
Oil cakes and meals	3	281	316	600
Animal Protein				
Fish Meal	47	–	111	158
Meat & Bone Meal	95	–	24	119
Milk Derivatives	6	–	–	6
Feathermeal	7	–	–	7
Other Protein				
Maize Gluten	–	57	–	57
Distillers' by-products	25	30	–	55
Field Beans	24	–	–	24
Dried Grass	25	–	–	25
Dried Poultry Manure	30	–	–	30
Urea	40	–	–	40
Others	2	–	–	2
Total	304	368	451	1,123

Source: JCO (1976)

THE CLASSIFICATION OF FEEDS

Feeds may be classified in numerous ways, and the system chosen is of less importance than a proper understanding of the correct definitions of the terms employed. One of the commonest errors in nomenclature is to regard a single 'straight' feed, such as barley, as a 'concentrate' merely because the nutrients contained in barley are more 'concentrated' than those contained in a succulent feed, such as fresh grass or a root crop. In this chapter, the definitions employed are those agreed, in 1977, conjointly between the Agricultural Development & Advisory Service (ADAS), the British Veterinary Association (BVA) and the UK Agricultural Supply Trade Association (UKASTA) and these terms are confidently recommended for wider adoption.

The three main divisions of animal feeds are: (1) Grass and forage crops; (2) Straights; and (3) Concentrated feeds.

Grass and forage crops

These are mainly dealt with in Chapter 29 but, in addition to grass products, maize silage and arable by-products are important.

Maize Silage

UK visitors to North America are always impressed by the large quantity and high quality of maize silage produced on that continent. Unfortunately, the UK climate is not so suitable for the growth of this crop, with the consequence that the quality cannot be so readily guaranteed. The most suitable varieties of maize for ensiling are the early-maturing ones and it is usual to harvest maize silage in October and November, depending on site and season. An average yield of over 4 t of dry matter/ha should be obtained, half coming from the cob and the remainder from the stem and leaves.

The energy value of maize silage is relatively high (usually over 11 MJ/kg of dry matter) but the protein content is always low (see Table 14.3). The consequence is that maize silage must be augmented with a protein supplement when this feed is provided as a large part of the total diet. In addition, minerals are needed to balance the low levels found in the maize crop.

Arable By-Products

The most traditional material falling into this category is cereal straw, an animal feed also employed as animal bedding. For every

Table 14.3 **The composition of fresh and ensiled maize**

	Fresh Crop	Maize Silage
Dry Matter %	27	26
pH (acidity)	6	4
Composition of DM %		
Cell walls (fibre)	46	47
Starch	19	18
Sugars	14	3
Acids	1	10
Ash	5	6
Crude Protein	8	9
Composition of Crude Protein Fraction %		
True protein–N	75	50
NPN	25	50

Source: Thomas & Wilkinson (1977)

tonne of cereal grain harvested there is another tonne of cereal straw, much of which is currently wasted. Of the 9 m. t of straw produced each year in the UK about 4–5 m. t of long straw are surplus to requirement. Indeed, the surplus straw is a major embarrassment hindering subsequent field operations and hence it is commonly removed by burning, a process which is potentially dangerous and, even when correctly supervised, releases large quantities of ash into the atmosphere.

Cereal straw can either be fed long, chopped or chemically treated. In its normal form, the feeding value of straw is low in energy (less than 6 MJ/kg of dry matter) and it is deficient in protein (indeed, the digestible crude protein can be negative, indicating that more protein is lost from the gut of the animal during digestion than is released into the bloodstream as an end-product of digestion).

Numerous other crop residues can provide potential feeds for farm livestock, although they are normally available on a highly seasonal basis. In the UK, pea-vining residues amount to between 2 and 3 m. t (on a wet-weight basis), sugar beet tops 6 m. t, sprout residues 0·7 m. t and other vegetable wastes just under a further 2 m. t.

In general terms, all these waste materials are capable of being utilised by farm livestock but the economics of so doing are often very doubtful, except in cases where the crop can be provided at very low cost. Animal enterprises situated close to intensive horticultural areas can sometimes use a variety of these arable wastes very profitably, one waste following another as the season progresses. However, it is seldom economic to transport such wastes over large distances, and artificial drying so as to render transport a better proposition is normally uneconomic.

Straights

These are single feedstuffs of animal or vegetable origin, which may or may not have undergone some form of processing before purchase. Single straights rarely provide the complete nutritional requirements for farm livestock, although the production of new genetically improved varieties, such as high-lysine corn, may change this situation.

In the case of the compound feed industry, the term 'raw material' is often used in place of 'straight', a word which is more commonly employed by farmers and their corn merchants. The various types of straights will now be considered:

Cereal Grains (e.g. barley)

Cereal grains are the most important constituents of animal feed in most developed countries. In the USA, maize (corn) is the most commonly used, whereas in the UK barley has traditionally provided a large quantity of animal feed, although in recent years the wheat tonnage has progressively increased. Most cereal crops are used both for human food and animal feed although a few crops, such as oats in the UK, were traditionally grown primarily for animal use.

Cereal By-Products (e.g. wheat offal)

Where cereals are grown primarily for human food they are normally processed and refined to some degree before being consumed. The outside husks or glumes are often rubbed off, and small grains removed by sieving. The resultant by-products of this physical processing are available for animal feed and it is likely that they will continue to be marketed for some time to come. It should, however, be noted that there is a trend towards a greater use of the 'natural whole cereal' for human food and, since the by-products are often rich in vitamins and fibrous, there are sound nutritional reasons why this practice may be extended.

Cereal grains are also indirectly used for human food because they are the substrate for the fermentation process leading to beer and spirit production. The 'spent grains' from breweries and distilleries have no further use by man and may then be available as an animal feed. Normally they are supplied directly onto farms in a wet condition, since the cost of drying is normally prohibitive.

Pulses (legumes) (e.g. field beans)

Although the seed from leguminous crops normally commands a more attractive price as a human food, nevertheless certain crops, such as field beans, are sometimes grown primarily for animal feed use. Additionally, legume seeds which do not come up to the standard required for human consumption are frequently diverted into the animal feed market. As the need for 'catch crops' increases and as the cost of nitrogenous fertiliser escalates it is probable that more pulses will be grown for direct animal feed use in the future.

Oil-Seed Residues (e.g. rapeseed meal)

The traditional use for oil-seed crops was for the production of either an edible oil (e.g. olive oil) or an industrial oil (e.g. linseed oil). The importation of oil seeds for this purpose has taken place over many centuries, but it is only comparatively recently that the residues have been used as protein-rich animal feeds. In recent times there has been a growing tendency for certain oil-seed crops, such as soya, to be available for animal feed use and to be sold entire for that purpose. A common term for such entire oil seeds is 'full fat' (e.g. full-fat soya).

Where the oil has been expressed from the seed by a simple rolling or crushing process the residue is known as 'expeller cake'. Such cakes can still be identified as one of the constituents of coarse feed preferred by some farmers for feeding to calves and, on occasion, dairy cows.

Where the oil has been extracted from the seed by the use of a chemical solvent the residue is known as a 'meal', and such materials are usually friable and more easily incorporated into crude mixtures than are the hard, large lumps of 'oil-seed cake'.

Certain oil-seed crops, especially groundnut, can have high levels of infection with the fungus *Aspergillus flavus*. This fungus and various other similar fungi produce a toxic material which can be very dangerous if consumed in large quantity by farm livestock. The toxin from the fungus *Aspergillus flavus* is known as 'aflatoxin' and is especially dangerous to young animals, particularly growing turkeys. The toxin is a powerful carcinogen, so there is a danger not only to the animal consuming the toxin but also a subsequent danger to man if the toxin is still present in meat or milk. The presence of such undesirable substances is now carefully regulated by law.

Root Crops (e.g. cassava)

Root crops are capable of producing very high yields of energy, in the form of carbohydrate, per unit of land area. The difficulty is that many root crops, particularly the tropical crop cassava, are difficult to harvest.

In tropical countries root crops can be air-dried prior to marketing, so that the material transported normally has no more moisture in it than would be present in cereal grain (approx 14–16%). In the UK, however, air-drying of root crops is not practical, hence the cost of storage and subsequent transport can often be prohibitively high.

During the last decade increasing amounts of cassava have entered world markets as an animal feed ingredient. Because of the very high yields of cassava in tropical countries, this trade is likely to continue unless artificial barriers to its free movement are introduced. Cassava enters the UK either as a very fine powder (which causes major difficulties when handling) or as a processed pellet. Most of the cassava imported is supplied to feed compounders but the material is also available, particularly on the Continent of Europe, for direct feeding by farmers as a straight.

Fibrous Materials (e.g. chemically treated straw)

This class of material covers a large number of crop by-products ranging from processed cereal straws to tropical products, such as bagasse pith (a by-product of the sugar cane industry).

In the past many of these materials were considered unsuitable for direct animal feed usage, because of their woody nature resulting from the presence of large quantities of lignin-like materials. However, in recent years a great deal of attention has been devoted to the chemical processing of these fibrous feeds, with a view to breaking down the integrity of the lignin structure and thus releasing the nutrients for animal use. A wide range of chemicals have been employed for this purpose, the most common being sodium hydroxide (used in factory processes for upgrading cereal straws) and ammonia, now being increasingly used for on-farm straw processing. The judicious use of these materials, some of which have to be handled very carefully, can increase the energy value by about 50%. Thus untreated cereal straws have an energy content of about 6 MJ/kg whilst processed straws have about 9 MJ/kg.

Animal By-Products (e.g. meat and bone meal)

These materials are by-products of the meat and fishing industries. A century ago, most of these materials would have been produced on a small scale and, since further processing is required before they are fit for animal feed use, they would have been wasted. The development of large-scale plants, now common in the meat and fishing industries, enables these by-products to be processed. Indeed, the economy of certain countries, such as Peru, is partly dependent upon a fishing industry almost completely geared up to producing fish meal for animal feed rather than fish for human consumption!

It is obvious that all these materials must be processed very carefully if they are to be completely safe for use as animal feeds. If the processing is deficient they can become a carrier for pathogenic bacteria (such as salmonellae and anthrax) or for viruses (such as the foot and mouth virus).

In the UK a 'Protein Processing Order' has recently been introduced with a view to ensuring correct processing procedures and strict hygiene measures where animal by-products are concerned. The State Veterinary Service has the duty to inspect the processing plants involved and also the end-product, whether the materials are home-produced or imported into this country from overseas.

Re-cycled Animal Waste (e.g. dried poultry manure)

Large quantities of these materials are available on-farm, especially on intensive pig and poultry enterprises. Many of these units are sited close to urban areas and the disposal of the animal manure becomes a major problem, costly to solve.

The consequence of these pressures has been the development of plant capable of processing the manures to a low-moisture material which can be transported relatively long distances. Such manures can be fed to ruminant animals as a source of non-protein nitrogen together with small amounts of energy and minerals. These materials, when properly processed, are quite safe but the obvious dangers of pathogenic bacteria and residual contamination on the one hand and of consumer resistance to their usage on the other, inhibit developments in this area. There is a reluctance on the part of the licensing authorities to give such materials a completely clean bill of health and there is a similar reluctance by animal feed compounders to incorporate them into their finished products. Many farmers would object to purchasing feed containing animal waste, even though it is a nutritionally sound material and economically priced. However, this may change in the future.

A further process has recently been developed in the USA and is used on a small scale in the UK. This is the feeding of 'soilage' to ruminants, the 'soilage' being a forage crop to which animal manure has been added. Cattle do not find such material unpalatable and the consequent feeding programme may prove economically attractive.

Synthetic Nutrients (e.g. urea)

This group of feeds comprises chemicals (such as urea) produced by the heavy chemical industry and fermentation products produced by the biotechnology industry (such as lysine and methionine).

This class of material currently contains relatively few products. However, it is highly likely that the list will enlarge in the future, as more micronutrients are recognised as essential ingredients in animal feed, so that large-scale production of such materials becomes economically attractive.

The first synthetic amino acid to be produced was methionine and, in its early days, its price was several thousand pounds per tonne. However, as world-wide demand has increased and the scale of production risen in consequence, the costs of production have tumbled in real money terms. Synthetic methionine can now be purchased at approximately one-quarter of the price originally required when production was on a small-scale basis.

At present many of these micronutrients, especially the essential amino acids, are destined primarily for use in pig and poultry diets. However, ruminant animals eventually require amino acids in precisely the same way as monogastrics, and it is therefore probable that in future these materials will be incorporated in some protected fashion so that they escape rumen breakdown, in diets for cattle and possibly sheep. Since the total market for cattle diets is about as large as the market for pig and poultry diets put together, it follows that any extension along these lines will double the market prospects for such materials. There is already some use made of synthetic amino acids in diets for the pre-ruminant calf and lamb; at this early age the young ruminant can utilise synthetic amino acids just as well as pigs and poultry.

Up to recently the two chief limiting amino acids produced in this way have been lysine and methionine. However, eight other amino acids are known to be essential for life, namely arginine, histidine, isoleucine, leucine, phenylalanine, threonine, tryptophan and valine. It is highly probable that some or all of these essential amino acids will be produced synthetically in the future by similar processes to those currently employed for the production of synthetic lysine and methionine.

Single-Cell Protein

The major oil and chemical companies have pioneered the production of this material, also known as biomass, using simple sources of feed energy which are themselves by-products of the oil-refining industry, such as methanol. The production of these microbial proteins requires large-scale fermenters and the capital cost involved in the construction of such plant is very great. Various micro-organisms are capable of being grown in this way. At the present time the main types employed are bacteria although fungi and filamentous algae could also be used.

Liquid Feed

The main material in this class is molasses, a by-product of the sugar-processing industries. About 300 kg of molasses are available for every tonne of sugar processed and since about half the contents of molasses are sugars, this represents a potential loss to the sugar industry of about 150 kg/t.

In the UK, the molasses sold comes from two sources: sugar cane and sugar beet. In the case of sugar beet molasses, some processing plants incorporate the molasses back on to the sugar beet pulp, and sell the combined product as 'molassed pulp'.

Molasses is not an easy ingredient to mix into concentrated feeds and in the UK it is usual practice for the molasses storage vessels to be heated in order to allow a more freely flowing liquid to be accurately metered into the mixers. Although large quantities of molasses are bought for on-farm use, the sticky nature of the material makes it very difficult to mix properly unless special equipment is employed.

Oils and Fats

These two terms are interchangeable. When a fat is in the molten state it is an oil. Thus butter, at normal room temperature, is a fat but, if heated, it turns into a yellow-coloured, clear liquid, butter oil.

The chief material in this class is tallow and since this material is derived from slaughtered animals processed through abattoirs, it can also be regarded as an animal by-product. Similar considerations apply to fish oil.

Just as oils and fats can be manipulated chemically to form high-quality materials suitable for incorporation in margarine and cooking oil, so it is similarly possible to modify oils and fats to render them more useful as animal feeds. In the case of the ruminant animal, the modification is usually by some form of fat protection, which enables the resultant product to pass through the rumen of the animal unchanged, so that the material can be digested in the mid-gut. In the case of pigs and poultry it is possible to alter the degree of saturation, and the length of the carbon chain of the component fatty acids, so that the proportion of 'essential fatty acids' is increased and the corresponding proportion of less useful material decreased.

To achieve higher levels of performance, there will be a natural requirement to increase the energy density of diets. This is most readily done by incorporating oils and fats because they have approximately $2\frac{1}{2}$ times as much energy per unit weight as do non-fatty feeds mainly comprised of carbohydrates.

For each livestock class there is, however, a definite limit to the amount of fat which can be added to the diet. If larger quantities are fed appetite can be adversely affected and excess amounts of fat can inhibit the digestion of other components of the diet, particularly fibrous components in the case of ruminant feeds.

Concentrated feeds

The agreed definitions of the main types of concentrated feeds are given below.

Compound Feeds

Compound feeds are made up of a number of different ingredients, mixed and blended in appropriate proportions, to provide a properly balanced diet. They are supplied in a wide variety of physical forms, of which the two main types are meals and pellets. The pellets can be very large, as in the case of cobs made of a size suitable for feeding to cattle and sows at grass, or they may be very small, as in the case of crumbles, or broken-down pellets, suitable for feeding to very young chicks or baby piglets.

Most compound feeds are produced in feed mills operated by the compound feed trade. Approximately 10 m. t of compounds are sold in the UK each year, but a further quantity, of unknown magnitude, is also produced in small farm plants for on-farm consumption.

Protein Concentrates

These are proprietary products specially designed for further mixing before feeding, at an inclusion rate of 5% or more, with planned proportions of cereals and other feeds, either on the farm or in a compound mill. Protein concentrates contain blended high-protein ingredients, such as fish meal and soyabean meal, fortified with essential nutrients, such as minerals and vitamins. Where the rate at which protein concentrates are used in the mix is as high as 50%, they will contain cereals or cereal by-products. Some protein concentrates are formulated so that, after further mixing with cereals in a ground or rolled form, the resultant product is then suitable for balancing farm forage, such as hay or silage. In practice, therefore, an on-farm mix of a protein concentrate together with a rolled or ground cereal is equivalent, in nutritional terms, to a compound.

Supplements

These are proprietary products for use at less than 5% of the total ration in which they are included. They are formulated to supply planned proportions of vitamins, trace minerals and non-nutrient pharmaceutical additives. To facilitate adequate mixing into the total diet, the active materials are normally present in supplements in a diluent carrier such as finely ground cereal.

Supplements may be supplied to compounders so as to save them the trouble of setting up their own dispensaries, or they may be supplied direct to farmers to enable them to home-mix their own diets based on home-grown or bought-in straights.

Processing of Concentrates

Concentrates are usually processed before feeding to livestock. Sometimes the processing takes place several times. Thus straights may be heat-treated; the straight may then be mixed with other

Table 14.4   List of processing methods for raw materials

Dry Processing	Wet Processing	Chemical Processing
Grinding	Soaking	Alkali treatment
Pelleting	Steam rolling	Acid treatment
Dry rolling or cracking or	Steam processing and	Formaldehyde treatment
kibbling	flaking	Mould inhibition
Extruding	Pressure cooking	Oil extraction
Micronisation	Exploding	Anti-oxidant addition
Roasting		Fatty acid manipulation
Popping		Chemical enhancement of
Sterilisation		wettability and dispersability
Spray and roller drying*		

* Mainly applicable to milk powders used in manufacture of milk replacer diets.

materials and then ground and the final product may then be pelleted. A list of the various processing procedures is given in Table 14.4.

PRODUCTION OF BALANCED DIETS FOR LIVESTOCK

In order to derive a nutritionally sound, least-cost diet the following steps should be followed:

Definition of type and class of livestock, weight (or age), and level of production required

A ration suitable for a dry cow will rarely be suitable for a cow yielding 7,000 l of milk. Similarly, the ration required for a large Friesian cow giving 5,000 l of milk need not be so concentrated as the ration needed for a much smaller Jersey cow giving the same milk yield. It is therefore essential to know very full details of the animals for which a balanced diet is being devised, since an economic ration for one group of animals may not be economic for a different group of different average weight or operating at a different level of productivity.

In practice, a decision has to be made concerning the number of different diets required throughout the life span of a meat animal, or at the different stages of lactation in the case of cows or egg-laying in the case of laying poultry.

In the case of broilers, which complete their life-cycles in about seven weeks, it is normal to feed three different diets – a starter, a grower and a finisher. Both nutritional and economic considerations dictate when one makes the change-over point. To delay will probably give slightly better growth but at a very much higher cost whereas to bring the change-over point forward will almost certainly reduce the growth rate so that the animals are not ready for slaughter within the seven-week period planned for them.

In the case of dairy cows it used to be common practice to change the rations, primarily by adjusting the feeding rate of the compound feed, at weekly or monthly intervals. Research has now shown that such frequent changes in diet are not necessary, at any rate in the case of average-yielding cows giving about 5,000 l of milk. It is, however, still common practice for pedigree breeders aiming at high milk yields to change the feeding programme regularly throughout lactation, although in the large group-fed herd only about three such changes may be made throughout the 10 months of lactation with a further change for the remaining two-month dry-period.

Predicted feed intake

Before a balanced diet can be drawn up, it is important to know in advance the expected feed intake of the class of animals for which the diet is designed. This is because nutritional requirements are most properly expressed in terms of weight of nutrients required per animal per day. As an example, a given amino acid may be required for a certain class of animal at a level of 10 g per day. If the animal is eating 1 kg of feed, the percentage of that amino acid in the diet will need to be 1%, but if the animal is to be fed, say, 2 kg per day, then the proportion will fall to 0·5%.

Where animals are fed *ad lib*, as is commonly the case with most types of poultry, the intakes of feed are quite critical. If they are higher than expected, a greater intake of nutrients will result and the animals will be overfed. The consequence will be a diseconomy in the feeding system. Conversely, if the appetite is lower than anticipated the nutritional requirements will not be met in full and a lower level of performance will result.

Although the composition of the diet must eventually be made up in terms of percentage inclusion of the different ingredients, it is important to realise that the nutritional requirements for the animals in question must always be calculated in terms of daily intake of each nutrient on a weight basis.

Nutritional specification

A specification is a description of the nutritional requirement which must be provided in a given balanced concentrated feed. In the case of in-wintered beef cattle, for example, the nutritional factors involved may be few in number, possibly restricted to digestible crude protein, energy and a few of the more vital vitamins and minerals. In the case of more specialised diets, for instance those designed for breeding poultry, some 30 or so nutritional factors may be listed including a large number of vitamins and minerals and both essential amino acids and essential fatty acids.

In the case of a nutritional specification used by compound feed manufacturers, the raw materials which may be employed in making the diet will be listed, together with the maximum and minimum inclusion rate for each.

Diet formulation

A formulation is a statement of the raw materials and inclusion rates used in a given concentrated feedstuff. The formulation is best worked out by means of a linear programme on a computer, in order to meet the nutritional specification at 'least cost'. Simple formulations, as may be used to compile complete diets for use by dairy cattle, are often worked out by hand on the farm. Such simple calculations can rarely achieve true 'least cost'.

With the development of on-farm computers and computer terminals linked to a main-frame by telephone line, it is possible to bring the advantages of least-cost formulation on to the farm at relatively low cost. Many advisory agencies, such as the ADAS, and several animal-feed manufacturers, offer such a service to farmers. In order to achieve the greatest economy, the programmes should be run fairly frequently, as raw material stocks change and as raw material prices alter. Because of rapid movements in world prices for straights, a formulation which is a least cost formulation at one time may be several pounds per tonne uneconomic a month or so later.

Feed additives

Various types of additives are administered to farm livestock, particularly pigs and poultry, and the use of such additives is increasing. Therapeutic drugs, prophylactic additives and growth promoters are dealt with in Chapter 15.

Chemical additives

A material which can be included in this group is sodium bicarbonate, a 'chemical buffer' which may help to counteract excess acidity in the rumen of cattle and sheep.

Palatability enhancers

It is a traditional practice to add spice to compound feeds, especially dairy cow feeds, and to add various flavours to milk replacers and dry diets for calves, lambs and baby pigs. There is little scientific data indicating whether or not the use of many of these substances is justified, but farmers tend to evaluate the quality of animal feed somewhat subjectively by 'nose' and they prefer feeds which smell pleasantly of aniseed, fenugreek, honey or caramel.

Non-nutritional additives

Various additives may be added to manufactured feeds for the prime purpose of improving physical quality. Such additives may have little or no nutritional significance – indeed some may be potentially harmful to the animal especially if administered in excessive quantities.

Examples of nutritionally neutral additives, which properly come into this class, are the various proprietary 'binders' whose function is to improve the physical quality of feed pellets; dispersal agents and surfactants, whose purpose is to improve the mixability of milk replacers; and anti-oxidants which are employed to stabilise the chemical integrity of fats and oils, especially by preventing the production of free fatty acids due to rancidity.

Feed quality

Many of the feeds and feed-additives mentioned above come under the provisions of various laws aimed to make animal feeds safe for both farm livestock and the ultimate consumer of animal products – man. The chief pieces of legislation regulating animal feeds and feed-additives are the Agriculture Act of the UK and the Straights Directive and Compound Feedingstuffs Directive of the EEC.

Although these various laws aim to safeguard the livestock producer and the animals under his care, nevertheless they do not remove the responsibility of the farmer to ensure that all feeds offered to his animals are of high quality and, therefore, 'safe'. All feeds are subject to contamination with crop sprays, pathogens, fungi, insect infestation and breakdown due to bad storage conditions, particularly high moisture and high temperature.

It is especially important for the farmer who mixes his own feed to be familiar with the various laws affecting what may and what may not be lawfully fed to his livestock, and it is even more important for him to exercise good quality control when purchasing feed. It is a prudent practice for the farmer to have his feed regularly analysed by a competent analyst, both to safeguard him from the inherent dangers in the use of substandard materials and also to enable him to put accurate nutritional values on the various feeds used for home-mixing so that the nutritional specification is correctly matched by an accurate feed formulation.

15

PESTICIDES AND ANIMAL MEDICINES

Pesticides are chemicals used to control pests in the widest sense of the word including weeds and diseases (caused by fungi, bacteria and viruses) as well as insects, mites, nematodes and other invertebrates, and also vertebrates such as birds and mammals. Although most pesticides are intended to kill pests, the objective is to prevent or reduce damage to an acceptable level at an economic cost and with least effect on non-target organisms and the environment. Thus, chemicals such as repellents, attractants and anti-feedants, which affect behaviour, are classed as pesticides, and because of similarity of use and manufacture, growth regulators affecting the physiology of the plant are also included. Some insecticides and fungicides used on crops may also be used against ecto- and endoparasites of livestock and for convenience veterinary products are included in this chapter.

Chemicals, at present the main means of controlling pests, must be seen in perspective with other methods (some of which may be essential accompaniments), such as: mechanical, e.g. hoeing of weeds, netting to protect against birds; physical, e.g. radiation, hot or cold treatment of plant material; cultural, e.g. use of resistant cultivars, rotational cropping; biological, e.g. introduction of parasites and predators; and legislation to reduce spread or the chances of introducing new pests.

It was not until the nineteenth century that any real progress was made in controlling pests and diseases and at the same time there was increasing understanding of the causes of attacks. Weeds had long been controlled by fallowing, hand-weeding and hoeing. Much fundamental progress was made in the early part of the twentieth century, particularly during the inter-war years, on the more valuable horticultural crops, especially fruit. However, it was the discovery of the insecticides DDT and gamma BHC (now HCH), and the herbicides MCPA and 2,4-D just before and during the Second World War that opened a new era, stimulated the systematic search for new compounds and led to the establishment of agrochemical manufacture as an important multinational industry often linked with the pharmaceutical industry with which it had much in common.

Nomenclature

Chemical names are usually inconveniently complex and common names for the active ingredient (ai) are essential for practical purposes. These are first agreed nationally (in the UK by a committee of the British Standards Institute) and then internationally by a technical committee of the International Standards Organisation, to which most countries belong, with the notable exceptions of the USA and the USSR, though in general American and international names are the same. When the ai is formulated for use, it is given a proprietary name by the makers and with some of the earlier, common chemicals, such as MCPA, which are long out of patent, there may be scores of identical or similar products with different names. Regulatory schemes for most countries now require that the name of the ai and its percentage should appear on the label and in the UK the current booklet of Approved Products (MAFF, 1983) provides a guide through the maze of names. The ai, especially of herbicides, is often an organic acid but it may be more convenient to use the sodium or some other salt, or an ester, and the choice may have some effect on formulation or performance. To avoid complications only the primary ai is mentioned in the text and Appendix Table 15.1 (see pp. 777-91).

Formulation

The most convenient and common diluent for spray application is water and, as few pesticides are easily soluble in water, a method of formulation must be selected which is practicable for the particular pesticide and circumstances. This requires a suitable solvent and various adjuvants to make the solution emulsifiable with water. The concentrated formulation may be an emulsion itself or miscible, i.e. it forms an emulsion when mixed with water. Another method is to prepare the ai as a paste, a wettable (dispersible) powder, or more recently, as minute granules or minute droplets, a process known as micro-encapsulation. For some purposes, especially aerial spraying, a concentrated solution in a mineral oil may be used. Other adjuvants may be required to improve the wetting, spreading or adherence properties to the leaf or other surface, and yet others to enhance stability or reduce corrosion of containers.

Dusts in which the ai is mixed with an inert powder are less frequently used than formerly, but granules of various sizes are used extensively especially for soil treatments. The ai may be incorporated in the granule or sprayed on to it after formation. Granules may provide a safer method of application for more poisonous chemicals, provided they do not easily break down into a dust.

Some liquid and solid pesticides are volatile and act partially or completely as a gas and there are a few which are gases at normal temperatures and pressures. Fumigants may be applied to enclosed spaces from gas cylinders, by volatilisation from solids or liquids distributed in various ways and sometimes with adjuvants to increase efficiency or act as a warning. Some pesticides have a partial fumigant action on the pest on the crop in the open.

Formulation is thus an important aspect of pesticide manufacture and requires specialised knowledge and skills to ensure that it does not unduly reduce efficiency or increase hazards and that the product remains stable and effective for at least two seasons without corrosion of containers.

Mixtures

Because one pesticide is unlikely to deal with all pests present at one time and to save costs of application, it is necessary to use mixtures. This has been a longstanding practice in orchards where, for over half a century, one or more insecticides have been applied with one or more fungicides, but the increasing complexity of herbicide mixtures has led to some problems. In the early days the failure of MCPA or 2,4-D to control cleavers (*Galium aparine*) and chickweed (*Stellaria media*) soon suggested the need to supplement them with other herbicides and, later, the combination of three or more became quite common. These mixtures may be incorporated in the formulation and the firm would have to ensure that the mixture was efficient and safe. Frequently, farmers prepare a 'tank-mix' and unless the mixture has been officially cleared by the firm or firms concerned, it could give rise to risks of incompatibility, reduced efficiency, increased phytotoxicity or possibly increased hazard to the operator. Sometimes the application of different pesticides within a few days has resulted in damage to the crop and instructions on time intervals should be observed. Although the risk of increased hazard is emphasised, the reverse is possible and mixtures can reduce phytotoxicity.

Application

Most pesticides are applied as sprays dispersed in water under hydraulic pressure but, because of the costs of transporting water, volumes per hectare have decreased for field crops from the earlier

1,000 l/ha down to 20–40 l/ha or even less with new methods. For fruit and bush crops, up to ten times these quantities have been used. A spraying machine may be hand-held or, more usually, mounted on a tractor. It comprises essentially a tank, pump, stirring or circulating device, preferably a pressure gauge, and one or more nozzles or other devices to convert the liquid into droplets. Variations on this theme range from simple knapsack sprayers, worked by hand or a small engine or compressor and fitted with a lance, to large tractor-mounted machines with booms of 12–20 m or more; high-clearance machines to treat tall crops such as oilseed rape are now available. The use of wide booms and low ground pressure machines permit application during autumn and winter. For fruit or other specialised crops the nozzles would be arranged to give a spray pattern appropriate to the crop. Air pressure may also be used to transfer the pesticide to the target and, in enclosed spaces, e.g. potato stores or under glass, fogging machines producing very fine droplets are often used. In these, as with other special applicators, appropriate formulations may be required. Individual weeds, if few in number, may be treated with special 'hand-wipers' containing herbicide, e.g. on wild beet or wild oats.

It so happened that crops were fairly tolerant to the earlier pesticides, and variations in the amount applied were less important than with some modern ones, especially herbicides, where too little may be ineffective against weeds and too much will harm the crop. Thus, much greater precision is needed in machines and their use. Most existing machines produce droplets with a broad size spectrum, of which the smallest and largest drops may be wasted and some may drift on to other crops. Modern tendencies are to produce droplets of optimum size for the required purpose (controlled droplet application or C D A) by using more refined nozzles or a spinning disc. Because the volume of a sphere varies as the cube of its diameter, halving the diameter of a drop reduces the amount of liquid to one-eighth, which can lead to large reductions in the quantity applied. The general practice has been to maintain the amount of ai applied per unit area constant but, against some pests, experiments suggest that concentrations could be reduced by up to a factor of ten; however, much more work needs to be done on this subject.

The principle of charging spray or dust particles electrostatically so that they will be attracted to the target has been under consideration for a long time but only recently has the subject advanced to a stage where practical application seems within reach and rapid progress may be expected.

Although quantities of ai of the order of 1–2 kg/ha seem remarkably small to have such a great biological effect, it must be remembered that these are far in excess of what would be needed if it were possible to limit the application to the target organism. It has been calculated that the proportion of foliar spray taken up by insects or crop plants is only about 0·02% and only 2–3% of a systemic pesticide applied to the soil is taken up by plants. Thus, there appears to be great scope for reducing the amounts of pesticides used, if more efficient application methods can be devised.

Dust and granule applicators consist of a hopper with gravity feed regulated by a mechanical device operated according to the speed of the machine, or the control of the operator for hand-held apparatus. The product may be applied overall, in a row or band, on or in the soil before or after sowing, after germination, or to individual plants.

Space fumigation (e.g. in warehouses or silos) and fumigation of grains, other foodstuffs or seeds, is usually done by professional operators, where necessary in special chambers; a simpler procedure is to pour some less toxic volatile liquid over the commodity in silos or heaps which are then covered to prevent the escape of gas. Soil fumigation or sterilisation also requires special machinery or special skills: it is normally only economic in glasshouses where high-value crops can be affected by serious disease or other problems. These were controlled formerly by steam sterilisation but this costly and laborious process has been replaced largely by soil sterilants or by fumigation with methyl bromide. Smokes, i.e. minute solid particles dispersed in air, are comparable with fumigants and are also used in confined spaces, especially glasshouses. Smoke generators are a sort of firework in which the ai is mixed with a combustible material in such a way that combustion is controlled carefully and does not reduce appreciably the efficacy of the pesticide.

Seed treatment (the word is preferred to dressing, now reserved for seed cleaning) is of long standing but first became widely used with the organomercurials against certain seed-borne diseases of cereals. Subsequently, insecticides were incorporated, first against wireworms and then other pests, and many different seeds can now be treated. Originally the chemical in the form of a dilute dust was mixed with the seed with shovels on the barn floor, or in a machine resembling a butter churn, but nowadays seed is treated by the merchant with either dusts or concentrated liquids in a special apparatus designed to apply precise amounts safely. There is a limit to the amount of deposit which can be retained by a particular seed and, if it is necessary to apply larger amounts, a sticker is used or the seed is pelleted.

Encapsulation is another method of formulation which may be used in special circumstances particularly for slow or delayed release of the ai. Volatile compounds like dichlorvos may be incorporated into a plastic block for use as domestic or industrial killers of flies and other insects.

Some insecticides, molluscicides and poisons for killing mammals may be mixed with a bait which may be generally distributed over a wide area or, if the bait is attractive, located at suitable intervals with precautions to prevent it being taken by non-target organisms.

Aerial spraying by fixed-wing aircraft and helicopters is a special form of low-volume application of value in treating vast areas of uniform crops quickly. It is also useful on trees and other tall dense crops and when the soil is too wet or too rough for ground machines. About 400,000 ha are treated annually with pesticides in the UK. Most are in the form of sprays but granules and pellets, e.g. slug baits, can also be used.

TYPES OF PESTICIDES

Pesticides can be classified in various ways according to what they kill, how they act or according to their chemical grouping. This is usually associated with their biochemical action although our knowledge of this is limited. Although some groups include compounds affecting more than one type of organism, for practical purposes it is convenient to deal with the generalities and practical aspects of herbicides, fungicides and insecticides separately; the last include chemicals to control other invertebrates, such as acaricides against mites, nematicides against nematodes and molluscicides against slugs and snails. The control of vertebrates, the use of growth regulators and of animal chemotherapy are also dealt with separately. A few of the more important chemical groups with examples are mentioned under each heading but more detailed information is summarised in Appendix Table 15.1 (see pp. 777-91).

Herbicides

Weedkillers were first used on crops towards the end of the nineteenth century. Copper sulphate and sulphuric acid could both be used selectively on cereals because of the difference in morphology of the cereal and broad-leaved weeds. Sulphuric acid was used quite extensively in the 1930s and 1940s and was followed by DNOC (dinitro orthocresol) but because it was very poisonous, especially in warm weather, it caused many deaths of humans and animals and was phased out as new and safer materials were developed. The discovery of the effective and relatively safe growth-regulator or 'hormone' herbicides, MCPA and 2,4-D, during the Second World War led to chemicals gradually replacing mechanical cultivations in many situations, especially in advanced countries. This encouraged changes in crop production, such as extended runs of cereals, and further advances led to minimum cultivation techniques and direct drilling. However, the change to more and earlier winter sowing resulted in new or increased weed and disease problems. Herbicides are also used as desiccants to kill the foliage of potato crops and of seed crops to facilitate harvesting. One advantage of herbicides is the reduction of inter-row cultivation, which, in itself, reduced the yields of some crops such as potatoes and shallow-rooting fruit crops such as raspberries and currants.

The effect of a herbicide depends on the chemical and its formulation, the method of application, the morphology and physiological state of the plant, and the weather. Some or all of these affect penetration and retention, and influence translocation (if it occurs) and accumulation at the site of action.

Entry into plants may be through the roots, foliage – including leaf axils as well as the leaf laminae – and also the coleoptile. Entry and retention on foliage depend on the size, number and trajectory of spray droplets; on the shape, angle and arrangement of leaves; and on the nature of the leaf surface – especially the presence and texture of wax particles and hairs. For soil-applied herbicides, toxicity depends on the mobility and availability of the chemical, which are affected by soil type and condition, particularly moisture and pH, and the relative position of the herbicide and plant roots. In addition to the physiological state of the plant, its age is also important and many herbicides at normal rates are effective only on germinating weeds or small seedlings. The plant may also have the ability to enhance the toxicity of the herbicide or detoxify it. All these considerations apply to the crop and the weed and selectivity is generally a combination of a number of small differences between the reaction of the crop and the weed. Selectivity is not absolute and herbicides may have an adverse effect on the crop which can be detected in the absence of weeds and particularly in unfavourable circumstances such as drought.

Herbicides may be applied to the surface of the soil or be incorporated into it to kill weeds already germinated or as they germinate i.e. pre-sowing; pre-emergence of the crop; or post-emergence. The last is usually soon after crop emergence but, if much later, herbicides may be applied between the rows of row crops. Post-harvest treatments are a fairly recent innovation.

Treatments may be overall or localised; they may be directed along or between the rows of a widely spaced crop, around the bases of trees or bushes, or to patches of weeds or individual weeds. Some herbicides are very persistent and have a residual effect up to a full season or longer, in which case they may affect subsequent crops. Soil type has a profound effect on the action of many herbicides (and other pesticides) which are adsorbed on to clay particles or to organic matter and are thus rendered less effective.

The development of herbicides started with the control of broad-leaved weeds in cereals, next to grassland and then to other arable crops, taking advantage of pre-sowing and pre-emergence techniques. The introduction of new compounds to control wild oats, blackgrass and other weed grasses was a major advance, as was the use of persistent herbicides in established orchards, on fruit bushes, nursery stock and forest trees, and some arable and horticultural crops. Some compounds, alone or in mixtures, could be used for sward destruction as a preliminary to reseeding or direct drilling. Most major weed problems on most crops can now be controlled by one or more herbicides, and without these, productivity would decline or costs escalate. Moreover the use of herbicides is often closely integrated with other technological advances and sugar beet provides a good example of how herbicides have become an essential part of crop production when linked with monogerm or pelleted seed, drilling to a stand and mechanised harvesting.

Because there are so many different groups of herbicides, only a few examples can be given here.

The phenoxy-acids, or 'hormone' weedkillers, are still amongst the most widely used herbicides and form the basis of many mixtures. The phenoxyacetics include MCPA, 2,4-D and 2,4,5-T; the phenoxybutyrics take advantage of the differential ability of plants to oxidise MCPB and 2,4-DB to MCPA and 2,4-D. Certain legumes cannot do this and can survive the treatment which kills many weeds. The phenoxypropionics include mecoprop, a useful complement to MCPA and 2,4-D, and dichlorprop, especially valuable in controlling *Polygonum* spp. Diclofop-methyl and trifop-methyl are effective against grasses in broad-leaved crops.

The dinitroanilines, long available as dyestuffs intermediates, were a logical development of the dinitrophenols, and include several useful compounds such as trifluralin, effective against grasses and broad-leaved weeds and used extensively in the USA on soyabeans and cotton. Nitriles are represented by dichlobenil, chlorthiamid and ioxynil.

The versatile carbamates include propham, useful against a wide range of germinating weeds; chlorpropham, also used to inhibit sprouting in stored potatoes; and asulam, which has proved of particular value in bracken control. The dithiocarbamates, diallate and triallate, are two of the many useful chemicals for controlling wild oats. There are numerous representatives of the ureas such as monuron, linuron and diuron, very widely used to control young weed seedlings. The persistent triazines, e.g. atrazine and simazine, are also used extensively as total weedkillers at higher doses, and selectively on certain crops at lower rates.

The bipyridyliums or quaternary ammonium compounds, diquat and paraquat, are absorbed rapidly by foliage, killing a range of annuals and the non-woody shoots of perennials. They are rapidly inactivated in the soil which can be sown again almost immediately.

The halogenated aliphatic acids, such as T C A and dalapon, are valuable for the treatment of many grass species. Dalapon is also an aquatic herbicide and is used like paraquat for the destruction of grass swards prior to sowing another crop. The amides, diphenamid, propanil and propachlor, are used widely in the U S A but have only a limited use in the U K. Glyphosate is one of the most promising of the newer chemicals as it is translocated to underground rhizomes and is effective against deep-rooted perennials.

Insecticides and related compounds

Insecticides are often divided into stomach and contact poisons and fumigants. This was useful when most of the few available chemicals had predominantly only one type of action and it was essential to use a stomach poison such as lead arsenate against leaf-eating caterpillars and a contact poison such as pyrethrum against 'sucking insects', e.g. aphids. Incidentally, although the term 'sucking insect' is familiar and convenient, aphids and other insects probing plant vessels and tissues have the sap forced into them as it is under pressure.

The earlier insecticides were either inorganic substances, which were fairly persistent, or complex compounds such as derris, pyrethrum and nicotine, derived from plants. Originally the part of the plant containing the poison – the roots for derris and the flowers for pyrethrum – was dried and powdered to produce a crude dust but, later, extracts were prepared and formulated as dusts or sprays. Much work was done to identify the active principles and for pyrethrum this ultimately resulted in the development of the synthetic pyrethroids. Although these early plant products were short-lived on the crop, pyrethrum (and the pyrethroids) had the valuable property of a quick 'knockdown' effect which is unusual in synthetic insecticides.

As with herbicides, the big advance came during the Second World War with the introduction of D D T and gamma B H C (now gamma H C H) or lindane. These organochlorines (O Cs), later followed by others such as aldrin and dieldrin, affected a broad spectrum of insects and were very persistent. At first these properties appeared to be advantageous, especially in reducing the number and range of applications and in controlling soil pests such as wireworms, but it gradually became apparent that adverse effects on beneficial insects, the development of resistance by many pests and persistence in the environment, might outweigh their benefits in the long term and the use of O Cs, except for the less persistent lindane, is now banned or restricted in most countries.

The O Cs were followed closely by the organophosphorus compounds (O Ps). Most are based on the general formula

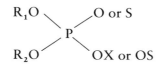

where R_1 and R_2 are usually ethyl or methyl groups. Several hundreds have been synthesised, of which 50 or more are still marketed. They have very varied properties; some have a high mammalian toxicity, others intermediate or low. Some have a broad spectrum of activity, others are relatively selective. Some, like chlorfenvinphos, can persist in soil for some weeks and can be used as seed treatments, yet others, like the highly poisonous mevinphos, are rapidly broken down in the presence of moisture and treated crops can be consumed after 24 hours. Most have contact and stomach action and some act as fumigants but a number are also systemic. (The term systemic is generally used for insecticides and fungicides, whereas translocation is generally used for herbicides.) Some have translaminar activity i.e. when applied to one side of a leaf they can kill insects on the other side. These two latter properties are of particular value in the control of aphids which are often protected by the curling and distortion they produce on leaves and which favour the growing point where systemic insecticides tend to concentrate.

All O Ps act by inhibiting cholinesterases and other enzymes both in mammals and insects and their selectivity depends on the specific enzymes concerned, the site of activity and the means and speed of breakdown. Many compounds become effective through the conversion of the —P S group to —P O in the animal or plant. One of the most widely used O Ps in the U K is demeton-S-methyl, often known by its proprietary name of 'Metasystox'. Others widely used are azinphosmethyl, chlorfenvinphos, chlorpyrifos, diazinon, dimethoate, fenitrothion, phosalone and triazophos. Phorate and disulfoton are generally used in granular form to control aphids and dipterous pests of vegetables. In this way they are less toxic to bees as well as to operators. Many O Ps have replaced the O Cs as sheep dips and to control ectoparasites of other livestock, and some are used to control internal parasites like warble fly larvae.

The carbamate group of insecticides of the general formula $RO.-$ or $R.NO.-CONHCH_3$, where the first group may be a phenol, an oxime or a hydroxy heterocyclic compound, are also inhibitors of cholinesterase. Most have a broad spectrum of activity and some are systemic. Some, such as aldicarb and oxamyl are effective nematicides and methiocarb is a molluscicide. Pirimicarb is mainly aphicidal and other examples include bendiocarb, carbofuran, a general soil insecticide and nematicide, and propoxur. Many are useful in the control of animal parasites.

The development of the synthetic pyrethroids has already been mentioned. The earlier ones, such as allethrin and resmethrin, were used mainly as alternatives to the pyrethrins as fly sprays, but the more recent ones such as permethrin, cypermethrin, fenvalerate and deltamethrin are broad-spectrum insecticides with moderate persistence and there is already speculation that they might give rise to some of the problems of the organochlorines in destroying beneficial insects. Their chemical structure is complex and their action both in insects and mammals is on the nervous system.

Apart from these four main groups, there are a few other types of special interest. Diflubenzuron inhibits the development of chitin and as such has a unique effect. Chemicals simulating the effects of juvenile hormones, e.g. methoprene, are also of promise, as are anti-feedants. Chemosterilants have also been considered but have a high mammalian toxicity and might be unacceptable. A number of synthetic pheromones, mainly sex attractants, are now available for many species; although so far they have been used mainly in traps to ensure optimum timing of sprays, they might be used to reduce total numbers of pests if combined with insecticides, or applied over a wide area to disrupt the finding of a mate. Repellents, though theoretically desirable, have not been sufficiently persistent in practice to be of much value on crops though they have some limited use for pests of livestock.

Acaricides

Although many insecticides, especially the OPs, are also effective against mites, because these important pests of fruit crops all over the world have many generations a year and are prone to develop resistance, it was necessary to find more or less specific acaricides. Some affect all stages of the mite, whereas others are mainly ovicides or only affect the active stages. Fortunately, several mildew fungicides, including dinocap and binapacryl, commonly used on apple, give a useful control of tetranychid mites and prevent the population rising above an acceptable level in normal circumstances, though there is a swing now towards fungicides without acaricidal action. More specific acaricides include amitraz, cyhexatin, dicofol, quinomethionate and tetradifon and these may be used on a wide range of fruit and glasshouse crops. One of the special problems in glasshouses is the need for frequent picking of crops such as tomatoes and cucumbers, so residues must be low and of short persistence. Chemicals which control tetranychid mites are not necessarily effective against eriophyid mites such as those causing 'big bud' of blackcurrants, and for these the O C endosulfan has proved useful. Various flour mites feeding on a range of stored products may require special treatment, often fumigation, and other mites and ticks ectoparasitic on animals present special problems, though some can be controlled by standard insecticides or acaricides.

Nematicides

Plant nematodes spend all or part of their life in the soil and proved very difficult to control until the introduction of the carbamates. Previously various soil fumigants had been used, including chloropicrin and dichloropropene, the latter originally in the form of D—D mixture. Dazomet, which, like metham-sodium, acts by producing methyl isothiocyanate, is a useful general soil sterilant as well as a nematicide. This, like methyl bromide, is used mainly in glasshouses. It should be mentioned that although many of the carbamates have been referred to as nematicides, in fact they are nematostats, i.e. they act by preventing feeding or development rather than by direct kill.

Molluscicides

Slugs and snails can be important crop pests in wet areas or seasons and are not easy to control, especially those species of slugs which rarely come to the surface and can seriously damage root crops such as potatoes. Metaldehyde and methiocarb, usually in the form of pellets containing a bait, can be effective against species feeding or moving about on the surface. Some snails are also intermediate hosts of various parasitic worms causing human and animal diseases, such as bilharzia in the tropics and liver fluke of cattle and sheep in the UK. Molluscicides may be needed to control the intermediate host as well as therapeutants for the mammalian host.

Control of vertebrate pests

These present special difficulties: first, man and domestic animals are likely to be more susceptible to poisons affecting vertebrates than to insecticides; secondly, in advanced countries the killing of certain mammals, e.g. rabbits and badgers, and of nearly all birds, however destructive, is unwelcome if not unacceptable; finally, both birds and mammals quickly re-invade cleared territories and re-establish themselves. This applies particularly to birds with their exceptional powers of dispersal. Rats and mice are less subject to sentimental attachment and are controlled mainly by poison baits containing anti-coagulants such as warfarin and difenacoum. This causes internal haemorrhage but in some areas rodents are now resistant to warfarin. The chemical control of rabbits is by the introduction of cyanide into their burrows, and strychnine is still used under licence for the control of moles by inserting poisoned worms into their runs.

Most birds are protected by law and in the UK poisons can be used only against certain species and then only under licence. Baits containing narcotics, such as alphachloralose, have been used with limited success. The idea is to collect and kill the pests, mainly wood pigeons, whilst still narcotised, and allow the other birds to recover before releasing them, but in practice there are difficulties. Repellents such as anthraquinone have been tested with varying success depending on species and circumstances. Undoubtedly an effective repellent would be of great value, especially against birds such as bullfinches which strip the buds off fruit trees and bushes. Although bird problems are often locally serious in the UK their total effect is not great, whereas they can have disastrous effects in the tropics where even drastic remedies have little permanent effect.

Fungicides

Fungal diseases can spread rapidly both within and between crops under suitable conditions and protection of the crop before the disease gets a hold is essential. Fungicides are therefore often applied on a routine basis as preventatives and thus some form of forecasting for epidemic diseases, such as potato blight, can be extremely valuable. Although fungicides based on copper, mercury and sulphur were among the earliest pesticides to be used extensively, the development of synthetic compounds lagged behind that of herbicides and insecticides.

Copper compounds are very effective but can be phytotoxic. One of the earliest fungicides was Bordeaux mixture, consisting of copper sulphate and lime, most efficient as a fresh tank mix. It, or comparable proprietary mixtures, is still used occasionally and organo-copper compounds form the basis of many wood preservatives. Mercury compounds still play a useful role but, even though the more poisonous types have been replaced by less poisonous ones, there is a trend in most countries to phase out mercury because of pollution of the environment. The organomercurials are still used as seed treatments and have played an important part in controlling seed-borne diseases of cereals such as smuts, bunts and leaf stripe, though this use is likely to decline as effective alternatives become available. Although these earlier metallic compounds are declining in use, more recently triphenyl salts of tin such as fentin acetate or hydroxide have been developed, especially for the control of potato blight. Elemental sulphur in the form of flowers of sulphur was one of the earliest chemicals for mildew control. Lime-sulphur, in which the sulphur is present as polysulphide ions, was also once widely used but both have now been replaced largely by colloidal sulphurs. The dithiocarbamates and related substances evolved from the use of sulphur and were some of the earliest synthetic fungicides, but are still widely used. They include thiram, especially useful as a seed treatment, zineb and mancozeb which are used in combination with each other and other fungicides for the control of a very wide range of diseases world-wide, though mainly used for potato blight control in the UK.

The success of systemic insecticides and herbicides stimulated interest in the development of systemic fungicides and during the 1960s rapid progress was made; by the end of the decade many candidates were available. Exploitation was rapid, partly following realisation of the importance of cereal diseases, especially barley mildew. Systemics were very effective against powdery mildews and other surface fungi and many had a wide spectrum of activity, though some were fungistatic rather than fungicidal. The major group comprises the benzimidazole compounds; the best known being benomyl, effective against a wide range of diseases and with a low toxicity to animals and plants. This substance and thiophanate-methyl act on the fungus by forming carbendazim or MBC in the plant, and this chemical is also formulated as a fungicide. Thiabendazole was first introduced as an anthelmintic and it was some years before its fungicidal properties were recognised. Like benomyl it affects a wide range of fungi, but its main use in the UK is for the control of storage rots of potatoes. Other important systemics are ethirimol, tridemorph and triforine, all used for the control of cereal diseases. Benodanil and oxycarboxin are very effective against rusts, but on cereals they have been supplanted largely by chemicals with a broader spectrum, such as propiconazole, fenpropimorph and triadimefon. Carboxin has a useful action against smuts and bunts when used as a seed treatment.

Some non-systemics worthy of mention are captan, for the control of apple scab; captafol for potato blight and *Septoria* on wheat; dichlofluanid, iprodione and vinclozolin for the control of Botrytis on soft fruit and glasshouse crops. As with many pesticides, mixtures are being used more widely to combat different types of disease simultaneously, e.g. mildews and eyespot on cereals.

Bactericides and anti-viral agents
Compared with fungi, bacteria are of minor importance in temperate crops though some cause problems on fruit and vegetables,

particularly root vegetables in store. Some copper fungicides are effective and some antibiotics such as streptomycin have been used experimentally. The chemical control of virus diseases has so far been directed at the vector but recent experimental work suggests that some substances can have a direct effect on plant viruses, though this is a long way from practical application.

Plant growth-regulating substances

These can be defined as chemicals which modify plant growth to the benefit of the grower. Their activities include:

(a) Retardation of growth, e.g. cereal straw length, where chlormequat is widely used to prevent lodging; of ornamentals, especially houseplants, e.g. daminozide on chrysanthemums; and of grasses and hedges to reduce frequency of cutting.

(b) Encouragement of root growth, especially of cuttings; this was one of the first uses of analogues of natural substances such as 4,indol 3, butyric acid.

(c) Thinning of fruit, e.g. carbaryl, induction of fruiting without fertilisation, prevention of fruit drop and hastening of ripening after picking, e.g. ethephon producing ethylene, a natural growth regulator.

(d) Defoliation to assist harvesting, reduction of side shoots, disbudding e.g. of chrysanthemums.

(e) Improvement of quality or yield (excluding fertilisers and trace elements).

(f) Resistance to drought at critical stages by restricting transpiration.

The chemicals used are many and varied: some are naturally occurring regulators such as abscissic acid or gibberellins or their analogues; others are well-known pesticides such as maleic hydrazide or carbaryl. This is a subject of increasing interest and many more compounds with various effects are likely to be produced in the near future. Some regulators have been shown to enhance or reduce the effects of pests and diseases in the crop.

CHEMOTHERAPY OF ANIMAL DISEASES AND PARASITES

Chemotherapy is the treatment of disease by using substances that have a specific antagonistic effect on the organism causing the disease. Chemotherapeutic agents are now widely used for treating and preventing diseases; some also for improving growth and food conversion. Examples include antimicrobial substances which are used against bacteria, fungi, viruses, etc, and parasiticides active against both external and internal parasites.

Since 1971 all medicinal products have been subject to the licensing requirements of the Medicines Act of 1968. This has the effect that no medicinal product can be marketed in this country unless it is licensed. A product licence is not issued unless the application satisfies the criteria of safety, quality and efficacy. Products that were already on the market before the operative date of 1971 were given licences as of right and are subject to review. In the context of the Medicines Act, safety of veterinary medicines includes not only safety for the target animal but also for the operator, the environment and the consumer. In this case the need, where appropriate, for the wearing of protective clothing and methods of safe disposal will be specified as well as a withdrawal period during which the product must not be administered to animals before slaughter, to ensure that meat, milk, eggs, etc, are safe for human consumption.

Antimicrobial substances

These consist of synthetic chemicals as well as antibiotics used for the treatment and prevention of animal as well as human infectious diseases. Antibiotics are complex substances produced initially by certain micro-organisms as metabolites and which are active in minute quantities against other micro-organisms. Most antibiotics produced by moulds are active against bacteria but some have activity against fungi, rickettsiae, viruses and, more recently, helminths.

It was discovered early on in the antibiotic era that certain hitherto sensitive bacteria could develop resistance to an antibiotic and this could seriously curtail its usefulness. Such resistance can often be mediated by plasmids (R factors) which can be transferred from resistant to sensitive cells. Because of the potential dangers that could arise from the development of resistance, a Joint Committee on the use of antibiotics in animal husbandry and veterinary medicine was set up by the British government and reported in 1969 (the Swann Report).

One of its many recommendations was the division of antimicrobial substances into 'feed' and 'therapeutic' categories. The former should only be allowed to be used off prescription if they were of economic value in livestock production under UK conditions, had little or no application as therapeutic agents in man or animals, and would not impair the efficacy of a prescribed therapeutic substance through the development of resistant strains of organisms. All therapeutic antimicrobial substances should be available only through a prescription.

Antibiotics can be either bacteriostatic, i.e. they prevent growth/cell division, or bactericidal, i.e. they kill the organisms. At all times an adequate therapeutic dose must be given by the appropriate route; if two antibiotics are given together they must be chosen with great care since some combinations are less effective than either singly. Whenever possible a narrow or medium-broad spectrum antibiotic should be used except in certain vital situations where a broad-spectrum antibiotic is used initially, pending the results of sensitivity testing. In animal use, advice must always be given on the need to observe withdrawal periods.

Penicillins and cephalosporins Penicillin is a narrow-spectrum antibiotic effective against gram-positive organisms and it must be administered parenterally. It is still widely used in intramammary preparations either alone or in combination with another antibiotic for treating mastitis. It is also widely used for treating gram-positive infections such as Erysipelas, Anthrax, Strangles and Joint Ill. It has no activity against gram-negative organisms.

Because of this and because of the development of strains resistant to penicillin, mediated through penicillinase production, search has continued for other penicillins with a broader spectrum of activity and uninfluenced by the production of penicillinase. This has resulted in penicillins that are stable to penicillinase (e.g. methicillin, cloxacillin) and those with a broader spectrum of activity especially against gram-negative organisms (e.g. ampicillin, carbenicillin, amoxycillin). Another recent approach has been the use of clavulanic acid which is a powerful inactivator of betalactamases (penicillinases). Although itself not active as an antibiotic, when incorporated with penicillin it protects it from being inactivated.

Cephalosporin resembles penicillin in having a betalactam ring and acts in a similar way to it but is much less susceptible to inactivation by penicillinase although some organisms produce a cephalosporinase.

Tetracyclines This group includes chlortetracycline, oxytetracycline and tetracycline.

The tetracyclines were one of the first groups of broad-spectrum antibiotics to be introduced. They are also effective when given orally. They are bacteriostatic and effective against a wide range of both gram-negative and gram-positive organisms as well as mycoplasmas, rickettsiae and some protozoa. Their widespread use has caused problems because of the development of resistance, especially as cross-resistance exists within the group. Caution is required when they are used orally in ruminants because they disturb the normal ruminal flora. Topical administration should be avoided because of the dangers of causing sensitisation to sunlight.

Aminoglycosides These include streptomycin, dihydrostreptomycin, kanamycin, gentamicin, amikacin and neomycin.

Streptomycin is still used in intramammary products in conjunction with penicillin to enlarge the spectrum of activity. A synergism can also occur when penicillin and streptomycin are combined. It is also used for treating systemic conditions caused by organisms sensitive to streptomycin, such as coli septicaemias. Both streptomycin and dihydrostreptomycin can give rise to ototoxicity, especially after prolonged administration, and widespread resistance has developed to both.

Other antibiotics which can be used for therapy include the macrolide group of antibiotics (tylosin, erythromycin, lincomycin) as well as chloramphenicol. Chloramphenicol is a broad-spectrum antibiotic with activity against gram-positive and gram-negative organisms as well as against rickettsiae and some viruses. It was also the first antibiotic to be synthesised. Because of the value of chloramphenicol in treating typhoid fever in man its use in animals is restricted to certain conditions. These include suspected systemic salmonellosis, particularly in cattle; respiratory infections, especially in calves; and panleucopaenia and ophthalmic use. Certain antibiotics are also available for treating fungal infections, for example griseofulvin against ringworm infection, and nystatin against candida infections.

Feed additives and growth promoters　Following the implementation of the recommendations in the Swann Report only certain antibiotics were authorised for inclusion without veterinary prescription in feed for growth promotion; these included zinc bacitracin, flavomycin and virginiamycin. In addition, a few preparations of copper and arsenic could also be used for certain animal species at specified inclusion rates. Therapeutic substances were not authorised for use as growth promoters.

At present the inclusion of substances in feed for growth promotion is controlled by an E E C Directive. Substances included in Annex I can be used in any member state whereas those in Annex II can be used by individual member states but are subject to further scrutiny before being included in Annex I or being withdrawn altogether. Copper for use at up to 200 ppm in feed for pigs is included in Annex II but authorisation to use arsenical compounds for growth promotion has been withdrawn.

Hormones have been widely used for growth promotion, usually being administered as implants in the ear, which is then discarded at slaughter. The natural hormones oestradiol, testosterone and progesterone are used as well as 'stilbenes' (diethylstilboestrol, hexoestrol, dienoestrol) and other synthetic hormones such as trenbolone and zeranol.

Under another EEC Directive the use of 'stilbenes', as well as thyrostatic substances (which are not used in Britain), is prohibited, but the natural hormones as well as trenbolone and zeranol are permitted for use for growth promotion pending a review.

Animal parasiticides

These are now widely used both for endoparasites (internal) and ectoparasites (external). Endoparasites of the animal body may be divided broadly into protozoa which are unicellular, and helminths, which include tapeworms, flukes and roundworms.

Protozoa   In Britain the commonest protozoal diseases are caused by coccidia but in areas where the tick vector (*Ixodes ricinus*) is common, bovine babesiosis or Redwater, due to a blood parasite *Babesia divergens*, may cause losses.

> Coccidiosis, due to species of *Eimeria*, occurs in most vertebrates but assumes significance in animals under intensive conditions, especially poultry, some sheep and occasionally in cattle. The parasites are highly host-specific and coccidia found in one host-species will not occur in another.
>   Poultry especially are vulnerable and most broilers are given drugs in the food for prevention. A range of compounds is available including amprolium, dinitolmide, clopidol, methyl benzoquate, halofuginone, arprinocid and the ionophore antibiotics such as monensin and lasalocid.

Treatment of established infections generally utilises various sulphonamide formulations some of which incorporate a pyrimidine potentiator.

Helminths   Different chemicals or anthelmintics are needed to treat and control different types of worm infestation.

Trematodes or flukes have complex life-cycles involving one or more intermediate hosts. The best-known example is *Fasciola hepatica*, the liver fluke of cattle and sheep. One approach to control is to destroy its intermediate host, the snail *Limnaea truncatula*, either by pasture improvement such as drainage or by using molluscicides.

> It is often more expedient, however, to treat infected animals with anthelmintics such as oxyclozanide, nitroxynil, rafoxanide, brotianide or diamphenethide. These drugs are highly effective only against mature (12-week-old) liver fluke infections. Rofoxanide, nitroxynil and brotianide have higher activities against six-week-old fluke infections than oxyclozanide; while at therapeutic dose levels diamphenethide is highly effective against flukes less than four weeks old. It is therefore possible to select a drug appropriate to the degree of control required.

Nematodes or roundworms are important parasites of the alimentary tract and lungs. Their life-cycles are direct, infection being transmitted from one group of animals to another by means of infective larvae on the herbage. Control of nematode diseases is commonly achieved by anthelmintic drugs used either by themselves or in combination with grazing management. In the case of lungworm infection of cattle a vaccine is also available. Most of the anthelmintics currently available belong to the benzimidazoles.

Thiabendazole was the first member of the group to be introduced, in 1961, and has been followed subsequently by parbendazole, cambendazole, mebandazole, oxybendazole, flubendazole, fenbendazole, oxfendazole and albendazole. Two other compounds, thiophanate and febantel, are broken down in the body to benzimidazole compounds and are therefore included in the group. Benzimidazoles are safe and highly effective against the majority of gastro-intestinal nematodes of cattle, sheep, horses and pigs. The range of activities of the more recently introduced benzimidazoles has also been extended to lungworms, tapeworms and liver fluke. Other compounds effective against a wide range of gastro-intestinal worms include levamisole, morantel and pyrantel. Levamisole is also effective against lungworms. The antibiotic avermectins, recently introduced, are highly effective against a wide range of gastro-intestinal nematodes, lungworms, warble fly larvae and ectoparasites such as lice and mites. Anthelmintics with narrow ranges of activity include bephenium hydroxynaphthoate, diethylcarbamazine citrate and several O P compounds such as haloxon and dichlorvos. Bephenium and diethylcarbamazine are generally indicated in the treatment of *Nematodirus* spp. and lungworms respectively while the O Ps are useful in the treatment of gastro-intestinal nematodes.

The cestodes or tapeworms have life-cycles which involve intermediate hosts. While adult tapeworms are probably of more importance in dogs and cats than in ruminants, *Echinococcus granulosus* and *Taenia multiceps*, both parasites of dogs, have cyst-like intermediate stages which are responsible for important diseases of man and sheep respectively, namely hydatid disease and gid or sturdy.

Niclosamide, nitroscanate, dichlorophen and bunamidine are all effective against adult tapeworms of dogs and, with the exception of nitroscanate, for sheep. Praziquantel is highly effective against adult *Echinococcus granulosus* in dogs.

Anthelmintics can be administered as drenches, by injection, incorporation in the feed or by means of medicated blocks. In general, drench and injectable preparations allow a greater degree of control over the amount of anthelmintic administered.

Ectoparasites   Livestock pesticides need certain characteristics which differ from others in general agricultural practice. Many will require to be persistent in the fleece or hair, so that repeat treatments can be kept to a minimum. Low mammalian toxicity is especially important when pesticides are applied directly to livestock, reducing the risk to young or sick animals and minimising residues in food and adverse effects on operators during treatment.

Arthropod ectoparasites may be obligate parasites, e.g. warble flies and blowflies. Some live almost entirely away from the host, visiting it only for blood meals which are nonetheless vital to the completion of the development cycle. Such blood feeders are significant in the spread of animal disease. Dung-breeding insects, especially flies, which can proliferate in enormous numbers in animal accommodation, can also be instrumental in the spread of disease.

Livestock pesticides are used principally in sheep (against blowfly strike, sheep scab, ticks and other fleece parasites) and on cattle to protect them, for example, against the effects of warble fly attack (in Europe) and against ticks, especially in the tropics. Mange in pigs and other animals is another widespread problem. In most cases pesticides are applied most effectively by the use of dipping or spraying techniques, although for certain purposes the use of systemic formulations (for internal parasite stages), or of dusts, is more appropriate. Most sheep dips now contain O Ps since the withdrawal of O Cs such as D D T and dieldrin. The reappearance of sheep scab led to the more general use of gamma H C H in many dips as a compulsory treatment; this has now to some extent been replaced by diazinon.

With the introduction of a Warble Fly Eradication Scheme the use of certain specified systemic O P 'pour-on' treatments, such as fenthion, phosmet, famphur and trichlorphon in the autumn, is generally recommended. Their use in the spring on affected cattle is mandatory.

Fly problems in animal houses can be prevented both by sprays, mainly based on pyrethrin aerosols, and residual treatments on surfaces, based on various residual O Ps and the newer persistent pyrethroids.

RISKS AND PROBLEMS OF PESTICIDES

As pesticides are necessarily toxic to some form of life they are likely to be toxic to a greater or lesser extent to other forms. The risks are (a) to the operator who applies the material and may have to handle the concentrate; (b) to the consumer who eats treated crops on which harmful residues may be present; (c) to wild life, which includes not only mammals, birds, fish and bees – creatures about which concern is usually expressed and for which standardised tests exist – but all other organisms in the environment (i.e. other than humans, the crop and the pest); and (d) to third parties, e.g. humans accidentally contaminated by walking through recently treated crops or sprayed from aircraft.

Of all these, risks to the operator are the most important, especially if he is working for long periods with some of the more toxic chemicals. For this reason users of certain chemicals included in the Health and Safety (Agriculture) (Poisonous Substances) Regulations are required by law to observe certain precautions, including the use of specified protective clothing when handling the concentrate and for some pesticides when applying the chemical. Some protective clothing can be uncomfortable to wear, especially in warm weather, and some pesticide solvents, notably xylene, may penetrate some types of gloves after a time: there is therefore plenty of scope for improvement. Even so, there have been no deaths due to the normal use of pesticides in the UK for many years and very few cases of illness. This contrasts markedly with deaths and accidents from farm machinery.

Risks to the consumer tend to be exaggerated, as, if pesticides are used according to instructions and normal practice, residues on food are nearly always minute even if they are detectable at all with the most sensitive methods, and are very unlikely to have any adverse effect. The most significant residues are likely to be found on crops such as lettuce which are eaten raw and may need to be treated shortly before harvesting. Even so, there are normally large built-in safety factors.

The problem of wildlife and the environment generally is complex and difficult to assess. In practice it is only possible to examine key representatives, if such can be identified, in a few important groups or situations, and even this demands costly and time-consuming effort, yielding results that are often difficult to interpret. Nevertheless, a great deal of work is going on and much progress has been made in developing techniques. These include careful observation of bird and mammal populations on more than one occasion before and after treatment of large areas, and the collection and analysis of residues in any corpses found and also in survivors representative of various groups including invertebrates (Bunyan & Stanley, 1980). Minimum requirements for most authorities include information for one or more species of birds, for one or two species of fish and often of aquatic invertebrates, bees and earthworms and possibly other soil fauna.

Risks to third parties are similar to, but much less than, risks to operators, though aerial spraying does create special problems; in practice these are more of a nuisance and an inconvenience than a serious hazard and only pesticides of lower toxicity are permitted to be applied from the air.

All developed countries have regulatory schemes designed to reduce these various risks by excluding or limiting the uses of products according to their known or potential hazard and by making recommendations for their safe use. Risks to humans are assessed primarily by animal experiments in which acute oral and dermal toxicities are determined for one or two species, invariably starting with rats. The standard test is the amount of the ai, and later the formulation, to kill 50% of the sample under standard conditions (LD 50); this is expressed as mg of pesticide per kg of the test animal (mg/kg) and the lower the figure, the more toxic is the chemical. Dermal toxicity, usually much lower than oral toxicity, is a particularly valuable measure as it relates more closely to the risk to the operator, and many pesticides can pass through the skin. These acute toxicities give only a preliminary guide and need to be followed by longer-term tests using lower doses, usually given in food, over 90 days and up to two years to study chronic effects, especially carcinogenicity, i.e. the risk of cancer. Since the thalidomide disaster, teratogenic, mutagenic and reproductive studies over three generations may be required. Other tests may cover metabolism in plants and animals to determine the fate of the original product, and any metabolites of, or impurities in, the original product may also require testing. Possible irritant effects on skin and eyes need to be investigated and also the possibility of sensitisation. Ideally 'no effect' levels are established for the major tests. Although most work is done with rats, mice and rabbits, acute toxicities are often obtained for a number of species and, if wide variation is found, the product is regarded with greater suspicion, and calculations are based on the assumption that man is at least as susceptible as the most susceptible animal.

The interpretation of animal experiments to assess the risk to humans is a matter of judgement and experience based on the amount operators are exposed to during application and the residues likely to be taken up by consumers of treated commodities. A large safety factor, usually about a hundredfold, is allowed. Practical use of a pesticide over many years without any detectable effects is the best demonstration of safety, though carcinogenic effects may not come to light for 20 years or more.

The Pesticides Safety Precautions Scheme (PSPS) regulates the introduction and use of pesticides in the UK. Briefly, firms wishing to market products have to supply detailed relevant information on all aspects of the use and potential hazards of their product and this information is considered, first by a Scientific Sub-Committee, and then by the Pesticides Advisory Committee which makes recommendations on safe use. Products are cleared in stages providing for trials, limited field experiments, provisional commercial clearance and, finally, full commercial clearance. More information is usually required at each stage. Although the scheme is not compulsory (as in many countries), in practice, by agreement with the industry, few if any products escape the net. In many countries, efficacy is part of the safety arrangements, and though the Agricultural Chemicals Approval Scheme in the UK is at present voluntary, proposals for linking the two schemes are under consideration.

Ecological effects on wildlife are extremely difficult to forecast. Though some effects of the persistent OCs were exaggerated, there seems little doubt that through concentrations in food chains, they resulted in the decline of certain birds of prey through complex physiological effects on the reproductive system. Biological persistence is taken into account in assessing the ecological hazards of new chemicals.

Residues on the aerial parts of plants are dispersed or broken down by volatilisation, the action of rain and sunlight, including ultraviolet light, and are also diluted by growth. Those which remain end up in the soil with that part of the application which was not retained on the crop. Here, most are broken down either chemically or biologically though some become attached for varying periods to organic or clay particles. The range of synthetic chemicals broken down by micro-organisms is quite remarkable, the OCs being exceptions. Although the possibility of pesticides affecting basic microbiological processes in the soil, e.g. nitrogen fixation, must always be borne in mind, present evidence is that pesticides in much larger amounts than are normally applied have little effect on important micro-organisms.

In addition to general ecological effects, there are particular ones which might affect the abundance of actual and potential pests in the way that red spider mites became a serious problem all over the world on fruit trees after the introduction of broad-spectrum and persistent insecticides. Even herbicides can have a profound effect on insect populations by destroying the hosts of plantfeeding species, which in turn reduces predator populations. Pesticides may also affect the physiology of plants and animals, making plants more or less susceptible to pests and sometimes affecting the behaviour or reproductive potential of pests or other organisms living on plants.

Resistance

Another problem is that of pests developing resistance to pesticides as a result of the selection of individuals which happen to possess resistance to a particular chemical. Such resistance is usually inherited, with the result that the proportion of resisters gradually increases, the rate depending on the rate of increase of the pest and its mobility, and on selection pressure, i.e. the proportion of the population exposed to the chemical. There are only a few recorded examples in weeds, but some are now resistant to the triazines in the USA. There is a potential problem with fungi, especially with systemic fungicides used against mildews and botrytis, but it does not appear to have developed as rapidly as was at first feared. The main concern is with insects and mites, and resistance to one or more insecticides has been noted for well over 200 species throughout the world, including a dozen or so in the UK. Insects and mites become resistant not only to the chemical to which they are exposed but to related chemicals and sometimes to chemicals of other groups through a process known as cross-resistance. Opinions differ about the future, some suggesting that before long all major pests will be uncontrollable by chemicals, whilst others point out that many pests have not developed resistance and that where it has occurred it is not universal. As usual, the truth probably lies between these two extremes but it seems to be only a matter of time before many more important pests become difficult to control. Despite a great deal of research, little progress has been made in solving the problem and regular changes of pesticide groups (ringing the changes) do not appear to hold out much hope except perhaps where large areas are controlled by one authority. Unnecessary use of pesticides aggravates the situation by increasing selection pressure, but it is not always easy to decide in advance when applications are necessary. Fortunately, the chemical industry has been able to find new chemicals with a different mode of action and has kept pace with many difficult situations, but the cost of developing new pesticides is becoming so prohibitive that it is unlikely that the pace can be maintained. Resistance has undoubtedly been a potent factor in encouraging the acceptance of biological control, notably in California and in the UK glasshouse industry.

Phytotoxicity

Phytotoxicity is, by definition, a feature of herbicides and, as explained earlier, selectivity is a matter of degree and a combination of several factors. Any pesticide may have an adverse effect on plant growth especially at certain critical stages or in particular weather conditions. Such effects may be obvious or detectable only by careful experiments and observations over many years. Apart from the treated crop, nearby crops may be affected by drift of droplets, or, in some cases, of vapour. Sometimes the effect may be the deposition of undesirable residues near to harvest, but the main risk is that of herbicides affecting a susceptible crop, usually adjacent but, in some meteorological conditions, crops some distance away may be affected. Present improvement of application methods should reduce, though not eliminate, the problem.

Different cultivars may react differently to herbicides and other pesticides. It has long been known that certain fruit varieties are 'sulphur shy' and that many varieties of ornamentals are very sensitive to particular pesticides, but, more recently, varietal susceptibility has been noted for cereals and other agricultural crops. This adds to the existing complications of testing procedures. A related problem is that of taint and because this is a very subjective quality, tests are based on statistically designed experiments involving taste panels. Tomatoes can be affected by vapour from chlorcresols at an appreciable distance and gamma HCH (and particularly the other isomers) can affect the taste of potatoes and blackcurrants. It often happens that taints not detectable in the fresh product are noticeable in the product when canned. Water authorities always have to bear in mind the possibility of pesticides tainting potable water though the amount of dilution is so great, even with aquatic herbicides, that an effect is very unlikely.

Risk benefit assessment

All these various problems tend to be exaggerated and many become an emotional rather than a scientific issue, so that some people advocate the banning of some or all pesticides and the substitution of other control methods. There is no doubt that pesticides have often been used unwisely and unnecessarily, and that cultural and biological methods have had some outstanding successes and many more will be achieved. Pests have a way of overcoming new obstacles and the rapidity with which new strains of fungi have broken down the resistance of new cultivars is a salutary reminder that we are never likely to find a permanent and perfect answer to any pest problem and that chemicals will almost certainly be necessary for the foreseeable future.

In order to assess the benefits of controlling pests, one must first consider the losses they cause. Losses due to pests of all types in the production and storage of crops worldwide are impossible to calculate with any accuracy but are accepted generally as being of the order of 30% (Cramer, 1967), probably rather more in the tropics and rather less in temperate and more advanced countries. Such calculations are based on the extrapolation of a small number of careful observations, surveys and experiments and a greater number of informed guesses. One of the more reliable estimates based on over 10 years of statistically designed surveys on cereal leaf diseases in England and Wales suggests that they cost the country about £100 m. annually, most of this due to mildew for which routine treatment can be justified (Cook et al., 1981).

Although the benefits of pesticides are easy to see where treated crops survive and untreated crops fail or are poor, it is difficult to quantify overall benefits, though in the USA it is claimed that for each dollar spent on pesticides there is an increase in a farmer's income of four dollars. Various detailed studies which attempt to quantify the benefits of pesticides have been summarised by Gough (1977) and it is reasonable to assume that, overall, benefits probably outweigh costs and unquantifiable risks by an appreciable factor. It is certain that pesticides will continue to be the main means of controlling pests for a long time and that, with more rational use, including integrated control, many of the problems could be reduced.

One point must be emphasised. All pesticides should be handled with care and the golden rule is to read carefully and follow the instructions on the label and not exceed the recommended rate. Empty containers should be rinsed out into the tank and disposed of safely and machinery should be cleaned after use. It is easy for familiarity to breed contempt and commonsense precautions should be observed even with what appear to be the most innocuous chemicals.

The agrochemical industry

The agrochemical industry has been described as convergent in that it is unified by the manner of use of its products and not by the starting point, though many products are derived from petrochemicals. Only a small number of firms, and these mainly multinationals, have the capacity to develop new pesticides, since the cost of one is now about £10 m. and it may be ten years or more before the costs of a successful product are recouped.

So far nearly all pesticides have been discovered by chance or by the systematic screening of thousands of chemicals, chosen more or less at random, for biological action by a range of relatively simple tests. If any of these indicate useful activity they are subjected to further tests and related compounds are synthesised and also tested. Thus after the discovery of the phenoxyacetic acid herbicides MCPA and 2,4-D, phenoxypropionic and phenoxybutyric acids were natural subjects for investigation. If preliminary toxicological studies indicate that further development is justified, larger field trials and more detailed toxicological work is undertaken. Finally, the difficult and irrevocable decision has to be made to go from the pilot plant stage to full-scale commercial production. Only one in many thousands of chemicals originally screened is likely to become a commercial success and the proportion has declined steadily over the years as regulatory schemes have become more demanding.

There is an increasing interest in a more selective approach as biochemical and physiological knowledge of the mode of action of pesticides is better understood but the possibility of designing pesticides for particular purposes seems a long way off. The cost of development is such that only pesticides useful on major crops or livestock over vast areas are likely to be profitable and the prospects of developing more or less selective pesticides for limited uses or limited areas like the UK are very remote unless they are subsidised.

Pesticide output and sales. World output of agrochemicals increased from about 100,000 tonnes in 1945 to 1·8 m. tonnes in 1975. Total sales of these and animal-health chemicals were estimated in 1979 at US $ 14,500 m. of which about two-thirds related to agrochemicals. The animal-health chemicals would include vaccines and vitamins as well as the therapeutants and additives covered in this chapter. Of the agrochemicals, herbicides accounted for over 40% and insecticides for about 35%.

For the UK the value of crop agrochemicals in 1980 was £393 m. slightly more than in 1979. The tonnage produced, amount exported and area of arable crops treated are given in Table 15.1.

Table 15.1   UK pesticides: production and use on arable crops (1980)

	Herbicides	Insecticides	Fungicides	Growth regulators	Others
UK sales £m.	124·8	22·1	38·7	3·5	5·5
Exports £m.	111·2	61·6	7·1		5·6
Area treated ('000 ha)	6,357	776	3,274	517	368

As each treatment is recorded as a unit hectare and many fields are treated more than once, the total can be larger than the total arable area involved. The value of ectoparasiticides was £4·3 m.

The value of UK pharmaceuticals in 1979 (supplied by the Association of the British Pharmaceutical Industry) was £13 m. for medicinal feed additives, and £48 m. for pharmaceuticals and biologicals. An estimated £10 m. can be added to these figures, making a total of £71 m., if sales from manufacturers who did not participate in the survey are included.

Production in all sectors has increased fairly steadily since the Second World War though there was a slight reduction in 1980 associated with the recession. It is worth mentioning that research costs are higher than in industry generally and are around 7–10% of total costs.

16

MACHINERY AND ITS USES

TRACTORS AND POWER

Although the development of the farm tractor has, from its beginnings in the steam engine, been evolutionary rather than revolutionary, the diversity of designs has increased rapidly in the 1960s and 1970s. The *general-purpose tractor*, with two driven wheels at the rear, still dominates and is available from 10 to 100 kW to suit farm size and tasks. For large arable enterprises larger, usually four-wheel drive, *cultivating tractors* provide power in the range 75–300 kW. These are best suited to draught operation, are less manoeuvrable despite often being of articulated frame design and wasteful of fuel in low power tasks. Specialised *tool carriers* for rowcrop work with implements mounted within the wheelbase give the operator improved view and hence more precise control. *Transport tractors* with wheel suspension permit road speeds of up to 80 km/h and improve economy where much materials handling over roads is involved; they particularly suit some contractor tasks.

Power from the engine is consumed within the tractor to operate hydraulic pumps for hydraulic lift, steering and other services. Useful power to the user is that available at the p.t.o. or for draught operation. The optimum power needed will depend on farm size, soil and crops (Table 16.1) but calculations suggest operating costs on high-power tasks are altered only by a few per cent over a wide power range. For example, with 200 ha of cereals, operating costs may vary only by 5% between a 50 kW and 150 kW tractor. Matching of tractor power and implement size is vital, with a mismatch likely to reduce workrate by 20%.

Table 16.1 Factors influencing optimum size of tractor

Factor	Effect on tractors
Farm size	Generally larger for large farms
Soil type	Heavier soils require larger tractors
Crops grown	All cereals – larger tractors
	Cereals/roots – smaller tractors
Forage methods	Larger tractors for silage than for hay
Cultivation methods	Smaller tractors for direct drilling and reduced cultivation

With high efficiency and good life, diesel *engines* are almost universal. Fuel efficiency varies by 30% from most economical to least, so that the specification should be studied. The specific consumption varies greatly over the operating range of the engine, which implies that the tractor should be driven in as high a gear as possible to minimise engine speed and maximise torque. So that the engine may be able to negotiate local tough areas when ploughing or cultivating, a good reserve of torque (> 12%) should be available above that at maximum power (Fig. 16.1).

Ground drive may be through two, four or more driven wheels, or through tracks. Generally the tracked tractor (a) has 20% higher tractive efficiency; (b) is driven nearer to its maximum power; (c) damages soil structure less; (d) costs less in maintenance than tyres

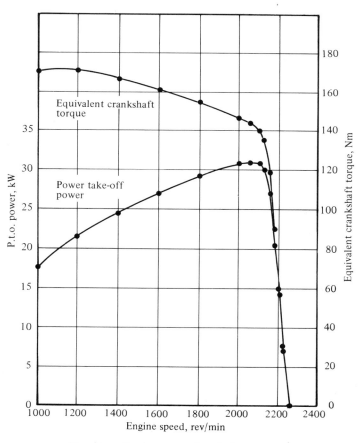

Fig. 16.1   Engine torque in relation to speed

Table 16.2 Comparative tractive efficiencies of two- and four-wheel drive tractors

Tractor type	Typical tractive efficiency %
Two-wheel drive	59
Four-wheel drive with smaller front wheels	63
Four-wheel drive with large equal wheels	67

on flinty soils but more on abrasive soils; and (e) presents transport problems between fields. Difference in tractive efficiency between two- and four-wheel drive is significant (Table 16.2), although a two-wheel drive model may be improved by dual wheels where practicable. For draught operation, ballast should be added to optimise efficiency. Figure 16.2 shows that for cultivation at 6 km/h, the driven wheels should carry a load of 100 kg for each kW of power. Soil compaction may affect both yield and the power needed for tillage; it is minimised by fitting larger or dual wheels, reducing inflation pressure to the lowest permitted by the load and travelling as fast as possible.

Implement coupling may be by the clevis drawbar, pitch-up hook

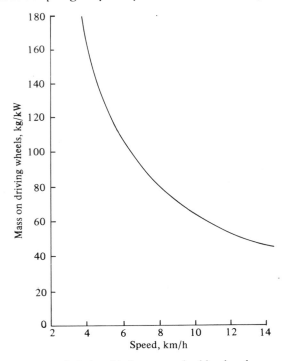

Fig. 16.2   Relationship between wheel load and power for highest traction efficiency

or three-point linkage. Automatic couplers for the three-point linkage permit rapid and safe coupling, even of heavy implements, by one man. Increasingly, the economic advantage of carrying out two implement tasks at the same time is being realised and the fitting of a linkage to the front of the tractor is permitting many useful combinations (Fig. 16.3). Three categories of linkage dimensions

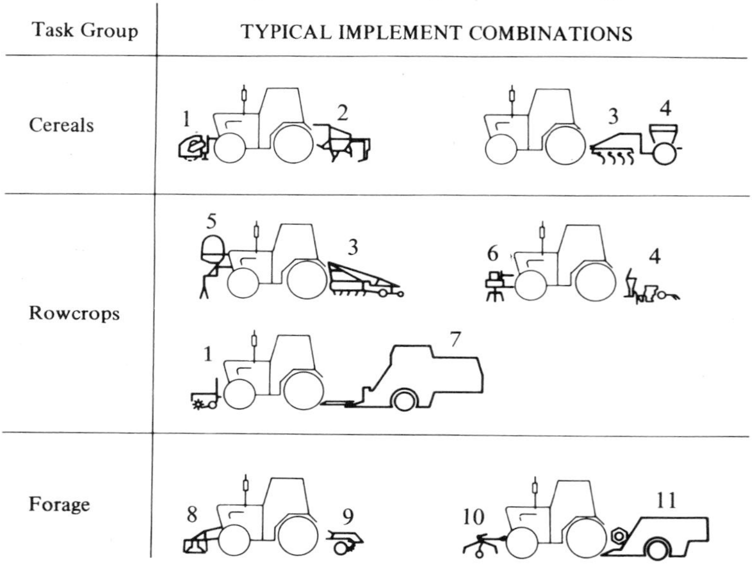

Task Group	TYPICAL IMPLEMENT COMBINATIONS	
Cereals		
Rowcrops		
Forage		

Fig. 16.3   Implements shown are 1, topper; 2, cultivator; 3, harrow; 4, drill; 5, sprayer; 6, tiller; 7, lifter; 8, mower; 9, conditioner; 10, tedder; 11, baler

cater for varying tractor sizes; similarly the power take-off (p.t.o.) shaft is categorised according to power (Table 16.3), with many tractors offering both Type 1 and Type 2 drives. Hydraulic power for operation of rams and motors external to the tractors is usually limited to 5–10 kW with an increasing number of auxiliary valves being optionally available. In matching tractor to implement, the lifting force available on the three-point linkage must be con-

Table 16.3   ISO classifications of three-point linkage and power take-off

Three-point linkage		Power take-off	
Category	Maximum drawbar power, kW	Type	Maximum p.t.o. power, kW
1	35	1	48
2	75	2	92
3	over 70	3	185

sidered, as must stability of the tractor when carrying a lifted implement.

Ergonomics and *safety* features have assumed increased prominence. In the UK all new tractors and all tractors driven by employees must be fitted with an approved protective cab or frame to prevent crushing in roll-over accidents. Noise level at the driver's ear is also covered in law with a maximum of 90 dB A. Seats mostly incorporate a sprung suspension with damper and, together with the noise limit, these encourage faster work and reduced fatigue. Other features of the workplace which should be considered at purchase are listed in Table 16.4. For use with toxic chemicals, cab

Table 16.4   Ergonomics features of importance on tractors

Seating and postures	Noise
Control design and layout	Vibration
Control identification	Cab climate
Guarding of moving parts	Visibility
Access to workplace	Protection from overturning

sealing and air filtration are vital. Full air conditioning is not normally considered necessary in the UK provided that windows can be opened.

Performance of individual tractor models is available from independent tests carried out by a station working to the OECD code.

For machines originating in the USA, a test may have been made to the Nebraska state code which also carries independent status. The OECD test will include measured values of p.t.o. power, draught performance, hydraulic lift capacity, hydraulic power available, brake performance, noise levels, turning circle, centre of gravity position and a brief check of transmission durability. There will be some fall-off in performance throughout life, but with correct maintenance, particularly in servicing of air filters, fuel injectors, engine and transmission lubricants and tyres, the loss of power should be no more than 5–10%.

Table 16.5   Main approximate costs of tractor operation

	%
Fuel and oil	40
Depreciation	30
Interest	15
Repairs	10
Tax and insurance	5

TILLAGE

The *purposes* of soil-tillage processes are to create a soil environment in which plants can easily grow roots and shoots and be provided with nutrients, to remove or restrain competing weeds and to permit crops to be harvested easily and in good condition. For cereals, particularly those planted in the autumn, successful application of direct drilling has shown tillage to be unnecessary on certain soils if weeds are controlled chemically. The choice is then, where appropriate, of traditional tillage with primary cultivation to perhaps 20 cm depth, followed by treatments to produce a seedbed to 10 cm depth, of 'reduced' cultivation with less treatments and to a lesser depth, and of direct drilling. Trials have shown that yields can be similar and costs reduced by the newer techniques, but farms will probably need to revert to more traditional tillage on occasions when soil or weather demands, so that most are likely to require a range of implements. Both for primary tillage and for secondary or seedbed preparation, a choice of draught or p.t.o.-operated machines is available. The former generally have the advantage of simplicity, although typically 20–30% of the tractor power is lost in wheelslip or rolling resistance in hauling a draught load. P.t.o.-operated machines may thus be more efficient at a greater capital cost. For primary cultivation the choice includes mouldboard plough, rotary cultivator, rotary digger, heavy disc harrow, chisel plough, 'flexitine' or heavy cultivator.

The *mouldboard plough*, although slowly being displaced by reduced cultivations, is the most effective implement for burying weeds and relevelling a rutted surface. Alternative body types and the features of their work are given in Table 16.6. Plough design

Table 16.6   **Mouldboard plough types and their characteristics**

Lea bodies	Turns an unbroken furrow slice, using long mouldboard
Digger bodies	Shorter mouldboard turns furrow more sharply and breaks up furrow
Semi-digger bodies	Intermediate, normally with some breakage of furrow
Shallow or 'stubble' bodies	For working only 75–100 mm deep, normally 250–300 mm wide

variants include *reversible ploughs* which, by having two sets of bodies with reversed attitude, permit all furrows to be layed the same way even though created in to-and-fro motions across the field; *semi-mounted ploughs*, normally with five or more bodies which incorporate a depth control wheel at the rear with an hydraulic ram to lift the plough out of work; and *disc ploughs*, having advantages in hard dry soils, with large concave discs scooping over the soil. Principal plough parts are shown in Fig. 16.4.

The *furrow press* has regained some earlier popularity where consolidation is needed to maintain moisture and soil contact

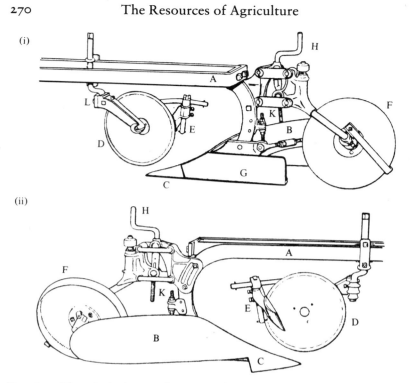

(i)

(ii)

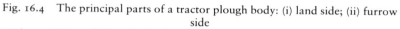

Fig. 16.4   The principal parts of a tractor plough body: (i) land side; (ii) furrow side
(A) beam, (B) mould-board, (C) share, (D) disc coulter, (E) skim coulter, (F) rear wheel, (G) landside, (H) rear wheel adjustment, (K) pitch adjustment, (L) coulter tilting adjustment

between the upturned furrow and soil below to ensure roots grow down.

Rotary cultivators are of two main types, the higher rotation speed 'Rotavator' which relies on the combined actions of cutting soil and of mutual impact of clods to obtain a fine tilth in a single pass, and the low rotation speed rotary digger which cuts and shears clods and, after inversion, leaves a rough tilth for weathering. The digger also incorporates tines to further disturb the soil below the level of the rotor. The rotary cultivator controls weeds by chopping them into small pieces. It also reduces crop debris such as Brussels sprouts' stalks to small pieces which may be incorporated in the tilth. The rotary digger, whilst not inverting surface soil and plant matter as effectively as the mouldboard plough, does incorporate some loose straw and other plant matter.

Disc harrows may be employed for primary cultivation or for seedbed preparation. Heavy types with two rows of discs of 50–60 cm diameter are used for primary cultivation on straw stubbles, where they function more effectively on dry soils than on wet. In hard conditions weights are added to aid penetration; scalloped discs also increase penetration. With a two-row set the most typical setting is for the front row to be 20° and the rear gang 15° on the opposite inclination to travel direction.

Tined harrows (often called cultivators) for primary tillage, cover the range from heavy rigid-tined implements (*chisel ploughs*) to spring-tined models (often called *flexitines*) which are normally used in two or three successively deeper passes. Tines are mounted in two or three rows fore to aft and staggered to permit debris and larger clods or stones to pass without blockage. Typical lateral spacing is given in Table 16.7. Wear of tine tips is economically

Table 16.7 Typical lateral spacing of tines on cultivators and harrows

Implement	Typical spacing of tines (mm)
Subsoiler	1,000
Chisel plough	300
Heavy Cultivator	200
Spring-tine cultivator	100
Drag harrow	50

important and replaceable – in some cases reversible also – tips are fitted. Tip life varies from 50 to 500 hours depending on soil. For sufficient break-up of soil in primary cultivation, two or three passes are usually necessary. Soil inversion and weed burial is still not as complete as with a mouldboard and energy needed for 2–3 passes will equal that of a mouldboard plough working to the same depth. A 'stubble cleaning' process with tines working to 10 cm, sometimes prior to the main tillage, causes weeds to germinate so that they may be destroyed in the later process.

For seedbed preparation lighter rigid or spring tines may be used in single or repeated passes. Both high tractor speeds and the spring action of the tines aid clod disintegration by impact. As tines penetrate less deeply and into already disturbed soil, these implements may be wider, sometimes with hinged sections for road transport. Final preparation of a fine seedbed or later covering of seed may be by *drag harrows*; normally made up of 6–9-m wide sets of flat zigzag frames with staggered 10-cm long vertical spikes. Recent developments have been of hydraulically operated frames to lift harrows or of hydraulic means of aiding penetration and soil levelling (*pressure harrow*). For the firm level seedbeds needed in beet growing, a heavy wooden-framed *dutch harrow* with vertical spikes is often used.

Powered harrows include rotary and vibratory types. Either the spiked rotary cultivator with a basic chassis similar to the 'rotavator', but blades replaced by 12-cm long straight spikes, or the rotary harrow with vertical tines rotating about a series of vertical shafts across the machine, are recommended when a seedbed is needed in one pass following primary tillage, particularly where clod disintegration is difficult. Vibrating implements vary from those with heavy tines which are vibrated at frequencies of 10–30 cycles per second to break more soil (at the same time normally also reducing draught by 10–50%), to light spiked harrows in which the spikes are transversely reciprocated over 10–20 cm at 1 cycle per second or less to increase soil stirring. Not powered from the p.t.o. but rotated by passage over the soil, implements such as the *turbotiller* have four, six or eight horizontal shafts fitted with bladed rotors similar to fans, which pulverise surface layers, penetrate well and are suited to higher speeds.

For *rowcrop* work, implements are available for creating and maintaining ridges and for surface weeding and cultivation between rows. Ridge creation for potatoes is commonly by ridging mouldboards attached to a toolbar behind, amidships or in front of the tractor; concave discs may also be used and more recently p.t.o.-driven rotary ridgers have been introduced to reduce ridge surface smear and hence clods. For inter-row weeding, wide-bladed tines are common and, although both p.t.o. and soil-driven rotary machines are sold for working between rows, their advantages are small.

Rolls are employed for consolidating soil, crushing clods and smoothing the surface. Two main types are 'flat' rolls and 'Cambridge' or ring rolls. The former consist of steel or cast-iron cylinders, 50–80 cm in diameter. In the heavy roll, they may be filled with concrete and provide sufficient weight to improve the smoothness of pasture surfaces for forage work, particularly by pushing stones into the soil where they are less likely to damage machines. Cambridge rolls incorporate a row of ribbed cast-iron segments 5–8 cm wide and loosely free to turn on a common axle. The ribs increase disintegration of clods and the looseness of individual rings helps avoid the adhesion of soil. All but the heavy rolls are normally used in gangs of three to make up a width of 6–9 m. Means exist of automatically arranging the gang for field work or transport, without heavy manual effort. One specialised type, the *crumbler roller*, which consists of bars arranged in cylindrical form, may be used to break soft clods or provide a depth reference.

Subsoilers, designed to break up a compacted layer below normal cultivation, consist of one or two tines capable of working to a depth of 60 cm and requiring tractors of high draught capacity. The 'paratyne' or 'paraplow' versions with angled tines are designed to lift and open a greater section of the soil without creating a ridged surface.

Combinations of tillage implements, often including seed drills, are recommended as a means of reducing labour and fuel by fully loading the tractor and reducing the number of passes. One pass tillage and planting may be achieved by combining disc harrow or cultivator with a cereals drill; the same combination may also follow a primary cultivation. Powered cultivators or harrows may be combined with drills whilst a whole range of tilth-producing implements such as cultivators, disc harrows, drag harrows and crumbler rollers may be combined under suitable coupling frames.

PLANT ESTABLISHMENT

Cereal drills are almost entirely of the random-distribution type although limited precision drilling of wheat and barley has taken place with some evidence of increased yields – generally following spring sowing. The main components of a cereal drill are the hopper, a metering device, tubes to a coulter and a suspension for the coulter to provide for its even penetration into the soil.

The most common metering or 'feed' mechanisms are the fluted roller (external force feed) in which adjustment of seed rate is by sliding the shaft carrying the rollers axially so that a greater or shorter length of flute is aligned with the outlet and the studded roller with which rate is generally varied by adjusting rotational speed. Both types are suitable for a wide range of seed sizes. A resilient roller, of 'Neoprene', which runs against a fixed plate, and into which seeds penetrate, is also used successfully. Coulters may be of 'Suffolk' or shoe type or of single disc design, the latter normally providing more effective penetration in hard soils and being also less prone to blockage in wet soils and by plant debris. Traditional spacing of coulters is 18 cm but there has been an increased adoption of 12 cm spacing on evidence that a small yield advantage is possible.

Drill layouts differ between the traditional design with a linear hopper over the width of coulters, gravity feed to the coulters and large wheels at either end, and the 'pneumatic' drills in which a central hopper is mounted on the tractor three-point linkage and seed is transported to the shoe coulters pneumatically through flexible pipes. Combine drills with two hoppers and feed mechanisms permit fertiliser to be applied during sowing.

Broadcasting of seed, generally restricted to autumn sowing, has been adopted by enthusiasts who claim faster work and satisfactory yields. With optimum soil moisture and temperature, the process is successful despite greater variability in depth and need for more careful covering.

For *direct drilling* a more penetrating coulter is needed and the most common is the triple-disc type in which the first disc cuts a slot which the following angled discs open to permit seed insertion. Greater weight is needed also for penetration and hence also a more powerful tractor.

For monitoring application rates, an electronic instrument to measure the area covered (*Acremeter*) is valuable. Another aid to precision in later fertiliser application and spraying is the establishment of '*tramlines*' along which tractors can be driven for these tasks. They are established most conveniently during drilling when appropriate coulters are blocked off to leave unplanted strips.

Precision drills are normally built as units for individual rows and ganged on a tool frame to suit the tractor or plot. They employ either an endless belt incorporating spaced holes to transport single seeds from the hopper to where they can drop to the ground, a 'cell wheel' on which a number of pockets carry the seeds, or a wheel on which seeds are carried over small holes to which they are attracted by vacuum. A refinement, capable of correct spacing at higher speeds, is the double cell wheel design in which seeds are accelerated by the second wheel to be dropped at a speed corresponding to that of the passing soil. Precision drills are fitted with press wheels to firm the surface after planting. Electronic monitors are available to warn the driver of planting faults.

Potato planters may be divided into semi-automatic and automatic types. The former rely on an operator to place tubers in cups which then carry them to the soil. These are suitable for chitted seed since damage is low, but are restricted to a maximum of 3 km/h, with each operator dealing with 120 tubers per minute. Automatic types, while capable of working at up to 10 km/h, can damage chitted seed. They sometimes employ one operator to fill in gaps as tubers pass on a belt. Most planters may be fitted with facilities for adding fertiliser or granular insecticide.

Transplanters are used for brassica crops. Although completely automatic types are under development, choice for the user is between 'hand fed' and 'mechanised'. With the former the machine opens a slit into which plants are manually inserted by an operator riding on the unit; the machine closes the slit and firms the soil. With the mechanical unit, the operator bends less to place the plants in grips on a rotating wheel, which carries them down to the soil. Units are for single row work and are combined on a tool frame which may take up to eight units. Watering attachments may also be added.

Grass seed may be sown into a fine seedbed by a general purpose 'cereals' drill with narrow coulter spacing, particularly the studded roller type. For the reseeding of pastures a direct drilling method is available in which the implement dispenses herbicide to kill competing grass in the vicinity of the row, cuts a slot and sows seed into the turf.

PLANT NUTRITION AND PROTECTION

Fertiliser distributors for solids may be (a) full-width type with the hopper straddling the bout, between a pair of wheels. Metering is by rollers or agitators, the latter combined with variable sized apertures to vary rate; (b) bulk hopper types combined with a wide boom along which delivery points are pneumatically fed from the hopper. Deflector plates at each point aid evenness of distribution. Typical widths are 10, 12 or 15 m and the pneumatic conveying may require 6–8 kW from the tractor p.t.o.; (c) bulk hopper types with spinning disc or oscillating spout mechanisms to distribute the material over bouts of 6–15 m. With this latter type, correct overlapping to ensure evenness of distribution from the addition of two individual distribution patterns is important (Fig. 16.5).

Liquid fertiliser may be applied by the larger, trailed types of boom sprayer, but requires expenditure in special on-farm storage tanks. Below-ground placement is often associated with drilling or planting with transport tanks on the tractor and attachments to the implement. Specialised direct injection of liquid or gaseous ammonia requires medium to high tractor power and involves low workrate.

(For *farmyard manure* spreading, see later section on livestock equipment.)

Spraying for crop protection includes *herbicides* to reduce competition from weeds, *fungicides* to reduce the risks of fungus diseases and *insecticides* to kill pests. In addition chemical applications may be made to shorten and strengthen straw and to destroy potato haulm prior to harvesting. Different techniques apply to ground crops such as cereals and rowcrops and to orchard or bush crops.

Spraying machine components include a *tank*, increasingly of plastic to permit more convenient shapes and resist corrosion; *mixing* either by paddle or pumped circulating flow; a *pump* of either diaphragm, centrifugal or low pressure rotary design; and a *boom* to which *nozzles* are attached at regular intervals. The boom is in sections which fold for transport, manually on small machines and hydraulically on large ones, which may incorporate five sections.

Distribution is affected by motions of the boom which, being long, is prone to amplify the motions of the carrying vehicle. Thus booms are

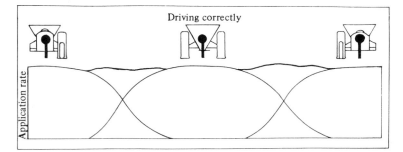

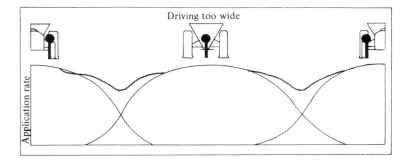

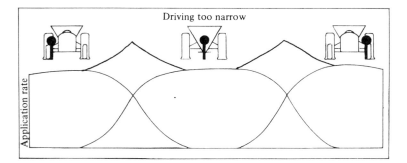

Fig. 16.5   Correct and incorrect overlapping of fertiliser distributor bouts

often attached through spring suspensions. Nozzles generally incorporate filters and vary in design according to operating pressure and volume. On wider sprayers nozzles on individual boom sections may be switched off to complete narrower strips of field.

Sprayer types divide into mounted tractor sprayers, usually having booms of 5–12 m attached to the frame also carrying the tank, trailed types with tank and a 10–18-m boom on a chassis fitted with large diameter narrow pneumatic-tyred wheels for good crop clearance, and self-propelled sprayers. The latter are based on a tractor chassis with the operator's cab at the front for best vision and with large narrow wheels for a crop clearance of 60–80 cm. Tank capacity may be 2,000 l, and boom width up to 24 m, with cost and workrate making these suitable only for large farms or contractors.

Low ground pressure sprayers are specialised vehicles suitable for running over cereals in late autumn or early spring when conventional tractors would cause excessive soil damage. The vehicle, incorporating a 6–8-m boom may weigh as little as 300 kg unladen and be fitted with tyres with an inflation pressure of only 0·2 bar or with low ground pressure tracks.

The most common *orchard sprayer* type is tractor hauled with a 1,000-l tank and 100 l/min. max. output fed through nozzles on both sides of the machine into the airflow generated by a large-diameter axial flow-fan to propel the droplets into the trees. For small-scale work the knapsack sprayer with a hand pump or small combustion engine or the electrically powered spinning disc hand-sprayer are available.

In *operation* of sprayers it is vital to (a) thoroughly wash all components as many chemicals used are corrosive; also, traces applied to the wrong crops could have disastrous effects; (b) ensure filters are regularly cleaned if spray is to be effectively applied continuously; (c) wait for suitable weather, remembering that fine particles can drift several hundreds of metres in a light wind; and (d) observe safety precautions, bearing in mind that many chemicals used have either an immediate or an insidious toxic effect.

FORAGE HARVESTING

Forage may be conserved as hay or silage, which is also made from other crops, notably maize. This section deals jointly with the cutting and early treatment of grass for either end product, but separately with later processes in hay or silage making. The making of conventional and large bales is included although the latter are more common with straw than hay.

Although three fundamentally different designs of *mower* – the reciprocating knife, rotary and flail types – have been used extensively, the rotary type is now dominant. *Drum mowers* (see Fig. 16.6) normally incorporate two rotating drums, each with several pivoted knives at the lower edges and driven by a shaft and gears or, exceptionally, by a V-belt at the top of the drum. Widths of cut available are from 1·4 to 3·0 m. Cutting height is maintained by skid discs or plates at the bottom of the drum shafts. *Disc mowers* have knives of similar type mounted on smaller rotors – four for a

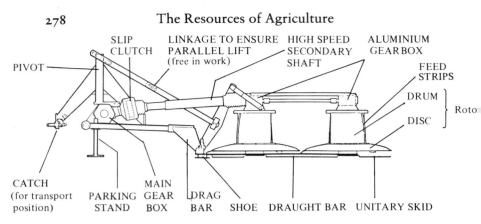

Fig. 16.6   Two-drum mower showing components

1·7-m machine – with the drive through an oil-bath geartrain below the discs. Both types function at up to 15 km/h. *Flail mowers* employ swinging cutting knives attached to a horizontal shaft. They are less susceptible to damage than drum or disc machines and are thus employed for rough grassland work. They do give a conditioning treatment to a crop but are generally more power demanding and slower than the other rotary types.

Conditioners for use together with or after mowing may be of flail, crusher, crimper or spoke rotor type. The combination of a rotary mower and a spoke or flail conditioner designed to scrape the crop surface lightly is becoming common. Crusher and crimper conditioner incorporate a pair of rollers running much faster than the forward speed of the mower. The lower roller lifts the crop and the roller pair squeezes it to break cells, so aiding drying.

After initial cutting and forming of swaths subsequent treatments are by tined implements. For hay in particular, frequent treatment by a *tedder* aids drying. Available implements fall into the classes of specialised tedder or general-purpose haymaking machinery with the additional capability of turning, scattering and windrowing. P.t.o.-driven tedders which lift, separate and fluff the swath, with deflectors allowing it to be moved sideways if needed, probably give the most effective treatment.

Combined haymaking machines offer a variety of actions. The main types are machines with near-horizontal rotors, usually p.t.o. driven, or finger wheel machines with independent obliquely fitted wheels driven by ground contact. For creating a windrow simple and multiple unit rotary rakes are available in addition to finger wheel rakes. Most are capable of speeds in excess of 10 km/h.

The *forage harvester*, whose performance is vital in the making of good silage, has evolved to form three main classes. The *flail-*

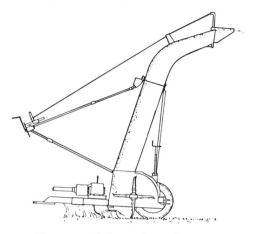

Fig. 16.7   Flail-type forage harvester

type (Fig. 16.7) consists of a high-speed flail rotor on a horizontal shaft and passing closely to one or more shear plates which cause crop laceration. The treated crop is guided by a hood to a chute about 2·5 m from the ground to be impelled into a high-sided trailer. Harvester and trailer may be towed in line with, or offset from, the tractor, care being needed to ensure stability on sloping land. A second tractor takes filled trailers to the silo. A flail harvester is capable of a 16 tonnes/h output from 15 kW p.t.o. power. A *double-chop* harvester also incorporates an auger to move the crop by the flail unit to a second rotor from where, after a second, more precise chopping, it is impelled through the chute. The crop is more finely chopped but, due to randomness of presentation, variation in chop length is still high. An output of 16 tonnes/h requires 25–30 kW p.t.o. power.

Metered chop harvesters (Fig. 16.8) incorporate a pick-up and feed mechanism so that the grass is fed uniformly and at a con-

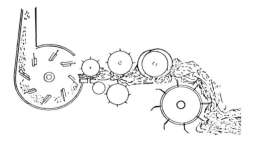

Fig. 16.8   Metered chop type

trolled velocity into a chopping cylinder. Chop lengths are uniform, aiding mechanical feeding to animals. Power need is similar to the double chop type for the same length but increases with finer chopping which may be to as little as 5–10 mm. Metered chop machines may be adapted for whole crop maize and be of *self-propelled* type.

Carted forage will be made into silage in a clamp or a tower silo. For the former a tractor-mounted *buckrake* will be needed to move silage onto the clamp and spread it evenly. The tractor is likely to be used also for consolidation. A *dump box* and *silage blower* will be needed for filling a tower silo. The dump box allows the contents of simple trailers to be instantly off-loaded, and provides means of slowly feeding the crop into the blower. The blower, powered from a tractor p.t.o. or by a 30–50 kW electric motor lifts the crop to 25 m or more.

Once hay is dry enough – ideally 20% moisture content or less – its handling is usually by pick-up baler making rectangular bales of about 20–30 kg weight. Modern *pick-up balers* are p.t.o. driven and incorporate a 1·3-m wide tine-bar pick-up, a cross-feed which moves the hay sideways into the compression chamber, where the ram working at 75–100 cycles per minute forces successive slices of material into the bale chamber against the resistance of adjustable spring-loaded pressure plates. After a preset length has been compressed, the knotter is tripped and the bale tied with two lengths of sisal or polypropylene twine. Bale lengths may be adjusted to suit subsequent stacking; it is important to monitor and adjust bale density throughout the day as crop moisture and toughness change.

If the hay is dry enough it is possible to carry the bales from the field right away, in which case a specialised *bale trailer* with pick-up and endless-chain conveyor carrying bales upward to a four-layer spiral may be used. However, it is more common for bales to be stacked in the field for further drying and for this the first step is generally a bale accumulator.

A *bale sledge* attached to the baler collects bales until the tractor driver trips the release; in this way bales are bunched for manual stacking. The 'flat-8' or 'flat-10' *bale accumulators* guide bales to lie in a regular array. Accumulators need to be paired with impaler- or gripper- type *tractor loaders* which are used to stack the grouped bales in the field, normally six or seven layers high. Later the loader can transfer bales to a trailer. Also for handling from the field, groups of 40–80 bales may be carried by (a) tractor rear-mounted bale carriers with hydraulically operated gripping 'gates'; (b) high lift carriers with hydraulic gripping on a rough terrain fork-lift; and (c) specialised transporters with a hinged body, which may be slid under and around a bale stack and then tilted horizontally for transport.

Another option is the use of the *bale packer* behind the pick-up baler; this implement ties together groups of 20 bales before they are released as a field stack.

Big bales have been developed so that farmers can take advantage of mechanised handling. The press-type *big rectangular baler* produces bales of 2·4 × 1·5 × 1·5 m weighing with hay about 500 kg. The machine requires about 45 kW from the tractor's p.t.o. and will pick up hay at up to 15 km/h, although the tractor and baler are stopped while the bale is tied. The *big roll baler* produces denser bales of varying size up to 1·8 m diameter and 1·5 m width, weighing up to 600 kg with hay. Speed of pick-up is similar to the rectangular baler and again the outfit must stop to allow the bale to be tied and subsequently ejected.

To handle large bales, specialised tractor-loader attachments and long, low trailers are available. If high stacking is needed, a fork-lift is a necessity. For traditional small bales a *bale elevator* is important if bales are to be manually stacked into barns.

Artificial or *barn drying* of hay is practicable although high energy costs make it unattractive economically. In view of the large number of possible methods, the equipment is not described here.

CEREAL HARVESTING

This section includes field harvesting and grain handling and drying equipment. Straw may be collected by the kind of balers described in the previous section: otherwise, straw spreaders and choppers, described here, are two options.

The *combine harvester* (combine) is used mainly to cut and thresh cereal crops, although it may also be used to pick up and thresh previously cut and swathed crops.

The principal parts of a combine (Fig. 16.9) are the reciprocating knife *cutter bar*, the *reel* which moves cut crop to the *feed auger* which moves it to the centre of the *table* to the *feed elevator* conveying it upward to the *threshing drum*. The drum by which most of the separation of grain from straw is achieved consists of a cylinder of beater bars rotating within a *concave* with a small clearance decreasing from the front, where grain enters, to the rear. Additional *beaters* may be sited before and after the drum to assist feed and separation. From the concave the straw passes to *shakers* (or walkers) which oscillate to separate remaining loose grains. The grain passes to *sieves*; at the uppermost an air stream separates chaff and light fractions. Material which does not pass through the first and second sieve but contains unthreshed ear fragments is returned to the drum for rethreshing. Straw is discharged to the rear from the shakers whereas grain is conveyed to a typically 2,000-kg-capacity *grain tank*. From the tank it may be transferred to a trailer by the *unloader auger*.

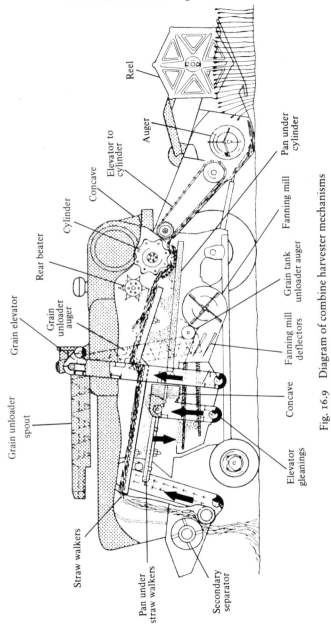

Reel

Auger

Elevator to
cylinder

Concave

Pan under
cylinder

Cylinder

Rear beater

Fanning mill

Grain elevator

Grain
unloader
auger

Grain tank
unloader auger

Fanning mill
deflectors

Grain unloader
spout

Concave

Straw walkers

Elevator
gleanings

Pan under
straw walkers

Secondary
separator

Fig. 16.9   Diagram of combine harvester mechanisms

Although tractor-operated combines are used in some countries, the self-propelled machine with cutting widths from 3 to 7 m is almost universal in the U K. Diesel engine power, from 50 to 100 kW is usual, and most modern machines are provided with continuously variable transmission. *Cabs* often incorporate air conditioning and air filtration to protect against dust.

Combine harvester *capacity* is difficult to define, being more dependent on straw throughput than grain. Figures quoted are normally for ideal crop and weather conditions, so in calculating seasonal capability with some margin it is reasonable to reduce these figures by half and assume that 150 hours of combining weather are available during the season.

> In the *rotary* combine, the traditional threshing drum and straw shakers are replaced by a longitudinal cylinder in which all threshing and separation takes place. Tests have shown that this design may have a larger throughput for a given size of machine. Specialised combines are available with bodies which are self-levelling for *slope working*. Alternative *headers* (cutting and collecting tables) are available for maize. Particularly for oil-seed rape and for peas *swathers* or *cutters* are available to permit further drying of the crop in a windrow.

Straw burning or incorporation in the soil in subsequent cultivation is aided by spreading or chopping. A *straw spreader* may be attached to the rear of the straw walker housing and driven from the engine without significantly slowing combine operation. A *straw chopper*, however, requires 20–50 kW. It is thus more usual to chop straw from the windrow in a separate action using a flail or purpose-designed chopper.

To assist in operating the combine at the highest throughput without unacceptable grain loss, an electronic *loss monitor* may be fitted. A group of detectors are mounted usually under the straw walkers and linked to an indicator in the cab. Other optional aids are *automatic table height control, area meters* and '*tank full*' *indicators*.

Grain is harvested at a moisture content (m.c.) normally within the range 15–25%. A range of electronic *moisture meters* are available, the most common and, in general, most satisfactory being types in which the electrical resistance of a ground sample of grain is measured and converted by meter reading or chart to % m.c.

Before storage, at 14% m.c. if in bulk or 16% in sacks, the grain must normally be dried. *Batch driers* dry a fixed load, normally in a tray, where grain is cooled before the tray is emptied and refilled with damp grain. An automatic batch drier, when fitted with moisture-content control, may continue the sequence unattended.

Ventilated bins may be used both for drying and long-term storage, although they do require that the grain is not too moist (generally < 21%) at harvest. Because drying may continue over a long period, heating of the air is much less than in other driers. Some heat is necessary, however, if the relative humidity of the air is to be reduced sufficiently to allow drying down to 14%.

On-floor drying involves similar principles with the building floor area also used for storage. To ensure sufficiently even airflow a system of main and lateral ducts must be arranged on the floor with the laterals spaced not more than 1·5 m apart. Grain may be piled to a height of 2·5 m.

Continuous-flow driers dry grain in a 3–30-cm layer, at the highest air temperatures which will not damage the germination, malting or feeding properties. Temperatures of 120–160°C are most common with grain in the final section of the drier being supplied with unheated air to cool the grain. Most driers are oil-fired and the most common are tower and cascade types, choice being largely to suit space available. Usually grain will take about one hour to pass through the drier for moisture-content reduction from 21 to 15%. The throughput and hence degree of drying is controlled by adjusting the grain outlet conveyor. This control involves moisture content measured at frequent – preferably hourly – intervals, but electronic moisture-content controllers will effectively monitor the grain and control the drying continuously.

Principal auxiliary equipment for a larger grain drying and storage installation will include a *cleaner* or '*dresser*', normally employed after drying, a *weigher* for yield recording or dispatch control and *conveyor* systems. The latter will vary according to the size and performance of the installation with *bucket elevators* and *belt* or *chain* and *flight* conveyors most suitable for vertical and horizontal movements respectively in large installations and the *auger* best for flexible use on small farms or with on-floor drying.

ROWCROP HARVEST

In this section equipment for harvesting potatoes, sugar beet and the more important vegetable crops will be mentioned. Space does not permit review of equipment for cleaning, grading and packing of produce.

Potato harvesting may involve machine lifting, followed by manual picking from the ground, manned harvesters, and harvesters with mechanical or automatic separation of potatoes from soil and stones.

Elevator diggers are for one or two rows with the former normally fully mounted and the latter semi-mounted. A wide share lifts potatoes and soil from the windrow onto a sloping metal elevator with agitating parallel bars which reduce clods and allow soil to fall through the elevator first so that the potatoes fall at the top of the conveyor on the ground surface. Work rates can be up to 0·3 ha/h, but area harvested will normally be limited by the capability of the picking gang. *Manned harvesters* are also generally for one- or two-row use, with from two to six operators supplementing the separation of the lifting elevator by moving good potatoes from the rubbish on one conveyor and stones or clods from the potato conveyor. Work rates depend on the stone and clod population but a harvester manned by four persons may typically deal with 5 tonnes of potatoes per hour. In suitable soils *unmanned harvesters* may lift, separate and convey to trailers by mechanical means only; further separation is often needed however at the store. *Electronic stone* and *clod separators* are available and use X-rays to identify items so that automatically controlled fingers linked to the monitor may be withdrawn in response to a stone or clod, allowing it to drop back onto the ground. Although tractor-hauled models still predominate, *self-propelled harvesters* of both manned and unmanned design are increasing.

Sugar beet harvesters may also be tractor-hauled or self-propelled. They must remove tops from the beet, lift it from the soil without breaking the tap-root too high to leave valuable sugar and, after separating maximum soil, convey the beet into a hopper or trailer.

Two principles are employed: the first a two-stage process in which the first machine tops and lifts two or three rows of beet, placing them in a windrow to be collected, cleaned on a web elevator, and loaded in a trailer alongside; the second a complete harvester which carries out all processes on one or two rows. An alternative is a harvester which needs to be preceded by separate topping. On the Continent of Europe, three-stage harvesting (topping, lifting and cleaning/loading) is common, using five- or six-row machines. Two- and three-stage systems are appropriate over 25 ha and single-stage harvesters below. As losses, resulting from inaccuracy in topping, failing to lift enough root and subsequent damage, may represent 10% of the crop, machine developments and feature variants are many. Topping is being developed for higher speeds, lifting may be by shares or 'Oppel' wheels, cleaning by spinners, rotating cages, agitators or web elevators. *Cleaner loaders* are used to load beet from clamp to road transport when, after some drying, further soil may usefully be removed.

Specialised harvesters for vegetables are limited by the small number of enterprises growing such crops in quantity. However, mechanisation is essential for many crops used for canning, freezing or other processing. Peas are harvested by cutting and windrowing followed by threshing or by shelling from the vine or by trailed or self-propelled *viners*.

MATERIALS HANDLING AND TRANSPORT

Much handling of farm inputs and outputs is covered in the specialised equipment discussed in the other sections. That described here is mainly of more general-purpose use, particularly the two most important types – tractor trailers and loaders. The importance of materials handling is indicated by the fact that tractors are considered to spend up to 40% of their time on this, and in this case the *all-terrain vehicle* such as the Land-Rover is important. Its use to haul empty trailers and implements between fields is underexploited. *Conventional tractors* with trailers, specialised trailed containers or mounted carriers are the most common options. Higher-speed transport tractors increase the economics of road journeys. *Trucks* are little used compared with the USA, since their traction on wetter soils is limited, and even for cereals they can be immobilised in wetter seasons. *Rough-terrain loaders* fitted with buckets or fork lifts are based mostly on tractor-type chassis with either rear wheel or articulated chassis steering. Their advantages are high bucket and fork-loading capacity – typically 1,000 kg per load compared with 200 kg for a tractor loader – and high reach/high stacking of bales. They may tow trailers, but are used most efficiently as loaders of trailers, manure spreaders, etc., hauled by tractors.

Trailers for tractor haulage are of vast variety. Most common are unbalanced two-wheel or, with tandem axles, four-wheel box types with part of the load carried on the tractor, so increasing its tractive ability and reducing rolling resistance draught of the trailer. These types, available from 3 to 20 tonnes capacity, are built for the transport of cereal and root crops as well as for general-purpose use. They have hydraulic rams for tipping to the rear, upper hinged rear doors and, for grain, must, if made of wood, be free from gaps at corners and between planks. Extensions to front and sides make them suitable for chopped grass for silage. Versions are provided with high lift and tip for discharging into higher heaps. For manures, slurry tankers and farmyard manure spreaders are also built for weight transfer onto the tractor and fall into the same handling class as unbalanced trailers. Self-loading models for one-man pick-up of hay or grass and self-unloading designs for feeding are also available.

On these designs advantage can be taken of the tractor *pick-up hook* which can be guided under a ring on the trailer drawbar. Operation of the hydraulic lift then lifts the trailer drawbar until it is automatically locked mechanically at the tractor. Release is also controlled from the tractor seat.

Flat-bed trailers for bales, sacks and pallets may be unbalanced with two wheels or four tandem wheels, but are also common as four-wheel designs with a steered front axle and fixed rear axle. Because of the low density of bales the beds may be up to 10 m long. They may be fitted with detachable 'ladders' at the front or rear to help stabilise high loads. The advantage of such trailers compared with unbalanced types is that they may easily be left loaded without jacking the drawbar. They are, however, less manoeuvrable, particularly when reversing, and need greater draught.

For safe road and field use *trailer brakes* must be provided, preferably so that they may be operated simultaneously with those of the tractor. (Note: Legal requirements are anticipated and should be checked.) The ratio of laden trailer weight to the weight of the towing tractor must also be limited and a check made that the vertical load on the towing hook or drawbar is within the capacity of the tractor tyres. Trailer *tyre equipment* is important to reduce draught and avoid soil damage. Generally larger tyres and more wheels will give worthwhile benefits, with slurry and manure equipment, which is most frequently taken on wet fields, the most important.

Implement-carrying trailers for drills, wide implements and tracklaying tractors have low chassis, sometimes with hydraulic lowering rams, for easy loading.

The tractor-mounted loader is available in two designs. The *front* or *fore-end loader* is attached to a frame built onto the sides of the tractor engine-transmission unit and is raised hydraulically to 3–4 m.

> It may be fitted with various handling attachments including a bucket for grain or soil, an open bucket for beet, a manure fork, or devices for lifting bales or tipping pallet boxes. Tilt is also controlled hydraulically from the tractor seat as are functions such as bale grips. The tractor must, of course, be driven to and fro to move materials which usually makes work slower than with a rear-mounted *slew-loader* with which the tractor is stationary and the bucket or grab is on a hydraulically swivelled, extended and lifted arm.

A *fork-lift* may be fitted to the tractor three-point linkage, where a lifting capacity of one tonne or more is available with suitable tyres and front ballast, compared with a limit of about 500 kg on the front loader. Lift heights of up to 3·5–5 m are available with tilt and sometimes side shift also possible. Some larger farms, and particularly vegetable growers, operate industrial types of *fork-lift truck* but these are limited to working on level floors free of mud.

Elevators are available in a variety of designs for bales and sacks within buildings. They are fitted mostly with small wheels for manoeuvring and with jacks or rams for altering the lift height and angles of individual sections. Most are powered by small two-stroke combustion engines.

IRRIGATION AND DRAINAGE

With the exception of very limited trickle irrigation, all watering in the UK is by overhead irrigation. Its importance has increased through the bids to increase cropping on limited land, and, in part, through the development of water reservoir constructional techniques including butyl linings. Irrigation equipment is normally obtained for adding over a season water equivalent to 10–30 cm of rainfall. The four main types are fixed spray lines, fixed rotary sprinklers, self-propelled sprinklers and large rotating spraylines attached to tractors or specialised vehicles.

A centrifugal *pump*, powered either by an electric or diesel motor in a fixed installation or from a tractor p.t.o., when water is drawn directly from a river or stream, is used to supply water to the irrigator through either 5–10-cm metal pipes with automatic couplers or flexible plastic piping. The *fixed spray line* is used normally for smaller areas of high-value vegetable crop. *Fixed rotary sprinklers* have similar application, but also find use in the grassland spraying of slurry liquid components. The *self-propelled sprinklers* or *rain guns* which use water flow through a hydraulic motor to propel the carriage carrying jets and sprinklers across a plot can be applied on field vegetables, sugar beet or soft fruit. The larger type of irrigator such as the *tractor-mounted rotating boom* is clearly for the largest areas. Coverage may be over a circle of up to 100 m in diameter and this equipment may add 2·5 cm of water to 1 ha in 8 h.

Farmland drainage covers the maintenance and digging of ditches, the installation of pipes at intervals across fields as main drainage channels, and the more frequent drawing of mole channels at a lesser depth to conduct subsoil water to the pipes. Piped drains can typically serve effectively for 20–50 years, whereas moles may need renewing at 5–10-year intervals.

Ditching is carried out mostly by a tractor-mounted hydraulic digging bucket. This is attached to the rear of the tractor, has two hydraulically actuated stabilising legs and provides for the operator to face rearward from a height from where he can see into the ditch.

Trenching machines for pipes will be chosen according to the length to be installed and whether clay tile pipes or continuous plastic pipe are to be laid. In larger projects, specialist contractors are normally employed and this may even be so for smaller installations where a *back-acter* trencher, a very narrow bucket attached to a slew loader straddling the proposed line of the trench, is used. Such machines, which may work to depths of 2·5 m are also recommended where rocks are present. Work is relatively slow.

Rotary wheel and *chain trenchers* work while moving continu-

24　Cubicle housing and self-feed silo for a 90-cow dairy unit

25　Sheep shed

26 An experimental aviary for poultry

27 Medium-power tractor with four-wheel drive for improved traction and handling on slopes

28 Five-furrow reversible plough with digger bodies

29   Contractor's high-capacity vehicle for fertiliser, fitted with low ground-pressure tyres

30   Slurry distribution tanker on 'flotation' tyres

31   Rotary cultivator for seedbed preparation and plant debris incorporation

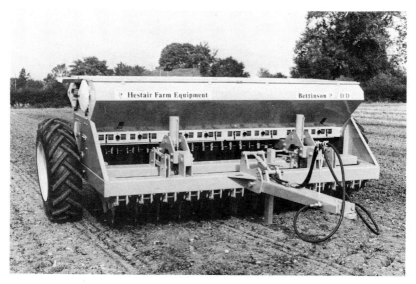

32   Triple-disc direct drill for planting cereals without soil tillage

33   Tracklaying tractor for use on easily compacted soils

34   Ultra-low ground-pressure spraying vehicle for use on wet soils

35   Two-row self-propelled sugar beet harvester

36   Manned two-row potato harvester

37 Smallford blackcurrant harvester modified for picking small bushes

38 Disc mower fitted with crop conditioner for rapid drying

39 Metered chop forage harvester working with a wilted crop

40 Big baler for round bales, working in straw

41 Automatic batch-type grain drier

42 Trenchless drainage machine working with continuous plastic pipe

43   Hydraulic digger-loader with ditching bucket

44   Hedge-laying: the stock- and trespass-proof finished article—Midland style.
This hedge will need no more than light trimming for the next two decades

45-6 The same fields before and after reclamation

ously digging to a depth of 1·5 m. The depth is hydraulically adjusted and to maintain a uniform gradient, depth may either be set by an operator making reference to sighting targets or controlled automatically by an optical or *laser beam* controller. The trencher incorporates a chute through which clay pipes can be fed continuously to leave them sited end-to-end at the trench bottom. Smaller machines may be carried on wheels, but tracks are more usual.

Trenchless drainage machines employ a large share to open a trench just wide enough to allow the pipe – normally continuous plastic – to be installed. In view of the high draught requirements winches are often employed with an anchor on a smaller tractor which is moved to be ahead of the drainlayer and a winch on the machine itself.

Porous *backfill* is added frequently above the pipe to form a coupling for water-flow from mole drains which are drawn across it. Chutes may be provided on drainlaying machines for filling the trench. As drainlaying is often carried out on wet soils, transport of the large quantities of backfill over soft surfaces poses a traction and potential soil-damage problem so that hopper trailers with very large tyres are common.

Mole drainage is more often a farm task. The mole plough consists of a single vertical leg carrying the bullet-shaped share through the soil at depths from 30–60 cm, preceded by a disc to cut the surface turf and leave a less ragged finish. Moles are drawn at gradients of about 1:200 at intervals of 2–6 m.

LIVESTOCK HUSBANDRY EQUIPMENT

More specialised equipment, such as milking machines, shearing equipment and poultry battery cages are dealt with in later chapters. This section, therefore, includes only brief descriptions of equipment employed for feeding, animal weighing and effluent collection and disposal.

Feed preparation involves milling or crushing of grain, mixing of ingredients and perhaps cubing or pelleting. Three types of *mill* are used on farms: plate mills, hammer mills and roller crushing mills. For coarse grinding, a plate mill is satisfactory, but hammer mills are the most popular and one attractive system employs a small hammer mill with an automatic control system so that a time-switch may start the mill during off-peak electricity hours and stop the mill when the batch is completed.

Feed mixers incorporate a vertical cylinder having a hopper bottom and containing a large auger raising material from the bottom to sprinkle it over the surface of the contents. In the most popular types material for mixing is added via an auxiliary hopper and mixed material removed from a low outlet. A 500-kg mixer requires 2–4 kW.

Cubers and *pelleting machines* employ a rotary-extrusion mechanism combined with a cutter. Power depends on the size of cubes or pellets but a 3-kW machine can produce 67 kg/h.

Mechanised feeding may be by bulk delivery vehicle or conveyor system. *Bulk delivery vehicles* which may be divided into more than one hopper section incorporate a long auger for transfer to hoppers attached to individual animal houses. As well as the larger models on a wheeled chassis, smaller tractor-mounted models are available.

Within the building a wide variety of feed conveyors is available. These include *auger, chain, belt* and *pneumatic conveyors*. Chain or pneumatic conveyors are most satisfactory for curved routes. Conveyors may be automatically stopped and started by pressure switches when dispensers are filled.

Feed dispensers are generally of volumetric types with the simplest types for manual operation by levers. Dispensed quantities can also be adjusted mechanically. Both milking parlour and piggery feed dispensers can be actuated electrically or pneumatically. In the parlour this actuator can be controlled from sophisticated controllers, including computer-based systems. In a piggery the dispensers may be actuated at intervals by a programmed controller, feed normally being consumed off the floor.

Liquid-feeding equipment is used for pigs, dairy cattle and calves. The advantages, particularly for pigs, are simplicity of installation and saving of a dry feed mixer on the farm. Mixing takes place, usually by a special auger within a 1,000–6,000-l tank, combined with re-circulation around the feed line.

The electric motor drives may vary from a single 1·5-kW motor for mixer and pump to separate 4-kW motors on each. Pipes of steel or polythene may be 38 mm diameter for smaller installations or 50 mm for larger, and dispensing into troughs may be manual or automatic. An automatic feeding system is normally programmed to mix, with complete re-circulation through the feeding pipes, before the dispensing valves are opened automatically in turn on a preset time basis. For calves, various types of teat feeder incorporate mixing, frequent batch dispensing and, in one case, a feeder travelling between pens.

Livestock weighers may be purely mechanical or incorporate electronic weighing elements and instruments.

Mechanical types for cattle usually incorporate a crush, which is also used for restraining the animal for veterinary treatments, a weightscale with damping to reduce the influence of animal movements and a lock-out device to protect the scale during transport. Cattle and pig weighers may be fitted with attachments for transport on the tractor three-point linkage. Electronic types with electrical circuits to provide accurate weight of a moving animal, digital read-out of weight and remote or automatic operation of gates, make work in large herds quicker and more accurate.

Animal effluents are collected for later fertilisation of fields either as farmyard manure or as a slurry.

Farmyard manure spreaders are of two main designs, those with trailer body incorporating a chain and slat conveyor in the floor which feeds the contents to a transverse shaft cutter-ejector at rear, or the side spreader, in which a cylinder type body contains a series of flails rotating at high speed on a longitudinal shaft. Both types are driven from the tractor p.t.o. and are capable of spreading at 8–50 t/ha. The cylindrical type of spreader is able to handle very wet materials. With a farmyard manure-storage and distribution system the only other equipment generally needed is a tractor loader with fork attachment and means of scraping up manure within the buildings.

When animals are kept in buildings with little or no straw, for example on slats or in cubicles with dunging passages, their dung is handled as a slurry. This conserves nutrients efficiently and is either stored under the slats or removed mechanically by scrapers that are tractor mounted or drawn along the passage by chains. Slurry cannot be spread on land throughout the winter because of damage to the soil so it is stored in either above-ground tanks, usually steel, or in lagoons excavated in the soil.

A *slurry storage tank* must be chosen to hold all slurry produced during the winter period when running with the tanker on the fields would damage grass or crops. The installation will require a *slurry pump*, normally driven from a tractor p.t.o. and often requiring up to 50 kW, or on some larger installations provided with its own electric motor. To keep slurry well mixed and able to be pumped, a stirrer may also be needed in the tank.

For field distribution, a *slurry tanker* is the most common method. The tanker may be filled by the pump incorporated in the storage system, by its own installed pump or by creating a vacuum within the tanker by p.t.o.-driven air pump and sucking up the slurry. Tanker capacities of from 1,000 l are available. Emptying may be by gravity onto a spinning disc or, preferably, to maintain spreading rate independent of level in the tank, by pump. The alternative *pipe distribution* method will normally require dilution of slurry and also prolongs the period of objectionable odours as it is slower.

To further facilitate pipeline distribution or to allow liquid to be returned to rivers, and to provide an easily handled solid nutrient, a *slurry separator* may be used. This will employ screen, centrifuge or separation settling with the former most common. One type consists of a concave screen with paddles on a rotor sweeping the inside. Treatment of the effluent to purify the liquid or to reduce odour is by *aeration*. A variety of methods exist of increasing the slurry-air surface, the most common being a compressor with nozzles to produce streams of small air bubbles. Others include stirring or pumping the slurry through a floating aerator which provides a series of liquid jets over the tank surface.

17

FARM BUILDINGS AND FIXED EQUIPMENT

REQUIREMENT FOR BUILDINGS

It may seem to be stating the obvious to say that farms need buildings but requirements for buildings are not always clear. Buildings are primarily an aid to management but more specifically they may be required to (a) shelter animals; (b) get stock off the ground to improve grass production; (c) provide an artificial environment for stock or crops; (d) improve working conditions of staff; (e) protect machinery from the elements; (f) protect materials from weather and vermin; and (g) house processing machinery and equipment.

Some buildings may encompass more than one of the above points but it should be clearly in the mind of the farmer and the designer why the building is required.

PLANNING NEW BUILDINGS

It takes at least a year from first discussion to completion of a farm building project of any size and this time can be reduced only by building without proper planning and design. Buildings are expensive and last a long time so mistakes are costly and not easily rectified. An error in the timing of cultivations or spraying may reduce this year's cereal crop but can be remedied next year. A poorly designed building will be with you for 20 or 30 years.

The farmer contemplating a new building should seek advice from government advisers, private consultants, neighbours, manufacturers and particularly from his management adviser. He should go and see buildings in use and talk to the man who operates the building as well as the farmer. He would be wise to employ an architect or surveyor on large projects except where the major part of the work comprises factory-built structures. Having taken advice and obtained information, the farmer should then make up his mind as to what he requires and properly brief those who are going to carry out the work.

There are a number of regulations that need to be complied with in the UK and it is prudent to consult the authorities as early as possible to save abortive work. The main requirements are:

(a) Building Regulations: all new buildings require Building Regulation consent from the local authority, although farm buildings are partially exempt. Consent must be obtained before work starts.

(b) Town and Country Planning: buildings over 465 m² in floor area or within 25 m of a classified highway require planning consent. Most other farm buildings are exempt but the regulations should be checked prior to starting work.

(c) Water Authority Consent: where any foul drainage is discharged to a ditch, stream or river the consent of the Water Authority must be obtained.

There are a number of other regulations, particularly those concerned with conservation, and ignorance of these is no defence.

Proper plans and specifications should be prepared for building contracts and competitive estimates obtained. The lowest estimate may not necessarily represent the best value and contractors should be chosen carefully. Such matters as terms of payment, starting and completion dates and approval of additions and variations should be set out and agreed.

CONSTRUCTION AND MATERIALS

British Standard 5502 gives comprehensive guidance on the design and standards of construction of farm buildings and should be used wherever possible.

The majority of farm buildings are of framed construction, i.e. the roof is supported on a framework and is independent of the walls and internal divisions. This method of construction is economical and flexible, enabling the building to be adapted to changes in methods or enterprises. Buildings are manufactured in standard sizes generally with bay lengths of 4·8 m or 6 m and widths in multiples of 1·5 m. Steel is used most commonly for the framework, being especially suitable for wide, clear spans. Concrete is used where freedom from maintenance is important and timber is economical for the smaller spans and especially suitable for on-farm construction.

Corrugated asbestos cement sheeting is used widely for roofing as it is maintenance free. Corrugated galvanised steel sheeting is also used but requires painting for a long life unless plastic-coated steel is used. Steel is more suitable for side cladding as it is less fragile than asbestos cement. Corrugated aluminium sheeting combines the strength of steel with the maintenance-free life of asbestos but is more expensive.

Concrete blocks are used commonly for walls and should be reinforced where a load is to be imposed on the walls. In livestock buildings concrete blocks are often built to 2 m high with a lighter cladding material such as galvanised steel or timber used above. Bricks are equally suitable for walls but are more expensive in most parts of the country. Concrete is generally used for floors, roads and yards and is best purchased ready mixed when consistency of mix is guaranteed. Properly laid, concrete has an indefinite life but it is important to use the correct mix based on BS 5328.

Timber is the most versatile material available for farm construction as it is easily worked, adaptable and resilient. If pressure impregnated with creosote or one of the proprietory timber preservatives it will have a long, maintenance-free life, and hardwoods are available which are stronger than softwoods and are naturally durable without treatment. Timber products such as chipboard, oil-tempered hardboard and particularly plywood are useful in partitioning and cladding. It is essential to use weatherproof, external grades of these materials and the strength of plywood makes it a valuable material for retaining structures such as silos and grain stores.

BUILDINGS FOR LIVESTOCK

Whilst an arable farm can operate with a minimum of buildings, the livestock farmer needs to invest a considerable amount of capital in buildings and fixed equipment. Pig keeping today is primarily an indoor operation and the dairy farmer requires buildings for housing the stock, conserving and feeding fodder, milking, handling and holding cattle and slurry storage. As the majority of the stockman's time is spent in and around buildings, careful design and efficient layout will pay dividends.

Effluent storage

In planning livestock buildings one of the first considerations should be effluent handling and storage. In very few cases is it economical to treat livestock wastes to a standard where the effluent may be discharged to a watercourse and so they must be spread on the land. To obtain maximum benefit from the manurial value of the dung or slurry, storage is necessary so that it may be applied to the land at the best time. Above-ground slurry stores constructed of steel or concrete are increasingly used particularly where the effluent has a high water content and where pollution risks are high. In many cases a lagoon which is excavated to a depth of 1–1·5 m and the excavated material used to build walls around the edge will provide economical storage in areas where there is no pollution risk. Where the effluent can be produced in a semi-solid or solid form storage problems are minimal but there may be a problem in disposing of polluted liquids.

Cattle housing

Traditionally cattle have been kept tied in stalls with a manger for feed at the head of the stall. Whilst economical on building space this system is laborious, does not provide ideal conditions for the cattle and is inflexible, with the result that it has been largely superseded by loose housing. In its simplest form loose housing consists of a covered yard with a feeder on one or more sides and the floor bedded with deep straw. This design requires a large straw usage and it is more common to find a strawed lying area occupying two-thirds of the floor area with the remaining one-third a scraped concrete feeding area alongside the feeder. This design saves straw and overcomes the problem of adjusting the feeder as the straw bed builds up. The best arrangement is for a long narrow building with the feeding area along one long side with continuous access from the bed. Restriction of access from the bed to the feed area tends to increase straw consumption through treading. To minimise capital cost the feeding area may be left uncovered but this will increase the slurry problem and may lead to food wastage.

The most common design of covered yard consists of a central tractor passage with a feed manger on either side backed by a concrete feed strip and strawed lying area. The planning dimensions are shown in Fig. 17.1 and on the basis of one fully grown beast requiring at least 600 mm of manger space the length of the building in metres can be determined by multiplying the number of cattle to be housed by 0.3. The dimensions shown are suitable for feeding from a forage box but the width of the centre passage may be reduced to 2 m excluding mangers where feeding is from a tractor and transport box. An alternative is to arrange the feeding on the outside of the building saving the cost of the feed passage.

Covered yards for milking cows do not need subdivision but

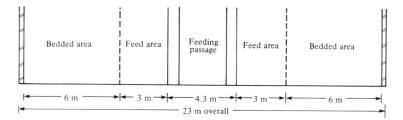

Fig. 17.1

where cattle are being fattened they perform better in small groups. Provided that the building has been properly planned it may be divided anywhere according to pen size required and still retain the right balance of feeding and lying areas. The positioning of the water trough poses problems but it is best sited on the edge of the bedded area adjoining the feed area with access from the feed area only. There are a number of designs of feed barrier available and the common ones are shown in Fig. 17.2.

An alternative to the straw yard is the cow cubicle where the animal has free access but lies in a stall instead of on a communal bed. The advantage of this system is that the cubicle is so designed that the cow can lie comfortably without interference from other

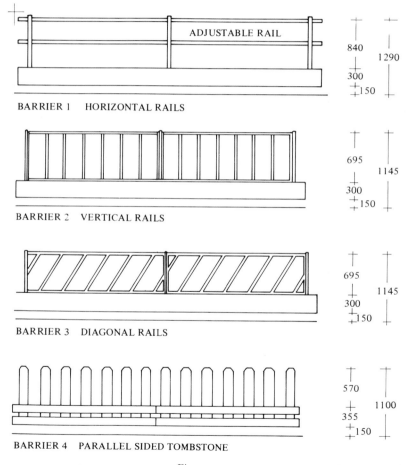

BARRIER 1   HORIZONTAL RAILS

BARRIER 2   VERTICAL RAILS

BARRIER 3   DIAGONAL RAILS

BARRIER 4   PARALLEL SIDED TOMBSTONE

Fig. 17.2

cows but cannot dung in the cubicle, so reducing bedding costs and enabling the cows to keep cleaner. This is particularly important for milking cows, where dirty cows waste time in the milking parlour.

The design of the cubicle division, in steel or timber, is less important than getting the correct dimensions and providing a comfortable bed. For Friesian cows the cubicle should be 2·15 m long and 1·2 m wide with a headrail set about 400 mm from the head. The bed should be approximately 225 mm above the level of the dung passage which should be 2·1–2·4 m wide.

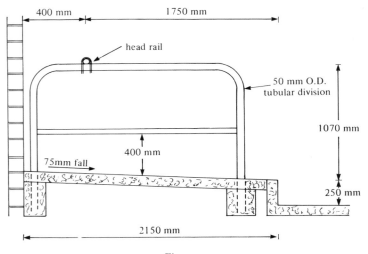

Fig. 17.3

Cubicles may be used for heifers and fattening cattle as well as cows but it is not easy to get the right dimensions for the growing animal. Topless cubicles, i.e. without a roof, have proved satisfactory for fattening cattle but the problems of maintaining a dry bed and the increased amount of slurry produced have restricted their development.

Slatted floors for cattle have always been popular in Scotland and are becoming more common elsewhere with the increasing cost of baling and handling straw. Slats are not suitable for milking cows as teats get damaged but can provide economical and labour-saving housing for fattening cattle. Although the cost of the building is high, the area required per beast is almost half that required on straw bedding and the overall cost per beast housed will be similar. Slats may be either high level, where the tractor can enter a cellar under the slats for cleaning out, or low level, where slurry is pumped from beneath the slats. The latter is more suitable for silage feeding where the slurry is more liquid. It is important to provide sufficient feeding space in a slatted house where stocking densities are high. A number of unroofed slatted units have been built recently and cattle performance appears to be unaffected.

Ventilation

Good ventilation is a prime requirement in any system of animal housing. Cattle require freedom from draughts and somewhere dry to lie and feed but they rarely require warmth as such. To reduce ventilation because it feels cold is a natural human reaction but it can be fatal to cattle. A constant supply of fresh air is required and is best provided by the use of spaced boarding about 1·2 m down from the eaves on both sides of the building and an open ridge which may be protected by an apron flashing piece (Fig. 17.4). The

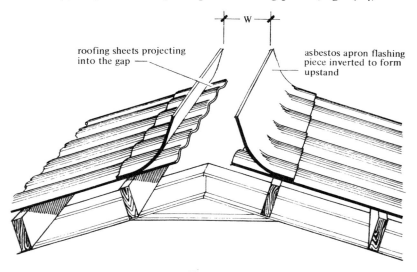

roofing sheets projecting into the gap —

asbestos apron flashing piece inverted to form upstand

Fig. 17.4

width of the opening (W) is related to the span of the building and the stocking density but, as a guide, 25 mm width of opening is required for every 3 m building width. The spaced boarding should have 15% gaps.

For very wide spans, multi-spans or dense stocking some form of roof ventilation should be used, e.g. spaced sheeting.

For pigs or poultry, where closer control of the environment is required, fan ventilation and artificial heating may be necessary and expert advice should be obtained when designing such ventilation systems. Although fan ventilation can provide better control, natural ventilation is easier to manage.

Milking parlours and dairies

The design of milking parlours is constantly evolving in an effort to increase the number of cows that a man can milk. A 1956 report in the farming press described a Somerset milking parlour installed at a cost of £680 which enabled the farmer to increase his herd to 30. Twenty-five years later parlours cost £30,000 and enable one man to milk 150 cows or more. The constant improvement in parlour design means that buildings for parlours and dairies should be capable of adaptation and the prefabricated parlour has much to commend it. The dairy must be large enough to house the bulk milk tank as well as the ancillary equipment and have a 150-mm-thick concrete apron outside for the bulk tanker to stand on.

The herringbone parlour is the most favoured design, accommodating 8–28 cows. One unit per cow is generally installed with automated feeding and cluster removal but for the smaller herd one unit per two cows with no automation can be considerably cheaper. The abreast parlour is satisfactory for the herd of up to 60 cows where individual cow attention is valued. Rotary parlours have a place where very large herds are milked and, when they can be made more reliable, their suitability for automation may make them attractive in spite of their higher capital cost.

Within the parlour area there should be sufficient space for collecting and dispersal yards, holding pens for AI and cattle-handling equipment. An area of 1·1 m² per cow is necessary in collecting yards which should either be long and narrow or circular with a backing gate. The circular yard has the advantage of combining collecting and dispersal but is more expensive and difficult to clean. As cows may spend four hours or more a day standing in collecting yards, they should be protected by a wall or windbreak, or, ideally, the collecting yards should be covered.

Calf housing

On most livestock farms calf housing is very much an afterthought and often provided in existing old buildings. Whilst these can be satisfactory, the very high mortality in calves reflects the inadequate housing so often found. Recent research and development work has shown traditional ideas on calf housing to be wrong and often dangerous. Healthy calves in the UK do not require heat provided that they are protected from draughts and the main requirement is for adequate ventilation. Fan ventilation and artificial heat are rarely necessary even in veal calf housing and a simple building with space boarding at the eaves and an open ridge is all that is required. Pens should be 1·8 m × 1 m with bucket holders at the front for food and water. Hurdle-type partitions are cheaper and more flexible than solid ones and if a double row of pens is placed in the centre of the building with a feed passage on each side mucking out will be easier and the walls of the building will not be fouled. It is important to provide proper facilities for the storage and preparation of food and the washing of buckets, etc. Considerable labour is involved in feeding calves and the feed room should be as near as possible to the housing. Housing in groups with automatic feeding will save labour.

On most farms calves will be moved at about 12 weeks old from the calf house to a covered strawed yard. Moving calves puts them under stress and the change in housing and feeding often leads to the development of respiratory disease at this time. This can be avoided by building a combined calf and follow-on house where the individual calf pen divisions are removed but the calves stay in the same building. Open-fronted pens about 6 m deep and 4·8 m wide will accommodate 10–12 animals from a week old to the time when they are turned out to grass.

Sheep housing

Housing sheep is a way of increasing stocking rates by reducing treading of the land and is an aid to winter management particularly at lambing time. But its justification must be established by management considerations alone as there is no evidence that sheep benefit from housing and hand feeding is increased as sheep get some feed from grazing throughout the winter. Given a choice, sheep will generally stay outside under all weather conditions.

The prime requirement of a sheep house is for adequate ventilation which means open sides above about 1·2 m and an open ridge or ventilated roof. The house may be protected from driving snow and rain by fixing a windbreak material to the open sides, either a proprietary plastic-based material or a double layer of chicken wire fixed at least 75 mm apart. There is little build-up of dung and the solid walls may be of timber or plywood as a cheaper alternative to concrete blocks. A simple timber-framed structure is adequate for sheep and home-grown poles can often be utilised to keep down the cost.

The internal layout should provide trough space of 500 mm per ewe and 1 m² for lying. This means that pens should be arranged with 2 m behind the troughs and a common layout is to have pens 4 or 5 m wide with a trough down either side. A 7·5-m-long pen will house 30 ewes. The trough design should provide for concentrate and hay feeding and be wide enough for the shepherd to walk down the centre. Early sheep houses provided wooden slatted floors but straw bedding is perfectly satisfactory so long as the floor is free-draining. Where slats are used they should be 50 mm × 30 mm with 19-mm gaps and constructed of hardwood. A constant supply of cool, clean water is essential and mountain sheep prefer running water which can be provided by guttering fixed along one wall.

Pig housing

Pig keeping is following the poultry industry by becoming a concrete-based factory enterprise with land required only for the disposal of effluent. These intensive enterprises use prefabricated buildings with mechanical ventilation and often automated feeding. Slatted floors are used to minimise cleaning and the units rely on high outputs to maintain profitability. The alternative is to use more adaptable buildings with straw bedding and natural ventilation, a system favoured particularly on large arable farms.

Feed is by far the biggest cost in pig production and housing systems must provide efficient feeding. Individual feeding of sows is essential and confining dry sows in stalls or tethers is increasingly popular. This is cheaper than yards with individual feeders but where the sow is confined the building must be well insulated with controlled ventilation. Boars should be penned alongside the newly weaned sow to encourage early conception.

Farrowing should always take place in a crate which prevents the overlying of piglets. Sows may spend up to three weeks in the farrowing house depending upon the weaning system and will then either move to a group-suckling yard or return to the dry sow house. Group suckling yards should provide room for 3–6 sows and their litters with a covered, heated creep for the piglets. At weaning the piglets can stay in the yard until moving to the finishing house at approximately 12 weeks of age.

Where early weaning is practised the piglets move from the farrowing house to either a flat deck or a veranda house. The flat deck consists of small pens with slotted floors in a building where close control of the environment can be maintained. After about four weeks the pigs move on to a veranda house which has a covered insulated kennel and a separate slotted floor dunging area. Three-week-old pigs may be weaned direct into veranda houses if underfloor heating is provided to the kennel area.

Fattening pigs over 35 kg in weight require controlled feeding and are moved at that time to a finishing house. The three main types of building are (a) the open-fronted unit where 15 pigs are housed in a low insulated kennel with a slatted dung area at the front. The pigman works outside and the unit relies on manually controlled natural ventilation, but it provides efficient fattening at a low capital cost; (b) the Suffolk-type house providing a kennelled lying area with straw bedding and a tractor-width solid dung passage with the trough alongside. This house is popular with the arable farmer as the main frame of the building could be adapted for other uses if pig production ceased; (c) the intensive house relying on fan ventilation, a high degree of insulation and slatted dunging areas. There are many proprietory designs but the houses, while offering potentially the most efficient conditions, are difficult to manage.

Fodder-storage buildings

A simple Dutch barn is all that is required for storing hay or straw and can be constructed cheaply of poles and a galvanised steel roof. Adequate height for loading is essential and the building should be sited to allow good access on all sides. If bales can be manhandled from the barn to the place of use, labour will be saved but once the hay or straw has to be loaded onto a trailer or transport box then distance from the covered yard or feeding area is not important.

Farmers are increasingly making silage instead of hay and require an airtight container for good quality. The tower silo provides the best conditions but is not popular because expensive machinery is required for loading and unloading.

Bunker silos are generally provided and may be either covered or uncovered. Not only do the bunker walls need to be airtight but they also have to withstand pressure from the silage and the weight of the tractor during consolidation. The design of thrust-resistant walls is a job for the engineer but there are many prefabricated wall units available which can be erected with local labour. Either concrete or timber is used for prefabricated units but walls may be cast *in situ* with concrete or constructed from reinforced concrete blocks provided that proper plans are used. Concrete block walls must be rendered to prevent deterioration from silage liquor. Where the silage is self-fed, walls 2·4 m high with a consolidated silage depth of 2 m are usual but if the silage is to be cut and carted, greater depths are possible provided that the walls are properly designed. Silage occupies 1·4 m³ per tonne.

The floor of the silo should be concrete of C 25 P mix laid to fall to an open channel at the front to carry the effluent to an adjacent sealed effluent tank. The tank may be constructed of rendered concrete blocks or precast concrete rings and large enough to take 2 weeks supply of effluent. The effluent should be diluted with equal quantities of water and applied to land at the rate of 50 m³/ha. On no account must silage effluent be allowed into a water-course or mixed with other effluent unless it is immediately to be spread on the land.

GRAIN STORES

Although the economics of centralised grain storage appear attractive, most grain is still stored on farms even when it will eventually be sold. The system of storage chosen depends primarily on whether the grain is for feeding on the farm or for sale. For feeding, the grain is best moist but grain for sale must be at a maximum 16% moisture and cannot be safely stored in bulk without ventilation at over 14%. Moist grain will go mouldy unless the air is excluded or it is treated chemically with propionic or formic acid and both systems are used for on-farm storage of feed grains.

Where grain is to be dried the choice is between storage in bins or in bulk on the floor. Floor stores are cheaper and require less fixed machinery but it is difficult to divide parcels of different grains. The grain may be dried through a continuous drier before being put on the floor or it may be dried on the floor provided it is not stored to a greater depth than 2·4 m. On-floor drying makes for quicker handling at harvest and enables grain to be stored at 16% moisture because it can be ventilated throughout the storage period. The main requirements are a waterproof concrete floor, thrust-resistant walls and main and lateral air-ducts. The main duct will be constructed as one wall or a central divider to the store with laterals either laid on the floor or constructed under the floor with perforated covers. Underfloor ducts ease loading and unloading but are more expensive.

With bins and fixed conveyors the grain store can be automated and a number of different varieties can be kept separate. Again, drying may be in-bin or continuous but some form of aeration is desirable in bins even if the grain is pre-dried. Square bins are most satisfactory indoors but large round bins erected outside can provide very economical storage. Provision should be made for cleaning the grain and the design of the store should minimise the risk of vermin entering. It is very difficult to keep out birds except by excluding all natural light.

When planning a grain store always allow for future expansion and good access from the public road is essential.

POTATO STORES

Whereas grain in store is inert, potatoes are living and are more demanding on the storage building. They similarly exert pressure on the wall but they also give off a large quantity of moisture and require to be maintained at a fairly constant temperature. The first essential is to keep the frost out and the simplest way to do this is to line the wall of the building with straw bales and place a layer of loose straw on top of the stack. Potatoes can be stored until March under these conditions but if storage is required through April and May when the price is advantageous, forced ventilation and better insulation are required. The insulation materials must have an effective vapour barrier and the wall insulation must take the thrust of the potatoes and resist mechanical impact.

A clear-span building at least 4 m to eaves should be provided, with a concrete floor, wide and high doors and good access from the highway. Where the potatoes are to be sorted and bagged on the farm, an additional 13·5 m should be added to the length of the building to provide a working area.

SERVICES

When planning new farm buildings few people adequately consider the provision and cost of services such as roads, drains, water and electricity. The importance of good access to buildings has already been mentioned and most new schemes require concrete roads and yards. These should be constructed with a C 20 P mix at least 100 mm thick and laid on a well-consolidated hardcore base. The strength of concrete is mainly dependent upon the strength of the base but the concrete should be reinforced or increased to 150 mm thick when it is to be used by heavy traffic. Roads and yards should be laid to falls of approximately 1 in 60 to surface water drains.

If the road is not to be used by livestock, hardcore alone may be adequate provided that the surface is properly bound with fine material and the water is kept out of the road by good drainage. Any potholes must be repaired immediately. An access road which is used by cars and lorries as well as farm traffic should be surfaced with macadam, either tar- or bitumen-based. This will require regular maintenance.

Practically every farm in the UK now has a supply of mains electricity but the cost of extending the supply to a new building site can be very high. Where powerful electric motors or heaters will be used, such as in grain drying, a three-phase supply should be provided. The wiring within buildings must be carried out by a competent electrician using durable materials and the installation should be checked regularly. Water- or vapour-proof fittings are required in many buildings and information is now available to enable the proper level of artificial lighting to be provided in all types of farm buildings.

A piped water supply is essential for any livestock enterprise and plastic-based piping is easily laid underground and within buildings. Piping should be protected from damage by livestock and troughs within buildings should be so placed to minimise fouling. Where there is a danger of freezing, low voltage electric wires may be attached to the pipes and heating elements placed in troughs.

Troughs where the inlet is on the bottom controlled by a floating ballcock are less likely to freeze.

For permanent boundary fencing woven wire with two strands of barbed wire on top fixed to timber posts at 2·8-m centres is best, but if only cattle are kept, a strained wire fence with droppers is more economical. Modern developments in mains electric fencing of high voltage offer the choice of substantial reductions in the capital costs of fencing, both permanent and temporary. Fencing posts must be of a durable species such as oak or pressure-treated with creosote or chemical preservative.

18

MARKETS AND TRANSPORT

Introduction

The UK food system includes a wide range of marketing channels associated with different commodities, and farmers may sell their output via a merchant or agent, marketing board, co-operative or group; by means of contracts or auction markets; to wholesalers, manufacturers or retailers, or directly to consumers through farm shops, pick-your-own, or processing and delivery services. Hence the market for farm products is not simply a place of exchange for disposal of output, but a series of linkages which involves farmers operating at many stages and with many of the other agents within the food chain.

A very simplified version of this food system is shown in Fig. 18.1. Despite the importance of the agricultural sector in the UK,

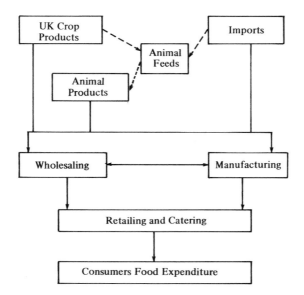

Fig. 18.1   Basic elements in the UK food chain

the value of farms' output for home consumption is surprisingly small, amounting to less than 25% of total household and catering expenditure on food in 1979. Partly this is explained by the crop products which are used in animal feeds, since it is only the resulting animal products which enter into this calculation of the agricultural sector's output, and partly by the food consumed in the UK which is derived from non-indigenous farm products (mainly tropical crops). Nevertheless, imported raw and processed food products amount to nearly as much, by value, as the output of UK farms, and are an important feature of our food chain. About one-third of indigenous-type food eaten by UK consumers is obtained from overseas agricultural production.

Most agricultural products reach the consumer in a different form from that in which they leave the farm. Current MAFF definitions indicate that over 70% of food bought by households is processed, and if meat slaughtering is included, the figure exceeds 85%, with eggs and fresh fruits and vegetables the remaining unprocessed items. About half the output of UK farms and half the food imports into the UK are channelled directly or through merchants and wholesalers into the food-manufacturing industry, which 'adds value' equivalent to about 25% of consumers' food expenditure to these inputs before passing the processed product on to the distribution stage. This latter, including wholesaling, retailing and catering, makes a similar addition to the value of the products before selling to consumers. Hence, considering the food chain in general terms, rather more than one-quarter of consumers' expenditure on food in the UK accrues to food distribution, about one-quarter to food manufacturing, and slightly less than one-quarter each goes towards the purchase of imports and the output of UK farms.

DEVELOPMENTS IN THE FOOD CHAIN

Between 1960 and 1980, in particular, the food system of the UK underwent many changes of importance to the marketing of farm products (see also Chapter 23). One general result has been a greater inclusion of post-farm marketing services in the food products that consumers' buy; and in most developed countries, including the UK, the farmers' share in the retail price of home-produced food products has been falling. Although some see this as unfortunate or even sinister, in part at least, it reflects some long-run trends in society. These involve more urban concentration of population, and the specialisation of farmers both in the production of particular crops or animal products and in specialist geographical farming regions, with the consequent needs for transportation to consumers. In addition, many commodities are produced only at specific times of the year, or vary in a production cycle, while consumers' food requirements need to be satisfied continually, subject to some seasonality in patterns of consumption. Hence, processing for improved preservation, storage and transportation is undoubtedly a significant element in the marketing services provided by the manufacturing and distribution sector. However, this is not the whole explanation.

Since, in the UK, most consumers have sufficient food to meet nutritional requirements, consumption levels in overall terms change slowly, and even at times when incomes have increased substantially little extra food has been bought. In fact, although food was still, at £26,781 m. in 1980, the largest single item of consumers' expenditure, its share was around 20%, compared with nearly 30% in 1960. With their generally satisfactory average energy intake, consumers look for other attributes in food, including convenience and ease of preparation, quality, novelty and variety, and desirable health or diet characteristics. Since, for example, convenience foods replace to some extent work in the home by further manufacturing and packaging of the product, as their consumption increases, so the food marketing bill rises. A further stage in this development occurs when catered foods replace home cooking. However, although more processed products have enjoyed relatively greater success than less processed foods over the long term, the value of convenience food purchases (as defined in the National Food Survey) has remained static at about 25% of expenditure on food eaten in the home during the 1970s. There is some evidence of quality shifts within the sector, as in 'trading up' when incomes rise in real terms, 'trading down' when they fall, and replacement of one convenience food by another, such as frozen for canned products. The net result, from the point of view of the farmer seeking a market outlet, is that the food processor is increasingly important as a buyer of farm products, and that the quality, timing and certainty of supplies become more significant elements in the terms of sale.

Changes in consumer life-styles have also been important in the developments in food retailing. The ownership of cars, refrigerators and deep-freezers has encouraged less frequent shopping, with bulk food buying for storage in the home. In turn, these factors have encouraged large unit purchases of meat, fruit and vegetables, some of which may be from farm shops and pick-your-own outlets. These trends are not only associated with improved standards of living but also relate to social factors, especially the high proportion, about 50% in the UK, of married women going out to work, and the large numbers of single-person families. These factors, in addition to others, such as the improvements in food packaging and transportation, along with various economies of scale and dynamic competition in retailing, have been instrumental in bringing about the striking changes in food distribution which have occurred over the past 20 years.

Currently, grocers (retailing packaged food, provisions and other foodstuffs) sell roughly 50% of food purchased by consumers in the UK; specialist food shops (butchers, greengrocers, fishmongers, bakers and dairies) about 30%; further retail outlets (including stores such as Marks & Spencer) another 10%; and other channels, such as street traders and market stalls, farm shops and other direct sales the final 10% (Tanburn, 1981). While grocers, generally, have been increasing their share of the total food market, taking, for example, about 30% of all carcass meat and 25% of all fresh fruit and vegetable sales (1979), it is within their own grocery sector that some of the most dramatic structural changes have taken place. The total number of grocery outlets has fallen from over 140,000 in 1961 to under 70,000 in 1979, while the share of the multiple grocer (with ten or more outlets) has risen from 27% in the former year to 52% in the latter (Table 18.1). The size of store has vastly increased. In March 1980, there were 176 stores of more than 2,500 sq m; in 1966 there was one (Tanburn, 1981).

Table 18.1 **Percentage distribution of sales by type of organisation and number of outlets**

		1961	1971	1978
Grocers:	Co-operative	21	15	15
	Multiple	27	44	52
	Independent	52	42	33
Total non-grocery food[1] retailers:	Co-operative	14	10	9
	Multiple	23	30	32
	Independent	63	60	59
All food[1] shops:	Co-operative	18	13	13
	Multiple	25	38	45
	Independent	57	49	42

[1] Includes off-licences (from Mordue, 1982)
Source: Annual Census of Distribution 1961, 1971 and Business Monitor S D M I Food Shops

Large retailers tend to deal directly with manufacturers or producers, specifying closely the characteristics and standards they require. While this is especially true for their own-brand packaged products, it also applies to many fresh foods bought directly from farmers or their organisations. Hence we again find pressures that require greater control by farmers over the quality, timing and preparation for sale of their products, and greater consistency of supply to meet regular demands.

When producing for, and selling to, the food-processing industry, farmers are faced with similar requirements. Both through pressures from retailers on manufacturers, and the desire to meet their own standards, food firms will stress their quality attributes and also price competitiveness in their input buying. Although farmers' trading relationships are often with first-stage processors using raw food materials, for example flour millers and beet sugar refiners, it must be borne in mind that to a large extent these industries rely on the second-stage manufacturers, for example, of bread, biscuits, cakes, sugar confectionery and soft drinks for the ultimate sale of the food product. These market-oriented companies have a long history of importing raw materials, and hence competition for farmers' markets is, in some sense, international, despite the degree of price protection afforded by the CAP.

In addition, the UK is remarkable for the number of multi-national and giant food firms which dominate production in many sectors. Table 18.2 indicates the percentage of various markets controlled by five firms, but in many cases one, two or three companies have a dominant share. Food firms are often diverse in their interests and many have significant links with animal feed-stuffs, drink and tobacco sectors, as well as more distant markets. Taking the 100 largest manufacturing firms (by net output) in the UK private sector, they control 43% of all manufacturing sales, but their food interests are responsible for 51% of the food-manufacturing market (Watts, 1982). It is against these changes in consumer preferences, and in the increasing concentration of ownership in manufacturing and destination, that the farmer needs to market his output. The larger organisations often seek a more controlled linkage of inputs into their production or distribution systems, and may well prefer contracts with farmers or co-operatives, rather than purchases through open market methods to obtain the quality and timing they require. Contracts may cover merely marketing arrangements such as quantity, quality, price and delivery, but some embrace detailed stipulations on production methods and inputs used. While farmers may benefit from the assured outlet, at least for the duration of the contract, and the provision of seeds, fertilisers and other inputs, some may find that their decision-making on-farm is reduced to frustratingly low levels, and mourn the loss of marketing flexibility when prices change unexpectedly.

The next part of this chapter considers developments in the main markets for agricultural products.

CEREALS

The cereals sector is a complex mixture of markets and end-uses (see also Chapter 23) served by both UK-grown and imported grains, but which has changed steadily over the past 20 years. The current (1980) situation is that animal feedstuffs make up the largest

Table 18.2 Five firm product concentration ratio groups in 1977

				Percentages			
Industry	0–49	50–59	60–69	70–79	80–89	90–99	100
	Poultry – unfrozen; Sugar confectionery;	Bacon & ham products; Sausages & sausage products; Meat – frozen; Poultry – frozen; Milk Fruit products – unfrozen	Fish & fish products – unfrozen Jam	Biscuits; Fish & fish products – frozen; Butter; Cheese; Fruit & vegetables – frozen; Soups and baby food	Flour – bread & biscuits; Bread; Yoghurt; Chocolate confectionery; Compound fat; Tea	Flour – household & self-raising; Cereal breakfast foods; Ice cream; Potato crisps; Malted drinks; Coffee	Sugar (8 firms); Margarine (7 firms)
Totals 30	2	6	2	6	6	6	2

Source: Department of Industry.

single outlet, using annually about 6 m. tonnes of UK barley, $3\frac{1}{2}$ m. tonnes of UK wheat and 3 m. tonnes of imported maize. The malting, brewing and distilling industries obtain nearly all their 2 m. tonnes of barley per year from UK farm supplies, while the 5 m. tonnes of wheat used in flour milling is supplied approximately equally from UK and imported sources. Of the wheat-flour end-uses, commercial bread making is the most important. Although this market has suffered long-term decline in consumer demand over many years, and a high proportion of imported hard wheats have been traditionally used in manufacture, developments both in UK-grown wheat varieties and bread-making processes have enabled the percentage of home-produced flour in breadmakers' grists to increase. Nonetheless, soft wheats grown in the UK remain especially suitable for biscuits and cakes.

COMPOUND FEEDS

Demand in the market for compound feeds is derived from the farm livestock population and its requirements within the UK, which in turn depends not only on food consumption patterns for animal products, but also international trade. The past 40 years have witnessed market changes in the pattern of meat demand. While beef has been stable, falls have occurred in the consumption of bacon and ham, and of mutton and lamb, with complementary increases in pork and poultry meat, both of which use feedstuffs intensively. Eggs and dairy products enjoyed periods of consumption growth, but have stabilised or declined over the 1970s. However, improvements in self-sufficiency, particularly for mutton and lamb, dairy products, and to a lesser extent beef, have counteracted poor consumption trends, and maintained almost constant feedstuff requirements in total over the past decade.

Nevertheless, this rising self-sufficiency has continued and accelerated the trend away from port locations and encouraged the expansion of feed compounders' capacity in farming areas which specialise in growing animal feed or rearing livestock. Thus, while traditionally the major half-dozen national feed compounders based their main processing in or near London, Bristol, Merseyside and Hull, the more recently developed sites have been in the country, and some supply large-scale bulk feeds. Hitherto, inland feed compounding had been the purview of the smaller companies producing smaller units, but many have been taken over by the national compounders, and much rationalisation has taken place. Hence, the distinction between 'port' and 'country' manufacturers is much less clear today.

Farmer relationships with feed compounders have normally been via agricultural merchants, although direct dealings with local smaller-scale firms are common. Merchants play an important role in this, and other cereals trades, especially in organising transport to and from farms, minimising the empty loads in the collection and distribution of grain, feedstuffs and other requisites.

Now, a good number of merchants are part of large feed company organisations, having been taken over in the acquisition of country compounding, and others are becoming integrated into the livestock industry. Another development is that farmers' co-operative organisations have slowly increased their stake in the feed sector to around 15–20% and, although they currently act principally as distributors for national companies' products, many of the largest are producers with strongly competitive potentialities.

Farmers play an important role in grain marketing in the UK, which arises from the fact that the bulk of grain storage is carried out on-farm, in contrast, for example, to most European countries where co-operative centralised stores predominate. There are signs that the UK is moving a little in this direction, especially with regard to grain exports (over 2 m. tonnes in 1980). The principal objective of storing is to obtain better prices by selling later in the marketing year, hopefully covering the storage costs; and in so doing farmers help in evening market supplies over time. Such transactions as these at the prevailing market price are termed 'spot' sales, but farmers may also enter into a 'forward contract' with a merchant, for example, selling the grain at an agreed price, the sale to be completed at some future date. Merchants are able to reconcile this method with unexpected changes in the spot prices through 'hedging' on specialised 'futures' contract markets. In addition to their 'insurance-type' function futures markets can help in establishing and maintaining appropriate market prices.

BREAD AND OTHER CEREALS

Similar general developments have occurred in the other cereals sectors, with a gradual switching of emphasis to home milling inland, and reduced reliance on imported wheats, although North American and other hard wheats remain significant in the bread industry. Economies of scale have led to high concentrations, with only three main wheat millers (two of them important feed compounders) and six major buyers of malting barley. In addition, two companies, as well as being major millers, supply together more than half the UK's bread, while two others control a similar proportion of the biscuit trade.

Again, merchants are important intermediaries between farmers and millers, continuing to be significant not only in transport, but in grading and the integration of like-quality lots. The merchant also plays an important role in price determination, especially in the higher quality cereal trade, where local markets are too small, and futures markets are not established on the same scale, and in home-produced feed grain. Grading is now carried out with many more scientific tests than previously. The results are particularly significant if grain may be sold into intervention under the CAP market support buying schemes, since wheat of bread-making quality secures higher intervention prices. However, merchants may find suitable milling outlets for rejected intervention grain, on the basis of their own and the miller's assessment.

The whole UK cereal chain appears to be becoming more vertically integrated, with major companies involved in several stages of production and trading, and in related activities, such as pet foods and frozen cereal products. However, smaller millers may have advantages in not having to plan so far forward in input supplies as major companies, and thereby take advantage of market movements. Hence, while the farmers' markets will continue to be dominated by the large grain users, smaller alternative organisations may continue to play a part in marketing.

MEAT

In 1980, total meat supplies in the UK – home-produced and imported – amounted to about 3½ m. tonnes. Approximately two-thirds was consumed as carcass meat, including bacon and ham, making up the largest category of consumer food expenditure at just over 20% of the total. The 1980 distribution of consumption is shown in Table 18.3.

Table 18.3 Distribution of expenditure and consumption of meat, 1980

	Total Expenditure by consumers on carcass meat £4,978 m. %	Total Domestic Consumption of carcass meat 2,385,000 tonnes %
Beef	34	29
Lamb	15	16
Pork	13	14
Poultry	15	23
Bacon/ham	20	15
Offal	3	3

Source: Economist Intelligence Unit (1981): Retail Business No. 283

Self-sufficiency in the UK varies between the different categories, but pork and poultry meat is almost entirely home supplied, beef approaches nine-tenths self-sufficient, lamb two-thirds, and bacon and ham two-fifths. The major sources of imports are: for beef, other EEC countries, particularly the Irish Republic; for lamb, New Zealand; and for bacon, Denmark.

Figure 18.2 summarises the main marketing routes between producers and importers, and meat retailers and manufacturers (see also Chapter 23). In 1977, almost 90% of pigs were sold dead-weight, while nearly two-fifths of cattle and one-third of sheep and lamb slaughterings followed this route (MLC, 1980). To some extent this reflects the intensive production methods commonly used for pigs, although the poultry industry, composed of a dozen or so major producers, is the best example of an intensive, integrated system in the meat trade.

Dead-weight sales on contract have advantages of an assured market for producers and certain supplies for abattoir owners, with published prices in terms of dressed carcass weight and quality known in advance on a weekly basis. In the case of pigs, the use of dead-weight and grade predominates, with 80% classified by the MLC Scheme, although this system is less well established for cattle and sheep.

Live-weight auctions, on the other hand, are subject to weekly supply and demand fluctuations, and hence price uncertainties, but many farmers maintain that a day at market provides good information to assist management and marketing decisions. In addition, if a live animal fails to meet the appropriate sale price, or the classification standard to receive government subsidy premiums, it can be withdrawn, and subsequently marketed. However, few animals are in fact taken out of the sale, as transport costs are the farmers' responsibility at this stage. Livestock auction markets have declined in number to about 350 in Great Britain in 1980, but have become larger and take stock from a wider region, as roads and transport methods have improved. Prices should, in these circumstances, become more stable and reliable indicators of overall supply and demand in the region, and less influenced by individual buyers and sellers.

Abattoir ownership has undergone various changes, but recent developments involve a decline in both public local authority-owned facilities in urban areas and those run by retail butchers. In 1980, out of an estimated 1,270 abattoirs in Great Britain, only 39 were publicly owned, although some of the 1,231 private operators were leased from local authorities (EIU, 1981). The private abattoirs handle about 90% of cattle units slaughtered, and vary in size from small retailers to those owned by the major vertically integrated concerns.

In general terms, the small slaughtering operations located near consumers are being replaced by medium and large-scale concerns, privately owned by wholesalers and located in or near production

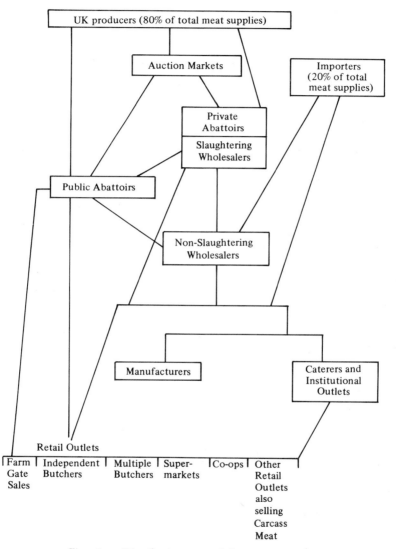

Fig. 18.2   Distributive network for carcass meat
Source: The Price Commission, *Prices and Margins in Meat Distribution,*
HMSO, London, 1975, p. 8

areas. The transport advantages of moving meat rather than live-stock favour the country site, especially as improved vacuum wrap-ping and temperature controlled vehicles preserve desirable meat qualities better after slaughter. Also, many urban premises have inadequate hygiene and effluent control to meet current regulations, and are too old to be worth the expense of modernisation. Develop-ments at the retail level have encouraged these changes, and parti-cularly as the specialist independent butcher has slowly declined in numbers and market share, and multiple butcher organisations and supermarkets have gained a larger proportion of meat purchases by consumers. The meat-processing industry is now often asked to produce primal and retail cuts, which require only limited further butchering and packaging in store, rather than the traditional carcasses and sides for the retail butcher. This development not only encourages cutting and packing at the abattoir, but also the processing of less desirable cuts and by-products which are often a very significant element in overall slaughtering profitability.

MILK

The organisation of the milk market (Fig. 18.3) is the responsibility of the five Milk Marketing Boards (M M Bs) (one for England and Wales, three for Scotland, one for Northern Ireland), which are producer-controlled monopolies with statutory rights embodied in the legislation of 1933/4, and subsequently confirmed, with modi-fications to meet E E C requirements, in 1978/9. In many respects this organisation has removed the need for marketing from the individual dairy farmer, although in the liquid sector some may become producer-retailers, undertaking their own pasteurising and delivery by authorisation of their Board, or manufacture farmhouse cheese, butter and cream under special contract with their Board. For the majority of producers, the M M Bs organise the collection of milk from farms and its delivery to first destination. Virtually all milk in Great Britain is collected from bulk tanks on-farm, and carried by road to town processing dairy or creamery. The Boards' own fleets deliver about half U K milk to its first destination, with hauliers, under contract to the M M Bs, handling the rest. Milk is owned by the Boards until it is unloaded, and producers bear the cost of this primary haulage. Since, with few exceptions, the trans-port cost is averaged over the country, milk production is encour-aged in areas remote from major centres of population and con-sumption, for example, in the west of England, and Wales, and hence location is determined by production rather than marketing considerations.

In England and Wales, 90% of liquid milk is doorstep delivered, although this proportion is slowly falling as more retail purchases,

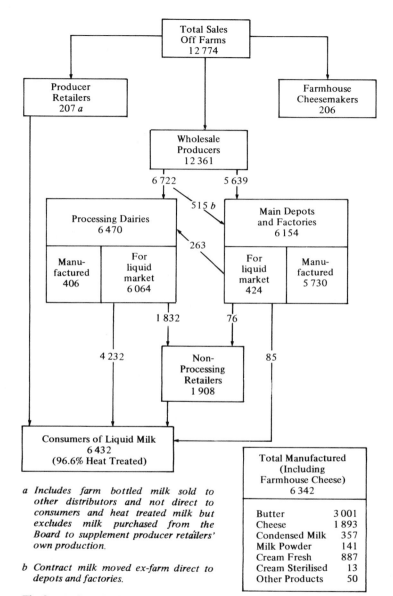

The figures shown (m.l) are in respect of milk produced in England and Wales and include milk transferred to other Boards for manufacture.

Fig. 18.3 Milk flow, England and Wales
Source: MMB Dairy Facts and Figures (1980)

particularly through supermarkets, take place, encouraged by some price competition and the provision of other types of milk, for example semi-skimmed.

While liquid milk has hitherto been supplied by home producers, although any European Court decisions to outlaw the UK's import regulations based on hygiene controls may change this, the milk products sector has traditional links with foreign exporters, particularly for butter and cheese from New Zealand and Europe for lactic butter. The UK is now (1980) nearly 60% self-sufficient in butter and over 70% self-sufficient in cheese, butter imports coming still from New Zealand by arrangement up to a specific quota, with remaining EEC countries providing the balance of butter and almost all cheese imports. Home processing of milk products is dominated by the MMBs, and in England and Wales about half of manufactured milk is handled by the Board, which has a 75% share of home-produced butter and a 60% share of home-produced cheese (Williams, 1980).

EGGS

In some respects eggs are a similar product to milk in that storage is difficult and fluctuations in supplies can lead to depressed prices. However, the marketing board solution has not been successful in this market, and since 1971 when the Egg Marketing Board was abandoned, control over free market supplies has been carried out by major producers, using Eggs Authority information to monitor market developments. Currently about half UK eggs go through packing stations, which tend to be supplied by the smaller producers, and half through co-operatives or directly to the trade.

Difficult market conditions, arising from falling consumption and increased cereals costs, caused many smaller producers to leave the industry, and the remaining larger-scale producers encourage direct marketing to the larger retail outlets.

FRUIT AND VEGETABLES

Horticultural supplies and potatoes in the UK form a large market, in 1980 approaching that of cereals by value (which it had, exceptionally, far exceeded in 1976, when potato prices were sharply raised in response to poor yields). The distribution system for potatoes tends to be different from that of other UK-produced fruit and vegetables, while imports, particularly of non-indigenous fruits such as bananas, also have specific arrangements.

Potato-growing areas and market utilisation are subject to regulation by the Potato Marketing Board (PMB), although the outlawing of UK import restrictions on main crop potatoes by the

European Court of Justice in 1979 has reduced the P M B's effective control. Consumption of 'raw' potatoes has been slowly declining, but accounts for about three-quarters of total usage by consumers, while in the processed potato products market, French fries have taken over from crisps in 1979 as the major sector. Raw potato distribution has a separate system of country and town merchants, before passing to the retail sector, while processors mainly contract with growers or merchants to obtain specific potatoes for their particular end-uses.

Other fresh fruit and vegetables were brought traditionally by producers or country merchants to local primary wholesale markets, sold on to secondary wholesalers and eventually greengrocers, or bought by large local retailers. A good deal of shortening and integration has taken place in this horticultural distribution chain (see Chapter 23).

Although 60–65% by value of fresh fruit and vegetable sales are through greengrocers, the number of their outlets is decreasing rapidly, with only about half of those operating in 1971 still in business in 1978. Supermarket grocery stores have gained a quarter of the market according to some estimates (Tanburn, 1981) and have simplified distribution by having direct delivery from grower or wholesaler into their own central depots, for subsequent delivery to individual stores. The multiple retail chains have tried to improve quality and reliability of supplies, and some have successfully adopted prepacking, especially with permeable shrink-wrap films. Others have used free-flow systems where customers select and bag their own choice of produce. Over-all, it may prove to be the impact of the larger retailers which is most effective in forming improved market efficiency in the horticultural trade, not least because of the chains' ability to import when U K produce is unable to meet their standards and long-term requirements.

The fruit and vegetable processing industry is wide ranging in size and activity, ranging from frozen, canned and dehydrated products designed to store and directly replace fresh in consumers' use, to more distinctly separate products such as jams, fruit juices, soups and baby foods. In order to obtain precise requirements and maintain continuous supplies when factories are working, processors of, for example, canned or frozen vegetables, offer farmers contracts which specify precise growing and marketing arrangements. Transport is arranged to minimise the time lag between harvesting and processing.

PART III

THE PRINCIPLES OF AGRICULTURE

19

THE PRINCIPLES OF
CROP PRODUCTION

The primary objective of crop production is to obtain as large a quantity as possible of a desired product per unit of the most limiting resource, consistent with the further objective in cash economies of leaving the largest financial return. The desired product may be some part of a crop, such as a seed or tubers, or it may be the whole of the above-ground part of the crop, such as grass for animal feed. Where only part of a crop forms the main product, the less important fractions are known as by-products. The desired product does not have to be a food for man or feed for livestock, but could be a raw material for industry such as timber, fuel or fibre for clothing.

Output is expressed usually as the weight of some specified plant product per unit area of land; this is usually called the crop yield, and that this refers to a yield per unit area is often implicity assumed. It is, of course, perfectly valid to discuss crop production in terms of yield per unit of some other resource, such as labour, or water or energy, depending upon which factor is regarded as being the most limiting. The importance of the fossil fuel contribution to crop production serves to remind us of the importance of using the most appropriate measurements in any discussion of crop production.

In this chapter discussion is centred on the more conventional use of the term yield as the weight of produce per unit area, partly because land is usually the most limiting resource, not unconnected with the fact that light, the ultimate limit of crop production, is received by crops on an area basis.

THE DETERMINANTS OF YIELD
PER UNIT AREA

The yield of useful product, sometimes referred to as 'economic' or 'commercial' yield, is only a fraction of the total organic matter synthesised by the crop, or the 'biological' yield as it is often called. Biological yield is difficult to estimate accurately as it must include the root system, and roots are very difficult to recover from the soil, where they commonly extend to depths of 1–2 m. For this reason, the term 'recoverable biological yield' may be used which is restricted to all the plant material contained in the above-ground portion of the plants, together with any below-ground storage organs, such as tubers and storage roots.

Farmers also commonly refer to economic yield in terms of the fresh weight of the product, such as tonnes of tubers, or of grain, or of turnips. This poses problems when discussing the biological principles determining yield, since the water content of these commodities may differ substantially. In seeds harvested at maturity, the water content is usually low, often between 12–16%, whilst vegetables, tubers and fleshy fruits may contain 70–80% water. It is therefore convenient when comparing the growth of different crops, or the effects of the environment on growth, to discuss yield in terms of dry weight.

All the components of a plant necessary for growth, development and maintenance can be derived from about 17 elements, of which carbon, hydrogen and oxygen, obtained from carbon dioxide and water, make up the bulk. The remaining elements are conventionally referred to as mineral elements and are normally absorbed by plant roots from the soil (see Chapter 11). Although plants may absorb many additional elements, only this relatively limited number have been proved to be essential. To be considered essential, the absence of the element must directly cause abnormal growth or failure to complete the life-cycle, or premature senescence and death.

The combination of carbon dioxide and water, which provides the starting point for all the compounds found in plants, requires a source of energy, which is provided by visible light (380–750 nm). The light energy is absorbed by pigments in the plants, known as chlorophyll, which give plants their green colour. Photosynthesis, then, is the process by which plants synthesise organic compounds from the inorganic raw materials mentioned, using the energy from sunlight to fuel the reaction, as shown in the following equation:

$$6CO_2 + 6H_2O \xrightarrow{\text{light energy}} C_6H_{12}O_6 + 6O_2$$

In this process, light energy is converted into chemical energy, in the form of carbohydrates, proteins and other plant constituents, and oxygen is evolved as a waste product. The manipulation of crops in order to maximise the utilisation of light dominates all other issues in crop production and consciously or unconsciously dictates both the way crops are grown and their genetic improvement.

The relationship between light interception and yield

The total amount of radiation intercepted by crops can be estimated by the use of solarimeters, instruments which convert the radiation which falls upon them into an electric current. The easily measured voltage produced is directly proportional to the radiation intercepted. By means of a solarimeter placed above the crop and one at ground-level below the crop canopy, the amount of radiation actually intercepted by the crop over its life-span can be measured.

In Fig. 19.1, the total dry matter yields of sugar beet and potato

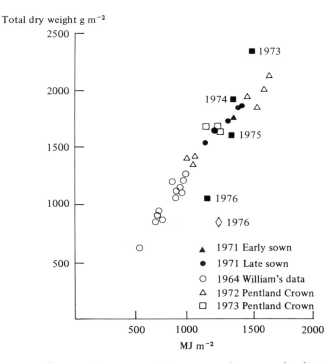

Fig. 19.1   Relation between total dry matter at harvest and radiation intercepted by foliage throughout the growing season: potatoes (open-) and sugar beet (closed symbols). The low efficiency with which radiation is converted into dry matter in 1976 is associated with drought
Source: Scott & Allen (1978); Williams, R.A.S. (unpublished data)

crops grown with different treatments (e.g. variety and time of sowing) and in different years have been plotted against the total amount of radiation intercepted.

With the exception of 1976, a dry growing season, total dry matter yield is very closely related to intercepted radiation. If the dry season is ignored, light energy is converted into dry weight at an average rate of about 1.4 g dry matter per megajoule (MJ) of radiation intercepted. Similar relationships have been shown for crops as dissimilar as barley (maximum standing dry weight) and golden delicious apples (Monteith, 1977).

Using this sort of analysis of the growth of crops it can be seen that the total dry matter yield of a crop will depend upon (a) the total amount of radiation intercepted by the crop canopy; and (b) the efficiency with which intercepted radiation is converted into dry matter.

As the farmer is not necessarily interested in total dry matter yield but in some fraction of it, he will also be concerned with the way dry matter is distributed within the crop.

The total amount of radiation intercepted by crops

This will depend upon the pattern of light receipts in a particular locality and season, which is less variable than most climatic parameters, and upon the development of the crop canopy – the above-ground light-intercepting structures of the crop, principally the leaves. A useful measure of leaves, relevant to their ability to intercept light, is the leaf area index (L), defined as the leaf area per unit area of land. It expresses the number of complete layers of leaves displayed by the crop, although of course leaves do not form unbroken layers one above the other. Leaves occur in different shapes and sizes, are held at different angles and may occur on long or short stems – all of which affect the 'structure' of the crop canopy. In the cereal crop, leaf sheaths are included in L, and it should be borne in mind that other structures such as the ears of cereals may be important photosynthetic organs. Under British conditions it would seem that values of L between 4–7 are required to intercept most of the incident radiation. In Fig. 19.2, the total light energy receipts, the development of L in a sugar beet crop and the resultant amount of light intercepted are shown to illustrate these relationships.

A knowledge of the temporal patterns of light energy receipts, the development of L, and the resultant interception of radiation by plants is fundamental to an understanding of the biological basis of crop production.

The efficiency of conversion of intercepted radiation into dry matter

It was suggested earlier that the average efficiency of conversion of intercepted radiation into dry matter (E) was about 1.4g/MJ. Several factors affect E, but few are within the control of the farmer or the plant breeder.

The efficiency of photosynthesis of individual leaves is greatest at low light intensities (Fig. 19.3a) because, at higher light intensities, carbon dioxide availability limits photosynthesis. In crop canopies, however, because of the geometrical arrangement of leaves

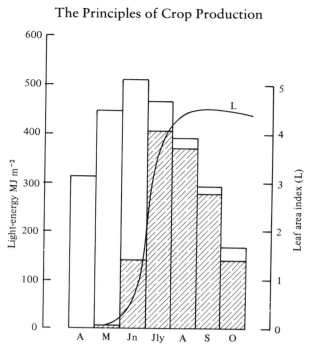

Fig. 19.2   Light energy receipts □ and the amount of light energy intercepted ▨ by a sugar beet crop at Broom's Barn Experimental Station (Suffolk), and the development of leaf area index (L)
Source: Jaggard & Scott (1978)

and the effects of mutual shading, most of the leaves are intercepting light at relatively low intensities, therefore in crop canopies E remains unaffected over a wide range of light intensities (Fig. 19.3b).

There are important differences between species in the efficiency with which light is utilised. The most efficient group contains many of the high-yielding tropical and subtropical graminaceous species such as sugar cane, maize and sorghum. These are referred to as C4 plants, as the first stable compounds formed during photosynthesis are acids containing 4 carbon atoms. The less photosynthetically efficient plants are known as C3 plants, as the comparable first stable compound contains 3 carbon atoms. A major difference between the groups is the greater efficiency with which C4 plants utilise carbon dioxide. However, this greater efficiency is evident only at high temperatures and light intensities; therefore under temperate conditions these differences will be minimal and offset by the higher threshold temperatures required for the growth of C4

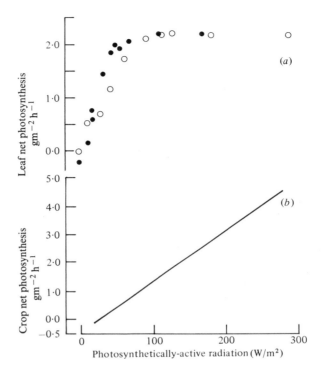

Fig. 19.3
(a) The change in net photosynthesis and radiation for two flag leaves of winter wheat from a crop with (●) and without (○) the addition of nitrogen fertiliser
(b) The relationship between net photosynthesis and radiation for a barley crop during mid-June
Source: Biscoe & Gallagher (1978).

plants, which restricts the period over which such plants can intercept radiation.

Plant breeding has had little or no effect on increasing E in plants and there is evidence that the photosynthetic rate has fallen during the evolution of wheat (Evans, 1975).

The photosynthetic efficiency of individual leaves declines with age: E will therefore be affected by the ability of a crop to replace old leaves with new ones. For some crop species (e.g. wheat), this ability is limited by the termination of the growing point in an inflorescence, which effectively limits the number of leaves which can develop. This type of growth is described as 'determinate' and can be contrasted with crops such as peas and beans in which flowers develop in the axils of leaves and new leaves can be produced throughout the entire life-cycle: this is known as 'indeterminate' growth.

While carbon dioxide is being taken up and oxygen released in the process of photosynthesis, the reverse process – known as respiration – is taking place. E therefore represents the net effect of photosynthesis and respiration. Respiration accounts for a loss of some 30–50% of the dry matter produced by photosynthesis. It is needed to release energy for the synthesis of new compounds required for growth and for the maintenance of the existing plant structure, the latter requirement increasing as the crop grows. Respiration rate is strongly affected by temperature, doubling for every 10°C increase: temperature, however, will only affect maintenance respiration as growth responds in a similar way to a rise in temperature.

Water stress is the most important environmental factor affecting E (see Fig. 19.1). The holes in the leaves (stomata), through which carbon dioxide diffuses to reach the chloroplasts, close in response to water stress, reducing the supply of carbon dioxide and hence the rate of photosynthesis.

The amount of water lost by plants (transpiration) is related closely to the amount of radiation intercepted by them, hence transpiration is often related closely to yield. There may even be a closer relationship than that between intercepted radiation and yield, as the amount of transpiration takes into account the reduction in E associated with the build-up of water stress in crops (see Penman, 1972).

There is some evidence that the capacity of a crop to accept the products of photosynthesis, or the 'sink capacity', may limit photosynthesis, as also may the ability of a plant to transport the products of photosynthesis from the 'source' (e.g. leaves) to the 'sink' (e.g. storage organ, such as grain, root or tuber). A limitation to cereal yield, for example, may be the number of grain sites in which assimilates can be deposited – the potential number of these sites being determined before grain filling has started.

Distribution of dry matter

The proportion of the total recoverable biological yield formed by the economic yield is often referred to as the harvest index (HI), thus: economic yield = recoverable biological yield × harvest index. HI will vary considerably between species: for example, maincrop potatoes may have an HI of about 85%, modern winter wheat varieties 40–50% and oil-seed rape about 15–25%. It is obvious that improvements in HI can lead to increases in economic yield without there necessarily being any concomitant increase in recoverable biological yield. Indeed, much of the improvement in cereal yields due to breeding is attributable to improvements in HI.

Changes in the environment or in the way in which crops are grown may have important effects on HI through the way crops grow and develop. Growth may be defined as an increase in size, and is measured normally in terms of dry weight (although it might be noted that, in seedlings, growth is better defined as an increase in fresh weight, as it is largely related to the uptake of water). Development may be defined as the progress towards maturity of a plant. Some of these changes are abrupt and easily recognisable, such as germination, flowering and senescence; and clearly alter the growth pattern of a plant. Other changes such as those taking place on the growing point of a cereal plant which determine whether leaves or flowering parts are being initiated, can be seen only by the microscopic examination of growing apices.

It is evident, even to the casual observer, that plants grow and develop in an orderly manner. Developmental changes are under the genetic control of the plant, but they may need to be triggered by signals picked up from the environment.

> For example, the latest 'safe' drilling date of most winter wheat varieties in Britain is the middle to end of February; winter wheat drilled later may remain vegetative and not produce any flowers and hence grain, and the HI will be zero. On the other hand, sugar beet drilled too early in the spring will be induced to flower in the same growing season and the plants are said to 'bolt'. In this case, flowering is undesirable and will reduce HI and have other undesirable effects, such as increasing harvesting difficulties and giving rise to potential 'weed beet' problems in subsequent crops. These crops are in effect 'biennials', and the environmental cues needed to induce flowering are periods of low temperature and/or short days. The choice of appropriate drilling dates enables the farmer to treat these crops as annuals or biennials for the particular component of yield he requires. Of course, the sugar beet seed producer will treat the crop like winter wheat and induce it to flower by drilling the crop in the early autumn.

The biological mechanisms bringing about such changes in development are many and complex, but one of the important systems controlling development is provided by chemical substances produced within plants known as 'plant growth substances' or 'plant hormones'. An endogenous plant growth substance may be defined as 'an organic substance which is produced within a plant which will, at low concentrations, promote, inhibit or qualitatively modify growth, usually at a site other than its place of origin.' Its effect does not depend on its place of origin.

Before leaving the question of development, two important factors may be noted. Within a crop species, different genotypes may differ in their developmental responses to environmental signals such as day-length, depending on their place of origin. This places some restrictions on the usefulness of the transfer of varieties developed in one latitude for use in areas further north or south, where the day-lengths are different. An historic example is provided by the potato, where the original crop, adapted to low latitudes in the Andes of South America, had short-day requirements for tuberisation and tubered very late in the long days of temperate latitude summers. It was nearly 200 years before the crop became sufficiently well adapted to these conditions to make any important contribution to food supplies.

The second point is that endogenous plant-growth regulators or chemicals closely related to them can be manufactured and applied to crops in ways which may affect their growth and development. One such chemical is chlormequat which is widely used to shorten straw and decrease the risk of lodging in cereals. A considerable research effort is being put into the development of such chemicals for use in agriculture.

Changes in the crop environment which affect growth rather than development may also affect the harvest index: for example, heavy rates of fertiliser and high plant densities often decrease HI, although the absolute increases in growth achieved by such practices may outweigh any reduction in HI. The approximate economic balance has to be determined empirically in field experiments.

Potential crop yield

The basic principle on which crop production is based is that of the large-scale conversion of light energy into the chemical energy contained in crops.

This principle can be used as a basis for action and is useful in estimating potential yield of a crop in a particular environment with which comparisons of actual yields achieved can be made. Such estimates are useful to show how big a discrepancy there might be between the best yields, or of average yields and the potential yield, to consider the reasons for such discrepancies and by implication to find out what scope there is for improvement.

The simplest calculation is (a) to take the total amount of energy which falls on an area of land over a year and (b) to divide this by the energy contained in a known unit of crop dry matter, say a gram, and convert this into practical units of tonnes of dry matter per hectare. In Britain (a) is about 3.3 GJ/m^2 and the amount of

Table 19.1   Average economic and recoverable biological yields of three major crops and the proportion of the total annual light energy their yields represent

| | | Economic Yield | | | | Yield as % 1890 t/ha | |
Crop	Component	Fresh Weight t/ha	Dry Weight t/ha	Harvest index %	Recoverable biological yield t/ha	economic	biological
Potatoes	tubers	30	6·0	85	7·1	0·32	0·37
Sugar beet	sugar	5·6	5·6	50	11·2	0·30	0·59
Winter wheat	grain	5·25	4·5	40	11·2	0·24	0·59

energy stored in dry matter (b) is about 17·5 kJ/g (Monteith, 1977). This would give rise to a yield of 1890 t dry matter/ha! Average total crop dry matter yields have been calculated and are presented in Table 19.1. Clearly on this calculation, with less than 1% of the radiant energy being transformed into dry matter, the process of converting radiation into crop production cannot be regarded as being very efficient.

Before considering how much of this inefficiency can be attributed to the shortcomings of the farmer, it should be made clear that much of this inefficiency is unavoidable.

Only 50% of the incident radiation is photosynthetically active and of this 50%, some 15% is reflected from or transmitted through the leaf. The maximum biological efficiency with which photosynthetically active radiation can be converted into glucose is 22·5% and this is only the case at low light intensities; finally, an allowance must be made for the loss of dry matter due to respiration. The effect of these losses on the conversion efficiency of photosynthetically active radiation into crop dry matter in a C3 plant is given in Table 19.2, where it is shown that about 5% of the photosynthetically active radiation (or 2·5% of the total radiation) should be converted into crop dry matter.

Given that the total amount of radiation received per day in June is about 170 MJ m^{-2} (Fig. 19.2) and that 2·5% of this is converted into crop dry matter with an energy content of 17·5 kJ per gram, then the rate of crop growth would be 243 kg/ha/day. Short-term maximum growth rates of many crops in NW European conditions reach about 200 kg/ha/day (Fig. 19.4), which agrees well with this calculation if allowance is made for the fact that in the estimates of crop growth rates in the field, roots were not included.

Since crops convert intercepted radiation into dry matter at a rate of about 1·4 kg/MJ total radiation, we can estimate that a crop in mid-summer receiving 17 MJ of total radiation per m^{-2} day^{-1} would give

Table 19.2 **An account of the conversion of photosynthetically-active radiation (PAR) into stored energy in crop dry matter by a C₃ crop (Heath & Roberts, 1981)**

	Relative Yield
1 Photosynthetically active radiation (PAR)*	100
2 The crop absorbs only 85% of PAR because of reflection and transmission	85
3 Maximum efficiency of conversion of PAR into glucose, 22·5%	19
4 Less 40% to allow for leaves in the crop canopy receiving light at a higher radiant flux	11
5 Less 30% to allow for photo-respiration	8
6 Less 40% to allow for dark respiration, i.e. 5 units of energy stored in the crop dry matter	5

* about 50% of the total radiation

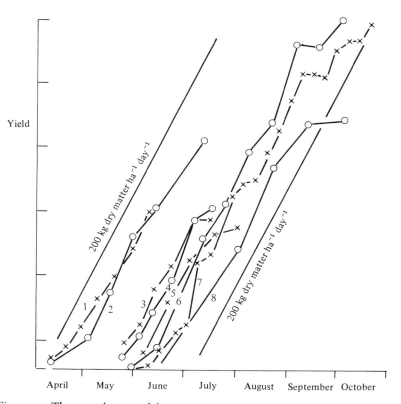

Fig. 19.4 The growth rates of the main agricultural crops in The Netherlands under (near-) optimal growing conditions, compared with a growth curve corresponding to 200 kg ha⁻¹ day⁻¹. 1 Grass; 2 Wheat; 3 Oats + peas; 4 Peas; 5 Barley; 6 Potatoes; 7 Sugar beet; 8 Maize
Source: de Wit, Laar & Keulen (1979)

rise to an increase in yield of $17 \times 1.4 = 23.8$ g m²/day or 238 kg/ha/day, which again agrees very well with the foregoing estimates and measurements.

It is possible to estimate the potential yield of a crop if we know the pattern of radiation receipts over the year, the pattern of leaf development (i.e. the % of radiation intercepted by the crop) and the proportion of dry matter yield found in the economically useful part of the crop.

In Fig. 19.5 diagrams are shown of three crops of major import-

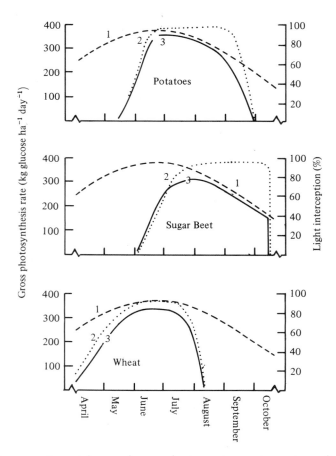

Fig. 19.5   Potential gross photosynthesis rate (curve 1), percentage light interception (curve 2) and calculated actual gross photosynthesis rate (curve 3) for potatoes, sugar beet and wheat

Source: Sibma (1977)

ance in western Europe: potatoes, sugar beet and winter wheat
(Sibma, 1977). In each diagram, line 1 is the gross production rate
(dry matter/ha) for a green, closed crop canopy optimally supplied
with water and minerals, which is determined by the amount of
radiation and how efficiently this is converted into dry matter.
Curve 2 shows the percentage of light intercepted by the three crops
and curve 3 shows the total production of dry matter in kg/ha/day.
As this line represents total dry matter, allowance must be made
for respiration, loss of dead leaves, and the harvest index in estimating economic yield. Assuming a respiration loss of 25% of dry
matter, the calculated potential yields of these three crops, for
radiation received at latitude 52°N is 96 tonnes of tubers, 77 tonnes
of sugar beet (approx. 12 tonnes sugar), and 11 tonnes of wheat per
ha per annum. These compare with average yields of 30 tonnes of
tubers, 35 tonnes of sugar beet and 5·25 tonnes of wheat per ha.

It is apparent that yield potential should increase from north to
south, as the potential length of the growing season becomes less
and less restricted by temperature, and the intensity of radiation
increases. In Fig. 19.6 potential gross production per annum is

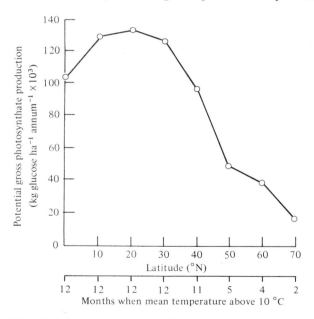

Fig. 19.6   The relationship between latitude for a transect along longitude 20°E
and estimated potential gross photosynthate production assuming that the
length of the growing season can be approximated to the time that the mean
monthly temperature is 10°C or above
Source: Heath & Roberts (1981)

plotted against latitude and shows that potential crop production in the subtropical latitudes is more than double that of the latitudes in which Britain lies (Heath & Roberts, 1981).

The main scope for increasing yield lies with extending the period over which there is a complete crop canopy, in order to increase the amount of radiation intercepted, and with increasing the harvest index. Inspection of Fig. 19.5 shows that only with winter wheat did the peak of light interception coincide with the peak of potential production (i.e. midsummer), but with the potato crop there would seem to be scope for intercepting more radiation earlier in the growing season, and even more so for the sugar beet crop where there is scarcely any radiation intercepted before the beginning of June. It has been estimated that if all the radiation in April, May and June were to be intercepted then the potential yield of sugar would be about 20 t/ha. This would require the crop to be sown in the autumn, which in turn would require some method of preventing the crop from running to seed in the following summer.

These considerations of potential crop yields are useful both to the practising farmer and to those engaged in trying to raise the yield potential, whether through techniques aimed at modifying the plant environment or the genotype.

THE PRACTICAL ACHIEVEMENT OF HIGH CROP YIELDS

The strategy for achieving yields that are close to a crop's potential, using economically justifiable methods of production, is relatively simple. It is to establish a crop canopy capable of intercepting most of the incident radiation, as rapidly as possible; to maintain that canopy, assimilating efficiently, for as long as possible (or desirable); and, where yield is a fraction of the recoverable biological yield, to maximise the harvest index.

The means by which the farmer may seek to implement this strategy are (a) the genetic material at his disposal (choice of species and cultivars); (b) the manipulation of this material (affecting emergence or density); (c) the modification of the crop environment; and (d) the control of negative biological agencies (weeds, pests and diseases).

In addition to producing high yields, these must be harvested and in most cases the produce stored, either for a matter of days or, in some cases, months, although seldom, by farmers, beyond a year. As the farmer's returns are determined by the quantity and quality of the product sold, he must pay due attention to the efficiency of harvesting and storage, and to any effects his production practices may have on them and on the quality of his produce.

Establishment of the crop

The establishment phase is a very vulnerable period in the life-cycle of a crop. The requirements for germination may not be met by the seedbed conditions, while the seeds and germinating seedlings may form a welcome food supply for animals, birds, fungi and insect pests. After germination the seedling must establish a root system and the shoots must find their way above ground before they can begin to intercept radiation and photosynthesise. Until this point, the plant is dependent upon the reserves of the seed or vegetative material – which, if depleted before the crop has emerged, will result in the death of the seedling. It is clearly of the utmost importance that the establishment phase should be completed as rapidly as possible, both in order to reduce the risk of death and to establish the crop as a photosynthesising unit.

Genotypic effects

There may be genetic differences within a crop species to be exploited with regard to the base temperature at which the crop will begin to germinate, but the major differences here are between crop species. Generally, little growth of temperate species takes place below soil temperatures of about 5°C, and for species, such as maize or phaseolus beans, native to warm temperate or subtropical environments the base temperature is about 10–15°C.

Manipulation of the seed

The germination capacity of the seeds sown is clearly an important determinant of how many seeds may be established.

> Legislation lays down minimum standards for germination and tests are carried out, in carefully defined conditions, to indicate the potential capacity of the seed to germinate under ideal conditions. Conditions in the seedbed are, of course, often not ideal, and there is some evidence that different seed lots, all of acceptable germination standards, may perform differently in the field, especially under adverse conditions. The ability of a seed lot to perform well under adverse conditions is termed 'seed vigour' and is defined as 'a physiological property determined by the genotype and modified by the environment, which governs the ability of a seed to produce a seedling rapidly in soil and the extent to which that seed tolerates a range of environmental factors' (Perry, 1970).

Treating seed to speed up the rate and improve the uniformity of emergence is also a possibility which may be attractive with species which are slow to germinate and emerge above ground.

> These treatments include partially imbibing the seed with water, but with insufficient to allow the seed to germinate. The seed is then dried

and sown in the normal way. This procedure is said to initiate the first phase of germination which includes the mobilisation of enzymes, which make the food reserves available to the embryo and also assists in the repair of damaged cells. The process may be taken further, by soaking seeds in aerated water and allowing germination to take place. Such seeds would easily be damaged by conventional drilling, and are protected by a viscous gel in a process known as 'fluid drilling'. The technique is not normally practised with commonly grown farm crops.

Seed may also be selected on the basis of size, and, in general, larger, heavier seeds are more likely to be viable than small, light seeds; such seeds are often removed by sieving or other means by the seed merchant.

The conditions under which seed is harvested may also affect the subsequent performance of the crop: for example, sugar beet seed harvested under cool damp conditions is more likely to 'bolt' or run to seed in the following year than seed harvested under warm, dry conditions.

The storage environment of the seed may affect viability. In general, the higher the temperature and moisture content at which seed is held, the faster the seed will deteriorate; hence seed should be stored under cool, dry conditions before use.

In British farming, seeds are the normal means by which crops are propagated, the exception to this being the potato crop, which is propagated vegetatively from tubers. With this crop, it is much easier to speed up development after planting, because growth of the tuber can be encouraged in the store before planting; this is standard practice when rapid establishment of plants grown for an early market is required.

Modification of the seed environment

The physical environment of the seed   The positive requirements of the seed for germination are water, air and a suitable temperature. Soil temperature is largely outside the farmer's control, depending largely on latitude, altitude and aspect. Sowing date will, therefore, largely dictate the soil temperature conditions met by the seed. In general, farm crops are not valuable enough to merit the expense of protecting the crop artificially, e.g. by plastic cloches, to encourage early growth.

The supply of water and air to the seed are related, under given climatic conditions, to the physical condition of the soil in the vicinity of the seed.

The basic problems facing the farmer, which will vary in severity according to the soil type and climate, are to provide the seed with sufficient air and water, and to minimise the risk of erosion due to rainfall or wind, which may remove both soil and seed. As the soil lacks vegetation at the time of sowing, the bare soil surface is particularly vulnerable to erosion due to the impact of raindrops or from wind.

Before considering what steps might be taken to optimise the physical environment of the seed it is important to be aware of the complexity of the soil environment, which may change rapidly both horizontally and vertically and with time. The soil can be broken up physically into inorganic constituents of various size particles, the proportions of which, together with the organic matter, determine the soil texture (Table 19.3).

Table 19.3 Classification of soil particle sizes

Soil Fraction	Size (microns)
Gravel	2,000 – 20,000
Course sand	500 – 2,000
Medium sand	200 – 500
Fine sand	100 – 200
Very fine sand	20 – 100
Silt	2 – 20
Clay	less than 2

These particles, together with the soil organic matter, may be further combined into aggregates which give the soil its structure. Between the particles and aggregates of particles are pore spaces of varying size: some of these are small enough to hold water by surface tension against the pull of gravity; above a certain size, water cannot be held and the spaces will be filled with air. Superimposed on this 'fine' structure are fissures and channels which arise from many causes, such as shrinkage and expansion on drying and wetting, channels formed when roots decay, and worm channels.

Owing to the structural characteristics just described, the soil can supply both oxygen and water to the seed and growing plant. It is logical to assume that there is an 'ideal' combination of particles and their arrangement into aggregates which will give rise to the most favourable air and water environment for the germinating seed. It has, for example (Russell, 1973), been stated that the ideal seedbed should consist of soil crumbs not much finer than 0·5-1·0 mm and not much coarser than 5-6 mm in a fairly firm packing. Finer crumbs will block the coarser pores needed for drainage and coarser crumbs are likely to give rise to open seedbeds in which the seed may be in poor contact with films of water and the soil will also be prone to dry out in the vicinity of the seed: it is particularly important that a germinating seed should not run short of water and become desiccated.

Pore size is also important with respect to root penetration. If roots have to exert even small pressures to enlarge the pores into which they are growing, their growth is considerably reduced. The majority of roots exceed 60 μm in diameter and the existence of a sufficient number of continuous pores into which roots can freely enter is an important requirement for their growth.

No simple solutions can be given to the problem of achieving an optimum arrangement of soil particles and aggregates or of how to maintain the stability of the soil structure. Traditionally, cultivations, especially mouldboard ploughing, have been used to achieve alterations in soil structure, while the stability of the soil structure has been favoured by the maintenance of organic matter through the application of organic manures and the operation of 'ley' farming systems where grass crops of several years duration alternated with arable crops.

Recent years have seen marked changes in attitudes towards cultivations, and particularly with regard to the basic cultivation process of ploughing. Reduced, minimal or zero tillage methods are being explored, not only with a view to reducing the cost of numerous tillage operations, but also with a view to improving the soil environment from the point of view of the emerging and emerged crop.

Table 19.4 The effects of cultivation systems on labour, energy requirements and yields of winter wheat (Bullen, 1977)

	1971–6		
Cultivation system	man hrs ha^{-1}	energy MJ ha^{-1}	yield t ha^{-1}
Plough (20 cm): disc or cultivator: drill	3·5* (2·2)	330 (179)	5·92 (4·91)
Chisel plough (13 cm) × 2 disc or cultivator: drill	3·4 (2·3)	310 (210)	5·78 (4·81)
Plough (20 cm): combined cultivator and drill	3·0 (1·9)	340 (178)	5·82 (4·76)
Shallow plough (10 cm): combined cultivator and drill	1·6 (1·2)	193 (104)	5·71 (4·65)
Sprayer: direct drill	0·5 (0·6)	27 (36)	5·62 (4·53)

* Clay loam (silty loam)

In Table 19.4 the effects of various cultivation systems on labour, energy requirements and yields of winter wheat are given for two soil types. They indicate that under the conditions of the experiment, yields were relatively insensitive to the various cultivation systems, but that these could have a profound influence on the time they involve, the energy required to perform them and cost. Because of the reduction in time some systems increase the flexibility or choice of planting date, which may have important effects on yield.

Waterlogging is the prime cause of anaerobic soil conditions, which occur when the rate at which oxygen enters the soil from the atmosphere is less than that at which it is used in the respiratory processes of plant roots, bacterial fungi and other soil organisms.

Because oxygen diffuses some 10,000 times faster in the gas phase than in solution, filling the pore spaces with water has a major influence on the availability of oxygen for respiration. Lack of oxygen is not, however, the only injury plants suffer in anaerobic conditions, for many other harmful changes, both chemical and physical, contribute to plant damage. For example, ethylene gas is formed: this is an endogenous plant growth regulator and can induce biological effects at very low concentrations, effects which include the stunting of roots and premature senescence. Ethylene will persist in poorly drained soils in the field for weeks at levels high enough to have appreciable effects on root development. It is possible that the ethylene content of many poorly drained soils is a more common source of damage to crop growth than lack of oxygen or a high concentration of carbon dioxide in the soil.

The injury caused by transient waterlogging can vary considerably with stage of growth, but the germinating seed is particularly prone to injury. With shoots above ground, the possibility of transfer of oxygen through the shoots to the roots exists, through the development of intercellular air spaces in the cortex of roots (*aerenchyma*), particularly well developed in rice and aquatic plant species.

Drainage is obviously important in humid climates when, for large parts of the year, rainfall may exceed evaporation. In Britain, for example, with spring-sown crops when the soil is bare at the time of drilling and also near to field capacity (i.e. the soil is holding all the water it can against the pull of gravity), little water is lost from the soil by evaporation once the top few centimetres of the soil are dry, and none is lost by transpiration (assuming there are no weeds) since there is no leaf cover. Rainfall will therefore fill up the large air spaces if for some reason, such as the presence of a clay subsoil, the water in excess of field capacity cannot drain away.

Remedies for such problems lie in draining the land artificially by means of an underground system of pipes, often combined with the formation of continuous channels in the clay by means of 'mole-ploughs'. Drainage problems associated with high water tables, as in river valleys, or on land below sea-level as in the Dutch Polders, or parts of the Fens, present more difficult drainage problems and may need expensive pumping equipment to keep the surface soil drained.

Soil erosion is particularly likely to occur when there is no vegetative cover of the soil surface. Water erosion can occur on sloping land and particularly under the impact of intense rains such as occur frequently in the tropics. Under the impact of the rain-drops, the soil aggregates are broken down, the pores become blocked, the rate of infiltration of the rain into the soil drops dramatically and water then flows over the soil surface, taking with it particles of soil. While the erosion of soil by water is continually taking place in temperate countries such as Britain, the rainfall intensity is such that the effects are rarely dramatic and its serious incidence is localised in time and space. Solutions to the problem are sought by ensuring that no large area of bare soil is exposed to rainfall, and that the velocity of water running over the soil surface is reduced by techniques such as cultivating and cropping along contour lines.

Factors predisposing soil to wind erosion are the soil texture, fine sands (100–200 μm) and organic soil aggregates (up to 1 mm) being particularly at risk, and the absence of any wind-breaks or vegetative cover. Solutions include the provision of wind-breaks and the retention or provision of vegetative trash on the surface soil.

The chemical environment of the seed   The only chemical re-quired for germination is water, and this has been discussed in the previous section. However, shortly after germination, given an adequate supply of water, the major limitation to growth will be the nutrient supply – although the reserves in the seed (or vegetative propagation unit) are likely to be sufficient to provide the emerging seedling with sufficient nutrients for the shoot to appear above ground. Of the utmost importance is the pH of the soil.

A neutral soil has a pH of 7, soils are said to be increasingly acid as the pH falls below this value and increasingly alkaline as the pH rises above it. The pH scale is logarithmic and a decrease from 7–6 implies a tenfold increase in acidity, and from 7–5 a hundredfold increase. Over a wide range, probably from pH 4–8, the harmful effects of acidity are indirect and are mainly related to the effect of pH on the concentration of different ions in the soil solution and hence their availability to the plant. A high aluminium content is the main cause of crop failure on acid soils and pH has an important influence on the incidence of trace element problems – all except Mo tending to become less available to crops as the pH rises.

Plants vary in their tolerance of soil pH and can roughly be divided into three groups: those adapted to and which only grow well on alkaline soils (calcicoles), those adapted to and which grow well only on acid soils (calcifuges) and those that tolerate a wide range of soil pH values. Table 19.5 shows the pH below which

Table 19.5   Critical soil pH values (Davies, Eagle & Finney, 1977)

Crop	pH	Crop	pH
Apple	4·9	Lucerne	6·1
Barley	5·8	Mangolds	5·5
Bean	5·8	Mustard	5·3
Beet	5·8	Oats	5·0
Brussel sprouts	5·6	Onion	5·6
Cabbage	5·3	Pea	5·8
Carrot	5·6	Potato	4·8
Cauliflower	5·5	Rape	5·5
Clover, red	5·8	Rye	4·8
Clover, white	5·5	Ryegrass	5·0
Cocksfoot	5·6	Swede	5·3
Kale	6·0	Timothy	5·2
Lettuce	6·0	Turnip	5·3
		Wheat	5·3

crops may be expected to fail. A slightly acid soil (pH 6·5) is probably best for a wide range of crops.

Acidity generally arises when soils are deficient in calcium. This is removed from soils by leaching and is therefore affected by the amount of drainage and also occurs on soils where there is little negatively charged colloidal material (i.e. sands as opposed to clays) to bind on to the positively charged calcium ions. It is corrected by the addition of calcium carbonate (lime) (see Chapter 11).

Problems arising from alkaline soil conditions are relatively rare in temperate agriculture – but may occur when land has been flooded with sea water. Under hot, arid conditions, soluble salts accumulate in the surface of soils whenever the ground water comes within a few feet of the surface. During dry periods, the surface of the soil is covered with a salt crust, which dissolves in the soil water each time it is wetted. The salts are usually the sulphates and chlorides of sodium and calcium, though they may occasionally be nitrates; the soils are light in colour and the pH, whilst high, is usually below 8·5. Under some conditions, sodium carbonate may accumulate; this raises the pH to 9 or 10. If sodium carbonate is predominant the humic matter in the soil may be dispersed and the soil assume a black colour.

Saline soils may arise naturally or through faulty irrigation and in arid regions salt control on irrigated soil is essential. The irrigation water always contains dissolved salts. The salt content of irrigation water and soil solution may be specified by its electrical conductivity and used as a guide to the frequency with which additional water must be added to leach the salts down the profile and as a guide to which crops to grow, as some are more tolerant of alkaline conditions than others. The harmful effects of soil alkalinity may be due to the specific toxicity of the ions contained or to the general effects of high concentrations of salts on the osmotic pressure of the soil solution – which can affect nutrient uptake, transpiration, photosynthesis and respiration. The general effect of this is to give dwarf, stunted plants and dull-coloured bluish-green leaves, which are often coated with a waxy deposit. The amount of soluble borates is important as boron is toxic to plants above certain minimum levels. Irrigation water containing more than 0·3 ppm B must be treated with caution, and 2–4 ppm restricts cropping to boron-tolerant crops, e.g. sugar beet, lucerne and some brassicas.

Because of the importance of leaching with excess irrigation water in controlling salt concentrations in arid land irrigation, it is essential that irrigation schemes should be complemented by adequate drainage facilities. Failure to ensure this may result in crop production having to cease because of salinity problems.

It is in the seedling stage that crops are most vulnerable to the effects of soil salinity. Some of the harmful effects at this stage can be avoided by drawing the land into ridges having flat sides either coming to an apex or having a flat top. Salts accumulate in the apex, or the centre of the flat top as this is where the wetting front of irrigation applied in the furrows will meet. Seeds planted on the sides or the edge of the top will be able to develop in soil with a salt content close to that of the irrigation water.

In temperate regions problems of salt concentration on the development of seedlings are most likely to occur from the use of fertilisers applied to the seedbed. Phosphates present no problem as they are relatively insoluble in the soil solution and hence have little influence on the osmotic pressure of the soil solution, but salts carrying nitrogen, potash or sodium (applied in the form of salt for sugar beet) may, if the levels applied are high and the rainfall light after drilling, have harmful effects on germination. In general the timing and placement of fertilisers can be manipulated to avoid harmful effects of these chemicals on emergence.

Another hazard for the germinating seed may arise from the use of soil-acting herbicides to control weeds in the early stages of growth. Faulty spraying, or environmental factors which reduce the tolerance of the crop or increase the concentration in the vicinity of the germinating seed, may have adverse consequences on germination.

Negative biological factors

The seed or germinating seedling may be attacked by a wide range of birds, small mammals, fungal and bacterial diseases. One strategy relevant to this stage of growth is to treat the seed with a fungicidal and/or pesticidal seed dressing: occasionally the seed may be dressed with chemicals with the object of repelling birds.

The other major biological hazard at this stage is provided by the potential competition from unsown plants, or weeds. Many arable soils contain a vast reservoir of weed seeds, often of the order of 5,000–10,000 viable weed seeds per m², but not all these seeds will germinate at any one time. Figure 19.7 illustrates some

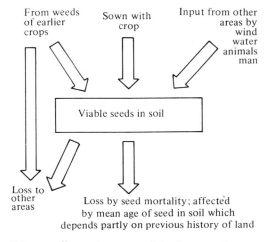

Fig. 19.7   Diagram illustrating some of the factors influencing the total population of viable seeds in the soil
Source: Hill (1972)

of the factors influencing the total population of viable weed seeds in the soil. In addition to seeds, some weeds may persist and regenerate from vegetative parts of the plant, for example from underground stems (rhizomes) in the case of couch grass (*Agropyron repens*).

Before the fairly recent influx of a galaxy of chemical herbicides, permitting the selective control of weeds in crops, horse- or tractor-drawn hoes provided the only major, rapid method of achieving selective weed control in the early stages of crop growth.

A major principle of crop production, which may be stated in relation to weeds, is that unless crops are kept practically free from weed competition, especially in the early stages of growth, yields will be reduced drastically.

Any weed control strategy should aim not only to eradicate weeds, but also to prevent new weed seeds being added to the soil. A major contribution to this objective is to sow seed which is free from weed seeds. Seed certification schemes which ensure certain minimum standards of weed contamination make a major contribution to this objective. Seed certification schemes are also important in ensuring freedom from diseases, and in the case of the vegetatively propagated potato especially, freedom from virus diseases.

The development and maintenance of an efficient crop canopy

A major principle which should guide practice is to maximise the interception of radiation by the crop as rapidly as possible and then to maintain the leaf canopy operating efficiently at this level of light interception for as long as possible or desirable, although this overall objective may need to be modified in particular cases.

Genotype

The choice of crop species and to a lesser extent the choice of cultivar will have marked effects on these objectives and will also interact with the other techniques at the disposal of the farmer. For example, leguminous crops (e.g. peas, beans, lucerne) will in general be unresponsive to additions of nitrogen fertiliser, due to their symbiotic association with bacteria which enable them to fix nitrogen from the atmosphere. Crops such as lucerne are far less affected by drought than others, such as grass and potatoes.

In adverse conditions, such as the semi-arid tropics, where crops are often short of water and plant nutrients, it appears that growing two or more crops together (intercropping) often makes better use of the limited environmental resources than growing crops in pure stands.

Manipulation of the crop

The major factor of importance here is the manipulation of plant numbers and their spatial arrangement. Plant density, sometimes referred to as plant population, is the number of plants per unit area of land, and their spatial arrangement concerns the geometrical relationships between the plant units.

The greater the number of plants per unit area and the more evenly they are spaced, then the greater the speed with which the ground surface will be covered by leaves and complete interception of radiation attained (Fig. 19.8). However, this does not mean that the optimal plant densities are those which result in the most rapid attainment of complete light interception, or that the optimal plant spacings approximate to square planting patterns, for the reasons discussed below.

Figure 19.9 shows the relationship between total biological yield and plant density at successive intervals of time. At the first harvest (time

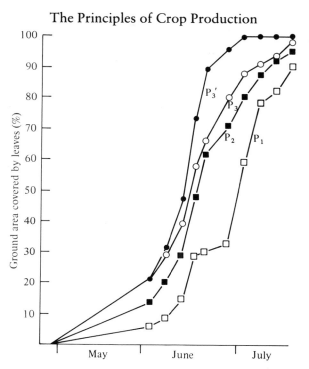

Fig. 19.8 The effect of plant density and spatial arrangement on the proportion of the soil surface covered by leaves when viewed directly from above.
(□ = 32,000 plants/ha 56 cm rows, ■ 86,000 plants/ha 56 cm rows, ○ 133,000 plants/ha 56 cm rows, ● 141,000 plants/ha 26 cm rows)
Source: Harris (1972)

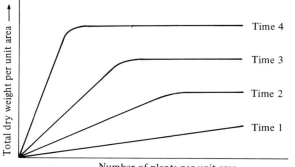

Fig. 19.9 Diagrammatic representation of the relationship between plant population and total weight per unit area on four occasions
Source: Bleasdale (1973)

1), taken so soon after the crop has emerged that the individual plants have been unaffected by competition (e.g. for light) with their neighbours, then yield will be directly proportional to plant density. As the plants grow they begin to compete with their neighbours and this will begin first at the highest density which leads to mutual shading between plants. This reduces the quantity of light received by an individual plant and growth rate is therefore reduced. Competition is delayed between plants grown at wider spacings and the individual plants will continue to grow at a faster rate for some time. Thus the plant density at which maximum biological yield is attained gets smaller as time progresses. Note that the maximum biological yield does not usually fall at high plant densities, for when light interception has been attained, the reduced growth of the individual plants is compensated by the greater number of plants; crop growth rate, as opposed to the growth rate of individual plants, therefore reaches a constant largely fixed by the level of incident radiation.

Plant density may affect E by altering the structure of the crop canopy, although this effect is probably less important than the indirect effect of increasing plant density on the rate of water loss from the soil, which increases the possibility of plant water stress reducing E.

In terms of economic yield, plant density has two important effects. In the first place, increasing interplant competition may affect the distribution of dry weight in the crop. In response to shading, stems tend to grow longer and leaves larger and thinner – mechanisms designed to enable the crop to intercept more light. This tends to result in a reduction in the supply of assimilate for other plant parts such as roots, which may eventually offset any gain in the total amount of dry matter produced in crops where the economic yield is in storage organs such as tubers or roots, e.g. potatoes and sugar beet. Secondly, it will be appreciated from what has been said, that the maximum biological yield may be attained from the collective contribution made by many rather small individual plants. But in certain crops, the size of the individual plant is important in terms of its saleability and hence economic yield. Thus the market for very small carrots or potatoes may be restricted to low-value stock feed. There may also be important premiums either for very large-size grades (e.g. baking potatoes) or grades of closely defined sizes (e.g. carrots). The manipulation of plant density is of the utmost importance where these considerations apply (Figs. 19.10 & 19.11).

There may be important limitations in the flexibility of the spatial arrangement at which crops can be grown. Most farm crops are drilled in rows in preference to broadcasting, as this usually

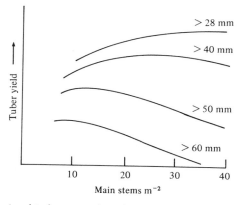

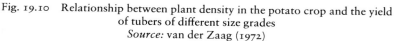

Fig. 19.10   Relationship between plant density in the potato crop and the yield
of tubers of different size grades
Source: van der Zaag (1972)

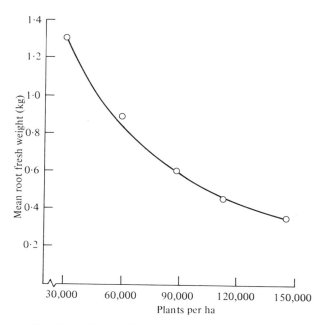

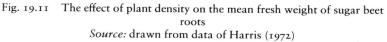

Fig. 19.11   The effect of plant density on the mean fresh weight of sugar beet
roots
Source: drawn from data of Harris (1972)

gives better crop establishment and there are practical limits to the closeness with which rows can be drilled.

Until the recent advent of a wide range of herbicides, it was often imperative to leave a fairly wide space between rows, to allow access for cultivation. Even with the availability of herbicides, they may be used in conjunction with inter-row cultivations for reasons of economy or for the control of weeds for which there may be no suitable herbicide treatment. For any given plant density, the wider the row, the smaller the distance between plants within the row: the geometrical arrangement is often expressed by dividing the longest distance between plants by the shortest, the resulting ratio being referred to as rectangularity. Thus plants arranged on a square planting pattern would have a rectangularity of 1, while a crop (e.g. sugar beet) grown on 50-cm rows with a mean spacing of 25 cm between plants within the row would have a rectangularity of 2. Some crops can tolerate a fairly high rectangularity: for example, cereal crops have been sown traditionally on 18-cm row spacings – given a plant establishment of 300 plants/m² this would give a mean spacing of 1·85 cm and a rectangularity of 9·7; decreasing the row spacing to 10 cm and the rectangularity to about 3 in this crop might be expected to increase yield by a small amount.

Before leaving the subject of plant density, it should be pointed out that the unit in which density is described may give rise to some difficulties. For example in cereals, as tillers (plants derived from buds developed in the axils of the lower leaves) are produced, the original plant loses its identity, the tillers produce roots and become independent plant units, thus making the tiller the logical unit of plant density. In potatoes grown from seed tubers, the ultimate plant density may best be described in terms of the stems arising directly from the seed tuber (main stems), but these are difficult to estimate and are not known before the crop has been established. Other crops such as turnips and carrots do not pose these problems, as one seed gives rise to one plant.

Manipulation of the crop environment

The two major components that can be manipulated are the mineral nutrient and water supply.

A shortage (or unavailability) of one or more essential mineral elements in the soil is one of the most frequent reasons for the failure of crops to develop sufficient leaf area fully to intercept all the incident radiation.

Figure 19.12 shows the effect of three major nutrients on the development of L in potatoes grown in an experiment at Rothamsted Experimental Station. Clearly this soil is fairly well supplied with phosphorus and potassium but not with nitrogen and supplying the latter as fertiliser would be expected to have large effects on the amount of radiation intercepted and on yield.

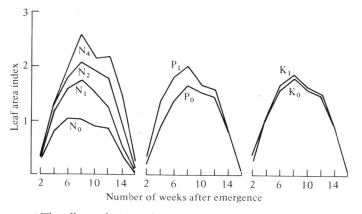

Fig. 19.12 The effects of N, P and K on changes with time in the leaf area index of potatoes, variety King Edward. ($N_0 = 0$, $N_1 = 94$, $N_2 = 188$, $N_4 = 376$ kg N ha^{-1}; $P_0 = 0$, $P_1 = 156$ kg P ha^{-1}; $K_0 = 0$, $K_1 = 82$ kg K ha^{-1})
Source: redrawn from Dyson & Watson (1971)

Nutrients appear to have little effect on E except perhaps where deficiencies are acute (see Fig. 19.3a).

It has been estimated that in the UK, between 36–55% of the production of our four main arable crops (wheat, barley, sugar beet and potatoes) may be attributed to the use of fertilisers. Of the mineral elements N is probably the one most frequently likely to limit yield. In other parts of the world, fertiliser may be unavailable or too expensive for the small farmer to purchase. There is therefore an increasing interest in alternative sources of nitrogen, either through the recycling of waste materials, or from biological fixation, principally through the symbiotic fixation of nitrogen in the nodules of leguminous plants.

The role of the green plant parts in trapping radiation is facilitated by the structure and by the arrangement of the leaves to intercept as much radiation as possible. An inevitable consequence of this arrangement is that a complete canopy of leaves attached to a root system well supplied with water will lose water at about the same rate as an open water surface.

The rate of water loss is fixed by the amount of energy available to vaporise water together with the rate at which water vapour is carried away from the surface of the leaves. For a short green crop which completely covers the ground with leaves and is liberally supplied with water, the amount of water loss can be calculated from standard weather data recorded at meteorological stations. Some allowance must be made for the fact that the stomata of the plants close during the night, which serves to cut down the rate of water loss to some extent. If these conditions are met, then the potential transpiration of crops can be calculated (Penman, 1949).

Potential transpiration is expressed in mm of water per unit of time (day/week) and would be expected to increase as temperature rises and radiation income increases with the advance of the growing season in temperate latitudes. In arid conditions, the loss of water from irrigated crops may be enhanced through the influx of hot dry air from adjacent unwatered areas, an effect often referred to as an 'oasis effect'. In the growing season rates of potential transpiration are typically 1–3 mm water/day: at the other extreme in very hot arid regions the rate may be 10–12 mm/day.

Knowing the rainfall and potential transpiration it is possible to build up a daily, weekly or monthly budget showing the balance between income in the form of rain or irrigation, and output in the form of potential transpiration. If the result is positive, it shows that more water has entered the soil, from rain or irrigation, than has been lost by transpiration. However this surplus water should be lost by drainage (assuming this to be adequate), and any positive value will be regarded as zero. If, however, more water has been lost than gained, a 'soil water deficit' (S W D) of that magnitude will have built up.

The soil acts as a reservoir of water, buffering the crop to some extent between short term fluctuations in supply and demand. The upper limit of water that can be held by a soil is fixed by the 'field capacity', or the water content of the soil at which it is holding the maximum amount of water against the pull of gravity. Plants can utilise the water in the soil profile, although often with increasing difficulty as the water is held in increasingly smaller pores and consequently at high tensions as the content falls, down to a lower soil water content which is described as the 'permanent wilting point'.

The tension at which this occurs is about −15 bars. At this moisture content plants will wilt even when their leaves are placed in a water-saturated atmosphere. The amount of water held between these two limits is said to be the 'available water holding capacity' of the soil. The amount of available water per unit volume of soil is strongly dependent on the soil texture and organic matter content.

Under conditions where the rates of transpiration are extremely high, e.g. on clear days with high temperatures, crops may wilt, even though the soil is close to field capacity. This is because, once the soil has dried within the immediate vicinity of the root, water has to move to the root across a layer of dried soil.

In the absence of any input of water, the diurnal water stress in the plant increases and its ability to recover during the night decreases. Such stresses on the plant inevitably affect growth. In the first instance, the efficiency of conversion of radiation into dry matter is affected, principally through the closure of stomata which cuts down the inflow of carbon dioxide, but as deficits increase, metabolic activities become increasingly disrupted.

The plant also reacts in other ways to reduce stress which also curtail growth, and this is primarily to decrease the amount of radiation intercepted, which as we have seen directly governs the rate of water loss. This takes the form of a decrease in the rate at which growing leaves expand, and an acceleration in senescence of leaves which are eventually shed. Thus water stress affects yields by decreasing the amount of radiation intercepted and the efficiency with which the intercepted radiation is utilised. There is therefore a strong positive relationship between transpiration and yield, and for crops to yield at their maximum potential, they must transpire at the potential set by the environment.

Two other important principles need to be noted here. If there is no crop (or weed) cover, water will be lost from a wet soil at the same rate as from a crop or from an open water surface. Once the surface of the soil is dried, however, the rate of loss will decline rapidly. As a crop cover develops, the rate of water loss will gradually increase and once a leaf cover of about 60% has been reached, water is being lost at rates not far below the potential. Use of this principle is made in dryland farming by keeping the soil free from any crop cover for a year, thus using two year's water supply to grow one crop, and by growing crops at lower densities in low rainfall situations.

The second factor of importance is that plant growth may be severely curtailed at soil water deficits which still allow crops to transpire at about the potential rate. The reason for this is that soil water is depleted from the superficial, mineral-rich layers of the soil progressively downwards, while roots may still extract sufficient water for growth from the mineral-deficient subsoil. As water is essential for the uptake of minerals, which enter the plant in solution, depletion of water in the upper layers of soil may result in a severe depression of growth due to a water-induced shortage.

A major concern of research into crop water stress is to determine when to irrigate and how much water to apply. To achieve this attempts have been made to define 'limiting deficits', or the soil water deficits at which growth is severely restricted, for the range of agricultural crops grown in Britain.

In general, for crops considered economic to irrigate, soil water deficits are maintained at less than 5 mm if possible.

Negative biological factors

Most of the negative influences on yield exert their effect through reducing the ability of the crop to intercept radiation, that is through reducing the leaf area of the crops. Traditionally, reasonable levels of pest, weed and disease control were achieved by growing plants in rotations of contrasting crops, but in modern intensive agriculture there has been a shift in some countries, such as Britain, to increased specialisation with a concomitant reduction in the opportunity to rotate different crops. For example, increasing areas of cereals, and particularly winter cereals, provide greater opportunities for the carry-over of pests and diseases from one crop to another, and certain weeds become more difficult to control.

The steady introduction since about 1945 of a wide range of both selective- and non-selective herbicides has made it possible to control most weed problems in most crops. In developed countries, where labour is relatively scarce and expensive, herbicide use has become a normal input on most arable crops.

Non-rotational control of pests and diseases has relied on two principal methods, the breeding of disease- and pest-resistant crop varieties and the development of crop-protection chemicals. Both methods of control suffer from the phenomenal rate of reproduction of many disease and pest organisms, which often contain sufficient genetic diversity for the development of new strains resistant to pesticides or capable of attacking hitherto resistant cultivars. This requires increased rates of chemicals and eventually the discovery of new chemicals and the breeding of new varieties in a race in which ultimate victory is rare or impossible. Chemicals suffer from two additional disadvantages: they may be harmful to organisms beneficial to crops, and some chemicals resist breakdown and may give rise to insidious long-term harmful effects. More rigorous testing of new compounds for such effects has had the effect of reducing the flow of new compounds into agriculture as well as increasing their development costs.

Perhaps the most outstanding 'bad' example of these problems is the use of chemical control of the cotton bollworm (*Heliothis zea*) which is practically immune to all available insecticides, following a spiral of more and more frequent applications of chemical, in extreme cases up to 60 applications in one growing season. 'The high costs entailed, and the ultimate failure to control the pest has in many instances made it uneconomic to grow the crop, has elevated formerly harmless organisms to pest status, has affected the quality of the environment and posed serious hazards to the health of the agricultural workers' (Luckman & Metcalf, 1975). Such problems have not reached this acute stage in Britain, mainly because pests are much less of a problem in cool temperate climatic conditions. However, the development of resistant aphids – an important crop pest – to certain insecticides has occurred.

More attention is now being paid to the possibility of controlling pests and weeds by biological rather than chemical methods. Biological control has been defined as 'the purposeful use of an organism or organisms to reduce a plant or animal population that is inimical to man' (Samways, 1981). This principle is illustrated in Fig. 19.13.

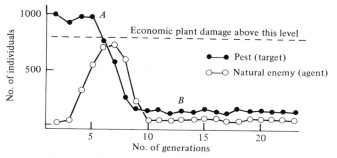

Fig. 19.13   Diagram of successful biological control. The increase in the population of the natural enemy (○) causes a decrease in the pest (●) from a damaging level (A) to a non-damaging level (B). With now fewer pests to sustain it, the natural enemy population declines. Both populations then continue at a low equilibrium level

Source: Samways (1981). From 'Biological Control of Pests and Weeds' by M. J. Samways. *Studies in Biology*, No. 132. Edward Arnold. The figure is modified after Varley, G. C., Gradwell, G. R. & Hassell, M. P. (1973). *Insect Population Ecology*. Blackwell, Oxford

The biological control of weeds has generally been most successful with perennial weeds, and especially introduced weeds. While most attention has been given to the insect control of weeds, a wider range of agents, including mites, nematodes, plant pathogens and parasitic plants are currently the objects of research.

There have been several examples of the successful control of insect pests by biological methods, particularly, for example, on relatively sedentary insects such as scale insect pests and mealy bugs in citrus fruit crops. In Britain, perhaps the most widespread application is the control of the glasshouse white fly (*Trialevrodes vaporatiorum* Westw) by the introduction of a parasitic chalcid wasp (*Encarsia formosa* Gahan). The small scale of such operations and the close environmental control has obviously facilitated implementation in this particular case. However, the lack of commercial incentive to develop such methods has restricted the research done on these problems, and a greater input of resources into such research may be expected to extend the range of applications of biological control.

The philosophy of pest control which appears to be emerging is neither chemical nor biological control but 'integrated control', in which appropriate chemicals are used alongside biological control, or 'pest management'.

Little has been said about the application of ecological thinking to the solution of disease problems. In general, the more diverse the cropping system in a given environment, the more stable it is likely to be and the less prone to suffer from major disease epidemics. Modern cropping situations are often very far removed from such stable ecological systems and therefore need constant inputs of artificial resources to maintain reasonably stable production. The understanding and harnessing of natural methods of maintaining stability from diseases is a challenging problem for which solutions become increasingly pressing.

Completion of the production cycle, harvesting and storage

The end of the production cycle may come when the crop has naturally senesced, and has no green plant parts with the capacity to intercept radiation and photosynthesise, as exemplified by the cereal crop. Other crops may senesce naturally by the time of harvest, but in some instances may still have a functioning crop canopy, depending on variety, season and the market outlet for the crop. This situation is exemplified by the potato crop: early varieties tend to senesce within the normal growing season if allowed to develop to maturity, but they are usually harvested as soon as an economic yield has been obtained, and the crop canopy is destroyed chemically or mechanically to allow the crop to be harvested. The canopy of a late-maturing variety may persist until killed by frost, or artificially. In the biennial sugar beet crop, the leaves persist all through the harvesting period (September to December), and are capable of photosynthesising, although the crop growth rate in the late autumn is negligible owing to low temperatures and low incident radiation. Some yield will be sacrificed by lifting in September and October. At the other extreme, the whole crop canopy may form the harvested product – as, for example, in grass cut for silage. The particular condition in which a crop is harvested and whether some yield has been sacrificed is therefore dependent upon the crop species and the purpose for which it is grown.

Method and ease of harvesting are important considerations which have to be taken into account when deciding certain details in the method of production. For example, the adoption of very wide row spacings in the potato crop is due largely to the need to reduce the production of clods which interfere with sorting the potatoes and lead to damaged tubers. Ease of harvesting and reduction of damage is also associated with the production of a deep, stone-free tilth in the same crop.

The need to spread the date of harvest will be affected by the choice of crop, e.g. early potatoes, winter barley, winter wheat, spring beans, maincrop potatoes and sugar beet could give a sequence of crop harvests which span a period from June to December in Britain. Within any one crop species, the choice of variety can influence the date of harvest, although this does not constitute the only way in which the plant breeder has influenced the harvesting operation. For example, the production of short, stiff-strawed varieties of cereals has considerably reduced the tendency of these crops to lodge and reduced the difficulty associated with harvesting such crops and the attendant loss of yield due to failure to collect all the grain. And, of course, long before plant breeding was a consciously organised science, selection of cereals for a non-brittle rachis was a vital factor improving the harvestability of the cereal crop. Breeders can also affect the degree of determinacy exhibited by naturally indeterminate crop species, such as peas and beans, which can condense the age range of the economic yield component and improve the harvestability of the crop. In the potato crop, the breeder may improve the ease of harvesting by concentrating the tubers in the ridge, by reducing the length of stolons on which the tubers are borne, by improving the shape of the tubers and by increasing the resistance of tubers to mechanical damage.

A weed-free crop at the time of harvest is also a vital prerequisite to the smooth running of this operation. Clearly, examples of the interaction between methods of growing the crop, its genetic make-up and the success of the harvesting operation can be multiplied, but enough has been said to indicate the importance of such interactions, and the need to be aware of them and to take them into account when selecting the crop species, variety and method of production.

Once harvested the crop is frequently, although not invariably (e.g. zero-grazed grass or forage), stored or conserved before use or sale at some later date, which can be a matter of days, months or even years.

Biologically, the major causes of deterioration in store are natural senescence, microbial degradation and non-vertebrate degradation. Natural senescence is associated with the hormonal balance in plant organs.

Microbial degradation is associated with saprophytic organisms which accumulate on crop tissues during growth, but which fail to colonise it: during storage however, given appropriate conditions, they will multiply and decompose the product. In store, it is therefore necessary to eliminate the decomposers, or retard their activity.

The particular strategy adopted will depend upon the nature of the stored product, its end use and cost.

For seed crops, water content and temperature are the two most important variables. Bacteria require very high relative humidities (approaching 100%) for optimal growth, while storage fungi operate best at somewhat lower relative humidities (about 85%).

The storage moisture content of most crop seeds in equilibrium, with R H between 65 and 70% (which should successfully eliminate microbial growth), is between 13–15% but is 8% for crops such as oil-seed rape and linseed. Storage temperature has an important bearing on whether moulding takes place, a marked rise in temperature leading to an increase in equilibrium R H to the advantage of the fungi. Fungal growth leads to a rapid rise in temperature, as the respiration of fungi is much higher than that of stored grain. Whilst most seed crops are stored under cool dry conditions, there may be some advantages in storing grain intended for animal feed at high moisture contents (e.g. 18–20%): this may involve storage at very low temperatures, in airtight conditions or with the addition of a chemical preservative such as propionic acid. These conditions must be imposed within a few hours of harvest.

To store vegetative material, in its natural state, without processing for human consumption, poses additional problems, because the material usually has a high water content (e.g. potatoes about 80%).

The main product of this type stored on farms is the potato and here the main causes of weight loss are respiration, evaporation, sprout growth and disease. The overriding requirement is to store under cool conditions (i.e. between 4–8°C) and to maintain an R H of between 95–98%. Again, the end use may affect the storage requirements; low temperatures, for example, give rise to reducing sugars which bring about unacceptable darkening in crisps and chips; storage temperatures in the range 5–10°C are required here with the use of chemicals to suppress the growth of sprouts, which is favoured by high temperatures.

Materials with a high water content, such as potato tubers, are much more prone to damage than are low moisture content grains. At the start of storage, use is made of the principle that wound healing is favoured by high temperatures, and in the first 10 days of storage potatoes are allowed to heat up to a temperature of about 18°C, from their own respiration, by withholding ventilation.

The main invertebrate pests of storage are the grain mites and storage insects. In cool, damp climates all grain is potentially at risk from mites which can survive at the normal moisture contents at which grain is stored. The storage insects (saw-toothed beetle, *Oryzaephilus surinamensis* L and the grain weevil *Sitophilus granarius* L) are of tropical or warm temperate origin and need temperatures of 12 °C or more in order to breed. Control is by cooling, drying and store hygiene.

Vegetative material, such as grass or lucerne, is conserved mainly as hay or silage.

Most farmers are concerned with storage of products for up to about one year. There is, however, major concern by plant breeders that, due to the widespread use of improved varieties and the resultant disappearance of local varieties of crops, important sources of genetic variability may be irrevocably lost. This has led to the setting up of so called 'gene banks' in which seeds of such local varieties are stored for future use.

Concluding remarks

The first choice a farmer must make is concerned with the selection of the species and the varieties he is going to grow, a choice dictated by many factors, including the prevailing climatic and soil conditions, other resources available to the farmer and his ability to manipulate them, and not least the market for his produce. It is possible to estimate over periods of time what yield advantages have taken place and it is further possible to estimate what proportion of this is due to the genetic contribution made by new varieties, and what to improvements in the environment in the widest sense of this term. Figure 19.14 shows such an estimate for wheat in England and Wales. The scope for improvements and the genetic variability available to realise them will vary with crop species but in wheat, especially, the improvements brought about by the introduction of new varieties in recent years has been very substantial. This is not only true of the UK, but of the world as a whole, where the introduction of new genetic material in wheat and rice, together with improvements in the inputs associated with their use, has given rise in some instances to spectacular increases in yield.

The basic biological principle on which plant breeding relies involves the processes which take place in sexual reproduction, which in the first instance leads to the division of the genetic information in the male and female 'gametes' and the subsequent recombination of the information after fertilisation. This process allows, via the selection of parents and offspring, the incorporation of 'new' characters into improved varieties. Techniques such as the transfer of desirable genes from one species to another are becoming established, and may at some future date provide a potent means of continuing the substantial advances that have already been achieved by the plant breeder.

However, while the broad principles on which crop production is based are reasonably well understood, the outcome of any par-

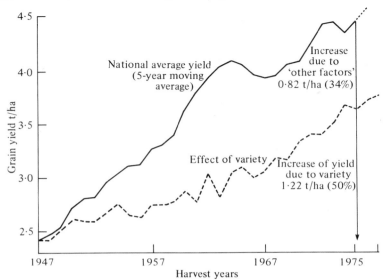

Fig. 19.14 The increasing trend in the national average yield of winter and spring wheat in England and Wales and the estimated effect of variety in achieving the increase, 1947–77
Source: Silvey (1978)

ticular course of action adopted is often very uncertain, due to the overriding influence of variables such as the weather and the incidence of pests and diseases, which are not only outside the farmer's control, but also largely unpredictable.

20

THE PRINCIPLES OF
ANIMAL PRODUCTION

Introduction

The primary purpose of animal production systems is to convert plant products into animal products usable by man. The most important animal products are considered to be those used as food by man, namely milk, meat and eggs, but other animal materials, such as wool, hair, fur and hides are also important. An equally important, although intangible, product is the use of horses, cattle and other species for draught purposes, and a further example is the use of dogs to control and guard other livestock. (Animals also serve man in roles not related to agriculture, as suppliers of pharmacological products such as hormones and antisera, as collaborators – or quarry – in sporting activities, and as ornament or companion in home and garden.)

Over the centuries since animals were first domesticated, their role has extended from that of ensuring man's survival to that of enhancing his enjoyment of life. The extension process has, however, varied greatly from one country to another. A man in one part of the world will kill a deer because he has nothing else to eat, while a man in another country will kill a deer so that he can display its antlers above his fireplace. More prosaically, the Indian farmer relies on the power of his cattle to till his cropland, on their milk to keep his children alive, and on their dung for fuel. In developed countries such as Britain, we have alternative sources of power and fuel, the technology to provide ourselves with a nutritionally adequate diet entirely from plant products, and the skills to manufacture synthetic alternatives to wool and leather, yet our use of (if not our dependence on) farm animals is much greater than that of the people of India. Table 20.1 shows the contribution of animal products to the national diet. Table 20.2 calculates the plant input to these animal products. These tables show first that the principle used to open this chapter – that animals are used to convert plant products into animal products usable by man – does not have a sound biological basis. But the tables also show that the principle is economically correct, for the British farmer annually converts £3,350 m. of crop products (some of which are scarcely

Table 20.1 Daily nutrient consumption per person in households in the UK (and the percentage provided by animal products)

Energy (MJ)	9·5 (40)	Calcium (mg)	964 (66)
		Iron (mg)	11·0 (35)
Protein (g)	73·4 (64)	Thiamin (mg)	1·22 (31)
Fat (g)	106·4 (64)	Riboflavin (mg)	1·90 (70)
Carbohydrate (g)	268·0 (10)	Vitamin A (μg)	1353 (65)
		Vitamin D (μg)	2·72 (47)

Source: MAFF (1979) Household food consumption and expenditure, 1979

Table 20.2 Approximate quantity and value of the major feeds used for animal production in the UK

	Quantity (Mt of dry matter)	Value (£m; 1981)
Pasture herbage		
Grazed	20	500
Conserved	12·5	625
Other fodder crops		
Root crops*	1·7	150
Kale etc.	0·6	40
Straw	1·5	35
Cereals		
Home-grown	8·5	850
Imported	3·0	400
By-products	2·0	250
Protein concentrates		
Oilseed residues	1·8	350
Fish and meat meals	0·6	150
		3350

* includes sugar beet by-products

saleable anyway) into £6,000 m. worth of animal products. The relative importance of crop and animal products in British agriculture is illustrated in Table 20.3.

Thus, although the contribution of animal production to the citizens of a developed country is complex and debatable, its contribution to the agriculture of such a country is simple and unequivocal.

ANIMAL CHEMISTRY AND ANIMAL NUTRITION

The general objective of animal nutrition is to convert feeds of plant origin into animal products. As we have already seen, animal

Table 20.3   **Annual value of UK agricultural output (1978)**

	£m
Cereals	1013
Other farm crops	484
Horticultural crops	691
Total crops	**2188**
Cattle	1262
Sheep	299
Pigs	687
Poultry	441
Other livestock	33
Total livestock	**2722**
Milk and milk products	1619
Eggs	379
Wool	32
Other products	10
Total livestock products (other than meat)	**2040**

Source: MAFF (1979) Output and utilisation of farm produce in the UK, 1972–8

products include the actual bodies of animals (meat, hides and wool), secretory products (milk and eggs) or intangible products (draught power). Both animal feeds (Chapter 14) and animal products are constructed from the same gross chemical components, as illustrated in Fig. 20.1, but they differ considerably in their relative proportions of the gross components, and also in the constituents that provide the fine structure of each component. The animal takes feeds apart and uses many of the constituents to build its products. Other constituents are oxidised like fuels as sources of energy, and constituents which cannot be utilised are excreted. The scientific principles of animal nutrition are therefore concerned first with determining the constituents (or nutrients) required by the animal, and second, with providing these nutrients in the correct quantities and proportions.

Water

If an animal feed such as fresh grass is placed in an oven at 100 °C for 24 hrs it will lose about 80% of its weight in the form of evaporated water. The greatest part of both plants and animals

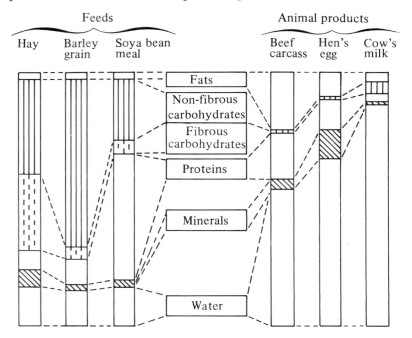

Fig. 20.1 Chemical composition of typical animal feeds and animal products.
(Note that animal products contain no fibrous carbohydrates.)

consists of water. In the animal body, water, protein and ash are found in approximately constant proportions (73 : 22 : 5) and the water content of the whole body is therefore mainly dependent on its fat content; as animals mature and become fatter, their water content falls (but rarely to less than 35%). Milk and eggs both have high water contents. The animal's requirement for water is generally high in relation to its requirement for food dry matter; a common ratio is 4 l/kg dry matter, but this is increased by a hot environment and by the demands of lactation. The most succulent plant feeds contain water and dry matter in the required ratio, but animals must often subsist on the drier feeds which they must balance with drinking water.

Dry matter

The components of a feed remaining after oven drying are known collectively as dry matter. With some feeds, particularly fermented materials such as silage, drying to remove water also removes volatile constituents such as alcohols and some organic acids; the dry matter content of such feeds may therefore underestimate their nutritive value. If the dried residue of a feed is raised to a temperature (600°C) at which it ignites and burns, the so-called organic matter is removed and an ash fraction remains. Most feeds contain 90–95% organic matter (hence 5–10% ash), although the dry matter of some animal products contains a higher proportion of ash (e.g. fish meal, 24%). The ash fraction of both feeds and animal products consists principally of metallic oxides, chlorides, phosphates and carbonates. The metals, together with sulphur, phosphorus and chlorine are known collectively as inorganic nutrients or minerals, and are discussed later.

Organic matter

Organic chemistry is the chemistry of carbon compounds. The basic building material for the construction of plant tissues (and hence of animal tissues) is the carbon of atmospheric carbon dioxide (CO_2), which in plant photosynthesis is combined with water (H_2O) to give organic compounds with the general formula $(CH_2O)_n$; note that oxygen (O_2) is released during photosynthesis. Organic compounds with the approximate general formula $(CH_2O)_n$ are known as carbohydrates. In the simplest members of the group n has a value of 3–6; such compounds are known as sugars, a typical sugar being glucose ($C_6H_{12}O_6$). From materials such as glucose the plant can build larger and more complex carbohydrates (see below). It can also modify the proportions of C, H and O and can add a wide variety of other elements such as nitrogen (N), sulphur (S), phosphorus (P) and the metals referred to earlier. The variety of substances synthesised in plants (and animals) is virtually limitless, but at this stage we need to consider only the carbohydrates and two major groups of their derivatives: (a) amino acids and proteins; and (b) fatty acids and fats.

Carbohydrates

Some examples of carbohydrates are listed in Table 20.4. The simplest members of the family, the monosaccharide sugars, have 5 or 6 carbon atoms and are named accordingly (pentoses and hexoses). For both of these groups the chain of 5 or 6 carbon atoms can carry the accompanying hydrogen and oxygen atoms in a variety of configurations. This means that the formula for hexoses $6(CH_2O)$ or $C_6H_{12}O_6$ can describe several quite different sugars; three examples are listed in Table 20.4.

Monosaccharide sugars are water-soluble and sweet-tasting. Although widely distributed in nature, they rarely occur in high

Table 20.4 Examples of major carbohydrates

Class	Subclass	Formula	Examples	Sugar units
Monosaccharide sugars	Pentoses	$C_5H_{10}O_5$	Arabinose	–
			Xylose	–
	Hexoses	$C_6H_{12}O_6$	Glucose	–
			Fructose	–
			Galactose	–
Disaccharide sugars	–	$C_{12}H_{22}O_{11}$	Sucrose	Glucose + fructose
			Lactose	Glucose + galactose
			Maltose	Glucose
Polysaccharides	Glucans	$C_nH_{2n-2}O_{n-1}$*	Starch	Glucose
			Glycogen	Glucose
			Cellulose	Glucose
	Fructans	$C_nH_{2n-2}O_{n-1}$	Inulin	Fructose

* where n is > 10

concentrations. For example, the dry matter of young, leafy grass might contain less than 5% of monosaccharides.

Monosaccharide molecules can be linked together, with the elimination of a molecule of water at each linkage, to form disaccharides. Table 20.4 shows that two glucose molecules can link together to form the disaccharide, maltose. Alternatively, glucose can link up with galactose or fructose to form, respectively, lactose or sucrose. Like the monosaccharides, disaccharides are water-soluble and sweet-tasting, but unlike the monosaccharides they occur in quite high concentrations in some plant and animal products. For example, the stem of the sugar cane plant and the root of the sugar beet plant contain 50–70% of their dry matter as sucrose. In cow's milk the disaccharide lactose comprises about 35% of the dry matter.

The process of linking monosaccharide molecules together can continue, to give trisaccharides, tetrasaccharides, etc. Eventually the number of linked molecules rises into the hundreds or thousands, and the products are given the general name of polysaccharides. In animal nutrition the most important polysaccharides are starch and cellulose. Starch is familiar to most people as a white powder which does not readily dissolve in water, and cellulose is recognised as the major constituent of the paper on which these words are printed. Table 20.4 shows that both of these compounds are formed solely from glucose molecules, and yet they obviously differ in their properties. The reason for the difference is quite simple, the linkages between successive glucose molecules being formed in a different way in each compound, but is of fundamental importance.

Both starch and cellulose are constructed (and dismantled) by the action of the biological catalysts known as enzymes. The enzymes responsible for constructing what are called the alpha linkages of starch and the beta linkages of cellulose are widely distributed in the plant kingdom. Animals also possess enzymes for creating alpha-linked polysaccharides, although the major product is not starch but the related polysaccharide, glycogen (sometimes called animal starch). Both plants and animals are well provided with enzymes for dismantling alpha-linked polysaccharides, but enzymes capable of breaking beta linkages are much less common. They do not occur in animal tissues, and are in fact more or less confined to bacteria and fungi. The significance of these differences is that starch and glycogen may be readily constructed and dismantled (and thus provide carbohydrate reserves in plants and animals), whereas cellulose once constructed by plants tends to remain in existence until attacked and dismantled by micro-organisms. Cellulose therefore provides the permanent structure of plants and is aided in this role by two important modifications. First, the linkages of cellulose encourage the formation of long chains of glucose molecules which eventually crystallise into semi-rigid fibres. Secondly, cellulose fibres are able to link up with other carbohydrates and related substances to form aggregates that are strong enough to resist even microbial enzymes. The most important compound associated with cellulose is lignin. Lignin occurs in high concentration in wood (20–30% of dry matter) hence the rigidity of this material, but it also occurs in cereal straws (approx. 10%) and indeed in most plants used as animal feeds.

From the foregoing it will now be clear that food plants contain two types of carbohydrates, the sugars and alpha-linked polysaccharides on the one hand and the beta-linked cellulose and its associated compounds on the other. Furthermore, it should also be clear that the latter group are inherently less useful as nutrients for animals, because animal tissues do not contain the enzymes needed to take them apart. This distinction has been long recognised and has resulted in the development of analytical procedures designed to divide the carbohydrates of feeds into fibre and what are loosely called soluble carbohydrates or nitrogen-free extractives (the reason for the last name will be made clear later).

Amino acids and proteins

Amino acids are relatively simple organic compounds; for example, the simplest, glycine, has the formula $C_2H_5O_2N$. The single nitrogen atom is combined with two hydrogen atoms to form the group NH_2, which is known as the amino group and is typical of all amino acids. Plant and animal tissues contain about 20 different amino acids. Plants (and also bacteria) possess the enzymes needed to synthesise all 20, but animals are unable to synthesise 10 of them. This group of 10 must therefore be provided in the food of animals and the 10 are known collectively as the essential or indispensible amino acids. Some amino acids, both essential and non-essential, contain an additional element, sulphur.

Just as the monosaccharide sugars can be linked together to form polysaccharides, so can amino acids be linked together to form proteins. Like the linkage of sugars, the linkage of amino acids is achieved by the elimination of a molecule of water at the junction, and the processes of forming or disrupting these linkages are again aided by enzymes. But the process of protein synthesis is much more complicated than that of polysaccharide synthesis, because instead of many molecules of the same sugar being used (e.g. glucose → starch), protein synthesis may require 20 different amino acids. Furthermore, the amino acid molecules must be linked together in a predetermined sequence. Faced with the problem of linking together a total of perhaps 500 components, comprising 20 different types, an engineer would use a plan or blueprint. The plant or animal uses as a guide, or template, substances called ribose nucleic acids (RNA). The RNA are stringlike molecules which may be envisaged as having various spaces cut into them, each type of space being of a size and shape to accommodate a specific amino acid molecule. In protein synthesis, amino acid molecules locate themselves along a strand of RNA in a precise order, and are joined together in the same order to produce a protein molecule.

In plants, proteins provide the 'working parts' as opposed to the structural contribution of cellulose and the storage components of starch and other carbohydrates. All enzymes are proteins, including the chlorophylls that are involved in photosynthesis. Thus young green plants and feeds derived from them contain a high concentration of proteins (typically 15–25% of the dry matter). As plants age and deposit more cellulose, their protein content falls; for example, the mature grass that is made into hay is likely to contain only about 10% protein in dry matter.

Animal tissues contain high concentrations of protein, there being no cellulose and very little storage carbohydrate (glycogen). The proteins of animals therefore have to provide the structural element, as well as the working parts. Muscles, which have structural and working roles, consist mainly of protein. Skin and similar internal tissues are known collectively as connective tissues, and consist largely of special proteins linked with polysaccharides. Bone consists of a protein matrix or framework made rigid by the deposition of a kind of 'concrete' composed of mineral salts. Wool and hair are composed almost entirely of proteins.

The concept of energy: fatty acids and fats

Starch was described earlier as a storage element of plants (and likewise, glycogen was described as a storage element of animals). These polysaccharides serve mainly as stores of energy. If an animal is cold, for example, it may 'burn' (i.e. oxidise) glycogen to keep itself warm. The glycogen is first converted back to glucose and the glucose is oxidised to yield a specific quantity of energy as heat:

$$C_6H_{12}O_6 \ (180\,g) + 6\ O_2 \rightarrow 6\ CO_2 + H_2O + 2 \cdot 82\ M\,J$$

The figure of $2 \cdot 82$ M J is the quantity of heat produced by the complete oxidation of one mol (i.e. the molecular weight in grammes) of glucose. In practice it is easier to express energy in relation to unit weight of a substance; the heat produced by oxidation of 1 kg of glucose is $15 \cdot 6$ M J. All carbohydrates have approximately the same energy content (see Table 20.5). The reason for this is that the energy content of organic compounds depends on their degree of oxidation, that is, the ratio of oxygen to carbon plus hydrogen. As we have already seen, this ratio is approximately the same for all carbohydrates.

Both plants and animals are capable of synthesising compounds known as fatty acids. These consist of a chain of carbon atoms, each with two hydrogen atoms, and an acid grouping —COOH at one end of the chain. The simplest acid of this type is acetic (or ethanoic) acid, which has in total only two carbon atoms, but higher members of the series have as many as 20 carbon atoms. These longer chain acids have a low ratio of oxygen to carbon plus hydrogen and therefore have a high energy content (Table 20.5).

To return to the short-chain acids, in their pure form these are liquids of low boiling point with rather distinctive smells. Acetic

Table 20.5   Energy content of feeds, animal products and their constituents (M J/kg dry matter)

Feed constituents	
Sugars (glucose)	$15 \cdot 6$
Polysaccharides (starch and cellulose)	$17 \cdot 5$
Proteins (casein)	$24 \cdot 5$
Fats (vegetable oils)	$39 \cdot 0$
Animal tissues	
Muscle	$23 \cdot 6$
Body fats	$39 \cdot 3$
Typical feeds	
Cow's milk	$24 \cdot 9$
Maize grain	$18 \cdot 5$
Hay	$18 \cdot 9$
Grass silage	$20 \cdot 5$
Linseed oil meal	$21 \cdot 4$

acid is the principle constituent of vinegar. With the next two members of the series, propionic and butyric acids, acetic acid is frequently produced through bacterial fermentation of carbohydrates. These three acids are produced in large quantities in the rumen of cattle and sheep, and they also occur in the fermentation product of grass, namely silage.

The longer-chain acids are solids at normal temperatures; for example, stearic acid has a melting point of 70°C. It is worth noting, however, that if long-chain fatty acids are slightly modified, by the removal of hydrogen atoms:

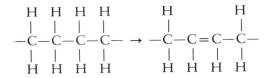

then their melting point falls. If stearic acid is modified as indicated above, the resulting acid (known as oleic acid) has a melting point of 10·5°C. The two types of acid (all hydrogen atoms present or some missing) are known respectively as saturated and unsaturated acids.

Fatty acids have the ability to link together, but unlike sugars and amino acids they do not link directly but through a linking substance known as glycerol (Fig. 20.2).

```
G   – Fatty acid
l
y
c   – Fatty acid
e
r
o
l   – Fatty acid
```

Fig. 20.2   Schematic representation of a triglyceride (fat) molecule

The resulting substance is known as a fat. Each glycerol molecule can link with up to three fatty acid molecules, and the latter commonly differ from one another. Fatty acids contribute their physical properties to the fats they form. Thus shorter-chain acids, and also unsaturated long-chain acids, will form a fat of low melting point (e.g. a vegetable oil such as olive oil) whereas the saturated long-chain acids will give rise to a fat of high melting point (such as lard).

Fats and oils occur in plants and animals mainly in a storage role, and because their energy content is high they provide a concentrated store of energy. Whereas starch, for example, contains 17·7 M J/kg, a typical fat contains nearly 40 M J/kg. Most animal feeds of plant origin contain a low concentration of fats. Grasses and cereal grains, for example, usually contain less than 5% fat in dry matter. The exceptions are the oil-accumulating plants such as soyabeans, groundnuts and rape. The seeds of these plants contain up to 60% of oil in the dry matter, but this oil is generally extracted from the seed for other uses and the oil-extracted residues (now containing less than 5% oil) are used as animal feed.

The ability of animals to store energy as fat is well recognised. They also synthesise fat for milk, which contains about 40% of its dry matter in this form and thus has a high energy value (25 M J/kg dry matter).

Other organic nutrients: the vitamins

In addition to carbohydrates, proteins and fats, animals contain many minor constituents. If the animal can synthesise these substances from the common food constituents, the actual substances do not need to be present in the food, and they thus pose no special problems for the animal nutritionist. There is, however, one particular group of substances which cannot be freely synthesised by animals. These are the vitamins, a chemically diverse group of 15–20 organic compounds which are generally present in the animal body in extremely low concentrations, but which are essential to the normal functioning of the animal body and must be supplied via the feed or by some special method of administration (e.g. injection). Fortunately, most of the vitamins are present in plants, where they also have a role. Those that are not present in plants in the exact form in which they are required by animals may be found in plants as closely related compounds which the animal is capable of converting to the required form; these related forms are known as vitamin precursors.

The small quantity of vitamin required by animals is illustrated by the following example. The pig's daily requirement for the vitamin cobalamin in about 20 microgrammes (1 microgramme, or μg, = 0.001 milligramme) or about 1 part in 100 million of the diet. The total feed requirement of Britain's 7 m. pigs amounts to 5 m. t per year, or enough to fill 250,000 lorries, whereas the total requirement of cobalamin, 50 kg per year, could be carried by one man.

Most of the vitamins are required in the animal in some regulatory role. For example, many of them are components of the enzymes referred to earlier. If the vitamin component of an enzyme is absent, the enzyme cannot function correctly, and the chemical reaction that it catalyses cannot take place. This explains why vitamins are vital to the normal functioning of animals, and why the deficiency of any one vitamin leads to the production of clear and characteristic signs of disease.

It is not possible here to describe in detail the individual vitamins and the diseases induced by their deficiency, but Table 20.6 provides a summary. Note that vitamin C (ascorbic acid) does not appear in the table, because farm animals – unlike man, monkeys and the guinea pig – can synthesise this substance.

Table 20.6 **Characteristics of some vitamins of importance in animal nutrition**

Common name (and chemical name)	Precursor	Role in the animal	Signs of deficiency	Feed sources	
				Good	Poor
A (retinol)	Carotene	Protects mucous membranes; vision	Night blindness; roughened skin	Green plants; fish liver oils	Cereals and other concentrates
D (calciferol)	Animal and plant sterols	Absorption of calcium; bone mineralisation	Rickets and other bone deformities	Fish liver oils; sun-dried forages	Cereals
E (tocopherol)	None	Anti-oxidant	Muscular dystrophy	Green plants; grains	Animal products; root crops
B$_1$ (thiamine)	None	Enzyme co-factor	Nerve degeneration	Most feeds	Refined cereals
B$_2$ (riboflavin)	None	Enzyme co-factor	Nerve degeneration; skin lesions	Green plants; yeast	Cereals
Nicotinamide	Tryptophan	Co-enzymes	Dermatitis	Most feeds	Maize
B$_6$ (pantothenic acid)	None	Co-enzymes	Loss of hair; dermatitis	Most feeds	None
B$_{12}$ (cobalamin)	None	Co-enzymes	Growth check; dermatitis	Animal products; bacteria	Plants

The vitamin requirements of ruminant livestock are simpler than those of pigs and poultry, because all the so-called B group vitamins are synthesised by the bacteria of the rumen. The B vitamin, cobalamin, however, contains the element cobalt, and if the ruminant's diet is deficient in this element, its rumen bacteria cannot manufacture cobalamin. Furthermore, all ruminants are susceptible to B and K deficiencies in early life, before their rumen flora has become established. Thus deficiencies of the water-soluble vitamins (B group) occur in ruminants only under rather special circumstances. Of the fat-soluble vitamins the precursor of vitamin A, carotene, and also the tocopherols known collectively as vitamin E, occur in relatively high concentrations in green plants. Ruminants are therefore unlikely to be deficient in A or E during the summer grazing season, but winter feeds such as hay, straw, roots and cereal concentrates are commonly deficient.

The contrast between the high carotene content of fresh grass (e.g. 250 mg/kg D M) and the much lower carotene content of hay (e.g. 25 mg/kg D M) illustrates an important feature of the vitamins, namely that they (or their precursors) may easily be destroyed during the processing or storage of feeds. Much carotene is destroyed when grass is bleached by sunlight during the hay-making process (thus barn-drying of hay minimises the loss of carotene, and the alternative form of grass preservation, namely ensilage, may entirely prevent loss of carotene).

Another principle of vitamin nutrition is that an animal's requirement for a particular vitamin may be affected by the presence of other substances in its diet, and this principle is illustrated by two facets of the vitamin E requirements of ruminants. The major function of vitamin E is as an anti-oxidant, and it is therefore able to prevent the oxidation of unsaturated fatty acids to potentially damaging substances known as hydroperoxides. This means that if, for any reason, the animal consumes and absorbs an unusually large quantity of unsaturated fatty acids its requirement for vitamin E will be increased. For example, calves given synthetic milks containing oils with unsaturated fatty acids may suffer from vitamin E deficiency unless an additional quantity of the vitamin is added to their food. The second illustration of interactions between vitamin E and other food constituents is entirely different. The mineral element selenium, which is a natural constituent of feeds, also has a role in the animal of preventing undesirable oxidations, and this means that in some circumstances, vitamin E and selenium are interchangeable.

Vitamin D also provides an example of an interaction between nutrients. This vitamin plays a major role in the absorption and metabolism of the bone-forming mineral elements, calcium and phosphorus. If diets are deficient or unbalanced in these elements, the role of vitamin D becomes critical, and the animal must be assured of a greater supply. Vitamin D is formed in the skin of the animal by the action of sunlight on its precursors, which are substances known as sterols. Like the other fat-soluble vitamins, A and E, vitamin D deficiency is therefore most likely to occur in winter, when animals are deprived of sunlight.

Pigs and poultry are likely to spend all their lives indoors and to be fed on diets containing no forages, thus their diets need to be supplemented with the fat-soluble vitamins. As they have no rumen, with its micro-organisms synthesising B vitamins, they generally require supplementation also with many of the B vitamins, particularly riboflavin.

From the foregoing, one might assume that vitamin deficiencies occur frequently in Britain's farm livestock, but in practice they are rarely seen. Ruminants are protected by their rumen microflora, by their ability to build up reserves of the fat-soluble vitamins when supplies are good and to utilise them in times of scarcity, and by supplementation on a relatively limited scale. Pigs and poultry are protected by much more extensive supplementation, the choice of supplements being based on detailed knowledge of these animals' requirements and on the ability of feeds to supply them.

The vitamin supplements referred to above were at one time based largely on naturally occurring materials unusually rich in vitamins. For example, some fish liver oils contain high concentrations of vitamins A and D, artificially dried forages are rich sources of carotene, and yeast is rich in B vitamins. More recently, synthetic forms of many vitamins have replaced the natural constituents of supplements.

Inorganic constituents of the animal body

It was explained earlier that ignition of animal products and tissues leaves a residue of ash. This ash fraction contains mineral elements, mainly in the form of oxides. Some of these elements, such as calcium and phosphorus, are present in relatively high concentrations and are known as major elements; others, such as iron and copper, are present in much lower concentrations and are known as trace elements. The members of each group, and their approximate concentration in the animal body, are listed in Table 20.7.

In animals, and also in plants, the mineral elements have a variety of roles. Some may have a structural role, the best example being the combination of calcium and phosphorus in bone. Others have what may be termed a regulatory role; for example, sodium and potassium are dissolved in body fluids and determine the osmotic pressure of these fluids, and also acid-base balance. Many elements, especially the trace elements, occur as constituents of complex organic compounds. Thus sulphur is a constituent of several essential amino acids, iron is a component of the haemoglobin of blood, and magnesium is a component of the chlorophyll of green plants. As components of enzymes, some elements have roles similar to those of vitamins; selenium, for example, is a constituent of an enzyme known as glutathione peroxidase.

The ash of animals contains many minerals with no known function. These are regarded as being non-essential to animals, although research scientists continue to find roles for them and hence transfer them from the category of mere contaminants to

Table 20.7   Characteristics of some mineral elements of importance in animal nutrition*

Element	Concentration in animal body	Role	Signs of deficiency	Feed sources	
				Good	Poor
Major elements					
Calcium	15 g/kg	Bone; nerve function	Rickets; paralysis ('milk fever'); soft-shelled eggs	Green plants; milk; fish and meat meals	Cereals; roots
Phosphorus	10 g/kg	Bone; co-enzymes	Rickets; depraved appetite (pica)	Animal products; cereals	Hay; straw
Magnesium	0·4 g/kg	Bone; co-enzymes	Tetany ('staggers')	Protein concentrates; legumes	Spring grass
Sodium	1·6 g/kg	Osmotic pressure; acid-base balance	Poor growth; depraved appetite	Fish meal	Plants
Trace elements					
Iron	50 mg/kg	Blood haemaglobin	Anaemia; poor growth	Green plants	Milk; cereals
Copper	3 mg/kg	Blood; enzymes	Anaemia; hair depigmenta-tion; nerve degeneration ('swayback')	(Soil dependent)	Milk
Cobalt	0·1 mg/kg	Cobalamin (vitamin)	(As for cobalamin deficiency)	(Soil dependent)	
Zinc	30 mg/kg	Enzymes	Growth check; skin lesions	Most feeds; yeast	None
Iodine	0·5 mg/kg	Thyroid hormone (thyroxine)	Enlarged thyroid (goitre)	Marine feeds	(Soil dependent)
Manganese	0·3 mg/kg	Bone enzymes	Bone deformities	Green plants	Animal products
Selenium	1·7 mg/kg	Anti-oxidant; enzyme (glutathione peroxidase)	Muscular dystrophy	(Soil dependent)	

* essential elements not listed, because they are rarely deficient, are potassium, chlorine, sulphur, molybdenum and chromium

that of essential elements. Some non-essential elements, such as silicon, aluminium and titanium, are relatively harmless, but others, such as lead, are potentially toxic. Even some of the essential elements can be toxic if present in excessive concentrations. The most important example in UK agriculture is probably provided by copper poisoning in sheep; sheep need about 5 mg of copper in each 1 kg of food dry matter, but a concentration of 20-30 mg/kg food causes an accumulation of copper in the sheep's liver which eventually proves fatal. Another trace element, fluorine, is needed in minute quantities by animals, but concentrations in food as low as 20 mg/kg may be toxic.

As with the vitamins, dietary deficiencies of the mineral elements bring about characteristic diseases in animals, and these are summarised in Table 20.7. In addition, there are a number of general principles governing the requirements for minerals of animals and the supply of minerals in feeds. Requirements depend principally on the rate at which minerals are lost from the body, and the quantities in which they are deposited in the body or its products. Some elements are carefully conserved by animals; for example, the iron of haemoglobin is salvaged when the red cells of the blood die, and is re-utilised to form new haemoglobin. Others, such as sulphur and sodium, play an essential role in excretion and are therefore lost from the body. The rate of deposition is obviously related to the growth of the animal; a young animal with growing bones has high requirements for calcium and phosphorus. These are met by milk, which is rich in both elements, and the lactating animal in turn has a high requirement for calcium and phosphorus. Another factor determining requirements is the ability of animals to store minerals. The bone-forming minerals, calcium and phosphorus, can be stored in quantities sufficient to carry the animal through times of scarcity (or excessive demand). Thus the lactating animal transfers calcium from bone to milk, and restores its reserves later, when milk secretion has diminished or ceased.

The chief determinant of the minerals supplied to animals by feeds of plant origin is the soil type. Plants obtain their minerals from the soil, and their composition tends to reflect that of the soil. If the soil is poorly supplied with minerals required by plants, then plants may grow slowly and show signs of disease indicative of mineral deficiency; examples of minerals in this category are phosphorus and manganese. In other instances, minerals required by animals may not be required by plants, and the deficiency may not be diagnosed until the plant (or soil) is analysed. Iodine and cobalt are two elements in this category. Finally, some plants may accumulate dangerously high concentrations of mineral elements; for example, there are 'seleniferous' plants which accumulate selenium.

Soil/plant/animal relationships for mineral elements are particularly important for animals such as hill sheep, which obtain all their food from the same farm and soil type. Thus the soils derived from Old Red Sandstone in Scotland are often deficient in the trace elements copper and cobalt, and sheep kept on the pasture, and fed on hay (and possibly cereal grain), grown exclusively on those soils, are therefore liable to be deficient in these elements. Conversely, animals such as dairy cows, pigs and poultry, which receive feeds purchased from other areas, will be less susceptible to such deficiencies.

Before an animal can make use of the mineral elements in foods, it must digest and absorb them. In some cases, an element may be present in the food in a quantity apparently sufficient to meet the animal's needs, but fails to be absorbed by the animal. Low 'availability' of elements in feeds may be due to several causes. The element may be present in the plant in an organic compound which resists breakdown by the animal's digestive enzymes. An example is phosphorus, which is present in cereal grains in compounds called phytates, which partially escape breakdown in the gut of pigs and poultry and so cannot supply these animals with phosphorus (in ruminants, phytates are broken down by bacterial enzymes). In other cases, elements may react with one another in the gut to form insoluble compounds. If calcium and phosphorus are present in feeds in a ratio very different from the normal 2 : 1, the one present in lesser quantity may be rendered unavailable by precipitation with the other. Another example is provided by copper, which in the gut of ruminants is liable to be made insoluble by high dietary concentrations of sulphur and molybdenum.

The first lines of defence against mineral deficiencies in animals are to identify the individuals likely to have high requirements, such as lactating animals, and to identify feeds which are likely to be deficient. Deficient feeds may then be balanced by including good sources of minerals in the diet (see Table 20.7). Thus calcium-deficient cereal grains may be balanced by calcium-rich roughages. Most foods of animal origin, such as fish meal and meat and bone meal, are well supplied with minerals. The second line of defence is to use special mineral supplements. These range in composition from simple substances supplying only one element (e.g. calcium carbonate) to complex mixtures containing all the elements likely to be deficient, and they may be given with other feeds or offered as blocks that are licked by animals. A further form of defence is to rectify the deficiencies of foods derived from plants by applying minerals to the soil; this is often neglected, although some of the supplementary minerals given to animals should – with proper husbandry – eventually reach the soil.

PHYSIOLOGY OF FARM ANIMALS

The anatomy of farm animals was outlined earlier, in Chapter 13, and the purpose of this section is to describe the functions and activities of the organs and tissues of the body. Because many parts of the body have several functions it is impossible to categorise each one exactly, but the major functions are outlined in the following pages.

Support and containment

Bone and cartilage support the animal, while skin and internal connective tissue hold it together. The skin also protects the animal from its environment. For example, sweat glands may control water loss; hair, wool or feathers control heat flow; the hooves provide hardened contact points with the ground.

Bone consists of an organic (protein) matrix of cells served by blood vessels, in which are deposited inorganic salts consisting mainly of calcium phosphates. In a mature bone, about one-third is organic matter and two-thirds, ash. The structure of a typical long bone, such as the femur (thigh bone) is shown in Fig. 20.3. In the foetus the bone appears first as uncalcified cartilage, and this becomes bone as the mineral material is deposited in it. In the young, growing animal, areas of cartilage known as epiphyseal

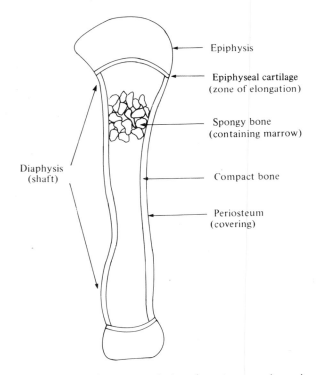

Diaphysis
(shaft)

Epiphysis

Epiphyseal cartilage
(zone of elongation)

Spongy bone
(containing marrow)

Compact bone

Periosteum
(covering)

Fig. 20.3   Typical structure of a long bone in a growing animal

plates continue to enlarge, thus lengthening the bone; when these are fully ossified (a stage known as 'closure') lengthening can no longer occur. In addition to providing support and protection, bone has the secondary functions of providing a mineral reserve and of housing (as bone marrow) the tissues responsible for manufacturing the red cells of blood.

Although bone is an essential component of the animal, its value as an animal product is low in relation to muscle or fat. A typical beef carcass, for example, contains about 15% of bone, and although some of this is saleable (as an integral part of joints of meat) much has to be 'recycled' in the form of mineral supplements for animals (or fertilisers for crops). In the improvement of meat animals there is therefore a tendency to minimise the proportion of bone by selecting compact, 'blocky' animals with short legs. Bacon pigs provide an exception to this: because their most valuable joints are those containing the backbone, such pigs have been selected for length of back, even to the stage of having additional vertebrae.

Skin consists of a tough layer of connective tissue (the epidermis), the outermost part of which is dead and cornified (i.e. hardened). The skin contains two types of gland, the sebaceous glands which secrete a soapy or waxy fluid, and the sudoriferous or sweat glands. In places, the epidermis grows down into the dermis to form pits known as follicles, and from the base of these follicles may grow hair or wool. Both of these fibres consist largely of a special type of hard protein known as keratin; the main difference between them is that hair fibres are solid, whereas wool fibres have a hollow centre. In sheep, the diameter of the wool fibres varies from $15\,\mu$ (0.015 mm) in the fine-woolled breeds, such as the Merino, to $50\,\mu$ in the coarse-woolled breeds such as the Scottish Blackface. Mountain breeds have a mixture of wool and hair fibres, the latter being known as 'kemp'. Another typical characteristic of wool fibres is the waviness or 'crimp'. In early summer, the 'suint' produced by the sebaceous glands moves out along the fleece, and its lubricating properties aid the shearing of the sheep. The feathers of birds are keratinised epidermal derivatives which, like wool, are formed at the base of follicles. Moulting (i.e. loss of feathers) is preceded by the growth of new feathers in the follicles.

The horns of cattle and sheep are based on a bone projecting from the skull but consist mainly of modified skin tissues. The interior of the horn is known as the corium, and consists of the dermis with an exceptionally good supply of blood vessels. Overlying the corium is the cornified outer layer. When cattle are dehorned at birth, the corium is destroyed by heat or caustic chemicals; it is important that all the corium should be destroyed.

The hooves of horses, cattle, sheep and pigs are similar in construction to horns, the bones at the extremities of the digits of each limb being covered by a corium, which is itself covered by the hoof tissue proper. When hoof growth is excessive or deformed, the excess tissue (which has no nerve supply) must be removed by clipping, paring or filing.

Movement

The animal must be able to move within its environment; in addition, as a complex organism, it must be able to move materials from place to place within its body. The first type of movement is carried out by muscles under the animal's voluntary control, such as those of the limbs, trunk and head; because these muscles are generally attached to bones they are also known as skeletal muscles. Many internal movements, such as the circulation of blood and the passage of food along the gut, are carried out by involuntary muscles (i.e. muscles not under the voluntary control of the animal).

Under the microscope, skeletal muscles are seen to have a striped appearance and are said to be striated. The individual muscle cells are elongated and may have several nuclei (Fig. 20.4). They are

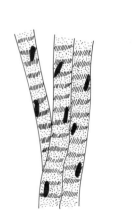

(*a*) Striated (skeletal)   (*b*) Unstriated   (*c*) Cardiac

Fig. 20.4   Muscle shells

grouped into bundles kept together by elastic connective tissue. In contrast, the tendons that attach muscles to bones are composed of an inelastic type of connective tissue. Within each muscle cell are fibrils composed of interleaved bands of two types of protein, actin and myosin. The contraction of muscle is brought about by links between the actin and myosin bands, which draw them closer together and thus shorten the cells. This process is stimulated by nerves (as described later).

The number of cells in a muscle is fixed at birth, but the length and width of individual cells increase as the animal grows. In addition, fat deposits appear between the muscle bundles (intermuscular fat) and even between the individual cells of a bundle; the latter is called intramuscular or 'marbling' fat and occurs only in animals carrying a considerable amount of fat.

The second major type of muscle is that known as unstriated or smooth muscle. It is found especially in the organs of digestion, excretion, respiration and reproduction, but occurs also in many other tissues such as the skin and the eye. Whereas striated muscle occurs typically as bundles of elongated fibres, smooth muscle consists of spindle-shaped cells, often arranged in sheets rather than bundles (e.g. in the wall of the stomach). Like the cells of striated muscle, those of smooth muscle contain actin and myosin fibrils, but these are less regularly aligned than in striated muscle. The typical contraction of smooth muscle is slow and prolonged (i.e. contrasting with the rapid and brief contractions of striated muscle). Contractions may occur rhythmically along the length of tube-like organs such as the intestine, thus propelling their contents.

Cardiac (i.e. heart) muscle is intermediate to the two types already discussed, being striated but involuntary. It shares with smooth muscle the capability of working continuously without tiring.

Sensing and control

The animal receives simple signals from its environment, most obviously through its special senses of sight, hearing, smell, taste and touch. But some messages are more complex; for example, chilling induces shivering, and a seasonal change in the light : dark proportions of the day can induce responses in the reproductive system. Yet other signals are produced within the animal itself; infection or injury may cause pain, and an empty stomach signals hunger. Most of these stimuli, whether external or internal, are conveyed by the nerves to the brain, where they are evaluated and used as a guide to action. In some cases, action follows the receipt of the stimulus so rapidly that we refer to the reaction as being involuntary. The cow that has its heel nipped by a dog kicks out almost immediately. One message (pain) passes from the foot to the spinal cord along one set of nerves, and the spinal cord passes a second message along another set of nerves to the leg muscles (action). In other cases the evaluation of the message is a conscious one and may not necessarily lead to action. Thus the cow that

The Principles of Agriculture

experiences hunger may delay a search for food because she knows that food will shortly be delivered to her, at milking time.

The fundamental units of the nervous system are specialised cells known as neurons (Fig. 20.5). The distinctive features of a neuron are

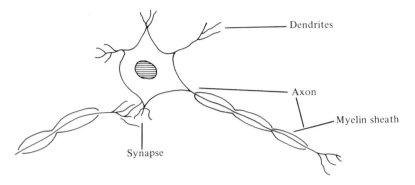

Fig. 20.5   A typical nerve cell or neuron

its long projection, the axon, and its numerous short projections known as dendrites. The axon of one neuron terminates near the dendrites of another. The intermingling dendrites of adjacent cells form what is known as a synapse, and this allows messages to be passed from one neuron to the next. Neurons located in the central nervous system (the brain and spinal cord) may have axons 1 m in length, terminating in a sense organ or muscle. Axons are generally protected by a fatty covering known as the myelin sheath.

Messages are conveyed from one neuron to the next by means of a chemical agent, and within each cell, by changes in electrical potential.

The neurons of the central nervous system are also protected by bones of the skull and spine. The brain consists of massed bundles of neurons, the main part of these cells (grey matter) lying around the outside of the brain and the axons (white) being collected together in the interior. The anatomy of the brain is complex and includes divisions into the forebrain (or cerebrum) which is responsible for higher mental activities, and the mid- and hind-brain which are more concerned with the routine regulation of body activities. Further characteristic features of the brain are its fluid-filled cavities known as ventricles. The spinal cord may be regarded as an extension of the brain. In it the cell bodies of the neurons, forming the grey matter, are in the centre, and the axons, or white matter, are on the outside of the cord (Fig. 20.5).

From each of the vertebrae of the spinal column, pairs of nerves – each consisting of numerous neurons – pass out to the skin and muscles. The dorsal (or uppermost) member of each pair is termed a sensory nerve and conveys messages from the periphery of the animal to its central nervous system. The ventral (or lower) member of each pair is a motor nerve, which sends impulses to a muscle. Thus the reflex reaction of the cow whose foot is bitten by a dog involves a sensory message from nerve endings in the skin that promotes a motor message back to the muscles of the leg. At the same time, the sensory message is transmitted along the spinal cord to the brain, where it is registered as pain.

Like the spinal cord, the brain has a peripheral network of what are known as cranial nerves. There are 12 pairs of such nerves, and several of them have specialised functions in serving the sense organs of the head.

This description of the peripheral nervous system has so far been mainly restricted to nerves controlling the voluntary functions of animals, and hence the striated muscles. The body, however, possess a secondary nervous system, the automatic system, whose function is to control involuntary (smooth and cardiac) muscles. The autonomic system is linked to the central nervous system, and is itself divided into two networks. The first, known as the sympathetic system, originates from the spinal cord in the thoracic and lumbar regions, and the second, or parasympathetic, network originates from the brain and sacral (pelvic) region of the cord. The actions promoted by stimulation of the sympathetic system are those which prepare the animal against danger or attack; hairs are erected and sweat secretion is stimulated. The parasympathetic system acts in the opposite way.

The animal's nerve network may be likened to a telephone system, which permits rapid receipt and delivery of messages. But the animal also possesses a second system of control, known as the endocrine system, which may be likened to a postal service. Messages are carried from one part of the body to another in the form of the chemical substances known as hormones. Hormones are mainly manufactured in either specialised organs (such as the pituitary gland below the brain) or in specialised parts of organs having other purposes; for example, the ovary, in addition to producing eggs, also secretes several hormones. The hormones are carried by the body fluids (especially blood) from the organ producing them to the organs or tissues upon which they act. Some hormones influence the activities of many parts of the body, while others have a very restricted and specific action. Some hormones of major importance in animal production are listed in Table 20.8.

Hormones are produced in response to a variety of stimuli. The endocrine system is linked with the nervous system, and a hormone may be produced in response to a nerve impulse; this association between the two systems is closest for the pituitary gland, which is located at the base of the brain. Another important source of

Table 20.8   Summary of the origin and action of the major hormones involved in animal production

Secreting organ	Hormone	Target	Main effects	Controlling factors
Pituitary gland (anterior lobe)	Somatotrophin (growth hormone)	General	Increases protein synthesis; decreases fat synthesis	Low blood glucose concentration stimulates
	Thyrotrophin	Thyroid gland	Stimulates secretion of thyroxin	Thyroxin feed-back
	Adreno-cortico trophic hormone (ACTH)	Adrenal gland cortex	Stimulates secretion of glucocorticoids	Glucocorticoid feed-back
	Follicle-stimulating hormone (FSH)	Ovary	Promotes egg development	Oestrogen feed-back
		Testis	Promotes semen production	
	Luteinising hormone (LH)	Ovary	Stimulates shedding of eggs	Progesterone feed-back
	Prolactin	Mammary glands	Initiation and maintenance of milk secretion	Oestrogen stimulates: progesterone depresses
Pituitary gland (posterior lobe)	Oxytocin	Uterus or male reproductive tract	Stimulates motility	
		Mammary gland	'Let down' of milk	Suckling stimulates
Thyroid gland	Thyroxin	General	Stimulates many metabolic reactions	Thyrotrophin (see above)
	Calcitonin	Bone	Inhibits release of calcium and phosphorus	Blood calcium feed-back
Parathyroid glands	Parathormone	Bone	Stimulates calcium release	Blood calcium feed-back
		Digestive tract	Stimulates calcium and phosphorus absorption	
Adrenal glands (cortex)	Glucocorticoids (e.g. cortisol)	General	Increases blood glucose Controls water excretion	ACTH (see above)
	Mineralo-corticoids (e.g. aldosterone)	Kidneys	Controls sodium and potassium excretion	Blood sodium and potassium feed-back
Adrenal glands (medulla)	Adrenalin	Numerous	Preparation for emergencies (increases heart rate, blood pressure, blood glucose, etc)	Brain (responding to pain or stress)

Table 20.8 cont.

Secreting organ	Hormone	Target	Main effects	Controlling factors
Adrenal (medulla)	Noradrenalin	Numerous	Often antagonistic to adrenalin	As for adrenalin
Pancreas	Insulin	General	Stimulates utilisation of glucose	Blood glucose feed-back
	Glucagon	General	Stimulates release of glucose into blood (antagonistic to insulin)	Blood glucose feed-back
Ovary (follicles)	Oestrogens (e.g. oestradiol)	Uterus	Preparation for pregnancy	Progesterone inhibits
		General	Stimulate growth	
Ovary (corpus luteum)	Progesterone	Uterus	Preparation for pregnancy	Uterus (see below)
		Mammary gland	Preparation for lactation	
		Pituitary	Suppresses FSH and prolactin (see below)	Pituitary gland
Testes	Androgens (e.g. testosterone)	Penis and scrotum	Stimulate sexual development	Pituitary gland
		General	Stimulate protein synthesis	
Uterus and placenta	Prostaglandin F_2	Ovary	Regression of corpus luteun	Foetal hormone activity

stimuli for hormone production is the hormones themselves. The endocrine system provides several examples of 'feedback' control, whereby organ (A) produces hormone (a); hormone (a) acts on organ (B); organ (B), as part of its response, produces hormone (b); and hormone (b) causes organ (A) to produce less of hormone (a).

A further type of stimulus for hormone production is provided by other chemical substances circulating in the body. For example, the hungry animal, in addition to receiving a nerve message that its stomach is empty, also responds to a reduction in the concentration of the sugar, glucose, in its blood. The response takes the form of the secretion of the hormone, glucagon, by the pancreas (see Table 20.8).

The interactions of the hormones involved in the control of reproduction (and lactation) are particularly complex – and later in this chapter they are explained in more detail than in Table 20.8.

Many hormones can now be synthesised in the laboratory; alternatively, compounds having the same action as, but differing in structure from, the natural hormones may be produced by various means. Both the synthetic hormones and their analogues may be administered to animals to control growth (see Chapter 15) and reproduction.

Digestion, absorption and metabolism

Food consumed by animals passes along the alimentary tract, or gut. Strictly speaking, food in the gut has not entered the body proper; it cannot be utilised by the animal until it has been absorbed into the blood and carried to the tissues. As we saw earlier, most food constituents consist of relatively large molecules, and before the food can pass through the wall of the gut into the body proper, its large molecules must be broken down to smaller and simpler molecules. This process of breakdown is known as digestion; it is accomplished mainly by the action of enzymes secreted into the gut either by the animal itself or by micro-organisms residing in the gut compartments. Some food components fail to be digested, either because the animal does not possess the enzymes needed or because food and enzyme fail to make contact; such components pass right through the gut and are excreted in the faeces.

The products of digestion absorbed by the animal are used as fuel to power its muscles, or as building materials for the growth of tissues and secretion of milk. The reactions required are again brought about by enzymes and are referred to collectively as metabolic processes or metabolism. Metabolism, however, is always incomplete in the sense that the reactions eventually yield substances which either cannot be further broken down, or cannot be used as building material. Such substances must be excreted by the animal. Some are passed back into the gut and excreted in the faeces, others pass through the kidneys and are excreted in urine, while some gaseous products (principally carbon dioxide and water vapour) are excreted via the lungs.

Digestion and absorption

The gut of a simple-stomached animal, such as a pig, consists of three main compartments (see Fig. 38.5). The stomach acts firstly as a storage organ and secondly as a reaction vessel for the animal's digestive enzymes. From the stomach the food passes into the small intestine, where more enzymes are added and many food constituents are absorbed. The last compartment, the large intestine, allows further absorption and also serves as an organ of excretion. In addition, the large intestine often harbours bacteria which digest the components of the diet, such as cellulose, for which the animal secretes no enzymes, although bacterial digestion at this late stage tends to be inefficient because its products have rather limited opportunities for absorption. Some animals, including the rabbit, are able to improve the efficiency of bacterial breakdown of food in the large intestine by consuming part of their faeces and thus recycling incompletely digested components through the stomach and intestines (a practice known as coprophagy).

Ruminant animals (e.g. cattle, sheep, goats and deer) have evolved an even more efficient form of microbial digestion. In addition to housing bacteria in their large intestine, they have these organisms and also protozoa (microscopic animal species) in a grossly enlarged compartment of the stomach known as the rumen (see Fig. 13.2). Thus, in ruminants, the food is digested *first* by micro-organisms and later by the host animal's own enzymes. One advantage of this arrangement is that in foods which consist largely of plant cell walls (such as grass and straw), the cellulose is broken down at an early stage and the more digestible cell constituents are exposed to digestion (by either bacterial or animal enzymes).

Earlier in this chapter, the formation of complex organic compounds from their simpler constituents was shown to be the result of condensation reactions, in which, for example, glucose molecules were linked together with the loss of one molecule of water at each linkage. Most of the reactions brought about by the digestive enzymes involve the opposite of condensation, hydrolysis. Water molecules are re-inserted into the linkages and the complex molecules are split into their components. The main reactions and the enzymes involved, are summarised in Table 20.9. As illustrated in Table 20.9, the breakdown of large molecules is often accomplished

Table 20.9   The major digestive enzymes

Enzymes	Origin and site	Substance digested	Substance produced
(a) Acting on proteins			
Rennin	Stomach (calf)		
Pepsin	Stomach	Proteins and peptides	Peptides and amino acids
Trypsin	Pancreas & small intestine		
Peptidases	Small intestine	Peptides	Amino acids
(b) Acting on carbohydrates			
Amylase	Pancreas & small intestine	Starch	Maltose
Maltase	Small intestine	Maltose	Glucose
Sucrase	Small intestine	Sucrose	Glucose and fructose
Lactase	Small intestine	Lactose	Glucose and galactose
(c) Acting on fats			
Lipase	Pancreas & small intestine	Triglycerides	Monoglycerides and fatty acids

stepwise, with one enzyme catalysing the initial stages, and another, the later stages (e.g. starch → maltose → glucose).

The appealing simplicity of the chemistry of digestion may cause one to overlook the complex physiology involved. Each enzyme requires a medium of defined acidity or alkalinity in which to work, and this is controlled by acid secretions (hydrochloric acid in the stomach) and alkaline secretions from the liver (as bile) and from the lining of the small intestine. Complex mechanisms involving hormones are required to control the flow of these secretions, and also of the enzymes themselves. Proteolytic enzymes (i.e. those attacking proteins) are capable of breaking down the animal's tissues as well as its food, and are therefore secreted into the gut in an inactive form which is activated by contact with a second enzyme.

The end-products of carbohydrate digestion (monosaccharide sugars) and of protein digestion (amino acids) are water soluble and thus readily absorbable through the wall of the intestine and thence to the bloodstream. The end-products of fat digestion (monoglycerides or glycerol and long-chain fatty acids) are insoluble in water and therefore present a problem of absorption. This is solved by converting these end-products into colloidal solutes with the aid of special chemicals (acting rather like detergents) that are secreted in the bile.

The minor nutrients, vitamins and minerals, must also experience digestion and absorption. The water-soluble B vitamins are small molecules, whereas the fat-soluble A and D vitamins (or their precursors) are larger, and have to be absorbed in a manner analogous to the fats themselves. Absorption of minerals cannot occur until the elements have been released from organic or insoluble inorganic forms. Solubilisation may be incomplete, or may be hampered by reactions among the elements themselves. Much of the phosphorus in plant materials is in the form of organic compounds known as phytates, and these are broken down in the gut by the action of microbial, rather than animal, enzymes. Copper in feeds is often rendered insoluble in the gut contents by reaction with sulphides to form insoluble copper sulphide.

Microbial digestion

The rumen provides an ideal medium for microbial digestion of food. The animal maintains the rumen at a constant temperature, and controls acidity by secretion of alkaline saliva. Food enters the rumen at frequent intervals, and the products of digestion are removed by absorption or by outflow to the remainder of the digestive tract. Although the micro-organisms of the rumen are valued for their ability to digest α-linked polysaccharides such as cellulose, they do in fact attack all feed components; it is usual for

70–80% of the breakdown of the food to be accomplished in the rumen.

The enzymes of micro-organisms begin their work like those of the host animal by breaking down complex molecules to simpler compounds. In the case of carbohydrates and proteins, however, breakdown is carried beyond the stage of simple sugars and amino acids, by means of a process known as anaerobic oxidation or fermentation. When a substance such as glucose is fermented by bacteria, a rearrangement of the molecule takes place which may be summarised as follows:

$$C_6H_{12}O_6 \rightarrow 2CH_3COOH + CO_2 + CH_4$$

The reaction releases some of the energy of the glucose, which is captured by the bacteria and used to fuel their own growth. The end products of the fermentation are the gases carbon dioxide and methane (CH_4), and a short-chain fatty acid (acetic or ethanoic acid). The gases are excreted by the animal (by belching) but the acetic acid is absorbed.

Other fermentations, which involve glucose and other sugars, lead to the production of additional fatty acids, propionic and butyric. The relative proportions of the three fatty acids depend on the diet; with diets rich in cellulose (e.g. roughages), acetic acid predominates; but with diets rich in starch (e.g. cereals), propionic acid approaches acetic in concentration. The overall effect of the fermentation of carbohydrate is that about 80% of its energy is transferred to the short-chain fatty acids, which are absorbed (through the rumen wall or from the intestine) and subsequently metabolised by the host animal.

The amino acids released from food proteins in the rumen are de-aminated by the bacteria. In de-amination the typical amino (NH_2) group of the acid is split off as ammonia (NH_3). The NH_3 is then either taken up by the bacteria and re-formed into the amino acids they require for their own proteins, or, if present in excess, it is absorbed from the rumen and excreted by the animal. The remainder of the amino acid molecules derived from the feed are fermented in the same way as carbohydrate. The net effect of microbial fermentation in the rumen is that the nitrogen of the feed proteins is transferred to bacterial proteins. The latter pass from the rumen to the small intestine and are subsequently digested by the animal's enzymes and absorbed. The extent of the transfer is, however, quite variable. Some of the feed proteins escape rumen digestion but are broken down in the intestine. The proportion of 'de-aminated' nitrogen that is recaptured may be low if the feed is rich in protein. However, if feed protein (and hence rumen ammonia) levels are low, they may be augmented by the addition of a simple nitrogen-containing compound such as urea. Urea may be added to the ruminant's feed by the farmer, but the animal itself possesses some capability of supplying urea by collecting it from body tissues and 're-cycling' it to the rumen. The important outcome of this utilisation of urea is that the quantity of actual protein passing out of the rumen (and subsequently being made available to the animal) may exceed the protein of the food.

A few additional features of microbial digestion may be noted briefly. Fats are hydrolysed in the rumen and the glycerol is fermented, but the long-chain fatty acids escape fermentation (although they may be converted from the unsaturated to the saturated form). Water-soluble vitamins are synthesised in the rumen.

Quantitative digestion: the concept of digestibility

If a food of known, standard composition, such as barley grain, is given to an animal, a predictable proportion of it will be digested and absorbed and the rest will be excreted in faeces. The proportion digested (i.e. not appearing in the faeces) is known as the digestibility of the feed, and may be expressed as a percentage or a coefficient (e.g. of the dry matter of barley, 80% or 0·8 might be digested). Digestibility coefficients may also be calculated for individual nutrients. For example, the starch of barley, being readily hydrolysed and absorbed, would have a digestibility coefficient close to 1·0, whereas the cellulose and hemicellulose (i.e. fibre, requiring bacterial degradation) might be digested only to the extent of 0·5. The digestibility of the dry or organic matter of starchy feeds is generally high and varies over a small range (0·7–0·9), whereas the corresponding values for fibrous feeds are lower and more variable (0·4–0·8). As one would expect, the digestibility of the fibrous feeds is higher for ruminants than for other animals.

Metabolism

This term embraces all the chemical reactions that take place in the animal's body. These reactions are carried out by enzymes produced in the cells of the animal; thus nearly all cells are involved in metabolism, although concentrations of cells in organs such as the liver and mammary gland may have special roles to play. The reactions of the body are numerous and complex. Furthermore, the construction of product Z from substrate A may involve successive steps in which A is converted to B, B to C and, eventually, Y to Z; the substances B to Y are known collectively as intermediates or intermediary metabolites. Here we can only outline the main functions of metabolism.

Some reactions are designed to break down chemical compounds derived from food or body tissues with the specific purpose of providing the body with energy; these processes, by which chemicals become fuels, are called *katabolic*. Others produce new tissues or repair existing tissues; these are synthetic in character and are called *anabolic*. Katabolic and anabolic reactions are linked by energy-carrying intermediary compounds. These usually contain phosphorus, a typical example being adenosine triphosphate (ATP).

A typical cycle of katabolic reactions is that used to oxidise glucose via a series of nearly 20 intermediates to carbon dioxide and water (i.e. the reverse of photosynthesis; see pp. 326 & 367). We saw earlier that this oxidation yields 2·82 MJ of energy per molecule of glucose. In the body about 70% of this energy is transferred to ATP and other energy carriers, and 30% is released as heat. A part of the same cycle is used to oxidise fatty acids, again with the transfer of energy.

A typical anabolic cycle is that used to synthesise protein from amino acids. Energy is required to forge the links between amino acid molecules, and this is provided by ATP. For protein synthesis to take place smoothly and efficiently, the amino acid molecules must be delivered to the DNA template in the correct forms and numbers, but there is no guarantee that amino acids absorbed by the animal from its gut will meet this requirement. The animal therefore possesses mechanisms for creating some of its own amino acids by transferring the distinguishing amino (NH_2) group from one compound to another (a metabolic process known as transamination). But about 10 amino acids cannot be formed in this way by the animal and must therefore be present in the mixture absorbed from the gut. These 10, of which important examples are lysine, methionine and threonine, are known as the essential or indispensible amino acids.

Other important synthetic reactions in the body are those involved in the formation of fats, of glycogen and of the sugar of milk, lactose; this last synthesis takes place in the mammary gland. These syntheses, and that of protein, differ in one important respect, the stability or permanence of the product. Glycogen, as a short-term energy store, has a brief life and within hours of its synthesis may be required for katabolism as a source of energy. Fats are longer-term energy reserves, but even they may be withdrawn from store within a day of synthesis (for example, in animals given one large meal a day). Lactose is in a different category, for after being synthesised it is soon removed from the body in the milk. Proteins, as the 'working parts' of the body, have lives that vary in length according to their activity. The protein of an active tissue such as that forming cell membranes in the intestine, wears out rapidly and must be replaced at frequent intervals. At the other extreme, the protein of wool is virtually without activity and is never replaced or 'turned over'. Proteins withdrawn from tissues are katabolised to amino acids, many of which are re-used to synthesise new proteins. The overall picture of protein metabolism in the body is one in which proteins are synthesised from a 'pool' of amino acids to which they eventually return these components. Re-utilisation is incomplete and there is a 'leakage' from the pool which must be balanced by an inflow from the food; in a growing animal, inflow of amino acids exceeds leakage, and protein synthesis exceeds protein katabolism.

Respiration and excretion

The katabolic reactions of the body are dependent ultimately on a supply of oxygen, the carbon and hydrogen of the compounds katabolised being oxidised to carbon dioxide (CO_2) and water. The respiratory system of the body, centred on the lungs and using the blood as a distribution network, provides oxygen to the tissues and removes most of the CO_2 and some of the water. The excretory system of the body, centred on the kidneys, removes some CO_2 and much of the water, and is also responsible for removal of other waste products discussed later.

Blood

This consists of a fluid medium, the plasma, which carries several types of cells and also a number of extracellular proteins. The three main types of cell are the erythrocytes (red cells), whose role is to carry oxygen, the leucocytes (white cells) which provide the body's first line of defence against infective agents, and the thrombocytes (platelets) which are involved in the clotting of blood. The erythrocytes, which are formed in the bone marrow, house the iron-containing protein, haemoglobin, that binds with oxygen. Normal blood contains 100–150 g haemoglobin per litre (anaemic animals have less) and 1 g of haemoglobin can carry 1·36 ml of oxygen. Erythrocytes have a life of 3–4 months; when they are broken down their iron is re-used to synthesise more haemoglobin, and other parts of the haemoglobin molecule are excreted via the liver as the bile pigments.

Of total blood volume, about 60% is plasma, and of plasma about 7% is plasma proteins. The latter include albumin, globulins (involved in protection against infection) and fibrinogen (involved in blood clotting). The blood also carries many other substances such as glucose and other nutrients, hormones, and inorganic elements such as sodium and potassium.

The principles of blood circulation are illustrated in Fig. 20.6. One branch of the circulation (pulmonary) takes blood to the lungs, where it is brought close to the air in fine, thin-walled vessels

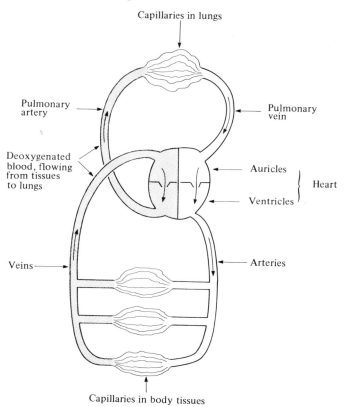

Fig. 20.6 The principles of the circulation of the blood

known as capillaries; here it collects oxygen and gives up CO_2. After returning to the heart, the blood is pumped into a second circulation (peripheral) which delivers it to the tissues; here – again in capillaries – it gives up oxygen and collects CO_2. The peripheral circulation also collects other waste products from the tissues and delivers them to the kidneys.

Kidneys

Each kidney contains one million or so minute filtration devices known as nephrons. The end of each nephron is a double-walled cup known as the glomerulus, through which blood capillaries flow. Water and small, soluble blood constituents such as glucose, urea and inorganic salts pass from the blood to the glomerulus and then through a series of tubes served by further capillaries. Here, the composition of the glomerular filtrate is adjusted to meet the needs of the animal. Much water, virtually all the glucose and most of the sodium are returned to the blood, but urea and other waste products are eventually passed from the kidney to the bladder as urine. Thus the kidneys act, not just in a passive, excretory role, but more positively as regulators of blood composition, and as the blood constituents interchange with the rest of the tissues, the kidneys' regulatory function extends over the whole body. For example, excretion of positively or negatively charged ions ensures the correct acid/base balance of the body, and excretion of a mineral such as potassium, which is abundant in foods and readily absorbed from them, ensures that its concentration in the blood and other body fluids does not become excessive. To function correctly, the kidneys must have a sufficient throughput of water.

Although urine (and faeces) may present problems of disposal to the farmer, they ensure the re-cycling of valuable nutrients to the soil and plants. For example, a dairy cow consuming 15 kg of grass dry matter a day, containing 3 kg of protein and 480 g N, might pass only 80 g N into its milk and excrete 150 g N undigested in faeces and 250 g N as urea, etc, in urine.

Reproduction

The male and female sex organs do not become fully functional until animals are well on their way to achieving their mature size. The attainment of sexual maturity occurs at what is known as puberty. In cattle, puberty occurs at a fixed body size regardless of age; for British Friesians it occurs at a live weight of about 270 kg. In sheep, puberty is again determined by body size, but its occurrence is complicated by the fact that sexual activity in the female of this species is restricted by daylength (to a period from September to February in the UK for many breeds of sheep). Thus a lamb born in April may, if well grown, achieve puberty in the following October, but a delay of two months may cause it to be held over until the following September. Pigs achieve puberty at a fixed age (about 200 days) rather than fixed weight, although females (gilts) may be jolted into sexual activity by sudden contact with males. The fowl lays its first egg at about 160 days of age, but this event is controlled by daylength; a long day (14 h of light) or an increasing daylength will advance puberty, while short days or decreasing day length will delay it.

The female

In the female, puberty is marked by the start of oestrous cycles. The immature ovary contains large numbers of special cells known as ova, each ovum being surrounded by a layer of follicular cells. Under the influence of the follicle-stimulating hormone (FSH) produced by the pituitary gland, one or more follicles will enlarge, each forming a cavity filled with fluid between the ovum and the follicular cells. In the cow, for example, a mature follicle may be 16–18 mm in diameter, and projects from the surface of the ovary sufficiently to be felt by a hand passed into the rectum. As the follicle grows it secretes oestrogens, which act on the anterior pituitary gland to stimulate the release of luteinising hormone (LH). The oestrogens also increase the vascularity of the uterine wall (in preparation for anticipated pregnancy) and provoke the signs of 'heat' or sexual receptivity.

Heat periods in the cow are marked by a discharge of watery mucus from the vulva, by excitability which includes mounting other cows, and – most particularly – by a willingness to stand when mounted by other cows. The ewe shows no generally recognisable signs of oestrus. In the sow there may be swelling and reddening of the vulva, but the most reliable indication is the animal's willingness to stand when pressure is applied to her back.

Towards the end of the heat period, the follicle ruptures and the ovum is released. In its place on the surface of the ovary grows the corpus luteum, which produces progesterone, the hormone that inhibits FSH production. If the ovum is fertilised and the animal becomes pregnant, the corpus luteum persists and prevents the growth of more follicles. If there is no pregnancy, the corpus luteum breaks down, FSH is again produced and the oestrous cycle is repeated. The timetable for these events varies between species, as shown in Table 20.10.

Table 20.10   Reproductive cycles for female domestic animals

	Cattle	Sheep	Pigs
Length of oestrous cycle (d)	21	16·5	21
Length of heat period (h)	16	36	48
Ovulation (h after start of heat)	30	36	36
Typical number of ova fertilised	1	2	12
Gestation period (d)	282	147	113
Interval from parturition to first heat (d)	40–70	–	4–7

In cattle, sheep and pigs it is now possible to influence the occurrence of oestrus (and the numbers of ova produced) by the use of exogenous hormones (i.e. hormones not produced by the animal itself). In sheep, the insertion into the vagina of pessaries impregnated with progestagens (hormones that help to prolong the life of the corpus luteum) blocks the oestrous cycle. If a flock of ewes so treated has its pessaries withdrawn simultaneously, the ewes will come into oestrus simultaneously and can then be mated together (two days after removal of pessaries) in order to lamb together on a prescribed date. This procedure, known as oestrus synchronisation, allows the shepherd to plan his lambing programme more precisely. A similar technique has been used in pigs, but with the progestagen included in the feed. In cattle, oestrous synchronisation can be achieved by an injection of a different kind of hormone, known as a prostaglandin, which causes an active corpus luteum to regress.

Another exogenous hormone preparation, known (from its origin) as pregnant mare's serum gonadotrophin (PMSG), possesses FSH and LH activity. When given to sheep or cattle it increases the number of ova shed at a single oestrous. Carefully controlled use of PMSG can increase litter size in sheep to three or four, and induce twins in cattle.

The male

The gonadotrophins, FSH and LH, have roles in the male which parallel those in the female. The production of spermatozoa in the tubules of the testis is stimulated by a combination of gonadotrophins (FSH and LH) and androgens; the androgens are produced by the so-called interstitial cells (lying between the tubules) in response to stimulation by LH. Androgens, such as testosterone, also promote the growth of (and maintain) the penis, scrotum and secondary sexual characteristics, and regulate the secretion of the glands associated with the testes, namely the prostate and the seminal vesicles.

Each spermatozoon consists of a head, containing the cell nucleus, a midpiece, containing the mitochondria and their enzyme systems, and the tail. The mitochondria provide the energy which moves the tail and gives the spermatozoon its motility. When spermatozoa leave the testis they are stored in the epididymis, and they are subsequently diluted with, and activated by, the secretions of the accessory glands, to form semen. The semen is expelled from the penis by the muscular contractions constituting ejaculation. In the bull, each ejaculate is 4–8 ml in volume, and contains about 1 m. spermatozoa per mm^3 (i.e. 6×10^9 per ejaculate). In the ram the volume is less (0·5–2·0 ml) but contains 3 m. spermatozoa per mm^3. In contrast, the ejaculate of the boar is large in volume (150–300 ml) but contains only 100,000 spermatozoa per mm^3.

Insemination

In natural mating, the ejaculate is delivered by the penis into the vagina or uterus, and the spermatozoa move up the fallopian tubes to meet the descending ova. In artificial insemination, this process is interrupted; the semen is collected from the male and examined, diluted (and possibly stored) before being introduced into the female. The advantages of artificial insemination are that a single ejaculate can be used to inseminate 20–200 females (thus extending the influence of superior sires on the next generation), and that the transfer of diseases is minimised. Cattle semen can be frozen and stored (at $-196°C$ in liquid nitrogen) for many years, and it is therefore possible to delay the extensive use of a bull until his value has been assessed by recording the meat or milk production of a small number of his offspring. The use of frozen semen also allows a bull to be mated with cows located in any part of the world. At present, these advantages are not fully obtainable with sheep and pigs, because their semen cannot be so extensively diluted, and their spermatozoa are damaged by the freezing techniques currently available.

Ovum, embryo and foetus

The fusion of an ovum with a spermatozoon is known as fertilisation. The single cell resulting immediately begins to divide, to form an embryo. In the cow the cells of the dividing ovum remain undifferentiated for the first 12 days, but during the period 12–45 days, primitive forms of all the major tissues and organs are laid down. The heart, for example, begins to function as early as day 21. Also formed are the membranes which surround and nourish the embryo. The amnion holds the embryo suspended in fluid, the allantois collects the excretions of the embryo and also carries blood vessels towards the lining of the uterus, and the serosa (or trophoderm) envelops the embryo and the other membranes. The allantois and serosa eventually fuse to form the chorion, and the latter becomes attached to the lining of the uterus (at 30–5 days) to form the placenta. The function of the placenta is to bring the blood vessels of the embryo and the mother in close proximity, so that nutrients and other substances may pass from one to the other. In the cow and sheep the conjunction of embryonic and maternal tissues occurs at a number of isolated spots, called cotyledons; in the pig, the placenta provides continuous contact.

At 45 days of age the bovine embryo weighs only 2·5 g and is about 3 cm in length. From then until birth it is known as a foetus. The foetus tissues grow rapidly, but at different rates, to produce the characteristic proportions of a newborn calf weighing 40 kg.

Pregnancy diagnosis

In the cow, a veterinarian passing his hand into the rectum can palpate the uterus and detect a developing foetus from about 60 days after conception. In the ewe and sow, the reflection of ultrasonic waves can be used to detect pregnancy from about days 70 and 50 respectively. In the cow, and also in other species, the corpus luteum remains active during pregnancy and continues to secrete progesterone, which circulates in the blood, and also reaches the milk, at a relatively high concentration. If a milk sample taken from a cow 20-23 days after mating is analysed for progesterone, a low concentration will indicate the recurrence of oestrus to be expected at that time, whereas a high concentration is indicative (although not conclusively so) of pregnancy.

Parturition

When the foetus reaches maturity, its pituitary gland secretes ACTH (see Table 20.8) which stimulates the adrenal cortex to produce a glucocorticoid hormone. The latter acts on the placenta to alter the balance of hormones produced there, and this in turn initiates the separation of the placental tissues and the muscular contractions of the uterus. Parturition can be induced artificially, by injecting the mother with a synthetic glucocorticoid.

Embryo transfer

In all domestic mammals it is now possible to transfer ova or embryos (fertilised or not) from one individual to another. Embryo transfer is now used routinely in cattle; for a reasonable success rate surgical methods are required for both collection and implantation of embryos, but non-surgical methods (analogous to those used in artificial insemination) are now being developed. The technique is used by breeders to multiply the offspring of outstanding females, such animals being 'superovulated' to produce up to 20 ova, which, after fertilisation, are transferred to nondescript cows. Embryo transfer may also be used to give twin pregnancies in cows.

Lactation

During pregnancy, oestrogens and progesterone, together with other hormones, promote the growth of the mammary glands, and a few days before parturition the glands fill with the secretion known as colostrum. The main constituents of colostrum are special proteins known as immunoglobulins, which, when consumed by the young animal, provide protection against bacterial infection.

After 3-4 days, colostrum is replaced by normal milk (see Chapter 35) having the composition described in Chapter 13.

Egg production in poultry

The hen has only one ovary (the left) and at ovulation a whole follicle (i.e. the chromosomal disc plus the 'yolk') passes into the oviduct. This tract, which is 80 cm long, consists of five sections, each with a specific function. The funnel-shaped infundibulum collects the follicle, and fertilisation may take place there. The next section, the magnum, is responsible for covering the follicle with albumen (i.e. egg 'white'), and this is followed in turn by the

isthmus, which adds a double membrane covering. The fourth
section, the uterus or shell gland, adds the shell, and the final
section, the vagina, passes the egg to the exterior via the cloaca.

Each follicle takes about 70 days to grow to the size (30–40 mm
across) at which it undergoes ovulation. From ovulation to oviposition (i.e. laying) takes about 26 hours, of which about 20 hours is
required for shell formation. The completed hen egg weighs about
57 g and contains (g/kg): proteins, 120; carbohydrates, 10; lipids,
105; mineral material, 110.

GROWTH AND DEVELOPMENT OF FARM ANIMALS

During their progression from a minute single-celled egg to a large
multi-cellular organism, farm animals show a clear pattern of
growth in weight (Fig. 20.7). The growth of the ovum and embryo

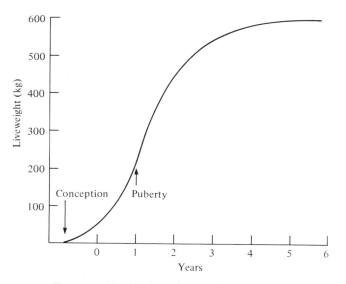

Fig. 20.7   Typical growth curve for a dairy cow

is slow, but the foetus shows an accelerating rate of increase to the
time of parturition. A further spurt of growth occurs at puberty,
and thereafter the rate declines until the mature weight is reached.
Before it is born, the animal's growth rate (and hence its birth
weight) is determined in the first instance by its genetic make-up
and by the size of its mother. The next major influence is the
competition it faces from brothers and sisters sharing the uterus;
for example, in breeds of sheep that produce single lambs weighing

5 kg at birth, twins will weigh only 4 kg each, and triplets, 3 kg. The nutrition of the dam may also affect birth weight, although the dam will use her reserves to protect her offspring from this influence.

After the animal is born the rate of growth is again determined by genetic factors but nutrition becomes the dominant influence. For example, it is quite common for cattle to grow from a birth weight of 40 kg to a slaughter weight of 400 kg over a period of two years and hence at a rate of about 0·5 kg/d. With a high plane of nutrition, however, the same cattle can reach 400 kg at one year old, after growing at 1 kg/d. In addition, the growth curve may not be as smooth as Fig. 20.7 shows. Where animals are subjected to seasonal fluctuations in food supply (hill sheep, for example), they may stop growing when food is scarce (or even lose weight), and then spurt forward again when food becomes plentiful.

Even in an animal growing at a uniform rate, the various parts of its body will grow at different rates. The body proportions of the animal therefore change as it grows, and to these changes we give the name 'development'. Post-natal development can be illustrated by comparing the weights of body components of the new-

Table 20.11   Weights of selected body components at birth and maturity in cattle (kg)

Component	(a) Birth	(b) Maturity	(b)/(a)
Total liveweight	50	600	12
Head	3	15	5
Hind leg	7	80	11
Heart	0·5	5	10
Bone	16	66	4
Muscle	20	240	12
Fat	1·5	80	53

born animal with those of the adult (Table 20.11). At birth, for example, the head of the calf is relatively large, and thereafter it grows slowly in relation to a part that is less well developed at birth, such as the hind leg. We therefore call the head an early maturing part of the body, and the hind leg, late maturing.

With regard to body tissues, bone develops earlier than muscle, and muscle earlier than fatty tissues. The same trend is shown by chemical components, protein being deposited in advance of fat. The general picture is one of a series of growth 'waves', with nerve tissues growing rapidly at first, followed successively by bone, muscle and fat. In practice there is a considerable overlap between the successive waves, the intervals between them being reduced by high planes of nutrition. Thus a fast-growing steer of 150 kg live weight will accumulate protein and fat at approximately equal rates of about 170 g/d.

There are also growth priorities within each of the major tissues. Fat, for example, is deposited first on the abdominal organs (e.g. the kidneys) and beneath the skin. Later, fat is deposited around the muscle bundles and – last of all – within the muscles themselves (i.e. as intramuscular or 'marbling' fat).

For each of the domestic species it is possible to discern a typical pattern of growth, in which the weight of a particular organ or tissue of the animal comprises a predictable proportion of the total weight. Nevertheless, the pattern for each species is likely to vary according to breed type, sex and nutrition. In cattle, for example, a small, early-maturing breed such as the Aberdeen Angus will have a mature weight of about 400 kg, whereas a larger, late-maturing breed such as the Charolais will have a mature weight of 600 kg. This means that if the two breeds are compared at 400 kg, the Aberdeen Angus – being mature – will be fatter than the Charolais. Within these breeds, females will be fatter than steers of the same weight, and the latter will be fatter than bulls. Males also differ from females in other growth characteristics; for example, males have greater development of the head and shoulders. But the major variations in growth patterns are due to differences in fat deposition; in other respects – for example, in bone-to-muscle ratio – differences between breeds and sexes are relatively small.

The same is true of the influence of nutrition on growth. In general, animals on a high plane of nutrition, hence growing rapidly, will be fatter at a given weight than animals that have grown slowly to that weight. This effect is seen most clearly in the pig, and with bacon pigs it is usually necessary to restrict food intake and growth rate to prevent over-fatness of the carcass.

Growth and meat production

The shape and form of today's meat animals is a compromise between the needs of the animals themselves – to eat, move about, reproduce, etc – and the requirements of man for meat. In species that reproduce rapidly, and hence can be changed in form more readily, modifications to meet man's requirements may sometimes seem excessive; for example, the selection of turkeys for massive breast muscle development has produced birds that have difficulty in copulating. Extreme as this example may be, it illustrates the importance of consumer demand for muscle (i.e. lean meat). Earlier generations of consumers also demanded fat in meat (to provide energy for physical labour) but the demand for fat has fallen to the point where today's consumer wants little or no visible fat (e.g. subcutaneous fat) on his meat but still values the flavour and juiciness imparted by intramuscular fat. As the latter is the last fat to be laid down it cannot be provided without fat being deposited elsewhere. There are also other contradictory requirements. The consumer's primary criterion of meat quality is tenderness; this is best achieved by slaughter of young animals, but meat from young animals lacks flavour (compare veal with beef, or lamb with mutton). A further complication is introduced by the requirement of the butcher for carcasses that can be cut up with minimal waste (e.g. as bone or trimmed-off subcutaneous fat).

The strategy adopted by the farmer to meet the requirements of the consumer will vary according to the animal species. With poultry, he uses specially selected hybrids (breeds) that can be grown rapidly to a slaughter weight that yields the carcass desired. With pigs the breed type and slaughter weight are again important, but nutrition (as indicated earlier) is a critical determinant of carcass fatness. For the bacon market in particular, a few extra millimetres of backfat can make the difference between top-grade and poorer (hence lower-priced) carcasses. With cattle and sheep the farmer's problems are complicated by numerous breed types (each suited to its particular environment) and by the influence of fluctuations in feed supply on optimum time of slaughter. For each individual, the farmer will have in mind some target degree of development (as judged mainly from subcutaneous fat cover) that is likely to suit market requirements.

MEETING NUTRIENT NEEDS: PRINCIPLES OF RATIONING

The scientific rationing of feed for animals requires knowledge of their nutrient needs and of the ability of feeds to supply nutrients. Consider, as a simple example, a 600 kg cow producing 20 kg milk per day; how much protein will she need per day, and how much feed will be needed to supply the calculated quantity of protein? The purpose of this section is to outline the procedures used to answer questions of that kind.

Energy requirements

As described earlier in this chapter, the animal requires dietary sources of carbohydrate and/or fat, proteins, minerals and vitamins. Because animals can inter-convert the organic nutrients, carbohydrate and fat (and, to some extent, protein), requirements for these are expressed in the common currency of energy. For example, 1 kg of milk might contain the following nutrients and energy:

Table 20.12

	(g)	×	(kJ g)	=	(kJ)
Fat	40		38·1		1,524
Lactose	47		16·5		776
Protein	34		24·5		833
Total					3,133

From this we may calculate that to produce 1 kg of milk the cow will need at least 3133 kJ ($= 3\cdot13$ MJ) of energy in its diet (and at least 34 g of protein). The qualification 'at least' is needed for reasons illustrated for energy in Fig. 20.8. The cow cannot be expected to transfer energy from feed to milk with perfect efficiency, for there will be losses in digestion and metabolism.

The example shows that of 258 MJ of energy in the feed $68 + 16 + 20 = 104$ MJ was lost in solid, liquid and gaseous excreta, leaving 154 MJ of what is termed 'metabolisable energy'. Of this, about one-third (54 MJ) was used for the maintenance of the cow (i.e. to maintain the essential bodily functions of the animal, such as respiration,

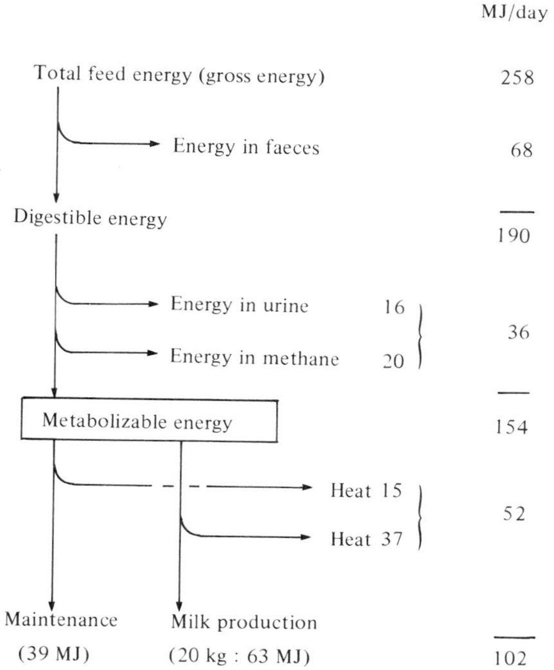

MJ/day

Total feed energy (gross energy) 258

Energy in faeces 68

Digestible energy 190

Energy in urine 16
Energy in methane 20 } 36

Metabolizable energy 154

Heat 15
Heat 37 } 52

Maintenance Milk production 102
(39 MJ) (20 kg : 63 MJ)

Fig. 20.8 An example of energy partition in the dairy cow

circulation and other essential muscular activities). These maintenance activities were achieved with an efficiency of about 72%, 15 out of 54 M J being wasted as heat. The rest of the metabolisable energy (100 M J) was used for milk synthesis, 60% (63 M J) passing into the milk components and the rest (37 M J) being lost as heat. Thus of the 258 M J entering the cow, only 63 M J (24%) eventually reached the useful product, milk.

From calculations of the kind illustrated in Fig. 20.8 it is possible to draw up tables of the energy requirements of dairy cows (and also of other livestock). For reasons that need not be discussed here, energy requirements are usually stated in terms of metabolisable energy. The maintenance requirements of animals will depend on their size (i.e. live weight), and their production requirements will depend on their rate of production (e.g. kg milk produced per day, or kg liveweight gain per day). As an example, the energy requirements of dairy cows are shown in Table 20.13.

Table 20.13 Metabolisable energy requirements of dairy cows (M J/day)

Live weight (kg)	Milk yield (kg/day)			
	0	10	20	30
400	42	92	142	192
500	48	98	148	198
600	55	105	155	205

Careful reading of Table 20.13 shows that maintenance requirements (i.e. requirements for zero milk production) are not directly proportional to live weight. Thus the requirement of the 600-kg cow is 32%, not 50%, greater than that of the 400-kg cow. In fact, maintenance requirements are commonly found to vary with live weight raised to the power 0.75 (i.e. $W^{0.75}$, commonly called 'metabolic live weight'). Another complication, not shown in Table 20.13, is that requirements will change if the composition of the milk, and hence its energy content, changes.

Protein requirements

Until recently, the protein requirements of pigs and poultry were expressed simply as weights or dietary concentrations of crude protein. For example, the protein requirements of a broiler chick might have been stated to be 180 g C P/kg diet D M. For ruminants, requirements were expressed in terms of *digestible* crude protein, since the protein in feeds for ruminants varies considerably in digestibility. Both of these measures have now been modified to take into account the variations between feeds in the nature or quality of proteins.

Non-ruminant animals (pigs and poultry), in addition to having a general requirement for protein, also have specific requirements for each of the 10 essential amino acids that they cannot synthesise themselves (see p. 369). The requirement for each amino acid is related to its concentration in the product being manufactured by the animal. For example, the proteins of poultry flesh contain about 70 g lysine per kg, and the broiler therefore needs approximately this quantity (actually about 60 g) of lysine to be present in each kg of its dietary protein. Maize contains 28 g lysine/kg protein (i.e. less than required), but soyabean meal contains more, 70 g lysine/kg protein. The proportion of these two feeds in a diet can therefore be adjusted to give the required proportion of lysine to protein. We can therefore amplify our previous statement of the broiler's requirement for protein (180 g/kg diet D M) by adding a figure of 10 g lysine/kg diet D M. Similar calculations can be made for the other essential amino acids, but a frequent outcome is that diets supplying sufficient lysine will also supply sufficient of the other acids. In practice it is usually necessary to specify requirements only for lysine, methionine and one or two other amino acids.

For non-ruminants, then, the ideal feed protein is one that contains amino acids in exactly the proportions they require. For ruminants, however, protein quality is much less dependent on amino acid proportions. As discussed earlier (p. 391) rumen micro-organisms re-fashion feed proteins into microbial proteins, whose amino acid proportions are more or less invariant. The microbial proteins thereby ensure that the host animal digests and absorbs amino acids in constant proportions. To synthesise protein efficiently, however, the micro-organisms must be presented with a steady supply of feed protein that can be degraded to ammonia in the rumen. The quantity of degradable protein they need will be determined by the quantities of energy-containing nutrients they ferment in the rumen; a simple expression of the relation between the protein and energy requirements of rumen micro-organisms is that for each 1 M J metabolisable energy in the diet they require 1·25 g N, or 7·8 g degradable protein. Thus, grass silage containing 10 M J M E/kg D M should contain 78 g degradable protein per kg.

An important measure of protein quality for ruminants is therefore protein degradability in the rumen. Green forages, hence grass silage, contain a high proportion of their protein (0·7–0·9) in a soluble and therefore degradable form. If the silage mentioned above had 0·8 of its protein degradable, then it would need to contain 78/0·8 = 98 g C P/kg D M to meet the needs of rumen micro-organisms.

If the protein produced by rumen micro-organisms was sufficient to meet all the protein requirements of the host animal, a high degree of protein degradability would always be desirable. For ruminants with relatively low protein requirements, such as adults kept at a maintenance level or growing slowly, this is commonly the case. But for more productive animals, with relatively higher protein requirements, the microbial protein must be augmented by feed protein that escapes rumen degradation but is digested in the lower gut. The scientific rationing of such animals requires the striking of a balance between the degradable protein requirements of the rumen and the additional undegradable protein requirements of the host animal. This is illustrated in Table 20.14, which shows that increasing milk yield in dairy cows calls for a greater proportion of undegradable protein in the diet.

Table 20.14   Energy and protein requirements of a 600-kg cow fed on a diet containing 11 MJ ME/kg DM

Requirement	Milk yield (kg/d)			
	0	10	20	30
[1] ME (MJ/d)	55	105	155	205
[2] Dry matter (kg/d)	5.0	9.5	14.1	18.6
[3] Rumen degradable protein (g/d)	430	815	1,210	1,615
[4] Undegradable protein (g/d)	—	65	320	565
[5] Total protein (g/d)	430	880	1,530	2,180
[6] Total protein (g/kg DM)	86	93	109	117
[7] Optimal degradability	1.00	0.93	0.79	0.74

Notes: [1] From Table 20.13
   [2] Line (1) values ÷ 11
   [5] Line (3) + line (4)
   [6] Line (5) ÷ line (2)
   [7] Line (3) ÷ line (5)

Methods for achieving the right balance of rumen degradable and undegradable proteins are described elsewhere (Chapter 14). It should be noted that for both non-ruminants and ruminants failure to supply protein of the desired *quality* means that a greater *quantity* will be needed. If a protein low in lysine is given to hens a greater quantity will be needed to meet both protein and lysine requirements (and excesses of amino acids other than lysine will be wasted). Likewise, a ruminant feed containing protein more degradable than required will need to be given in greater quantity (and the nitrogen of excess degradable protein will be absorbed as ammonia from the rumen and excreted as urea in urine). For example, from Table 20.14 we may calculate that if a cow yielding 30 kg milk per day were given protein 0.8 degradable (i.e. 0.2 undegradable) it would need 565/0.2 = 2825 g/d (or 30% more than if the protein had the desired degradability of 0.74).

Requirements for minerals and vitamins

These may be stated in simple terms as the total quantity or concentration of the nutrient in the diet (see below).

An example of rationing

The requirements of a typical dairy cow are summarised in Table 20.15. The list of nutrients should include trace elements and vitamins, but these have been omitted to simplify the example.

Table 20.15   Daily nutrient requirements of a 600-kg cow yielding 20 kg milk (g)

	Maintenance	Production	Total
Energy (M J M E)	55	100	155
Rumen-degradable protein	430	780	1,210
Undegradable protein	—	320	320
Calcium	15	33	48
Phosphorus	12	32	44
Magnesium	11	14	25
Sodium	4	13	17

Let us suppose that the feeds available for the cow are hay and a simple, farm-mixed concentrate based on cereals and soyabean meal. The composition of these feeds is shown in Table 20.16.

Table 20.16   Composition of feeds for a dairy cow (g/kg dry matter)

	Hay	Concentrate
Energy (M J/kg D M)	9	12
Rumen-degradable protein	70	95
Undegradable protein	20	40
Calcium	4·5	1·0
Phosphorus	2·0	4·0
Magnesium	0·9	1·5
Sodium	3·0	1·2

The calculation of the quantities of the two feeds required might adopt the following sequence:

(a) Assume that sufficient hay is given to meet the maintenance requirement for energy.
(b) Deduct from the total requirements of the cow the quantities of nutrients supplied by the hay.
(c) Calculate the quantity of concentrate needed to meet the remaining requirement for energy.
(d) Determine the quantities of nutrients supplied by the concentrate, and compare them with requirements.

These four steps are illustrated below:

	Requirement	Supplied by hay	Remaining requirement	Supplied by concentrate	Excess or deficit
Dry matter (kg)	—	6	—	8·4	—
Energy (M J)	155	54	101	101	0
Rumen-degradable protein (g)	1,210	420	790	790	0
Undegradable protein (g)	320	120	200	340	+140
Calcium (g)	48	27	21	8	−13
Phosphorus (g)	44	12	32	33	+1
Magnesium (g)	25	5	20	12	−8
Sodium (g)	17	18	−1	10	+11

The final ration (6 kg hay dry matter and 8·4 kg concentrate dry matter) has a rather wasteful excess of undegradable protein, and a relatively large (but harmless) excess of sodium. Calcium and magnesium are in deficit, and the ration could be supplemented with 33 g calcium carbonate (containing 400 g Ca/kg) and 21 g magnesium oxide (containing 520 g Mg/kg).

In this example, feeding standards (Tables 20.13, 20.14, 20.15), and data (Table 20.16) were used to derive a ration for an animal with a predetermined level of production; this procedure is known as *ration formulation*. The same tables can also be used in a different way, to predict the performance of animals given a predetermined ration. For example, from Tables 20.16 and 20.13 we calculate that a 500-kg cow given daily 8 kg hay dry matter and 10 kg concentrate dry matter would have enough energy (172 M J ME) to produce about 25 kg milk (although this ration would be deficient in other nutrients); this procedure is known as *performance prediction*.

In practice both types of calculation may be more complex than illustrated here; further details are given in standard textbooks on animal nutrition.

PRINCIPLES OF GENETIC IMPROVEMENT

Comparisons of today's farm animals with those of former times, 50, 100 or 200 years ago, disclose considerable differences in outward appearance and in productivity. For example, today's dairy cattle tend to be black and white in colour, to have no horns and to give an average milk yield of about 5,000 kg per lactation. Their predecessors of 50 years ago were generally red and white in colour, carried horns, and yielded about 2,000 kg per lactation. These changes have been due to a variety of factors. The change in coat colour is due to the replacement of one breed, the Dairy Shorthorn, by another, the British Friesian. The loss of horns is due to the routine use of chemical agents to destroy the horn bud in calfhood. The increase in milk yield is due partly to the breed substitution, partly to improved feeding and management and partly to the selection of superior breeding stock in each successive generation.

These factors which have operated in the past are available today to the farmer who wishes to improve his livestock in some way. Basically he may impose external changes in the *environment* of the animals (e.g. improve their nutrition), or modify the nature of the animals themselves. To achieve the latter type of change he may introduce animals of the desired quality from elsewhere. Such a change need not be as drastic as the replacement of one breed by another, however, and this section is concerned with the alternative method of improvement, namely genetic improvement of livestock by selection of superior individuals for breeding. The advantage of this method is that once achieved, improvements are maintained without further effort, whereas an improvement obtained by changing the external environment of animals is likely to require continued implementation. Today's British Friesian calf, even if its ancestors have been chemically dehorned for the past 10 generations, will still grow horns if left untreated.

The mechanism of inheritance

The calf grows horns because it possesses in its body cells what is called a 'gene' for horns. A gene is a strand of the string-like molecules known as nucleic acids. We saw earlier (p. 370) that *ribose nucleic acids* (R N A), which occur in the fluid content of cells, act as templates for protein synthesis. Genes, however, are composed of *deoxyribonucleic acids* (D N A) and are found in the nuclei of cells. They are joined together to form structures known as chromosomes which are arranged in pairs. In cattle there are 30 pairs of chromosomes, carrying many thousands of genes.

The D N A acts as the fundamental template for protein synthesis, because it provides the template for construction of R N A. Thus we may envisage the development of horns in cattle as depending on the presence of one or more specific proteins (perhaps as enzymes), which can be synthesised with the aid of R N A, provided the cell nuclei contain the appropriate gene, or strand of D N A. One of the fundamental properties of D N A is that it can reproduce itself, by acting as a template for another molecule to be produced; this ensures that all the cells of the calf contain identical D N A, because they all originate from one cell, the fertilised ovum. The only exception to the rule of every cell in an individual carrying the same genes occurs in the formation of the gametes (ova and spermatozoa). Fertilisation depends on the fusion of the nuclei of an ovum and a sperm, and if each of these two nuclei had its full complement of DNA, the nucleus of the fertilised ovum would contain twice the normal complement. This is avoided by a halving of the DNA in the gametes which is achieved by the segregation of the pairs of chromosomes; thus the gametes of cattle contain 30 chromosomes.

Each pair of chromosomes is identical in the sense that each strand of DNA that we call a gene has a corresponding strand on the other member of the pair. The two genes may be identical, in which case the gametes produced by an individual must always carry the same gene; the individual is then said to be homozygous for that gene. Alternatively, the genes may differ in some way. For example, one chromosome might carry the gene for horns while its pair mate carries a gene for 'no horns' (i.e. a DNA strand which fails to promote the synthesis of the protein required for horn development). Such an individual is said to be heterozygous for the horns gene and its gametes could contain either the horns or the no-horns genes.

In a population having an equal distribution of horns and no-horns genes an individual has a one-in-four chance of having two horns genes, a one-in-four chance of having two no-horns genes and two chances in four of having one horns and one no-horns gene. It seems likely that the homozygous animals will have horns or have none (i.e. be naturally polled). But what will the heterozygous animals have: will they be intermediate to the extent of having small horns or even one horn? In fact the no-horns (or polled) gene is dominant to the horns gene (which is, conversely, said to be recessive) and the heterozygotes are therefore polled. For other characters, however, neither gene may be dominant, and the heterozygotes will be intermediate in character to the homozygotes. In cattle an example is provided by the Shorthorn breed, in which a red-coated animal crossed with a white-coated animal (i.e. the homozygotes) produces heterozygotes with a mixture of red and white hairs that is described as a roan colour.

The segregation of chromosomes in the formation of gametes is largely a random process, and the phenomenon of heterozygosity ensures a great many possible assortments of chromosomes and hence of genes. Thus two brothers will each inherit half of their genes from each parent, but the random segregation will ensure that they do not inherit the same assortments of genes, and therefore that they will not be genetically identical. Only twins derived from the same fertilised ovum (called monozygous twins) are genetically identical. Nevertheless, the likeness between brothers, or between parent and offspring, is strengthened by the fact that many genes are linked on the same chromosome. The effect of this linkage is that characters are inherited in 'packages' rather than in a completely random fashion.

One of the most obvious effects of linkage is the association between the sex of an individual animal and its other characters. One pair of chromosomes does not obey the rule described above, of one member of the pair matching the other. In male mammals the chromosome called the X chromosome is paired with a distinctly different Y chromosome; a male therefore produces sperm half of which have an X chromosome and half, a Y chromosome, and is said to be heterogametic. Female mammals, on the other hand, have two X chromosomes; their ova all have X chromosomes and they are said to be homogametic. An ovum that is fertilised by a sperm carrying an X chromosome becomes a female, and an ovum fertilised with a Y-containing sperm becomes a male. Thus:

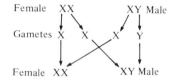

Fig. 20.9   The determination of sex by chromosome combination

The Y chromosome carries very little genetic information, but the characters it does carry can be expressed only in the male. In addition, recessive genes carried on the X chromosome have a greater chance of being expressed because they are not paired with a dominant partner.

Although characters such as horns and coat colour provide some striking examples of single-gene inheritance, the important production traits of farm animals – such as growth rate, milk yield or litter size – are determined by the combined action of many genes. Another important distinction between, say, coat colour and milk yield is that whereas the former is unaffected by the environment, the latter is susceptible to many environmental influences. A cow with a genetic potential for high milk yield may fail to achieve her potential because she is not given enough food or because she is affected by some disease such as mastitis. While a farmer may have no difficulty in identifying cows carrying the gene for a particular coat colour, it is much more difficult for him to select cows carrying the gene combination required for high milk yield. Furthermore, with the multigenic inheritance of milk yield, the chances of obtaining the same combinations of genes in successive generations are extremely small.

We can now distinguish between the genotype and phenotype of a character. The genotype refers to the genes present in the individual and the phenotype to the way the character controlled by the genes is actually expressed. In the coat-colour example referred to earlier, genotype and phenotype correspond exactly, the heterozygotic roan being recognisable as an intermediate to the homozygotic red or white. In the horned/polled example, however, we cannot distinguish by eye the homozygote polled animal from the heterozygote which is also polled, and phenotype does not correspond fully with genotype. In the case of the production traits, the correspondence between genotype and phenotype is disturbed also by environmental influences. This last correspondence is often expressed in terms of what is called *heritability*. The heritability of a character is measured on a scale from 0 (least correspondence) to 1 (perfect correspondence). Heritability is a measure of the proportion of the difference between an animal's own performance and the population average that will be inherited by its offspring. Some examples of heritability for a selection of production traits are given in Table 20.17.

Production rates, such as milk yield and live-weight gain, have low to moderate heritability, while reproductive traits have very low heritability. Product quality characteristics, such as milk and carcass composition, tend to be more highly heritable.

Table 20.17 Approximate values for the heritability of important production characters in farm animals

Heritability range	Species			
	Cattle	Sheep	Pigs	Poultry
0–0·2	Calving interval	Litter size	Litter size	Egg production
0·2–0·4	Milk yield	Fleece weight	Live-weight gain	Yolk colour
	Live-weight gain			Live-weight gain
0·4–0·6	Milk fat percentage	Wool fibre diameter	Bone length	Egg shell quality
0·6–0·8	Milk protein percentage	Birthcoat quality	Backfat depth	—

Principles of selection

In any population of farm animals, a herd of dairy cows for example, there is a considerable range in individual productivity. The range may be defined by a very small group of exceptionally good producers, an equally small group of exceptionally poor producers, and a large mass of intermediate producers. A frequency distribution for individual milk yield in a herd of 100 dairy cows might appear as in Fig. 20.10. Yields range from 2,500 to 7,500 kg per lactation, with a mean of 5,000 kg. The aim of genetic improve-

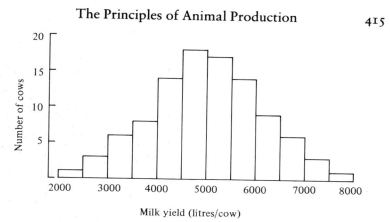

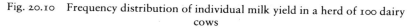

Milk yield (litres/cow)

Fig. 20.10   Frequency distribution of individual milk yield in a herd of 100 dairy cows

ment is to identify the superior animals, to use them in preference to the rest as the parents of the next generation, and by doing so to raise the average level of production of the next generation.

This simple concept of genetic improvement is in practice complicated by several factors. The first has been mentioned already, namely that phenotypic performance is an unsure guide to genotype. For milk yield, heritability is only 0·25–0·40. A second problem illustrated by the example above is that the scope for selecting females may be quite small. For example, the average herd life of a dairy cow is about four lactations, so one in four of all calves born must be retained as herd replacements. As only half the calves born are females the proportion selected is reduced to 1 in 2 (=50%), and may be further reduced by loss of herd replacements by death or illness. Genetic improvement in milk yield will also depend on the selection of superior males, since these will provide half the genes of the new generation, yet males have no phenotype for milk yield.

A further factor to be borne in mind is that selection of cattle on milk yield alone may lead to deterioration in other important traits. In Britain, dairy cows provide calves for beef production, so their offspring must be able to grow rapidly and yield a satisfactory carcass. Another important characteristic of dairy cows is that they should conceive at regular intervals of a year.

Genetic improvement in milk yield of dairy cows will therefore depend primarily on the heritability of milk yield and the difference in yield between the population average and the average of the individuals used for breeding. The latter difference, known as the selection differential, will in turn depend on the efficiency with which superior animals (of both sexes) can be identified, on the extent to which breeding can be concentrated on a small proportion of the population, and on the extent to which characteristics other than milk yield must be taken into account in selection. Finally, the *rate* of genetic improvement will depend on the generation interval, which in cattle is about five years (in poultry, in contrast, the generation interval is less than one year, and the rate of improvement is correspondingly faster).

Selection methods

The simplest method of selecting animals for breeding is to base selection on individual performance. Such *performance testing* is feasible only if the character(s) used for selection are expressed in the appropriate sex (e.g. males may be performance-tested for meat production but not for milk production). Furthermore, it is effective only if the heritability of the character(s) is high (>0.25). If heritability is low and/or the character is not expressed in the sex to be selected from, then selection must be based on the performance of close relatives of the individuals concerned. Such *family selection* may involve preceding generations (*pedigree selection*), contemporaries (*sib testing*; sib being a corruption of sibling, a sister or brother), or the offspring of the individuals concerned (*progeny testing*).

The selection differential for males is generally greater than that for females because only a small proportion of all available males need be used for breeding. Even with natural mating, one bull may serve 100 cows a year, and with artificial insemination he will be mated with several thousand females. Most programmes of genetic improvement therefore place emphasis on the selection of males.

Performance testing

A typical test of this type is carried out at a special test centre to which males are sent by individual breeders. The males are fed on a standard ration and their growth rate over a predetermined age or weight range is recorded. The growth rate of each individual is then compared with the group average. If large differences between breeds may be expected, as in cattle for example, the test is confined to one breed. An example of a test conducted on bulls by the UK Meat and Livestock Commission is shown in Table 20.18. The four bulls tested varied quite considerably around the breed average, especially in fat depth.

By standardising some major features of the environment (housing and feeding) the performance test gives the breeder a better chance of identifying *genetically* superior individuals than he would get by testing his animals on his own farm. Nevertheless, the degree

Table 20.18 The results of a performance test of Hereford bulls, carried out by the Meat and Livestock Commission

Bull No.	Withers height (cm) at 365 days old	Live weight (kg) at 400 days old	Daily gain (kg)	Feed conversion efficiency	Fat depth (cm)
1	116·5	514	1·3	6·7	7·8
2	110·5	513	1·3	6·9	5·2
3	116·0	499	1·2	7·0	3·5
4	110·5	480	1·1	6·5	5·4
Breed average	114·4	507	—	—	5·7

of standardisation is usually insufficient to justify the comparison of a male's performance with that of individuals tested at another time or place. Another drawback to the performance test is that it does not allow the quality of the animal product (i.e. carcass quality) to be assessed directly. This really requires a progeny test but there are some simpler alternatives. When boars are performance tested they may be accompanied by some of their littermates, who are slaughtered for carcass appraisal at the end of the test (i.e. a sib test). It is also possible to make an assessment of carcass quality of males on test by the use of non-destructive methods such as ultrasonic probing of fat and muscle.

Progeny testing

This type of testing can be illustrated by the tests currently carried out on dairy bulls by the Milk Marketing Boards. Each year, about 130 bulls selected on the milk production of their mothers are each mated by artificial insemination with a randomly selected group of 300 cows in seven or more commercial herds. The daughters of these matings eventually calve and complete a lactation. Their milk yield and composition is then compared with those of unrelated heifers in the same herds to give what is known as an improved contemporary comparison (ICC). The ICC of a bull might be expressed as +250 kg milk, +12 kg fat, +8 kg protein, for example, and this would indicate the extent to which he would be likely to improve the lactation milk yield, milk fat yield and milk protein yield of his daughters when mated to unselected cows. In practice about 18 bulls are selected, from the 130 tested, for general use by artificial insemination.

The attraction of the progeny test is that it provides a direct estimate of the bull's genotype and hence of his breeding value. The disadvantages of progeny tests are the numbers of animals required (which are inflated by some inevitable losses) and the time required. A bull's ICC may not become available until six years after he is first mated, by which time he may be dead. In practice, this problem is overcome by collecting and storing semen by deep-freezing while his progeny test is in progress, and by either releasing it for use or discarding it once the result of the test becomes available.

Pedigree selection

As mentioned earlier, the initial selection of young bulls for progeny testing is likely to be based on the performance of the bulls' mothers. This is a form of pedigree selection, in which the records of ancestors are used to identify both promising families and individuals. The individual breeder of large farm animals generally relies on pedigree selection, because he has insufficient animals to conduct his own progeny tests. Pedigree selection was used by the eighteenth-century breeders who established many of our well-known breeds of livestock. It has the advantage of allowing rapid selection from records accumulated previously, and such records may be more detailed than those provided by a progeny test (e.g. milk records for several successive lactations of the same cow).

Selection indexes

When selection is based on more than one criterion it is often desirable to combine the measures of each character into a single or index value for each animal. An example is provided by the Meat and Livestock Commission's performance-testing scheme for young boars. Two boars – plus a castrate and a gilt from the same litter – are grown from 27 to 91 kg live weight at a central testing station. The boars are assessed for growth rate, food conversion ratio and backfat depth. The castrate and gilt are slaughtered and their carcasses are dissected. All measures are weighted for heritability, economic value and correlation with other measures, and the total score is expressed as a deviation from the current average score for boars of the same breed, which is maintained at 100 points. Below-average boars (<90 points) are slaughtered. Boars well above average (<200 points) are likely to be used for artificial insemination. In between these extremes, the better boars are likely to enter specialised breeding herds, while boars nearer the average will be used in commercial herds.

Breeding plans

Although the selection of breeding animals is often the most difficult part of genetic improvement it must, to be fully effective, be accompanied by a breeding plan.

Inbreeding

The interbreeding of close relatives may sometimes be used to give a concentration of desirable genes. For example, a bull may be mated to his own daughters and granddaughters, or brother may be mated with sisters. Inbreeding was used by some of the breed pioneers, who wished to establish uniform flocks or herds carrying distinctive traits. The problem with inbreeding is that it can concentrate undesirable, as well as desirable, genes, which means that improvements in the desired production characters may be accompanied by a decline in other traits, such as resistance to disease; in addition, inbreeding is likely to increase the frequency of inherited physical defects.

Breeding plans based on inbreeding are rare today, but may be used to produce contrasting families which are then interbred (i.e. used for crossbreeding).

Linebreeding

Like inbreeding this involves the mating of close relatives, but the relationships are less close. The purpose of linebreeding is to try to preserve in successive generations the combination of genes responsible for an outstanding individual.

Crossbreeding

In its most extreme form this type of breeding can involve the mating of different species, as, for example, when European-type cattle (*Bos taurus*) are crossed with the native cattle of tropical countries (*Bos indicus*). More commonly, it involves the crossing of contrasting breeds; an example is the crossing of Border Leicester rams with hill breeds of sheep to produce 'half-breds' that are suitable for use on improved pastures. Crossbreeding within a breed may, as suggested above, be used to combine the merits of inbred lines.

Much crossbreeding is carried out for reasons other than genetic improvement. Thus Hereford (beef) bulls may be crossed with Friesian (dairy) cows to produce offspring better suited to meat production than is the purebred Friesian. The crossbred males are not used for breeding and so have no influence on subsequent generations. The crossbred (Hereford × Friesian) females might be used as breeding cows in beef herds, where they would be mated with another beef bull to produce offspring that were three-quarters beef type.

The choice of a suitable breed of bull to cross with crossbred cows illustrates one of the main problems of crossbreeding. Crossbred animals are normally not bred together because their offspring are likely to vary in type between the extremes of the grandparent breeds. It is sometimes appropriate to 'backcross' the crossbreeds to one of the parents (e.g. Hereford × Friesian heifers crossed with a Hereford bull). Alternatively a bull of a third breed, such as a Charolais, might be used. In sheep breeding the Border

Leicester × hill breed 'half-breds' are commonly crossed with a ram of a breed selected for its carcass quality, such as the Suffolk. A ram used to produce animals for slaughter is known as the 'terminal' sire. For replacement half-bred ewes, the flock master must return to the original Border Leicester cross.

An example of crossbreeding intended to give genetic improvement is provided by what is called 'grading-up'. Nondescript cows are mated to a bull of a recognised breed, their offspring are mated to another bull of this same breed, and later generations are similarly crossed. Eventually the crossbreds are genetically so close to the breed selected that they are indistinguishable from members of that breed, and may be admitted to the breed themselves.

One of the main purposes of crossbreeding is to gain the advantages of what is called 'heterosis' or 'hybrid vigour'. If two separate breeds or lines within a breed are homozygous for many genes (i.e. each line has genes on its chromosome pair mates which are the same, but differ from the genes carried in the same positions by the other line), the interbreeding results in a high degree of heterozygosity. Such heterozygous animals often perform better (by as much as 20%) than the average of their parents. Thus the hybrid females that the pig farmers may buy from a specialist breeding company can be expected to produce litters at weaning containing more piglets, each individually heavier, than would be expected from their parent lines.

Conclusion

It is hoped that this very brief account of the principles of animal breeding will have given the reader some insight into breeding practices on British farms. No great emphasis has been placed on individual breeds, although the development and maintenance of distinct breeds has played a most significant role in genetic improvement. Individual breeders, often relying on somewhat subjective methods of selection, still make an important contribution to animal improvement, but the future lies more in the hands of breeders who collaborate – in national schemes or smaller groupings – by submitting their breeding stock to performance and progeny tests. A relatively new development is the group-breeding scheme, in which several breeders contribute their best females to a nucleus herd which is used to breed sires. Both sexes can be tested under standardised conditions, and tested sires can be returned to members of the group for use in their own herds. For poultry, and to an increasing extent for pigs also, genetic improvement is now carried out by commercial companies, who supply stock to farmers.

As mentioned earlier the rate of genetic improvement in poultry is rapid. Growth rate in broilers, for example, is estimated to have been increased through breeding by about 2·5% per year over the past 20 years. In dairy cattle breeding, detailed progeny testing and intensive selection of bulls could theoretically increase milk yield by about 2% per year, but in practice the improvement has been <1% per year. For beef cattle and sheep, genetic improvement is probably even slower. A recent development that is likely to increase the rate of improvement in ruminants is the use of techniques for multiple ovulation and embryo transfer (see p. 400). By allowing genetically superior dairy cows, for example, to produce as many as 20 daughters a year, the selection of females can be considerably intensified.

PRINCIPLES OF DISEASE CONTROL

The first principle of maintaining the health of farm livestock is that prevention is better than cure. To maintain health, however, farmers and stockmen must be familiar with the more common diseases of livestock, the circumstances in which they occur, and the agents responsible for them. Farmers and stockmen must be able to recognise ill health when it does occur, to make a preliminary diagnosis, and to apply remedial measures, either on their own initiative or on the advice of a veterinary surgeon. Although the veterinary surgeon must be called in to deal with severe or prolonged illness, he does not expect to treat minor conditions of ill health; it is generally the farmer who must distinguish between the major and minor cases.

Signs of ill health

The first manifestation of ill health in an animal may be a change in its behaviour. It stands apart from the herd, adopts an atypical posture, is reluctant to move, or shows no interest in food. Ruminants may stop chewing the cud; there may be sudden reduction in productivity (which can be detected quickly if it is a fall in milk yield but not so quickly if it is a reduction in growth rate).

Close examination of the animal will show changes in appearance. Head, wings or tail may droop, eyes may be dulled, the coat (of cattle) may be dull and rough rather than shining and smooth. Swellings or sores may be seen. The nature of excretions and secretions may change. Thus the faeces may be liquid (diarrhoea) or contain blood, urine may contain blood or become cloudy, milk may contain clots or blood. Abnormalities of movement, ranging from shivering to uncontrollable convulsions, may be seen.

As in man, some simple tests may help to establish the nature of the disease. A high respiration rate may indicate fever and/or lung infection, and elevated pulse rate and body temperatures are also indicative of fever. Normal values for these measures are shown in Table 20.19.

Table 20.19   Normal values for respiration rate, pulse rate and rectal temperature in farm animals

	Respiration rate (per min)	Pulse rate (per min)	Body temperature (°C)
Horse: adult	8–16	28–42	37·5–38·5
: foal	10–15	40–58	37·5–39·0
Cow : adult	26–30	60–90	37·5–39·5
: calf	30	100	39·0–40·0
Sheep or goat	10–20	68–90	39·0–40·5
Pig	10–20	60–90	38·0–40·0
Fowl	15–48	120–160	41·5–42·5

Ill health at birth

An animal may be abnormal at birth, because either it has a genetic defect which has affected its early development, or it has been damaged *in utero* by some maternal inadequacy or through competition from its littermates. An abnormality acquired *in utero* is described as congenital. Some examples of genetic defects are the misshapen head of the 'bulldog' calf, or the imperforate anus of piglets; they cannot be treated, but recording their occurrence and selecting against them will reduce their future incidence. Some congenital disorders such as 'swayback' in lambs (muscular incoordination resulting from maternal copper deficiency – see Table 20.7) are also irreversible.

Genetic and congenital defects, however, account for only a minor proportion of the animals that die at or shortly after birth. In pigs, for example, of all such *neonatal* deaths, these defects are responsible for only about 12% (i.e. they account for losses of about 0·2 pigs per litter). Attending animals during parturition will save many young from death. Early recognition of those that are poorly equipped, because of size or lack of reserves, to adapt to extra-uterine life is also very effective in preventing losses. Animals disadvantaged at birth may be temporarily removed from their dams (e.g. lambs to a warmer place) and given special treatment.

Injuries

Poorly designed or constructed housing can be a major cause of injuries to livestock. Slippery or broken floors, projecting metalwork, and unsafe feeding yokes or neck chains must be avoided. Animals may often be injured by their mates. A badly designed farrowing crate allows the sow to crush piglets when she lies down. The close confinement imposed in some modern systems of production may be responsible for injurious vices in livestock; some examples are tail-biting in pigs and feather-picking (perhaps leading to cannibalism) in poultry. Injuries may be prevented by modifications or mutilations of animals, such as de-horning in cattle and partial beak removal in poultry. A common cause of injuries due to fighting is the mixing of animals that are strangers to each other. A final point to remember is that physical injuries may predispose animals to diseases due to other causes. For example, a crushed teat (sometimes a self-inflicted injury) may allow the invasion of the mammary gland by bacteria which cause mastitis (see below).

Diseases due to micro-organisms

The principal pathogens in this category are the bacteria (i.e. minute unicellular plant organisms) and the viruses. The latter are extremely minute particles, consisting largely of DNA, which can multiply only inside the cells of other organisms. Two further types of pathogen in the same category are protozoa (i.e. minute unicellular animal organisms) and fungi.

These pathogenic organisms commonly gain access to animals through their respiratory, alimentary and urinogenital tracts, and may be transmitted to other animals by direct contact (e.g. venereal diseases), or via food, water or air (e.g. the virus causing foot and mouth disease may be carried by wind for long distances). Some organisms may always be present in animals but become harmful only if their numbers are excessive and/or they move to a more sensitive part of the animal (e.g. invasion of intestinal tissues by bacteria living in intestinal contents). Other pathogens may be capable of surviving for long periods outside animals; an example is the bacterium causing anthrax, which forms spores capable of surviving in soil for many years.

Some diseases may be transmitted between farm animals and man and are known as *zoonoses*. In Britain the most important examples are tuberculosis, brucellosis and a group of intestinal and other disorders caused by *Salmonella* bacteria.

When pathogens invade animal tissues they commonly stimulate the production of what are known as *antibodies*. These are proteins which react with pathogens and suppress or kill them. Pathogens are also contained, and thus inactivated, by the white cells of the blood.

The production of antibodies offers several methods for control of diseases. After an animal has been infected by a particular pathogen, and has combated the infection by production of antibodies, the latter may remain in its blood and confer immunity to further infection. Antibodies may also be passed from one animal to another. This transfer occurs naturally from mother to offspring, either via the placenta or via the initial secretion of the mammary gland known as colostrum (see p. 400). During the first day or so of life, the young animal is able to absorb intact proteins (the immuno-globulins of colostrum). In adult animals, antibodies may be trans-ferred by collecting blood serum from an individual that has ex-perienced infection in the past and injecting it into an individual recently exposed to infection. Animals which acquire immunity *passively*, at birth or by transfer of antiserum, receive protection that is immediate but which is also short-lived, because no produc-tion of antibodies has been stimulated.

Animals may be *actively* immunised by injecting them with pathogens that have been killed or modified but are still capable of stimulating the production of antibodies.

Control of diseases caused by micro-organisms

Having established the background to diseases caused by micro-organisms, we can now consider the strategy used to control them. If pathogens fail to make contact with animals there will be no infection, so the first means of control is *isolation*. A policy of isolation may be exercised at several levels. An island country, such as Britain, can prevent the importation of livestock (or require them to spend time in quarantine to ensure they are free of disease); this policy has (so far) proved particularly successful in keeping out many diseases caused by viruses, such as rabies and cattle plague. In the case of diseases spread mainly by close contact between animals, an individual farmer may be able to isolate his flock or herd if it is known to be free of the disease. This strategy is used to control virus pneumonia in so-called 'minimal disease' pig herds. Again, on the individual farm, a single animal that contracts a disease may be isolated from the herd until it recovers. Many farms have a special building for this purpose, and it may be used also to quarantine purchased stock.

A second type of strategy used to control disease requires the slaughter and hygienic disposal of infected animals. This strategy may be used to control diseases which occur as sporadic outbreaks but which can spread rapidly if given the opportunity. In Britain, foot and mouth disease is controlled by immediate slaughter of all susceptible stock on a farm once cases have been diagnosed on that farm. A slaughter policy may also be used to eliminate diseases that occur in a relatively small number of individuals. In Britain, bovine tuberculosis and brucellosis have been reduced to a very low incidence by a policy of routine testing of animals for these diseases, and slaughtering those found to have them.

In the case of diseases caused by organisms widely distributed in nature, and which can survive apart from animals (for example, in faeces), a strategy of routine hygiene is required.

The routine used to control infection of calves by organisms such as *Salmonella* bacteria, which cause intestinal upsets and scouring (diarrhoea) may be used as an example. Young calves are housed separately from older animals (to prevent infection from elsewhere) and are kept in individual pens with individual feed buckets (to prevent one calf infecting another). Their buckets are cleaned frequently (to prevent bacterial growth in feed residues) and the calves are inspected frequently (to allow detection and isolation of infected animals). A further precaution, which is not always practicable, is to rear calves in batches; in between each batch, the calf house is clear of animals and can be thoroughly cleaned and disinfected.

Intestinal disorders in calves are caused by a wide variety of organisms, and it is at present difficult to protect the animals by vaccination. This technique is used, however, to control diseases caused by specific organisms. In lambs, for example, there is a small group of diseases (pulpy kidney, lamb dysentery and tetanus) caused by bacteria of the *Clostridium* group, which can be prevented by vaccination at birth. In poultry, vaccines are used to control Marek's disease (a viral infection which affects the nervous system and causes paralysis), infectious bronchitis and several other diseases.

If a vaccine is used routinely for a long period, the incidence of the disease concerned may be reduced sufficiently to justify a change to an alternative control strategy. An example of such a disease is brucellosis in cattle, which was at one time prevented by vaccination but is now controlled by the slaughter of infected animals.

The last line of defence against disease is the use of therapeutic agents or drugs. For some livestock, principally poultry, drugs are given routinely, as a constituent of the diet, an example being the use of dimetridazole to control the protozoal disease, blackhead, in turkeys. Drugs may also be used routinely (i.e. to prevent infection) but intermittently; thus dairy cows may have their teats injected with antibiotics at the end of lactation, in order to treat existing infection and prevent infection during the dry period. Finally, drugs may be used selectively, to treat infections.

The use of antibiotics to treat animals has been questioned because they may encourage the production of resistant strains of bacteria that may subsequently infect man. Antibiotics used in human medicine are not now used routinely for farm animals, but may still be used for selective treatment. The treatment of mastitis in dairy cattle, for example, is based on the use of antibiotics.

Diseases due to parasites

The major internal parasites of livestock in Britain are helminth worms, such as the roundworms that inhabit the stomach and intestine (and also the lungs), tapeworms of the small intestine, and flatworms exemplified by the liver fluke. The major external parasites are small invertebrates such as ticks, mites, lice and flies.

Worms

The nematode roundworms found in the stomach and small intestine of sheep and cattle damage the lining of the gut, thus interfering with absorption and also causing loss of blood. The adults lay eggs which are excreted on the pasture and may hatch to produce larvae within a few days or lie dormant over winter. The larvae are ingested during grazing and develop into adults to complete their life cycle. Lungworms have a similar cycle but the adults live in the respiratory tract, where they cause bronchitis and predispose their host to pneumonia.

The life cycle of tapeworms and the liver fluke is complicated by the need for an alternate host. The adult fluke, for example, lives in the duct system of the liver, its effects ranging from chronic 'unthriftiness' to severe liver damage and death of the host. Its eggs pass through the bile duct and thence to the faeces. The larvae derived from them invade a type of small snail found commonly on ill-drained pastures, and subsequently develop into a cyst which is ingested by the grazing animal.

Ideally, pastures would be kept completely free of infective larvae, but in practice this is difficult to achieve because a newly sown pasture is soon re-infected by the stock brought on to it. Nevertheless, helminth infestation of livestock can be kept to an acceptably low (i.e. non-injurious) level by the use of husbandry practices that break the life cycle of the worms. Life cycles can also be broken by the destruction of alternate hosts. For example, the draining of pastures discourages invasion by snails and hence reduces fluke infestation.

Like bacteria and viruses, some helminths stimulate the production in the host of antibodies. Older animals are therefore less affected by worms than are their offspring, and rotational grazing strategies are often designed to give young animals the 'cleanest' areas. Young cattle, when turned out to graze in the spring, are particularly susceptible to a lungworm infestation which causes the coughing condition known as 'husk'. For this parasitic disease, however, it has proved possible to produce a vaccine which is now used routinely in cattle.

During the past 30 years there has been a remarkable development of anthelmintic drugs, most of which are administered by oral 'drenching' but the most recent of which can be given by injection. Although the first lines of defence against worms are the husbandry practices suggested above, anthelmintics provide valuable reinforcement, especially when stocking rates are so high as to present special problems of control.

External parasites

As a group, these parasites cause ill health in farm animals through three major effects. First, they may irritate the animal to such an extent that its feeding behaviour is affected and its productivity reduced (e.g. biting flies). Secondly, the parasites may have a direct damaging effect on the animal's tissues (e.g. ticks may suck blood and cause anaemia, warble fly larvae burrow through the skin of the animal). Thirdly, external parasites may act as *vectors* (i.e. carriers) of diseases caused by micro-organisms. The most serious example of insect transmission of disease is trypanosomiasis (sleeping sickness) of ruminants (and also man), which is caused by protozoa carried by tsetse flies. While this does not occur in Britain, there are several diseases here (e.g. tick-borne fever and louping ill in sheep) which are transmitted by external parasites.

Parasites that can survive only on selected host species can be eliminated by routine treatment of their hosts with insecticides. The disease known as sheep 'scab', which is caused by mites, was eliminated from Britain (and other countries) by compulsory 'dipping' of all sheep in insecticide solution, but has recently recurred. The warble fly of cattle, which spends part of its life cycle in animal tissues, is currently the subject of an eradication campaign.

Nutritional diseases

Three types of disease may be considered under the heading of nutritional diseases. The first group are the deficiency diseases discussed earlier in this chapter (e.g. Tables 20.6 and 20.7). The second group are the diseases caused by ingestion of poisons. Thirdly, there are the so-called 'metabolic' disorders which are characterised by abnormalities in the absorption, metabolism and excretion of nutrients.

Poisons

The range of substances toxic to livestock is extremely wide, and only the major classes of poisons can be indicated here. Grazing animals may consume weeds, growing in or around their pastures, that contain toxins (e.g. bracken and ragwort). Normal crop plants may contain substances that become toxic if they accumulate in plants, or are otherwise consumed in excessive quantity.

> Many members of the *Brassicae* (i.e. cabbage family) contain an unusual amino acid derivative that causes anaemia in ruminants. Grasses may accumulate toxic quantities of nitrates, and pasture legumes (like red clover) can contain phyto-oestrogens, which interfere with the reproduction of grazing animals. The toxicity of a normally safe plant or feed may sometimes be due to fungal infestation. The classical example of this is ergotism, the nervous disorder of man and animals that is caused by consumption of grains infested with ergot fungus, *Claviceps purpurea*. A more serious threat today are the carcinogenic aflatoxins produced by fungi found on groundnuts.

Poisoning may also be caused by contamination of feed. In the field, airborne pollutants such as fluorine from aluminium smelters, may make crops toxic to livestock. In the feed-mixing plant, pig feeds fortified with 200 mg/kg of copper as a growth promoter may contaminate feed intended for sheep, which are sensitive to diets containing >20 mg Cu/kg. Animals may also ingest toxic substances from sources other than feed; for example, lead ingested by licking painted surfaces is a common cause of poisoning in livestock. The many chemicals used on farms to control weeds, plant diseases and pests represent a continuous hazard to animals.

Methods for preventing poisoning are largely self-evident. The farmer must be aware of the hazards present on, or likely to arrive at, his farm and must be prepared to protect his livestock accordingly.

Metabolic disorders

The complexity of the chemical transformations taking place in the animal is evident from earlier parts of this chapter. Nutrients absorbed from the gut are used continuously to construct new or replacement tissues and products, or are shunted into or withdrawn from stores; nutrients derived from tissue breakdown are salvaged for re-use, and unusable materials are excreted. Occasionally something goes wrong. We have seen already that a dietary shortage of a particular nutrient can be responsible for serious ill health. But even when there are no dietary deficiencies, an enzyme deficiency, for example, may mean that some vital product fails to be delivered in sufficient quantity to the site where it is required. The disease which follows such a failure of metabolism is known as a metabolic disorder.

Metabolic disorders generally arise from some overloading of the animal's systems. The high-producing animal, such as the high-yielding dairy cow, is particularly susceptible. The likelihood of metabolic upsets is also very dependent on the speed at which the increased load is imposed. At calving, the dairy cow moves rapidly from a state of relatively low nutrient intake and nutrient demand, to the more intense metabolism imposed by the high milk yield of early lactation. She is therefore particularly susceptible to metabolic disorders in early lactation. It would be wrong to assume, however, that the low producer is unlikely to suffer from metabolic problems. In an animal fed at a maintenance level, and with few reserves, a slight change in nutrient demand can upset metabolism as disastrously as can a much greater change in a normally well-fed animal. Although the dairy cow is used here to illustrate the main types of metabolic disorder, these diseases are found in all forms of animal production.

When a dairy cow begins to lactate, the nutrient demands of milk secretion rise faster than does food intake and thus nutrient supply. The cow therefore uses her reserves. Body fat is transformed into milk fat and may also be used for maintenance purposes; tissue proteins might also be transferred to milk, but scope for this is limited by the absence of any substantial protein store; minerals, especially calcium, are moved from bone to milk.

To metabolise fat effectively, the cow needs an adequate supply of carbohydrate in the form of glucose or propionic acid. If the balance is disturbed by too much fat, the latter is imperfectly metabolised to substances which may be toxic to the animal; some of these substances belong to the class of chemicals called ketones, and the resulting disorder is called *ketosis*. A further manifestation of too rapid a removal of body fat from reserves may be an undesirable accumulation of fat in the liver.

If calcium cannot be drawn from bones fast enough to meet the drain in milk, then the calcium concentration of body fluids is reduced. Nerve and muscle functions depend on adequate calcium, and a fall in blood calcium concentration (hypocalcaemia) causes muscular spasms and incoordination; this condition is commonly called *milk fever*. A related condition is associated with low blood magnesium (hypomagnesaemia); it occurs commonly in grazing cows and is called *grass staggers*.

Additional metabolic disorders arise from imperfection in the metabolism of rumen micro-organisms. Excessive production of gas, often trapped in a foam, causes bloat. Too rapid fermentation of diets rich in soluble carbohydrates to lactic acid and the subsequent absorption of this acid by the animal may cause *acidosis*.

As metabolic disorders are commonly associated with high levels of production, isolated cases may be accepted by the farmer as part of the cost of increased productivity, and are treated as they occur. Thus, cows with milk fever may be given an intravenous infusion of calcium in the form of calcium borogluconate, which relieves the tetany and may maintain blood calcium concentration at a normal level for sufficient time for calcium supply and demand to return to balance. First-aid measures, however, are inadequate if the incidence of milk fever or other metabolic disorders is severe, and some means of prevention must be sought.

Although metabolic disorders are not generally due to dietary deficiency, they can often be prevented by modifying the diet. For example, the dairy cow can be prepared for the calcium requirement of lactation by feeding her on a *low* calcium diet before calving. This has the effect of mobilising bone calcium, and ensures that the mobilising machinery is operating before a large demand is suddenly placed on it. A rather different example is provided by the ewe in late pregnancy that may suffer from ketosis (in the form known as pregnancy toxaemia) because she is drawing heavily on her fat reserves to provide for growth of foetuses. A small supplement of a starchy concentrate given at this time will help to restore the balance in metabolism between fat and carbohydrate.

As production levels rise, and as other diseases are eradicated, metabolic disorders (sometimes called *production diseases*) assume greater importance in animal production. A device that is sometimes used to diagnose them when they are still sub-clinical (i.e. manifested in a form too mild to be recognised by the farmer) is to carry out what is called a *metabolic profile* test. In a dairy herd subjected to such a test, blood samples would be taken from several animals having different levels of milk production, and the samples would be analysed for a series of 10 or so metabolites, including calcium, glucose, etc. The extent to which the overall profile of these metabolites differs from normal is used to indicate where preventive measures are needed. Although interesting in principle, this test has not proved to be as effective in practice as it was hoped.

Preventive medicine

The first principle of animal health control, that prevention is better than cure, is now enshrined in the important branch of veterinary practice called preventive medicine. Instead of merely treating sick animals, the veterinarian advises the farmer on the procedures he himself can use to prevent infections and disorders, and he visits the farm at regular intervals to inspect the stock and apply any specialised tests or treatments that are required (e.g. pregnancy diagnosis). The veterinarian may also discuss with the farmer the wider aspects of animal husbandry such as nutrition, ventilation of buildings and pasture management.

21

THE PRINCIPLES OF
GRAZING MANAGEMENT

Introduction

A grazing system can be considered as a chain of processes involving the production of herbage from populations of plants, its consumption by populations of animals, and its conversion into animal product. The processes of herbage production and utilisation interact directly with one another so that management decisions which improve efficiency at one stage of the production chain may result in a reduction in efficiency at another. Thus an understanding is required of the factors influencing efficiency at each stage, the degree to which they are amenable to manipulation, and the form and extent of their associations with one another.

THE SWARD

Temperate grasslands are populated primarily by relatively low-growing perennial plants, most of which can perpetuate themselves by the development of new tillers as well as by seed. In forage plants adapted to grazing the buds capable of sustaining new growth are sited close to the base of the plant where they are unlikely to be removed except by the most severe defoliation (see Fig. 12.9). Perenniality is a consequence of continued tiller development, because few individual growth sites have an effective productive life of more than 12 months. In perennial plants the supply of new buds is virtually inexhaustible, and their development into tillers is limited primarily by competition for light, water or nutrients. Growing tillers develop their own root systems at a relatively early stage and thus become capable of sustaining growth independent of the parent plant, and it is usual to consider the sward in terms of populations of tillers in the case of grasses and of their nearest analogues, rooted nodes, in the case of the stoloniferous legumes.

Leaf production is virtually continuous over the lifetime of a vegetative tiller, individual leaves being produced in sequence from the stem apex which is borne close to the base of the tiller and encased in the sheaths of older leaves. Each leaf will extend for a period of 7-21 days during the main growing season, depending upon climatic conditions, but for a much longer period of time in the winter. Once it is fully extended an individual leaf has a relatively short period of existence before irreversible degenerative changes begin which lead rapidly to senescence and death. Thus a vegetative tiller of perennial ryegrass characteristically carries only three live leaves, one of which is in an active phase of extension, one fully expanded or approaching full expansion, and one mature. This pattern of development and death proceeds irrespective of whether individual leaves are defoliated or not.

Under appropriate conditions individual tillers may develop from a vegetative to a reproductive state in which the stem apex elongates to form a flowering head. In many species of grass the stimulus to flowering is increasing daylength in the spring, and in some species, such as *Lolium perenne*, an initial cold stimulus is required. In other species, such as *Poa annua*, tillers may develop rapidly to a reproductive state at almost any time of the year. Reproductive changes involve progressive elongation and lignification of the main stem, resulting in a rapid increase in the height of the sward canopy. Once the seed head borne at the top of the flowering stem has ripened, leaf production has effectively ceased and the tiller proceeds through the final stages of senescence and death. After a phase of development of reproductive tillers, plant growth can be sustained only from any vegetative tillers which remain, by the development of new shoots from the bases of the old reproductive tillers, or eventually by the development of new plants from shed seed.

Plant growth is dependent upon the supply of assimilates resulting from the process of photosynthesis in active leaves exposed to the light. Since leaves develop from growing points situated close to ground-level they must penetrate through the overlying sward canopy in order to reach the light, and successive leaves tend to over-top those earlier in chronological sequence. Characteristically the proportion of live leaf increases from the bottom to the top of the sward canopy and the proportions of stem, sheath and dead material decline. The bulk density (mass per unit volume) of foliage also declines progressively from the base of the sward upwards. As a consequence of stem elongation, reproductive swards are usually taller than vegetative swards and have a lower bulk density, individual tillers tend to adopt a more erect habit of growth, and there is a more heterogeneous admixture of stem and leaf material. All of these factors may influence both herbage production and the herbage intake of grazing animals.

Young, actively growing leaf tissue contains relatively limited amounts of fibre, and is therefore easily digested by ruminants. The ratio of structural to non-structural material increases with increasing leaf maturity, particularly once the degenerative changes of senescence begin and the more soluble components are progressively lost by respiration and remobilisation. Thus the digestibility of leaf tissue may be as high as 0·9 during elongation, declining to 0·7 in senescence and 0·5 in dead leaves in the litter layer. The digestibility of reproductive stem declines rapidly with advancing maturity as a consequence of its high structural content and degree of lignification, so that the digestibility of herbage in reproductive swards falls progressively with increasing maturity and increasing ratio of stem to leaf tissue (Fig. 21.1).

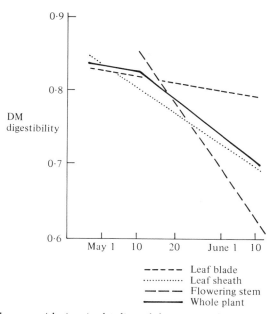

Fig. 21.1 Changes with time in the digestibility (DM) of the whole plant and its components in S24 ryegrass during primary spring growth

Herbage growth and utilisation

The processes of growth and senescence occur simultaneously and virtually continuously in temperate swards. Conventional measurements of herbage 'production' – strictly net production or accumulation over a finite period of time – depend upon estimates of the balance between the rates of growth and senescence. Successful management involves optimising this balance, so it is important to know how the two rates can be manipulated, but neither is particularly easy to measure in the field.

One approach is to measure the rates of carbon uptake in photosynthesis and release in respiration, and the balance between them, the net carbon assimilation. These rates are not strictly analogous to the rates of herbage growth, senescence and net accumulation, but cumulative net carbon assimilation must set an upper limit to production because carbon forms such an important component of plant tissue and plant energy supply. The net rate of carbon assimilation per unit of ground area (P_{nc}) increases progressively to a maximum with increasing leaf area index in a developing sward as the efficiency of light interception increases, but eventually will start to decline as mutual shading reduces the rate of carbon uptake per unit of leaf area and as the losses attributable to respiration increase (Fig. 21.2).

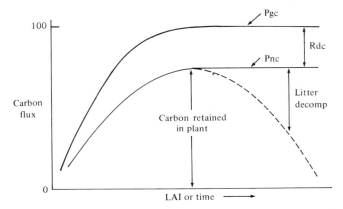

Fig. 21.2 Diagrammatic representation of changes in the rates of gross photosynthesis (Pgc), net photosynthesis (Pnc) and respiration (Rdc) in a grass sward over a period of regrowth following defoliation. Values shown are rates of carbon flux ($gm^{-2}\,h^{-1}\,CO_2$) relative to maximum $Pgc = 100$

Herbage production can also be viewed as the product of the number of growth sites (i.e. of grass tillers and clover nodes) per unit ground area and the rate of production of plant tissue per site, both of which can be measured. A reduction in the herbage mass maintained under continuous stocking management towards 1,000 kg O M/ha results in a reduction in the size of individual tillers but an increase in tiller population density, an increasingly prostrate habit of growth and an increase in the ratio of young to old leaf tissue. These adaptive changes help to maintain both the efficiency of light interception and the photosynthetic efficiency of leaf tissue, but with further reduction in herbage mass the tiller population may decline and the rate of growth falls away rapidly (Fig. 21.3). The rate of senescence of mature tissue declines progressively as herbage mass is reduced and, when the balance be-

tween the rates of herbage growth and senescence is taken into
account, net herbage production per hectare is usually maximised
at a herbage mass of 1,200–1,500 kg O M/ha (L A I 3–4) under continuous stocking, though the variation in net production is in fact
relatively small over the range from 1,000 to 2,000 kg O M/ha (L A I
2–5) (Fig. 21.3).

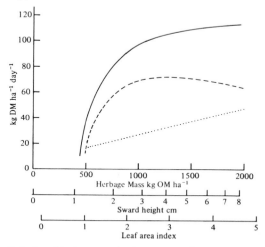

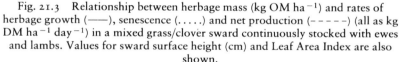

Fig. 21.3   Relationship between herbage mass (kg OM ha^{-1}) and rates of
herbage growth (——), senescence (.....) and net production (- - - - -) (all as kg
DM ha^{-1} day^{-1}) in a mixed grass/clover sward continuously stocked with ewes
and lambs. Values for sward surface height (cm) and Leaf Area Index are also
shown.
Source: Adapted from Bircham and Hodgson (1981)

Intermittently cut or grazed swards are in a continuous state of
change and therefore have less opportunity to adapt, but the effects
of management can be described in the same terms.

THE ANIMAL: HERBAGE CONSUMPTION

In many circumstances the daily intake of nutrients from herbage
is the major determinant of the productive performance of grazing
animals, and has a substantial influence on the efficiency of conversion of ingested herbage into animal product. The process of digestion in ruminant animals is relatively slow and the consumption of
food may be limited by the physical volume of digesta in the
alimentary tract or some part of it. The rate at which food dis-
appears from the tract, and thus allows more to be consumed,
reflects its digestibility, the rate of digestion and absorption of the
digestible components, and the rate of passage of indigestible com-

ponents down the tract. These three characteristics tend to be correlated in long forages, and are positively related to forage intake (Fig. 21.4). In ruminants fed on mixed diets of forages and

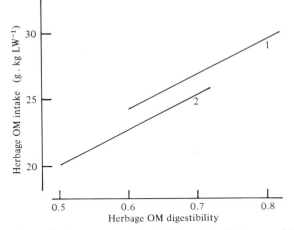

Fig. 21.4 Relationship between the digestibility of the diet ingested (OMD) and the daily herbage intake (g OM kg LW⁻¹) of weaned calves strip-grazing swards of perennial ryegrass. 1: primary growth; 2: regrowths

concentrates, intake may also be limited at high nutrient concentration by the level of circulating metabolites in the blood, but this limitation is seldom if ever important in animals eating all-forage diets, particularly under grazing conditions.

The bulk density of the herbage within a sward canopy is relatively low, particularly in the surface horizons of the sward, and grazing ruminants must take up to 30,000 or more individual bites of herbage and spend up to 12 hours or more grazing per day in order to satisfy their appetites. Factors which limit the weight of herbage in individual bites are therefore likely to limit daily herbage intake unless the animal can compensate by increasing the rate of biting or the time spent grazing. Both changes are possible, but they are seldom of sufficient magnitude to compensate fully for changes in intake per bite which usually appear to be the major determinant of daily intake. In homogeneous vegetative swards intake per bite is influenced by variations in sward height and in the bulk density of herbage in the grazed horizon (Fig. 21.5), and animals appear to exhibit little deliberate selection between the alternative botanical and morphological components. In more heterogeneous swards, selective grazing will work to the animal's advantage by increasing the nutrient concentration of the diet but this advantage is likely to be offset to some extent (and, in extreme cases, completely) by the

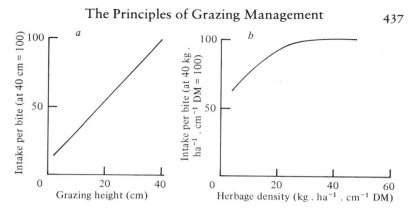

Fig. 21.5   Relationship between intake per bite (mg OM kg LW^{-1}) and (*a*) the height of grazing (cm) and (*b*) herbage bulk density (kg DM ha^{-1} cm^{-1}) in the grazed horizon

reduction in intake per bite and/or in biting rate which is likely to accompany selection.

Herbage intake per animal is therefore likely to be influenced primarily by the stage of maturity of the sward, the vertical distribution of herbage, and the disposition of the component species and morphological parts within the canopy. It will be seen that the factors which characterise a well-grazed sward – a dense population of tillers bearing relatively young leaves – are likely to be as beneficial to herbage intake as to herbage production. However, the close grazing necessary to maintain a sward in this condition is unlikely to maximise herbage intake per animal because it will result in a relatively shallow sward canopy.

EFFICIENCY IN GRAZING SYSTEMS

Examples of the efficiency of each stage of the production process are given in Table 21.1, taking the ratio of $\dfrac{\text{energy out}}{\text{energy in}}$ as a common basis for measurement. The examples refer to temperate conditions. Management is concerned with the interrelationships between the efficiencies of the three stages and the factors influencing overall efficiency.

The efficiency of conversion of the sun's energy into plant tissue is always low but, because of the enormous quantities of light energy striking the earth and the consequent benefits which should accrue to even a limited improvement in efficiency, there has always been great interest in the possibility of manipulating the sward to increase herbage production in a predictable fashion. Under a cutting management the amount of herbage harvested is almost

Table 21.1 Estimates of the energetic efficiency of (a) herbage growth; (b) herbage utilisation; and (c) conversion of ingested herbage into animal product: temperate grassland

Stage of production:	Ratio $\dfrac{\text{Energy Output}}{\text{Energy Input}}$
(a) Herbage growth*	
$= \dfrac{\text{Energy in plant tissue}}{\text{Photosynthetically-active radiation}}$	0·02–0·04
(b) Herbage utilisation**	
$= \dfrac{\text{Energy in herbage consumed}}{\text{Energy in herbage grown}}$	0·4–0·8
(c) Conversion to animal product***	
$= \dfrac{\text{Energy in product}}{\text{Energy in herbage consumed}}$	0·02–0·05

* Mean value for year; during active growth efficiency may be 0·04–0·08
** Grazing animals
*** Sheep. Estimate based on annual food input to ewe and lamb(s); during grazing season, efficiency on a daily basis may be 0·05–0·15

invariably higher at low than at high harvesting frequencies (Fig. 21.6), but there is less unanimity about the effects of severity of defoliation on the amount of herbage harvested. The relatively high average herbage mass and LAI on infrequently cut swards should result in efficient light interception and a high rate of carbon uptake

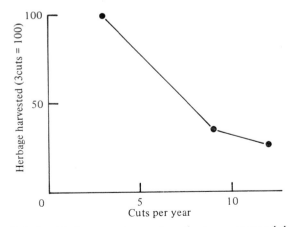

Fig. 21.6 Relationship between the number of cuts per year and the herbage harvested (value at 3 cuts per year = 100)

(Fig. 21.2), but some of the advantages of infrequent cutting may also be attributable to improved harvesting efficiency.

There appears to be much less scope for improving the efficiency of herbage production by modification of grazing management, though herbage growth rates may still be depressed by prolonged, severe defoliation (Fig. 21.3).

The efficiency of utilisation of the herbage grown is in general substantially higher than the efficiencies of the other two stages in the production process, but this is probably the stage at which most unobserved loss of efficiency occurs. In continuously stocked swards the rate at which herbage is harvested by grazing animals (kg/ha/day) appears to be maximised at a herbage mass and LAI at which herbage growth rate is already depressed, but at which the ratio of consumption to growth is high. Maximising herbage consumption per hectare therefore depends upon avoiding managements which are severe enough to depress herbage growth seriously at one end of the range and lax enough to increase herbage losses (or depress the consumption/growth ratio) at the other. Fortunately, it seems that the effective zone of approximate balance between these extremes under grazing conditions gives considerable scope for manoeuvre (Fig. 21.3). It is worth noting in passing, though, that most farmers still tend to err too close to the upper limit of the range.

Herbage consumption per unit area is the product of intake per animal and stocking rate (SR). It almost invariably increases with increasing SR, despite some reduction in intake per animal as a consequence of the associated decline in herbage mass. In qualitative terms, the diet consumed from an efficiently utilised sward is likely to be just as good as that from a laxly defoliated sward, and this in itself will help to maintain nutrient intake.

The efficiency of conversion of ingested food into animal product is also inherently low, due partly to the relatively high demands made on nutrients for the maintenance of body function, including the digestion process. It is directly related to the level of herbage intake per animal and, for a population of animals, is influenced by the proportion of productive individuals and their genetic potential for growth or lactation.

Overall efficiency is commonly measured as the output of animal product per unit area of grassland, and the management variable likely to exert the greatest impact upon output per unit area is the stocking rate (Fig. 21.7). In meat animals, performance (LWG) per animal declines steadily with increasing SR over virtually the full range studied and there is very little indication of 'plateaux' of animal performance at low SR, except perhaps at the very lowest levels. This reflects the progressive reduction in intake per animal with increasing SR. The relationship between output (LWG) per

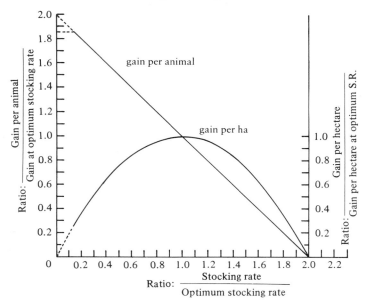

Fig. 21.7 Relationship between stocking rate, weight gain per animal and weight gain per unit area (all expressed relative to value at maximum weight gain per unit area)
Source: Adapted from Jones and Sandland (1974)

ha and SR takes the form of a rectangular hyperbola, being zero (predictably) at zero SR and at the SR giving zero LWG, and a maximum at the SR midway between the two. This defines the point of maximum biological efficiency for the system; it represents a compromise, because none of the three main stages in the production process are likely to be operating at maximum efficiency at this point. The SR maximising gross output, gross margin and profit per ha will be progressively to the left of that maximising biological efficiency and, although this generalised relationship forms a very useful basis for making comparisons between levels of management inputs, the assessment of stocking rates appropriate to particular farm enterprises requires more detailed information on input costs and output values, and on anticipated output responses.

22

THE PRINCIPLES OF
FARM BUSINESS MANAGEMENT

Introduction

This chapter is about the procurement of resources and the organisation and use of those resources across the whole farm business in an endeavour to achieve certain predetermined objectives.

Ultimately, perhaps the most important objective for any farm manager, is that his or her business will survive – and for that to happen it is essential that over a period of time sufficient profits are generated to reward and motivate owners and managers and to provide at least part of the funds from which future expansion and development are made possible.

This is not to suggest that profits are the only thing that matters, but without them insolvency may follow. This is most likely to be avoided if the nature of the task and certain general principles of management are understood and applied to farm businesses just as to other businesses and organisations.

This is in no sense to belittle the importance of good husbandry: it is very important to 'do things well' and this will seldom be in conflict with doing things profitably. It is merely being argued here that *too much* preoccupation with the *purely* technical and with day-to-day matters could be dangerous if it is at the expense of the more strategic, commercial and human aspects of management.

Before proceeding to discuss the component parts of the farm manager's job it may be helpful to offer a definition of what the job entails.

Management is a comprehensive activity, involving the combination and co-ordination of human, physical and financial resources, in a way which produces a commodity or service which is both wanted and can be offered at a price which will be paid while making the working environment for those involved agreeable and acceptable in financial and non-financial terms.

This definition has been worded carefully in order to draw attention to certain key aspects of management whether it relates to a large or a small organisation. The majority of farms – but not all of them – fall into the small-firm category and this definition of management is applicable to them. It stresses that – especially in an industry such as farming, where opportunities for the delegation of management are fairly limited – management is a *comprehensive* activity and top management has the ultimate responsibility for all that goes on. It also stresses that management is concerned with the *combination* of different kinds of resources, with the need to be market oriented, and also with the human factor. Spelt out in more practical and diagrammatic terms the farmer's or farm manager's job – for the purpose of this discussion the two can be regarded as virtually synonymous – can be depicted as in Fig. 22.1.

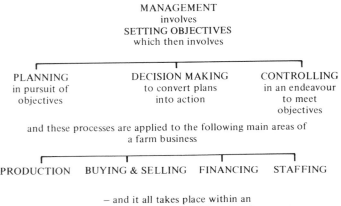

Fig. 22.1   The farm manager's job

It must be appreciated that a diagram of this kind reflects a very personal view of management and it should not be assumed that the various processes and activities depicted are as independent of each other as the diagram may suggest. Decisions about what to produce, for instance, will have obvious repercussions on market-ing arrangements, capital requirements and on the number and quality of staff to be employed. In a similar way, decisions which relate primarily to such questions as the supply of capital or to the availability of labour can have an important influence on produc-tion possibilities.

It is also appreciated that this diagrammatic view of the farm manager's job may represent more or less than the whole story for many farms. For most farming situations, however, it is believed that this fourfold division is appropriate. Assuming that the need to set, and, when appropriate, to quantify objectives has been accepted and acted upon, managers will then be responsible for *making plans, taking decisions* and *exercising control* across the four sectors of the farm business that have been identified. Indeed, it is in the exercising of responsibility in this way that a manager earns his rewards: a salary in the case of a hired manager, and profit, with the final risk for the capital that is at stake, for the self-employed farmer.

In the rest of this chapter each of the processes and business sectors discussed in this Introduction will be looked at in more detail.

SETTING OBJECTIVES

It is important for any organisation to take a periodic look at its objectives for a limited number of years ahead – say three, four or five years. This helps to give the organisation itself a purpose; it provides the individuals in the organisation with a common goal and it minimises the risks of going off at too many tangents. Farming, like other industries, has its yardsticks by which the financial results for any particular trading year can be assessed and judged. Here we are concerned with something more fundamental: what a farm business has to offer; its potential; what the individuals concerned want out of it, and what path forward it is going to follow.

In an informal sense farmers have done this, but it is doubtful if many of them have put themselves and their businesses through the more formal approach to this exercise offered, for instance, by such techniques as *Management by Objectives*.

This comprehensive approach to thinking about the whole management process involves discussions over a period of time by an appropriately chosen small group of people concerned with the management of a business, including perhaps somebody from outside the business, such as an adviser. It involves a careful consideration of the resources available to the business, including, most importantly, the manager himself (all of these known as the internal environment); the many and various factors and influences that impinge directly upon the business from outside (the external environment); clear statements, quantified where possible, about where the business is aiming to get to and how it intends to get there; the identification of the critical so-called 'key results' areas; the necessary performance levels and appropriate control systems – and finally, a willingness, periodically, to reconsider, replan, restate and re-dedicate.

This might call for a farmhouse meeting on rather a formal basis for long-term objective setting. With the small family unit, discussions may involve only the farmer and his wife whereas with the large company the directors and manager would ideally be joined by unit managers as well as consultants. The eventual outcome should be an agreed statement of what is to be achieved and how and when it is to be achieved.

It should not be pretended that this approach to management – be it on a farm or any other kind of business – is either easy or a panacea for all problems; but a periodic and rigorous questioning of the kind that has been described here can only be beneficial. The task will be time consuming and may need to be repeated. But the difficulties should not be an excuse for not thinking seriously, from time to time, about long-term objectives. Translating that thought into achievement then calls for effective planning, decision-making and control – topics to which we now turn.

PLANNING

If objectives are going to be achieved, then plans have to be drawn up in advance of the desired events. Planning is precisely about that: about taking the time *before* something happens in an endeavour to arrange and influence events when they *do* happen.

Almost inevitably, in the farming scene, the word *planning* is synonymous with the word production, or with the *farm plan*. It is certainly true, of course, that the planning of production is very important, whether it relates to the long term or to the short term. However, the planning process is likely to be equally applicable and necessary across the whole spectrum of a farm business and the whole range of a farmer's problems. It may be just as important, for instance, to spend time planning such matters as the cash flow for the next six months, or the recruitment of a new member of staff, or the marketing of some particular commodity, as it is to spend time on the farm cropping plan.

The framework of action provided by a well-thought-out plan, in any sector of the business, provides the stability against which considered and careful adjustments can be made – and 'management by crisis', as it is derogatorily called, is avoided.

DECISION MAKING

Some would say that management is essentially about decision making of one kind or another, but whilst this contributor cannot see management in such limited terms, there is an element of truth in this view – to the extent that decision-making is one of the most persistent and, therefore, the most nagging aspect of management. Decisions are often associated with change, from one activity or method to another, but essentially they are about choice; and sometimes, but not always, they are about problem solving.

Decisions can be categorised into two types: the tactical and the strategic. The so-called tactical decisions relate to short-term matters where, often, a fairly quick decision is required. In farming, for instance, weather conditions will invite day-to-day decisions about how to deploy labour and machinery in the face of conflicting demands for their use. Strategic decisions, as the name implies, are about the longer-term issues and often need more careful thought. An inherent difficulty in the process is that decisions have to be taken in the present, often based on information drawn from the past, about events that will happen in the future. It is also the case that wherever a choice exists, it is only the outcome of the chosen course of action that is known. To that extent it can never really be claimed that the *best* decision has been taken; only the one that appears to be best before the event.

For strategic decisions, the farmer may find it helpful to follow these steps: identify the issue/choice/problem in hand; assess its significance; consider the alternative choices open; collect information/data about those choices; evaluate the data, e.g. with budgets; make and implement a choice; check actual results; and, finally, accept responsibility for the success or failure of the decision.

It is not suggested here that in every situation or wherever a choice of any kind has to be made by a farmer or a farm manager he will or should go through this formal and apparently laborious process, but something like this sequence is probably required – and where major investment is involved it should certainly not be by-passed.

CONTROL

It is essential that setting objectives, planning and decision making are followed by the appropriate control mechanism to ensure that what is supposed to happen does, as far as possible, happen. In this sense it could be argued that 'control' comes closest to the heart of practical farming itself, i.e. getting things done and getting them done well. It is very easy for inadequate performance in one sector to be masked by superior performance elsewhere: budgetary control techniques can provide an effective mechanism for avoiding that masking effect.

There are, in fact, two essentially different elements to control: the *physical* and the *financial* – but while they entail different approaches to the task they have a similar and single aim: to achieve what was intended. To consider the physical element first, realistic performance levels must be determined and then careful supervision at the practical workaday level must follow if those levels are to be achieved. The endeavour will be to grow crops and produce livestock products in an efficient way, and close collaboration and discussion with staff will be an important ingredient in this if errors are to be identified, checked and corrected.

Supporting this physical approach can be the operation of budgetary control techniques: very much an office job but with the same aim, i.e. to identify, check and correct error. In essence what is involved is that budgets should be prepared for each item of output and each item of cost, expressed in sufficient detail (i.e. physical and financial) so that, after the event, the discrepancy between actual results and the budget can be identified, measured and explained (see Appendix Table 22.1, p. 792).

There are, of course, many influences at work in farming that may be quite outside the farmer's control; the weather and many other 'external' influences immediately come to mind. Budgetary control is concerned with the things which are controllable and the actual application of control. The essence of control is that something should be done.

It should be added that farm enterprises and farm costs vary in the extent to which they lend themselves to immediate or longer-term action being taken. The so-called 'factory enterprises', for instance, offer the opportunity of continuous control and numerous costing schemes appropriate to this need are offered to the farming community from advisory and commercial services. Other enterprises, such as arable and livestock rearing, do not lend themselves so readily to continuous checks and control. A farmer might conduct an annual comprehensive examination of every item in the farm economy, or examine the whole financial structure of his farm more regularly than once a year, in which case cash flows on a monthly or quarterly basis, setting budgets against actual figures, can help to provide the answer. The precise methods and approaches adopted in this context are less important than the recognition of the value of budgetary control and the physical improvements it can help to bring about. Each farmer or farm manager can adopt those techniques and methods that he finds suit him. What is probably true, however, is that there is more scope for improving profits on many farms by tighter control of what is happening than there is by radically altering what is happening itself, i.e. the farm system.

PRODUCTION

At first sight the concept of agricultural production seems straightforward enough: the creation of commodities that shortly will be turned, either directly or indirectly, into food or drink for human consumption. It is only when one begins to consider precisely how that is actually brought about that the concept gets a little more elusive. It is not always easy to see production happening or even to measure it. It results from managerial effort in combining and co-ordinating a collection of inputs – land, labour and capital – in a way that transforms them into something else, i.e. the commodity that is required, and can be offered at a price that will be paid.

There are three basic production questions to which every farmer has to find an answer: *what* will he produce, *how* will he produce it, and *how much* will he produce? The answers to those questions revolve around the concept of marginal adjustment. In theory, adjustments will continue, and 'equilibrium' will be reached only when there is no further financial incentive to change in any of the areas implied by the three questions, i.e. the choice and combination of enterprises, the methods of production and the overall scale of production. Until that point has been reached increasing the use of resources in the production of one commodity rather than another, in the use of one input rather than another, or increasing the overall scale of operations will be beneficial in economic terms. In practice, of course, all kinds of considerations prevent this final degree of equilibrium from being reached, but somehow or another, farmers *do* have to make decisions about such matters. They do have to decide on the allocation of land, labour and capital between the different enterprises on their farm; they do have to explore ways and means of producing each commodity in an economical way, and they do have to consider how intensive they should become, recognising when *increasing returns* give way to *diminishing returns*. Armed with those pieces of economic thinking which seem relevant and combined with their own commercial instincts and organisational ability, farmers and farm managers (sometimes aided by outside advice) do have to plan, make decisions about and control their production.

A convenient way of thinking about the production process is to divide it into three separate parts: (a) the building of a production plan, having due regard to market opportunities and the availability of fixed resources; (b) acquiring the necessary additional resources and employing them in the appropriate combination; and (c) operating the plan, with due regard to required levels of performance and appropriate supervision.

Formulating a production plan can be approached in a variety of ways, ranging from the simple subjective decision of the kind made by many farmers in the light of their resources, experience and interests (e.g. 'I intend to milk 100 cows on half of my farm and put the rest into corn') to the use of computerised planning methods, such as linear programming, where the complexity of the available choice between enterprises and the range of constraints on resource-use is recognised, along with the impossibility of reaching planning solutions by hand-operated methods. In between these two extremes there is a whole collection of techniques, themselves varying in their level of sophistication, but all based on the gross margin concept.

A *gross margin* is the difference between the output generated by an enterprise minus the *variable* costs associated with that enterprise. The variable costs are those costs which vary in direct proportion to changes in the scale of an operation, such as concentrated feed for livestock and seed, fertiliser and sprays for crops. The gross margin does not, of course, measure profit but it does indicate the contribution that an enterprise makes towards covering the *fixed* costs of running a farm (i.e. those which do not vary directly with every change in the scale of individual enterprises, like rent, regular labour, machinery and general overheads) and then providing something towards final profit. Appendix Table 22.2 (p. 792) shows example layouts for calculating gross margins. Expressed per unit of the most scarce resource (in this example, and often in practice, land) they can provide a hierarchy of enterprises which provides a helpful guide to planning on the individual farm. The more simple planning techniques which use this approach are sometimes referred to as 'ordered budgeting', whilst the more complex of them carry the broad label of 'programme planning'.

There is a good case for the use of such 'middle of the road' methods, which are more objective than 'back of the envelope' budgeting but do not call for the full rigour of the computer. Based on the selection of enterprises according to their gross margin, such methods would have in common something like the following separate steps:

(a) A detailed appraisal of the resources available.

(b) A review of the available marketing opportunities.

(c) A listing of all enterprises that could be included in the plan.

(d) Calculation of the 'normalised' gross margin per chosen unit for each of these enterprises (i.e. taking account of known performance over a number of years and anticipated future costs and prices).

(e) A determination of the physical 'limits' to each enterprise.

(f) The selection of each enterprise in descending order of gross margin per unit of scarcest resource to its predetermined limit.

(g) A check on the availability of other scarce resources.

(h) The consideration of other enterprises making no demands on scarce resources.

(i) The calculation of total gross margin for selected system.

(j) The subtraction of fixed costs to determine profit.

Finally it is necessary to emphasise here that plans seldom, if ever, exist in a vacuum. In practice the planning of production programmes very often means adjusting the existing plan and the manager's task may be simplified in this respect, if he accepts that there are, in essence, only four kinds of possibilities. First, he can try to improve the gross margins from existing enterprises by husbandry or commercial means. Secondly, he can alter the farming system, perhaps substituting enterprises at the margin in favour of those yielding higher gross margins, perhaps by making changes in a reverse direction with a reduction in fixed costs, or by introducing new enterprises. Thirdly, he can attempt to reduce fixed costs without changing the pattern of enterprises. Or, finally, he can, if circumstances permit, add to the economic base of the business by the introduction of fresh land, new enterprises not requiring land, or new non-farming activities. Every possible change to a farming system can be considered under one or other of these four manœuvres and the periodic and systematic examination of any farm business in this way could be valuable.

The second and third parts of the overall production process will be touched upon later in the chapter so will not be elaborated here, other than to say that, having developed a suitable plan in terms of utilising the major resources of land, labour and capital, management has two other essential responsibilities before that plan can become effectively operative. First, it must make available whatever range of additional resources are required, including the basic materials required by each enterprise; and, secondly, these resources must be so organised, the necessary jobs scheduled and appropriate supervision given, that things do actually get done as required.

BUYING AND SELLING

Farmers are frequently urged to 'improve their marketing'. It is a plea that often leaves them perplexed, asking themselves 'What does that mean, and what can I do that I am not already doing?'

It is important to remember that farmers operate *in the market* when they *buy* as well as when they *sell*. It is important that any commodity or requisite that is purchased should meet the need it was intended to both in quantitative and qualitative terms. Hopefully purchase will be made at a competitive price, with whatever guarantee of continued service is appropriate, along with the maintenance of goodwill on both sides and, of course, without an undue expenditure of time and effort.

So far as selling is concerned, there are three basically different kinds of decisions that confront a farmer. First, there are the strategic decisions about *what* commodities to produce: these decisions are, in effect, about what will be sold and to whom. When a decision is made about *when* to produce (e.g. summer or winter milk) a decision is also being made about when to sell.

Secondly, farmers must decide what kinds of outlets they will use for their produce. Sometimes the commodity itself will determine this for them (e.g. milk via the Milk Marketing Board) but more often than not a choice will exist.

Thirdly, each farmer will have to decide how closely and independently he wishes to be involved in the marketing process, in addition to physical production. The dividing line between production and marketing can be a fine one, but some individuals may wish to limit their involvement in marketing and concentrate their efforts on the production side of things. Others may wish to get nearer to the consumer, if not right at the farm gate, then in some form of commercially integrated activity. Some may prefer to operate on their own in these matters, or, alternatively, to accept the advantages and the disadvantages of co-operative effort.

It is worth stressing the fact that selling produce is the natural extension of producing it in a physical sense, and it would be poor management to expend valuable effort on one part of the job, only to ignore the other. And yet it seems that many farmers do feel in something of a dilemma in this area of their business. Whilst they may recognise the inevitability of their involvement – they do, after all, have to buy and sell – many of them tend to see marketing as a highly professional commercialised activity in which it may be difficult and even not sensible for them to compete. In the interests of a good division of labour they often argue that they should concentrate their effort on 'the farmer's real job': physical production.

The truth of the matter is that both of these attitudes to marketing are tenable. In a narrow professional sense it is probably true that it is best left to the professionals. It is also true, however, that in the sense that marketing is part and parcel of the overall production process, and because it involves buying as well as selling, a commitment to this aspect of his business should be an important part of any farmer's thinking and activity, alongside all other aspects of management.

FINANCE

As the rest of this chapter will have demonstrated, business management is not simply about financial management, but it is an especially important part of any manager's task. It is a complex part of management and a convenient way of viewing the subject is with the help of Fig. 22.2.

The diagram suggests that financial management can be subdivided both in *content* and in *time* into two parts.

So far as records and accounts are concerned it is important, whether a farmer is responsible for keeping his own 'books' or whether this is done for him by a secretary or, increasingly these days, with the help of a computerised system, that he personally

	The past and the present	The future
Trading considerations	RECORDS AND ACCOUNTS	BUDGETS
Capital considerations	BALANCE SHEETS	INVESTMENT APPRAISAL

Fig. 22.2   What financial management involves

understands the data and the 'profit and loss account' that is presented at the end of the year. It is probably even more important, however, that he is capable of understanding and interpreting the kind of management analysis that can be derived from such accounts – indicating the annual output from each enterprise, the level of the major inputs, and, where appropriate, the relationship between the two. With initial help from a trained adviser/consultant the systematic examination of input/output of this kind (whether it is cast in gross margin form or not) can help to indicate where improvement and/or adjustment is desirable. Unfortunately this process is too often reserved for situations in which profits have disappointed, when it could usefully be applied as a routine tool of financial management, using appropriate yardsticks of performance, and following something like the sequence shown in Fig. 22.3.

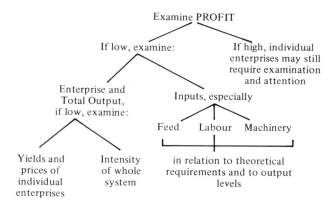

Fig. 22.3   Systematic management analysis

Turning from the past to the future, it is important for farmers and farm managers to understand, and preferably to be able to prepare for themselves, simple budgets which will help them to test the likely financial outcome of changes that they might wish to consider. The main types of budget that will be required will be the so-called 'complete' and 'partial' budgets. As its name implies, the complete budget should include every item of input and output that can be anticipated for a given farm system, as far as possible relating known levels of physical performance to the best possible estimates of future prices and costs. This kind of budget will be most useful when a new farm is being tendered for or taken over, when an existing farm system is being radically altered, or as a basis for budgetary control along the lines discussed earlier in the chapter. The partial budget, also as its name implies, relates only to part of any farm economy: that part of it (the outputs and the inputs) that will alter as a consequence of any contemplated change. The principles which govern the calculation of the individual items are the same as those used in complete budgeting, but to the extent that farmers will often need to contemplate modest change – but seldom radical change – the partial budget will be the tool they most frequently need. Their 'back of the envelope' calculations will frequently be of this nature. The framework for both the complete and the partial budget is shown in Appendix Tables 22.3 and 22.4 (p. 793). In Appendix Table 22.5 (p. 794) the framework for a cash flow budget is shown. This form of budget – setting out the flows of cash into and out of the business – can be drawn up to suit the needs of its user. It can relate to trading items only or (more usually) can include capital and personal items as well; time periods may be monthly, quarterly or longer, as required. It may stretch as far ahead as is required although it may begin to get unreliable beyond about two years. The technique, which can be used to help negotiate loans, to help exercise budgetary control, to permit investment appraisal, and simply to help 'manage' the cash balance, is by no means new; but it has been given a new prominence in recent years, as inflation and increased uncertainty about prices have given rise to new liquidity problems.

If the 'profit and loss account' and future budgets are usually concerned with the trading aspects of financial management, it is to the 'balance sheet' and to the field of investment appraisal that a farmer must turn when considering capital questions. A balance sheet provides a statement in the form of a *balance* between the assets employed in a business – ranging from the very liquid one of cash, to the most *fixed* of all assets, the land – and the debts and loans that have been incurred (the liabilities) to make those assets available. Hopefully the assets substantially outweigh the liabilities, in which case the business could be sold up and the owner remain *solvent*, keeping for himself the difference between the assets and liabilities, known usually as the Net Worth, or the Owner's Equity. In the setting out of a balance sheet Net Worth is actually shown as a liability (i.e. the business's liability to its owner) so that the two sides of the statement automatically balance. Traditionally farmers have 'strong' balance sheets, i.e. they own a high proportion of the assets they employ. This may not be so in every case, however, and banks from which loans are being sought will be especially interested in the relationship between the farmer's own capital (the Net Worth) and the amount he has borrowed. Where the latter is high in relation to the owner's capital the business is said to be 'highly geared' and there is a riskiness associated with further borrowing, due to the first call that existing interest charges will have on profits. The balance sheet will provide the necessary insight into this and other aspects of a farm's financial status – and it is important therefore that in this context, the asset values depicted in it are kept at a realistic level.

Finally, a brief word about investment appraisal. To the extent that much of the capital in farms represents a past commitment and is very often irreversible in its use, the element of farming capital that should come under the closest scrutiny is that small share of it that is being freshly invested – when a choice still exists. There are various methods of appraising such investment – ranging from the simple to the complex. At the simple end of the range is the expression of the extra profit likely to be generated by the new investment as a percentage of that investment, i.e.

$$\frac{\text{extra profit} \times 100}{\text{extra capital}}$$

Another simple method of judging new investment is to calculate the so-called 'pay back period', measured in years, by dividing the expected annual profit (before depreciation allowances) into the total extra investment. Both of these methods are relatively crude but may help a farmer to make his decision in an area where there will usually be non-financial as well as purely financial considerations to take into account. When major investment projects are involved which can be expected to yield a return over an extended period of time with unequal returns between the years, appraisal methods which embrace a discounting of future yield will offer an advantage over the simpler methods – but they will also involve most farmers in seeking specialised help in their application.

Not infrequently, farmers, like other businessmen, may follow a particular line of investment because instinctively they feel it to be right – and where past investment has proved to be sound, such intuitive approaches may be justified. There can be no guarantee of this, however, and, at the very least, some form of partial budgeting would be wise – relating the extra profit, if at all possible, to the extra capital to be injected. It is important also that fresh investment is directed towards the enterprise or enterprises that are known to be successful already. That assumes, of course, that fresh investment is needed in the first place. There may be other ways to increase profits, if that is the objective, involving the better management of existing resources, and if that opportunity exists, it should never be overlooked.

STAFFING

The number of employees working on many of Britain's farms is small; the working relationship between employer and employee is very often a close one and labour relations in the industry are generally claimed to be good. None of that, however, should blind farmers and farm managers to the need to give careful and continual thought to their staffing needs and policies. Essentially they should be concerned to employ the correct number of staff, with the necessary skills to perform those jobs that will help predetermined objectives to be achieved – in a working environment that is acceptable and rewarding. They should keep an eye on the future as well as the present and there may be no more important aspect of management to which they should devote their attention.

Assessing the correct number of workers required to run a farm can be done, especially for small enterprises, without too much fuss by experienced judgements. In more complex situations, however, it is perhaps best done with the help of 'standard man days'. This involves multiplying the hectares of each crop and the headage of each different livestock type by the known number of standard man days (i.e. 8 hours) that each requires. Fifteen per cent is then normally added to the total to allow for maintenance work. A 'labour profile' can be prepared to identify seasonal peaks and troughs and consideration given to choice of enterprise, as well as the use of machinery and casual and contract labour in order to minimise the possibility of overstaffing with costly regular labour.

Recruiting the necessary labour can be subdivided into three main parts. First, there is the need to clearly identify the job (or jobs) to be filled as well as the kind of person most likely to fill the vacancy satisfactorily. This entails preparing a so-called *job description* and a *man description*. Secondly, there is need for a well-designed and placed advertisement, followed by the careful selection of a short-list of applicants to be interviewed, followed by the interview itself. Interviews should not be rushed and there should be adequate time to show applicants the farm, the local housing situation and schools and introduce them to the existing staff. Ideally, candidates should bring their wives (or husbands) and be able to obtain all the information they require in order to make a decision should they be offered the job. Thirdly, when a choice has been made and the offer accepted it is essential that an adequate induction takes place. A good start is essential and no new member of staff should begin work without feeling that he is welcome to the farm team, informed about facilities and procedures, and thoroughly clear about his own responsibilities and authority.

The correct induction of a new worker is, of course, just the beginning. If each employee is to give his best and contribute to the overall success of the farm business, preferably over a lengthy period, management has the ongoing responsibility of training, motivating and remunerating each employee in an appropriate way. An employee's training may take place on or off the farm according to needs and circumstances; his motivation will depend upon good working conditions, good leadership, good communications and on the extent to which his endeavours, physical and mental, are appreciated; his remuneration will be financial and non-financial; both are important. And if at the end of the day the right man – properly skilled, well motivated and satisfied with his recompense – is kept on the farm, he will be a priceless asset.

Conclusions

This chapter has tried to meet three particular objectives. First, to emphasise the comprehensiveness of the farm manager's task. In many small businesses (which, relatively speaking, most farms are), opportunities for the delegation of managerial responsibility may be rare, so that many farm managers have a wide-ranging and complex task. Jack of all trades, master of none, they may sometimes feel, but, in management terms, they certainly have to be responsible for a wide range of 'trades'.

Secondly, it has tried to reflect the job as it is. Farm management is primarily about setting objectives and meeting them – with all of the day-to-day and longer-term demands which that implies. This chapter has, therefore, as far as possible avoided techniques and figures and concentrated on the tasks and the problems which occupy a farm manager's thoughts and activities most of the time. The books listed for further reading will provide all the guidance that is normally necessary on business management techniques and their application.

Finally, although the subject has been discussed under various headings no attempt has been made to suggest that those subheadings represent watertight divisions. The dividing lines, for instance, between setting objectives, planning and making decisions are narrow ones, and decisions taken within any one sector of a farm business have immediate repercussions on the other sectors. All of these topics are present; all of them are important – and all of them interdependent on each other.

23

THE PRINCIPLES OF MARKETING PRIMARY PRODUCE

THE MARKET ECONOMY

In the UK we have what is known as a *market economy* for the distribution and sale of most agricultural and horticultural produce. The essence of a market economy is that prices are determined by the interplay of supply and demand. If the supply is short and demand strong, then clearly buyers will have to pay a relatively high price for their requirement. By contrast, if supply is plentiful and demand weak, then sellers will have to accept a relatively low price to dispose of their produce. In a market place, prices are constantly changing, reflecting the different pressures of demand and supply.

To see what happens in practice, we may consider the position of a wholesaler standing in New Covent Garden early in the morning and trying to make up his mind what price to ask for his apples. In the first place he knows whether demand on the previous day was strong or weak. He knows too, whether growers have sent him overnight more or fewer apples than usual. If he looks around at the neighbouring stands, he will be able to see whether his competitors' stocks are greater or less than normal. He bases his first asking price early in the morning on this information.

Soon the buyers start arriving. Some stop at his stand, and inquire the price. It is not usual for a buyer to buy immediately, because he will want to compare both prices and qualities in a number of stands before committing himself. After asking the price, the buyer will therefore move on. If he comes back to our wholesaler fairly quickly and makes a purchase, then the wholesaler learns that he may have set his price a little too low. If the buyer does not return at all to make a purchase, then he realises he may have set his price too high. As the evidence accumulates one way or the other, he will almost certainly alter his asking price for the apples appropriately. In this way buyers and sellers, through their negotiations, reach what is known as the *equilibrium price*, that is to say the price which will clear the market of its available supply given the present level of demand.

The market economy is by no means the only system for the

distribution of food. In very primitive societies (*subsistence econ-omies*) each family hunts for or collects the food it needs for its own use, and the only 'marketing' which takes place is interchange of produce between families. In other societies the food available is distributed according to the status of the receiver; thus the chief of the tribe will have first choice, the young hunters second choice and so on. However, the principal alternative to a market economy in developed countries is the *planned economy*.

Planned economies are usually also socialist economies, and prices do not as a rule depend on supply and demand but are fixed at levels which the government believes to be socially desirable. Thus, in the Soviet Union the prices consumers paid for primary foods did not change appreciably for some 20 years, although as farm prices have regularly been increased, this has only been pos-sible by a system of massive subsidies. In a planned economy the national requirements for any commodity are calculated by the government planners, and state farms, or farmer co-operatives, are assigned production targets which, if met, result in the target quantities being available for the population.

However, there is no such thing as a completely planned econ-omy; given the millions of transactions necessary to feed the popu-lation, this would be an impossibility. Equally, there is no such thing as a perfect market economy. Virtually every country in the world intervenes in its agriculture to some extent for a variety of reasons which may be economic, social or political. Nor is the distinction clear cut between countries with left-wing ideologies and right-wing ideologies; for example, there is a great deal of intervention in South African agricultural marketing, whilst both Hungary and Yugoslavia, although socialist countries, allow part of their economy to operate on market principles.

At first sight the planned economy may seem to have much to recommend it. However, the historical record of planned econ-omies in feeding their populations has not, on the whole, been impressive. Furthermore, the market economy is much less of a hit-and-miss system than it may appear at first sight. This is because prices are not simply the amount of money a buyer pays for his produce – they are also signals which convey messages. In a market economy there are usually quite a number of middlemen – mer-chants, dealers, wholesalers, etc – who are looking for opportuni-ties to make money by buying and selling. A low price in, say, market X may send the message to one of these people 'Buy here and transport the produce to market Y, where the price is higher', or it may signal 'Why not buy now, and put the produce into store to sell back later when the price is higher?' To the farmer it may signal 'Don't produce so much of this next year'. To be noted is the fact that if we were able to plan our economy in every detail, these are exactly the instructions which we would be sending out from our central office.

What this means is that if the market economy is operating efficiently (an important proviso) the system is *self-adjusting*. Farmers hear the messages, and adjust their production as required, traders hear the messages and, in pursuit of profits, move produce about, or into or out of store in response to the true needs of the market. At its best, the market economy is an instrument of great flexibility which is in sharp contrast to the relative inflexibility of centrally planned economies.

It will be evident from this account that price is central to the efficient operation of a market economy. Unless accurate and timely information about prices is available, then the signals which generate the process of self-adjustment convey the wrong messages or no messages at all. Even when good information is ostensibly available, it may not be precise enough. For example, published statistics would suggest that there is a fortune to be made by transporting top-quality Friesian calves from one market in East Anglia to another no more than 20 or 30 miles away. This is not, in fact, the case – it simply happens to be the practice to sell them rather older and heavier in one market than in the other, and in consequence they fetch a somewhat higher price there. Again, research showed quite a difference in the officially reported prices of apples in the markets when compared with prices actually received by a sample of growers. In other words, the UK may have a complex and sophisticated marketing system, but one can be far from happy that the critical price information necessary to make the system operate properly is always available.

THE MARKETING SYSTEM FOR FRESH PRODUCE

In this context the word 'system' must be read as meaning rather more than just 'the way things are done'. It is used to mean an assembly of people and organisations which are *inter*-dependent and not independent of each other. Because of this interdependence, if things change at one point in the system, there are very likely to be consequential changes which are necessary at many other points in the system. For example, we shall see, in later paragraphs, how the development of supermarkets in the retailing sector created the need for substantial changes in practice throughout the whole system.

A modern agricultural economy shows a high degree of specialisation, with farmers concentrating on the particular enterprises which are most suited to their local conditions. Thus, cereal production is concentrated into eastern and parts of southern England, livestock production more into the west and north. Some crops are quite heavily concentrated: virtually all our very early potatoes come from the counties of Cornwall and Pembrokeshire, and something like one-third of all our Brussels sprouts from the single county of Bedfordshire. Further, unlike, say, the motor industry, for most agricultural commodities there are many thousands of suppliers. Economists talk about agricultural production being 'atomistic' in nature.

On the demand side there are some 17 m. households waiting to be fed, so that demand, too, is atomistic in nature. Furthermore, the population is concentrated into large towns often quite far removed from where produce is grown. It follows from this that there are two essential stages in a marketing system. The first is that of *assembly*, that is to say collecting all the very early potatoes from Pembrokeshire and Cornwall and the sprouts from Bedfordshire, and taking them somewhere where retailers can buy them. For the moment we can call this a wholesale market. Here the second, or *distribution*, stage begins, but this time there is a difference. Whereas the lorries coming into a market will be filled with bulk supplies of a single product, the lorries going out from markets to retail shops will be filled with the wide range of products which people want to buy when they go shopping.

Before describing a marketing system in more detail there are two preliminary points to be made. The first one is that the last 20 years have been a period of considerable change in marketing systems in the UK. These changes have not, however, applied to the whole of the system, but only to a part of it, while the balance continues to be served in substantially the same way as it has been for many years. In other words, we have at the moment two significantly different systems existing side by side. In the rest of this chapter we shall first of all describe the traditional system, and then go on to look at the factors which have created the need for change, and show how the marketing systems have responded in consequence.

The second point is that, although the two stages of assembly and distribution apply to all agricultural products, the details of the system differ quite considerably from one product to another. This is not surprising. Obviously, a livestock and meat marketing system must contain a slaughterhouse, which will not exist in a horticultural marketing system. With some products such as apples or potatoes, storage is an important ingredient in the system, whereas with lettuce and tomatoes it is not. Meat, on the other hand, can be stored, but to do so is expensive and tends to reduce the value of the product. Some produce is consumed fresh, while other produce moves through processing factories. Some products have characteristics which make expensive packaging desirable, which would be uneconomic with cheaper, more staple, foods.

In these and many other ways products differ one from the other, and these differences inevitably affect the details of the marketing system. There is, therefore, no single system. The description which is given in the following paragraphs relates to the way in which the majority of horticultural produce is marketed, and the way in which the systems for some of the other major product groups differ will be indicated in later paragraphs. However, in real life, even within a single sector such as horticulture, the patterns can be very complex as people adapt the basic system to their own particular needs.

An example: the horticultural system
(see also Chapter 18)

The 'main line' of the assembly stage concerns three categories of people in three kinds of location. First of all there is the farmer, probably living in a village. He sends his fruit or vegetables to a *country merchant* who has a business in the farmer's local market town. This merchant may either buy the farmer's produce or sell it for him, getting the best price he can for it and returning to the farmer the achieved price, less transport and commission costs. A number of farmer co-operatives act, in effect, like country merchants. The third person involved can be called a town wholesaler, whose premises will be in a *primary market*.

There is no very precise definition of a primary market, but it can be taken as meaning one where a significant proportion of the produce wholesalers handle is sold on to wholesalers in other towns. The primary wholesale markets are those in the very large cities – London, Birmingham, Manchester, Glasgow, etc. Imports also often come into these primary markets, so that there is a very wide selection of all possible kinds of produce on sale here.

A *secondary market*, on the other hand, is one in which the wholesalers buy most of what they want from primary markets. Here we are talking about towns such as Leicester, Coventry, Brighton, etc. Sometimes these secondary 'markets' are not really markets at all, but simply a street or two where a number of wholesalers' businesses have congregated.

The distribution stage begins in the primary market with the town wholesaler. The produce which has come into him from the country is displayed on his stands, and potential buyers come into the market to inspect it. The town wholesaler has (at least in the simplified version of the system being described) two kinds of buyers. Because he is situated in a very large town, there are many greengrocers within easy reach of the market, and they will come in early in the morning to buy what they need. However, there are only about a dozen primary markets in the country so that a very large number of retailers are too far away to make this daily trip economic. So the second type of customer, in the primary market, is a secondary wholesaler, who takes the produce he has bought on the next stage of its journey. This is to the secondary market, where most of the rest of the retailers in the country will buy their requirements.

However, even this is not the end of the chain. While retailers near the primary markets and retailers near the secondary markets can buy at those markets, there are still the needs of retailers in outlying areas. These people are served by what is known as a *travelling wholesaler*. This is, at its simplest, a man with a lorry who goes either to a primary market or, more usually, to a secondary market and buys the requirements of the local shopkeepers whom he serves. Finally, of course, the consumer buys the produce from a choice of retailers.

As noted, however, this is only the main line of the system and there are a number of variations which confuse the pattern. In the first place, of course, there may be no marketing system at all – the farmer may simply sell direct to consumers, either through a farm shop or by means of a 'pick your own' enterprise. Some farmers – particularly perhaps egg producers and occasionally milk producers – deliver produce direct to retail shops in their own local area. However, the quantity they can dispose of in these ways is usually much less than their total production, so that they have to send some produce through the normal marketing system as well.

Another point of interest is that, logically enough, the country merchant and the secondary or the travelling wholesaler are often the same people. A country merchant will bring lorry-loads of the Brussels sprouts which he has assembled from his home in Bedfordshire into London, will sell them to the primary wholesalers and will then load up with the wide range of produce he needs to service his local retailers in his wholesaler function.

Not infrequently, it is possible to miss out a stage in the process. Large farmers in the home counties, for example, may well send their produce direct into the London primary markets – indeed, some of them even have small wholesale businesses of their own in those markets. The farmer co-operatives, and indeed sometimes country merchants, may miss out the primary market and sell direct to wholesalers in secondary markets. Imports may similarly go direct to secondary markets or even to the depots of the large retail chains. One of the dangers of drawing diagrams to illustrate marketing systems is that this tends to make them look neat and static, whereas they are very complex and high dynamic.

MARKETING SYSTEMS FOR OTHER PRODUCTS

Livestock and meat

In the livestock and meat marketing system, the assembly stage concerns livestock, the distribution stage deals with meat, and the slaughterhouse is the borderline between the two. There are three ways a farmer can sell his livestock (see also Chapter 18). The first is to a *dealer* who comes round and buys on the farm. In this sense the dealer is largely equivalent to the country merchant. The second way he can sell his cattle is at the *auction*, and there are facilities for auctions in most of the larger country towns in the livestock areas. Here, too, his stock may be bought by a dealer, or by a local butcher or by a representative of a slaughtering wholesale company.

Most dealers nowadays are either employees of large meat companies or working regularly as independent agents for them, and so buying to meet a known demand. Dealers may be buying for a number of different customers, so they often have some land of their own to which the animals they have bought are taken so that they can be sorted out into various consignments.

There used to be a time when butchers bought a lot of cattle in auction markets and took them home and slaughtered them in abattoirs at the back of their shop. However, the regulations covering health and hygiene in abattoirs have become increasingly stringent over time, so more and more butchers, unable to face the costs of adapting their premises, have had to give up slaughtering their own animals, and indeed it is a comparatively rare practice nowadays.

The third way the farmer can sell his animals is direct to a *slaughtering wholesaler*. Historically, slaughtering has been seen as a public rather than a private responsibility, and the municipal authorities in most towns of any size were responsible for setting up and running an abattoir. Many of these have, however, now become something of an embarrassment to their owners. They are often in old premises, inconveniently placed in the middle of what have become busy industrial towns, and – for reasons to be described in a later section – a large number of much newer, larger, better-equipped, privately owned slaughterhouses have grown up in recent years either in the country areas or in convenient smaller towns just outside the major urban centres.

The existence of these municipal slaughterhouses, however, means that not all meat wholesalers have their own slaughterhouses – some get their slaughtering done at the municipal abattoir. The wholesaler may buy his animals from a dealer or direct from a farmer or from an auction market.

The farmer selling to a slaughterer may sell in one of three ways. He may sell live weight (but does not now often do so) which means that he is either offered £x for the live animal, or x pence per kilo of its weight alive. Alternatively, he may sell dead weight, which means that the animal is slaughtered and the carcass 'dressed' (i.e. skinned and the bits not suitable for human consumption removed) before being weighed. Most common, however, is to sell 'dead weight and grade'. This means that the wholesaler offers a different price per kilo according to the grade an inspector assigns to the carcass. These prices are usually published weekly by the wholesalers so that farmers can know in advance what they are likely to receive.

There are primary markets for meat just as there are for horticultural products (Smithfield is the best-known example), but they are rather less important by comparison. One reason for this is that while there are relatively few fruit and vegetable markets in the country, there are many more slaughterhouses, so that more retailers have one in their immediate locality from which they can buy their supplies direct. The second reason is that there are fewer types of meat; the range of fruit and vegetables in our diet is very large, so that one needs a large market to provide buyers with a full choice. With only cattle, sheep and pig meat being bought, a butcher is likely to be able to find what he wants even in a fairly modest-sized abattoir.

The primary markets for meat have, therefore, two main functions. The first is to handle imported meat. The second is to act as a 'clearing house'. Slaughterhouses will sell as much as they can of their meat locally (or direct in bulk to retail chains or processors), but if they are in livestock-producing areas they are likely to have a larger volume of supplies than they can dispose of locally and they will send their 'surplus' to the primary markets.

Milk

The central point of the milk marketing system is the processing dairy where milk is 'treated' (e.g. pasteurised or ultra high-temperature treated) in bulk and then bottled or put in packs. Most of the milk in the UK is treated by a very small number of very large firms, although there are still a few farmers who run their own, very localised processing and retailing businesses. There are Marketing Boards for milk (see also Chapter 18). Milk is, of course, produced every day of the year, so that a centrally organised assembly system is both more desirable and more feasible than for the product groups described above.

Cereals

Cereals are not, of course, usually sold to households. For the most part the eventual buyers are large manufacturing companies – bakers, millers, animal feed companies, brewers – who buy in tens of thousands of tonnes. Even the other major buyers, livestock farmers, will be buying in at least hundreds of tonnes a year (see also Chapter 18).

The assembly stage is performed by local corn merchants or co-operatives who fulfil much the same role as the country merchant in horticulture. The end buyers negotiate with the merchants, building up their large order in a series of 'deals'. Farmers will buy their more modest requirements from the merchants.

The cereals market is known characteristically as a 'commodity' market, a term which is used about products (metals or rubber are other examples) which either occur naturally in a fairly standardised form or can readily be standardised. Because of this, judgement by the buyer of the quality of the produce on offer plays very little part in his buying decisions. A consequence of this is that prices vary almost wholly in response to supply and demand.

Cereals also differ in being a genuinely international market. Some countries – notably the USA and Canada – are suppliers of cereals to the whole world and there are a handful of very large international traders. These organisations ship vast quantities of cereals, often changing the destination of a ship in mid-ocean in response to what may appear to be extremely small changes in price. The quantities involved are, however, so large that to do so can be highly profitable; and, of course, occasionally highly unprofitable.

Cereals trading, therefore, tends to be a high-risk business in which knowing the requirements of buyers is less important than having an intelligence network to keep one informed of price changes, not just at a UK or even at a European level, but at a worldwide level. The Soviet Union, a very large producer but one which has very variable harvests, is a major influence, exporting large quantities some years, importing heavily in others.

MARKETING BOARDS

Marketing Boards in the UK differ from those in many other countries in that they are producer-controlled rather than state-controlled organisations. At one time it was the policy of the National Farmers' Union that Boards should be set up for all products. Various factors have weakened this policy. In the first place the government increasingly resisted the idea of setting up more Boards. Indeed the producers themselves became disenchanted with one or two (for tomatoes and eggs, for example) and disbanded them.

More recently their status has become very doubtful in terms of EEC legislation, which holds them to be contrary to 'fair competition'. While it is still too early to say that Boards are only of historical relevance, their roles and structures at the time of writing are certainly changing unpredictably. The Milk Marketing Board, for example, by far the most powerful and highly regarded, has become a voluntary, commercial organisation without the strong, statutory backing it used to enjoy (see also Chapter 18).

In a sense the term 'Marketing' in their titles is somewhat misleading, since they are not for the most part actively engaged in buying and selling. They sometimes, however regulate prices or margins (milk and eggs) or the quantity of production (potatoes) as well as providing valuable services to their industries, such as research and development, generic advertising, etc.

MARKETING SYSTEMS ADAPTING TO CHANGE

As noted above, marketing systems are highly dynamic and are subject to virtually continuous adaptation to meet the needs of their changing environment. The participating organisations will expand or contract; take on or abandon certain functions or product lines; develop new supplies or outlets. They will do this in response to factors such as changes in the economic or political environment, to developments in technology; to changes in consumer taste; and, since the constituent parts of a system are interdependent and not independent of each other, to changes elsewhere in the system.

In the 1960s and 1970s, for example, a significant change took place in the pattern of food retailing when the supermarkets began to gain ground at the expense of the smaller, independent retailers such as the greengrocer, the butcher and the grocer. These new stores had several points of difference. They were large; they dispensed with shop floor assistants; they stocked a very wide range of foods, thus permitting what is known as 'one-stop shopping'; and, perhaps most important, they brought a more sophisticated, businesslike management into retailing.

It soon became clear that these characteristics led them to have a very different type of demand for farm produce. The small retailers buying in markets look for 'bargains'; they can accept produce of a wide range of quality, but they are looking for produce which stands higher in the range of quality than it does in the range of price, i.e. a bargain.

Supermarket demand, however, is very specific with respect to quality, and something 'almost the same' will not do, even if the price is slightly lower. There are several sound economic reasons for this, of which only two need be quoted. The first is that their reputation among housewives depends in large measure on their providing produce of unvarying quality consistently. The second is that, like sensible businessmen, they allocate space within the store to a product appropriate to its capacity to produce a profit. If sent, say, apples of a different quality from that specified, they could sell them all right – but they will be unhappy because the shelf space allotted to them might have brought a higher return if filled with jam or breakfast cereals.

A primary market is a place where produce of all qualities is on display so that buyers can choose whatever is appropriate for their own trade. A supermarket chain's buyer, wanting hundreds of tonnes, cannot go to such a market and build up his order by picking and choosing in this way; he wants large quantities of the specified quality delivered to him. The primary markets were unable to meet this need.

To meet it, clearly somebody had to sort out the produce suitable for the supermarket's needs. In the horticultural sector this meant central grading stations out in the country where produce could be graded to the required uniform standard. Country merchants were slow to see this opportunity, but a number of farmers responded by setting up co-operatives for this purpose. In the livestock field this same need resulted in the setting up of a number of private abattoirs whose owners consciously bought animals which had a high probability of providing meat of the required standard.

If one is sorting out produce to meet supermarkets' specific quality demands, it is pointless to send it through a primary market; it can travel direct. Thus both wholesalers and supermarkets themselves set up depots in smaller towns to which produce of various kinds, already selected to the required quality, could go in bulk, to be broken down into the individual loads necessary to provide a particular store with all it needed. Thus the position of the central markets weakened perceptibly.

Nor was this the end of the consequential changes. To set up a co-operative grading station, for example, costs money, and farmers would be unwilling to undertake such investment without the certainty of a sale. Equally, the supermarket itself wanted an assured supply. Increasingly, therefore, contractual arrangements developed between co-operative and wholesaler and supermarket. A result of this was to stabilise prices more, because produce was less affected by the very short-term fluctuations in demand and supply which markets display, and could respond only to 'real' long-term trends.

This example shows how a recognition of the dynamic nature of marketing systems is essential for success. Too often marketing is seen as simply receiving produce on one side and selling it on the other and taking a margin for doing so. Successful marketing demands a sensitive awareness of changes in the environment and a willingness to adapt one's product range and one's systems so that they remain relevant through all the changes which take place around one.

PART IV

CROP PRODUCTION SYSTEMS

24

CEREAL PRODUCTION SYSTEMS

Introduction

Cereals are the most important arable crop in the UK, occupying over 3·0 m. ha; about 79% of the total arable acreage. Table 24.1 shows that barley and wheat are the major cereals, oats being grown on less than 5% of the cereal area. While the total barley area has changed little in the past ten years (Table 24.1), the proportion of winter barley increased to account for one-third of the barley area in 1980. A major reason for this increase has been new varieties with yields comparable to those of modern winter wheat varieties, particularly on the thinner and poorer soils, e.g. in the Cotswolds and in Norfolk. These varieties flower and mature earlier than winter wheat and are less affected by the drought that occurs on these soils at ear emergence – a major reason for poor wheat yields. Ripening 2–3 weeks before wheat, the pressures of a limited harvesting period are eased and cultivations for the next crop can begin earlier. This has been important where oil-seed rape is grown, because early sowing of rape greatly improves the yield expectation. Winter barley is grown principally to produce large yields of feed barley and a specialist grower expects about 7 t/ha.

Unlike barley, the wheat area has greatly increased between 1970 and 1980 (Table 24.1) and currently winter wheat accounts for over

Table 24.1 Areas, yields and total production of wheat, barley and oats in the UK

		Area ($\times 10^3$ ha)	Yields (t/ha)	Production ($\times 10^3$ t)
1970	Wheat	833	4·04	3,364
	Barley	2,413	3·59	8,664
	Oats	381	3·43	1,308
		3,627	3·68	13,336
1980	Wheat	1,371	5·21	7,143
	Barley	2,343	4·08	9,559
	Oats	136	3·96	539
		3,850	4·48	17,241

90% of the area. New varieties were again a major contributor to this change, because during the 1970s varieties such as Huntsman and Hobbit became available with yields that were bigger than existing varieties grown under the same conditions. Also, these varieties enabled larger amounts of fertiliser to be applied without increasing the risk of lodging, because of shorter and stiffer straw. Concurrent advances in the agrochemical industry produced a range of chemicals that were extremely effective for the control of fungal diseases and weeds. Membership of the EEC increased the price of grain and, in northern Europe during the 1970s, several integrated systems for growing large wheat yields were developed and widely publicised in the UK. The combined effect of these developments has been a 30% increase in wheat yields between 1970 and 1980, with specialist growers expecting yields of between 7 and 8 t/ha. The present world record yield of wheat is 14·07 t/ha at 16% water content, which was grown in Scotland in 1981. Initially, the new varieties were feed wheats, but some of the modern varieties now combine high yields with good bread-making quality; a trend that will probably continue, if the large quantities of imported 'hard' wheats are to be decreased.

The first stage when considering cereal production must be a clear definition of its objective, which can be expressed in several ways including maximum profitability, largest grain yield or highest quality. Then, differences in soil type, climate and seasonal weather patterns must be accounted for, but in the UK these are extremely variable and so a single, rigidly defined and timed series of cultural operations, i.e. a 'blueprint', for cereal production is totally inappropriate. The only way to cope with the regional and seasonal variations relies on a knowledge of the physiological principles of crop production (discussed in Chapter 19). For cereals, grain yield is related closely to the total growth of the crop and hence maximum grain yield can be obtained by encouraging maximum growth. Crop growth is proportional to the amount of sunlight intercepted by healthy green tissue, mainly leaves but also stems, ears and awns, which can be maximised by ensuring rapid and uniform ground cover and by maintaining the green area with proper nutrition and protection from pests and diseases. Grain yield also depends on the availability of water; about 50 mm of water are required for the production of 1 t of grain, which must come from rainfall and water stored in the soil. Only by understanding how the physiological processes governing crop yield are influenced by the natural environment and how husbandry practices affect these processes will it be possible to manipulate crops to consistently achieve the defined objectives. Regular and detailed inspections of crops are necessary throughout the season so that changing circumstances, e.g. disease infestation, are quickly identified and appropriate action taken. This contrasts markedly with previous attitudes, particularly for winter wheat, which was left throughout the winter and only during late spring, when it began to yellow, was it given attention.

GROWTH STAGES

The production of cereals depends on using a wide range of chemicals, e.g. fertilisers and crop protection chemicals that are applied to the crop on a number of different occasions throughout the season. It is essential that these chemicals are applied at the correct time to ensure maximum effectiveness coupled with minimum crop damage; the time of application referring to the stage of development that the crop has reached and not to a calendar date. The terms growth and development should not be confused and the distinction can be best appreciated by examining their definitions. Growth is an increase in size, e.g. weight or area. Development is the progress of a crop from germination through its life-cycle to maturity. The period between germination and flowering is the same well-defined period in the development of all cereals, but during this period there can be very large differences in the amount of crop growth.

For many years the Feeke's-Large scale was the only one available to enable the 'growth' stage of a crop to be determined. This system relies on a comparative assessment of plant size and leaf arrangement, particularly when the crop is young and before rapid stem extension. However, the use of this system can be confounded by differences in growth habit during the winter, caused by varietal differences or growing conditions, so that crops with very different structures can be at exactly the same stage of development. This system, therefore, tends to be most unreliable at the time when critical decisions have to be made, for example, the application of hormone-based herbicides in the spring.

Recently a more comprehensive decimal scale has been introduced by Zadoks, which describes the morphological characteristics of plants in terms of the number of leaves, tillers and nodes. A complete description of this scale is given in Appendix Table 24.1. The number of leaves that have appeared on the main stem has been found to be closely related to the development of the crop. However, it is important that a regular check is made of the leaves appearing, because during the winter the first-formed leaves can die without trace, and a count of existing leaves in the spring will be in error. Recent work at the Weed Research Organisation has shown that, in part, this problem can be overcome by measuring the length of the main stem from the ground to the ligule of the youngest, fully expanded leaf, because this is also related to the

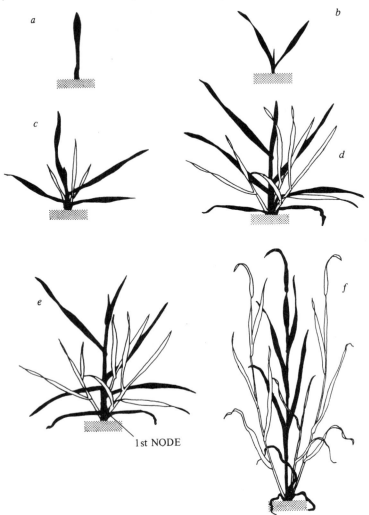

Fig. 24.1 Diagrammatic representation of winter wheat plants showing selected
Zadoks stages

a) First leaf emerging from the coleoptile (stage 10)
b) Two leaves unfolded, the third emerging (stage 12)
c) Four leaves unfolded, two tillers emerged (stages 14, 22)
d) Six leaves unfolded, three tillers emerged, pseudostem erect (stages 16, 23, 30)
e) Seven leaves unfolded, three tillers, first node just detectable (stages 17, 23, 31)
f Main shoot with more than nine leaves unfolded, three tillers, four nodes
   detectable, boot swollen (stages 19, 23, 34, 45)

crop's development. Otherwise the Zadoks scale is capable of providing an accurate indication of crop development that is independent of the growth habit or posture of the crop. (See Fig. 24.1)

Modern cereal production requires that crop development must be reliably and consistently determined. There are many dangers associated with trying to substitute the calendar as an alternative approach and these are illustrated in Table 24.2. This shows the

Table 24.2 Variations in time taken from sowing to reach a particular development stage (Zadoks 16) and mean air temperature

Sowing Date	Mean Air Temperature (°C)	Date of reaching development stage
4 October 1975	7·1	14 January
11 October 1977	6·2	7 February
30 October 1974	6·4	4 March
31 October 1976	4·5	8 April
16 November 1973	5·4	2 April

date on which the correct developmental stage for the application of hormone-based herbicides occurred for the same variety of winter wheat sown in different years. While sowing dates varied by only 43 days, the developmental stage occurred between 14 January and 8 April, a period of 84 days. Consequently, the use of a calendar to determine the time of application is quite unreliable and could lead to serious crop damage and yield losses.

VARIETIES

The choice of variety is a major decision for the cereal grower because it influences the quantity, quality and profitability of grain produced. The NIAB produce a 'Recommended List' every year which summarises the main characteristics of wheat, barley and oat varieties recommended for England and Wales.

Varieties are listed in one of four categories, for general use (G), for special use (S), provisional recommendation (P) for a variety on which further trials are in progress and seed supplies may be restricted, and becoming outclassed (O). Results from the NIAB trials sites are used to calculate a 'control' yield for each crop, which is the mean of a few established varieties, and an expected yield of each variety, which is expressed as a percentage of the control yield for that crop. Other important characteristics, e.g. disease resistance and grain quality, are usually expressed on a 0 to 9 scale, the large numbers indicating that the variety shows the character to a high degree. All new varieties must spend at least two years on the provisional list.

Although very good, the NIAB ratings have drawbacks that relate to the level of input of crop protection chemicals and nutrients used on the trials. In 1982, for the first time, NIAB published some ratings from trials where the main diseases had been well controlled. While the ranking order of yield for most varieties was similar, whether or not they were treated with fungicide, some varieties performed relatively better when kept free of disease. The more resistant varieties tend to be largely free from disease when untreated, so there is only a small yield increase when the disease is controlled. The more susceptible varieties tend to be more infected when untreated, hence there is a larger yield increase when fungicide is applied. Thus, Hobbit, with a yield just above the control, despite having the lowest resistance to both mildew and yellow rust, responded to a fungicide programme by becoming about the highest-yielding variety. There is a similar difficulty with nitrogen. NIAB use relatively small amounts in their trials, partly to avoid excessive lodging which could cause difficulties in harvesting and in interpretation of the results. There is an indication, however, that some varieties respond to larger amounts of nitrogen than others; which has to be remembered when selecting a variety.

After yield the most important difference between varieties lies in their different resistance to, or tolerance of, diseases. Varieties with moderate or high resistance are available for all the most important diseases except take-all. Where particular diseases are a local problem varieties with a high resistance should be chosen, but even then new races of fungal diseases arise periodically and may overcome the resistance of a previously safe variety. This risk can best be reduced by using the varietal diversification scheme published by NIAB to select at least three varieties for every farm so that each has a different resistance factor. Even more benefit can be obtained if the same approach is used to select varieties for adjacent fields. The concept has even been extended to sowing a mixture of varieties carefully chosen for quality and agronomic compatability especially for ripening uniformity. These mixtures can often so reduce the spread of disease that fungicide applications are unnecessary.

As previously indicated there are a range of agronomic differences in varieties, most of which are published in the NIAB list. These range from tolerance to some herbicides (notably those to control grass weeds) and insecticides, through different growth characteristics, e.g. straw length and standing ability, to harvesting characteristics, date of ripening, grain shedding, the risk of grain sprouting in the ear. The relative importance of these depends on the grower's needs, e.g. straw for bedding; resources, e.g. time taken to harvest; and the environment, e.g. a sprouting-prone variety would be less acceptable in the wetter west than in the drier east.

Lastly, the selection of a variety determines the potential market for the grain. Thus, the selection of a barley variety not recommended by the Institute of Brewing as a malting variety means that the grain will have to be sold for feed at a lower price. However, the selection of a malting variety does not necessarily mean a premium will be obtained, as growing and harvesting conditions may prevent the variety achieving its potential. This also applies to wheat, where some varieties may obtain a small premium for being suitable for inclusion in bread-making grists and some a larger premium as being 'bread-making' wheats. Although some wheats are of 'biscuit-making' quality these seldom achieve any premium over feed price.

Overall, the grower is interested in net cash return and must consider in the choice of variety the growing costs as well as the likely revenue. Thus, varieties with low disease resistance will have higher fungicide requirements and hence cost more to grow. Similarly, the seed of newly introduced varieties is usually more expensive and may offset most of the expected extra revenue. As the yield and the agronomic features of the newer varieties improve, the grower needs to change his varieties to obtain the maximum returns. However, he should not abandon, too soon, varieties with which he is very familiar and has done well.

ROTATIONAL ASPECTS

Rotations have been used to provide a series of crops that enable specific problems caused by diseases, pests and weeds to be minimised or controlled. Modern advances in agricultural chemistry have enabled many cereal growers to simplify their rotation because a wide range of chemicals is now available to control many of these diseases, pests and weeds. It is not uncommon in intensive cereal-growing areas, such as East Anglia, to find large areas where the only rotational consideration is the choice between winter barley or winter wheat. However, much of the cereal crop is still grown in a rotation with at least one break crop, e.g. oil-seed rape, potatoes, sugar beet. Obviously, the choice of break crop and the position of the cereal crop in that rotation significantly influence the yield expected, the economic return, and the management system chosen to achieve them.

The previous crop influences the time and type of soil cultivations required to drill the cereal crop. Following another cereal, direct drilling or minimal cultivations are commonly used, which will enable relatively early drilling but particular attention would have to be given to the control of grass weeds (including cereal volunteers) and some pests, e.g. slugs. If cereals frequently occur in the rotation then take-all becomes a problem. Winter-sown crops are more susceptible than spring-sown and wheat more than barley. Consequently, the safest position for winter wheat is immediately after a break crop; second and successive wheats are increasingly susceptible. It has been shown that if wheat is grown continuously take-all damage reaches a maximum after about the third year and then declines, the yields never recovering to achieve those of first wheats. Winter barley is considered an ideal entry for oil-seed rape because the earlier harvest allows early drilling of rape. However, the crop residues left after the rape harvest seem to encourage slugs, which can be severely damaging to the following cereal crop. Potatoes can also provide a good entry for winter cereals but wheat bulb fly can be a problem if early lifting is followed by late drilling. Sugar beet is likely to delay drilling a cereal and, because of crop residues, ploughing is essential for the preparation of a seedbed. Many modern winter wheat varieties can be sown as late as February, so, if most of the cultivations are done during the autumn, drilling can take place when conditions allow and the expense of ordering seed of a different variety or crop (e.g. spring barley) is avoided, although yields are likely to be slightly lower than with the spring variety. Grass weeds can be effectively eliminated from sugar beet crops and combined with the late sowing it should be possible to reduce the herbicide costs for the cereal.

CULTIVATIONS

The choice and timing of cultivations needed to provide a seedbed and establish a crop are governed by soil conditions and the previous crop. Maximum productivity depends on roots deeply penetrating the soil to reach the largest possible reserves of water and nutrients. This is impossible if the soil is waterlogged or there is a mechanical barrier (a pan). Therefore the soil must be drained and if pans exist they must be broken by cultivating rather deeper than the pan. Except for a few soil types, pans are created by poor soil management, especially by cultivating or allowing vehicles to be used when the soil is too wet, loose soil being particularly vulnerable. There is very little evidence that subsoiling, in the absence of problems with drainage or compaction, gives any increase in yield. Every effort should be made to prevent soil damage by keeping traffic on the field to a minimum, using tramlines to limit the extent of damage and keeping grain trailers to defined haulage tracks.

To drill seed uniformly at a controlled depth the soil surface must be reasonably level, have a good structure and be relatively free from the residues of the previous crop. The soil structure should be crumbly, with the crumbs in the size range 1–5 mm, which will be large enough not to blow away but still allow good contact with the seed to aid water absorption and hence germination. The traditional cloddy autumn seedbed, intended to provide shelter during the winter, is no longer fashionable.

Crop residues cause blockages of the drill coulters and, even when some are removed, for example when straw is baled and carted, the remaining stubble and chaff pose some mechanical difficulty in drilling and encourage slugs. The problem caused by cereal straw is often resolved by burning, which gives further benefits by returning nutrients to the soil, removing weeds and destroying their seeds, and killing spilt grain. If burning is to be practised then the NFU code should be followed, which specifies a cultivated, straw-free firebreak around the perimeter of fields, the attendance of staff and fire-fighting equipment and the notification of neighbours and emergency services. Burning is unpopular with the public and if cereal growers are not very careful the risk of the practice being banned by legislation is considerable. If burning is impossible, e.g. sugar beet tops, then the residue must be buried by ploughing, which is also used to control difficult weeds, e.g. sterile brome grass. The incorporation of large amounts of organic matter will increase microbial activity in the soil, temporarily locking up nitrogen and releasing chemicals such as acetic acid which are harmful to seedlings. Consequently, residues should be evenly distributed and mixed thoroughly into the soil below seed depth and additional nitrogen should be added to ensure that some is available for the crop.

Following residue incorporation, subsoiling or wheel damage, a further surface cultivation will be necessary to get a good seedbed. Discs are good in hard, dry conditions where their innate compacting action will not cause further damage, otherwise shallow tines should be used. A good tilth often exists after a cereal harvest and, when there has been a good burn, drilling can be performed with no cultivation at all, i.e. direct drilling. This is only possible on certain, well-structured soil types and permits substantial savings in time and money. However, where acid conditions or a pan at the surface develop after prolonged periods of direct drilling, these must be rectified.

The advent of herbicides has rendered the harrowing of cereals uneconomic as the cost, in terms of the mechanical damage to the crop, is high. Pre-emergence harrowing may be necessary to cover the seed adequately or sometimes to incorporate volatile herbicides, e.g. tri-allate, into the soil. Some growers roll their cereals in the autumn to provide a level surface on which to apply residual herbicides. Rolling in the spring may be necessary to press stones into the ground or to counteract 'heave' caused by frost; some think it encourages tillering.

ESTABLISHMENT

The number of established plants per unit ground area is a major determinant of yield in some crops, e.g. sugar beet and vegetables. Until recently this aspect of cereal production had been largely disregarded because of the apparent ability of cereal crops to compensate for a deficiency in one yield component by increasing one or more of the others. For example, if plant numbers are low there is a tendency for more ears per plant to be produced by increased tillering. However, recent integrated systems for cereal production have focused attention on the importance of plant density for consistently achieving large yields. Current recommendations for winter barley are 300 plants/m² and for winter wheat 250 plants/m², increasing to 350 plants/m² for late-drilled winter cereals and all spring-sown crops.

The first factor influencing plant population is the number of seeds sown. Traditionally, it has been standard practice to sow a given weight of seed per unit area of land, usually about 160 kg/ha for winter wheat. This approach is inadequate when trying to establish a defined population because a change in 1,000 grain weight influences the number of seeds sown.

The second factor affecting plant population is the percentage of seeds sown that produce plants. Surveys have shown that for winter-sown crops on average there is only 65% establishment. Percentage establishment depends on many factors including the type of seedbed that has been produced and the conditions at drilling; for example, deep or late drilling tend to decrease establishment. When considering a target plant population it is important, therefore, that seed rate also allows for the percentage establishment and this can be done using the following formula:

$$\text{seed rate (kg/ha)} = \frac{\text{target plant population (plants/m}^2\text{)} \times 1{,}000 \text{ grain weight (g)}}{\text{percentage establishment (\%)}}$$

For example, the seed rate required to achieve 300 plants/m² when the 1,000 grain weight is 50 g and the expected establishment is 65%

$$= \frac{300 \times 50}{65} = 230 \text{ kg/ha}$$

Uniformity of establishment is also important because the aim is to cover the ground quickly with leaves so that most of the available sunlight is intercepted and used for crop growth. This doesn't happen with irregular establishment or gappiness which can occur for a number of reasons, e.g. surface waterlogging or a blocked coulter, and tillering cannot fully compensate for these missing plants.

Cereal drilling in the UK takes place between September and April. Winter barley, which should be drilled before winter wheat, generally achieves a better yield the earlier it is sown. The evidence to support a similar policy for wheat is less clear, although whenever possible drilling should be completed by mid- to late-October. Obviously this is not always possible, for example when wheat follows sugar beet, but under these or similar circumstances it is even more important to calculate the seed rate accurately and to allow for the expected percentage establishment if an acceptable plant population is to be obtained. However, there are occasions when despite every good intention the weather and other factors conspire to produce a very poorly or irregularly established crop and a decision has to be taken to re-drill. Experience has shown that if establishment is reasonably uniform then, for winter-sown crops, re-drilling should be done only if plant populations are less than 50 plants/m². With very irregular establishment and where large areas of the field have no plants, then these areas must be re-drilled because the existing crop would be unable to compensate and there would be a serious danger of weed and pest infestations.

FERTILISERS

In cereals it is standard practice for all of the phosphorus and potassium fertilisers to be given in a single application before emergence. This can be done either before drilling by applying the fertilisers prior to seedbed cultivations or at drilling by using a combine drill, which is slower. The amounts of phosphorus and potassium applied will be influenced by soil fertility, previous cropping and the amounts likely to be removed in the crop at harvest; typically, a wheat or barley crop yielding 8 t/ha contains the equivalent of 100 kg K_2O/ha and 70 kg P_2O_5/ha in the grain plus straw. On soils of index 2 or above, the fertilisers should serve three purposes: first, to maintain fertility by replacing that removed by crops; second, to insure against unsuspected deficiencies; and finally, to provide some extra phosphorus and potassium that may stimulate root growth in an unfavourable season. On soils with lower indices the amount of fertiliser should be increased to ensure that satisfactory soil reserves are developed. In both of these cases the fertilisers need not be applied immediately prior to the crop and in many cases it is unnecessary for every crop to be fertilised. If potatoes or sugar beet are in the rotation the dressings given to these crops can be planned so that they leave enough for following cereals. However, on index 0 soils it is recommended that at least

30 kg/ha of the phosphorus and potassium fertilisers are applied using a combine drill because they will be immediately available to the young crop and are very effective in increasing yields. While the amounts of fertiliser applied will depend on prevailing circumstances, for a reasonably fertile soil and an expected grain yield of 8 t/ha the recommendation would be 55 kg/ha of both P_2O_5 and K_2O.

The principles for determining the required amount of nitrogen fertiliser are more complex because nitrogen is available from two sources, either fertilisers or released from soil organic matter. At present there are no reliable techniques for predicting the nitrogen availability from soil organic matter at each stage in the cereal crop's growth. Therefore the basis for current recommendations relies on estimating the expected requirement of the crop, allowing for soil type and rotation, with particular emphasis being given to the previous crop. The average amounts applied to winter wheat, winter and spring barley in 1980 were 145, 129 and 87 kg N/ha respectively. For winter-sown cereals large amounts of nitrogen fertiliser need to be applied, often up to 200 kg N/ha, hence there are generally two or more applications during the season.

Determining the amount and timing of nitrogen fertiliser is helped by an understanding of the nitrogen demand of cereal crops from sowing to harvest. For winter-sown crops the amount of nitrogen taken up between emergence and February is small, about 20 kg N/ha for wheat, although early-sown barley may contain about 40 kg N/ha. For both winter- and spring-sown crops most of the nitrogen is taken up when the crops are growing quickly between April and July: when maximum rates of uptake can be 4 kg N/ha/day with the average about 1–2 kg N/ha/day. In most seasons little nitrogen is taken up during grain filling. The total amount taken up depends directly on the yield. The amount of nitrogen in winter wheat grain is about 16 kg N/t at 85% dry matter and in the straw and chaff, removed by baling or lost by burning, it is approximately 4 kg N/t, making a total of about 20 kg N removed as grain. Similar figures apply for feed barley, so that a winter-sown cereal yielding 8 t/ha would remove about 160 kg N/ha from the soil.

WINTER WHEAT

Currently, more and more crops are being sown in September and as they are relatively large during the late autumn and winter, there has been a tendency to apply seedbed nitrogen. Surveys indicate that about 60% of crops receive seedbed nitrogen, averaging 25 kg N/ha, although a definite yield response has been found only for direct-drilled crops, because in the undisturbed soils little nitrogen is released from the organic matter.

The first top-dressing, approximately one-third of the total, should be applied at maximum tillering and when the young ear is forming (Zadoks 16), but the date for this application varies because of the effect of temperature on the rate of crop development. There is little evidence to suggest any advantage in applying this top-dressing before mid-February. The second, and usually the main top-dressing should coincide with the beginning of rapid stem extension (Zadoks 18). Even on relatively backward crops this application should always occur before the end of April, because if it is delayed there is a risk of a dry soil surface which means that the fertiliser will not dissolve and enter the soil so that uptake will be slow when demand is greatest. On high-yielding sites, such as the water-retentive boulder clays of eastern England, where wheat is usually the first or second cereal after a break crop of peas, beans or oil-seed rape every four or five years, yields consistently exceed 8 t/ha and the total amount of nitrogen fertiliser is usually about 200 kg N/ha. Under these circumstances nitrogen is applied at least three times during the spring, but again the final application will be before the end of April; the additional dressing being given during March.

Several of the modern winter wheat varieties have good bread-making quality, and the associated high protein levels depend on site, season and management. Medium or heavy soils which have large soil nitrogen residues, because of high organic matter or previous cropping, are most likely to produce high protein levels because grain protein contains 6% nitrogen. Usually, the maximum protein levels are obtained when all of the nitrogen is applied by early stem extension (Zadoks 30), although further increases in protein have been obtained by the application of an extra 20–40 kg N/ha at flag leaf emergence (Zadoks 39). When wheat is grown after another cereal, protein levels tend to increase up to a total application of 200 kg N/ha, although it is impossible to predict whether the final 50 kg N/ha is sufficient to increase the protein level above any particular threshold value.

WINTER BARLEY

While a major objective of winter barley production is early drilling and the production of a large, deep-rooted crop before winter, there is very little evidence to indicate that it responds either to seedbed nitrogen or to an autumn top-dressing. Recent surveys of farm practice suggest that approximately three-quarters of the crop receives autumn nitrogen, typically 25 kg N/ha in the seedbed.

The principles of timing top-dressings of nitrogen in relation to growth stage are similar to those for winter wheat. A split top-dressing has been found to give the best results and when correctly timed can give a grain yield increase of nearly 1 t/ha, compared to the best single application. As winter barley crops are usually larger and developmentally more advanced than wheats the first application is earlier. This application should be during early tillering (Zadoks 21–22), which coincides with the appearance of the fifth and sixth leaves (Zadoks 15–16), when half of the total amount of nitrogen should be given, but there seems little advantage from applications as early as the beginning of February. The second application should be at the beginning of stem extension (Zadoks 30), the same developmental stage as wheat but it will occur earlier in the year with winter barley. Feed varieties have been shown to respond to 180 kg N/ha when a full fungicide programme is used.

Malting varieties need a low grain nitrogen concentration of between 1·5 and 1·6% but respond to nitrogen timings and amounts in the same way as feed varieties. To achieve this requires a site that is inherently low in nitrogen, most likely a sandy or chalk soil in a predominantly cereal rotation. Over the range 50–150 kg N/ha applied the rate of increase in grain nitrogen is 0·1% for every 25 kg N/ha. Consequently an application of 90 kg N/ha instead of 140 kg N/ha will reduce grain nitrogen by about 0·2% but may also decrease yield by nearly 0·5 t/ha. However, there is no guarantee that grain nitrogen will be less than 1·6% even when grown on a suitable soil and the amount of nitrogen applied is limited.

SPRING BARLEY

The crop emerges when daylength and temperature are increasing, conditions which favour fast crop growth. The amount of nitrogen fertiliser needed to support this growth depends on those factors which effect the nitrogen reserves in the soil, which are previous cropping, the weather of the previous winter, which alters the size of the reserves left from previous crops, and soil type. All the nitrogen fertiliser can be applied at drilling but to ease the pressure of work in spring it has been shown that nitrogen fertiliser can be broadcast before drilling without decreasing yields. Where more than 90 kg N/ha is recommended then only half should go into the seedbed and the rest be given as a top dressing in April or early May. This allows for an adjustment to be made in the total amount if spring rainfall has been excessive or if the crop appearance warrants it. The second dressing should not be delayed too long because it has been shown that a yield loss occurs if it is delayed until the late tillering stage. The requirements for growing barley of good malting quality are similar to those previously described for winter barley.

WEEDS

Weeds can cause substantial losses in yield if uncontrolled, primarily because of direct competition with the crop for water, nutrients and light. Losses also occur because of the physical presence of weeds, such as cleavers, in the field at harvest, or weed seed contaminating the grain, e.g. wild oats, and because weeds allow a carryover of pests and diseases from one crop to the next. Therefore weeds must be controlled or even, depending on economic balance, eliminated.

Cultivations are usually inadequate to control heavy infestations although they may check light infestations. A dense, vigorous crop will suppress many weeds, but any gaps will allow weed growth which will, in turn, make it harder to establish the next crop uniformly. Thus, most growers rely on chemical control, particularly for heavy infestations. Chemicals, applied while the crop is growing, rely on a selective effect to kill the weed and not the crop. Under these circumstances crop damage may easily occur by applying the chemicals at the wrong stage of crop development; in the wrong weather, e.g. too cold; when the crop is under stress, e.g. drought; at the wrong concentration; in mixtures with other chemicals or at the wrong rate, e.g. due to spray boom overlap.

Before sowing is the only opportunity to use non-selective herbicides and the target is to have no live weeds on the surface at drilling. A general 'cleaning-up', especially of cereal volunteers if they have grown, can be carried out using paraquat. Couch grass can be well controlled with glyphosate provided that the leaves are at least 150 mm long and that the land can be left undisturbed for a week after application.

After sowing there is a choice between pre-emergence and post-emergence application, although sometimes the choice of chemical determines the timing. Pre-emergence is likely to be less damaging to the crop, but rather less effective and not so long-lasting. The balance of the choice changes during the season, with early-sown crops being best sprayed post-emergence, but late-sown crops benefiting from pre-emergence sprays, especially if temperature or ground conditions are likely to restrict late applications. The smaller the weed the easier it is to kill and the sooner the weed is killed the less the competition it provides. It is advisable to delay application until most of the weeds have emerged, unless the herbicide has a residual action. If it does, this will decline with time and an application too early in the autumn may risk weed development in the spring. However, the dose rate for residual herbicides is usually devised to give season-long protection and so if it is known that a second application will be made it may be possible to use lower rates than normal for the first application.

Whichever timing is chosen during the autumn a wide range of chemicals is available with the following as examples only.

Blackgrass must be controlled, if present, and this can be done with chlortoluron, if the cereal variety is tolerant, or isoproturon, which will also give good control of wild oats and broad-leaved weeds, with the notable exceptions of speedwells and cleavers. If blackgrass is not a problem then methabenzthiazuron or terbutryne at their lower recommended rates will give good control of broad-leaved weeds, including speedwells. Some winter-growing weeds are not controlled by the autumn sprays and will be too large for effective control in the spring. These should be controlled during the winter if ground conditions permit; examples are cleavers using mecoprop, speedwell using ioxynil and wild-oats using the lower rate of dichlofop-methyl.

Spring control should be needed only if earlier control has been inadequate and the crop has failed to smother the weeds. Many contact and hormonal sprays are available. Timing related to the stage of crop development and the label recommendation is very important. There is a high risk of crop damage, especially with the hormonal chemicals, if conditions are not quite correct, and this risk increases as crop development proceeds. The safest stage is when the distance between the soil surface and the ligule of the youngest, fully expanded leaf is between 5 and 12·5 cm.

In spring cereals the strategy for weed control is similar to that for winter cereals although there will be less problem from weeds which have low seed dormancy, e.g. blackgrass, or which tend to germinate in the winter, e.g. speedwell. However, it is imperative that overwintered, and hence large, weeds are killed before crop emergence in the spring, otherwise they will be difficult to kill.

If elimination of a weed species is required then it will often be necessary to remove individual plants (rogue), e.g. wild oats, or to make a later application of a chemical. Couch grass infestations can be virtually eliminated by spraying glyphosate when the crop is nearly ripe (grain moisture content 30% maximum, Zadoks 87) provided the couch grass is growing vigorously.

DISEASES

While the need for disease control has been clearly established, too many growers spend large amounts of money unnecessarily or to poor effect. In part, this is because the implications of other husbandry decisions on disease risk are either ignored or poorly understood.

Disease susceptibility may be increased by choice of varieties, crop hygiene or management practices. When varieties with low resistance are grown or when too large a proportion of the cereal area is occupied by one variety, this encourages disease. Some diseases will be able to bridge the non-cereal break in a rotation if crop hygiene is poor, thus defeating the objective of the break. For example, allowing cereals or grasses to be present in a break crop or having volunteers that persist in stubbles after harvest. Management practices such as early drilling, high plant populations and heavy applications of nitrogen may increase disease susceptibility. Many of these factors appear to be additive in their effects, so the more factors that are present the higher the disease risk and the more damaging the disease will be. Several of these management practices are used to obtain high yields and so, where it is planned to use them, it will also be necessary to be prepared to take more action against disease.

Once the major management decisions have been made the grower must decide, for each disease, whether to adopt a policy of either preventative fungicide use or treatment when disease appears. If the disease risk is high, considering the criteria previously discussed, then insurance treatment is generally favoured. Insurance spraying may also be used for diseases which (a) are very likely to appear, e.g. eyespot; (b) do much damage before they can be seen, e.g. yellow rust; (c) are favoured by weather that makes spraying difficult, e.g. Septoria and wet weather; and (d) are controlled by fungicides that prevent the disease entering the plant. Treating the disease when it appears involves more risks and demands that crops are regularly inspected. Unhealthy plant appearance can be caused by many factors, e.g. herbicide damage, drought and nitrogen deficiency and often these conditions resemble disease. Consequently, before money is spent on fungicides it is essential that the disease is recognised and accurately diagnosed and if the grower is uncertain then the diagnosis should be confirmed by experts. Once decided, it is important that the treatment is quickly applied and completed and certainly it should be possible to spray all of any one variety in a day.

The potential loss from a disease depends partly on which part of the plant is attacked and partly on when, during the season, the attack occurs. Seed-borne diseases on or in the grain may be controlled by chemical seed treatment. This is cheap and effective and by its constant use some diseases have been practically eliminated, e.g. bunt and leaf stripe. Where the infection lies deeply inside the grain, e.g. loose smut, then more expensive treatments are required and so are used mainly on valuable basic seed stocks.

Attacks on root and stem can be costly because they may kill the whole plant but not until late in the season when the opportunity for neighbouring plants to compensate is limited. Take-all attacks plant roots and unless it is a severe attack, the effects are not obvious until ear emergence. This disease can be controlled only by rotation, there is no effective chemical at present. Eyespot attacks the lower section of the stem causing lodging or even death. First wheats in the rotation are frequently affected while second and subsequent wheats are almost always attacked, usually in the spring. The disease is cheaply controlled using a benzimidazole and prophylactic application at Zadoks stage 30 is recommended. Sharp eyespot is a different disease affecting the stems, it is seldom serious and there is no chemical control currently available.

Foliar diseases attack the leaves, decreasing the area of green leaf available for intercepting light and causing significant yield losses, which are greatest when the youngest two or three leaves at the top of the crop are damaged. As the cereal approaches ripeness then the later the leaf area loss the smaller the likely yield loss. Examples are mildew, yellow rust and brown rust, with barley susceptible to *Rhyncosporium* and net blotch and wheat susceptible to both *Septoria nodorum* and *Septoria tritici*. These diseases need to be controlled through the crop's life. Generally, foliar diseases do not require insurance sprays in the autumn. However, treatment should be considered if there is substantial leaf area loss, especially on the youngest leaves or if the build-up is such that control would be difficult in the spring. In a mild autumn, mildew can develop to cause serious damage to barley and also to make it more susceptible to cold. The thresholds for spraying vary, but if 3% of the leaf is covered then potential benefits from spraying should be assessed carefully. As rusts cannot be seen for ten days after infection the presence in the locality of races of rust, to which the grower's varieties are susceptible, should be regarded as an indication of the need to spray. A wide range of fungicides are available to control these diseases and the choice should be made according to the circumstances at the time, but those fungicides that tend to give the greatest green leaf area persistence should be favoured. Spring-sown barley is so likely to get mildew, especially if potential inoculum sources are close by, e.g. winter barley, that prophylactic control is usually justified and can often be achieved using a seed dressing.

The ear can also be attacked by most of the foliar diseases, although if properly controlled on the leaves they will probably not be serious. Fungicide applications are rarely worthwhile after anthesis but can be made up to watery ripe (Zadoks 71) and may be beneficial in crops with large potential yields. A range of other, non-specific diseases such as *Fusarium*, *Botrytis*, *Alternaria* and *Cladosporium* also attack the ear and grain. These may decrease yields as well as cause loss of quality, partly by shrivelling of grains and partly by discolouration. A protective spray, e.g. a dithiocarbamate or a benzimidazole can be applied just prior to anthesis for the control of these diseases.

PESTS

Pest attacks, unlike many diseases, tend to be sporadic and irregular so that routine control measures are rarely possible. As with disease the damage done by pests depends on the plant part attacked and the time when the attack is made.

Of the bird and mammal pests the very largest, e.g. deer, Brent geese, tend only to be local problems, although often devastating and very difficult to control. Smaller birds can cause substantial losses of inadequately covered seed. Rooks and starlings frequently feed on seed from which shoots have just emerged, killing many of the plants in the process; this problem is most severe where a relatively small area of crop is at the susceptible stage and damage is concentrated. A similar pattern occurs when sparrows eat grain from nearly ripe ears of early-ripening areas of winter barley. Bird control may take many forms but provided they are not allowed to concentrate their activities then the damage birds do to cereals will be limited. Rabbits do most damage when grazing widely over winter cereals, especially when crops are poorly established or late-sown; yield losses of up to 30% occur. Control is by shooting or gassing wherever possible, but, if not possible, wire netting or electric fences can be economically worthwhile.

Soil-dwelling pests generally cause problems in specialised circumstances. Wireworms, which frequently attack cereals following grass, are easily and cheaply controlled by seed dressings. Slugs can cause a complete crop loss if a lot of residue from a previous crop is left on the surface together with loose soil, and they are particularly prevalent after rape or when cereals have been direct drilled. They are contained by hygiene and can be treated with chemical baits, although not always successfully. Baiting is also the main control for leather-jackets, which can be a problem in cereals following grass. Cereal cyst eelworm is the only significant eelworm and primarily attacks oats, especially on light land; it is best controlled by rotation.

Wheat bulb flies lay eggs into bare soil from mid-July until late August and wheat will be at risk if drilled under these circumstances, for example following early potatoes or a fallow. The larvae hatch in early spring and they will bore into a stem, killing the growing point and hence that stem, then moving to another stem. Seed dressings are often an adequate control measure and should always be used when crops may be at risk, but sprays may also be necessary when attacks are severe. Opomyza flies lay their eggs later in the autumn, usually in October, selecting a site near established cereal plants. The larvae attack the crop a little later than wheat bulb fly but only one stem is killed by each larva. Early-drilled crops are most at risk and control relies on an insecticide application timed to coincide with egg hatching, so that larvae are killed before entering the stems. While the control of Opomyza has resulted in up to a 17% yield increase, occasionally yield has been decreased.

Thrips and blossom midges lay their eggs in the developing florets, causing sterility and blind grains. Losses are seldom substantial, although very occasionally they are devastating, e.g. midges in 1926 caused a 30% loss in wheat yields. Accordingly a marked build-up of either pest should not be ignored, especially as control is both cheap and can usually be applied as a mixture with disease control sprays. Population explosions of other pests can suddenly occur, like the severe attack of the rose-grain aphid on flag leaves in 1979, so that in general any sudden development of pest populations should be carefully considered to assess their likely effect on the crop and the economic damage that would be caused if the population increased unabated.

Some aphids can transmit viruses while feeding, for example the bird-cherry aphid can transmit barley yellow dwarf virus (BYDV), which can also attack wheat and oats. This virus disease causes severe stunting, and sometimes death of the plant and can drastically decrease yield. The main spread of the virus occurs by wingless aphids in the late autumn, although the symptoms are not usually evident until the spring. Consequently treatment is impossible and prevention must be relied upon. An aphicide spray at the end of October or beginning of November will be effective, but the problem is knowing whether it should be applied. Crops sown before mid-October are most at risk and if 5% of barley plants (10% for wheat) have aphids present at the end of October then spraying is recommended. Finding aphids at the end of October can be very difficult and so, in areas where serious losses from BYDV regularly occur, it is advisable to spray all crops drilled before mid-October if any aphids are found.

INTEGRATED SYSTEMS

During the 1970s several different integrated systems for producing consistently high yields of winter wheat were developed in northern Europe. The aim of these systems was to provide farmers, usually in a relatively small, well-defined edaphic and climatic region, with a series of guidelines for crop management. Consequently, there are large differences between systems in the timings and amount of seed, fertilisers and other chemicals used. Not surprisingly, when directly applied in the UK, the results from these systems were very varied and sometimes disappointing in relation to the claims made for them. However, they significantly influenced cereal production in the UK by demonstrating, first, that large yields (e.g. 8 t/ha) could be produced and second, that regular and detailed attention to all aspects of husbandry is a necessary prerequisite for achieving these yields. The best-known systems are the Schleswig-Holstein, the Laloux and the I T C F.

Schleswig-Holstein

Named after the region in northern Germany where it was developed, this is regarded as a high-input system aiming for yields of 8·5 t/ha of grain with a high protein content. The deep silty soils, rich in organic matter, are relatively easy to work and sowing should be completed by mid-September. All crops receive about 100 kg P_2O_5/ha and up to 180 kg K_2O/ha depending on the position in the rotation. The target population in spring is between 300 and 400 plants/m^2, achieved from a seed rate of between 350 and 450 seeds/m^2 and an expected establishment of 90%. The emphasis on early sowing with a relatively heavy seed rate can be related to the autumn and winter weather. The warm soil during the autumn promotes fast development and leaf growth, so by the end of December plants have five fully expanded leaves, two or three tillers and roots penetrating to about 1 m. Little growth occurs during January and February because of near-freezing temperatures, and heavy winter rainfall means that the amount of mineralised soil nitrogen will be minimal. The potential for growth is large when temperatures and daylengths increase during March and so the first application of nitrogen fertiliser, 80 kg N/ha, is recommended at the end of February to stimulate fast crop growth.

The continental climate of this region means that the winter weather is invariably the same so that the timing of fertiliser applications can be based on calendar date, which cannot be done reliably in the UK because the weather is so very variable. A further 40 kg N/ha is applied at the beginning of April to maintain fast growth during the early summer, when there is bright sunshine and adequate rainfall. A final application of 80 kg N/ha is recommended at ear emergence, when there is enough rain to wash it into the soil, to improve grain protein content, for which a substantial premium is paid, and to aid straw decomposition (straw burning is impracticable).

The crop protection programme is intensive and expensive. Crops are sprayed routinely against foot-rot diseases, the chemical being mixed with C C C, which is applied as a split application in March and April. The aim is to keep the crops free from weeds and diseases throughout growth, which usually involves spraying on at least four separate occasions, because the dense crop and warm, moist climate are conducive to a rapid build-up of foliar diseases.

Laloux

Developed by Professor Laloux for the Limoges region of Belgium, with a loamy-clay soil, where sugar beet is the main entry for wheat in a predominantly three-course rotation; this system aims for consistency of production with minimum costs. A seed rate of 220 seeds/m² is recommended to achieve the target population of between 180 and 200 plants/m², as the expected establishment is about 85%. The timing of the first application of nitrogen fertiliser is considered to be very critical but because the weather in Belgium, like that in the UK, is very variable and unreliable, calendar date cannot be used as the basis for recommending the timing of husbandry operations.

> To overcome this a detailed scale of plant development was devised. The first application of nitrogen fertiliser, 30 kg N/ha, must be at the end of tillering (Zadoks 30). Applied at this stage the fertiliser will increase tiller survival, which is essential if a good yield is to be achieved from a relatively low plant population. It is recommended that the heaviest application of nitrogen fertiliser, 80 kg N/ha, should be applied at the beginning of stem extension and when the growing point is between 1 and 2 cm above the base of the plant. At this stage the plant has produced almost eight fully expanded leaves (Zadoks 18, 31), is intercepting most of the sunlight and should be growing at close to maximum rates.

The purpose of this nitrogen application, therefore, is to provide the crop's requirements and prevent any limitation to growth. At the same stage, CCC is applied to strengthen the straw and prevent lodging. Lodging is a major problem because of the frequent and severe summer rain storms and is the reason why low density crops succeed better in this region. A final application of 30 kg N/ha is given at flag leaf appearance and the summer rain means that it will be washed into the soil and available to the crop. The aim of the fungicide programme and the late nitrogen applications is to maintain an active, green leaf area throughout grain filling; if the late nitrogen is omitted, 1000 grain weight decreases by about 3 g, equivalent to about a 6% yield loss.

ITCF

The major objective of this system, developed by the Institut Technique des Cereals et Fourages in Paris, is to ensure that both the timings and amounts of nitrogen fertiliser are correct for the location, soil type and yield expectation. This is based on the assumption that the potential yield of a crop is primarily determined by nitrogen fertiliser and that a crop protection programme can only prevent the erosion of this potential and not cause it to increase.

A budget sheet is used to calculate the correct amount of nitrogen fertiliser. On one side the total nitrogen requirement of the crop is estimated, which is the amount of nitrogen expected to be removed in the grain plus that remaining after harvest. The amount of nitrogen available is calculated taking account of soil type, rotation, previous manuring and yield levels, winter rainfall, rooting depth and yield expectation. The recommended amount of nitrogen fertiliser is the difference between the requirement and the amount available.

Although this system recognises the importance of timing nitrogen application in relation to stage of crop development, the recommendation for the first application is given as a calendar date. For Northern France the average date is 5 March. The amount applied at this time is calculated using plant and tiller numbers and the visual appearance of the crop, i.e. pale green, dark green etc., the maximum recommendation being 100 kg N/ha. The remainder of the nitrogen must be applied when the growing point is 1 cm above the base of the plant, defined as the joint from which nodal roots appear and determined by the dissection of plants in the field. The amount may be increased to make up for any nitrogen that has been leached (estimated from soil type and rainfall data). To improve grain protein a third application of nitrogen is recommended when the flag leaf is appearing on 50% of the stems; the amount can be part of that recommended for the second application.

25

PULSES

All of the pulse crops are leguminous plants; they have the ability to fix their own nitrogen by means of the symbiotic nitrogen-fixing bacteria which live in their root nodules. The grain or pulse forms of legume have a high total protein content (average 20-26%) and they represent a considerable proportion of dietary protein in many developing countries. Most grain legumes are reasonably palatable and acceptable under conditions of home cooking, although some may cause flatulence, and some also contain toxic elements that are generally removed by cooking. On a world basis, legumes are second only in importance to cereal grains whilst grain legumes can be considered as a natural supplement to cereals. They are normally deficient in the essential amino acids, methionine and cystine, but contain adequate amounts of lysine, whereas cereals are deficient in lysine but contain sufficient methionine and cystine. They contribute about 7-10% of dietary protein in the UK.

In the UK, the major home-grown pulses for human consumption are peas, broad beans and runner beans whilst processed soya beans are imported, mostly from the USA, for use in human foods. As far as animal feeds are concerned, peas and beans are home-produced whilst locust-beans and soya-beans are imported.

PEAS

The majority of the peas grown in this country are harvested dry (about 120,000 tonnes per annum) for canning and packeting although a small proportion, usually of the Maple type, are sold as feed for domestic pigeons. Harvesting the peas is the most difficult part of their production. Timing of harvesting is critical for the maintenance of quality standards whilst handling and drying require much care. On average, peas yield 3t/ha but they have a slightly lower protein content than beans so protein production is about the same from either crop. Peas have some advantage in that, in general, their yield stability is better than that of field beans.

In the early 1970s, concern about the speed of throughput of pea harvesting (vining) equipment led to the development, from amongst the very wide range of naturally occurring forms of pea plants, of a leafless pea (Snoad, 1980). In this plant the leaflets are converted into tendrils and the stipules are much smaller so that there is far less foliage in the crop and vining or combining can be an easier, faster and more reliable process. More recent modifications of harvesting equipment have reduced the need for less foliage in the pea plant but, as a crop, the leafless pea has been found to yield well when grown at high densities and to be better adapted to field conditions than conventional types. Since leafless peas are also less prone to lodging and the diseases associated with lodging they appear to be better suited to high moisture areas where conventional peas tend to collapse and succumb to disease. It may also prove possible to develop deep primary rooting systems on pea plants to improve drought resistance. Such developments bring us nearer to a situation in which it will be possible to obtain 5 t/ha of peas. If home-produced, dried peas can be efficiently and reliably grown and harvested then the dried pea could become a much more important source of protein for animal feedstuffs and substitute for much of the soya which is currently imported.

The remainder of the pea crop, about 60,000 tonnes per annum, is harvested at a point when it is at optimum condition for immediate human consumption. Production of such peas is located within a reasonable distance of processing plants since the peas have to be harvested, transported and processed on the same day. Small quantities of these peas are marketed fresh in pods but the majority are taken to factories for shelling and freezing or canning. Most of the peas are grown under contracts from the processing companies who advise on all aspects of growing and harvesting the crop. Almost all of the crop is harvested by grower-owned mobile viners.

BEANS

Field beans (*Vicia faba*) are potentially a very good source of protein that would be more widely grown if yields were more reliable. At present yields are very dependent on the uncertainty of pollination by bees. There are a number of varieties of field bean which vary in size of seed from the large-seeded variety *major* (broad bean) through the horse bean var. *equina* to the small-seeded var. *minor* (tick bean).

In Europe, both winter and spring types are grown, whilst the ratio of winter to spring beans in England as a whole is about 20:80. Winter-sown beans tend to yield better but the timing of sowing is more critical and they tend to be more at risk from frost and disease (especially chocolate spot).

Broad beans are used for human consumption mostly in the fresh form but they are also canned and frozen. It is thought that the higher digestibility and sugar content of some of the broad bean varieties could, with suitable breeding, be transferred to varieties fed to animals. Traditionally, the medium- and smaller-seeded forms have been used for livestock although the horse bean forms a normal part of the diet in parts of Africa and Asia.

The importance of the field bean in the UK has varied in recent years depending on the availability of soyabeans which are preferred by feed compounders and food manufacturers. Soyabeans have a higher protein content than field beans but they cannot be grown successfully in temperate regions. Field beans can provide a useful break in intensive cereal producing systems but they might be even more widely grown if yields were more reliable. Yield instability appears to be due partly to adverse weather conditions and susceptibility to pests and diseases and partly to the uncertain activity of the necessary insect cross-pollinators. The long-tongued bumble bees visit the flowers to gather nectar and at the same time activate a tripping mechanism whereby the stigma and stamens are released from the keel petal. The pollination and fertilisation of individual flowers results in the production of pods and it cannot be achieved easily by other types of bees. Thus one of the major objectives of plant breeders working with field beans is to develop satisfactory self-fertile varieties which are not dependent on insect-pollination. If yield reliability were improved by breeding varieties that are self-fertile and less susceptible to adverse weather conditions then more use might be made of the crop, particularly if market outlets were improved. Much of the crop is exported (presumably for feed) and most of the rest is used as feed for livestock or domestic pigeons. Only a small proportion passes through the hands of feed compounders. However, there may be some interest in the crop for protein extraction and conversion into spun vegetable protein; the residual starch could then be used as an energy substrate for fungal culture.

Table 25.1 Yield of pulse crops

	Grain yield kg/ha	Protein yield kg/ha
Peas (UK)	2,750	550-825
Field beans (UK)	2,600	624-910
Soyabeans (USA)	1,836	459-955
Peanuts (USA)	2,360	566-802

In the food production and feed processing trades the soyabean *Glycine max* is preferred because it presents the fewest problems. It can outyield the pulse crops that can be grown in this country (Table 25.1) in protein terms but it has probably had more breeding development to improve it than our current varieties of peas and beans. Soyabeans have become an important industrial crop because they provide both oil and a protein-rich meal. The USA is the world's major producer of soyabeans and, because the plant requires relatively high temperatures, its production is restricted in the EEC to warmer, southern areas. The plant grows between 20 and 180 cm high and has clusters of white or purple, pea-like flowers in the axils of the leaves; these flowers give rise to up to 15 pods, each containing 2–4 seeds, with a high oil and protein content. These seeds can be processed into edible oils (for use in margarine, for example) and industrial oils for paints, resins and plastics. Extraction of the oils leaves a high protein flour, which can be used in foods such as sausages or processed meats, or 'textured' or spun into meat analogues. The residues of processing the soya beans are used as protein supplements in animal feeds.

The pulses are not at present dominant components of British agriculture but there are a number of promising varieties of peas and field beans which can be grown in this climate. They provide a useful break from cereal crops in many parts of the country and may contribute some residual nitrogen to subsequent crops. Some research and development is required in key areas before home-grown crops of peas and beans could take over soya's role in livestock feeds; soya's advantage will remain in its oil co-products and its disadvantage that it has to be imported. A similar import disadvantage relates to one of the important pulse components of the human diet. In the UK, more than 285,000 tonnes of canned baked beans are consumed each year. These are the Navy bean *Phaseolas vulgaris* for which our climate is marginally unsuitable. Research is in progress to develop suitable varieties of Navy bean for production in the UK in order to reduce dependence on imports.

26

OIL-SEED CROPS

CROP TYPES

The oils produced by plants and used by man fall into two broad types. The so-called essential oils usually have a strong taste and aromatic odour and are mostly benzene or terpene derivatives or straight-chain hydrocarbon compounds. They may occur in any or all of the tissues of a plant in amounts which rarely exceed 1-2%. Examples are cinnamon (extracted from bark), ginger (root), and mint (leaf). They have a variety of culinary, medical and perfumery uses but are of very localised production and are not further considered here.

The second and much more important group are the fatty oils, composed of glycerides, which are fatty acid esters of glycerol. They are widely used in cooking oils, margarines, soap and various industrial processes, and are derived from seeds and fruits in which the oil acts as an energy reserve. They are often associated with concentrations of protein, and after oil extraction can provide a valuable animal feed.

Most vegetable oils are mixtures of several glycerides. These differ in the length of the carbon chain of the component fatty acids and in the degree of saturation; an unsaturated chain has one or more double bonds in the carbon chain and carries two fewer hydrogen atoms for each double bond. The composition of some vegetable oils is given in Table 26.1. which also indicates the chain length and the degree of saturation. Saturated glycerides tend to be more common in fats of animal origin and have been associated with heart disease, so moderately unsaturated oils have something of a dietary premium, for example in margarine. They are also used in soap and detergent production. The manufacturer is not necessarily restricted to the oil in its natural state, the degree of saturation can be manipulated chemically. Oils like linseed oil which are high in unsaturated glycerides, for example linolenic, are often called drying oils because they form an elastic skin as a result of absorbing oxygen from the air. They are important components of paint and varnish.

The soyabean crop is the largest single source of the world's

Table 26.1 Fatty acid composition of some vegetable oils

	Palmitic C.16:0*	Stearic C.18:0	Oleic C.18:1	Linoleic C.18:2	Linolenic C.18:3	Eicosenoic C.20:1	Erucic C.22:1
Rape (*B. napus*) high Erucic	4·0	1·5	17·0	13·0	9·0	14·5	41·0
Rape (*B. napus*) low Erucic	4·7	1·8	63·3	20·0	8·9	1·3	0
Soyabean	11·5	3·9	24·6	52·0	8·0	0	0
Sunflower	7·2	4·1	16·2	72·5	0	0	0
Olive	14·6	3·1	76·2	5·5	0·6	0	0
Linseed	6·1	3·8	15·5	15·3	59·3	0	0

* Length of carbon chain and number of double-bonds
Source: after Bunting (1974)

supply of vegetable oil. Next in importance but at a considerably lower level of production comes a group consisting of oil-palm, sunflower, groundnut, cotton and rape-seed. Less important oil-seed crops include crambe, maize and safflower. Within the E E C, sunflower and olives are of some significance but oil-seed rape is its most widely grown oil crop. About 80% of the E E C's oil-seed and vegetable oil requirements are imported.

Only two oil crops are of consequence in Britain, oil-seed rape and linseed. These are discussed below. Homegrown rape-seed provides the equivalent of about 15% of the country's vegetable oil requirements. There is considerable scope for the development of alternatives, but although some, like sunflower and lupins, show potential, none are on the immediate horizon.

OIL-SEED RAPE

There has been a rapid increase in the area of oil-seed rape in the UK since 1970, and yields also have been rising (Fig. 26.1). The increase in world commodity prices and the price stability which entry to the E E C has provided are two factors which have enhanced

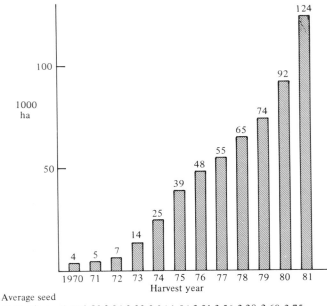

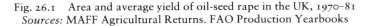

Average seed
yield, t/ha: 2.03 1.93 2.04 2.22 2.26 1.56 2.31 2.56 2.38 2.68 2.75

Fig. 26.1   Area and average yield of oil-seed rape in the UK, 1970–81
Sources: MAFF Agricultural Returns. FAO Production Yearbooks

the attractions of the crop. It is not, however, a new crop to the UK; from the seventeenth century it provided much of the local demand for vegetable oils and oil-seed meals until replaced by cheap imports after the middle of the nineteenth century.

The plant and the crop

Four distinct types exist: *Brassica napus* forma *oleifera* (swede rape) and *Brassica campestris* forma *oleifera* (turnip rape), both of which occur in annual and biennial forms. The annual type is suitable for spring sowing, the biennial for autumn sowing. Turnip rape is the hardier and earlier species and occupies a considerable proportion of the Canadian crop, but it has a lower seed yield and a lower oil content and in Europe has been largely replaced by swede rape.

The seed of all these types is small and globular with a brownish black testa. On germination it produces a single erect shoot bearing alternate hairy and lobed leaves. The plant remains short in the early stages but internode elongation causes a rapid increase in height from just before anthesis and for about three weeks after. At flower initiation the apical meristem produces an inflorescence which is a terminal and simple raceme. Usually 6-8 axillary branches develop from the uppermost nodes and produce secondary racemes. These in turn may give rise to a few tertiary racemes but in Britain they never contribute to seed yield. Flowering occurs over 3-4 weeks. The first flower to open is the most basal one on the terminal raceme and flowering proceeds acropetally on all racemes, the lowest axillary raceme being the youngest. The fruit is a stalked pod with a small distal beak and with the seeds arranged in two rows, one in each loculus.

Leaf area development reaches a maximum during the flowering period and then declines rapidly. Its major contribution to seed yield is therefore in producing an adequate number of pods. The upper leaves make some contribution to the assimilate contained in the seed, but the pods themselves and their subtending stems are the major contributors; at this stage the pods form a substantial canopy above the senescing leaves. The main period of seed growth in winter rape occurs between the end of May and early July when radiation is at its peak. The lower yield of spring rapeseed is partly due to that fact that this stage occurs a few weeks later when radiation is lower and when higher temperatures hasten maturity.

The plant carries some hundreds of flower buds, but many of the later buds abort before opening and many which do open do not produce seed-yielding pods. About 60% of the seed yield is derived from the terminal raceme and the first axillary raceme below it. There is a spread of 2–3 weeks in the growth and ripening of pods, longer in the case of sparse populations.

The different components of seed yield vary considerably between cultivars and between different management and environmental conditions. Typically a crop yielding about 2·5 t/ha might consist of 4,000–8,000 seed-containing pods per m², and 8–16 seeds per pod weighing about 4 mg per seed. This may be provided by a plant population at harvest of 40–100 plants per m², thus bearing between 80 and 200 yielding pods per plant. Total number of pods may be much greater, but many of the later-formed contain no seed and a few will have shed their seed before harvest. The number of seed-yielding pods per unit area is the component of yield most affected by management and growing conditions.

Climate, soil and farming system

In Britain, the growing season is sufficiently long to allow swede rape to be grown, and the winter is mild enough for the more productive autumn-sown biennial form. The main arable areas of Britain present no serious climatic limitations. Drought rarely occurs early enough to affect the main growth period of winter rape, and although low autumn rainfall can pose establishment problems and stormy weather can cause harvesting losses, these are more a function of season than of district.

The crop is adapted to a wide range of soils. Performance may be reduced if the pH is much below 6·0, and the crop shows some susceptibility to waterlogging partly because of the direct effects of anaerobic conditions and partly because of the importance of spring growth to an early flowering crop like winter rape-seed.

Oil-seed rape fits well into cereal farming because it uses the same basic equipment. Economies in tillage procedures have increased the potential area of arable crops which can be sown in the autumn and on some heavier soils rape has replaced sugar beet or even potatoes. Whilst it can be sown after winter wheat, more commonly the preceding crop will be winter barley or peas which allow more time for land preparation. The increasing use of winter barley has fortuitously provided a very suitable slot in the rotation for winter rape-seed.

Varieties

Rape is 30–40% cross-pollinated and the variability which this induces means that new varieties can be produced fairly rapidly. They can also be bulked rapidly because of the wide ratio between seed rate and seed yield. Recommended varieties in the 1981 list are shown in Appendix Table 26.1 (pp. 797–8); all are swede rape types.

The characteristics sought in an oil-seed rape variety can be separated into field factors and seed factors. Shortness of stem and resistance to lodging can both reduce harvesting problems. Early lodging must be avoided but should not be confused with 'leaning'. A heavy crop can absorb wind pressure by leaning with it, reducing the pod shatter which can occur in a crop which stands very firmly. In winter rape-seed earliness of maturity is not a particularly important character, but a range of varieties with different maturity dates can be useful in fitting the crop into farming systems. Disease resistance is also becoming an important requirement.

Seed factors relate to oil content, oil quality and meal quality. Oil content is more affected by species and subspecies than by variety. The most important aspect of oil quality is the content of erucic acid. Though valuable for some industrial lubrication purposes, it is an undesirable dietary constituent except in very small quantities. All currently recommended varieties are low (sometimes called 'zero') erucic types, with less than 2% erucic acid, and should be grown in isolation from high erucic types. Meal quality is primarily concerned with low glucosinolate content, because these compounds can be enzymically hydrolysed to form substances which show some toxicity. The Canadian spring-sown crop now consists largely of 'double zero' varieties, low in both erucic acid and glucosinolates. There are some double zero spring varieties in Britain but suitable types are not yet available in winter rape-seed. Meals containing glucosides can be safely fed if they are heat-treated to destroy the enzyme responsible for the hydrolytic action, but because the meal also contains some other potentially harmful or tainting substances, it is used only as one component in a compound feed.

Cultivations and sowing

Autumn-sown rape initiates its inflorescences in late winter. The potential seed yield is determined largely by plant size at that stage, so that adequate growth during the autumn is crucial for crop performance. In practice this means that winter rape-seed should be sown in the period mid-August to mid-September and preferably before the end of August.

Sowing in late summer means that soil water status can be a critical factor affecting both cultivation procedures and crop establishment.

Direct drilling or a suitable minimum cultivation technique are likely to give better establishment in a dry season than a more protracted sequence of operations. However, a compacted soil, the presence of a pan or of unburnt crop residues may require ploughing, though seedbed preparation can then be difficult in dry conditions. A fine and firm seedbed is essential for uniform depth of sowing, 1·5–2·0 cm, and hence for even and rapid germination and emergence.

The rape-seed plant is relatively adaptable to differences in plant population and good yields can be obtained from a wide range of population densities. It is common to aim at a population of 100–120 plants per m², but rather more than this for spring rape-seed which has less time for adequate axillary inflorescence development. A seed sample of rape will contain about 220,000 seeds per kg so that the usual seed-rate of 6–7 kg/ha will provide about 140 seeds per m². Allowing for establishment losses this can be expected to provide the required population. Broadcast, late-sown and spring-sown crops should be sown at 9–10 kg/ha.

Drilling in narrow rows (12–20 cm) probably gives a slight yield advantage because the reduced rectangularity delays the onset of interplant competition, but a more important benefit is that when the crop is to be windrowed at harvest it provides a better base than wide rows to hold the cut crop and facilitate drying and combining.

Fertiliser use

Fertiliser requirements for winter and spring rape-seed are shown in Appendix Table 26.2 (p. 799). Responses to phosphate and potash are not common and above index 0 the recommended rates of these two nutrients are maintenance dressings at a similar level to those used on cereals. Maintenance dressings can be applied at any time in the rotation. Oil-seed rape has a greater nitrogen requirement than most arable crops and its supply is a major factor determining yield.

Because of the importance of growth before infloresence initiation there is an appreciable need for a nitrogen dressing to the seedbed, usually 40–60 kg/ha. This autumn nitrogen can to some extent compensate for late sowing. The amount of nitrogen to be applied in the spring will depend on soil type and previous management but will usually be in the range 150–250 kg/ha. The early spring supply of nitrogen is more critical than for most crops so that an application should be made in the period late February to mid-March. Whether this is the only application or some is reserved for later will depend on circumstances. A split dressing may be advantageous if a large application is intended, or on light soil, but present evidence does not justify the extra operation. The expense of aerial application may be worthwhile if wet conditions prevent conventional methods.

Weed control

Weed problems in oil-seed rape include self-sown cereals, grass weeds such as black grass (*Alopecurus myosuroides*) and wild oats

(*Avena* spp.), and a variety of broad-leaved weeds. Some, like cleavers (*Galium aparine*) and Mayweed (*Anathmis* spp.), are becoming more common because of the increasing emphasis on autumn-sown crops in the rotation.

Grass weeds and volunteer cereals are best controlled by a pre-sowing application, for example of TCA, often incorporated into the soil. If a delay in sowing results in a flush of weeds before crop emergence, a non-selective, non-residual herbicide such as paraquat may be used. Autumn-germinating broad-leaved and grass weeds can be controlled in the emerged crop by a variety of herbicides applied in the period October to mid-November when the crop has at least three expanded true leaves and when the need for a herbicide can be assessed. In a mild autumn chickweed (*Stellaria media*) can be a problem; even if a late flush of seedlings is not competing with the crop, spraying may be desirable to avoid trouble in the spring. The weed-killing properties of herbicides are described elsewhere (see Ch. 15). Appendix Table 26.3 (p. 800) gives a list of those suitable for use in oil-seed rape.

Pest and disease control

The pest and disease problems of a crop are often relatively minor when it is first introduced but become more serious with time and as the area of the crop increases. This is not to say that all will become more severe, since natural control mechanisms often come into play, but it does mean that the occurrence and economic importance of particular hazards can change.

The major pests are pigeons (*Columba palumbus*), cabbage stem flea beetle (*Psylliodes chrysocephala*), blossom (or pollen) beetle (*Meligethes* spp), seed weevil (*Ceutorhynchus assimilis*), and the pod midge (*Dasineura brassicae*). Slugs can also be a problem in direct-drilled crops and baiting may be necessary especially in damp conditions.

Pigeons can be a major problem in winter rape. Their grazing greatly reduces performance and may even cause crop failure. TCA or Dalapon used for weed control can reduce the palatability of the foliage but the best control is a crop without bare patches which is sufficiently well grown by late October to deter pigeons from landing on it.

The common flea beetle (*Phyllotreta* spp.) is controllable by gamma HCH seed treatment, but control of the larvae of the cabbage stem flea beetle requires late autumn or early spring application of the same insecticide. At that time the larvae enter the stems and leaf stalks, causing leaf shedding, collapse of stalks and loss of yield. They develop from eggs laid in the soil by adults emerging from crops and volunteer plants the previous July, and occasionally it may be worthwhile controlling the adults if they cause appreciable damage to seedlings of the new crop.

Blossom beetles emerge from hibernation in the spring. They cause damage primarily to the buds of spring crops but backward winter crops can also be affected. The dark, shiny and oval-shaped beetle can be controlled by spraying at the green bud stage with one of several different insecticides but this is unlikely to be economic unless there are 15 or more beetles per plant. By contrast, spraying is economic if seed weevil populations exceed one or two per plant, and serious infestations can occur on both autumn and spring crops. Though the adults cause some damage, the main hazard is that eggs are laid in the pods and the larvae destroy some of the developing seeds. Control measures should attack the larvae but must be delayed until after flowering because of the danger to bees. Phosalone or Endosulfan can be applied from the air or ground, preferably late in the evening and only after warning local bee-keepers.

The pod midge lays its eggs into damaged pods where larvae feed on the seeds and can also cause premature shattering. Frequently it is seed weevil damage which allows entry to the pod so that control of the seed weevil is sufficient to suppress midge activity.

Three diseases are presently of general importance and are briefly described here. Others such as downy mildew (*Peronospora parasitica*), white leaf spot (*Pseudocercosporella capsellae*) and club root (*Plasmodiophora brassicae*) are so far of lesser consequence.

Canker (*Phoma lingam*, also known as *Leptosphaeria maculans*) is spread by spores from infected stubbles. The first sign in the crop is the presence in the autumn of 0·5–1·0 cm fawn spots on the leaves. The infection spreads to the stem where brownish black cankers appear during flowering at lower nodes. Lodging and premature ripening can result, both causing loss of yield.

Light leaf spot (*Cylindrosporium concentricum*, also known as *Pyrenopeziza brassicae*) produces pale green lesions on the leaves of young plants from November. These spots enlarge and the infection can spread by rain-splash to later leaves and to stems, flower buds and pods.

Alternaria (*Alternaria brassicae*) causes black or brown spots to appear on the leaves and later on the pods. Pod infestation causes premature ripening and shattering, particularly if they are infected at an early stage.

Harvesting, drying and storage

Harvesting is a more critical operation than in cereals because pod ripening is staggered and because of the greater risk of shattering. The spread of ripening is greatest in autumn-sown crops and thin crops because axillary inflorescences contribute a substantial proportion of the yield. Thus whilst spring-sown crops can often be combined direct, winter rape-seed is usually either windrowed first or a desiccant is applied before combining. Windrowing is done when the seeds in the middle of the terminal inflorescence are turning brown and those at the base are dark brown. By contrast, a crop to be desiccated would be sprayed about three days later than this. Combining will follow when the majority of all seeds are black, 1–3 weeks after windrowing, 1–2 weeks after desiccation. The windrowed crop should have a 20-cm stubble so that air can circulate around the windrows. The desiccated and direct-combined crop requires less investment in extra equipment and will usually show least shedding of seed if calm weather conditions prevail between the two operations, but windrowing commonly gives better results in more difficult weather.

Rape-seed is small and flows easily; the combine should be carefully adjusted and cracks and crevices in trailers sealed. The safe moisture content for rape-seed storage is about 8%. Seed which comes in at up to 15% moisture can be held for a time with low-volume ventilation but should be cleaned first. However, the uncertainty of the British climate suggests that the rape-seed grower should have drying facilities immediately available because excessively damp seed can deteriorate very rapidly.

LINSEED

Linseed (*Linum usitatissium*) is a minor annual crop in Britain, occupying less than 1,000 ha. Recently there has been some renewed interest in it because it receives an E E C subsidy of around £200/ha. The plant has a slender wiry stem with numerous small lanceolate leaves arranged alternately. The small white or blue flowers develop on branches near the apex. Each flower gives rise to a five-celled capsule, with two oval flattened and shiny brown seeds in each cell. Average seed yield is about 1·9 t/ha.

The crop is suited to temperate damp conditions and can be grown on most soils except the very light and those where the pH is below 6·5. Early spring sowing is desirable, using about 90 kg seed per ha and with a row-width not greater than 18 cm. Maintenance dressings only of phosphate and potash are required, with 40–80 kg nitrogen per ha broadcast in the seedbed. Herbicide use may be necessary because the crop is not very competitive with weeds. The main pests are rabbits, pigeons and flea-beetle all of which affect the crop in the early stages. Flea-beetle attack can be effectively controlled by seed dressing.

The crop is ready for combining in late August or early September when the seeds rattle in the golden-brown capsules. In damp seasons it may be useful to make prior use of a desiccant. The seeds flow easily and care is needed to avoid losses during harvesting and handling. Drying characteristics are similar to those of oil-seed rape.

27

ROOT PRODUCTION SYSTEMS

This Chapter deals with the two major cash crops: sugar beet and potatoes. (Minor root crops, swedes, turnips, mangolds and fodder beet are dealt with in Chapter 28.)

SUGAR BEET

Trends in production and processing

Sugar beet is a relatively recent crop in British agriculture. With the exception of Kidderminster and one near Telford all the factories are in eastern England and are amalgamated into the British Sugar Corporation. It was written into the Sugar Act of 1935 that there would be a statutory levy on the grower and the processor for each ton of beet grown, which would provide a fund for research and education.

The area grown increased from about 135,000 ha in the 1930s to 165,000 ha in the 1940s and 1950s, 175,000 ha in the 1960s and 215,000 ha at the end of the 1970s. Now there is evidence of a small cutback in area but home-grown sugar supplies about half the nation's requirements. Since 1960 the number of growers has declined steadily, halving between 1965 and the present (14,000). The average area per grower has doubled since 1966 and is now 15 ha. There seems every likelihood that, faced with increasing capital requirements and the increasing productivity of men, machines and land, the trend to specialisation will continue. Farms where there are few or poor alternative breaks and where harvesting problems are minimal can be expected to increase production of sugar beet. On the light soils the crop is well suited to provide the required break from cereals, and frequently these are the only other crops in the rotation. Holdings tend to be larger on the lighter lands and therefore expansion of contracts will doubtless continue. On the heavier soils harvesting can be a nightmare and oil-seed rape is in many ways more attractive.

Until about 1960 yields increased steadily, approximately doubling over 30 years, with improved varieties, earlier sowing, disease control and fertiliser usage. The changes in methods of establishment during the 1960s and early 1970s ended the trend of increasing yields because the first monogerm varieties were lower yielding than the then best multigerm varieties. Plant establishment problems caused yield losses which were previously compensated by the hoer and the newly introduced herbicides often slowed early beet growth. In the early 1970s the new monogerms started to outyield the best multigerms but in 1974-6 yields were severely depressed because of drought and disease. At present there is some concern that since then there has been no clear evidence of a return to the trend of increasing yields. Root yields of about 35 t/ha are similar to those of the early 1960s. There is a definite trend for growers on heavier soils to cease production and, with an increasing area on less water-retentive soils, the national crop becomes more vulnerable to a dry year, unless irrigation becomes more widespread. There is evidence of intense activity amongst plant breeders and several new high-yielding varieties are being recommended, the best of which outyield Sharpes Klein E of the early 1970s by about 10%. The breeder has made improvements in root quality; during the last four years sugar concentrations at the Bury St Edmunds factory have averaged more than 17%, a standard never previously achieved so consistently. Production of white sugar has now reached 1,100,000 tonnes per annum. As well as sugar the crop provides molasses, and important sources of animal feed, the pulp (about 650,000 tonnes per annum), which is the residual root material after the sugar has been extracted, and the tops which can be fed direct.

We are nearing the end of a revolution in methods of production which began in the late 1940s with the beginnings of mechanical harvesting (16% of the crop area was harvested mechanically in 1950, 95% in 1965). During the late 1950s and early 1960s it was clear that unless the mechanisation could be extended from harvesting into the spring period, the home sugar industry would be in jeopardy, particularly in the West Midlands. The crucial developments are charted by specifying the periods over which the adoption of components of the new system were first introduced, then became almost universal: precision drilling, 1958-66; herbicides, 1962-70; pelleted seed, 1965-71; monogerm varieties, 1966-73; drilling-to-a-stand, 1968-78. The man-hours involved in sugar beet production decreased from 310/ha in 1954 to 50/ha in 1980 and there were instances in 1981 where crops were grown with as few as 25 man-hours/ha.

Suitability of the crop for processing

The grower is paid on the basis of a fixed price at a standard sugar percentage (16) with increases or deductions per percentage unit above or below the standard. Formerly, yield was of overriding importance and while it is still true that returns depend heavily on yield, the pricing structure has changed so that low concentration incurs an increasing penalty. To a considerable extent sugar concentration is beyond the grower's control; it is governed primarily by a negative relationship with rainfall during the pre-harvest period. Excessive nitrogen usage and infection with virus yellows and downy mildew depress sugar concentration. At the factory the extraction of sugar is impaired if the roots contain excessive concentrations of impurities, compounds containing nitrogen, primarily, but also sodium and potassium. The grower's return is not affected directly by these factors although their concentration changes in parallel with that of sugar.

Because foliage and the crown, i.e. that part of the root above the lowest leaf scar, contain high concentrations of impurities, payment for each load is adjusted according to a top tare – that part of the weight of a sample of roots which is removed after hand topping to the prescribed standard. Top tare is increased where crops are gappy because it is impossible to set the harvester to deal properly with crowns of different diameter and which protrude from the soil to differing extents. Frost damage either in the field or in the clamp depresses sugar concentration and in badly frosted beet little sucrose remains, having been converted to invert sugar and gums, which at best slow down the factory slice rate and at worst render beet unprocessable.

The production of seed

About 1,000 ha of sugar-beet seed crops are grown annually in the UK. Some seed is exported, especially to the USA, and some is imported from south-west France and the Po Valley region in Italy. The proportion of seed which is home grown varies with the success of our seed houses in producing varieties for the recommended list relative to their competitors from the Continent. Originally seed crops were in Essex and around the Wash. Intermingling of the biennial seed crop and the annual root crop effectively maintained the disease cycle of virus yellows and downy mildew. Both aphids and viruses overwintered on the 'stecklings' and the seed crop provided a springboard for infection of the root crop in the following spring and early summer. It was discovered that raising sugar beet by sowing in spring under a barley cover crop was an effective means of protection from aphids and in districts where the root crop is grown this is the standard method of growing seed. In areas where root crops are few, notably Huntingdonshire, Northampton, Berkshire and Wiltshire, seed crops are direct-drilled in July without a cover crop.

In hot, dry years such as 1976 seed crops ripen in August but usually they are not ready to be swathed until September. It is known that when ripening is delayed in cool, wet summers such as 1968, 1972 and 1977, cool spells in September and October, when seed is still 'on the straw', can partially vernalise it, causing the succeeding root crop to be unusually prone to bolting. Partly for this reason European seed houses have transferred seed production from Sweden, Denmark, The Netherlands and Germany, to south-west France and the Po Valley in Italy. Ironically this has, in part, contributed to an increasing weed beet problem in Europe. In England, one-quarter of our beet fields have in recent years contained misplaced seedlings whose source is seed shed in previous crops – dormancy can last for 15 years. Extreme annual types have been introduced by contamination of seed crops with pollen of wild annuals which grow as weeds in those regions of southern Europe recently used by seed producers.

Seedbed preparation and sowing

Traditionally, many cultivations were made to prepare a fine, firm and level seedbed for the small seed. In recent years there has been a marked trend to minimise the amount of cultivation. The essence of the process is to plough uniformly avoiding an uneven surface, which on heavier soils may require some autumn or winter cultivation to produce a more uniform but coarse 'winter wheat type' seedbed, then in spring to do the minimum required to provide a fine tilth at seed depth for maximum seed/soil contact. Compaction below the seed and through the plough layer impairs plant establishment and inhibits root penetration and growth.

Increasingly, P, K, Na and Mg are applied, according to soil analysis, before ploughing and after autumn or winter levelling where necessary, it may be possible to create a seedbed with one pass, possibly with two sets of harrows in tandem. Where pre-sowing cultivations are kept to a minimum, nitrogen may be applied at or after sowing but not before, and directed between the rows to avoid adverse osmotic effects on emergence. A satisfactory alternative is to broadcast one-third of the required nitrogen at sowing, then to broadcast the remaining two-thirds when establishment is complete.

With seeds spaced 15 cm or more apart, as is common with drilling-to-a-stand, potential yield will be lost due to gappiness unless 75% of seeds produce established seedlings. Moreover the proportion of crop left in the field at harvest, which averages 8%, is likely to increase because harvesters work inefficiently in gappy crops. The target on mineral soils is 75,000 plants/ha but on organic soils fewer can be tolerated without loss of yield provided plant distribution is even. On light peat or sandy soils wind can cause seeds or small seedlings to be blown away with the surface soil and various measures (inter-row planting of straw or more rapidly growing species) are taken to stabilise the soil surface during April and May.

Surveys indicate that establishment is improving but it is estimated that overall 10% of potential yield is still lost because establishment is less than the target. There is potential to make emergence more rapid and uniform; seedlings emerge over a protracted period, commonly 40–50 and occasionally 80 days. Where fluid drilling techniques have been attempted with sugar beet establishment has been erratic and more needs to be known on carriers and techniques for accurate seed placement.

When sowing is earlier than about 20 March there is a probability that a gappy stand with many bolters will result. The target is to sow as much as possible before the 10 April because each day's delay thereafter incurs an ever-increasing yield penalty. The first sowings should be slightly more closely spaced and restricted to bolting-resistant varieties. In dry springs late-sown crops rarely establish well because of lack of moisture, which also inactivates pre-emergence herbicides.

Raising seedlings under glass and transplanting to the field at normal sowing time lengthens the growing season and usually increases yield by 10–20%, but the question is whether the costs of raising and planting can be justified; present indications are that the answer is, at best, marginal.

Fertilisers

As well as to the usual major nutrients, N, P and K, sugar beet responds to sodium on soils other than some peats and silts. In part sodium can replace potassium but it has unique properties and interacts positively with nitrogen in stimulating sugar production. It seems likely that profits are being lost by giving too much nitrogen and phosphate and not applying sodium; the average usage of nitrogen is currently 145 kg/ha (160 in the early 1970s) whereas experiments show that more than 125 kg/ha is rarely justified. Nitrogen increases yield by accelerating leaf production early in the year. However, when more than the optimum is given, foliage growth is stimulated later in the season without increasing sucrose content of the storage root. Juice purity and hence the value of the crop to the processors also decreases.

As yields and nutrient offtakes increased and less F Y M was used, many crops on lighter soils showed severe symptoms of magnesium deficiency in the early 1960s. Growers now regard magnesium as a major element and base decisions regarding its use on soil analysis. On organic soils the crop often shows symptoms of manganese deficiency and yield is lost when it is severe. Conventionally, manganese is applied as a foliar spray as soon as the leaf surface is sufficiently extensive but incorporating manganese within the pellet is giving promising results.

Irrigation

Crops use about 25 mm of water to produce each tonne of dry matter. On most soils, the root system can extend to at least 1·5 m below the surface and can extract of the order of 75 mm on sands, 125 mm on loamy sands, and 150 mm on clay loams. Irrigation up to the end of August does not decrease sugar percentage and can increase it. In September it is usually only crops on shallow or light soils that respond; irrigation then can decrease sugar percentage. Because of slower closure of the canopy and lower value, sugar beet take lower priority than potatoes in an irrigation programme: however, in dry summers irrigation of beet gives profitable increases. Almost half the crop is grown on sandy loams or coarser-textured soils and it is desirable that more of the national crop, currently less than 10%, is irrigated. One difficulty of assessing the true potential yield benefits from irrigating sugar beet is that fine dry summers are favourable to aphids and frequently virus yellows infection limits responsiveness to added water.

Diseases and pests

The principal disease of sugar beet, Virus Yellows, is caused by two viruses, Beet Mild Yellowing Virus and Beet Yellows Virus. The former is more common but the latter more markedly restricts growth and yield; severe attacks of yellows can decrease yield by 40–50%.

The peach potato aphid *Myzus persicae* transmits both viruses and control measures rely on killing the vector. Since 1973 some strains of the aphids have developed resistance to the aphicides then in general use. To prolong the useful life of the few materials now available, the spray warning scheme aims to restrict spraying times to when it is most effective. The scheme is based on monitoring of the overwintering of aphids and viruses, insect trap catches and daily crop inspections during May and June. Aphids overwinter on a number of hosts including sugar beet and brassicas and in glasshouses. The viruses overwinter on sugar beet, both seedcrops and groundkeepers, some weeds and clamped mangolds. Aphids and virus sources are most numerous after a mild winter. Granules applied in seed furrows give protection until about mid-June and are used as a routine in the south-east of East Anglia where crops are most likely to be infected early. Even in years of severe outbreaks there is considerable field-to-field variation in the severity of infection, an observation which might provide researchers with some clues for an improved strategy for control.

Previously, powdery mildew was regarded as of little economic importance but the disease was unusually prevalent in 1975 and 1976 and since then a single sulphur spray when the fungus is first spreading, usually in August, has consistently given worthwhile increases in yield.

In order to avoid the build-up of beet cyst nematodes, the grower's contract restricts cropping to fields that have not grown members of the beet and crucifer (e.g. cabbage, kale, rape, turnips) families in the two preceding years. This does not guarantee that the nematode will not increase to damaging populations and, because 20% of the national crop is grown after only a two-year break, growers are warned to be ready to widen their rotation at the first sign of trouble. Ectoparasitic stubby root or needle nematodes are the principal causes of the condition on sandy soils known as Docking disorder which is particularly prevalent in years when May is wet. This is controlled largely by routine application of aldicarb as seed furrow granules.

With drilling-to-a-stand techniques, pest depredations of seeds (notably by woodmice) and seedlings (notably by a complex of spring-tails, millipedes and symphylids, and skylarks at a later stage) become increasingly significant. Seed-borne blackleg (*Phoma*) is controlled by steeping seed in ethyl mercury phosphate solution.

Weed control

Because the crop does not form a closed canopy until mid-summer, late-germinating species such as fat hen, mayweed and the polygonums tend to predominate. Although hand-weeded experiments demonstrate that weeds usually do not have to be controlled until four weeks after emergence, early germinators are by then no longer susceptible to post-emergence herbicides. Pre-emergence herbicides (usually band sprayed) restrict the range of weed sizes with which post-emergence herbicides have to cope and thus pave the way for their effective use. The key to effective use of post-emergence herbicides is to apply them before weeds, especially the polygonums, have grown beyond the cotyledon stage.

Recently, two new techniques for herbicide application have been developed. The first involves a sequence of overall applications of some post-emergence herbicides at lower volumes than previously recommended. This is quicker than band spraying so more of the crop can be treated when the weeds are at the most susceptible stage. Because less herbicide is used the danger of crop damage is lessened. The other development is the application of non-selective herbicides to tall growing weeds, notably fat hen, potatoes or weed beet, using wick-applications.

POTATOES

Trends in area, production and utilisation

The total area of potatoes grown in the UK was at its peak of 560,000 ha during the Second World War and subsequently has decreased to the present 174,000 ha. Annual variation in area is caused partially by the Potato Marketing Board (PMB), who recommend an area related to expected demand. Their control is exercised by a quota for each individual registered producer, but growers frequently ignore the advice of the Board.

The number of registered producers has decreased more rapidly than the area and currently the average area per potato grower is 5·8 ha. Both in seed and ware production it is not uncommon for growers to have more than 400 ha, often including rented land.

The national average yield has, in general, increased over the last 25 years and was 37 t/ha in 1981. The only marked deviations from the trend were years of severe drought (1975 and 1976) and water stress has been the principal factor causing seasonal variation in yields. As an increasing proportion of the crop is now irrigated (23%), the impact of rainfall on national production is likely to be reduced.

There have been considerable changes in the utilisation of the crop in the last 25 years. The proportion of the crop processed (currently 22%) into chips and crisps as well as dehydrated and canned has increased steadily and even in the traditional ware market 10% is now pre-packed in some way.

Suitability of crops for outlets

For all outlets there are specifications for the required sizes of tubers and, for some, tubers of certain sizes receive a higher price (Table 27.1). There is a marked preference in certain areas of the country for ware potatoes with wholly or partly pink skins. This is probably a legacy of the widespread acceptance of the variety King Edward as a quality potato and is of commercial importance, as higher prices are paid for these coloured varieties. For the production of crisps and to a lesser extent, chips, a high dry-matter content and a low content of reducing sugars are required to ensure a bright

Table 27.1 Tuber size specifications for different outlets

Outlet	Size (mm)
Crisps	40–70 (Preferably 45–65)
Chips	50–80
Dehydration	Not specified
Canning	20–40
Diced	50–80
Pre-packs	45–80
Bakers	>70
Ware	40–80

yellow/brown colour in the fried product. The level of reducing sugars is influenced partly by growing conditions but mainly by temperatures at harvest and in store; temperatures below 8°C increasing the reducing sugar content.

For all outlets freedom from greening, damage and rotting diseases is essential and standards are laid down by the PMB.

Types of crop

Potatoes are harvested from early May (Isles of Scilly) through to November or even December. Crops harvested before the end of August are normally sold directly to markets or processors; those harvested from September to December are predominantly put into stores to provide supplies until the following May or June. The sale of one year's crop is, therefore, completed after the start of harvesting in the next year.

Earlies and second earlies

These crops are harvested while the tops are green and sold quickly for consumption. They are concentrated in the western coastal environments (Cornwall, Isles of Scilly, Pembrokeshire, Ayrshire) and some mild eastern coastal ones (Kent) in which early planting is possible and growth in the early spring and summer is largely uninterrupted by frosts. The western sites normally produce the highest yields and most of the first new potatoes come from these areas. From late June crops of so-called second earlies from Lincolnshire, Cheshire and the West Midlands begin to be harvested. These are, normally, different varieties and are harvested until August when harvesting of maincrops begins. Harvesting begins early in May in the Isles of Scilly but not until late May in other south-west districts. Yields are initially small (10-14 t/ha) but a very high price (£300/t or more) is paid for these 'new' potatoes. As the season progresses yields increase but prices normally decrease and by early July yields are 40-50 t/ha.

Maincrops

These are concentrated in the eastern parts of England and Scotland, although important production areas are also found in the west (notably the West Midlands, Cheshire and Lancashire). For harvesting in September for storage the crops are normally defoliated 2–3 weeks before intended harvest and produce the highest yields (up to 70 t/ha).

These divisions of the harvesting season are not rigid but represent the basic sequence. As yields increase with time the whole sequence is severely affected by any significant delay in the harvesting of the early crops. If weather conditions or a low price delay harvesting each remaining hectare has a substantially higher yield when subsequently harvested. The effect of serious delays in harvesting of the earliest crops is to increase yields of many crops intended for June/July harvesting and this usually results in low prices which are a further disincentive to harvest. Thus, over-supply may persist for the whole of the season. The impact of any interference with early harvests, e.g. imported early potatoes, is much more than simply on the return to the early grower.

Seed potatoes

High-grade basic seed is produced in upland parts of the UK but the industry is heavily concentrated in Scotland. Further multiplication of certified seed is practised in lowland England. As seed tubers are smaller than ware tubers seed crops are normally defoliated before ware crops, although not always harvested before them. Potatoes are subject to several virus diseases (the most important of which are spread by aphids) and before a statutory seed certification scheme was introduced, yields were often low as a consequence of virus infection of the seed tubers.

In the belief that the windier, cooler conditions of northern upland areas were inimical to the early establishment of large aphid populations the production of virus-free seed was established between the wars in such areas, especially Scotland and Northern Ireland. At present the multiplication begins with virus-free tubers produced on stem cuttings in a glasshouse and called VTSC (Virus Tested Stem Cuttings). These are produced by DAFF in Scotland and supplied to registered seed producers in lots of 10–12 tubers. The method of propagation ensures that these tubers are free from latent diseases (notably gangrene caused by *Phoma exigua* var. *foveata* and dry rot caused by *Fusarium solani* var. *coeruleum*). The small tubers are multiplied in fields as separate lots and inspected each year before being certified as meeting certain standards. As the quantity of tubers increases, the area of multiplication increases and eventually after 5–8 years the progeny tubers are sold to ware growers as basic seed. As multiplication proceeds, the grades are called V T S C (for four years), Foundation stock (up to four years) and A A (for as long as the stock meets the certifying standards). V T S C multiplication often occurs at higher altitudes than later multiplication and the disease standards are more rigorous than for succeeding grades of seed. Thus, A A allows slightly more diseased plants than F S and this more than V T S C. Any high-grade stock which fails to meet the intended specification is downgraded to its actual standard or, if very infected, excluded from the system by denial of a certificate. The major problems associated with certification are virus diseases and blackleg (*Erwinia carotovora* var. *atroseptica*).

The most important aphid-transmitted viruses are leaf roll and severe mosaic (virus Y). These may be transmitted late in the growth of the crop and as there is no requirement for growers to defoliate their crops shortly after their final inspection, infection may occur post-certification. In The Netherlands, a test on the harvested tubers is used to guard against such occurrences. The problem in Scotland is further exacerbated by the absence of any segregation of seed and ware crops and the occurrence of seed production at all elevations in the eastern part of the country. Consequently, the efforts of seed growers can be negated by the practices of their ware-growing neighbours.

It might be expected that seed producers would defoliate their crops very early in order to control tuber size (and also restrict opportunity for transmission of virus), but this is frequently not the case and seed crops often contain a substantial quantity of over-sized tubers. This is regarded as an insurance against a low price for seed, but clearly is contrary to the objectives of seed production.

Components of the production system

Sprouting

All early crops, most second earlies and approximately 38% of maincrops are sprouted prior to planting. Sprouting advances emergence and the whole cycle of growth of the crop; for earlies and second earlies there is no doubt that it increases yields and in many maincrops increases in yield also result. Sprout growth has been likened to an ageing process and its extent is an illustration of the physiological age of seed. The large effects of physiological age on growth patterns allow varieties to be used for harvesting outside their nominal classification if growers are prepared to control their sprout growth.

Planting and seed rate

Planting of early crops, especially in the west, may begin as early as January and is usually completed by early April. Maincrop planting does not usually begin in southern England until late March. Planting is normally later as one moves northwards and in northern Scotland seed crops are frequently planted in May.

Many of the early crops are hand-planted to cope with the well-sprouted, large seed but in all other crops machine planting is almost universal. There is considerable continuing debate as to the most effective size of seed for maincrops and the total quantity (seed rate) to be planted. For maincrop varieties the optimum seed rate (expressed in t/ha) for saleable yield is lower the smaller the seed and there is considerable evidence that yields are not affected by seed size if each size is planted at its optimum seed rate. As Table 27.2 shows, the optimum seed rates for specific seed sizes are

Table 27.2 Seed rates for maximum gross margins for different seed weights in two contrasting varieties (seed costing twice the value of ware)

Variety	Seed weight (g)	Optimum seed rate (t/ha)
Pentland Crown	30	2·1
	60	3·3
	90	4·2
King Edward	30	1·5
	60	2·4
	90	3·0

quite different in the popular varieties. Thus, growers should adjust their seed rates according to seed size and variety if they are to achieve the highest yields and avoid excessive seed costs. There is much evidence to suggest that many do not bother. In general, commercial crops have too low a seed rate.

The limited adjustment for seed size is surprising as close-graded seed (e.g. 35–45 mm, 45–55 mm) is essential to achieve accurate spacing with many planters. As seed prices do not vary negatively with seed size to any marked extent, the greatest financial returns normally result from the use of small seed. This is not the case in The Netherlands where the prices of seed more accurately reflect their value and seed size does not, therefore, significantly affect the return.

Fertilisers

Most fertilisers are applied either before or during the planting operation. For many crops fertilisers are broadcast prior to cultivation of the ploughed land and incorporated by the subsequent cultivations. Some crops receive their fertiliser (especially in liquid form) placed to the side and slightly below the seed. On light soils splitting the N application between planting and a top dressing just after tuber initiation has been shown to increase yields compared with one application at planting.

Rates of fertiliser application have increased steadily and in 1980 averaged 188 kg N, 188 kg P_2O_5 and 264 kg K_2O per ha. Thus, many crops must receive more than 200 kg N/ha. Experiments have frequently failed to show responses to the application rates currently used and very high yields (>75 t/ha) have been produced from much lower applications (<150 kg N/ha).

Weed control

Good weed control is achieved in most crops with the use of residual, soil-acting herbicides applied to the completed ridges; in most cases the herbicide is applied prior to crop emergence and in some cases may contain a contact herbicide such as paraquat to kill any emerged weeds. Metribuzin can be applied post-emergence up to a maximum shoot length of 15 cm on certain varieties. On organic soils, metribuzin is used widely as a herbicide, being incorporated prior to emergence and used successfully post-emergence on several varieties. Weed control is sometimes a problem in the Fens because of the frequency and diversity of weeds.

Water use

Experimental evidence suggests that for a tonne of tuber dry weight to be produced, approximately 25 mm of water must be transpired per ha. In many areas, the total contribution of stored soil water and rainfall during the growing season seems to be about 300 mm which suggests a maximum unirrigated potential yield of approximately 60 tonnes. In dry years and on light soils potential yields are much lower. Under irrigated conditions, yields of 100 t/ha have been recorded on soils of markedly different water-holding capacity.

The application of water is now increasingly via rain-guns and self-propelled systems which apply the water more quickly and are less laborious than conventional spray lines. There is, however, some doubt as to the uniformity of application, which influences the uniformity of tuber growth and may cause erosion and damage to the ridges through the increased impact of the droplets.

Pests and diseases

As a consequence of too frequent cropping, especially in the last war, large areas of light and organic soils became severely infested with potato cyst nematode. Such soils are so well suited to potato production that the wide rotations (no potatoes closer than 8–10 years) recommended for control of the pest were unattractive to growers. Fortunately, both nematode-resistant varieties (e.g. Maris

Piper) and nematicides have been introduced. In the Fens, close cropping with potatoes is again practised.

Blight is still a potentially serious disease for it may destroy the foliage prematurely and also infect tubers which are unsaleable after storage. Routine applications of fungicides are made and at the same time aphicides may be applied if high populations are expected. The use of fungicides normally controls the disease well.

Until recently the fungicides were protective and an adequate cover of the leaves had to be maintained through the season. This necessitated frequent spraying in the wet conditions which favour the development of the disease. Recently, systemic fungicides, which do not require applications more frequently than every 2–3 weeks, have been rapidly adopted by the majority of growers. However, although strains of the blight fungus resistant to metalaxyl have been identified, new systemic fungicides are in advanced testing and are likely to be widely used.

Harvesting and storage

For earlies and second earlies the green haulm is pulverised off with flails and the crop harvested shortly afterwards. For maincrops a similar sequence of operations occurs in those crops sold immediately from the field, but crops intended for storage are defoliated chemically, usually with sulphuric acid or diquat, some 2–3 weeks prior to the intended harvest.

For earlies much handpicking continues as the tubers are small and the severe slopes of the generally small fields are unsuitable for many machines. Gangs of pickers work behind elevator diggers which lift the tubers from the ridge and return them to the soil surface. Mechanical harvesters are now used widely in maincrops and the machines are unmanned on soils requiring little separation of tubers from extraneous material and clods. Where greater separation is required, some hand sorting on the machines is still needed.

It is imperative that tubers intended for storage are harvested with the minimum of damage; tubers may be bruised, cut or crushed during harvesting and damage reduces out-turn from store due to increased water loss, infection with gangrene and dry rot, and premature sprouting. Additionally, damaged tubers are unacceptable to processors even if predominantly sound, for the cost of hand-trimming of the damaged areas is too great. The avoidance of damage requires good setting of the harvester, so that the whole ridge is harvested and separated, the minimum number of drops and changes of direction for tubers during harvesting, and careful delivery of the tubers from harvester to trailer for delivery to store. Similar avoidance of damage must be achieved while loading into store. An increasing proportion of the crop is graded in some way before storage and thiabendazole is frequently applied as a protection against storage diseases. Considerable improvements in harvester design have occurred recently and with skilful operation mechanical harvesting can be done with little damage to tubers. Many conveyors and trailers are lined with plastic foam which minimises impact. Although the need for separation of tubers from other matter is mainly a function of soil type, the farmer's skill in cultivation is important for the ridges should contain no large clods. In the main, these are formed during cultivations in the spring. On many soils the presence of stones necessarily causes damage in harvesting and the continued production of the crop on many light soils has only recently been assured by techniques of stone separation which remove most of the stones from the ridged soil during planting. This is especially important in seed crops in Scotland where the damage caused by stones frequently resulted in gangrene.

Stores are insulated, frost-proof buildings which have some form of ventilation. Most stores still load the crop in bulk but a significant number use 1-tonne boxes; this latter system is especially attractive to seed producers. Once tubers are in store they must be cured for 10–14 days at approximately 15°C to ensure the healing of wounds. For crops harvested in September ambient temperatures are usually above 15°C and curing is both easier and more rapid than for later-harvested crops. Long-term storage requires suppression of sprouts, which is achieved by repeated doses of CIPC (or less commonly tecnazene) beginning 3–4 weeks after loading of the store. The temperature of storage is lower (5–8°C) for tubers for ware consumption than for processing (8–12°C), because reducing sugars increase below 8°C.

There is also an important consequence of any tubers or tuber pieces left in the field as some will survive the winter and become 'volunteer' potatoes in succeeding crops. These are a source of disease (both virus and fungal) and perpetuate the problems for further years.

28

FORAGE CROPS

Forage crops are highly digestible and palatable and are either very quick growing or very high yielding. They have another advantage in that, when either grazed or harvested and stored, they provide feed at times when grass growth is poor. They can be grown as catch or as full-season crops. Catch crops, such as rape, can be spring-sown to provide summer grazing or, like stubble turnips, can be sown in summer following early harvested crops, to provide autumn grazing. Full-season crops of kale, turnips and swedes can be grown to provide autumn and winter grazing whereas mangels and fodder beet can be harvested in autumn and stored for winter feeding. Cereal crops such as maize can be purpose-grown for ensiling or, like whole crop wheat and barley, can be ensiled on an opportunist basis when shortage of grass silage demands. Finally, winter rye can be sown in the autumn, overwintered and grazed in the spring prior to the onset of grass growth.

TRENDS IN PRODUCTION

Changes in the areas of forage crops grown over the last 50 years are summarised in Table 28.1.

Table 28.1 Areas of forage crops in England and Wales ('000 ha)

Crop	1939	1960	1965	1970	1975	1980
Kale & rape	na	176	114	77	74	43[1]
Turnips, swedes & fodder beet	160	83	58	43	51	38[2]
Mangels	85	51	23	10	7	6[3]
Maize	–	–	–	1	26	22[4]
Arable silage (England only) ('000 tonnes)	na	378	487	213	632	684

na = not available
[1] Includes cabbage.
[2] Excludes fodder beet
[3] Includes fodder beet
[4] Includes some maize for grain

The area of traditional forage crops has declined markedly, associated with their high labour requirements, a fall in livestock numbers in arable areas and improved grassland production. However, the introduction of precision drilling, improved herbicides and better harvesting techniques have reduced their labour requirements and slowed the rate of decline. In addition the area of catch crops (which do not appear in the above statistics) has increased.

The area of forage maize increased rapidly between 1970 and 1980, due to improved varieties, weed control and harvesting machinery, and peaked in 1977, since when it has declined slightly. The data for arable silage are less reliable than the other statistics, but indicate some increase in the quantity made.

RELATIVE YIELDS AND FEED VALUES

Comparative yields and feed values, based mainly on plot yield data from NIAB leaflets and analytical data from MAFF Technical Bulletin 33 are shown in Table 28.2. The yields shown are 10–20% higher than would be generally achieved in practice but comparisons between crops are still valid.

Table 28.2   Relative yields of forage crops

Crop	Dry matter yield (t/ha)	ME content (M J/kg D M)	Crude Protein (g/kg)	ME yield (G J/ha)	CP yield (kg/ha)
Kale (autumn)	8·6	11·0	153	95	1,316
Rape	4·2	9·5	200	40	840
Turnips (Continental white-tops)	3·3	9·2	192	30	634
Swedes (roots)	7·6	12·8	108	97	821
Fodder beet (roots)	13·5	12·5	79	169	1,066
Mangels (roots)	12·0	12·4	83	149	996
Maize silage	11·9	10·6	90	126	1,071
Arable silage (w. wheat)	8·6	8·4	78	72	671
(s. barley)	8·0	9·6	95	77	760
Grass (high D value)	9·7	11·3	168	110	1,630
(low D value)	15·0	9·7	102	146	1,530
Spring barley: grain	4·9	13·7	108	67	529
straw	2·9	7·3	38	21	110
total	7·8	–	–	88	639

Although mainly valued as sources of energy, forage crops also produce high yields of protein. Most compare favourably in yield with barley and grass, and those which yield less well are generally very quick growing.

FORAGE CROP UTILISATION

In order to realise their high yield potential, forage crops must be utilised efficiently. The majority of forage crops are grazed, usually by sheep but also by cattle. Wastage can be high, as shown in Table 28.3.

Table 28.3   Dry matter wastage (%) of forage crops grazed by lambs

	Stubble turnips	Swedes	Rape	Kale
Level of wastage	4–80	5–65	5–70	5–70

Source: Bastiman and Slade (1978) *ADAS Quarterly Review*, 28

Soiling is the main cause of rejection and wastage. Leafy crops are less prone to soiling than are root crops, which are particularly susceptible if easily uprooted due to variety or method of sowing having limited the proportion of root in the ground. Heavy soils and high rainfall conditions obviously make the soiling problem worse. Wastage can be reduced by increased stocking rates, drilling crops in rows (which provide walkways) and direct drilling into a surface mat which provides a firmer base for grazing. The provision of grass run-back areas is also beneficial. Grazing method, such as strip or block grazing, has little effect on utilisation. Efficiency of utilisation can be affected by animal factors, cattle, and lambs which are losing their teeth in late winter, making inefficient use of root crops. Harvesting forage crops for indoor feeding is not without loss. In most cases tops are discarded and root losses of about 15% have been measured, either as small roots or broken tips left in the ground. Losses also occur in store and, while averaging 15–17%, can be much higher in roots harvested in wet conditions, or in low dry matter roots bruised and damaged during harvest. In dry areas and on light soils storage losses may be avoided by harvesting roots as needed through the winter, but harvesting difficulties and soil damage make this impossible on heavy land and in wet areas.

The tops of mangels and fodder beet can be utilised. If fed fresh their oxalic acid content can cause scouring, however, and if ensiled there are problems with effluent and butyric fermentations. Losses are also incurred in ensiling maize or whole crop cereals. Field and effluent losses are negligible, because the crops are harvested direct at dry matter contents above 25%, and respiration and fermentation losses are usually low. However, these crops are prone to aerobic degradation, and inadequate sealing or slow use (meaning prolonged exposure to air after opening) can give dry matter losses up to 30%. These can be controlled to some extent by the use of additives such as propionic acid.

FORAGE CROP HUSBANDRY
Kale and rape

There are two types of kale: marrow-stemmed varieties, which are high yielding; and thousand-head varieties, which are leafier, more digestible and frost resistant. Bred hybrids between them combine many of the favourable characteristics of both. Kales generally out-yield rapes but the latter grow more rapidly and are more suited to catch cropping. They are, however, less winter hardy. Kale and rape can be grown in northern and upland conditions but, because they are usually grazed, heavy soils in wet areas should be avoided.

Position in rotation   Kale can be grown as a main crop, as part of a normal root break or, in southern areas, as a catch crop following early harvested crops or grass broken out in June. The hardiness of kale allows late utilisation which means that it must be followed by a spring crop. Rape is usually grown as a catch crop sown in April for use in July (when it can be followed by a winter crop) or sown in August for use in October to January (when it must be followed by a spring crop). Most rape varieties are susceptible to club root and so must not be grown in too frequent rotation with other brassica crops.

Establishment and plant population   These crops require fine, moist seedbeds which have been well consolidated. They can be drilled or broadcast and rape has been established successfully by aerial sowing into standing cereal crops. Both crops can be direct drilled with the advantages of quicker establishment and cleaner utilisation that this technique brings. Seed rates of 2–4 kg/ha for drilled crops should be increased to 4–8 kg/ha for broadcasting, and good-quality seed treated with insecticide and fungicide should be used. Main-crop kale should be sown in April or May, earlier sowing giving heavier yields but increased tendency for the crop to become woody. As a catch crop it should be sown in June. Rape as a catch crop can be sown in April or August, early-sown crops being susceptible to mildew attack.

Manuring   The N, P and K requirements of kale and rape are given in Table 28.4.

Utilisation

Kale used to be hand harvested for feeding indoors, and has been forage harvested successfully, but, like rape, it is now almost in-

Table 28.4 Fertiliser recommendations for kale and rape

Index*	Nitrogen (kg/ha N)		Phosphorus (kg/ha P₂O₅)		Potassium (kg/ha K₂O)	
	Kale	Rape	Kale	Rape	Kale	Rape
0	125	100	100	75	100	100
1	100	75	75	50	75	75
2	75	50	50	25	50	50
3	–	–	25	–	50	50
over 3	–	–	–	–	50	50

Source: ADAS Booklet 2191

* ADAS Index based on requirements for N and soil analyses for P & K (see Glossary).
For direct drilled crops the nitrogen should be increased by 25 kg/ha and extra potash should be applied through the rotation where crops are zero grazed rather than grazed *in situ*.

variably grazed. Utilisation can be improved by strip grazing especially when being fed to cattle. Shorter-stemmed varieties simplify the use of electric fencing, as do shorter, leafier crops produced by late sowing.

Feeding large quantities of kale, and to a lesser extent rape, can cause haemolytic anaemia as the result of their relatively high contents of S-methyl cysteine sulphoxide (SMCO). Other constituents of brassica leaves, such as nitrogen compounds and goitrogenic substances can also cause metabolic disorders. Consequently stock should be introduced to these crops gradually.

Turnips and swedes

Turnips are generally lower in dry matter content than swedes and have no neck, the hairy leaves arising straight from the bulb. They are divided into two main groups: white-fleshed varieties of about 8% dry matter content, and yellow-fleshed varieties of between 8 and 10% dry matter. The former, often called stubble or 100-day turnips, are frost-susceptible, prone to bruising and store less well than the latter, but have the marked advantage of very rapid growth, especially of leaf, and are valuable catch crops. Yellow-fleshed varieties are not suited to catch cropping but as whole-season crops produce high yields, mainly root, for autumn grazing.

Swedes have smooth leaves which arise from a neck above the bulb. They have dry matter contents between 8·5 and 10% and vary in skin colour and root shape. Shape is agronomically important, globe being best for machine harvesting but the intermediate shape is preferred for grazing. Swedes are slower growing than turnips and are grown as full-season crops. Having higher dry matter contents than turnips they are hardier, less prone to bruising and store better. However, very high dry matter varieties may be so hard that they require chopping to ensure high intakes.

Turnips and swedes grow best in cool, higher rainfall areas of the north and west, and are less suited to dry areas than are mangels and beet. Turnips can be grown on less fertile soils than other fodder roots.

Position in rotation Turnips and swedes can be grown as main crops within a rotation, usually following and followed by cereals. They should not be grown in close succession because of the danger of club root build-up especially in soils with low pH.

Turnips can be grown as a catch crop sown in either April or May, to provide grazing in July to October, or sown in July or August following early harvested crops, to provide autumn grazing. They can be used as pioneer crops in the improvement of upland grassland.

Establishment and plant population Although traditionally grown on ridges the development of improved herbicides and precision seeding has led to these crops being grown on the flat. Shallow tilths are adequate but seedbeds should be fine, firm and moist. Many crops are direct drilled, following herbicide use to give good weed control, which reduces labour input, improves germination in dry conditions (because it conserves soil moisture) and provides a firm base which reduces crop wastage under grazing. Where land is cultivated this usually involves autumn ploughing and cultivation, with minimum soil disturbance in the spring to retain moisture in the seed bed.

Full-season crops are sown between late April and mid-June according to latitude. Earlier sowing increases root yield but also increases susceptibility to mildew attack, although effective fungicides have been developed to control this. Very early crops may bolt which reduces feed value, but the greater problems are flea beetle or cabbage root fly attack if establishment is slow in cold springs. Plant population is not critical with turnips or swedes, but populations should be about 66,000/ha, on rows about 450 mm apart. Seed rates of 2·2–4·5 kg/ha are used for conventional drilling, reduced to 280–370 g/ha for precision drilling when it is advisable to use graded seed which has been treated with insecticides. Although plant population does not affect total yield it does affect top to bulb ratio. Low populations encourage large bulbs and high populations encourage top growth. Manipulation of the crop in this way can be useful with high top yields preferred for early grazing and large bulbs more convenient for harvesting. Large bulbs are prone to crown rot and splitting, however, and are less frost resistant than smaller bulbs.

Manuring Swedes and turnips are sensitive to acidity and grow best at pH levels of about 6·5. Overliming should be avoided because alkaline conditions cause boron deficiency and this leads to heart rot which can result in serious yield losses. This problem can be avoided by the use of specially formulated fertilisers or by borax application to the seedbed. The recommended rates of N, P and K are shown in Table 28.5. These levels can be reduced if slurry or farmyard manure has been applied. Excessive nitrogen use on swedes should be avoided because it reduces dry matter content and so reduces frost hardiness.

Table 28.5   Fertiliser recommendations for swedes and turnips

Index	Nitrogen (kg/ha N)		Phosphorus (kg/ha P$_2$O$_5$)		Potassium (kg/ha K$_2$O)	
	Swedes	Turnips	Swedes	Turnips	Swedes	Turnips
0	75	100	150	75	125	100
1	50	75	125	50	100	75
2	25	50	75	25	60	50
3	–	–	50	–	60	50
Over 3	–	–	50	–	60	50

Source: A D A S Advisory Leaflet 591

Utilisation

Turnips are usually grazed by sheep but are also suitable for cattle grazing. White-fleshed varieties tend to become woolly and unpalatable with maturity and should be eaten off quickly, especially in the autumn because of the additional risk of frost damage. Swedes can be grazed or harvested for indoor feeding to cattle or sheep. Harder, high dry matter varieties require chopping to achieve high intakes. The problems of storage are similar to those of mangels and beet.

Mangels and fodder beet

These represent a continuum of the same crop from low dry matter mangels to high dry matter fodder beet. They are usually grown in eastern and south-eastern areas, and are very high yielding. Traditional varieties produced multigerm seed from which several seedlings grew at the same site. This necessitated laborious hand singling to produce one large root per site rather than several stunted ones. Monogerm varieties have now been bred (especially of fodder beet) which can be precision drilled at the desired plant population, accepting that as few as 50% of seed sites may give rise to established plants, and hand singling dispensed with. Effective precision drilling requires seed pelleted in a clay coat to give it a regular spherical shape. The pellet may be impregnated with fungicides or insecticides. Varieties of mangels and fodder beet range in skin colour from red through orange to white, but this has no agronomic significance. A more useful grouping is on dry matter content into low and medium dry matter mangels (10–12% and 12–15% D M respectively) and medium and high dry matter beet (15–17% and 17–20% D M respectively). Dry matter content is associated with proportion of root in the ground, which ranges from about 36% for low dry matter mangels to 66% for high dry matter fodder beet. Less root in the ground means easier lifting and cleaner roots at harvest. Mangels tend to produce larger roots than beet, but about 10% less dry matter yield. Both crops yield best on deep, well-drained loams. Mangels are more suited to heavier soils but beet can be grown (and harvested more easily) on lighter, thinner soils. Both crops are deep rooting and yield well in dry conditions provided moisture is sufficient for good establishment.

Position in rotation   These are full-season crops grown in the root break in arable rotations, usually following and being followed by cereals. Due to their relative freedom from pests and diseases they can be grown frequently in rotations, and on favoured fields can be grown for several years in succession. However, care should be taken over rotational spacing where there is risk of beet cyst nematode infestation, especially in sugar beet growing areas.

Establishment and plant population   Fine, firm and moist seedbeds are required for satisfactory establishment. On medium or heavy land this means autumn ploughing to produce a weathered tilth in spring. Perennial weeds are best controlled in the previous cereal crop or by cultivations prior to ploughing. Crops are sown between early April and late May, according to soil conditions, with higher yields following earlier sowing. Maximum yields require fairly precise plant populations of 60,000 and 70,000 plants/ha for mangels and beet, respectively, obtained by sowing 10–13·5 kg/ha multigerm seed or 3·5–7·0 kg/ha of rubbed and graded or monogerm seed. Row width is not critical and about 500 mm is usual with plants about 240 mm apart in the rows. Drilling to a stand using low seed rates is risky and good-quality seed and fine seedbeds are essential. Where multigerm seed is used crops should be singled at about the 4 true-leaf stage, as interplant competition becomes severe beyond this stage.

Manuring   Mangels and fodder beet are sensitive to acidity and yield best at soil pH levels of about 6·5. Lime required to achieve this should be applied to the previous crop or in the autumn prior to drilling. These crops respond to sodium, applied as agricultural salt at 150 kg/ha in the previous autumn or in the spring, at least four weeks before drilling. The N, P and K requirements are shown in Table 28.6 for different soil indices. These levels can be reduced if farmyard manure was applied in the previous autumn.

Table 28.6   Fertiliser recommendations for mangels and fodder
beet

Index	Nitrogen (kg/ha N)	Phosphorus (kg/ha P₂O₅)	Potassium (kg/ha K₂O)
0	125	100	200
1	100	75	100
2	75	50	75
3	–	–	75
Over 3	–	–	75

Source: ADAS Advisory Leaflet 591

Utilisation

Mangels and fodder beet are seldom grazed but are mechanically
harvested in October and November to avoid severe frosting. Care
must be taken with mangels and low dry matter beet varieties
which are prone to damage that can precipitate rotting in store.
The crops should be topped at harvest, mangels above the crown
to avoid bleeding and beet below the crown, but topping too low
can cause rotting. All tops should be removed from crops stored
until spring to avoid carry-over of virus yellows. Roots can be
stored indoors or in clamps which should be on well-drained sites,
windproof and insulated with straw. Too rapid or deep insulation,
or failure to provide ventilation, may cause over-heating especially
if much soil or green material is included. Deterioration is also
encouraged by putting wet or over-dry roots into store. Mangels
should be left to mature until the new year, before they are fed, to
avoid the risks of scouring.

Forage maize

Forage maize can be grown on a range of soils provided that the
pH exceeds 6·0, but grows best on deep well-drained loams with
high organic matter content. Optimal growth requires high temperatures. Spring growth starts at soil temperatures above 10°C
and during stem elongation temperatures in south-east England are
nearly always below optimum. Consequently, factors, such as shelter, slope and aspect, which marginally affect soil temperatures,
can have a marked effect on yield or yield components, even within
a farm, and choice of field can have an important effect.

The maximum growth potential of maize occurs in July and
August when the evapo-transpiration rate is high, so despite being
deep rooting it can suffer from moisture stress, especially on light
soils. During vegetative growth this can cause stunting but during
flowering it reduces grain set and so reduces forage quality. Irriga-

tion is justified only in extreme cases, however, and moisture stress is best avoided by growing maize on heavier soils, but such soils warm up slowly in spring, giving poor establishment, and can give harvesting problems in wet autumns.

Position in rotation There are few rotational constraints on maize production. Its relative freedom from pests and diseases means that it can be grown repeatedly on the most favoured fields. Alternatively, being neither a host nor susceptible to pests and diseases of other crops, it can provide a useful break crop. Being tolerant to the broad spectrum residual triazine herbicides it can be regarded as a cleaning crop. It contributes little to soil structure or condition but can utilise heavy dressings of organic manures, so subsequent crops can benefit from phosphate and potash residues.

On mixed farms it usually follows a run of cereals where its late sowing provides opportunity for weed control, or it can be grown as part of the root break. Late harvest makes it an unsuitable entry for winter crops and it is usually followed by spring crops or grass. In grassland areas, when grown after grass, it is susceptible to frit fly and wireworm attack.

Establishment and plant population Maize should be drilled in late April to mid-May according to latitude.

Later sowing limits grain yield and so reduces forage quality. A deep, fairly fine, moist seedbed is required and the crop should be drilled about 50 mm deep. The optimum plant population is 110,000–140,000/ha, achieved by sowing 40–50 kg/ha seed. Spatial arrangement has little effect on yield or its components and row width can range from 450 to 750 mm.

Manuring Fertiliser recommendations are given in Table 28.7. Nitrogen requirements are low because low temperatures in UK limit responses. Phosphate and potash are applied to maintain soil status, yield responses have not been recorded in this country.

Table 28.7 Fertiliser recommendations for forage maize

Index	Nitrogen (kg/ha N)	Phosphorus (kg/ha P_2O_5)	Potassium (kg/ha K_2O)
0	60	75	75
1	40	40	40
2	40	40	40
Over 2	–	–	–

Source: A D A S Booklet 2191

Phosphate and potash can be ploughed down in autumn, or together with nitrogen, incorporated during spring seedbed preparation. Combine drilling should be avoided as emergence can be reduced.

Maize can exploit heavy dressings of organic manures. Up to 60 t/ha slurry can be applied over winter and incorporated in spring. This can supply sufficient phosphate and potash, but it is still worthwhile applying 40 kg/ha N.

Utilisation

Maize is usually ensiled prior to feeding, using specialised harvesting machinery, but conventional ensiling techniques, paying particular attention to sealing and limiting aerobic wastage after opening. Autumn grass can be augmented by zero-grazing maize.

Whole-crop cereal silage

On mixed farms the seasonal variation in grass silage production can be evened out by ensiling immature cereal crops. The crops most commonly grown and so most commonly ensiled are winter wheat and spring barley, the latter being preferred because of the higher profitability of wheat as a grain crop. Cereals are seldom grown specifically for silage, but are taken out of grain production as required. However, they are sometimes grown as nurse crops for grass or clover re-seeds with the aim of ensiling them to give the re-seed an earlier opportunity to grow away. Because the crops are grown initially for grain, cultivations, seed rates and fertiliser use are the same as those for conventional crops.

Whole crop cereals yield up to 10 t/ha dry matter and reach maximum yield in mid- to late July, well after the peak of grass silage making. At that stage (when the grain is milky/mealy ripe) changes in digestibility are slow and so timing of ensilage is not critical. Whole crops have dry matter contents of 30–35% in early July (suitable for clamp silage) or 35–40% in mid-July (suitable for tower silage). This obviates the need for wilting and so eliminates field losses and speeds up harvesting. It also eliminates effluent. The high dry matter and sugar contents mean that satisfactory fermentations can be achieved without the use of additives and conventional ensiling techniques are used. The disadvantages of whole crop cereals for silage are that they have lower digestibilities (58–60D) and crude protein contents (about 8% CP) than grass silage. Their susceptibility to aerobic breakdown means that sealing of the silo and careful feeding are necessary to contain in-silo losses.

Forage rye

Rye is the most winter-hardy cereal and being one of the earliest crops to start spring growth produces grazing about four weeks before Italian ryegrass. It yields best on free-draining fertile soils but tolerates acid conditions down to pH 4·9. On cereal farms rye can follow early harvested cereals and not being susceptible to the majority of cereal diseases it provides a good break. On grass farms it is usually the first crop after a ley. Being grazed off early in the season it can be followed by kale or swedes.

Highest yields are obtained from drilling in mid-August and decline with later sowing. Early sowing means using year-old seed on which germination tests are recommended. Forage varieties do not tiller readily and so seed rates of 220–250 kg/ha are required. Seedbed dressings of up to 50 kg/ha nitrogen and phosphate and 75 kg/ha potash are recommended, depending on previous cropping. A spring dressing of 75 kg/ha N should be applied, timing depending on forwardness of the crop and grazing requirement. Utilisation is often improved if spring nitrogen applications are staggered. Early-sown crops produce considerable autumn growth which is best grazed, but severe grazing can give poor recovery.

Earliness of spring grazing depends on sowing date, season and location, but sheep grazing can be available in February. Sheep can be set-stocked but cattle are best strip grazed. Grazing with sheep should start when the crop is 100 mm high and with cattle when 150–200 mm high. Delaying beyond this or failing to stock heavily enough leads to a stemmy crop and poor utilisation. If the crop outgrows grazing requirements the surplus is best ensiled.

Minor forage crops

As an arable by-product sugar beet tops have long supplied autumn grazing, following wilting to avoid metabolic disorders. Machinery developments have increased interest in ensiling them but as yet only a small proportion are ensiled.

Fodder raddish is a fast-growing catch crop suitable for sowing between June and August. Although resistant to club root its use is limited because it runs to seed after a short growing season and palatability falls off rapidly, so that it must be utilised within a 2–3-week period. It is also very frost susceptible and must be grazed off by October.

Cabbage is a more demanding crop to grow than other forage crops but produces heavy yields of easily grazed forage in November and December. Varieties sufficiently winter hardy for grazing in January to March, when lambs are losing their teeth and find root grazing difficult, are lower yielding.

New hybrid forage crops are being produced but as yet they have not been fully evaluated and occupy negligible areas.

PESTS OF FORAGE CROPS

Forage crops are subject to damage during establishment from birds and from such pests as slugs, leatherjackets, wireworms and cutworms. These can be controlled by cultural methods, seed treatment and pesticide application to the soil. None of these general pests is dealt with here. Specific pests of forage crops are outlined below. This is not meant to be a comprehensive list but to cover only those pests of major importance.

Cabbage root fly (Delia floralis)

This pest attacks swedes, turnips, kale and cabbages. There are two and sometimes three generations of flies per year emerging in April/May, June/July and mid-August.

Each generation lays eggs on or just below the soil surface which hatch within seven days and the maggots feed on root tissue. After about three weeks, when fully grown, the maggots move into the soil and form pupae from which the next generation of flies emerges. Pupae from the later generation overwinter in the soil and give rise to flies in the following year.

Conditions for egg laying are most favourable in spring and infestations from the first generation of flies are the most concentrated and severe. The maggots attack establishing crops causing the stunting or death of seedlings. The attack of later generations on more mature plants is less damaging.

Preventive applications of organophosphorus or carbamate insecticides can be given as soil-incorporated granules or as a band spray at drilling. These control maggots and reduce early root damage to seedlings, but do not protect throughout the season. Chemical treatment has little value in controlling an attack once the maggots are down at the plant roots.

Flea beetles

A variety of flea beetles attacks a range of crops including swedes, turnips, kale and cabbages and to a lesser extent mangels and beet.

Adult beetles overwinter under tree bark or in sheltered sites like hedge bottoms, emerge in spring and fly to invade host crop seedlings. Eggs are laid in the soil in May/June, larvae emerge and feed on leaves or roots and then pupate. Adults emerge two to three weeks later in July and August, feed on crop foliage and then overwinter. Damage is caused by the small adult beetles eating holes in the seed leaves and first true leaf. The worst attacks occur in April/May, especially in dry seasons when seedling growth is slow, and become progressively less damaging as plants grow bigger. Damage is minimised by cultural practices which encourage rapid seedling growth. Chemical treatment can be as seed treatment with gamma-HCH, usually applied as a routine to brassica seed (it should not be used on beet or mangels), soil incorporation or bow-wave application of carbofuran at drilling, or spraying the seedling crop with gamma-HCH if damage is seen.

Mangel fly (Pegomya betae)

The flies emerge in April and lay eggs from April to June on the lower surfaces of mangel, fodder beet and sugar beet leaves.

After four to five days these hatch into maggots which bore into the leaves causing blistering. After about 15 days maggots drop from the leaves, burrow into the soil and pupate. Adult flies emerge after 14–20 days. There are two or three generations per year, and pupae from the last generation overwinter.

Mining into the cotyledons or foliage leaves retards seedling growth and may cause plant death in bad growing conditions. Damage is worst in May and June but, provided the central shoots survive, crops can recover and yield loss is limited. Later attacks on outer leaves have little harmful effect.

The pest can be controlled by organophosphorus insecticides which should be applied only if infestations are severe at early stages of crop development. Maggots (not eggs) can be controlled and choice of insecticide should be related to aphid populations – some mangel fly insecticides can increase aphid numbers (by killing predators) and so increase the incidence of virus yellows.

Frit fly (Oscinella frit)

This is the only serious pest of maize in Britain. There are three generations each year, but only the first attacks maize. The second and third attack and then overwinter on grasses or cereals. The first generation flies emerge in April and May and lay eggs on young maize plants. After hatching the maggots bore into the shoots and cause severe damage if they invade the growing point.

Frit fly can be controlled by incorporating organophosphorus insecticide into the soil prior to drilling or by applying granules or spray at crop emergence to give more lasting control.

Beet cyst nematode (Heterodera schachtii)

This nematode overwinters in soil as egg-filled cysts formed from the bodies of females fertilised in the previous year.

Some eggs hatch each spring but others remain viable in the soil to hatch in future years. The nematode can be spread from field to field by anything which transfers cyst-infested soil, from tractor tyres to wind erosion. Nematodes invade and kill rootlets of host plants, and in severe attacks the whole tap root may be stunted or killed, with obvious reductions in yield.

Fodder beet and mangels are attacked as are brassica forage crops. The latter suffer little damage but can cause marked increases in nematode populations. Although damaging to forage crops the main concern over this pest is for sugar beet. There is no economical chemical control, the only practical control being by crop rotation. There are statutory regulations on the growing of host crops on soils infested above certain levels, and rotational controls on the growing of host crops are incorporated into British Sugar Corporation contracts.

DISEASES OF FORAGE CROPS

The following are the most damaging diseases of forage crops.

Club root (Plasmodiophora brassicae)

This serious soil-borne disease of turnips and swedes also attacks rape, cabbage and kale.

Soil-borne spores invade root hairs and stimulate root cells to swell. Infected plants wilt in dry weather and may die. If they survive they remain stunted and their roots show characteristic distortion and swellings which eventually rot and release new spores into the soil, which can remain viable for several years. Development of the disease is favoured by lime deficiency and poor drainage, and spread is via infected soil or plant tissue.

Control should be by cultural or rotational means. Adequate liming and good drainage limit its development and careful disposal of infected plant material can limit persistence and spread. Soil infection can be reduced by allowing at least five years between susceptible crops and resistant varieties of swedes and turnips should be grown in risky situations.

Virus yellows

This disease attacks mangels and fodder beet and is particularly serious in sugar beet.

Aphids (particularly Myzus persicae) introduce viruses to the crop and then spread them from plant to plant, giving typical yellow patches within the crop. The aphids overwinter on brassicas or host weeds.

Winged individuals are produced in spring and migrate to host crops. Cold weather between January and March reduces numbers but warm weather in April increases numbers and activity.

Aphids introduce viruses only if they have first fed on infected plants such as beet seed crops, sprouting mangels in clamps, groundkeeper beet or certain weeds. Infection causes leaf yellowing and secondary fungal infections may cause dead patches within leaves. Yield reductions may be as high as 30% especially following early infection.

Control is best achieved by minimising overwintering infected material from which aphids pick up their initial infections. Early sowing is beneficial because the older plants are before becoming infected the less they are affected, and closely spaced even stands are less likely to be colonised.

Aphids can be controlled with organophosphorus or carbamate insecticides. Granular applications to the seedbed give protection up to the 6 or 8 leaf stage but are recommended only in high risk areas and following mild winters. Later sprays should be applied when aphid numbers warrant it. Already there are strains of aphids resistant to these insecticides and their further development following incorrect or unnecessary use should be avoided.

Powdery mildew *(Erysiphe cruciferarum)*

This is a fungal disease of brassicas which can be particularly harmful on swedes and turnips, especially in warm dry areas and seasons. Infection is spread by spores into crops and between plants within crops.

It first appears as small round 'mealy' patches on leaves. Under warm conditions these spread to cover the whole leaf surface and severe infections can cause leaves to die and fall off, causing marked reductions in yield.

Early-sown crops are particularly susceptible and yield losses are heaviest if the disease develops by early August. Late-sown or quick-growing catch crops are rarely affected. Varieties differ in susceptibility but none is resistant. Benomyl, dinocap and tridemorph are effective fungicides and are provisionally accepted for use on swedes.

WEED CONTROL IN FORAGE CROPS

Weeds can be controlled in whole-season root crops and leafy brassicas by preparing seedbeds well in advance of drilling so that weeds can germinate and be chemically destroyed before the crops are sown. This technique is not always appropriate, and cannot be applied to early drilled crops or catch crops because the delay in crop establishment is unacceptable. In such circumstances control has to be by chemical treatment at different stages of crop development as shown in Table 28.8.

Table 28.8 Chemicals for weed control in forage crops

Chemical	Turnips/Swedes	Mangels/beet	Rape/kale
Pre-drilling			
chloridazon	No	✓	No
cycloate + lenacil	No	✓	No
glyphosate	✓	✓	✓
metamitron	No	✓	No
trifluralin	✓	No	✓
TCA	✓	✓	✓
Contact pre-emergence			
paraquat	✓	✓	✓
desmetryne	No	No	✓
Contact and residual pre-emergence			
nitrofen	✓	No	✓
Residual pre-emergence			
dinitramine	✓	No	✓
ethofumesate alone or in mixtures	No	✓	No
lenacil	No	✓	No
metamitron	No	✓	No
propachlor	✓	No	✓
propham/fenuron/chloropham	No	✓	No
chloridazon	No	✓	No
Post-emergence			
dalapon	No	✓	No
metamitron	No	✓	No
phenmedipham	No	✓	No

Source: ADAS Short Term Leaflet 163

There are no post-emergence herbicides approved for use in brassicas and so control by inter-row hoeing may be necessary. Weed control in maize is usually by atrazine, simazine or EPTC (with protectant) which are very effective soil-acting herbicides. Tri-allate can also be used pre- or post-drilling for the control of wild oats and blackgrass.

As new herbicides are approved they are listed in *Approved Products for Farmers and Growers* published annually by HMSO, and care should always be taken to apply them according to manufacturers' label recommendations.

29

GRASSLAND AND GRASS PRODUCTION SYSTEMS

Grassland occupies a greater area than all other crops grown in the British Isles and provides most of the feed for all our herbivorous livestock; some 60–70% of the needs of cattle, some 90% for sheep and a substantial contribution to the diet of the horse population. The crop is therefore a major resource within the agricultural industry.

ECOLOGICAL ASPECTS

Grassland contains a wide variety of plant species, mainly grasses, leguminous plants and other dicotyledonous forbs, all of which are able to regenerate themselves from growing points at or near ground-level. This characteristic distinguishes pasture plants from those which would replace them in the absence of successive defoliation, such as the shrubs, bramble (*Rubus fructicosus* L.), hawthorn (*Crataegus oxyacantha*), hazel (*Corylus avellana*) and blackthorn (*Prunus spinosa*).

These would in turn be replaced by the taller-growing woodland species such as oak (*Quercus* spp.), ash (*Fraxinus excelsior*) on the better-drained soils of the lowlands, with alder (*Alnus glutinosa*) in the wet situations; on the drier limestone soils, beech (*Fagus sylvatica*) would be prevalent. At the higher altitudes on the wetter peaty soils, birch (*Betula* spp.) would be the natural climax forest, and in the northern climates it would be one of pine (*Pinus sylvestris*).

In our climate with its cool wet winters and damp mild summers there is this natural ecological succession to the taller-growing species which, due to the shade provided by their foliage, will eliminate the shorter-growing plants. An oak tree can provide an area of foliage fifty times that of its canopy ground cover. Such shrubs and trees, as young saplings, quickly succumb to the removal of their terminal and unprotected growing points by grazing animals.

Grassland is therefore a dynamic association of plants occupying the first phase of vegetation in the colonisation of bare land to forest, and to this extent is semi-natural.

There are some 150 species of grasses in the British Isles and some 60 species of leguminous plants, together with several thousand species of flowering plants – of the dicotyledonous type, sedges (*Cyperaceae*) and rushes (*Juncaceae*) growing in various associations according to their soil habitat. Some meadows taken only for hay twice yearly and otherwise untreated are known to contain more than eighty species of plants. The habitat is influenced largely by its location in respect of climate, rainfall, temperatures and exposure to wind forces and by such edaphic factors as soil type, its depth, physical and mineral constitution, and the water relations within the soil. In turn, the plant associations themselves influence the type of habitat in which they grow.

As in the macro-vegetational changes from bare land to forest, so in the grassland ecosystem the taller species become prevalent where only lightly controlled. Certain fairly widely defined plant associations of semi-natural grassland can be distinguished as being typical of different situations.

Upland acid peat soils (300–600 m)

At the highest elevations on peaty soils in almost permanently stagnant water conditions there are the cotton grass (*Eriophorum* spp.) associations which are virtually useless agriculturally.

Overlying the mineral soils still with a peat formation of lesser depth of around 20 cm and at somewhat lower elevations there are the distinctive purple grass moors or 'flying bent' (*Molinia caerulea*) so called because of its deciduous habit.

The herbage is of limited value to grazing animals, since the Molinia is acceptable only in its early growth stage in late April or early May, and any improvement in changing the herbage must begin with control of the soil water.

Surrounding or adjacent to the Molinia moors, in areas of more freely moving water having a shallow peat surface, are large areas dominated by mat grass (*Nardus stricta*). In the more stagnant areas this grass species becomes tufted in habit with its semi-decayed closely knit leaf bases and surface rhizomes virtually excluding other species.

These semi-natural grassland communities are typical of large areas of hill land under an accumulation of highly acid peat soils with very poor drainage.

Where similar soil and climatic conditions prevail, but with good natural drainage which prevents a high water table developing in the upper 15–30 cm of soil, heather moors develop. Typically, such heather moors comprise three species: ling (*Calluna vulgaris*), heather (*Erica* spp.) and bilberry (*Vaccinium myrtillus*) and form such dense woody growth that few other species are able to survive, though this will vary with the control exercised by burning.

The stock-carrying capacity of these semi-natural grassland communities is very low. The herbage has a comparatively short season of growth and quickly becomes unpalatable to livestock. Moreover, the mineral status of the herbage is deficient in all the major elements.

Nonetheless, should economic conditions within the industry favour their further development, improvement is technically feasible and they represent considerable potential. Their complementary grazing value is realised on farms adjoining such areas and having the grazing rights. The additional summer grazing releases fields within the farm-fenced areas, for the conservation of winter fodder or fodder crop production.

Mineral acidic soils: fescue and Agrostis–fescue pastures

The fine-leaved fescues (principally *Festuca rubra* and *F. ovina*) are widely distributed from the highest areas of grass to that of the sea salt marches, where an ecotype of *F. rubra* is commonly found as a primary coloniser of the raised mud banks.

> They are often the dominant or co-dominant species on the naturally well-drained and shallow soils occupying the hillsides of enclosed land adjoining the acid peat areas, on lowland free-draining sand and gravel areas, and are quite prevalent on the chalk and limestone soils of the 'Downlands'.
>
> Under acidic soil conditions, both major species form a close matt of herbage at ground-level, which tends to accumulate, making the underlying herbage quite unpalatable to livestock.

In the deeper, more moisture-retentive soils, the Bent grasses (*Agrostis tenuis, A. canina, A. stolonifera*) become dominant or co-dominant with the fine-leaved fescues and together they form the more characteristic type of semi-natural grassland vegetation in Britain.

Being such ubiquitous species, not closely related to particular soil types, the Agrostis/fescue associations contain a plethora of other grasses and non-gramineous species.

On the drier and deeper soils in sheltered situations, bracken (*Pteridium aquilinum*) can rapidly colonise such grassland, and it remains a feature of many upland and hill areas. On the drier slopes of shallow soils, gorse (*Ulex europaeus, U. gallii*) can quickly invade and, where drainage is impeded, rushes (*Juncus* spp.) become dominant.

In unmanured hay meadows the taller species of grasses growing in association with *Agrostis* and fine-leaved fescues may consist of cocksfoot (*Dactylis glomerata*), tall oat grass (*Arrhenatherum ela-*

tius), Yorkshire fog (*Holcus lanatus*), bromes (*Bromus mollis*) and sweet vernal (*Anthoxanthum odoratum*).

In the grazed situations, the Agrostis/fescue components are more pronounced, but may contain scattered areas of crested dog's tail (*Cynosurus cristatus*) and, in damp situations, of meadow foxtail (*Alopecurus pratensis*) and of the meadow grasses (*Poa annua, P. trivialis*).

These areas of semi-natural grassland are marked by the general absence of leguminous plants, particularly of the white clovers, because of the acidic nature of the soils, with pH generally of the order of 4·5–5·5 and general deficiencies of phosphates, and potash. Their densely matted herbage is not conducive to the establishment and spread of wild white clover (*Trifolium repens*).

Basic soils: chalkland and limestone

Although similar grass associations are to be found on the limestone and chalkland areas, they are often distinguished in having an indigenous flora of leguminous plants, including wild red clovers (*Trifolium pratense*), wild white clover, black medick (*Medicago lupulina*), birdsfoot trefoil (*Lotus corniculatus*), yellow suckling clover (*Trifolium dubium*) and various vetch species (*Vicia* spp.).

Areas of chalk downland may typically, too, be dominated by Tor grass (*Brachypodium pinnatum*), which is almost entirely neglected by livestock.

Mineral soil: neutral grasslands

Agrostis with ryegrass pastures

On the more naturally fertile soils, with no serious impediment to drainage, more nearly neutral in their lime requirement (pH 5–6), ryegrass (*Lolium perenne*) becomes increasingly abundant, especially where fields are grazed regularly. The type of pasture formed is quite distinct from that of the Agrostis fescue type in that it is not matt forming. The ryegrasses produce an open-textured sward in which the stoloniferous legumes such as white clover can thrive. These species grown in association form the basis of our pasture plants in the most productive pastures. A few of these pastures were renowned for their ability to fatten mature bullocks and wethers, without the aid of fertilisers or feedingstuffs, notably in Leicestershire, and on the Romney Marsh in Kent.

The grasses: their structure

This is considered in some detail in relation to the principles of grazing management (Chapter 21) and in Chapter 12.

Characteristic of most grasses is the ability to develop tillers. Tillers which produce seed heads are known as 'fertile' tillers, those not producing seed heads are referred to as 'blind' tillers (Fig. 29.1). Since the nutritive value of the plant to the grazing animal lies largely in the leaf, those types of grasses which are more profusely tillering in their habit not only give more leaf production relative to stem and seed heads, but are better able to withstand continued defoliation.

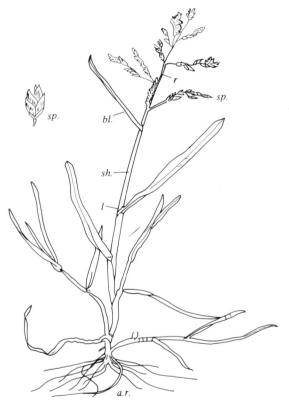

Fig. 29.1   A typical grass—annual meadow grass (*Poa annua*)
a.r., adventitious root system. *l*, ligule. *sh*, leaf sheath. *bl*, leaf blade. *r*, rachis.
sp, spikelet

Grasses such as smooth-stalked meadow fescue have other forms of vegetative reproduction: they do not tiller in the accepted sense, but produce above-ground creeping stems or stolons, which produce new leaves and adventitious roots from the nodes and tend to produce a matted turf. Others, of which couch grass (*Agropyron*

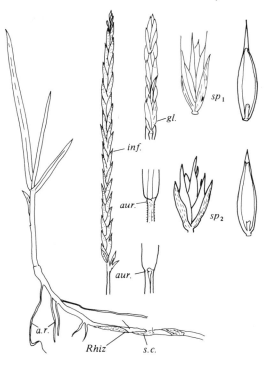

Fig. 29.2   Couch grass (*Agropyron repens*)
Rhiz, rhizome. *a.r.*, adventitious roots. *sc.*, scale leaves. *inf.*, ear. *gl.*, glumes.
aur., auricles. *sp₁*, *sp₂*, spikelet and seed of the awned and awnless forms

repens) is an example, produce underground stems or rhizomes
(Fig. 29.2), rendering them highly competitive and difficult to eradicate, especially in less competitive annual crops such as cereals.

HERBAGE SPECIES USED IN AGRICULTURE

Of the 150 or so species present in the British Isles, only a few are
of agricultural importance. Plant breeding has tended to concentrate on the more productive species, such as the ryegrasses, rather
than on other species of more limited value.

Ryegrasses

(*Lolium perenne, L. multiflorum, L. multiflorum* var. *wester-woldicum*)

Perennial ryegrass can readily be recognised by its overall dark green colour with shiny undersurfaces of the leaves giving the pasture a glistening sheen.

More particularly the plant is glabrous, the young shoots are flattened or oval in section and characteristically as they age have a pink or reddish base. The leaf blades (see Fig. 12.8) are slightly keeled with distinct ribs on the upper surfaces. The auricles are small and the ligule is short and blunt. The spikelets, each bearing 6–10 florets, are borne edgeways and recessed directly on the rachis, only the terminal one containing two glumes.

Perennial ryegrass is a pasture plant *par excellence*. Palatable to stock at all times it has a long season of growth but demands a high level of soil fertility in terms of good drainage and soils of pH 6 or above, well supported by available phosphates and potash. These requirements form the basis of maintaining the ryegrass/ white clover association in which the clover nitrogen contribution can be of the order of 150–175 kg/ha per annum. Such levels of nitrogen input together with those derived from stock residues, however, generally give less than half the production derived from perennial ryegrass swards which are supplied with some 375 kg/ha of fertiliser nitrogen per annum.

Yields of the order of 10,000–14,000 kg of dry matter per hectare may be obtained in this way, and as such are capable of providing the whole of the forage needs of the dairy cow from 0·4–0·5 hectares.

Though primarily best suited to grazing, perennial ryegrass swards are capable of giving very high yields of silage or hay.

Following the scientific breeding of varieties initiated in the first quarter of this century at the Welsh Plant Breeding Station, varieties or cultivars were primarily selected and bred according to intended purpose or use, that is, either grazing (e.g. S23), or grazing with hay production (e.g. S24).

The number of varieties now available has greatly increased following the enactment of Plant Breeders Rights. More recent introductions within each group include those of the tetraploids, obtained by chemical treatment during the breeding process, doubling the number of chromosomes.

Varietal introductions are to be found in the *Plant Varieties and Seeds Gazette* published by the MAFF, and the whole range of varieties is published in the *Classified List of Herbage Varieties* produced by the NIAB, who also produce an annual Recommended List in their Farmers Leaflets.

Italian ryegrass (Lolium multiflorum)

A biennial capable of giving the highest yields of herbage, of the order of 11,000–16,000 kg/ha. Being one of the earliest grasses to make active growth in spring, it is much used on dairy farms for this purpose, and because of its relatively late heading produces heavy yields of high-quality silage subsequently. Growth is extended well into the late summer.

Westerwolds ryegrass (L. multiflorum var. westerwoldicum)

An annual form of ryegrass. Quickly established from sowing, the Westerwolds varieties have limited use for the partial replacement of poorly established short-term ryegrass leys.

Cocksfoot *(Dactylis glomerata)*

A tufted perennial, it is a very widespread species, more tolerant than the ryegrasses of inherent low soil fertility. Where well fertilised it is capable of producing high yields (9,000–13,000 kg/ha) of fodder of lower quality. It is not suited to wet soils but is a highly persistent species, both as grazed or mown.

Timothy *(Phleum pratense)*

A glabrous tufted perennial generally of light green colour, with the blades rolled in the shoot.

There are several variants of the species, that used in agriculture being a hexaploid. All, however, favour damp, high organic-content soils, but as sown in mixtures they remain highly persistent on the heavier mineral soils. Timothy excels as a hay plant in conjunction with ryegrass and contributes substantially to total yield where only moderate levels of fertiliser are applied. As a pure sward, yields are of the order of 8,000–10,000 kg/DM/ha.

Meadow fescue *(Festuca pratensis)*

A loosely tufted perennial, glabrous, with leaf blades rounded in the shoot: the leaves, liked those of the ryegrass, are dark green with smooth shiny undersurfaces, and characteristically showing a restriction mark near the apex.

Occurring naturally in the more fertile, wetter areas of heavier clay soils, where it may replace perennial ryegrass, it provides both good-quality grazing and high yields for hay production. Compared with perennial ryegrass its yields are lower and it is now very little used in seeds mixtures.

Other grasses in agricultural use

Tall fescue (*F. arundinacea*), a more strongly tufted, bigger and more coarsely leaved plant than meadow fescue, is now mainly used on farms for grass-drying purposes.

Grasses such as rough-stalked meadow grass (*Poa trivialis*) and smooth-stalked meadow grass (*P. pratensis*), although highly palatable and valuable as constituents in old pastures, are not now used because of their lower overall production in comparison with the ryegrasses.

Clovers and other legumes

The most valuable and widely occurring species is white clover (*Trifolium repens*). It is used in all long-duration leys where the use of fertiliser nitrogen is restricted to less than 190–250 kg/N/ha per annum. Whereas, formerly, red clover (*Trifolium pratense*) was much used either alone or as the principal constituent of short-term crops for hay, its use nowadays is greatly diminished. Similarly, lucerne (*Medicago sativa*), with its more particular requirement of soils and climate, is being grown on a diminished scale. Sainfoin (*Onobrychis sativa*), though never grown extensively, is a crop now of very rare occurrence. Trefoil (*Medicago lupulina*), once used in a system of catch-cropping for sheep grazing on the alkaline soils of the limestones and chalk downland, is largely of historical interest only. Alsike (*Trifolium hybridum*) is a biennial now little used except as a pioneer legume on the wetter, more highly organic soils. The ability of the legumes to 'fix' atmospheric nitrogen in symbiosis with the rhizobacteria of their root nodules has long been recognised, as has their ability to raise the nitrogen status of soils and contribute substantially to animal requirements, especially for protein.

The higher content of calcium and magnesium of the legumes compared with the grasses assists in mitigating the incidence of animal disorders due to deficiencies of such minerals in the diet.

However, there are disease problems (such as bloat) associated with high contents of clover in the animal diet and their reliability as forage plants is affected by various pests.

The much higher use of nitrogen fertiliser with its more predictable and reliable response, both in arable cropping and on the more intensively farmed grassland, has largely been responsible for the decline in the use of the more specialised forage legume crops.

White clover derives its perennial habit from the production of stolons which readily root at the nodes on being trodden into the soil. In all closely grazed pastures where the pH of the soil exceeds 5 and there is a sufficiency of phosphate, white clover is indigenous.

There is a wide genetical range of variants within the species, from the large-leaved white clovers suitable for silage or hay to the small-leaved white clovers that are truly perennial in habit and withstand continuous and severe defoliation. Within this range are several varieties of medium leaf size of which S100, with succeeding introductions, and the widely used New Zealand Huia are notable examples. The ability of white clover to persist is much affected by varietal susceptibility to the soil-borne fungus known as clover rot (*Sclerotinia trifoliorum*). Widely endemic throughout the country, the effects of the disease are more pronounced in the drier areas, varieties of continental origin generally being the more resistant.

Red clover is characterised by its tall growing habit arising from a strong tap root, the whole plant being hairy to a greater or lesser extent. Introduced into Britain in the sixteenth century red clover was valued highly both as a forage plant for hay production, for sheep grazing, and as a means of improving soil fertility in alternate cropping systems.

There is a considerable range of types within the species, from those selected as ecotypes in different areas to the more recently bred tetraploid varieties. The latter are capable of producing high fodder yields comparable with those of grass receiving high nitrogen fertiliser treatment.

Within the range of genotypes, some are suitable only as annual or biennial crops, the broad red varieties, whilst others, possessing a higher number of stems and shorter and more numerous internodes with smaller leaflets, may survive productively for two or three years. These are the late- or extra-late-flowering varieties.

Yields from straight crops of red clover may attain 11,000 kg/ DM/ha, on twice or three times cutting, but this can be reduced seriously where clover rot or clover sickness caused by the stem eelworm (*Ditylenchus dipsaci*) are endemic.

Lucerne is a multi-stemmed glabrous perennial arising from a very strong tap root capable of growing to more than 1 m in length. Lucerne will grow satisfactorily only on high pH soils, 7 or above, throughout the upper 30 cm of the soil profile. The soil must be well drained at all times of the year and the crop succeeds best at altitudes less than 150 m. Its demand for both phosphates and potash is high since it is essentially a conservation fodder crop capable of yielding some 14,000 kg/DM/ha annually. The main disease is Verticillium wilt (*Verticillium albo-atrum*) which is soil-borne, contagious and reduces yields in the second and subsequent harvest years. Stem eelworms are often responsible for the death of areas of the crop. The most destructive disease problem is that caused by the bacterium (*Corynebacterium insidiosum*), which is devastating but of limited occurrence. Varieties showing a high degree of resistance to these diseases are available but have a lower yielding ability.

Lucerne is difficult to conserve as field-dried hay, due to high leaf loss, and for silage the use of additives is essential to obtain a satisfactory fermentation.

Alsike is a glabrous perennial, in growth-form rather like red clover; it is now little used because of its low persistency and yield.

Sainfoin is now almost defunct as a crop plant, because of its low yield (about 7,000 kg/D M/ha) and its high cost of establishment. It is a crop of low seed yield and high seeding requirement, and like lucerne is exacting in its soil type requirements. Sainfoin has the particular merit, however, of being one of the few legumes which do not cause bloat, due to its tannin content. There were two main varietal groups: common sainfoin, which persisted for some 5–7 years as a ley, and giant sainfoin which was a two-year crop.

Miscellaneous herbs

Formerly much valued for their ability to provide a choice of diet and a growth of green leaf during drought periods, together with their generally high mineral contents, such plants as narrow-leaved plantain (*Plantago lanceolata*), yarrow (*Achillea millefolium*), chicory (*Cichorium intybus*) and burnet (*Sanguisorba officinalis*) were included as herbal strips in grazing leys.

THE ESTABLISHMENT OF GRASSLAND

The object of grassland improvement is primarily one of increasing the stock-carrying capacity of the farm and secondly of improving the quality of the diet. This may mean the replacement of existing herbage which for a variety of reasons has been allowed to deteriorate both in yield and quality, or the sowing of high-yielding grass crops for grazing or as conserved fodder grown in an arable rotation. With developments in the use of herbicides, of fertilisers and machinery, there is now available a wide range of methods by which land may be reseeded in the many and varied circumstances ranging from hill land to the lowlands.

Reseeding from arable land

Where alternate systems of grass/arable farming are practised the undersowing of a spring-sown cereal crop is the customary method of ley establishment. Winter ploughing followed by the later spring cultivations provide an ideal frost tilth and produce a firm, fine seedbed for the smaller-seeded grasses and clovers. Competition from annual weeds in the early stages of establishment is reduced to a minimum by the cultivations.

Provided a stiff-strawed, early-ripening cereal cover crop is used at reduced seed rates and the grass mixtures are sown at the same time, good establishment can be secured, but the many problems associated with the undersowing of cereal crops which have caused poor ley establishment have led to an abandonment of the practice in the more specialised farming systems. Competition for moisture, sunlight and nutrients, in which the cereal crop has the advantage, can be critical in the early stages of establishment. Laid crops smother the undersown grasses, giving only patchy establishment. In wet seasons grass growth can be vigorous causing added difficulties at harvest and in early straw removal. Where there are special weed problems the use of herbicides is more limited, especially where clovers are being established.

However, where spring-sown cereals are cut for silage at the early-flowering stage in July, they provide a very good means of establishing undersown grass and clover leys. This technique is particularly appropriate where reliance is placed on securing good clover plant establishment, or that of lucerne, since all leguminous plants benefit from an early establishment during the growing season.

Reseeding without a cover crop in spring is the most reliable way of establishing the grass ley, and is greatly assisted by early grazing. Production is some 50–60% of that of a fully established crop. For these reasons reseeding the leys is now done commonly following the cereal harvest. The earlier the cereal crop is harvested the better are the chances of securing a full establishment of the grass ley. Whereas all sown grasses and clovers will normally succeed best when sown in July, perennial ryegrass may be sown well into August, and Italian ryegrass will succeed when sown as late as mid-September. The proportion of sown species surviving the winter months from later sowings is greatly diminished. Though drought may delay germination in July and August, both temperature and humidity in September are conducive to rapid germination and growth. The winter barley crop harvested in July affords better opportunities for an earlier ley establishment. Ploughing, wherever practicable, followed by the necessary cultivations, is the more effective means of providing a tilth free of shed cereals and other annual weeds. On the heavier, often drought-stricken soils it is more usual and practical to surface cultivate either the burnt stubble or one treated with grass-killing herbicides, such as paraquat or glyphosate, and sow in the more easily prepared tilth. Care must be taken wherever such herbicides are used that sufficient time elapses between their application and sowing to grass. Herbicidal contamination resulting from the organic residues of an old stubble has a lethal effect on new seedlings and such residues must be removed or effectively buried in the soil. These herbicides are particularly useful in controlling germinated shed cereals, black grass (*Alopecurus myosuroides*) and meadow grass. The required cultivations depend very much on soil type: in uncontaminated free-working loams, grass can be drilled using disc drills as a one-pass operation. Leys which are established well during the early autumn months give full production in the ensuing year, those based on Italian or the hybrid ryegrasses being particularly successful.

Normally, the residual fertiliser from that applied to the cereal crop is sufficient for July sowings, but where phosphate and potash are known to be low they should be applied together with some nitrogen in the seedbed.

Reseeding after grass using herbicides

Similar techniques of applying paraquat or glyphosate and either ploughing, rotovation, or direct drilling into the destroyed herbage may be used when short-term leys of Italian ryegrass are being renewed with the same type of ley. The direct drilling method gives a quick establishment regrowth and by retaining the sod surface prevents undue damage from animal treading.

Renewal of older ryegrass leys which may contain a high proportion of other species such as the meadow grasses, bent grass or fine-leaved fescues, present more formidable problems, particularly in the absence of cultivations. The more herbicidal-resistant species such as the fine-leaved fescues and, to a lesser extent, meadow foxtail generally require more than one spraying application for a complete kill, and much depends upon getting a uniform application at the first spraying. Repeated spraying may be necessary. Residues from both the old turf and from those surviving the herbicide can be highly inimical to establishment. Delayed or partial establishment renders the newly germinated seedlings very vulnerable to damage by insects, which are always present in older swards. Whilst insecticidal treatments are available they add greatly to costs.

Following the use of the grass-killing herbicides, the complete destruction of the old herbage by ploughing, discing or rotovation provides a more reliable means of securing good establishment.

Similar problems exist where techniques such as slot-seeding have been developed for the establishment of grasses and legumes injected by various means into existing swards as a method of improving their production whilst retaining a continuous output of grass. One of the major advantages of introducing an arable crop into the grass rotation lies in controlling grass pests, of which Frit fly (*Oscinella frit*), Crane fly larvae (*Tipula* spp.), wireworm (*Agriotes*) and snails are the most important.

Reseeding in upland areas

In upland and hill areas it is often necessary to build up a satisfactory soil mineralisation and fertility status well before the land is resown to high-fertility-demanding ryegrass swards which are intended to remain for the longest possible duration. Preventing the regeneration of such perennial weeds as bracken, from its deeply buried rhizomes, or of rushes, with their abundance of seed in the soil, often requires some years of competitive cropping despite good initial control from the use of herbicides. The use of pioneer crops such as turnips on peaty land, rape or a mixture of both on mineral soils, serves these purposes where other arable crops cannot be used. Three or more such crops may be required to attain the necessary soil fertility status before resowing to ryegrass leys. More than one application of lime may be necessary to obtain the requisite pH level of 6.

Conditions for establishment

Whichever method is used to prepare the land for ley establishment the principles of obtaining a friable yet firm, fine tilth in which the seed is buried not more than 1·5 cm must be the aim, followed by consolidation by rolling as the final operation. The use of compound fertilisers containing nitrogen, phosphate and potash, worked into the seedbed before sowing, is necessary to ensure a readily available supply of nutrients during the early establishment phase of the crop.

Where clovers are used, the nitrogen fertiliser should be restricted to less than 50 kg/ha since it reduces their establishment. All reseeds benefit from being grazed in their year of establishment.

TYPES OF LEY

The term 'ley' refers to the intended duration of the grass crop, whether for one year or a longer specified period: permanent pasture refers to land which remains solely as grassland, at least for a very long period. With careful management and by maintaining a high soil-fertility status, however, perennial ryegrass pastures may be retained as such for twenty years or more. These are referred to as long-duration leys since they are derived from sown seed.

Long-duration leys

These comprise the more persistent varieties of the species used and with few exceptions are based on perennial ryegrass with or without white clover. The trend has been towards simple mixtures and, taken to its extreme, only one variety of a species may be used. On dairy farms this might consist of one of the late-heading perennial ryegrass varieties serving all purposes of grazing, hay or silage. By choosing varieties of different heading dates for different fields a succession of grass crops can be taken for silage at the optimum stages of yield and quality. This practice is particularly relevant to the production of dried grass where a predictable quality is required. Lucerne grown for silage or dried grass is generally sown as a one-variety crop allowing the use of selective herbicides to prevent competitive weed invasion.

While single-variety grass leys are capable of giving high yields for grazing, hay or silage, each has some defects which can be complemented by others having different characteristics. Such factors as winter hardiness, disease resistance, seasonal growth and unpalatability, ability to withstand seasonal high grazing pressure and drought resistance are highly relevant in different climatic situations and particular farming systems. Appropriate combinations of complementary and compatible varieties as mixtures are therefore the more reliable as general-purpose, long-duration leys. On dairy farms the use of tetraploid varieties of perennial ryegrass greatly enhances the palatability and acceptance of the diploid ryegrasses at all times of year. This is especially relevant in cloverless leys. The addition of timothy to ryegrass leys performs a similar function and can make a substantial contribution to the hay crop.

The use of cocksfoot, formerly much despised in ryegrass mixtures because of its incompatibility and tendency to form clumps which were not easily controlled, merits a place on the dry, shallower upland soils under systems of mixed livestock rearing where only moderate quantities of fertiliser are used.

Mixtures without ryegrass are sometimes used on dairy farms where more reliance is placed on the clover contributions and less on applied nitrogen. These mixtures contain timothy, meadow fescue and white clover. All species in the mixture are relatively slow-establishing and should be undersown, preferably in an arable silage crop. This type of ley is uncompetitive against weeds and should follow a weed-controlled arable sequence of cropping. The special attribute of this type of ley is its high acceptability to dairy stock throughout the season. It is less suitable for sheep grazing since close defoliation allows rapid ingress of the meadow grasses.

Short-term leys

Capable of giving the highest production over periods of one to three years these are based on Italian ryegrass or the hybrid Italian perennial varieties. Being highly responsive to nitrogen applied in the early spring, they provide both an early grazing crop and high yields for silage. Their relatively late heading allows full leaf production with a high soluble carbohydrate content and digestibility up to the cutting stage in late May or early June. With added nitrogen there is quick recovery growth which may be grazed, or cut a second time for hay or silage. Growth continues well into the autumn, when a third cut may be taken, but better plant survival is obtained if the sward is grazed at this time of year. When combined with broad red clover, high forage yields can be obtained for hay or silage without added nitrogen.

THE MANURING, MANAGEMENT AND UTILISATION OF GRASSLAND

All methods of pasture renewal involve some loss of production during the process, and may be inappropriate due to problems of access, or inconvenience in the management of livestock.

Pastures of moderately productive capacity can be greatly improved by applying the appropriate fertilisers, provided other soil conditions are not limiting.

Sustained improvements can be made only where soil drainage is effective at all times of the year. The more obvious effects of poor drainage are seen by the growth of rushes, sedges and aquatic species. Effects are often less recognised where soil water impedance results in later growth and the diminution of the more productive elements of the sward. Heavily compacted soils resulting from animal treading restrict root extension and impede the uptake of fertiliser nutrients.

The gradual depletion of lime reserves in the naturally acidic soils by drainage loss is reflected in the diminution of the clover content and the ingress of the Bent grasses, Yorkshire fog and Fine-leaved fescue, with the eventual extinction of the ryegrasses. Efficiency in the uptake of other fertiliser ingredients depends upon the presence of an adequate lime content.

Phosphates are an essential requirement for all soils shown to have an analysis less than 2 on the ADAS scale, and are especially necessary for the establishment of grass swards, and the maintenance of clover content in existing swards.

Potash requirements vary as much according to the type of grassland management practised as to differences in soil type. The well-drained sands and gravels, limestone and chalk soils are naturally deficient, but the heavier clay soils or silty clay loams may well have enough for consistent high grass production under an all-grazing management. The potash in the herbage in these circumstances is largely recycled by the grazing animal. There is usually, however, some transference within fields from more distant areas to those where animal excreta and urine accumulate.

Where crops are cut and conserved, the whole of the potash content (some 1–2% of the dry matter) is removed and repeated conservation cuts quickly deplete the soil reserves.

Yields are depressed where serious potash deficiencies occur and become evident in a poorer leaf growth with discoloration in the grasses and white spotting and stunting in clover leaves. Only in the presence of an adequate potash supply can added nitrogen have its full effect. However, an over-supply of potash to pastures can result in hypomagnesemia in the animal. Such disorders can be fatal unless quickly diagnosed and treated. Due to the high intake by potash immediately following application, dressings are best applied to grazed pastures in mid-summer when the magnesium content of the herbage is at its highest.

Nitrogen is the fertiliser constituent having the most potent effect on grass growth, provided all other nutrients are adequately supplied.

Experimentally it has been shown that yields of the order of 14,000–15,000 kg of dry matter per hectare can be produced from straight ryegrass crops receiving some 450–500 kg/N/ha per annum with three to four times cutting. Together with other sources of nitrogen from animal residues, the soil and the atmosphere, the response to added fertiliser nitrogen is generally of the order of 21 kg/DM/ha per kg of nitrogen applied. Whereas there is a straight-line response in dry matter yields to 450 kg/N/ha in cut swards, under grazing management the indications are of a declining response above 200 kg/N/ha.

The best responses from applied nitrogen depend both on the timing and rates of application. Where solid nitrogen fertilisers are used the best response is derived from frequent applications supplying some 2·5 kg/N/ha per day of the growing season. Within the growing season, however, responses to added nitrogen vary considerably, the maximum response coinciding with the rapid extension growth of grass up to the preheading stage in May or early June. The difference in response during this spring growth period compared with that in the post-heading phase of grass growth is of the order of 3:1.

The ryegrasses make active growth when soil temperatures remain at some 6°C for five days but, without added nitrogen, soil temperatures of 9°C are required to give a similar effect. Growth in such circumstances may be delayed by some 14-21 days.

Nitrogen and clover

Applied nitrogen can affect the growth and contribution of clovers both in suppressing the development of the nodular bacteria and in promoting a vigorous grass growth overshading that of the clover plant. White clover will contribute substantially in areas of adequate rainfall where up to 185 kg/N/ha are used on grazed swards. The larger-leaved, longer-petioled varieties are the more competitive at this level of nitrogen application and are often used in long-duration leys for silage. Above this level of nitrogen use the white clover contribution is greatly diminished. In a red clover/ryegrass crop there are no beneficial effects from applications of nitrogen above 50 kg/ha because of the diminished clover contribution.

The changes brought about in the composition of the herbage from applied nitrogen may result in some lowered protein content where the clover contribution is affected, but that of the grasses is increased and maintained at a level often in excess of stock requirement. The carbohydrate content of the crops is slightly reduced where nitrogen is used in quantity, but the overall digestibility of the organic matter is not affected.

Pasture improvement

Enhanced soil fertility will in itself cause changes in botanical composition, favouring the more productive species, such as ryegrass if present, white clover, timothy, cocksfoot and the meadow grasses, which will gradually replace the inferior species (Bent grasses, fine-leaved fescues and broad-leaved weed constituents). The degree and rapidity of change which can be achieved depends upon the initial herbage composition. The presence of a deeply matted turf may take several years of such treatment before the semi-rotted organic matter is destroyed sufficiently for the better grasses and wild white clover to become established. Where some perennial ryegrass is present, however, the response to added fertilisers is both more immediate and productive. The higher production must be fully utilised by some higher level of stocking intensity, which in itself has two important influences on the sward, that of treading and the distribution of excreta. The effects of treading, with an adequate soil lime content, reduce and eventually prevent a mat formation, and the higher amount of animal faeces more evenly distributed adds to the soil fertility cycle.

Highly productive systems can be based on swards of only moderate quality, in terms of their botanical constitution, where high quantities of nitrogen are consistently used. The higher stocking rates accelerate the changes in favour of an increased perennial ryegrass/meadow grass content. Whilst such methods of obtaining high production and a gradual improvement in sward composition are feasible and practical, they are of greatest advantage in the higher rainfall areas and on the more intractable, less-well-drained, heavier clay soils, where effects of drought are rarely, if ever, experienced. They are much less applicable to swards, of like constitution, in low rainfall areas where limitations of soil moisture in the summer period restrict growth and nitrogen response.

The use of herbicides in pasture improvement

Normally, herbicides have only an ancillary role in weed control in grassland. The most effective means of controlling unwanted plants lies in the promotion of the crop itself, by attention to fertiliser needs and to soil drainage where necessary. The practices of alternate mowing and grazing will keep many weed plants in check, notable exceptions being infestations of docks (*Rumex obtusifolius*, *R. crispus*) and nettles (*Urtica diocica*). A range of herbicides is available for the selective control of docks. Both weeds may be effectively destroyed by spot treatment with glyphosate, or paraquat repeated when necessary.

Of more widespread occurrence are the various species of thistle, the more common being the creeping thistle (*Cirsium arvense*), in pastures which have been heavily stocked over winter, or heavily overgrazed in the spring period. Control can be effected by treatment with 2,4-D at the early-bud stage, but by this time much valuable grazing will have been sacrificed. More effective longer-term means of control lie in promoting a vigorous early spring growth of grass, either by resting the pasture at this time or by applying nitrogen. Similar herbicidal measures may be taken to control ragwort (*Senecio jacobaea*), a poisonous plant which inhabits land generally in a low state of fertility. Buttercups (*Ranunculus repens*, *R. acris* and, to a lesser extent, *R. bulbosus*) can also be controlled with MCPA or 2,4-D herbicides in their early-flowering period.

Whilst the control of some species of rush (*Juncus* spp.), particularly the soft rush (*J. effusus*), can also be obtained by the use of MCPA or 2,4-D applied at the pre-flowering stage, other species such as the hard rush (*J. glaucus*) are resistant and where these species occur in mixed populations control of the former can result in an increase in the latter.

Bracken (*Pteridium aquilinum*) has long posed special problems of control. Two herbicides in particular have proved effective, asulam or glyphosate applied during July at maximum growth stage. Some regeneration may occur in subsequent years and can best be prevented or reduced by establishing a vigorous densely grown sward in its place. The ensuing competition from the earlier-growing grass limits frond emergence, and those which succeed are often destroyed by the grazing animal.

Selective grass herbicides

Inevitably, with time, the meadow grasses, Bent grasses and other species appear in the sward, and in most unsown pastures they co-exist with perennial ryegrass. Investigations with selective grass herbicides have shown that some of the less desirable species can be controlled both during ley establishment and in existing swards.

The meadow grasses, particularly annual meadow-grass, black-grass (*Alopecurus myosuroides*) and meadow barley grass (*Hordeum secalinum*), can be controlled by ethofumesate in perennial ryegrass swards. The perennial ryegrass is unaffected, but the herbicide is harmful to clovers. Yorkshire fog is more heavily suppressed than perennial ryegrass by asulam, and, in the higher rainfall areas, dalapon used in July, at half the normal rates for total sward destruction, has given effective control of the Bent grasses. Although the growth of ryegrass is checked for some weeks after such treatments, provided it is uniformly distributed in the sward and in sufficient quantity, it quickly recovers following a nitrogen application. The clovers are unharmed. In lucerne crops the growth of annual meadow-grass is well controlled by very low doses of paraquat (0·35 l/ha) applied in early spring before the crop has commenced active growth. Carbetamex is selective in its control of grasses in forage legume crops.

Grazing systems

The principles of grazing management are dealt with in Chapter 21 and details of grazing systems are given in the chapters on milk, beef and sheep production systems (Chapters 35, 36 and 37). The main grazing systems are as follows:

Strip grazing with cattle

The area to be grazed is controlled by an electrified wire held by insulated, looped iron posts fixed manually in the ground. Allocations of grazing may be made twice daily or once every two days according to circumstances of herd size, grass growth, ground conditions and labour availability. When applied throughout the grazing season for dairy cows a back fence may be used to prevent undue treading and spoilage of the herbage previously grazed. The system is commonly used on dairy farms as a means of controlled grazing for the early spring growth of grass. Grass surplus to the grazing requirement in the peak growth period may be taken for silage or hay production.

Paddock grazing

The setting up of equal-sized paddocks, whether by temporary fencing or, usually, by more permanent fences, affords a means of rotational grazing, supplying a fresh growth of grass forward of the stock, and at the same time allowing an uninterrupted regrowth of the grazed area.

The number of paddocks may be as high as 22–30, each giving one day's grazing for the herd and a regrowth period of 3–4 weeks. Alternatively, half the number of paddocks may be used, each giving two days' grazing and thereby reducing the amount of fencing required.

Modified systems of paddock grazing have been developed for the more intensive stocking of younger beef animals and dairy heifers in the 6–12 months age group.

Initially, during the months of April/May, their grass requirement can be met from one-third of the total area, fenced as necessary into three paddocks, whilst the remaining area is taken for silage. The grazing area is then extended to include the aftermath in order to provide for the higher consumption of grass with increasing growth of the animals. Depending upon the amount of nitrogen applied, the initially grazed area may also be taken for a later cut of silage and afterwards used for grazing again.

Grazing systems for sheep may also involve strips or paddocks and, in addition, may incorporate a 'creep' to allow lambs access to areas not simultaneously available to the ewes.

Zero-grazing

Formerly known as 'soiling' the system refers to the cutting and carting of grass to animals continuously housed or confined to a steading area. Much used on the Continent and in hot climates it has been practised only on a very limited scale in Britain. It requires a much higher investment in machinery, which is in constant use, since only limited quantities of grass can be provided without deterioration in storage at any one time. Disposal of the effluent and manure pose special problems. The system provides for a high utilisation of the grass crop without the need for fencing or water supply points, but must be integrated with a conservation system such as silage making whereby the quality of the crop can be kept under control.

Set-Stocking

Field-by-field grazing to which this refers has long been the traditional practice in Britain. Stock are allocated to a particular field or number of fields where they remain for indefinite periods. It is the traditional practice on mixed livestock farms, concerned with stock rearing.

The intensity of stocking on any particular part of the farm is seasonally adjusted according to the growth of grass. During the winter months all fields may be grazed by sheep while the cattle are confined or housed. With the onset of spring grass growth, stocking levels are intensified during the lambing and calving period, together with the need to set aside fields for fodder conservation. Because of the high stocking pressures at this time of year, the closure of fields for conservation is often delayed, impairing their production of hay or silage. Access to hill grazing alleviates the situation on the more heavily stocked upland farms, hence the need for improvement to allow an earlier turnout. In some instances extra land is rented on a seasonal basis, but much can be achieved in the judicious use and timing of fertiliser applications to boost grass yields when most required.

THE CONSERVATION OF GRASS FODDER CROPS

Requirements

In the UK climate, conserving the grass crop is essential to cater for the winter needs of livestock, throughout some 5–6 months for cattle and as supplementary feed over a period of 3–4 months for sheep. The total requirement has to be assessed in the earliest part of the growing season and fields set aside for the purpose as an integral part of the grassland management system. Where hay is used solely as the long fibre requirement for dairy cattle the quantity needed is of the order of 1·5 tonnes per beast. When fed wholly on silage, the requirement is 6–7 tonnes. With well-manured high-yielding crops such yields can be obtained from 0·25 ha taken in one cut, but increasingly, with the more frequent use of nitrogen fertiliser, two or more cuts may be taken, at six-weekly intervals.

The requirement for young cattle, depending on the use made of other fodder and on their age, is of the order of half to two-thirds of these quantities. In sheep farming, more reliance is placed on hay production, being of a higher dry matter content, more palatable and more easily transported during the wetter, winter months. Grass silages are used, but these need to be of high dry matter content and well fermented. Sheep flocks wholly maintained on the open hill are rarely supplemented with hay (or silage).

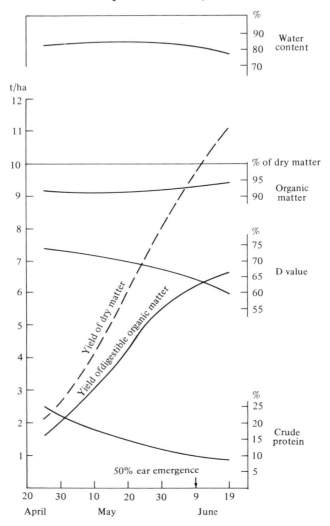

Fig. 29.3   Changes in yield and quality in relation to mean date of ear
emergence in Perennial Ryegrass, S.23
Source: Green, Corrall & Terry (1971)

Forage quality in conservation

The changes in nutritive value which occur as the grass crop
matures are seen in Fig. 29.3. Whilst the total yield of dry matter
increases up to the full-flowering stage, the digestibility (D value)

of the organic matter declines, at first steadily, then more rapidly following ear emergence. There is a marked decline in protein content and the moisture content is also reduced. This pattern of changes occurs in all grass species but they show considerable differences both in terms of their specific D values and in yields of digestible organic matter when cut at the same level of D value. Investigations have shown that the ryegrasses outyielded other cultivated grasses by some 15% in terms of their digestible organic matter at any stated D value.

Within the ryegrasses there are varietal differences in their digestible organic matter contents, largely determined by their growth patterns and time of ear emergence. The earlier-growing varieties of perennial ryegrass, Italian ryegrasses and hybrids tend to retain a higher level of digestibility during ear emergence.

Clover plants retain both a high level of digestibility and protein content to the full-flowering stage, but lucerne is exceptional as a forage legume, in that its digestibility is of a lower order throughout the season.

The D values of the grass crop are unaffected by high nitrogen fertiliser applications, except where this may influence the clover content within the sward. Due to the seasonal growth pattern, the time at which certain D levels are reached is delayed at higher altitudes and in more northerly latitudes by some 3–7 days.

Conserved herbages which have a D value above 60% contribute to animal production, those of only slightly lower value provide for bodily maintenance only. As there is some loss in digestibility of the herbage during conservation the aim must be to cut crops before the D value falls below 65%.

Frequency of cutting influences both yield and quality. Lower yields of higher nutritional quality are obtained from frequently cut herbage; the optimum output of digestible nutrients is obtained following re-growth periods of 6–7 weeks.

Methods of conservation

The principles used are either (a) reducing the levels of moisture content in the crop to those at which bacteria, fungi and other microbes cannot operate or (b) controlled fermentation of the fresh crop in an acid medium from which air is excluded.

Artificial drying

Commonly referred to as 'grass drying' this relies upon the burning of fossil fuels to remove the water content of the crop. One tonne of freshly cut grass with a moisture content of 75–85% requires some 360–550 litres of fuel oil, or the equivalent energy provided by other fuels, to reduce the moisture content to a level of 7%. Much development has taken place in designing drying machinery of high thermal efficiency, including the use of reheat systems in triple-pass drums, to give a high throughput of grass. Similar developments have taken place in the machinery used to cut, collect and transport the grass to the drier, and in that used in processing the product. Such is the investment in equipment and the rising costs of fuel that this method of conservation is now confined to relatively large enterprises.

The artificial drying of grass shows the least losses of nutrients of any conservation system and, being largely independent of weather conditions, is capable of reliably producing a product of high nutritional value. Nutrient losses are of the order of 5%. Dried grass has a high carotene content, varying from 200–400 mg/kg, and a crude protein content varying from 10 to 20%.

Hay making

Conserving the grass crop as hay in the traditional manner is wholly dependent on there being sufficiently long, fine and drying spells of weather. Three to four days of fine weather are normally required in the early part of the season and some 2–3 days when the humidity falls with rising air temperatures. Such seasonal spells of weather rarely occur before the end of May in southern areas, or 2–3 weeks later on high land and in the northern areas.

These periods coincide with the ear emergence of Italian and the late varieties of perennial ryegrass, when both yield and quality are at their optimum. At the critical stage a choice has to be made during inclement weather between delayed cutting, with the consequent decline in nutrient digestibility, or exposing the crop in the swath to high nutrient losses incurred by delayed harvesting. Total losses of the crop through prolonged exposure following cutting are not uncommon. The risk of such occurrences can be countered in a number of ways. The diploid ryegrasses are more easily field dried than the tetraploids, the inclusion of clovers gives some greater flexibility in arresting the decline of digestibility with delayed cutting, and risks are greatly reduced where mechanical treatments are used to increase the rate of drying and shorten the time of exposure to adverse weather.

The use of flail mowers which chop and bruise the crop increases the rate of drying appreciably but such are the losses of leaves that their use has largely been discarded. Machines which break the cuticle, particularly of the grass stems, either by pressure, as with the roller crushers, or by crimping at short intervals by crimpers, have a similar effect on increasing the drying rates without the attendant leaf losses. Crops so treated are no more vulnerable to nutrient losses in the field during recurrent spells of rain than those untreated. The main influence in speeding up drying rates lies in the frequent movement of the swath, exposing the maximum leaf surface to air movement by tedding the crop. This can be done within a few hours of cutting, and as moisture loss is rapid in the initial stages of drying, the more the crop can be exposed the more uniformly and quickly it dries in the raised and 'loosened' swath. There is little loss of leaf during this early treatment, while the leaves remain pliable.

Field-dried hay is normally baled at 25% moisture content, using the small rectangular baler, but the bales require several days of further drying in the field before they attain a level of 20% moisture, at which they can be stored under cover. Where large balers are used, the moisture content at baling should be 20% or less. Further drying occurs in storage until the hay reaches an ambient moisture content of 16%. Hay baled at moisture contents higher than 25% gives rise to mould development and heating losses which can be prevented only by means of forced ventilation. Crop nutrient losses in field-dried hay are rarely less than 30%.

Barn hay-drying

This is a system of grass conservation which is a compromise between the high-cost, low nutrient losses of grass drying and the converse in field hay-making. The crop is cut and treated in the normal way, as for field-dried hay, but is baled at a higher moisture content, generally around 35%. This shortens the period of lying in the swath to within two days. The size and weight of the bale are often reduced to facilitate handling.

Generally, the method of drying is by deep storage, which does not involve movement of the hay as in batch-drying or tunnel-drying systems which may be operated in the field. The barn consists of air-sealed walls to a height of 2 m with a raised floor consisting of strong steel mesh, 75 mm square, through which air is blown. The edges of the mesh are also sealed by sheeting for a distance of 1 m, so as to prevent air escape along the walls. Each square metre of floor area will take about 250 kg of dried hay, thus a barn measuring 33·5 m² will deal with a batch of 9 t. The bales are laid on edge and alternate courses have their long axes at right angles to each other. The depth of loading should not be more than 2 m at a time, and as each layer is dried another is placed on top.

Ventilating fans should be capable of forcing about 1·3–1·4 m³ of air per minute through every square metre of the floor. Initially, cold air is used since the higher moisture of the hay is more easily extracted, but as drying proceeds over several days, heated air is needed to extract the remainder to that of ambient air humidity.

The quantities of hay which can be dealt with in this way depend upon the floor area and the number of power-driven fans used, and, of course, on the initial moisture content of the baled hay. Whilst hay bales containing as much as 50% moisture can be treated in this way, more especially by the batch-drying system, over three times as much water has to be evaporated, doubling the cost of fuel required.

The advantages of barn hay-drying lie in its greater reliability compared with field drying, and despite its higher costs through extra installation and fuel, there are savings in field operations, and a reduction of field losses which can result in an extra 15% of collected material.

Ensilage

Conserving the crop in its green fresh state has many advantages, especially in its being much less dependent on weather conditions, and full use can be made of fertiliser applications, particularly of nitrogen, to secure the highest crop yields. The crop may be taken at younger stages of growth, when the crude protein content and the digestibility of nutrients are high. Earlier cutting promotes a quicker aftermath growth uninhibited by the cover of mown grass. The whole of the procedure from harvesting through to feeding can be mechanised.

Though much remains to be done in producing a conserved product from ensilage comparable in feeding value to that of the crop from which it is made (as in artificial drying), much has been achieved in terms of practical application over a wide range of farming situations. On a national scale, half the conserved fodder (on a D M basis) is now in the form of silage.

Essentially, the process entails storage of the green crop in silos from which air is excluded and temperatures are controlled, promoting the development of lactic acid produced as a result of the arrested breakdown of the soluble carbohydrate content of the crop. The exclusion of air is particularly important in preventing the development of moulds and the effect of other rotting microbes which, in the presence of air, result in total degradation of the ensiled material.

Many types of crop may be ensiled, including cereals, pulse crops, legume forage crops and maize, but chiefly they are of grass leys produced as an integral part of the farming system. Each type of crop requires special treatment in the ensiling process, largely dependent on its content of fermentable sugars, its dry matter when harvested, and the compaction it requires in the silo. Legumes are generally low in soluble sugar content and, where these constitute a high proportion of the herbage, additives, which increase the sugar content, thereby inducing the correct type of fermentation, or various acidifying agents are applied to inhibit microbial development. The additives used may be molasses to supply the extra sugars, or formic acid, sulphuric acid or formalin. Some additives contain deoxygenating agents.

The ryegrasses have the highest soluble carbohydrate content and are thus better able to produce silages of good fermentation. The Italian and hybrid varieties are especially suitable in this respect, with soluble carbohydrates often exceeding 30% in the dry matter.

Cutting and wilting

The optimum stage of growth for cutting is at pre-emergence of the flowering shoots, when the percentage dry matter has increased, together with that of its soluble carbohydrate content. Grass crops taken at an earlier stage of growth have very high moisture contents and, when cut and directly ensiled, often result in over-compaction which leads to a butyric type fermentation with its unpleasant odour. Additives are necessary when dry matter is lower than 20%. Low-dry-matter silages result in lower feed intake by animals and in unnecessary losses in the effluent exuding from the base of the silo. Silage effluents are particularly obnoxious as pollutants of water courses and must be contained in specially prepared earth sumps.

Crops cut at the pre-emergence heading stage will contain 80% or more of moisture and, for the reasons given, it is common practice to allow a wilting period in the swath over a period of 12–24 hours to reduce this to a level of 70–75%. Movement of the swath accelerates moisture loss. The swaths are then loaded into tipping waggons using the flail pick-up harvester, the latest of which have chopping mechanisms which deliver the material with a predetermined length of chop. Harvesting the grass at this lower moisture content means that a greater weight of dry matter is taken with each load, and chopping the grass disseminates the sugars throughout the mass of material to be ensiled.

Crops wilted to even higher dry matter contents may be ensiled in walled clamps, but these need to be quickly filled, well consolidated and rendered airtight, if secondary heating is to be prevented on being opened for use. Dry matter contents of 40–50% are required for air-sealed tower silos.

Type of silo

The 'clamp'

Originally clamps consisted of wide trenches sunk into the ground.

They had the merit of air sealing by the earth at one end and along the sides, a principle now followed by the use of polythene-lined and strongly supported walls inclining outwards. They may be roofed and should be covered by weighted polythene sheeting. A large concreted area is necessary, on which the grass is dumped for loading into the silo by means of fore- or rear-loading tractors. The method of filling the silo is by stacking as much grass as high as possible into the far end, bringing it forward as filling proceeds in a wedge pattern. This method provides for adequate consolidation and enables each day's filling to be completed with the minimum area to be covered with polythene sheeting. The capacity of silo required is $0.5 \ t/m^3$.

Stacks or field storage

Formerly used when transport was limiting, the grass was stored in rounded 'hillocks', or, where small quantities of silage were made, in 'draw-over' clamps without side walls: these methods of storage are now largely discarded due to the problems of filling, consolidation, adequate sealing and protection.

Tower silos

Air-sealed tower silos enable higher-dry-matter silages to be made with both fully mechanised loading and extraction. Though not of recent origin, this method of storing and delivery of the ensiled material to livestock involves a high investment and, as such, is more suited to larger cattle enterprises.

Baled silage

The silage is placed in polythene bags by balers specially adapted for the purpose: the bags are mechanically handled, stored in heaps and covered by extra polythene sheets. Care has to be taken to prevent vermin damage.

Feeding of silage

Self-feeding involves 24-hour access by the animals to the cross-face of clamped silage. Control of the forward movement of the animals and their access to the feeding face is controlled by an electrified wire or a bar, moved as necessary. A fully concreted base, sloping slightly away from the feeding face, is essential for the efficient removal of the slurry. The width of feeding face should be of the order of 150–200 mm for each mature animal (cattle) and the height of the clamp 1.5–1.8 m.

The self-feeding system is more relevant to the medium-sized herds of 40–80 cows in its economy of labour, but may be applied to herds both larger and smaller in numbers and to young cattle stock. Chopped material is more easily accessible and in well-managed systems the full ration can be taken without undue wastage.

In large herds, or where the silage ration is limited in the diet, the material is extracted mechanically, loaded into feeder waggons and distributed along the feeding face or mangers of the stockyard.

Feeding quality

The quality of silage is determined primarily by the crop and the stage at which it is cut. Dry matter should be of the order of 25–30% in clamps and 40–50% for tower silages.

The protein content of grass silage is of the order of 14%, but those containing high proportions of legumes may attain 20% or more. Losses in silage making normally vary from 15–25%.

Conserved feeds are now rated in terms of metabolisable energy, expressed as megajoules per kilogram of dry matter (M J/kg/D M).

Silages are generally	9·5–10
Field-dried hay	8·5–9
Barn-dried hay	9 –10
Dried grass	9·5–10·5

THE GRASSLAND CONTRIBUTION

Despite the much-reduced area of grassland during the past four decades, numbers of cattle and sheep have increased substantially. Fatstock are produced at younger ages and milk yields have more than doubled during the period. The area of forage root crops has been halved and a large proportion of the land listed as rough grazing has been transformed into productive pasture.

In all these developments improvements in the methods of grass production, utilisation and conservation have made a significant contribution.

Britain is for the most part ryegrass country, made possible by the introduction of high-yielding persistent varieties. As with all forms of crop production, both their yields and their persistency are influenced primarily by the provision of good soil conditions nurtured by the rational use of fertilisers.

The use of herbicides has greatly facilitated sward establishment and maintenance, and the control of major weed invasion. The range of herbicides is constantly being extended.

With more precise data on grass yields and response to nitrogen, stocking rates and requirements can be more accurately assessed.

Such information forms the basis of much of the financial planning of grassland enterprises. Each farm is its own particular business enterprise, however, and grassland can be farmed remuneratively at widely different levels of production.

30

FIELD VEGETABLES

More than twenty different vegetables are grown on a field scale in the UK and although they are very diverse in form, nevertheless several economic and marketing features are common to them all. Continuity of supply of closely graded produce to high-quality standards is now all important whether the vegetables are grown for processing, pre-packing for supermarket outlets or the traditional wholesale market. In these circumstances plant uniformity within a crop can be far more important in determining the *marketable* yields of the required size-grades of produce than the *total* yield. Similarly, reliability in producing a continuous supply of high-quality, perishable vegetables such as lettuce over a period of weeks or months is of the greatest importance in meeting an agreed marketing commitment.

Rapid changes have occurred in recent years in the systems of production brought about as a result of scientific and technological advances and the need to meet the increasingly high-quality and grading requirements arising from trends in marketing and consumer preferences. With those vegetables where the field operations can be mechanised there has been a marked trend towards large-scale production by farmers already experienced in arable crop production.

Several basic systems of production have evolved, each broadly suitable for crops of similar habit such as brassicas, legumes, root crops and salad crops. Each system is a sequence of individual operations many of which are common to all vegetable crops although others are largely specific to crop groups or individual crops.

OPERATIONS COMMON TO THE PRODUCTION OF MOST VEGETABLE CROPS

Plant establishment

A feature unique to vegetable production is the overriding importance of predictable and precise crop establishment in the field. It is crucial to obtain the correct plant population in order to control plant size (and hence that of the marketed product) and it is equally important to get rapid and uniform establishment in order to meet planned market requirements on time. There are two main methods used: by drilling *in situ* or by transplanting plants raised elsewhere.

Transplanting is generally preferred for widely spaced crops of high value, or when early crops are required, or when seed costs are high, or to enable a closer crop rotation to be practised, or to grow cold-susceptible crops which are not planted out before the risks of frost are past. This method of establishment gives precise control of plant spacing and is more reliable but it is also more labour-intensive and, therefore, a more expensive system of establishment than drilling the seed directly into the field. Plants for transplanting are raised in heated or cold glasshouses, cold frames, polythene tunnels, or in a seedbed outdoors. The grower can raise his own plants or buy them from specialist plant raisers. In recent years transplants have been raised in peat blocks singly, or several to a block, although for large-scale planting of such crops as brassicas 'bare-root' plants 'pulled' from the raising beds are still extensively used. The plants are transplanted by hand, or by hand-fed machines, but palletised plant-handling systems and automatic transplanting machines are now being used by the most progressive growers.

Successful establishment of drilled crops is critically dependent on seedbed conditions for many vegetable seeds are small in size and are vulnerable to adverse conditions. Everything is done to get rapid germination and emergence. Seeds are frequently treated with an insecticide and/or a fungicide before drilling to give protection against soil-inhabiting pests and fungal diseases, and more attention is now being given to the improvement of soil structure. The bed system of production is being used increasingly for vegetable production especially for drilled crops. There are many variants of this system in which the field is divided after ploughing into beds which are straddled by the tractor wheels. All subsequent operations are carried out using the same wheelings and, because there is less compaction within the bed, more even seed germination and more uniform plant growth are then obtained.

For most vegetable crops precision drills are used to achieve the desired plant population and spatial arrangement. Rows can be drilled as close as 3 cm with seeds spaced 2 cm apart within the row for crops such as carrots; the precision drills perform best with graded seed. To ensure that seeds are sown at a uniform depth the seedbed must be as level as possible. Some soils are prone to form a surface 'cap' or crust after rain which can prevent seedling emergence. Irrigating to wet, and thus soften the cap, may be necessary to ensure timely emergence. On a small scale with high-value crops soil conditioners can be sprayed on the surface soil to prevent capping or the seeds can be covered with additives such as peat or compost. Recently a new system has been developed for fluid drilling pre-germinated vegetable seeds in a protective carrier gel. Advantages reported for this system include more rapid, uniform and predictable seedling emergence, earlier maturity, and higher yields, the benefits varying with the crop. This is a high-technology system which, perhaps, presages the shape of things to come in the production of high-value vegetable crops.

Manuring

Most field-scale vegetable crops are grown on soils which do not receive bulky organic manures and nutrients are supplied from inorganic fertilisers. The different nutrient requirements for most vegetable crops have now been determined but the proportion of nitrogen, phosphate and potash and rates of application will vary with the nutrient status of each soil. To guide the grower the National Vegetable Research Station (NVRS) has developed a method for predicting the likely effects of freshly applied fertilisers on the yield of vegetable crops on most soils provided that the phosphate and potassium status of the soil is known. Copies of the predictor can be obtained from the NVRS.

Irrigation

Irrigation is of particular importance for vegetable crops in order to maintain growth and quality during dry periods. Where successional drillings of salad crops are needed to maintain continuity of supply, irrigation is essential to ensure rapid emergence of the crop. In addition, all transplanted crops are vulnerable to water shortage after planting and may need irrigation on many soils. Apart from improving quality and continuity of cropping, carefully timed irrigation increases the marketable yield of many crops. For vegetable crops which do not fully cover the ground it is inadvisable to use equipment which gives a large droplet size as damage may be caused to the structure of the surface soil.

Weed, pest and disease control

For the majority of vegetable crops weeds can be effectively controlled by pre-emergence and post-emergence herbicides so that close-row bed systems of production can be employed where appropriate. However, with some crops such as onions, where weed control by herbicides is not always completely reliable, rows must be wide enough apart to allow occasional cultivation when necessary.

The many economically important pests and diseases that attack vegetable crops must be controlled efficiently by approved methods if the marketable product is to meet the stringent statutory and other quality grades laid down for many vegetables. Whenever possible, insecticide/fungicide-treated seed should be used, or treatments applied at drilling, and varieties which are resistant to specific pests or diseases should be used. Growing crops should be examined frequently and when necessary treated for pest and disease control. The newer systemic insecticides and fungicides which are absorbed by the plant have many advantages over the older materials often providing a longer-lasting protection to most parts of the plant.

Harvesting and post-harvest operations

Each vegetable crop is handled somewhat differently at maturity depending on the way it is harvested and its market outlet. The simplest system is where the crop is harvested by hand and the produce put into bundles (salad onions, radishes), nets (Brussels sprouts) or crates (cabbage, cauliflower, lettuce) in the field where they are often loaded onto the lorry for the wholesale market. If, however, these crops are for supermarket outlets the produce will be brought to a packhouse and the 'field-heat' may be removed by recently developed refrigeration systems such as 'ice-bank' coolers or vacuum coolers.

Washing may be necessary (celery, lettuce and radishes), the produce then being graded closely for size, weight and quality and pre-packed into weighed units for direct sale to consumers. The pallets of pre-packs may be held briefly in cool store before being transported to the supermarket, in some instances maintaining a 'cold-chain' by refrigerated transport. Crops such as bulb onions are lifted in early autumn and dried, cured and stored to supply the packhouse at intervals throughout the winter. With other crops such as carrots and parsnips which keep satisfactorily in the ground, liftings are made at intervals, using a cold store to provide a buffer supply sufficient to run the packhouse for a few days if the weather prevents harvesting from the field. For crops such as tomatoes and white cabbage 'controlled-atmosphere' (CA) stores may be used to prolong the storage life.

With several of the crops the harvesting operation can be mechanised but subsequent operations in the packhouse are very labour intensive. Only with pea and bean crops for processing are all the operations from drilling to processing completely mechanised in a high capital/low labour production system which has to be planned meticulously from the programmed drilling schedule to the factory operation.

In contrast, in recent years there has been the 'pick-your-own' development where customers are provided with containers and are allowed to pick what they require. For this type of operation a range of vegetables such as peas, beans, salads, courgettes, sweet corn and tomatoes are grown to be available over as long a period as possible.

DETAILS OF CROPS

Brassica crops

Hybrid varieties of Brussels sprouts, cabbage and calabrese enable very uniform crops to be grown and machines are now in commercial use for once-over harvesting of sprouts. However with existing varieties of cauliflower a succession of selective cuts by hand is needed to clear this variably-maturing crop. Using a succession of sowings and varieties cabbage and cauliflower can be cut virtually throughout the year, Brussels sprouts from August to March and calabrese from July until October.

The majority of brassica crops are transplanted with spacings differing for each crop (Table 30.1). All benefit from high levels of nitrogenous fertilizers and, with the possible exception of Brussels sprouts, they all respond to frequent irrigation, especially after transplanting. The most important pests and diseases of these crops are listed in Table 30.1.

Peas and beans

Peas grown for processing are drilled in rows 9–20 cm apart from February to early June to a planned schedule based on 'heat-units' (day-degrees). Crops are harvested mechanically at precisely defined stages of pea maturity, as measured by equipment such as the tenderometer, from June to early August. French beans are drilled in rows 20–50 cm apart during May and June and are harvested during August and September.

Whereas peas and other legumes do not benefit from N fertilisers, French beans respond to dressings of nitrogenous fertiliser, as well as phosphate and potash, depending on the soil analysis. Yields are increased by irrigating during the flowering and pod-filling stages of both crops but this is not often practical on a large scale.

Broad beans (*Vicia faba*) and runner beans (*Phaseolus coccineus*) are grown on a small scale for the fresh market and for processing. Runner beans grown as 'pinched' or 'stick' crops are considered to have a better flavour than French beans but are more expensive to produce.

Further information on the pulse crops is given in Chapter 25.

Table 30.1 Statistics relating to the most important field vegetable crops grown in England and Wales

Crop	Area grown in June 1980 ('000 ha)	Season of supply (with storage)	Population density (plants/m²)	Important pests	Important diseases and disorders
Asparagus	0·40*	April–June	2–3	Asparagus beetle	Violet root rot
Beetroot	2·31	June–May	60–100	Mangold fly, Cutworms	Scab, Violet root rot
Beans, broad	3·64	June–Aug	15–20	Bean seed fly, Black bean aphid, Red spider mite, Pea and bean weevils	Anthracnose, Grey mould, Halo blight, Mosaic virus
dwarf	6·78	July–Sept	35–40		
runner	1·80	July–Oct	5–10		
Brussels sprouts	12·92	Aug–March	2–5	Cabbage root fly, Cabbage aphid, Caterpillars, Flea beetles, Turnip gall weevil	Clubroot, Cauliflower mosaic virus, Downy and Powdery mildews, Damping off and wirestem, Dark leaf spot
Cabbage	11·53	All year	2–30		
Calabrese		July–Oct	8–20		
Cauliflower	12·56	All year	2–4		
Broccoli, sprouting		Feb–May	2–4		
Kale, curly		Nov–April	3–5		
Carrots	12·45	All year	50–500	Carrot fly, Willow-carrot aphids, Cutworms	Carrot motley dwarf virus, Cavity spot, Violet root rot
Celery	0·98	June–Dec	4–10	Celery fly, Carrot fly, Cutworms, Aphids	Leaf spot, Blackheart
Leeks	1·75*	Sept–May	25–50	Onion fly, Onion thrips, Stem nematode	Rust, White rot, White tip
Lettuce (Butterhead, cos and crisp)	3·54	May–Oct	8–12	Lettuce root aphid, Foliage aphids	Beet Western Yellow virus, Downy mildew, Grey mould, Mosaic virus, Tipburn, Big vein
Marrow and courgettes	0·16*	June–Sept	2	Aphids	Grey mould, Powdery mildew, Cucumber mosaic virus
Onions, bulb	6·77	All year	80–300	Onion fly, Onion thrip, Stem nematode	Neckrot, White rot, Leaf spot
salad	1·03	All year	300	Carrot fly	
Parsnips	2·39	Aug–April	20–50	Carrot fly	Canker
Peas	55·09	June–Aug	90	Pea moth, Pea and bean weevil, Pea midge	Downy and Powdery mildew, Leaf and pod spot
Radishes	0·34*	April–Nov	100	Cabbage root fly, Flea beetles	Clubroot, Damping-off and wirestem, Downy and Powdery mildew
Sweet corn	0·60*	July–Oct	4–6	Frit fly	Damping-off, Maize smut
Tomato, outdoor	0·10*	July–Oct	3	Potato cyst nematode	Potato blight, Botrytis, Stem canker
Turnips	2·92	June–April	10–30	Flea beetles, Cabbage root fly, Turnip gall weevil	Clubroot, Dry rot, Powdery mildew, Mosaic virus
Swedes		Sept–May	10–30		

Onions and leeks

The area of bulb onions grown in the UK has risen from about 800 ha in 1960 to 6,770 ha in 1980 resulting from improved production and storage systems and better varieties. Home-grown onions can now supply the market throughout the year. The main crop is drilled from February to April, harvested from August, is dried, and can be stored until May. Winter-hardy varieties of mainly Japanese origin drilled the previous August will give mature bulbs from early June onwards. Few crops are transplanted although there is developing interest in the multi-plant block system of production.

Seed is treated with fungicide to control neck rot in store and the crop is often grown on a bed system. Weed control is by herbicides but later cultivations may also be necessary. Crops must be dried, cured and stored under closely controlled conditions to maintain bulb dormancy and quality.

Leeks can be harvested in virtually every month of the year but they are grown mainly for autumn and winter production. A combination of sowing times and varieties provides continuity and both drilling and transplanting systems are used. Plant spacing varies somewhat with season of production but for winter harvesting 40 × 15 cm is common.

Vegetable root crops

Although a significant proportion of carrots and beetroot are grown for processing most roots are grown for the wholesale market or pre-packs. Both for prepacking and processing maintaining continuity of supply of high-quality, closely size-graded roots is essential. This is done using successional drillings of different varieties and by varying plant populations according to the season of production. For example, for early carrots drilled in February and March on the lightest soils, seeds of the Amsterdam Forcing type treated with a suitable fungicide are sown to give a population of about 50 plants/m², while for mid-season and late production Chantenay types are grown at 150–500 plants/m² and for overwintered crops Autumn King types are usually preferred. With all root crops, the plant population and the timing of liftings enable root size to be controlled within limits. A wide range of row spacings is used to suit different harvesting machinery and many crops are grown on bed systems. Parsnips are drilled from February to early May at populations of from 20 to 50 plants/m² depending on the market outlet, the higher population producing smaller roots for pre-packing. Canker-resistant varieties such as Avonresister should be grown where this soil-borne disease is prevalent. Beetroot is drilled from March to July and early-sown varieties must be resistant to bolting.

Lettuce

Maintaining continuity of supply of high-quality heads is of paramount importance and as cut lettuce cannot be stored, this can only be achieved from successive drillings or transplantings. Both systems of production are used, transplants raised in peat blocks generally giving higher yields and being more predictable in their maturity but costing more to produce. Where market outlets are guaranteed by contracts this system is often preferred and the proportion of transplanted crops has increased in recent years. Production several weeks earlier can often be achieved on a small scale by the use of polythene film mulches, 'floating' mulches or tunnels.

Three main types of lettuce – butterhead, crisphead and cos – are grown at spacings of approximately 30 × 25 cm, often on a bed system, and harvesting is done by hand.

Lettuce crops can be attacked by four species of aphid. These are controlled by foliar sprays, preferably using systemic insecticides, not only to maintain the high quality of the heads but also to prevent the transmission of virus diseases such as lettuce mosaic.

Celery

Three main types of celery are grown in the UK: winter-hardy trench celery for winter production and frost-susceptible self-blanching and American green varieties for summer and autumn maturity. Almost all of the crop is transplanted with specialist propagators raising the plants usually in 2·7-cm-square peat blocks. The self-blanching varieties such as Lathom Blanching are closely planted 30 × 30 cm apart or 30 × 21 cm on organic soils, whereas the trench celery is planted 12·5 cm apart within rows 135–150 cm wide to allow for the successional earthing-up operations around the plants. Planting is done between the end of April and mid-July. Celery plants require a plentiful supply of nitrogenous fertiliser applied as base and top dressings and regular irrigation is essential to maintain growth and ensure high quality.

In addition to the crops described, specialist producers grow such crops as asparagus and radish, while recently, with the introduction of new varieties and production systems, the area of marrows and courgettes, sweetcorn, and outdoor bush tomatoes has increased substantially. The total area of field vegetables (excluding potatoes) grown in England and Wales in 1980 was 178,728 ha; the statistics for the most important vegetable crops are in Table 30.1.

31

FRUIT AND HOP PRODUCTION

Introduction

Fruits and hops grown in the UK (Table 31.1) cover nearly 71,000 ha many of which are south of a line between the Wash and Worcestershire within three main regions: Kent and Sussex

Table 31.1 **Fruits and hops in 1979**

Cropped area, gross production, yield and value of output marketed

Orchard Fruits	ha	'000 tonnes	Yield tonnes/ha	£'000
Dessert Apples	18,711	221·9	11·8	36,470
Cooking Apples	11,742	141·3	12·0	15,241
Cider Apples	4,900	25·5	5·2	1,507
Pears	4,599	72·7	15·8	6,924
Perry Pears	337	2·8	8·3	157
Cherries	1,439	6·2	4·3	3,620
Plums	4,763	48·6	10·2	5,174
Others and Mixed	1,100	3·2		460
Total Orchard Fruits	47,591			69,553
Soft Fruits				
Strawberries	7,582	54·6	7·2	40,676
Raspberries	4,154	20·5	4·9	16,212
Blackcurrants	4,091	21·7	5·3	11,223
Gooseberries	1,160	6·9	5·9	2,060
Red and White Currants	219	0·9	4·1	587
Logan and Blackberries	790	4·7	5·9	3,685
Total Soft Fruits	17,996			74,443
TOTAL FRUITS	64,848*			143,996

*Excludes area of soft fruits grown under orchard trees

Grapes	300	Not published		
Hops	5,708	10·0	1·8	20,700

Adapted from MAFF

(21,451 ha), East Anglia (10,037 ha) and Worcestershire and Herefordshire (6,882 ha), and there are three main climatic factors which influence this distribution. First, in these regions annual rainfall is less than 76 cm, giving relatively good light conditions and moderately dry conditions, the latter being unfavourable to diseases, and facilitating disease and pest control by spraying. Secondly, mean daily sunshine is between 4·5–5·0 hours, which is favourable for vegetative growth and fruit ripening. Finally, the incidence of air frosts after 1 May is low, so there is less chance of loss of crop due to frost damage to flowers and young fruits.

Within these regions fruits and hops are grown on the best, deep, well-drained soils which provide maximum water-holding capacity for dry periods, yet good drainage for wet ones since waterlogging can be a problem for all these crops. Calcareous and acid soils are avoided so a soil pH between 6·0 and 6·5 can be readily maintained. Choice of site is important for fruits and hops and, in particular, frost pockets and exposed sites are avoided, but where these crops have been poorly sited some frost protection may be achieved by water sprinkling, and wind protection may be provided by the use of living or artificial shelter.

Two other factors influencing the siting of fruits and hops are that they generally require ample water supplies for irrigation in dry years, for frost protection during blossoming and for disease and pest spray programmes throughout the growing season, and they also make heavy demands on labour, especially for casual workers for picking.

Vegetative propagation

Fruits and hops do not breed true from seeds, so they have to be propagated vegetatively.

Because this can lead to the spread of viruses, some of which are lethal, and many of which lead to yield reduction, particular techniques are used at the research stations to produce plants free of known viruses, which then can be multiplied by specialist nurserymen for sale as certified plants to fruit producers. For orchard fruits the E M L A scheme has been introduced by East Malling and Long Ashton Research Stations for the propagation and distribution of virus-tested trees, and for soft fruits individual certification schemes are available.

Fruit trees and grape vines contain two components: rootstock and fruiting variety. Rootstocks are propagated by layering shoots or taking hardwood cuttings and fruiting varieties are grafted on to the rootstocks, by 'T' buds, chip buds or dormant scion grafts. Soft fruits and hops grow on their own roots and are propagated in various ways depending on the particular crop.

Planting distances

With tree crops it is often necessary to plant close and then thin as the ground is covered and competition begins to occur. In general, close planting leads to small plants with low yields per plant, but high yields per unit area of land, although planting too close can lead to loss of yield as well as quality, due to competition for light, water and nutrients. The distance between rows is influenced by the width of the tractor and other machinery and the distance between plants in the row is influenced by the need for efficient management, especially with regard to weed, disease and pest control.

Pruning, flowering and pollination

The perennial nature of fruit plants makes it very necessary to prune and train most of them, otherwise they grow too large and unmanageable, thus making disease and pest control, and the production of quality fruit, difficult and costly. High fruit yields are dependent on abundant flowers which are formed during the late summer and autumn of the year previous to their opening, so an adverse effect of one year may influence cropping the next year. Seeds are generally necessary for the development of fruits but hops develop without seeds. For seeds to be formed in fruits, pollination is essential and there is a difference between fruits in the ease with which pollination will occur. The flowers of orchard fruits are self-sterile so that cross pollination is necessary and more than one variety must be planted in an orchard. In order to be certain of cross pollination the pollinator variety must be planted at regular intervals amongst the main variety and there must be a good insect population, and if there is not, hives of bees have to be introduced during the blossom period. Flowers of soft fruits and grapes are self-fertile so only one variety has to be planted to achieve satisfactory pollination.

Organic manures and fertilisers

Fruits and hops benefit greatly from the application of organic manure and where farmyard manure is available it is applied before planting at 150 t/ha.

After planting organic mulches are also beneficial especially early in the life of plants as they help to conserve moisture. In addition, there is considerable research interest in the use of black polythene mulches for young plants. The requirement for fertilisers varies with type and age of plant, climate and soil conditions, but general recommendations for fruits and hops, based on soil analysis, except in the case of nitrogen for fruit crops, are shown in Appendix Table 31.1 (pp. 801-2).

Irrigation
With fruits and hops, soil moisture deficiency during one growing season will affect crops in that season and in future seasons, but the deeper the soil and the more extensive the root system the better is the plant able to withstand a soil moisture deficit. Therefore, older plants are less affected by dry periods and trees less affected than, for example, strawberries. Irrigation during the growing season for all fruits and hops has been found to increase yields. Appendix Table 31.2 (p. 803) provides a guide to times of application and to recommended rates.

Crop protection
Crop protection is one of the most costly aspects of fruit and hop production. The crop plants occupy the land for from three to thirty years, depending on the crop, so that once perennial weeds, particularly couch grass (*Agropyron repens*), creeping thistle (*Cirsium arvense*) and bindweeds become a problem they are difficult to remove from the growing crop by chemical means and it is essential to remove most of them before planting, by means of a combination of fallow, application of a translocated herbicide and cultivation. Frequent movement of heavy machinery along alleyways between trees may lead to compaction of soil in orchards with the result that the technique of grassing down in orchards is practised.

> This leads to a particularly complex concept of a weed because if the grass sward gets out of control it competes with fruit trees, especially for water and nutrients. Therefore, in considering weed control in orchards, it is important to include control of the sward by mowing which has to be carried out every week in the growing season to restrict grass height to 10 cm. In addition, the plant components of the sward must be controlled, especially broad-leaved plants, such as docks (*Rumex* spp.), with selective, translocated herbicides.

Finally, all competing plants in tree rows have to be controlled by applications of herbicides; residual, translocated or contact as appropriate. Soft fruits, grapes and hops are generally grown free of all other plant competition and weed control is achieved mainly by herbicides, the deep-rooting nature of the established fruit plants making them particularly safe from phytotoxic effects of soil-applied residual herbicides against annual weeds and some perennial weeds, but spot applications of contact and translocated herbicides are also made.

The long-term nature of all fruits and hops makes the occurrence of soil-borne diseases and pests particularly serious because crop rotation cannot generally be used to control them. In soils in which fruit trees have been grown for a long period there is a particular problem that after old trees are grubbed newly planted trees grow poorly, a condition known as replant disease, and normal growth can be achieved only if the soil is treated with an appropriate fumigant. The perennial structure of fruits and hops enables diseases and pests to overwinter on the plants, thus providing a nucleus for multiplication of diseases and pests once conditions are suitable in the spring. The marketable portions of all these crops, except hops, are fleshy with relatively high sugar contents which makes them very susceptible to attacks by a wide range of diseases and pests. This not only reduces the weight of crop, but also may lower the quality of the crop, whilst the consumer demands a very high standard of quality of fruit. In order to maintain this high quality, the E E C has introduced Standards for strawberries and for orchard fruits when both are destined for the wholesale market. There are several reasons, therefore, why routine spray applications are applied to fruits and hops, but it is essential not to spray insecticides during flowering, otherwise pollinating insects are killed.

Picking and marketing

The picking of fruits and hops is very labour-demanding so there is a great need for the development of mechanical fruit pickers, yet the delicate nature of the harvested product makes the design and operation of fruit-picking machines very difficult. In spite of this some progress has been made, especially for picking fruits, except grapes, destined for processing, and for picking hops. In recent years one other approach to the problem of obtaining hand pickers at reasonable cost has been for growers to develop the 'Pick Your Own' trade (P Y O) where the consumers travel to the fruit field and pick their fruit, which they buy at a price below that obtainable at retail outlets.

Economics

Fruits and hops are high-cost crops; they are both capital and labour intensive. Orchard fruits and hops, in particular, involve high capital investment with a slow return on capital, as four to seven years must elapse before a crop of commercial value can be picked. Soft fruits, especially strawberries, are cheaper to establish and quicker to return a profit. Two to three times as much labour/ha is required for fruits and hops as for farm crops, although fruits and hops have higher outputs/man. In addition, these crops have high outputs/ha, which may be three to six times those of farm crops.

POME FRUITS

Two pome fruits are grown commercially in the UK: the apple, genus *Malus*, and the pear, genus *Pyrus*, of which the former is more important, being valued at £53,218,000 compared with £7,081,000 for pears (in 1979). They both contain varieties of three kinds which are used for dessert, cooking or fermenting, in the last case to make cider or perry, respectively. Other points of similarity are that the fruits of both are relatively hard and they can be kept for six to nine months if stored under appropriate conditions of low temperature and controlled atmosphere.

Apples

The apple is not only the most important UK fruit crop but it is second to the grape in importance amongst world temperate fruits. The apple is the most researched fruit crop in the UK with the result that techniques have been developed to enable the grower to exercise considerable control over tree shape and form by means of choice of rootstock, selection of pruning and training technique and the use of chemical growth regulators. Apples are grown in most counties in the UK but 84% of those for dessert and cooking outlets are grown in Kent, Sussex and East Anglia, where the climate is relatively dry and sunny and frost-free at blossom time, while 70% of cider apples are grown in Herefordshire and Somerset where the high rainfall favours cider production. Although apple trees will tolerate a wide range of soil types, the pressure of overseas competition is resulting in only the best, deep, well-drained loams being newly planted with dessert and cooking varieties. Apple trees are amongst the most susceptible of fruit crops to waterlogging.

Varieties and rootstocks Although more than 2,000 apple varieties are recognised, only a few are grown commercially in the UK, of which Cox's Orange Pippin (10,600 ha) and Bramley's Seedling (7,000 ha) are the most important dessert and cooking variety, respectively. Worcester Pearmain, ripening in September, before Cox's Orange Pippin, and Laxton's Superb ripening after Cox, have both been planted widely in the past. Several new varieties have been introduced recently and show commercial promise, including Discovery, Kent, Spartan and Crispin. Golden Delicious is also grown but does not yield as well in the UK as it does in warmer climates. There is a wide range of rootstocks available for apples including M27, very dwarfing; M9, dwarfing; MM106, vigorous; and M25, very vigorous. Layered shoots or hardwood cuttings are used to propagate rootstocks, to which the appropriate cropping variety is grafted.

Planting and training

Planting systems are termed extensive when trees are planted far apart and allowed to grow large, or intensive when they are planted close together and pruned to a small size. In recent years the tendency has been to plant intensive systems which can be more efficiently managed with the production of higher quality fruit, well exposed to light.

There is a wide range of possible planting systems, depending on vigour of variety, vigour of rootstock, planting distance and training system, but several trials have been carried out with the result that the two most popular systems are the bush and the spindle bush. Bush trees of Cox on MM106 are planted 4·5 m apart, in rows 5·5 m apart, giving 395 trees/ha. The trees are tied to stakes 70 cm high, so only the trunks are supported, and trained with open centres. Spindle bush trees of Cox on MM106 are planted 2·4 m apart, in rows 4·3 m apart, giving 969 trees/ha. These trees are tied to stakes 1·8 m tall and each is trained with a central vertical stem, which is tied to the stake as it grows, from which two or three tiers of horizontal branches are allowed to develop which are then trained along the row. In both systems laterals arise on the branches and are pruned by the renewal system which allows fruiting on long laterals which are cut out the following winter.

Flowering and fruit development

Flowers are initiated the year before opening, at various positions depending on variety; at the tip of current year's growth, on axillary positions on current year's growth or on one-year-old growth. As flowers open and set fruits in the following April and May they are very susceptible to frost, against which continuous sprinkler irrigation may be used. If fruit set is too great fruitlets should be thinned, either by hand within three weeks of full bloom, or by chemical thinning with carbaryl when the apples are 12 mm in diameter, but with both techniques very careful judgement is necessary to make sure that over-thinning does not occur. As the apple fruits approach maturity there is a tendency for pre-harvest drop to occur and to prevent this the chemical a-naphthalene acetic acid or a mixture of ethylene-releasing ethephon and 2-(2,4,5-tri-chlorophenoxy)-propionic acid may be applied.

Disease and pest control

Four important fungal diseases attack aerial parts of apple trees: canker (*Nectria galligena*) on stems, scab (*Venturia inaequalis*) and powdery mildew (*Podosphaera leucotricha*) on shoots and fruits and bitter rot (*Gloeosporium* spp.) on fruits.

Canker attack can lead to death of young shoots, older branches or even whole trees, and as it infects only through wounds it is controlled by spraying to protect leaf scars and by painting large pruning cuts with protective fungicide paint. Scab and powdery mildew are controlled by spraying with fungicide every 10–14 days throughout the growing season resulting in up to 12 sprays in a season. Bitter rot attacks fruits in stores and can be controlled by three or four sprays of fungicide in August and September or by dipping fruits in fungicide before placing them in store. Aphids, winter moths, sawfly (*Hoplocampa testudinea*), fruit tree red spider (*Panonychus ulmi*), codling moth (*Cydia pomonella*) and tortrix moths are all important pests which have to be controlled by spraying with insecticide on four occasions: pre-blossom, immediately post-blossom, mid-June and early July. Thus a complete spray programme for the control of diseases and pests may involve up to 16 sprays throughout the growing season costing 34% of the total growing costs. In the past these sprays have been applied as an insurance against possible disease or pest attack, but today there is a trend towards an assessment of threshold level of diseases and pests before spraying is carried out; this is known as supervised control which it is hoped will reduce the cost of the spray programme and cut down on the contamination of the environment by fungicides and insecticides.

Picking, storage and marketing

The apple fruit for dessert or cooking, because of its size, is the easiest fruit to pick by hand so there has been little incentive to develop machine picking, although the National College of Agricultural Engineering does have a prototype machine, the potential of which is being investigated experimentally. Cider apples are already picked mechanically, since some fruit damage can be tolerated.

In order to extend the marketing life of some varieties they are stored for 6–8 months in purpose-built, insulated and gas-tight chambers at a temperature between 0 and 3·5 °C and with a CO_2 level raised to between 1 and 10% and an O_2 level lowered to between 3 and 11%. The bulk of the apple crop is sold through the wholesale market and the rest of the crop is sold direct to the retail trade, at the farm gate or through the P Y O outlet.

Pears

Pears are often grown in conjunction with apples since they have similar requirements for warm, dry summers but pears are more tolerant of waterlogging.

Varieties and rootstocks   The most popular variety is Conference which represents 60% of the total pear area; it is a regular cropper and can even set fruits parthenocarpically following frost damage. Doyenne du Comice is the most important pollinator, with the best flavour of

English-grown pears, but irregular in cropping. The choice of root-stocks for pears is much smaller than for apples giving a smaller range of tree size and the choice lies between pear rootstocks which give vigorous trees, suitable only for perry pears, and quince, of which the selection Malling Quince A is most used, which gives relatively small trees.

Planting and training Pears grow with an upright habit and the central leader form of tree is generally grown either as a bush or a spindle bush. Trees to be trained as bushes are planted 3·0 m apart, in rows 4·5 m apart, giving 500 trees/ha and those for spindle bushes are planted 2·5 m apart, in rows 4·0 m apart, giving 1,538 trees/ha. A renewal system of pruning is generally practised but as pear trees age they have a tendency to produce complex spur systems, rather than extension growth, in which case spur thinning is necessary.

Disease and pest control Pears suffer less than apples from diseases and pests, so only a simple spray programme is necessary, involving not more than 10 rounds. Scab (*Venturia pirina*) is the most important fungal disease which is controlled by spraying with a protective fungicide at 10-day intervals from 10 days after bud burst until late June. Fireblight (*Erwinia amylovora*), a very serious bacterial disease in warmer climates, occurs sporadically in the UK depending on weather and variety, and if it occurs it must be reported to the Plant Health Board of the Ministry of Agriculture and stringent measures have to be taken to prevent it from spreading. Aphids are the most important pests and pear sucker (*Psylla pyricola*) occurs in some years in some areas, but a petal-fall spray of insecticide will control both, with a second application three weeks later if pear sucker is a problem.

Picking, storage and marketing Pears are picked by hand when they are ripe, but still hard, and assessment of correct time of picking is based traditionally on judgement, but today a starch-iodine test has been developed which improves the chance of picking at the right stage. They may be stored for up to six months at 0·5°C in 5% CO_2 and 6% O_2. To be sure of best conditions after storage pears are placed in conditioning rooms at 18 to 20°C with high humidity for two to seven days depending on variety. Pears are marketed similarly to apples.

STONE FRUITS

Two stone fruits are grown relatively successfully in England: the plum, *Prunus domestica* and the cherry, mainly the sweet cherry *Prunus avium*. For both, over the past 30 years, there has been a decline in area with a halving of output, due to three main causes. The first is that both crop irregularly since they are amongst the

first fruit crops to flower (mid-April), when conditions for pollination may be poor. Secondly, competition from overseas has
increased in the form of fresh and canned fruits, such as peaches,
and thirdly there has been a change in people's eating habits with
less demand for plum jam.

Plums

Plums, many of which are processed, are grown mainly in Worcestershire (35%), Cambridge and Isle of Ely (22%) and Kent (22%),
and jamming and canning factories have been established in these
regions. Although they succeed best on deep, heavy loam soils,
plums will tolerate less well-drained soils than other fruit crops.

Varieties and rootstocks   The two most popular varieties are Victoria
and Pershore Yellow Egg, representing 34% and 15% respectively of
the total area of plums. There is not a very wide range of rootstocks
available and not a true dwarfing stock, but the most-used rootstock
today is St Julien A, which gives a moderately sized tree replacing the
more vigorous Myrobalan B and Brompton, and the most promising
dwarfing stock for the future is Pixy.

Planting and training   Unlike other fruit trees most plums are self-
fertile so it is not essential to have more than one variety in an orchard.
Trees are generally planted 3·7-5·0 m apart, in rows 4·8-6·0 m apart,
giving up to 563 trees/ha. In the past, plums have generally been trained
as half-standard or as bush trees but for the future there is interest in
training them as spindle bushes or dwarf pyramids.

Disease and pest control   The two most important diseases, both of
which can be lethal, are silver leaf (*Chondrostereum purpureum*) and
bacterial canker (*Pseudomonas mors prunorum*). As they infect through
wounds in the stem it is important that most of the pruning is carried
out in the summer when wound healing is rapid and also that large cuts
are protected with a fungicide paint. Aphids are the most important
insect pests and are controlled by routine spraying at white bud, and a
suitable insecticide at this time also controls caterpillars. If red spider
and sawfly (*Hoplocampa flava*) occur they are sprayed post-blossom,
giving four spray rounds altogether. Birds, especially bullfinches, cause
damage in February and March when they eat developing flower buds
and if they are abundant they have to be controlled by trapping or
shooting.

Picking and marketing   Plums for the dessert trade are picked by hand
as they ripen, but those for processing may be mechanically shaken
from the trees and caught on collecting frames. Plums for canning

are picked under-ripe while those for jamming are picked in a riper condition.

Cherries

Cherries are mostly grown for the fresh market. They are more demanding than other UK fruit crops for a dry climate, being particularly susceptible to high rainfall in April and June, and for a deep well-drained soil; Kent, where 80% of the crop is grown, provides such conditions. In this county, too, there is a long tradition of growing, resulting in sufficient casual labour willing to pick from high ladders.

Varieties and rootstocks Cherry varieties are cross-incompatible so several varieties are generally grown in each orchard to ensure satisfactory pollination. In addition, it is necessary to plant a number of varieties in order to spread the ripening period, and Early Rivers, Waterloo, Roundell, Merton Glory and Stella are a few of the varieties which are grown. Compared with other tree fruits, there is a narrower range of rootstocks for cherries and trees have been trained as tall standards using the vigorous F 12/1 rootstock, but recently the more dwarfing Colt has been introduced which could lead to smaller trees.

Planting and training Trees are generally planted on the square 10.0–14.0 m apart, with an interplant in the centre of each square which is thinned when overcrowding occurs, so at planting there are up to 200 trees/ha, thinned to 100 trees/ha. Pruning is generally minimal, to achieve a delayed open centre head with wide-angled branches in the formative years, and to remove crossing branches during cropping.

Disease and pest control In order to reduce infection by the diseases bacterial canker and silver leaf, pruning is carried out in early spring in young trees and immediately after fruiting in older ones and large pruning cuts are protected with fungicide paint. Spraying with copper fungicide at bud break and at leaf fall will also reduce infection by bacterial canker. The most important insect pests are aphids (*Myzus cerasi*) and caterpillars, against which spraying is essential at white bud giving five spray rounds altogether. Birds, particularly starlings, consume large quantities of fruits and are difficult to control, but they can be scared away and nets can be used to protect the trees.

Picking and Marketing Cherries are picked by hand, as they ripen, and graded at the same time, and the delicacy of the fruits makes mechanical picking for the dessert trade unlikely in the near future.

BUSH FRUITS

Bush fruits include black, red and white currants and gooseberry, all belonging to the genus *Ribes*, and all are characterised by having a bush habit which is pruned to about $1\frac{1}{2}$ m high and $1\frac{1}{2}$ m across. The blackcurrant, *Ribes nigrum*, is the most important bush fruit and 80% of the English crop is used for processing, especially for juicing because of the high vitamin C content of the fruit. Of all soft fruits this is the least popular for the P Y O trade. Blackcurrants are grown in the main fruit-growing areas in England and thrive on a wide range of soils, even on the heavier types, and a high organic content is desirable for moisture retention, so applications of bulky organic manures are particularly beneficial. The blackcurrant is one of the first fruit crops to flower so it is essential to plant on sites which are least likely to have spring frosts.

Traditionally, a good supply of labour was required for picking, up to 60 people to pick 3 tonnes a day, but within the last 10 years great progress has been made in the development of picking machines; most of the crop is picked by this means today.

Varieties and planting material

The main variety, representing 70% of the area, is Baldwin, an old variety, which has been difficult to replace because of its high vitamin C content, and the next most important is Wellington X X X which ripens slightly earlier thus spreading the picking season. Blackcurrant bushes are generally cropped for 10 years and then grubbed because of infection by reversion virus and it is essential that new bushes are free of this virus, so certified stock from specialist nurserymen is planted. The blackcurrant is vegetatively propagated quite readily from hardwood cuttings taken from stool bushes in September.

Planting and training

Blackcurrants are grown mostly in continuous rows although multiple row or bed systems are sometimes planted. For continuous rows, bushes are planted 0.6–1.2 m apart in the rows, with 2.7–3.0 m between rows, giving from 2,778 to 6,173 bushes/ha. Length of row has to be considered in relation to picking and 90 m is recommended for hand picking, but for machine picking the ideal row length is one which, when picked, will just fill the capacity of the trailer being towed behind the picking machine. At planting, bushes are cut down to ground-level but in subsequent years, weak growth and some of the fruited shoots are cut out to encourage new growth which carries fruit in the following year. It takes 50–70 man hours/ha to prune mature bushes.

Disease and pest control

Two important fungal diseases, leaf spot (*Pseudopeziza ribis*) and powdery mildew (*Sphaerotheca mors-uvae*) are controlled by seven sprays of a protective fungicide applied at 14-day intervals from 'grape' stage (tight flower bud). The most important pest is gall mite (*Cecidophyopsis ribis*), the vector of reversion virus, which is controlled by spraying with an insecticide and which, if chosen correctly, will also control aphids. The complete spray programme generally involves seven rounds a year.

Picking and marketing

Fruit for processing should be uniformly ripe but firm, while fruit for the wholesale market should be less ripe so that crushing does not occur before reaching the consumer.

CANE FRUITS

Cane fruits include raspberry, blackberry and loganberry, all belonging to the genus *Rubus*. All three are characterised by having a perennial stem base and root system and a biennial shoot system, and the shoots (canes) need a post and wire system for support which makes these crops relatively expensive to establish. The raspberry, *Rubus idaeus*, is the most important cane fruit and is grown widely in the UK with a concentration of 3,300 ha in Perthshire and Angus in Scotland where the summers are cool and moist. Being late flowering, the raspberry seldom suffers from frost damage although planting in frost pockets must be avoided. In Scotland most of the crop goes for processing, and factories for this purpose have been established in the raspberry-growing areas. In England the fruit is sold on the fresh market, as well as for processing, and in recent years there has been an increased interest in the PYO outlet. Up to 38 pickers/ha are required during the short ripening period so a nearby centre of population is essential when a large area is being planted, and this requirement has stimulated research on machine picking at the Scottish Crops Research Institute.

Varieties and planting material

Breeding programmes at the Scottish Crops Research Institute and at East Malling Research Station have made a particularly important contribution to the productivity of raspberries. Malling Jewel has been the mainstay of the raspberry industry in recent years but is being replaced by the Scottish Glen Clova and by new varieties, such as Malling Delight. The raspberry is naturally vegetatively propagated by means of suckers developed on the roots and new plants are raised by specialist nurserymen.

Planting and training

Plants are spaced 0·4–0·6 m apart in rows, with 1·5–3·0 m between rows, giving from 5,555 to 16,600 plants/ha. Two growing systems are used. In Scotland separate stocks (stools) are maintained by removing all suckers which are produced in between the original plants, while in England suckers are allowed to grow between original plants, thus forming continuous rows. There are a number of ways of supporting the canes, but the most usual is with two wires 0·6 and 1·5 m above the ground, secured to posts 1·5 m tall and 14 m apart in the rows. After fruiting the old canes are cut out and 13 canes to the metre are tied to the top wire by means of a continuous ball of twine. Unwanted suckers between the rows are controlled by mowing, by flails, or by the use of one application of dinoseb-in-oil applied when the suckers are 20 cm high.

Disease and pest control

Grey mould (*Botrytis cinerea*) of fruit is the most serious fungal disease and for control, four sprays of protective fungicide are needed at 10-day intervals from early flowering through fruit development. The raspberry beetle (*Byturus tomentosus*) is the most serious pest: it can be controlled by one spray as the first fruits ripen and if the right insecticide is used, aphids will also be controlled. Four spray rounds are required altogether.

Picking and marketing

Plantations are picked-over every two or three days to obtain pink fruits for canning and for marketing fresh, but less frequently to obtain fully ripe fruit or over-ripe fruit for freezing and pulping, respectively. Some raspberries are marketed through a cool chain where the fruit is taken to a cold store at 2–4°C as soon as possible after picking and transported to market in a refrigerated van and finally sold from a cooled display counter.

STRAWBERRIES

The large-fruited, cultivated strawberry is a hybrid resulting from the crossing of two American species *Fragaria virginiana* and *Fragaria chiloensis*. The plant is a low-growing perennial which can be cropped for up to four years. The low growth of the strawberry makes it particularly susceptible to radiation frost damage but the long flowering period of each plant prevents complete kill of flowers from occurring. Picking of strawberries is particularly laborious and as 24 pickers are required per ha it is getting increasingly difficult to obtain sufficient pickers, and this has encouraged the development of a picking machine, a prototype of which has been produced by the National Institute of Agricultural Engineering.

Varieties and planting material

70% of the area of strawberries in the UK is planted with Cambridge Favourite and 20% is planted with Redgauntlet. The strawberry is vegetatively propagated naturally by runner plants which are produced at the ends of stolons, from parent plants, during long days of summer. As viruses and mycoplasmas spread readily from plant to plant by insect vectors and from parent to runner, every effort has to be made to minimise the spread of these diseases, and the Certification Scheme for strawberries was the first to be introduced by East Malling Research Station and the Ministry of Agriculture. Two systems of growing which are most used are the spaced plant and the matted row but with the development of mechanical harvesting there is increasing interest in the matted bed. In the spaced plant system plants are 0·3–0·5 m apart, in rows 0·9–1·0 m apart, giving up to 37,037 plants/ha, and all runners are removed as they develop so that the original plants remain separate. To produce matted rows, plants are placed at similar distances and runners are encouraged to root around them to produce a mat about 0·45 m wide. In the matted bed, runners are allowed to root between paired rows giving beds about 1·2 m wide with paths 0·45 m wide in between. Unwanted runners between the rows are removed by means of cultivation or by spraying with a contact herbicide.

Protected cropping

The strawberry is the one temperate fruit crop which it is economic to grow under glass or polythene to produce out-of-season fruit at a premium. Generally, glass or polythene cloches or walk-in polythene tunnels are used to produce fruit for the early market in May from early areas in the south and south-west. If the variety Redgauntlet is used for early production it may also give a second crop if protected from September to November.

Disease and pest control

There are three important diseases of strawberries two of which are soil-borne: red core (*Phytophthora fragariae*) and verticillium wilt (*Verticillium dahliae*) both of which are difficult to control; the third is grey mould which attacks the fruits, especially in wet seasons, and for which four sprays of protective fungicide are required at 10-day intervals from early flower. There are two important pests: aphids which carry viruses and red spider (*Tetranychus urticae*) both of which may be controlled by spraying once before flowering, so five rounds altogether are required.

Picking and marketing

Strawberries destined for the wholesale market are picked ripe and graded as they are picked, but those destined for processing are not graded and are picked firm with one- to two-thirds of their surface coloured. In recent years the strawberry has become the most popular fruit for the P Y O trade.

GRAPES

The grape, *Vitis* spp., is the most recent temperate fruit to be grown commercially in England, having been little grown since the dissolution of the monasteries, and it is grown primarily for wine production. The English climate is only just suitable as grapes flourish under long, warm, dry summers and cool winters, conditions which occur mostly below latitude 50°.

Grapes must be grown where there is little chance of spring frosts yet not above the 100 m contour, otherwise damage from wind will occur and, of the 200 ha grown in England, 70% are in the south-east or east of the country. Grapes tolerate a higher pH than other fruit crops.

Varieties   Those suited for English conditions have been selected from varieties grown in West Germany and the Champagne area of France; Müller Thurgau has been most popular in the past, but for the future Seyval Blanc, Madelein Angevine and Triumph d'Alsace show promise. The E E C has very strict regulations about grape varieties and wine categories, dividing them into 'Recommended', 'Authorised' and 'Unclassified' varieties, and only the first can be used for 'quality' wines, the other two groups being used for 'table' wines. The cropping varieties are generally grafted on to rootstocks of American origin which are resistant to the root pest *Phylloxera*.

Planting and training   The grape plant (vine) is a climber, needing a trellis for support, and there are two main types of trellis and training. The Guyot System has wooden posts 1·85 m tall and 6·0 m apart with four wires in one plane 0·6, 0·9, 1·4 and 1·8 m above soil level. Vines are planted about 1·5 m apart in the rows, with rows 2·0 m apart, giving 3,333 vines/ha. In this system the fruiting stem is tied along the bottom wire and the laterals which arise from it are trained vertically above. The Geneva Double Curtain System, which is best suited to vigorously growing vines, requires similar posts 7·3 m apart with a cross batten about 1·2 m long supporting two parallel wires 1·8 m from the ground and the vines are planted 2·45 m apart, with 3·7 m between rows, giving 1,120 vines/ha. As the vines grow they are tied alternately to opposite wires with two permanent stems (rods), totalling 4·9 m in length, to each vine and from these rods the laterals hang down forming the 'curtains'.

Disease and pest control   Two important diseases have to be controlled by routine sprays of fungicides; powdery mildew (*Uncinula nectator*) requires six sprays at 14-day intervals from mid-June and grey mould requires five sprays from early July at 14-day intervals and two sprays in October at 14-day intervals, giving eight rounds altogether. Aphids are seldom a pest and are sprayed only if an outbreak occurs, but wasps are generally serious and can be avoided only by growing varieties which ripen in October. Birds are always a problem and grapes must be netted against them.

Picking and processing   Grapes should be picked once they have reached a suitable sugar concentration, judged by experience and the use of a refractometer. The harvested grapes then have to be processed into wine which demands costly equipment and great skill; some growers produce the wine themselves while others send their fruit to a vinery. The average yield of grapes is 5 t/ha which make approximately 930 (700 ml) bottles of wine.

HOPS

There are several features of the hop, *Humulus lupulus*, and its cultivation which make this crop unusual. The plant is a climbing herbaceous perennial which requires very costly poles and wire-work for support. The marketable portion of the plant is the female inflorescence (cone) on the bracteole of which are carried lupulin glands containing resins and essential oils. The value of hops for brewing lies chiefly in the bitterness of the resins and the aromatic and flavouring properties of the essential oils, the amount and quality of which are varietal characteristics influenced by climatic and drying conditions. Hop marketing has been particularly well organised since 1932 when, to avoid over-production in England, the Hop Marketing Board was set up and sales quotas were introduced, but today the compulsory nature of the Board does not comply with EEC Regulation, so a voluntary Co-operative Society is being established as a replacement.

Hops are concentrated in the two traditional fruit-growing areas in England: the south-east and the west midlands where the growers and work force have particular knowledge of the crop and where the local commercial firms have specialised in supplying the hop growers' requirements. A most important climatic factor influencing the distribution of hops is temperature, especially during August when high mean temperature is correlated with high alpha-acid, the most important component of resin. Hops are particularly susceptible to wind and have to be protected against its effects by trees, tall hedges or netting.

Varieties

In the years between 1972-6 there was a distinct change in the varieties that were grown. In 1972 Fuggle, Brambling Cross and Golding represented 63% of the hop area but by 1976 these varieties represented only 30%, having been replaced by new high alpha-acid varieties: Wye Northdown, Wye Challenger and Wye Target.

Planting and training

New plants (sets) are vegetatively propagated by hardwood cuttings, layered shoots or even leafy cuttings under mist. Two main systems of planting are used: a square plant with plants 2·0 m apart, giving 2,575 plants/ha, and a rectangular plant with 1·0 m between plants in the row, with 2·4 m between rows, giving 4,025 plants/ha. The square plant is cheaper to establish and allows movement in two directions but requires a higher yield per plant to achieve equivalent yield per ha. There are a number of different post, wire and stringing systems for supporting the hops. As growth starts in early April removal of downy mildew shoots (spiking) is carried out and later the new shoots (bines) are twisted clockwise around the strings. Later still the bottom leaves, up to 1 m from the ground, are removed and when the bines reach the top of the strings some of the laterals are removed to allow easier movement. Hops respond to a high level of manuring so applications of organic manure, farmyard manure, straw or shoddy and inorganic fertiliser are required at intervals throughout the season.

Disease and pest control

The soil-borne fungal disease, progressive verticillium wilt (*V. albo-atrum*), is the most serious attacking hops, since once it gets into a hop field it causes great crop losses and it cannot be eradicated. Two airborne fungal diseases, powdery mildew (*Sphaerotheca humuli*) and downy mildew (*Pseudoperonospora humuli*), are controlled by seven routine sprays of fungicides applied at 14-day intervals from late May to mid-August. The three most important pests are flea beetle, aphids and red spider mite (*Tetranychus urticae*) which are controlled by four sprays of insecticide at 14-day intervals from mid-May. The complete spray programme involves seven spray rounds.

Picking and drying

Resin glands are fully developed when the bracts and bracteoles are crisp and yellowish green but if a large area has to be picked some hops must be taken early and some late. The traditional hand picking by holiday-makers from urban areas has been replaced largely by machine picking. The bines are cut and taken to static machines which separate the cones from leaves and other debris but the product is not as clean as hand-picked cones, and physical damage leads to 10% loss of resin. Ripe hops contain about 80% moisture when picked which has to be reduced to 10% quickly and efficiently to avoid loss of resins and oils. Most hops are dried in traditional oast houses, the process being highly specialised and labour intensive (Fig. 31.1).

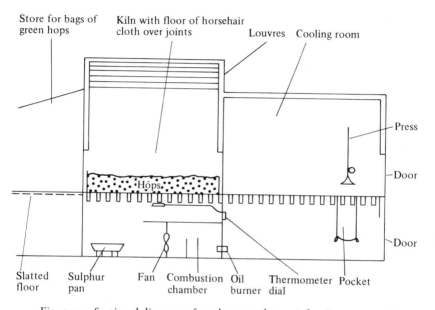

Fig. 31.1   Sectional diagram of modern oast house (after Burgess, 1964)

32

PROTECTED CULTIVATION

Protected cultivation is the second in economic importance of the four main sectors of horticultural production covering vegetables in the open, protected crops, hardy fruit, and nursery stock and bulbs. At 1980 values the gross output of protected crops is £200 M. and this sector accounts for 26% of the horticultural total. There are about 9,200 holdings with a significant area under protection. Just over half have an area of 0·1 ha or less. Three hundred and thirty holdings have more than 1·0 ha under protection and these account for 40% of the total area. It would be impossible for 0·1 ha under protection to provide a full-time living for one man. Most of these small areas form part of a larger enterprise and service the production of outdoor vegetables or hardy nursery stock.

The commercial production of crops under protection developed from the forcing of fruit, vegetables and flowers in the gardens of the landed gentry in the late eighteenth and early nineteenth centuries. From the middle of the nineteenth century production expanded following the invention of industrial methods of making sheet glass and the influence of Sir Joseph Paxton on the design and construction of greenhouses. The construction of the large conservatory at Chatsworth in 1837 and the Crystal Palace for the 1851 Exhibition provided the base from which modern glasshouse design developed. From 1870 there was a rapid expansion mainly in proximity to large urban centres such as London where production developed in the Lea Valley, the Hampton area of Middlesex, the Swanley area of Kent and around Worthing in west Sussex.

Up to the mid-1950s the main structural framework was built in timber and covered with glass. From 1955 timber was replaced by steel and aluminium and almost all greenhouses are now framed in galvanised steel or a mixture of steel and aluminium. From 1960 film plastic provided an alternative covering to glass. Film plastic requires a much lighter supporting framework than conventional glasshouses but the skin needs replacing every two years, the structure is more subject to wind damage and there can be problems caused by condensation forming on the inner surface of the plastic. For the production of vegetables the most widely used design is the 3·2 m-span Venlo. This is built usually in multi-span blocks up to

3 ha in area. For ornamental crops wider spans of 6·4 m or more are preferred with multi-span blocks smaller in size than for vegetables. Single wide-span houses up to 30 m wide were built in the 1960s but these are now uneconomic to erect and the widest span in general production is 14 m.

As a result of the relatively poor light integral in Britain in winter, light is a limiting factor in growing crops in greenhouses from October to March. Modern greenhouses are designed to transmit the maximum amount of winter light. The advent of aluminium and steel sections for glasshouse construction reduced the size of structural members and increased the light transmission. Orientation of greenhouses E-W rather than N-S also increases winter light transmission and the single wide-span houses built in the 1960s were particularly advantageous in this respect. It is now general practice, wherever the site permits, to build multi-span blocks with the ridges orientated E-W to obtain better transmission of winter light. Houses covered with film plastic tend to have poorer light transmission due to the deposition of condensation in discreet droplets on the inner surface. For a given area covered, the cost of building increases with the span; Venlo designs have the lowest cost per unit area. The basic 3·2 m span of the Venlo design has been adapted to make it more acceptable for ornamental crops by carrying two spans on a lattice girder to give a clear 6·4 m working area.

There are about 2,900 ha under protection in the whole of the British Isles with 2,130 ha in England and Wales, 410 ha in Guernsey, 250 ha in Eire and 90 ha in Scotland. Detailed long-term statistics are available only for England and Wales, where there was a peak of 1,919 ha in 1954 followed by a decline to 1,538 ha in 1965. Since 1965 there has been a steady increase until 1976 when the total reached 2,039 ha. The area has since been steady around this figure but provisional figures for the June 1981 Census suggest that the area may now be declining. Since 1960 film plastic has provided a less costly alternative to glass and the area under this type of protection has risen to 200 ha accounting for about 10% of the total.

In the British Isles the industry is fairly widely distributed but with major concentrations in Guernsey, the southern and eastern coastal counties of England, west Lancashire, Humberside and the Clyde Valley. Fifty per cent of the total area is in East Anglia and the south-east and 20% in Lancashire and Yorkshire. By county the most important areas are Essex (190 ha), West Sussex (188 ha), Humberside (187 ha), Lancashire (176 ha) and Lincolnshire (144 ha). Vegetable crops are the more important accounting for 72% of the area compared with 27% for ornamental crops. In value of output, vegetables (including £52·5 m. for mushrooms) account for £143 m. and ornamentals for £55 m. annually. Most of the ornamental crops are grown in the southern counties and the importance of flower crops decreases from south to north. Almost all the output from Guernsey and 25% of the output from Eire is marketed on the UK mainland. All the output from the UK is marketed fresh on the home market. None is exported or processed.

Compared with all other systems of crop husbandry, protected cultivation is characterised by the degree of control possible over the crop environment. The majority of modern greenhouses are equipped with automatic controls. It is possible to control aerial and root temperature, carbon dioxide concentration, relative humidity, nutrition and water supply. The control of all these factors can be integrated and optimised for particular conditions using micro-processor systems. It is practicable and economic to supplement solar radiation with light from discharge lamps in the early stages of growth in winter. Accurate control of the environment gives the grower a high capacity to time, programme and control the growth and development of crops and to achieve levels of productivity higher than in any other sector of agriculture or horticulture. Ability to control the environment also facilititates pest and disease control and allows the exploitation of biological control methods.

Seventy-five per cent of the total area under glass is heated compared with only 20% under film plastic. Of the area heated, 5% is heated by solid fuel, 6% by gas and 89% by oil. Half of the area heated by oil uses the lighter, more expensive distillate oils. Within the structure, heat is distributed by steam on 12% of the area, by low-pressure hot water on 37%, by medium-pressure hot water on 15% and by direct-fired, warm-air heaters on 36%. Warm-air heaters are most popular where only a small temperature lift is required and in film plastic houses.

Major disadvantages of cultivating crops under heated protection are the quantity and high cost of the support energy needed and the poor energy balance measured as an input/output ratio. Protected cultivation in the UK and Channel Islands uses about 500,000 tonnes of oil and 85,000 tonnes of coal annually for heating. In the UK the oil used represents 25% of the petroleum fuels directly used in agriculture. Though agriculture as a whole uses only 4% of the national energy budget, protected cultivation uses a very large and disproportionate share of that total. The cost of energy is the largest single item of cost in the production of heated crops and accounts for up to 40% of the total costs of production of crops requiring temperatures in the range 15–20°C. The current economic and energy situation is putting severe pressure on this sector of horticulture to increase the efficiency of its use of energy. Since 1973 the consumption of oil has been reduced by 21% without any loss of output simply by increasing the efficiency of energy utilisation, reducing heat loss from structures, by improved control systems and by optimising environmental factors. There is also a developing interest in alternative fuels, in the use of industrial reject heat and in combined heat and power generation systems.

Cooling in summer is provided normally by ventilation either by mechanically operated ridge ventilators or by electrically driven fans. Fans are now less popular because of the high cost of power but they are the only practicable method in large film plastic houses. Forty per cent of the total area is fitted with some form of automatic ventilation. For many crops it is common practice to raise the level of carbon dioxide in the atmosphere from the ambient level of 300 vpm to about 1,000 vpm to increase growth and raise quality in winter and spring. The carbon dioxide is produced either by burning propane, natural gas or low sulphur kerosene or by evaporating the gas from bulk liquid storage. Twenty-five per cent of the area is fitted with equipment for carbon dioxide enrichment. The automatic application of water and nutrients to crops is general and 75% of the area is fitted with suitable equipment.

Productivity of labour has been increased by work study, by the introduction of cropping systems with a lower labour content and by mechanisation. The training of cucumber and tomato crops has been simplified by high wire supports and by layering systems. Materials handling has been greatly improved by the use of self-propelled trolleys and crop-spanning gantries. Operations such as block making, seed sowing, container filling and planting have been fully mechanised. It is now possible to mechanise completely the production of lettuce from seed sowing to harvesting and production systems have been developed for pot plants which allow newly potted plants to be fed on to one end of a production line and marketable plants removed from the other.

Production systems in protected cultivation are essentially continuous monocropping. Where soil is used as the growing substrate it is necessary to 'sterilise' the soil between successive crops. This may be done by the injection of steam or by the application of chemicals. About 55% of the total production area is sterilised each year. Of the area sterilised 30% is steamed, 50% fumigated with methyl bromide and 20% treated by other chemical methods. Due to the high cost of energy and labour and toxicity problems with methyl bromide, production in soil is being replaced by peat or rockwool modules and by the use of hydroponics as in the Nutrient Film Technique. The growing modules are discarded after use for one crop and with culture solutions there is no requirement to sterilise.

Tomato (763 ha), cucumber (234 ha) and lettuce (1,200 ha)* are the most important vegetable crops. The situation is dominated by the tomato which accounts for half the area used for vegetable production in England and Wales. There is a continuing search for alternative vegetables to diversify production and lessen the major dependence on tomato. Sweet pepper and celery have increased in recent years. Small areas of many other crops are grown including radish, aubergine, Chinese cabbage, courgette and calabrese but it is difficult to find crops capable of sustaining any large area of production. Tomato and cucumber are grown normally as single, long-season crops planted from December onward and cropped from March to November. Lettuce is produced most frequently as a succession of short-term crops from October to May. There is, however, an increase in the demand for protected lettuce in the summer from the supermarket chains and there is now production on an all-the-year-round basis.

Chrysanthemum, carnation, rose and forced bulb flowers are the most important cut flower crops. Chrysanthemum is by far the most important single crop. It is produced partly from crops which flower naturally in the autumn and early winter and partly from crops grown through the year on a 12- to 14-week cycle in which flowering is programmed by the control of daylength and temperature. Carnation and rose production has declined since the mid-1960s largely as a result of competition from high-quality imports in the winter from Israel and Colombia. Forced bulb flowers have also declined but here the main cause is the high price of energy for heating. A very wide range of flowering and foliage plants is grown for sale in pots. Production has increased over the last 15 years and the crop is second only to cut flower chrysanthemums in value.

It is likely that the cost and availability of energy will continue to be the main problem in this sector. Pressure will be maintained to increase the efficiency of energy utilisation and to move from the direct use of fossil fuels to the use of industrial reject heat and renewable energy sources.

* This is derived from the number of lettuce crops multiplied by the area used for each. In practice, several crops may be grown on the same area within one year.

33

FLOWERS AND NURSERY STOCK

Although the title suggests a biologically well-defined group of crops, several flower crops are grown under protected cultivation and have therefore been considered in Chapter 32. Similarly, as bedding plants are usually produced under protection they are also included with the protected crops. The present account is limited to outdoor flowers (excluding bulbs), bulbs and bulb flowers produced in the field or forced under glass, and outdoor-grown nursery stock.

The total value of this sector of the British commercial horticulture industry is about £112 m. (1980 values). In addition to its economic importance as an industry it provides a valuable contribution to the quality of modern life both within the home and as part of the urban and suburban environment of the majority of the population.

OUTDOOR CUT FLOWERS (EXCLUDING BULBS)

The area devoted to outdoor-grown cut flowers has declined fairly steadily from a peak in 1951 of 3,000 ha to its present area of 1,000 ha. There are some signs of stabilisation at this level, which would be encouraged by the increasing costs of fuel used in producing cut flowers under glass. The value of the industry in 1972 was £6·1 m.; in 1980 it was about £10 m.

Much of the outdoor cut-flower industry is situated near large towns, which facilitates transport and marketing, but there are several specialised growing areas such as Devon and Cornwall (especially in the more westerly part), and in Cambridgeshire, Bedfordshire and Lincolnshire not close to the main markets. Changes in the location of production have occurred over the past 30 years or so because of urbanisation in the south-east, and as a result of a general increase in flower production in the eastern counties, whilst changes to increased production of other, more profitable crops, notably hardy nursery stock and bulbs, have been responsible for much of the general decline in the area of outdoor cut flowers. Factors contributing to this decline include a high labour input, high costs of materials, and the difficulties of marketing a product subject to seasonal variation in both quantity and quality.

A wide range of species is grown, but there is little information on their relative importance. Chrysanthemum represents about 31% of the area, with dahlia 17%, annuals 24% and biennials 5% of the total. Other important species, listed alphabetically, are Alstroemeria, Aster, Delphinium, Lupin, Marguerite, Peony, Phlox, Pink, Pyrethrum, Scabious, Stock, Sweet Pea, Sweet William, Violet and Zinnia. There is also a group of 'everlasting flowers' such as Helichrysum and Statice, dried before sale for indoor winter decoration. Some species grown are annuals and biennials grown from seed sown *in situ*, in a seedbed for transplanting, or under glass where protection is required. Most seed is sown in spring and early summer, but frost-hardy species can be sown in early autumn, and advantage can be taken of areas of milder climate and of sheltered locations to produce early crops sold at higher prices. Perennials are frequently given some protection by frames, polythene covers or mobile glasshouses to induce extra earliness, to improve flower quality and to prevent weather damage. Provision of wind-breaks is often an essential feature of outdoor flower growing, especially in spring.

BULBS (INCLUDING CORMS)

Bulb and corm plants are mostly members of the three monocotyledon families, *Amaryllidaceae*, *Iridaceae* and *Liliaceae*, but some of the miscellaneous bulb crops are from others including *Primulaceae*, *Ranunculaceae*, *Alliaceae*, *Begoniaceae* and *Gesneriaceae*. Within the UK the major crop is Narcissus (including Daffodil), which occupies three-quarters of the total bulb area; indeed, the UK is the largest producer in the world of Narcissus with slightly more than half of the world's true bulb area. In the UK the Tulip is a poor second with 12% of the area, followed, in decreasing order of importance, by Anemone, Gladiolus, Iris and Lily. A further range of miscellaneous bulbs is grown under glass (Hippeastrum, Begonia, Gloxinia, Nerine) or outdoors for cut flowers, pot plants or for sale as garden plants (Crocus, Snowdrop, Scilla, Chionodoxa, Grape Hyacinth, Autumn Crocus, Amaryllis and Nerine, as well as many others of even lesser economic importance). Narcissus and Tulip are grown for both flowers and bulbs; the other species are grown almost exclusively for flowers.

The total area of field-grown bulbs is about 5,400ha, and the output value of the industry is about £32 m. (1980) of which Narcissus is £18 m. and Tulip about £5 m. The area devoted to bulbs increased since the Second World War to a peak of 6,600 ha in the mid-1970s. The subsequent reduction in area is not a true reflection of the industry because higher planting densities are now used for Narcissus. Over the past decade, an export trade has been developed in both dry bulbs and flowers, with a present-day value of about £3·5 m. Total imports amount to about £7 m. (1978) made up, in descending order of value, from miscellaneous bulbs, Hyacinth, Iris and Gladiolus.

Bulbs are grown in the field by farmers rather than by horticulturists; the crops fit in with other farming operations and crops (broccoli and early potatoes in the south-west, arable farming in eastern England), and allow the long rotations which are essential for the production of healthy bulbs. Other growers of bulbs are specialist growers of individual crops, e.g. Lily, nurserymen who grow a range of ornamental stocks, and a few specialist bulb nurseries. Narcissus and Tulip are grown normally, like potatoes, in ridges in the field, and similar machinery is used for planting and lifting.

Commercial bulb growing started in the UK in 1880 in south Lincolnshire. Although bulbs are now widely grown commercially, there are two main production areas, one in eastern England (Lincolnshire, Cambridge and Norfolk) and the other in the far west of Cornwall and the Isles of Scilly. That in eastern England represents 70% of the total area, the south-west 14%, and Scotland (east coast) about 6%. The character of bulb growing differs markedly between the different growing areas, although these differences are less distinct than formerly. In eastern England many bulb growers are concerned with three aspects of bulb growing: production of bulbs, forcing flowers and harvesting outdoor flowers if this is justified by market prices. Narcissus, Tulip and some Iris and Gladiolus are grown, but the first two are by far the most important. The scale of operations is large, the bulbs form part of a highly mechanised arable farming on easily worked fertile silt soils which allow efficient production and handling and long rotations with cereals, potatoes and sugar beet. Bulb growing is capital-intensive, growers are equipped with facilities for bulb storage and hot-water treatment and with specialised machinery for planting, lifting and grading of bulbs. Many growers also force bulbs and therefore have the necessary specialised controlled temperature stores and glasshouses. In the south-west, the emphasis is more on outdoor early flower production, mainly from Narcissus, although there is a considerable trade in the bulbs produced as a result of flower growing in the field. Most of the Anemone-growing is in the south-west and there are some cut-flower growers of Iris, Lily and other miscellaneous bulbs.

The bulb-forcing industry involves flower production under heated glass (about 33 ha) in winter and early spring from field-grown bulbs artificially treated with low temperatures during summer and autumn. There is a rapid turnover of bulbs through the forcing glasshouse, with several 'rounds' of bulbs being forced in each season. The early rounds are Narcissus, later Tulips are produced as well. Cooling can be given to the bulbs either on 'standing-grounds' outside under straw or in controlled temperature stores. Flowers of Narcissus are available in quantity from early November and Tulip from early December, whilst Iris are available all year. About 60% of the Narcissus bulbs, and 65% of Tulip are forced. Approximately 40% of the Narcissus are sold as dry bulbs for export or the home market. For Tulip the corresponding figure is 35%.

HARDY ORNAMENTAL NURSERY STOCK

The present area of hardy nursery stock (excluding fruit stocks and that for outdoor flower production) in England and Wales is 5,500 ha (1980). It was only 3,500 ha in 1955, increased to a peak of 6,000 ha in 1974, and has since fallen somewhat irregularly to the present position. The current farmgate value is about £70 m., there being about 2,000 holdings.

In England and Wales the main production areas are in the south-east and eastern regions, with 35% and 21% respectively of the total production. Intensive methods are used for the high-value products and there are many growers with small areas of nursery stock. During the 1970s, several hundred producers with small areas gave up nursery stock production, but three-quarters of the holdings still account for only 16% of the total area. Many growers have areas of 20 ha or more; although numerically small at 2·5% of the total, they account for over 40% of the total area.

There is a wide range of products, but no information on the areas of individual crops grown except for roses. The range covers monocotyledons, dicotyledons and conifers, both deciduous and evergreen plants, of a range of habits: herbaceous plants, climbers, alpines, ground-cover plants, hedging plants, shrubs, bushes and trees. The plants are produced in a variety of ways; from seed, by budding, grafting or from cuttings etc., and sold as bare-rooted plants or as container plants to local authorities, landscape contractors, development corporations and to the gardening public direct or through garden centres.

There are two production systems – field and container growing. Plants of the former are marketed only when dormant, but the latter can be sold throughout the year. The high capital outlay and recurrent costs of container growing are balanced by the benefits of year-round sales. Although some heathers are sold when one year old, many of the plants sold are several years old, and some trees are 6–8 years old from propagation when marketed, so that there is a high capital investment in the crop.

Changes in the cropped areas of nursery stock are shown in Table 33.1.

Table 33.1 Cropped areas ('ooo ha) of nursery stock in England and Wales

	1973	1975	1978	1980
Roses	1·5	1·2	1·0	1·0
Shrubs	1·6	1·7	1·3	1·5
Trees	1·4	1·6	1·5	1·5
Herbaceous	0·3	0·3	0·3	0·3
Others	0·9	1·0	1·0	1·2
Totals	5·7	5·8	5·1	5·5

The clearest trend is the decrease in the area of roses (including stock for budding), which appears to be a direct response to falling demand. In the past decade or so there has, however, been an increase in UK rose rootstock production, and we now produce about 21 m. annually (1977 figure). The area of field-grown hardy ornamentals has fallen, reflecting reduced spending by public authorities, less house building and a tendency to smaller gardens. This change has been offset partly by increased numbers of container-grown plants each year, from 27 m. in 1976, successively to 31 m., 34 m, 47 m. and 53 m. in 1980.

For container-grown nursery stock, statistics are poor, but the results of a survey in 1976 indicate the following percentages of the total plants grown: alpines and heathers 5.2, azaleas 3·4, rhododendrons 4·0, climbers 6·2, roses 3·8, conifers 13·1, other shrubs 62·7 and trees 1·6.

Although UK growers can produce most of the species required, there has been a long-standing import trade from Europe, especially from The Netherlands and Belgium. Despite increases in home production, we still import almost two-thirds of our rose rootstocks. The annual cost of rose imports is £1·4 m. and for other hardy nursery stock £14·1 m. (1979 values). Some of the imports are young plants intended for growing to saleable size in this country.

34

FUEL CROPPING

Although some form of fuel cropping has been practised for many centuries, crops have not made a significant contribution to Britain's fuel supplies since wood became a relatively unimportant source. Renewed interest in fuel crops is essentially a recent phenomenon and large-scale applications in Britain a future possibility.

Ever since the discovery of fire, plant products have provided man with fuel. Prior to the Industrial Revolution Britain was fuelled almost exclusively by wood. In medieval times tenants had a right, known as *hedgebote*, to branches of hedgerow standards for use as firewood; this provided fuel for a significant proportion of the rural population of that time. Substantial amounts of firewood were also provided by the branches of coppice standards and the offcuts and wastes after coppice products had been made from the underwood.

During the sixteenth and seventeenth centuries wood consumption increased dramatically as the population grew and industry expanded; at the same time large areas of woodland were cleared to provide extra land for cultivation. This resulted eventually in a severe 'energy crisis' as supplies dwindled and prices rocketed.

With the Industrial Revolution and the large-scale exploitation of coal, wood diminished in importance in Britain. The twentieth century saw increasing reliance being placed on petroleum and natural gas, and wood and other plant-derived fuels are now of little importance. This is not the case in many other countries, particularly those of the Third World, where over 90% of the population are dependent on firewood as the main fuel source.

The 'energy crisis' of the seventeenth century was overcome eventually, after much readjustment, as wood was replaced by the more energy-dense fossil fuel, coal. The 'energy crisis' of the 1970s, brought about by rising prices and an awareness of the limitations of supply of fossil fuels, notably petroleum, has provoked a renewed interest in fuels derived from plant materials and, hence, in fuel-cropping.

Like the fossil fuels – coal, oil and natural gas – fuels derived from crops and other biological materials are the result of photosynthesis in green plants and, hence, contain energy fixed in chemical bonds; apart from the possibility of generating hydrogen by photochemical and photobiological means, the derivation of fuels from biological materials is the only means, at present, of providing solid, liquid or gaseous fuels and, therefore, of providing direct substitutes for fossil fuels.

Fuel crops are capable of providing storable forms of energy and are ecologically inoffensive, renewable and rely for their production on existing technologies. Disadvantages associated with fuel cropping include the large area of land required, possible competition for land use with food production, the low energy density of the feedstock and the associated transport and storage problems and the losses and inefficiencies associated with the conversion processes.

Where fuel crops compete with food production, fuel cropping is not likely to be implemented until it can compete successfully with other enterprises using the same land. However, some opportunities do exist and options outlined in this chapter will become increasingly attractive as the prices of conventional fuels rise.

As plant materials vary in their moisture and chemical content a number of crop-to-fuel conversion technologies are needed. Of most importance are the direct combustion of dry materials such as wood or straw, various thermo-chemical conversion processes (e.g. pyrolysis, gasification) yielding a range of fuels, fermentation to produce ethanol, anaerobic digestion to produce methane and the extraction of oil from oil-seed crops such as rape.

In this chapter energy yields are expressed in petajoules (P J) equal to $J \times 10^{15}$. All estimates were based on 1979 data and the UK primary energy consumption is taken as 8369 P J.

Strictly speaking, a fuel crop is one grown specifically and exclusively for that purpose, but a broader definition will be adopted in this chapter. Here, fuel cropping refers to the use of the whole or part of a crop for fuel and to the management of existing resources towards the production of fuel.

EXPLOITATION OF EXISTING RESOURCES

At present only a very small proportion of agricultural crop material is used for fuel. Some farm woodland is used to provide firewood for sale or use on the farm, and an increasing number of farms are installing straw burners to provide heat for domestic consumption, grain- and grass-drying, mushroom-house heating, glasshouse heating and the heating of pig buildings (MAFF, 1980a). Farm woodland, crop residues and areas of natural vegetation such as bracken and heather do, however, represent sources which could be further exploited to produce fuel, without adversely affecting food production.

Animal wastes also represent a significant energy resource within agriculture, providing a feedstock suitable for anaerobic digestion to produce methane. This is a very indirect form of fuel cropping, but the technology associated with it can also be applied to wet crop residues (i.e. sugar beet tops) and purpose-grown fuel crops. Anaerobic digesters have been set up on a few British farms to operate on livestock wastes. The potential energy yield as methane from cattle, poultry and pig waste in the UK has been estimated as 37 PJ/year or 0.4% UK primary energy consumption.

Crop residues

A number of crops grown in the UK provide dry residues suitable for direct combustion or biological or thermal processing to liquid or gaseous fuels, or wet residues most suited to anaerobic digestion to produce methane. Notable among these is cereal straw which is capable of supplying all agriculture's heating demand when burnt as a fuel (White, 1980). Estimates of the potential UK output of cereal straw and other crop residues are given in Table 34.1.

Table 34.1   Estimated yields of various crop residues for 1979. For dry residues, energy yields are gross energy in residue; for wet residues, as methane after anaerobic digestion

Crop	Yield of Residue $t \times 10^3$	Energy Yield PJ
Cereals	9,709	173
Dried Peas	113	2
Field Beans	126	2
Oil-seed Rape	554	10
Total Dry Residues	10,502	187
Sugar Beet	963	9
Potato	241	2
Vegetables	259	3
Total Wet Residues	1,462	14
Total Residues		201
%UK Energy Consumption		2.4%

The yield of straw in cereals is determined by the species, variety, by environmental factors and the cutting regime. Straw yields could be improved by breeding and selection or a change to existing longer-strawed varieties. Lowering the combine cutter bar or stubble harvesting with a forage harvester would increase straw yield and it is claimed that the technique of whole crop harvesting recovers 100% more straw, as well as the light chaff fraction, than conventional handling (Lucas, 1978).

Cereal straw and other dry residues can be utilised as fuel on the farm in straw burners, which have been used successfully in a number of European countries and which have, more recently, increased in popularity among British farmers. A wide range of designs (and prices) are available operating on big or small bales or chopped straw, and capable of providing heat for a number of different applications.

Farm woodland

In 1979 there were 190,854 ha of woodland in agricultural holdings in England and Wales (MAFF, 1980b). Assuming an average output of 4t DM/ha/year of fuelwood this represents a resource of about 15 PJ/year wood energy or 0·18% UK primary energy consumption. Of course, not all this is easily exploited, but where the woodland is small, single-stem trees or coppice, easily accessible and fairly young, a farmer with sufficient time in winter, a chain saw and a ready market may produce firewood quite profitably. Growing trees for fuel may also provide a more profitable enterprise for many areas of farmland in the UK, particularly the uplands.

Natural vegetation

The exploitation of areas of natural or semi-natural vegetation to provide feedstocks for biofuel production has some attractions; feedstock costs are likely to be low and competition with other land uses minimal. Approximately 40% of Britain is covered by natural or semi-natural vegetation and it is considered that of this area heather moors ($1\cdot491 \times 10^6$ ha) and bracken fells ($0\cdot322 \times 10^6$ ha) offer the most suitable areas available immediately for cropping (Lawson et al, 1980).

Other types of natural vegetation which may be exploitable include various heaths, moors and bogs, saltmarsh, mire and waste sites such as stream banks, road-sides, field-corners, graveyards and sewage farms where existing vegetation often shows very high productivites.

Estimates of the potential UK energy output from heather and bracken are 10·74 PJ/year and 32·46 PJ/year respectively, a total of 43·2 PJ/year, or 0·52% of UK primary energy consumption. In many situations heather and bracken are managed already, at some expense; harvesting these crops for fuel may prove a more profitable use of the land.

LONGER-TERM OPPORTUNITIES

There are a number of opportunities which, although requiring higher financial inputs than the options described in the previous section, would not compete with food production and could be implemented in response to moderate rises in the price of conventional fuels. These include the growing of fuel crops on derelict land, dual-purpose crops providing both fuel and food or animal feed, and catch crops, grown between the harvest and sowing of main crops.

Derelict land

The area of derelict, despoiled and degraded land in Britain suitable for growing energy crops has been estimated as some 0.25×10^6 ha capable of yielding 1×10^6 t DM/year of coppiced woodland (Dennington and Chadwick, 1979). This amounts to an output of 20 PJ/year gross energy in wood or 0.24% UK energy consumption.

Dual-purpose crops

Crops which provide food or animal feed and a residue suitable for use as a fuel are, in effect, dual-purpose crops; the potential of residues from existing crops has already been discussed. In the longer term it may be possible to replace current crops by species that produce both a food/feed and a fuel component. The growing of large plantations of multi-purpose crops that can be fractionated into the basic components of human food, fuel and other raw materials may also be possible, but would require major changes in technology, land use and human diet.

The most promising dual-purpose crops are oil-seed rape, Jerusalem artichokes, fodder beet and sugar beet. Oil-seed rape provides vegetable oil capable of mixing with or replacing diesel, protein-rich meal used as animal feed and straw suitable for feed, bedding or for fuel. Replacement of crops grown for animal feed with oil-seed rape would provide 93.2 PJ of energy in oil and straw equivalent to 1.11% of UK primary energy consumption.

Fodder beet and sugar beet produce tops which can be fed to animals; the roots can also provide feedstocks for anaerobic digestion to produce methane or for ethanol production in addition to their traditional use as animal feed or in sugar manufacture. Jerusalem artichoke tubers provide a suitable feedstock for ethanol or methane production and can be used as animal feed or human food; the tops could be harvested wet, for animal feed or for methane production, or dry as straw for direct combustion or thermal processing. The residue from ethanol manufacture may also be suitable as an animal feed or even a feedstock for anaerobic digestion.

Catch fuel crops

Catch crops are grown in the time period available between the harvest and sowing of main crops and would provide a feedstock suitable for anaerobic digestion to produce methane. Traditionally, catch crops have been used to provide animal feed in the autumn/ winter period, but with the decline in mixed farming many areas of arable land lie fallow at this time of year. The main advantages of using this opportunity to produce fuel are that the system can be introduced into current cropping regimes without displacing food production, and that some of the cost can be borne by the main crop.

The potential UK output of catch fuel crops is dependent on the area, duration and time of land available, the yield of crops grown on that land (largely a function of sowing date and location) and the efficiency of the anaerobic digestion process to produce methane. One estimate places the potential energy output for catch crops grown on all the UK tillage area at 214 P J/year as methane or 2·6% of UK energy consumption.

DEDICATED FUEL CROPS

As fuel prices rise fuel cropping will become increasingly profitable and compete successfully with other enterprises, thus stimulating the transfer of land used for food or animal feed production to fuel crops. Dietary change away from animal products towards the more energy- and land-efficient vegetable foods would also release land for fuel crops.

Initially, fuel crops may be annuals such as oil-seed rape or sugar beet grown as alternatives to conventional break crops. In the longer term, fuel crop plantations might be established, consisting either of crops for fuel or of dual- or multi-purpose crops.

Where dedicated fuel crops are grown a number of possibilities are available; these will produce feedstocks for different conversion processes and, hence, different fuels, and include coppiced woodland, short-rotation single-stem trees, perennial agricultural species and wild or naturalised plant species. The strategy adopted will depend on the type of land and the value of the fuel produced.

If a fuel cropping opportunity is exploited the fuel produced is determined largely by the nature of the opportunity. Where land is released to dedicated fuel crops the fuel required will determine the crop sown. It seems likely, therefore, that crops yielding liquid fuels (e.g. methanol, ethanol or vegetable oils) capable of substituting for diesel or petrol will be most valuable.

Trees

Trees represent one form of fuel crop plantation which could be established on agricultural land. On suitable land in the lowlands coppiced woodland may achieve yields of up to 20t/ha/year; single-stem trees on rotations of 12–20 years could yield up to 12t/ha/year and may be more suitable for upland regions. Trees yield a feedstock suitable for conversion to a range of gaseous and liquid fuels, notably methanol.

Agricultural crops

Agricultural species suitable for dedicated fuel cropping include starch crops such as cereals and potatoes, sugar beet and fodder beet suitable for ethanol production; and wet, green crops such as ryegrass, forage maize and lucerne producing high yields of a feedstock suitable for methane production via anaerobic digestion.

Wild species

Certain species growing naturally in the UK exhibit very high yields and could be considered as dedicated energy crops. The naturalised alien species Japanese knotweed (*Reynoutria japonica*) seems capable of annual production in excess of 20t/ha in small stands; *Reynoutria sachalinensis* and *Impatiens glandulifera* are also promising.

ECONOMIC CRITERIA

There are three main ways that fuel crop material produced on the farm can be used: (a) the crop and the fuel can be produced and used on the farm, in which case the cost is that associated with production and conversion and the profit is the value of the conventional fuel saved; (b) the fuel can be produced and sold and the selling price will include the farmer's profit; and (c) the crop material can be sold to a central processor and is, therefore, treated in the same way as any other saleable commodity.

Implementation of fuel cropping systems is largely dependent on profitability. The use of residues is, in many situations, already an economic proposition. As fuel prices rise, opportunities such as catch cropping and dual-purpose crops could be implemented. Substantial rises in fuel prices and probably government incentives would be needed to effect the implementation of dedicated fuel crop systems.

Many of the systems referred to in this chapter are still in the research and development stage and exact costs are not yet available. Fuel cropping in the U K is not likely to make a major impact on total energy supplies, but has the potential to more than provide the energy needs of agriculture.

PART V

ANIMAL PRODUCTION SYSTEMS

35

MILK PRODUCTION SYSTEMS

Introduction

The dairy cow is one of our most valuable domesticated animals, producing high-quality human foods from a wide range of feeding-stuffs, especially grass and forages. Milk is the major product from the national dairy herd, although cull cows are an important source of meat, particularly for manufacturing purposes – and surplus calves a vital input into beef production.

The aim of most dairy farmers is to achieve a high level of profitability from their herds, although in practice, a wide range of physical and financial performance is achieved. Some of the variation is due to farm location, climate and soil type, some to managerial ability of the farmers themselves, but a considerable amount to the fact that the behavioural pattern of dairy cows responds markedly to the care they receive from stockpersons. There is an art in cow care; it is difficult to define but it involves such aspects as close observation and the taking of timely action. Cows are creatures of habit, responding well to routine and to an environment with minimum stress. Such a situation can be provided with relative ease in a small family-operated herd but is more difficult to achieve in a larger highly mechanised unit involving a number of employees.

Mechanisation has in recent years considerably reduced the chores involved in dairying. It is now almost universally adopted for the milking process and for the making of conserved fodder. It is widely used in manure handling but has a varied involvement in feeding, depending largely upon herd size. Automation and electronics in milking and in feeding are more recent, although increasing, thus adding to the considerable range of skills required by the staff of a modern dairy farm.

Before describing the numerous types of dairy enterprise and the production systems in use it is necessary to consider briefly some general aspects of the UK dairy industry and the reasons why farmers have milk production enterprises.

It can be seen from Table 35.1 that in recent years dairy cow numbers have remained virtually static and the number of produ-

Table 35.1   U K dairy farming

	1960	1970	1980	1981
Dairy cow numbers '000	3,165	3,244	3,226	3,224
Registered milk producers	151,625	100,741	56,247	53,525
Average size of herd	20	31	53	55
Average annual milk yield per cow (litres)	NA	3,840	4,720	4,810
Sales of Liquid Milk (m.l)	7,036	7,458	7,291	7,136
Milk for manufacturing as % total sales	27·3	38·4	51·9	53·1

Source: M M B Dairy Facts & Figures 1981

cers has decreased markedly with a corresponding increase in the average size of herd. Yield per cow continues to increase and, associated with the fall in retail sales of liquid milk, the influence of the manufacturing outlets has increased.

With the E E C market for milk and dairy products in surplus and with rising costs of production, the profits of U K producers have for several years been under pressure. Schemes aimed at reducing the number of dairy cows in the community have been in operation but their influence upon the U K cow population has been minimal. A number of dairy farmers have discontinued production, some being older farmers, many already having made plans to retire, and others who were able to change their farming system by introducing or expanding a viable alternative enterprise, such as cereal production.

For the majority, milk production continues to be potentially the most profitable enterprise. The cash flow advantages are considerable, with payment for milk being received on a monthly basis. Considerable sums of capital are invested in a dairy enterprise, some in the livestock which, if required, could be recovered, but much in specialised fixed equipment such as milking parlours, which have a minimum resale value. The intensity of a milk production enterprise can be varied according to the availability of resources. On farms with limited land area it is possible to increase output by using higher levels of inputs such as fertiliser and purchased feedingstuffs. Where labour is a limited resource, capital can be utilised by providing mechanisation and automation.

DAIRY HERD TYPES

Milk production enterprises are to be found in a wide range of situations in the U K from small upland holdings to large arable-based farms. Hill farms are the only exception; they do not carry dairy herds due to the problems of producing adequate forage supplies and to the lack of access, especially in the winter months. The number of producers in upland marginal areas has fallen in recent years due to economic pressures but also by an enforced change in the milk collection system from churns to refrigerated bulk tanks. Some producers with farm roads unable to take tanker lorries have been able to stay in business by utilising portable tractor-drawn bulk tanks for the transfer of dairy supplies to a main road collection site.

Table 35.2   Regional dairy cow numbers and average size of herd (1979)

Region	Cow Numbers ('ooo)	Average cows per herd
Northern	297	51
North-Western	576	53
Eastern	97	72
East Midland	135	64
West Midland	260	62
North Wales	120	38
South Wales	211	39
Southern	147	85
Mid-Western	420	74
Far-Western	308	46
South-Eastern	163	82
England & Wales	2,734	56
Scotland	247	81
Northern Ireland	265	26
United Kingdom	3,246	52

Source: MMB Dairy Facts & Figures 1981

Small family herds

Despite the increasing size of the average herd, small herds continue to be of considerable importance in the dairy industry. This is particularly the case in Northern Ireland and in Wales but less so in Scotland. In England, the northern and far western regions have the highest proportion of small herds but even in the eastern region, one in three herds carries less than 40 cows (Table 35.2).

On small family dairy farms, milk production is usually the only enterprise, although due to a general interest in breeding and in individual cow performance, heifer replacements are frequently reared. Intensive pig and poultry enterprises are sometimes found where capital and labour are less of a limitation than land. A sheep enterprise such as store lamb fattening, can be complementary to a dairy herd when capital is more limited than land, although there is often competion between a ewe flock and a dairy herd for spring grass, especially in a late season.

The general features of small family dairy farms include a conservative attitude to business expansion, to bank borrowing and to unnecessary complication of the farming system. The herd is cared for by the farmer and his family with the labour input per cow being higher than with any other type of herd. Long working days, infrequent time off and minimum holiday periods are common, but the attraction and the major reasons for success are the satisfying life obtained. Detailed care and attention are given to the cows, and involve individual feeding and management. Capital is often an additional limiting resource, so that investment in machinery and equipment tends to be low.

With a general trend in the agricultural industry towards machines with higher output, it is possible for the small farm to be over-equipped in physical terms but, if second-hand equipment is purchased, excess investment can be prevented. Machinery is usually well maintained as it is predominantly owner-operated. As a means of reducing the capital input into mechanisation, some small farmers have joined machinery syndicates, whereas others are increasingly utilising the services of local contractors for such operations as forage conservation and manure spreading. Breeding performance of cattle in these herds is generally good due to the close and accurate observation of oestrus. Artificial insemination is widely used, especially as it is less expensive than having a bull.

Some marked differences occur in the systems of feeding and housing according to location. In the cooler, higher rainfall areas, as in the northern counties, the land is predominantly in permanent pasture. This provides reliable summer grazing and conserved fodder, traditionally in the form of hay, and, due to the climate, of varying quality. Buildings are substantially constructed of stone, proving difficult and expensive to adapt to changing technology. They comprise cowsheds, hay storage, dung compounds and effluent storage tanks.

In the milder southern counties, and especially in the south-west, a wider range of forage crops can be grown but there is still considerable dependence upon hay. Buildings are less substantial, allowing for more flexibility, herd expansion and in some situations outwintering is practised. Although the Friesian breed is predominant, there are many Channel Island cattle to be found on family farms in the southern counties. They respond well to individual management and, being smaller, can be stocked at a higher rate than other breeds. Spring calving is common, as peak fat production is appropriately matched to the needs of the holiday trade in this region.

Large family herds

These are some of the most progressive and profitable dairy enterprises in the country, having usually been developed from successful small herds. Care and attention to detail are maintained with family workers still much involved and working alongside employees. Management input to these herds is considerable, with effective communications, animal identification and data recording systems operating. Alternative enterprises are rare, apart again from heifer replacement rearing, as milk production is so well suited to these farms. Relief milking and feeding is frequently carried out by the farmer himself, thus assisting him to keep in close contact with operational detail. At other times, he is free to leave the farm, to attend meetings or conferences, a situation in contrast to that of the owner of a small herd.

Intensive grassland management is a feature of these farms as many are located, perhaps not by coincidence, in good grass-growing areas such as Cheshire, Somerset and south-west Scotland. Both permanent grassland and leys receive considerable quantities of slurry in addition to high inputs of fertiliser nitrogen and stocking rates are high. Winter feeding is based almost entirely on silage, except in Scotland where root crops and distillers grains are important feeds. Silage is usually self-fed but in some situations it is handled mechanically. The latter could be the case where a member of the family has undertaken a period of training and brings to the business the necessary skills and interest in mechanisation. Cubicle housing is used widely as it fits in well with these predominantly grassland holdings, using minimum bedding material and allowing for expansion of cow numbers with few difficulties. Slurry stores tend to be of high capacity, avoiding the need for winter spreading on the heavy soils. Application to the grassland normally takes place following a silage cut.

Large, high-throughput milking parlours, such as 20-stall, 20-unit herringbones, are common, so that milking which involves two people can take place in a relatively short period. By allowing a wide interval between milkings, which is advantageous to the cows, a useful block of time is available during the day to carry out other duties. Friesian herds are again the rule although, with an increasing interest in the feeding of high levels of forage, Holstein bulls are being used increasingly to breed cattle with higher intake potential. Extra milk is being produced from these animals and when coupled with the more efficient feeding, this is considered to more than compensate for the reduced returns from the bull calves which are not as popular with beef producers.

One-operator herds

Another numerically important type of herd is that operated by one employed stockperson. The herd is often the only dairy enterprise on a mixed farm or it may be one of a number of such dairies in common ownership, located together or scattered over a wide area. The main feature of these herds is that, apart from relief milking and holiday periods, the herd is milked, fed and managed by the one person. In theory, this should be an ideal arrangement with the cows getting used to the stockperson, as on a small family farm. In practice, physical and financial performance are extremely variable, depending upon the skills but more upon the attitude and motivation of the person in charge. Operating such a herd singlehanded is generally a lonely existence and one which can have many frustrations. Tasks such as assisting a cow at calving tend to be relatively easy if help is on hand but can be a considerable struggle when carried out alone. A highly motivated person obtains impressive results and considerable satisfaction, not only from the physical work but also from having responsibility for the herd. Others, with personalities unsuited to the environment, often feel dissatisfied with the employment and therefore achieve low performance levels. Standards of care and maintenance of equipment vary widely, as does the quality of record keeping. Relief work for an efficient stockperson is straightforward but can be a considerable problem for others, when, for example, the breeding records have not been brought up to date. Arranging for one person to act as regular relief for a number of herds is a common and satisfactory system, either for herds in the same ownership or on a co-operative basis for a group of owners.

Systems of production also vary in detail according to location. On the southern chalklands, typified by Wiltshire and Hampshire, outwintered herds are to be found on farms with large cereal acreages, utilising kale or silage from simple field clamps. Notable developments of this type of unit took place in the 1920s and 1930s by such pioneers as Hosier and Paterson. At that time, the milking facility comprised a simple portable bail which was moved regularly over a grass field. Milk was cooled and transported in churns but, with the more recent change to bulk tanks and therefore the need for an electricity supply, milking now takes place on a permanent site. Some farms have two sites, one located adjacent to the summer grazing area and the other in the farmyard for the winter months.

On arable farms, one-operator units are to be found, utilising by-products or areas of grassland which are in rotation or on parts of the holding unsuited to arable cropping. The feeding arrangements normally include a manger, which is filled by a self-unloading forage box operated by the stockperson or, more often, by a member of the arable staff. Loose housing is common, utilising the readily available supplies of straw, increasingly from big bales, and producing manure which is of considerable value, especially on light sandy farms. Milking systems vary widely in type and sophistication, depending to a large extent upon the size of herd. Simple abreast parlours are used with small herds and where the stockperson is content to spend extended periods involved in the milking process. Higher-throughput herringbones fitted with automatic cluster removers are more common in situations where a keen person is content to manage a larger herd single-handed. Relief milking is often a particular problem in arable areas due to there being few available people with the necessary interest and skills of dairy farming. Another specific problem is that of the reduced quality of forages which are presented to the herd at certain times during the winter months; a situation frequently outside the control of the stockperson. Such feeds directly reduce cow performance but also have an additional effect in demotivating the person who is supposed to be in charge of the herd. The arable farmer, and particularly the tractor drivers, have their major involvement in the cropping enterprises. In the interest of carrying out timely subsequent cultivation and planting, they tend to collect arable by-products from the field with speed as their key objective rather than producing quality fodder free from soil spoilage.

Large herds

These are units operated by two or more employees and they seldom directly involve the herd owner in the day-to-day work or organisation. These herds may have been established as large units or been expanded from small units with the aim of overcoming some of the problems discussed above. The frustrations associated with single-handed tasks can be minimised, the varying skills of individual staff members used to advantage and a self-relieving rota arranged. Improved opportunities are available for staff training both off-farm but more particularly on-farm, with a new entrant being able to work alongside skilled stockpeople.

Operational responsibility is usually delegated to a foreman or head stockperson and care has to be taken when this person is off duty to ensure effective further delegation. Second and third stockpeople can soon lose motivation if they are always given the unpleasant jobs and where there is a lack of encouragement for them to take an interest in the physical and financial performance of the enterprise.

A wide range of housing, milking and feeding systems are in use but as many of these units are in arable areas, the factors affecting choice of system are influenced predominantly by location. There are, however, particular problems associated with herds of this type in respect of the large volumes of materials to be handled. The removal of slurry or manure from buildings is a prime example, often being carried out by junior staff and with equipment that has a tendency to suffer from lack of maintenance.

Producing silage of an adequate quantity and quality proves difficult, particularly when the dairy staff have little or no involvement in the operation. It is common for the arable staff to be given responsibility for grassland conservation, especially on larger holdings where there may be more than one dairy enterprise.

Very large herds

Having experienced some of the problems associated particularly with two- and three-operator units, a number of farmers have followed examples of herd management from parts of the USA and Israel and increased cow numbers still further to establish herds with 300 cows or more. The economies of scale from such large units have proved to be small but it has been possible to create a system of management which provides a good environment for the cows as well as for the staff. Physical and economic performance of the better-managed units approaches that of the best large family units, although a number of others soon went out of business due to basic failures in planning or, more often, due to problems in staffing and in the control of the business.

A herd of this size justifies the employment of a highly skilled manager, together with a small team of specialised staff. An organisational structure can be set up so that at all times there is at least one person on hand to make decisions and to take any necessary action. The herd is usually subdivided, especially in the winter months, into manageable groups of 60–80 cows according to yield, date of calving or age. Forage of varying quality can be utilised by appropriate groups, although with silage feeds, having several clamps in use at any one time, the problem of deterioration at the silage face is accelerated. Complete-diet feeding is used in many of these herds, involving the mixing of concentrates with forage and allowing the animals to have *ad lib* feeding of the mix.

Materials-handling problems are potentially considerable in such large herds although, by using strongly constructed, high-output machines, as used by contractors, and having one member of the team in charge of maintenance, these can be controlled. For milking, large high-throughput parlours such as rotaries or herringbones are widely used but in some situations, capital is saved by utilising smaller one-operator facilities for extended periods. Shift-milking fits well into such a system, providing each operator with a reasonable length of day but allowing the cows a longer working day. An efficient system of record keeping is essential with a herd of this size and many have introduced computerised data control, using farm-based mini-computers, or are in contact with a central bureau.

With all dairy herds, irrespective of size or location, there are alternative feeding and management systems which can be operated. These will now be considered for the summer and winter periods.

SUMMER FEEDING AND MANAGEMENT

The importance of grassland in UK agriculture and particularly in milk production has been discussed in Chapter 29. Well-managed grass is a very economical source of food for the dairy cow but considerable problems arise in matching seasonal growth to the changing nutritional needs of a herd of cows. Very young or over-mature forage results in inefficient digestion, so that the general aim is to graze grass before it passes its best feeding stage and to conserve surplus material. Supplementation of an all-grass diet is necessary at certain times in the season. High-magnesium minerals need to be fed at all times: fibrous feeds, such as hay or straw, may be needed in early spring and energy supplementation is often necessary during dry periods, as hay, silage or early-planted kale. The feeding of concentrates to grazing cows, although widely practised, has seldom proved to be economic as, in most cases, replacement rather than supplementation of nutrients takes place. Concentrate feeding may be essential to the high-yielding spring-calved animal and is used by many as a convenient method of mineral supplementation.

The majority of UK dairy cows are turned out to pasture by day and by night during the summer months, although a few herds are housed after the evening grazing so as to be handy for morning milking. In exceptional cases, as for example where a farm is subdivided by a busy road, zero- grazing is practised. This involves the demanding process of cutting and carting fresh grass, on a daily basis, for the herd which is kept in confinement. An alternative system, but again rare due to high costs, is that of feeding conserved forages throughout the year.

For grazing, three main systems of control are in use, each aimed at minimising wastage but at the same time allowing the grass to be consumed at the optimum stage of growth.

Strip grazing

A limited new area of the field is provided once or twice each day by moving an electrified fence. This is the most efficient system of rationing grass but to operate satisfactorily it requires a stock-person with considerable skills and dedication. Areas of surplus material can be taken for conservation with relative ease, but poaching along the fence line is often a problem following heavy rain. In large fields it is necessary to use a back-fence to prevent the grazing of regrowth but this often creates problems in providing access to water troughs. Strip-grazing is also a satisfactory method for the grazing of kale, turnips or early-bite rye.

Paddock grazing

The area of grassland is subdivided into a number of permanent or semi-permanent paddocks. Each is grazed in rotation, nitrogen fertiliser applied, then follows an extended rest period for regrowth to take place. Ideally, each paddock provides one day's grazing but without adjusting cow numbers this is possible for only part of the season. Two- or three-day paddocks reduce the need for fencing and are ideal for a small herd. At the peak of spring growth, paddocks may be subdivided, whereas in late season, one paddock will provide grazing for only part of the day. Conservation of surplus material is usually an inefficient process due to the small size and often awkwardly shaped areas to be cut. The system is tending to lose favour due to the increasing cost of providing fences and water supplies.

Set-stocking

The herd is given free access to a fixed area which is grazed continuously over the season. It is simple to operate, with fertiliser applications taking place in rotation over parts of the area, thus avoiding problems with taint of the herbage. Stocking rates tend to be slightly less than with paddocks and although cows are generally content, they do have to spend more time grazing. It is therefore a more appropriate system for autumn-calvers, with their lower in-take requirements, rather than spring-calvers. The herd needs to be introduced into the area in early spring, receiving supplements of winter rations until full grass growth occurs. Additional areas can be added in mid-season from conserved aftermaths. The one main disadvantage reported by stockpeople is the additional time required to collect the herd from the larger area for milking.

In all grazing situations the need for an adequate supply of clean drinking water is important. Herd expansion on many farms has not been matched by an appropriate increase in the supply of water to grazed areas. Average consumption is 45 kg per day, with an increase in periods of hot weather. Cows tend to drink as a herd at particular times, such as 1–3 hours after milking, so that to provide adequate supplies, large-capacity troughs need to be available.

WINTER FEEDING AND MANAGEMENT

One of the most important decisions to be taken by a dairy farm is the type or types of bulk feed to be used as the basis of winter rations. A wide variety of feeds are in general use but the choice available to an individual farmer is limited by location and climate of the farm, as well as by such factors as capital availability, building layout and skills of the farm staff.

Hay

This is a traditional feed for dairy cattle and one that is still widely used, particularly with smaller herds. The capital requirement for field machinery and storage buildings is less than with silage, and manure or slurry handling is usually less of a problem with hay-fed cows. The handling of small bales is a reasonable task and, where building layout allows, large bales can be utilised. Hay does not have the smell problems often associated with silage but there can be a health hazard to staff handling mouldy material. Dry matter losses in field curing are much higher than with silage, even if the latter is wilted. The major problem with hay is that quality is so dependent upon weather conditions.

Artificial dehydration produces a high-quality material with low field loss, but in view of the high capital cost of the equipment, and the level of support energy required, it is unlikely that dried forage will have any significant future in dairy diets. Barn conditioning, however, which utilises cool air, may continue to have a place, especially on units where quantities to be handled are relatively small and where the buildings are suited to retaining bales which shrink on drying. The mechanical handling and storage of loose chopped hay has been tried on some farms but not taken up, due mainly to the low bulk density of the material.

Straw

Barley and oat straw are widely used in dairy units, although in small quantities, as a fibrous constituent, often as an alternative to hay. Chemical treatement is being increasingly used to improve digestibility by either on-farm processing or in specialised factories. This material can be used in greater quantities in diets to replace silage, but adequate protein supplementation is essential when the straw is treated with sodium hydroxide rather than with ammonia. As straw is a widely used bedding material, both in yards and in cubicles, many cows take the opportunity after new bedding is supplied to consume some material.

Silage

With the weather conditions generally prevailing in the UK, but particularly in the major dairying regions of the north and west, ensiling offers the greatest scope for maximising the output from conserved grass. Capital requirements for machinery and storage containers are higher than for hay but methods of reducing this commitment include the use of contractors or involvement in a machinery syndicate. The milk-producing potential of silage is determined by its nutritive value and by its potential intake, both affected by its digestibility and by such factors as wilting and the use of additives. Wilting, although of considerable benefit in materials handling and in avoiding effluent production, appears to be of minimal value to the cow. Additives, on the other hand, have been shown to increase intake and milk yield but some, i.e. the acid types, are particularly corrosive to forage harvesters. The digestibility (or stage of growth) of the crop at ensiling is the most important factor affecting the milk production potential from silage. Adopting a system of frequent cutting to increase digestibility lowers annual yield and therefore involves a reduction in stocking rate.

Alternative crops for ensilage are grown with success in certain regions of the country. Lucerne, if wilted to at least 25%, short-chopped and with a high rate of additive applied, can make good high-protein silage. Maize is a crop which, being harvested late in the season, spreads the use of harvesting machinery. It is high in energy content and therefore does not require an additive to obtain satisfactory fermentation. It is low in N content and therefore requires careful balancing, especially of protein quality, if optimum digestion is to be obtained. Many farmers prefer to feed maize silage to cows in mid- and late-lactation as an aid to live weight gain.

The feeding of silages can be arranged in numerous ways, with the particular system selected depending upon method of storage, building layout, herd size, capital and labour availability, but especially upon the interest and skills of farm staff in machinery.

Self-feeding is a very common method adopted, especially when the silage is stored in a concrete-floored clamp. Cows are held back from the face by an electrified wire or by a form of barrier (as described in Chapter 17). Each cow requires 150 mm of feeding face, or rather less if 24-hour access is provided. Difficulties arise with bullying, especially in large groups and particularly when there is a spread in calving dates. Heifers may suffer considerably unless they calve and settle to the system some time before the older and bigger animals. Wastage is sometimes difficult to control, especially in years when material of varying quality has to be ensiled in the same clamp. In addition to the self-feed face, farmers have found it to be advantageous to have, in the building layout, an alternative site for feeding other forages and concentrates.

Manger feeding of silage is gaining popularity, especially in the larger herds where subdivision of the herd is practised and where self-feeding space is limited. Tractor-mounted block-cutters are of increasing use for the cutting, and transporting to the feeding site, of blocks weighing up to 500 kg each. These are consumed direct from a facility such as a circular manger or they can be forked out along a straight feeding fence relatively easily by hand. Block-cutters leave a smooth undisturbed surface at the clamp which minimises deterioration, especially when compared to the more common method of unloading, which uses a tractor fore-loader. Silage made in large round bales and stored in individual plastic sacks is gaining popularity on some farms, the bales being transferred for feeding to circular mangers.

Self-unloading forage trailers are used to transport and dispense silage into mangers. These have to be loaded with care in order to avoid large lumps which easily damage the unloading mechanism. They are well suited to use with tower silos, as the material from this type of container is dispensed in a loose, even flow from bottom or top unloaders. Although some manufacturers claim that their forage trailers will reliably handle silage of varying chop length, all machines operate more effectively when the material has been chopped short by a precision type harvester. Forage trailers are relatively large and cumbersome machines and in order to avoid damage to them and to dairy buildings, skilful driving is essential, this being aided by the availability of wide passageways and by adequate turning space at the ends of buildings.

In a number of large herds, i.e. in excess of 200 cows, complete-diet feeding is practised. This involves a high capital outlay in a machine which has a V-shaped body fitted with three or more longitudinal augers, which mix together the various constituents of a diet. The machines are fitted with electronic weighing equipment so that accurate loading of ingredients is possible, as is the dispensing of known quantities of the mix for a specific group of cows. Even greater care and attention is required of the driver who also needs to have skills in stockmanship and a basic understanding of nutrition, in order to operate the system satisfactorily. Unless individual concentrate rationing is possible in the milking parlour, it is necessary to be able to subdivide the herd into several groups if over- or under-feeding of some cows is to be avoided.

Arable by-products

A wide range of farm-grown feeds are utilised by dairy cattle including manufactured by-products such as brewers' grains and sugar beet pulp. Potatoes, carrots and forage-harvested kale are suitable for mechanical handling, but kale and sugar-beet tops are more often strip-grazed. High yields of dry matter can be obtained from root crops such as mangolds or fodder, but considerable management input is required, especially in the winter months, to ensure efficient harvesting, storage and chopping of the roots.

Concentrates

The reasons for concentrate feeding of dairy cattle, especially in early lactation, were outlined in Chapter 20. Concentrates have traditionally been fed at milking time and, in a cowshed with the animals tied, rationing creates few problems. There is no limitation on feeding time and milk let-down is thought by many farmers to be improved. With the widespread move to loose housing and parlour milking, however, some changes to the method of feeding concentrates have been necessary. In a modern parlour, with a milking unit at each stall, high throughputs are being achieved so that the time available for feeding is restricted. Unless three-times-a-day milking is practised, which is rare, it is not possible for high-yielding animals to eat sufficient concentrates in the parlour, unless throughput is to be sacrificed. Additional quantities are therefore fed on one or more occasions during the day in an outside manger, this being achieved more accurately if the herd is subdivided into yield groups. Following the introduction of electronics into agriculture, the development of out-of-parlour concentrate dispensers has taken place. It is possible, although not widely practised in cubicle housing arrangements, to dispense small quantities of feed at frequent intervals to the front of each cubicle. More common is the use of special stalls in the housing area, connected to a bulk cropper, which have a dispensing mechanism activated by keys which are carried on a neck chain or collar on the cows which justify extra feeding. Problems with bullying have been experienced on many farms and some training of cows to the system has been worthwhile. The more sophisticated models ration the cows individually as each animal carries a key with a different signal. A digital display facility enables individual daily consumption to be easily checked and this has proved to be a valuable aid to health management and especially to oestrus detection (as the cow tends to eat less when on heat).

Alternative rationing strategies for concentrate feeding are in use, with the farmer's personal preference being largely the factor which determines choice on a particular farm. The most common system is for a regular adjustment of feeding levels following weekly, fortnightly or monthly milk recording. More liberal allocation is provided in early lactation, to assist high peak yields, with reduced inputs per litre of production in late lactation. An alternative system, particularly suited to late winter and early spring calvers, is 'flat rate' feeding which involves few changes to the level of concentrate input. All cows in a group receive an identical quantity with variation in dry matter intake being compensated by individual intake of silage or other background feeds. For optimum results, it is advantageous to arrange for block calving of the herd, within a limited number of weeks, so that more appropriate yield groups can be formed. There is always the danger that staff motivation will suffer when the decisions on individual cow rationing are removed from the stockperson's duties.

SEASONALITY OF CALVING

The season of calving affects the milk yield of a herd as well as both the costs and returns of production. Autumn calvers tend to have a higher lactation yield than spring calvers, since, just as their yields are beginning to decline, they are turned out to grass, which usually gives a fillip to their lactation curves. Autumn and early winter calvers generate a higher income as they produce a greater proportion of their milk during the winter months when prices are higher. Feeding costs are, however, considerably higher for autumn calvers as they consume more expensive concentrates and conserved feeds rather than grass. Stocking rates tend to be lower in spring-calved herds, especially if they are located in drier areas, although the majority of spring-calved herds do tend to be in the wetter western counties. Up to now it has proved much easier to achieve higher levels of technical and economic efficiency in the use of forage with spring calvers, although promising developments are under way with high forage-based feeding systems for autumn calving animals.

Dairy farmers need to match their season of calving to the supplies of forage, i.e. spring calving in the grassy south-west, or to match forage supplies to calving season, i.e. high-quality silage for autumn calvers in the drier south and east. The majority of producers have no concentrated period for calving, with cows calving throughout the year. This may be advantageous or even necessary to maintain supplies for a retail outlet but if unplanned, as in many cases, it leads to inefficiencies in feeding and frequently to a long calving interval.

BREEDING SYSTEMS

Artificial insemination is the most common method in use for the breeding of dairy cows. Semen from a wide selection of beef and dairy breeds is available from national as well as private organisations. Most operate a scheme for the testing of young bulls so that a high proportion of the inseminations are from bulls which are progeny tested. It is normal for an inseminator from a breeding centre to travel to the farm but increasingly, especially in large herds, semen is stored at the farm for insemination by a member of the farm staff who has undertaken specialised training. Where proper housing and penning are available, and where staff time is adequate, there are advantages, in a large herd, in having a bull at the farm. A small proportion of animals appear to conceive better to natural service and a bull is often a valuable aid to identify the animals which are in heat but are not seen to be mounted.

Bulls of beef breeds usually have a sufficiently docile temperament to be allowed safely to run with a milking herd. It is necessary to separate cows in the first 42 days of lactation to avoid early conception, short lactations and loss in yield. A bull running with a herd can easily be over-worked and it may be necessary to segregate cows which have been served. The use of a marking device strapped to the chin of the bulls is a valuable aid to identification of service dates, which is essential to efficient management.

REPLACEMENT MANAGEMENT

The average herd life of a dairy cow is about four lactations so that to maintain numbers in a 100-cow herd, 20 replacement heifers are required each year. Replacement rate varies considerably between herds and is influenced by such factors as the prevailing economic circumstances.

Short-term cash flow can be improved in difficult times by retaining in production, older or lower-yielding animals that would normally be culled. It is also possible to delay or even avoid the purchase of replacement heifers or, in herds where rearing takes place, to raise cash by selling a number of animals in the rearing stage. A period of high cull cow prices can be utilised to replace a higher than usual number of cows or to temporarily reduce herd size until the next batch of heifers calve. In many situations, high levels of enforced culling take place as the result of health management problems, especially with animals failing to breed satisfactorily. Farmers in these situations have a reduced opportunity to make decisions on culling for low yield or poor conformation, unless exceptionally high replacement rates are accepted. Costings clearly demonstrate that the returns from utilising farm resources to rear heifer replacements are much lower than for milk production so that high replacement rates are an additional cost burden on many farms. Another factor concerning replacement policy is that overall output may be reduced due to the lowering of staff motivation as a result of the retention of difficult animals, such as the one that will stand quietly for milking only if she is last of her batch to enter the parlour. Allowing for mortality and infertility, 50% of the average herd needs to be bred pure so as to supply the required heifer calves.

A choice of replacement policy can be selected from three major options, i.e. home rearing, purchase or contract rearing. Home rearing minimises the risk of disease, allows for tighter control, especially of the animals before calving, but utilises land, labour, capital and management resources. Contract rearing avoids the use of land and labour but it does require an investment of capital and it may be difficult to obtain a competent rearer. The purchase of down-calving heifers, although requiring considerable lump sums of ready cash, enables animals to be obtained with an appropriate calving date, age, weight and particularly conformation. Whichever system is selected on the individual farm, according to the availability of resources, the overall farm profitability is influenced by operating an efficient system which is well planned and controlled.

Age at calving is an important consideration and, although very early calving (i.e. less than 22 months) is not recommended, calving at around two years has proved to be a satisfactory method of minimising herd replacement costs. Although first lacatation yield is lower than that of older heifers, lifetime yield has proved to be greater. There are seasonal constraints on the age of calving due to the fact that the average calving interval is nearer to 13 months than to 12. In order, therefore, to maintain a calving season, heifers need to calve before the older cows. In herds where block-calving, e.g. within a three-month period, is practised, heifers can be calved only at two or at three years of age. A wider range of options exist for a herd with a widespread calving pattern but then $2\frac{1}{2}$ years is usually favoured.

In order to calve the heifers in a given month and at an optimum precalving weight for the breed, a definite plan of rearing has to be followed. It is important to achieve target growth rates at all stages of the rearing period. Growth rates which have to be accelerated in a subsequent period, usually increase cost and can lead to over-fat animals which can have an increase in calving difficulties.

The choice of bull for use on heifers is influenced by the risk of difficult calvings associated with some breeds, the number of dairy calves required and the relative value of beef-cross calves of the different breeds. Large beef breeds are generally avoided but the

Hereford and Aberdeen Angus are widely used, as are pure dairy breeds, particularly individual proven bulls which have no reported cases of dystokia.

CALF REARING

At calving, the cow and calf are, ideally, housed in an individual box or pen. Immediately after birth, the calf's most vital need is to suck colostrum, which is the milk secreted during the first few days after calving. Colostrum has a high nutritive value and is a mild laxative but, above all, contains valuable antibodies which help to build up the resistance of the calf to the infective organisms on the farm. The rate of absorption of antibodies into the bloodstream from the digestive tract is known to be at a peak some 3–6 hours following birth, so care needs to be taken to ensure that the calf sucks satisfactorily at that time. Leaving the cow and calf together for 24 hours is sound practice but, after then, stress is increased at separation and difficulties are often experienced when calves have to be trained to drink from a bucket. At this stage the calf should be moved to a straw-bedded individual pen in a well-ventilated but draught-free building. The availability of a number of smaller-capacity units is preferable to one single building, because this allows an all-in/all-out system to operate in each and therefore considerably minimises health problems. Dehorning and the removal of surplus teats, when necessary, takes place in the early weeks of life.

During the initial stages of rearing, the calf depends on single-stomach digestion and so requires its food in a liquid form. The practice of suckling up to four calves at a time on a nurse cow produces fast-developing animals but is a system better suited to the rearing of bull calves destined for beef production. The feeding of whole milk is expensive and is therefore normally replaced by milk substitutes. The feeding of surplus colostrum in a sour form or following deep-freeze storage, has proved to be satisfactory. The rumen of young calves develops rapidly when access to dry foods is provided thus allowing 'early weaning' to take place.

Numerous systems exist for the feeding of milk substitutes, although bucket feeding is the most widely used method. It allows easy recognition of any calf not drinking and enables full control of the level of intake of individual animals. 'Twice-a-day' feeding of milk at blood temperature is traditional but alternative systems include cold milk i.e. at tap water temperature, and 'once-a-day' feeding. The latter is a most suitable system for smaller dairy and one-man farms, allowing calf feeding to take place during the day, between milkings.

Ad lib milk feeding systems have been developed which use

automatic dispensing machines or lower-cost bulk containers which supply the milk by plastic tubes to a bank of artificial teats. These systems save labour and provide an ideal little-and-often supply of milk, but the consumption levels and therefore feed costs per animal are much higher than with rationed feeding. Special long-life acidified powders have been formulated for *ad lib* feeding systems, the liquid staying fresh for up to 3 days. Such systems are well suited to the rearing of spring-born calves at pasture, utilising low cost temporary shelters.

After weaning, concentrates and high-quality forages need to be fed to maintain target growth rates. Calves are usually turned out to grass at about six months of age and it is during this first grazing season that most difficulty arises in maintaining adequate growth. Supplementary feeding may be essential at both ends of the season and parasite control is of major importance. Stocking rates are very dependent upon the level of fertiliser nitrogen applied to grass. It is possible to allow lax grazing with the calves and to have a group of older animals following in rotation to clear up lower quality forages, a system known as 'leader/follower'.

Weight at service is a key factor in obtaining heifers of adequate size at calving. As service takes place at 15 months for calving at two years, concentrate supplementation of conserved feeds may be necessary in the early months of the second winter-feeding period in an autumn-calved situation. Rations for the in-calf animal can be based predominantly on forage feeds, which will involve grazing in the case of the autumn calves. Animals need to approach calving in a fit but not fat condition, in order to avoid difficulties.

HERD HEALTH

Considerable progress has been made in recent years to control and, in some cases, to eliminate disease from the national dairy herd. Nevertheless, disease continues to add to the cost of milk production both directly for preventive and curative measures and, indirectly, in terms of losses in production. Good stockmanship is a key factor in minimising this cost, by early diagnosis of disease symptoms and prompt corrective treatment, together with closer cooperation with the veterinary profession, particularly in terms of preventive medicine.

Outbreaks of highly contagious diseases, such as Foot and Mouth and Anthrax, are fortunately rare. These are notifiable diseases and therefore dealt with using strict regulations. Tuberculosis and Brucellosis have now been virtually eradicated but research programmes continue to investigate the diseases which infect such organs as the lung, liver, alimentary tract, uterus and udder. Readers are referred to appropriate texts for a comprehensive description of cattle diseases so the comments that follow are only related to the common disorders of dairy cows, their prevention and control on the farm.

The three diseases which most seriously affect production efficiency and limit longevity in the dairy cow are mastitis, infertility and lameness. Mastitis is an inflammation of the udder, caused predominantly by microbial infection but also by injury, secretory malfunction or physiological change. It is difficult to recognise in the initial stages, the first symptoms being clots or discoloration of the milk and/or a hot swelling of the infected quarter. These are followed by a reduction in yield and even the complete loss of the quarter. The disease can be present undetected in the udder, so laboratory tests are needed to identify such cases. Milk tests are also used to type the strain of bacteria so that appropriate antibiotics can be prescribed by the veterinary surgeon: the most widely used materials are penicillin, streptomycin and cloxacilla. Great care needs to be taken to identify the cows which receive treatment so that milk can be rejected for the specified period, to prevent any traces of antibiotic reaching milk to be sold for consumption. Heifers calving for the first time are usually free from infection, with the normal spread of the disease being at milking time. It is therefore important to have an efficient hygiene routine, especially one which includes post-milking teat disinfection as well as having a regular maintenance programme for the milking plant. Contamination of the udder during the dry period has also been shown to be an important factor with mastitis, so it is common practice to treat animals with long-acting penicillin at drying off. Records should be maintained of all clinical cases and treatments as an aid to making culling decisions, which may be necessary with chronic and incurable cases.

Infertility is another major production disease of dairy cattle and is the most important reason for culling. Failure or delay in breeding due to difficulties in coming into oestrus, conception or abortion are costly in terms of lost production. Specific diseases such as *Vibrio foetus* and metritis contribute to infertility but major predisposing factors are nutrition and management. The effects of feeding on successful conception can start in the previous pregnancy and particularly critical is the management at calving. Cows with retained foetal membranes require close observation and in most cases antibiotic treatment, and cows with ovarian malfunction frequently require hormonal therapy before conception is possible. Pregnancy diagnosis is an important management aid and is normally carried out by a veterinary surgeon 40–60 days after service. As with mastitis, an efficient recording system is essential to the managment of herd fertility. The role of the stockperson cannot be overstressed in the effective observation of oestrus and in spotting pus-like discharges from the genital organs following calving.

Foot problems are another major reason for the culling of dairy cattle. Lameness caused by inflammation of the hoof is an obvious symptom, with accompanying loss of weight and production. Infection enters through wounds in the skin near the coronet of the hoof and the problem is increased in wet and muddy conditions. Antibiotic and sulphonamide injections are usually effective but keeping the feet in sound condition by regular trimming and foot bathing in formalin or copper sulphate has been shown to be worth while. Overfeeding, followed by laminitis, can easily occur in high-yielding herds and a new concrete floor with a highly abrasive surface often causes severe foot troubles. A sound policy is to select bulls for breeding replacements whose progeny have above-average ratings for feet and legs, particularly in respect of depth of heel.

There are a number of fairly common metabolic diseases of dairy cows which are very much related to herd feeding and management.

Milk Fever (hypocalcaemia) is characterised by low levels of calcium in the blood. It normally occurs within a few days of calving. The symptoms include loss of control of the limbs, followed by a coma and finally by death if no treatment is given. However, the administration of a subcutaneous injection of calcium borogluconate using a flutter-valve is usually a successful cure if given in time. The feeding of high phosphorus minerals pre-calving and also an injection of vitamin D_3 have been shown to be valuable aids in preventing milk fever.

Acetonaemia (ketosis) is the result of inadequate carbohydrate metabolism due to incomplete oxidisation of fatty acids. The result is an abnormally high concentration of ketones in the blood, producing the characteristic symptom of breath smelling of acetone: the animal is off her food and there is a rapid drop in milk yield. The disease normally occurs in housed animals at the end of the winter feeding period, particularly in high-yielding cows. It is preferable to prevent the disease by providing high-energy feeds. Direct administration of glucose temporarily alleviates the condition but spring grass tends to have a magical curative effect.

Grass tetany (hypomagnesaemia) is associated with low blood magnesium levels. It commonly occurs in the first few weeks of grazing in the spring but also at other times, especially when the animal is under stress. The symptoms are quick to develop, starting with a general nervousness, staggering gait followed by collapse, convulsion, coma and death, all in the space of a few hours. The treatment, if given in time, is very effective, and consists of a subcutaneous injection of magnesium and calcium salts. Prevention is normally by feeding a daily supply of 55 g of calcium magnesite per head.

Bloat occurs when the gases produced during rumination cannot escape and is usually associated with the grazing of lush pastures, especially those with a high clover content. The most obvious symptom is a swelling on the left upper side of the cow. Emergency measures include piercing the rumen wall with a special instrument known as a trochar and cannula to release the gases. Dosing with detergent or oil is possible in the early stages but prevention by increasing fibre intake is a worthwhile practice.

External parasites cause animals discomfort and are often transmitting agents for disease. Lice are a particular problem during winter housing. Warble flies lay eggs on cattle in early summer. After hatching the larvae burrow beneath the skin and eventually damage the skin by creating breathing holes and by the emergence of pupae. Control is exercised by applying systemic dressings to the cow's back in October and November.

Ringworm is caused by a fungus and affects the surface cells of the skin and hair, more commonly in young cattle than in cows. Patches of hair fall off, leaving hard whitish areas of exposed skin. The disease spreads quickly, especially among housed cattle, and man can also contract ringworm. Disinfection of buildings helps in control but drugs are available which can be included in the diet of infected animals.

Internal parasites are of considerable economic importance in milk production. Liver fluke is more widespread in dairy cows than was earlier imagined. Losses include condemned livers at slaughter and an unknown reduced level of milk production and decrease in fertility. Drugs are available to dose infected animals but veterinary advice is needed to draw up a safe control programme which avoids discarding considerable quantities of milk.

Husk is a condition which, although easily prevented, causes severe coughing and loss of production in young cattle. Lungworm larvae picked up from infected pasture develop in the lung, larvae are produced on a lavish scale in the bronchi, eventually being excreted in the faeces. A very effective vaccine is available, but it is expensive so some farmers take a calculated risk by not vaccinating calves but grazing them on 'clean' pasture (ungrazed, at least by cattle, for a year or more).

Stomach worms are also common in young cattle, especially in intensively grazed systems. Heavy infestation causes scouring and loss of condition. As with husk, provision of 'clean' grazing is advisable but a wide range of drugs are available for use as a dose or injection.

36

BEEF PRODUCTION SYSTEMS

RESOURCES USED IN BEEF PRODUCTION SYSTEMS

There are great advantages to the farmer in using planned systems of production as a guide to achieving predictable performance. Beef production, more than most agricultural enterprises, lends itself to *ad hoc* methods of utilising feeds on the farm, because of the ready availability in markets of growing cattle of all ages which can be purchased for the purpose. However, such methods leave the farmer at the mercy of fluctuating market forces. Greater consistency of output and of economic returns over a period of years can be expected from defined systems than from unplanned methods of production. Systems of beef production are described in terms of their interacting components. These are primarily the cattle, feeds and management methods which are required to meet well-established targets for live weight and live-weight gains at defined points in the life of the animal. The performance of the farmer's own livestock can be compared with these targets by monitoring live weight and feed supplies. If the specified animal performance is achieved in relation to a well-defined market requirement, then economic returns can be predicted within a defined cost/price structure. This is the main justification for well-described systems, but they have the added advantage of providing a valuable link between research, development, advisory work and farm practice.

In the following sections the components of systems, i.e. the principal resources required for the beef enterprise in meeting the needs of the meat market, are described with particular reference to their importance in the systems considered. The source of the animal (dairy or suckler herd) is closely linked with breed; it also determines the method of rearing, and influences the system of production in which the calf is used. Sex, like breed, has similar effects, in general, across all systems. Feeds used in systems vary with the agro-climatic region, which has a considerable influence on the type of beef enterprise undertaken. Other resources, such as the size and type of farm, and the availability of buildings, capital or labour, have their effect on the choice of a beef production system, and the hazards of ill health vary with different systems. Finally, descriptions of the product, in terms of the breed, age and weight of the animal, and the weight and composition of its carcass, provide a most important link with the beef market and the economic returns to be expected from the system.

Level of fatness of the animal is an important determinant of the end-point of a beef production system since the deposition of fatty tissue (of high energy content) is expensive in feed energy intake by the animal. At the fattening stage in cattle, therefore, when a small increase in lean meat content in the body is accompanied by a high proportion of fatty tissue (e.g. Table 13.9), increases in live-weight gain beyond the optimum level of fatness neeeded to meet consumer requirements are uneconomic. However, given a slaughter point at equal levels of fatness, although large breeds take longer to reach this end-point, and therefore consume higher levels of feed, than small breeds, there are not large differences in efficiency of feed conversion (kg LWG per kg feed). The need, in beef production systems, is therefore to choose the breed and sex to suit the feeding system appropriate to the farm, and to ensure that the cattle are slaughtered at an optimum level of fatness. The optimum will be determined by the market requirement and its choice will be assisted by assessment of fat cover in the live animal and by carcass classification (see Chapter 13).

Sex effects and growth-promoting implants

In general the target live weights and gains described for beef systems are for steers. Many of the cross-bred heifers from the dairy herd are not available for fattening systems as they are purchased as replacements for culled cows in suckler herds. In general, the early maturity of heifers makes them less suited to high plane feeding systems with a large proportion of cereals in the diet, whereas the later-maturing bull is well suited to such diets. Bulls are less suited to out-door grazing systems than to indoor feeding, because of the higher cost of fencing required to meet safety regulations at pasture, and other management problems. The retention of uncastrated bull calves also poses management problems in the suckler herd because of the difficulty of managing mixed sexes after the calves reach five months of age. The young bulls then become capable of mating the heifer calves, which are intended for sale as fattening stock, and at a later stage may serve the suckler cows themselves.

There is much experimental evidence on the use of implants of anabolic (growth-promoting) substances in cattle. Increases in growth rate in steers of 10–40% have been obtained from hormones and other anabolic agents placed as a subcutaneous implant at the base of the ear. New regulations for the control of anabolic agents were introduced in 1981 within the EEC, when stilbenes and thyrostatic growth promoters were banned. The continued use was allowed of implants of three natural hormones, oestradiol, progesterone and testosterone, and of two substances which do not occur naturally in the animal, trenbolone and zeranol.

Sources and characteristics of feeds used in beef systems

The area of effective grassland in the UK in 1979 was estimated to cover about 8 m. ha when heath and moorland of little agricultural value are excluded. Most of this grassland is in the west of the country, where it comprises 70–90% of the agricultural area as compared with 17–40% in the east. Beef cattle are also more densely distributed in the west (50–90 beef cattle per 100 ha agricultural area) than in the east (17–37 per 100 ha). This broadly reflects the dependence of beef production on grassland, although appreciable numbers of beef cattle are also kept on arable farms to utilise by-products, and forage crops such as maize.

In the most widely used beef production systems, grass provides 75–85% of the energy requirements of beef cattle and up to 100% of their protein for much of their lifetime requirements. Overall, the contribution of grain and other concentrate feeds used in beef production in the UK does not exceed about 20%.

The predominance of grass in the agricultural area in response to the temperate climate and topography of the country, and the relatively low cost of energy in the form of pasture and conserved grass has resulted in a greater dependence on grass for beef production in the UK than in other countries of the EEC. The predominance of early-maturing breeds in the mainly cross-bred population of the beef industry in the UK is consistent with this dependence on grassland. Although the energy content of grass at very young stages of growth can be as high as that of some samples of barley grain and the protein content of the grass is much higher, the average nutrient content of grazed and conserved herbage, in practice, is low. Furthermore, the restriction of intake by the animal which often results from management practices imposed during grazing (Chapter 21) and winter feeding can limit the nutrients supplied to beef cattle from grassland. Under such restrictions, early maturity in cattle is an advantage. The larger, later-maturing breeds can fail to reach a suitable condition for slaughter on grassland unless management is carefully controlled to that end.

Economic assessment of beef systems

In assessing economic returns from beef production systems, most emphasis is placed on the costs of production which have to be met for individual cattle: namely the 'variable costs' of purchase of the livestock and feed and of the home-produced forage, bedding, marketing and veterinary charges. These variable costs are deducted from returns on sales to give gross margins. Costs which do not vary directly with livestock numbers; namely the 'fixed costs' of general farm overheads, rent, buildings, machinery and labour, are deducted from gross margins to give net profits. It is important to make optimum use of these farm resources, and the choice of beef production system for the individual farm will be guided by the availability of land, buildings, capital and labour as well as the supplies of crops or by-products which form the basis of the system.

SYSTEMS OF BEEF PRODUCTION

Systems are usually named by reference to their duration and the types of feed and cattle employed. Many systems encompass the whole lifetime of the animal on the same farm, but there is a great deal of trade in cattle which leads to partial systems on many farms. Descriptions of systems require a specification of seasons, specific types of feeds and management, live weights, and live-weight gains. When systems have been in use for many years and have been recorded in large numbers (as by the MLC), a good guide to performance of the cattle, or 'target', is provided by the average live weights or gains derived from a number of recorded farms. The MLC further analyses its records to examine the mean performance of the producers who attain the top one third of the range of gross margins per hectare. The components of success in the system can be identified by comparing the various factors involved for the average and 'top third' producers. Although results of systems operated on experimental farms can be used, these are usually at a higher level of performance than can be expected under commercial conditions.

SYSTEMS OF PRODUCTION FROM CALVES BORN IN DAIRY HERDS

Calves born in dairy herds are the most important single animal resource used for beef production in the UK since they supply some 60% of the prime beef from systems starting with calves (i.e. excluding culled cows and imported store cattle). These calves from milking herds are mostly sold at one to two weeks of age to specialist calf rearers or beef producers, and enter systems of production with an age at slaughter which ranges from ten months to over two years.

Grass/cereal and cereal systems

The duration of systems increases as greater dependence is placed on grass. On cereal feeding the lifetime of the cattle is 10–12 months. Three main grass/cereal systems can be distinguished in which age and live weight at slaughter become greater as more use is made of grass and less of cereals (Table 36.1). The quantity of concentrates used per kg live-weight gain declines from 5·4 kg in the cereal system to 2·3 kg in the 24-month system. This factor is important when the ratio of cost of concentrates to grass changes.

Table 36.1 Performance of cattle and use of concentrates in the cereal and three grass/cereal beef systems (means for 5 years, from Kilkenny & Dench, 1981 and MLC, *1981)

	Live weight at slaughter kg	Daily live-weight gain (DLWG) kg	Stocking rate cattle/ ha	Concentrate used tonnes/ head	Concentrate, kg per kg (DLWG)	Gross margin,* £ per head	Gross margin* £ per ha
Cereal system	392	1·0	–	1·8	5·4	53	–
Grass/ cereal systems							
15-month	433	0·8	7·2	1·2	3·3	–	–
18-month	471	0·7	3·2	1·1	2·7	175	539
24-month	498	0·5	2·4	1·0	2·3	189	399

* Adjusted for inflation to 1980 prices

Eighteen-month beef

One of the most widespread systems of using calves born in dairy herds is based on the use of grazed and conserved grass with limited supplements of cereals, to produce cattle fit for slaughter at eighteen months of age. Since a higher proportion of calves in the UK are born in the autumn than in spring, the system most commonly starts with calves born in July–December. These are reared indoors in their first winter, grazed during the following summer and 'finished' (i.e. brought to a carcass composition suitable for slaughter) during their second winter on a diet of conserved grass and concentrates.

The reared calf is fed from 2–3 months of age to about six months on hay or silage with a concentrate supplement limited to about 2·0 or 2·5 kg per head per day. With steers the aim is a daily live-weight gain (D L W G) of 0·7–0·8 kg per head, with a target live weight of 180–5 kg at the time the calves are turned out to pasture in spring (Fig. 36.1). Rates of gain higher than this can be achieved with a higher level of concentrate supplement, but if this is done, the rates of gain at pasture are relatively lower. Costs per unit of live-weight gain are usually lower on pasture than on winter diets. Rates of gain in the first winter which are much below the target, however, may result in calves which do not meet the levels of live weight required later in the system.

The grazing season, covering the period 6–12 months of age, depends on efficient grazing management methods which aim for high rates of gain by the cattle. Targets are 0·8–0·9 kg D L W G, and the cattle should weigh 300–50 kg at one year of age at the end of the pasture season (Fig. 36.1).

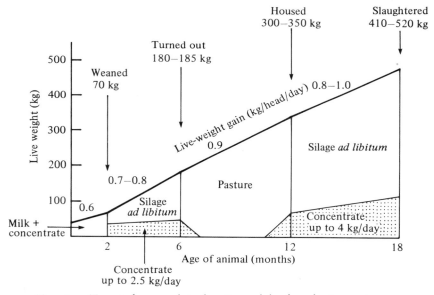

Fig. 36.1　Targets for growth in the 18-month beef production system
Sources: based on Baker *et al.* (1967) and MLC (1979)

An important aspect of grass management in this period is to supply adequate levels of fertiliser. Nitrogenous fertiliser is applied four or five times in the season, following cutting or grazing, to give a total application of 200–300 kg N per ha. About one-third of the grass area is grazed in spring and the remainder cut for conservation as silage or hay in late May or early June. The grass grown after cutting is grazed, with some further areas cut for conservation in late summer. Anthelmintic drugs are used, ideally in combination with a move for the cattle to previously ungrazed areas in July, to control infestations of intestinal parasites.

In the final winter period of six months the cattle are fed on the conserved silage with increasing quantities of a supplement of concentrates such as mineralised barley. The amount of supplement fed will depend on the quantity and quality of conserved feed available but will be about 2–4 kg per head per day. The target DLWG is 0·8–1·0 kg, leading to a live weight at slaughter of 410–520 kg per head and a dressed carcass weight of 180–270 kg. In practice the range of ages at slaughter of individual cattle may vary between 16 and 20 months.

Experiments have shown the possibility of omitting supplements of barley during the winter periods of this system. With cattle of an early-maturing cross-breed, the Sussex × Friesian, mean carcass weights of 235 kg were attained with steers on a forage-only system, compared with 244 kg on the normal (supplemented) system. The respective carcass weights for heifers were 217 kg and 223 kg. Feed costs were £20/head lower, and gross margins per head were higher, in the forage-only groups.

Breed effects The breeds most commonly used in the 18-month system are the Friesian and the Hereford × Friesian. The cross-bred Hereford × Friesian reaches a given level of carcass fatness somewhat earlier (at about 450 kg at 17-month age) than the Friesian (at 500 kg at 19-month age), but alternatively a lower daily quantity of concentrate supplement may be given to the cross-bred animal in the final winter than to the Friesian. Aberdeen Angus × Friesian cattle have a smaller size and fatten earlier than the Friesian or Hereford × Friesian. The Charolais × Friesian is less fat at a given weight, and grows to larger sizes (550 kg at 20-month age) than the Friesian or Hereford × Friesian.

Sex effects and growth promoters Steers are most commonly used in the 18-month system. When bulls are used, suitable fencing is required for safety, both at pasture and indoors. Additional cereals may be required in the final winter to raise the level of nutrition and ensure adequate levels of fatness in the carcass of bulls. Targets for daily live-weight gain are 0·1 kg higher for bulls than for steers of the same breed and live-weight targets will be some 10% higher than for steers at the time of housing and slaughter. This will result in live weights and carcass weights considerably higher than for steers of early-maturing breeds. Such targets have been achieved with small experimental groups of bulls of relatively late-maturing breeds. Adequate levels of carcass fatness (conformation classes U+, U and R, and fat classes 3–4 of the EUROP classification) were attained profitably, without the need for high levels of concentrate supplementation, at slaughter weights of 550–80 kg and carcass weights of 300–30 kg in bulls of Friesian and South Devon × Friesian breed at 18 months of age.

Field trials with implants of the growth-promoter trenbolone or zeranol in the ears of Friesian steers in an 18-month system gave an increase in growth rate respectively of 7% and 13% in a 4-month treatment period prior to slaughter.

Economic aspects   The beef enterprises on farms recorded by the MLC which make the highest gross margin show a higher DLWG, slaughter weight and stocking rate, but a lower use of concentrates. Most of the greater success of the producers with the top third of gross margins derives from the higher stocking rate (34% contribution to superiority) linked with higher applications of nitrogenous fertiliser, and lower usage of concentrates (48% contribution). At 1979 prices, fixed costs in the 18-month system comprised some 60% of the gross margin.

An 18-month period of feeding is less suitable with spring-born calves because the time of slaughter is in the autumn, when many cattle are finished at pasture and the seasonal price for beef tends to be low. The 15-month system of finishing indoors on an *ad lib* supply of cereal/protein concentrate avoids this, and is also used for calves born during winter.

Fifteen-month beef

This system applies to calves born between November and February, which are reared to gain at 0·8 kg per day up to the time when they are turned out to graze. Since they are fed in yards in their second winter on a diet almost entirely composed of concentrates, it is important to the profitability of the system that gains during summer are as high as possible, and a gain of 0·7 to 0·8 kg per day is the aim. Concentrates may be fed at pasture at high stocking rates to achieve this target. The best results, however, are attained when other stock are utilised in a 'leader-follower' system in which the young cattle have first access to fresh areas of pasture but are moved on when plenty of grass is still available, leaving older stock to clear up the remainder. Age at slaughter is at 13–17 months, at a live weight of 385–454 kg and carcass weights of 205–50 kg. The Friesian steer is more suited to this system than the Hereford × Friesian, which tends to fatten too rapidly on the all-concentrate diet. For the same reason, i.e. their suitability to high-plane finishing, bulls are well suited to this system.

Two-year beef

Steers from calves born in November–February are often kept for up to two years to be finished at pasture in their second summer.

Target gains and management in the first grazing season are the same as for the 18-month system. However, since there are cattle of two age groups on the farm in this system, a suitable method of grazing management is the 'leader-follower' one in which the young cattle graze fresh plots of grass first and are followed by the older cattle. This gives the younger age group advantages of high availability of grass, an opportunity for selection of high-quality herbage, and reduced exposure to nematode parasitic infections. These calves will have a target weight of 225–75 kg at the end of the grazing period. The second winter period is suitable for 'store' feeding, i.e. a rate of live-weight gain which will allow slow growth of skeleton and muscle, but not much development of fat. A diet of silage or hay with 1·5–2·0 kg of concentrate is suitable to achieve the required low growth rate of 0·5 kg D L W G. Such cattle, when turned out to graze, exhibit 'compensatory growth' i.e. they grow rapidly at pasture and reach about 1·0 kg D L W G. These animals are suitable for slaughter from pasture between 20 and 24 months of age between July and October at live weights of 450–550 kg.

The leaner, later-maturing breeds such as the Friesian or Charolais × Friesian are less suited to this system than the early-maturing crosses such as the Hereford × Friesian because the late-maturing animals may not all reach the required degree of fatness at pasture and would then need a period of feeding indoors. The late-maturing breeds will give carcasses in fat class 3 or 4L of the E U R O P classification, whereas the early-maturing breeds will classify in fat class 4H. Stocking rates will be between 2·0 and 2·4 cattle/ha.

Cereal beef

Cereal beef is a specialised system of production from dairy-bred calves which comprises less than 5% of total beef production in the U K. The cattle are kept entirely indoors to slaughter at 10–12 months of age and are fed on *ad lib* supplies of cereal grain (usually rolled barley) with supplements of protein, minerals and vitamins. Calves enter the system at 10–13 weeks of age with live weights of 100–110 kg, when the dietary crude protein requirement is 14% and the target growth rate is 1·2–1·3 kg D L W G. At 6–7 months of age, with a live weight of 250 kg, the dietary protein content is reduced to 12% and the target is 1·3–1·4 kg D L W G. Steers of the Friesian breed are taken to slaughter at 400 kg; bulls at 450 kg.

This system can be started at any time of year, but the buildings used must be of a design suitable to minimise the incidence of respiratory disease. Safety regulations, when bulls are housed, require pen divisions of suitable height and design, and specify the need for two people to be present if a pen is entered.

Early-maturing breeds and crosses such as the Aberdeen Angus and Hereford are not suitable for this system since they lay down excessive amounts of fatty tissue at the high plane of nutrition which results from *ad lib* cereal feeding. Late-maturing breeds and crosses are ideal, giving high carcass weights without excessive fat. This system will result in carcass weights of 210–240 kg, with a EUROP fat class of 3.

Disease hazards in cereal feeding are chiefly those concerned with the diet and life-time housing. Acidosis and bloat (ruminal tympany) can occur, and the feeding of 1·0–1·5 kg per day of long straw or hay will help to minimise these conditions. Disorders of the feet, and rumenitis, can be prevalent, and there is often an incidence of liver abscesses. Pneumonia can result from poor ventilation and humid housing conditions, and is aggravated by draughts and over-crowding. Veterinary advice and treatment should be sought for these conditions.

The mean gross margin per head in this system (1976–80) was £53, which compares with £175 per head for the 18-month system (Table 36.1). Producers with gross margins in the top third of the range (most of whom keep bulls rather than steers) achieve higher feed conversion efficiency (45% of their superiority), higher sale weights and prices (18% and 19% of their superiority) and lower feed costs (11% of superiority).

As alternative feeds in the cereal beef system, oats, wheat, maize and sugar-beet pulp can be introduced to a varying extent to replace barley as the main source of energy. Protein supplements can include dried grass or lucerne, and urea can be used as a source of dietary nitrogen at limited levels of inclusion in the protein supplement.

Roots such as turnips or swedes can be used to replace half of the cereals in the diet of cattle at 150–250 kg live weight or the whole of the grain above these weights. Since roots yield more energy per ha than cereals, this will lead to increased outputs of beef per ha of land used to grow the feed.

Silage systems

Although grass silage provides an important part of the winter diet in 18-month and two-year beef systems, silage can also be used as the major part of the diet in cattle which are housed for their entire lifetime. These systems lend themselves to the use of bulls in situations where it is more practical to control these animals indoors throughout their lives.

Maize silage beef

Maize is a crop which can give high yields in a single harvest and is easy to ensile to provide a feed of high energy content. Protein supplements are required to meet the needs of cattle in early stages of their growth, but non-protein nitrogen supplements such as urea can be used to cheapen the supplement at later stages of growth. A daily gain of 0·9 kg per day is the target for Friesian steers, with a live weight at slaughter of 450 kg. This is a high-energy system, well suited to bulls (slaughter weight 490 kg for Friesians) and late-maturing breeds.

Grass silage beef

Grass silage of high nutrient content (10·2 M J/kg D M) has been used successfully with minimum supplementation (2 kg concentrate to one year of age, 2–4 kg barley from one year to slaughter) in an indoor lifetime feeding system for Hereford × Friesian bulls (Hardy, 1981). At slaughter, a mean live weight of 406 kg and carcass weight of 220 kg resulted, with a E U R O P carcass classification of U or R for conformation and a mean of 3·9 for fat class.

SYSTEMS BASED ON CALVES BORN IN SUCKLER HERDS

The proportion of beef produced from calves born in the U K which is derived from suckler herds is about 40%. Suckler systems tend to produce cattle at lighter weights (Table 36.2) than dairy beef systems (Table 36.1), probably as a result of the higher proportion of beef breeding and more rapid growth rates in suckler herds. Con-

Table 36.2 Performance of cattle and use of concentrates in suckler systems (means for 5 years, from Kilkenny & Dench, 1981 and M L C, 1981)

	Live weight at slaugh- ter kg	DLWG kg	Stock- ing rate cattle/ ha	Concen- trate used tonnes/ head	Concen- trates, kg per kg DLWG	Gross margin,* £ per head	Gross margin* £ per ha
Suckler systems 18-month beef calf – autumn calving	418	0·8	1·7	0·9	2·5	–	–
2-year beef calf – spring calving	444	0·5	1·2	0·7	1·8	–	–
Lowland	–	–	–	–	–	145	267
Upland	–	–	–	–	–	173	257
Hill	–	–	–	–	–	170	199

* Adjusted for inflation to 1980 prices

centrate use tends to be lower, but gross margins are also lower. The margins are improved when calves are finished on the same farm as compared with selling them as weaned calves and stores. Several factors need good management if profitable production is to be achieved in any suckler herd.

Calving dates

A traditional pattern of spring calving has developed in over half of the total suckler herds, which allows the cow and calf to make optimum use of summer pasture. The weaned calves are mostly sold at 6–9 months of age at autumn sales, and these calves may be fattened at 12–15 months of age or kept as stores for sale in spring and fattened at 18 months to two years of age (Fig. 36.2). Autumn calving has been found to give higher gross margins than spring calving but choice of calving date will depend on specific farm situations and resources.

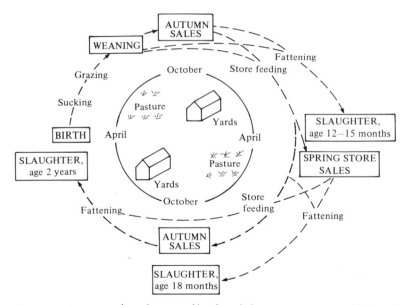

Fig. 36.2   Patterns of production of beef cattle born in spring in suckler herds
Source: Wilkinson & Tayler (1973)

Reproductive efficiency

Most of the feed used to produce a carcass from a single-suckled calf is consumed by the cow, so it is important that the herd should produce a calf per cow as regularly as possible every year with little loss of time between calving and conception, and with minimum mortality of cow and calf. A low spread of calving dates within the herd is a sensitive index of average reproductive performance, since it directly reflects a high rate of conception in the previous period of service. Short calving periods improve profitability by aiding herd management and producing more uniform groups of calves for sale. An important aid to the attainment of short calving intervals and high weaning rates is the use of body condition scoring to help ensure good condition at mating.

Breed and environment

Breed, and hence mature weight, of both sire and dam have major effects on the growth of the calf, derived from both genetic and nutritional sources. Birth weight of the calf is higher when the dam and/or sire is of greater mature weight, and higher quantities of milk are consumed by suckled calves of higher birth weight. Choice of breed for a suckler herd, however, depends not only upon the high mature weight required to achieve rapid growth in the calf, and hence increased weight and profit when it is sold, but on the suitability of the dam for the climatic and grazing conditions in which the herd is to be kept. On average, cross-bred cows of medium size (440–500 kg, e.g. crosses of Hereford or Aberdeen Angus with Friesian) produce a higher weight of calf at 200 days of age per 100 kg weight of cow than do cross-bred dams of higher mean weight (550–630 kg, e.g. crosses with Charolais, Lincoln Red or Sussex).

Feeds, feeding and economics

Feeding the suckler cow is the most expensive part of the production process. For half of the year the cow can be fed at the maintenance level on cheap feed supplies, but it is essential to raise the nutrient supply in preparation for calving, for lactation and for mating (Fig. 36.3). A considerable range of feeds, suitable for different environments and dates of calving, is available on farms, and only broad indications can be given for some major systems of production.

The value of beef herds as an effective means of utilising low quality feeds, and marginal land in the uplands and hills, has been recognised within the EEC, which provides subsidies to support beef production in this form. In 1980 a Suckler Cow Premium was introduced by the EEC which provides a payment of £12·37 per cow provided no milk is sold from the farm over the year. In the

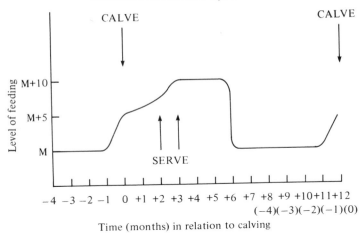

M = maintenance ; M+5 = maintenance + 5 kg milk per day ;

M+10 = maintenance + 10 kg milk per day.

Fig. 36.3   Level of feeding of suckler cows in relation to requirements
Source: Broadbent (1971)

hills there is an additional Hill Livestock Compensatory Allowance which in 1982 was raised to £44·50 per cow.

Suckler herds in the hills and uplands

Suckler herds in the hills are numerically important, with some 793,000 cows (56% of which were in Scotland) being in receipt of the hill cow subsidy in 1978. Spring calving is best suited to the hills and uplands where the transport of both bulky and concentrate feeds to the farm is expensive. Productive areas of upland associated with rough grazing have a main function of providing winter feed, but are also used for spring grazing before there is sufficient herbage on the poorer or higher land. Pasture is usually supplemented with concentrates in spring and autumn, particularly with a feed rich in magnesium in spring, to avoid hypomagnesaemia.

Targets in the hills and uplands (M L C, 1979) for autumn- or spring-calving herds respectively are: a maximum calving spread of 100 or 60 days; a maximum number of barren cows of 4–5% or 3–4%; a maximum calf mortality of 4–5% or 3–4%; 91–2 or 93–4 calves weaned per 100 cows; and, for each date of calving, 5·4–5·5 calves in the lifetime of the cow. Calf performance targets for autumn or spring herds are: 0·8–0·85 or 0·9–0·95 kg D L W G; 280–

295 kg or 230–240 kg weaning weights. Stocking rates are 0·9–1·5 cows/ha or 1·0–1·8 cows/ha.

Suckler herds in the lowlands

On lowland arable farms there are often ample supplies of cereal straw which can be used as part of the diet of a suckler herd, and cereal grain is also readily available to raise the energy content of a straw-based diet. Sugar-beet tops and sugar-beet pulp, root crops and potatoes can also be utilised by a beef cow herd. Grassland areas which are not suitable for ploughing are available on many farms. These feed sources, or silage where it can be made, provide the winter diet, and pasture usually provides the summer feed.

Autumn calving Autumn-calving herds can be given straw during the winter, but the need for nutrients is increasing to its highest level at this stage (Fig. 36.3). A basal diet of 10 kg barley straw needs to be supplemented with 3 kg daily of a protein-rich concentrate for two months after calving, rising to 5 kg per day in mid-winter. Urea can be used at up to 70 g per day, with barley and a supplement of minerals and vitamins, to cheapen the diet, but lower calf weights may result when NPN is used at this stage. Grass silage of medium to high digestibility, when fed *ad lib*, will be eaten in quantities of about 40 kg per day (at 25% dry matter) by a cow weighing 450 kg. This will supply the needs for maintenance and 5 kg milk per day, and a daily supplement of 1·5 kg barley will supply energy and protein for a further 10 kg milk per day.

Pasture alone will meet the needs of the autumn-calved cow in summer, but the requirements of the calf are high. The dam's milk supply dominates the performance of the calf up to three or four months of age. At about this stage, solid feed begins to predominate over milk in its contribution to the calf's performance, and creep feeding (i.e. allowing access to concentrates only to the calf, through a small entrance) is provided at this age and at pasture in mid-season. Leader-follower grazing allows for maximum nutrient intake by the calves and good utilisation of the remaining pasture by the cows. Calves are weaned in late summer, at least by ten months of age. Scouring in the calf, and hypomagnesaemia in the cow, are health hazards which require appropriate feeding and management for their avoidance.

Spring calving Grass growth meets the cow's lactation requirements, and calves do not require creep-feeding with concentrates until the second half of the grazing season, the calves being weaned when placed in yards for winter feeding, or when sent to the autumn sales. Since the feeding level needed for the cow is no more than maintenance, for much of the winter, straw and silage can make a greater contribution to the diet than for the autumn-calving cow.

Targets in lowland herds (MLC, 1979) for autumn- or spring-calving herds respectively are: a maximum calving spread of 100 or 60 days; a maximum number of barren cows of 3 or 2%; a maximum calf mortality of 3 or 2%; 95 or 96 calves weaned per 100 cows; and, for each date of calving, an average of 5·7 calves in the lifetime of the cow. Calf performance targets are: 0·9 or 1·0 kg DLWG, leading to weaning weights of 320 or 250 kg. Stocking rates are 2·0 or 2·3 cows per ha for autumn or spring calving.

Partial systems

There is much selling and purchasing of cattle which have been taken through part of a system and complete the production process on other farms. It is important that the feeding of these animals should be appropriate to their weight and age, and to the objectives of the system of production for which they are intended.

Lightweight spring-born suckler calves purchased at autumn sales cannot be finished in the following winter, but need to be fed to grow at 0·4–0·6 kg per day preparatory to a pasture finishing period. Compensatory growth then operates at pasture and optimum total gains can be obtained. Winter finishing of suckled calves and stores is most successful when well made silage of high digestibility is given with up to 2·75 kg barley/head per day. Stores which are purchased at the start of a grazing season and finished a year later in yards, or those bought in autumn and pasture fattened, make higher gross margins than those finished in six months, by some 40%.

Calf rearing and veal production

Most of the calves purchased from dairy herds for beef production are reared on the early weaning system. It is most important to the future health of the calf that it receives colostrum from its dam shortly after birth. Details of calf rearing methods are given in Chapter 35.

Veal production in the UK makes up less than 1% of the total production of beef and veal. Calves are reared on milk substitute to gain 1·2 kg per day and are slaughtered at 3½–4 months of age to produce carcasses weighing 95–120 kg.

37

SHEEP PRODUCTION
SYSTEMS

Sheep are kept in a variety of systems, depending on various factors. The natural environment, altitude, slope, weather and soil type govern the type of sheep likely to thrive; the political and financial environment influence which aspect of the sheep and its husbandry is likely to be profitable, and this can change radically. In medieval times wool was more important than meat and was used as a money equivalent with knights being held for ransom in exchange for sacks of wool. The change in importance from wool to meat has had a great effect on the type of sheep kept and on the method of husbandry. Politics influence the amount of financial support available for keeping sheep in the hills and also the season when prime lambs are marketed. Trade agreements affect exports and the destination of the product affects the type of product required. People may also keep sheep simply because they like them.

All these factors help to explain the multiplicity of systems that exist in the British sheep industry and a large part of this chapter will be devoted to classifying and analysing some of the more important of these in more detail.

SYSTEMS OF SHEEP PRODUCTION IN BRITAIN

Early lambing flocks

The object of lambing early is to market lambs when the price is highest, which is normally at Easter. This implies that the ewes should be mated in July or August to lamb down in December and January. As most breeds and crosses in Britain start their natural breeding season in late August and September, either some kind of hormone or light treatment has to be used to advance the breeding season or else a breed with a longer breeding season, such as the Dorset Horn, can be used. Because the ewes are being mated at the start of their breeding season the proportion of the flock showing oestrus over 16 days will be lower, which will result either in a longer period of lambing or else a lower conception rate, depending on whether the ram is left in for longer than the normal period. The use of synthetic progestagen sponges will avoid both these undesirable results but it will add to the cost. There should be no problem with ram fertility in July or August but the use of sponges to synchronise oestrus will mean that instead of the normal ram:ewe ratio of 1:40, a ram to ewe ratio of 1:10 will be required. This will necessitate the buying and keeping of more rams.

Finally, because the ewes are being mated at the onset of their breeding season, ovulation rate and therefore mean litter size will be lower than if the ewes had been mated in mid-breeding season. Selling percentages for recorded flocks are 120-40%, with a mean litter size of 1·5 to 1·6.

After lambing in December/January the farmer has alternative methods of fattening the lambs: he can either house the ewes for the last six weeks of pregnancy, lamb inside and then wean the lambs at six weeks of age onto a largely concentrate diet and fatten them rapidly indoors; or else he can opt for a cheaper and possibly less reliable system in which the ewes are kept outside on catchcrops or grass supplemented with hay and concentrates. The ewes lamb outside and the lambs are then reared on the ewe and finished on spring grass. This last system implies a mild winter and an early spring which makes it more suitable for the more sheltered areas of the south and west.

Rearing weaned lambs indoors on a controlled diet gives greater certainty of growth rate and therefore of meeting a target slaughter weight by a set date, but a balance has to be achieved between the cost of the ration and the speed of growth. Milk substitutes for rearing lambs artificially from one day of age are expensive and so are concentrate diets. Broadly speaking, the longer the lamb is kept on milk, whether natural or artificial, the faster the growth rate, but the more expensive the food. The nutritional policy will depend, therefore, on judging the trend in market price. Lambs receiving milk substitute *ad lib*, together with concentrates and hay, may grow as fast as 450 g/day. A more usual growth rate for artificially reared and early weaned lambs of Suffolk cross type offered a barley-based concentrate diet is 340 g/day, with a conversion efficiency of 3-3·25 kg of concentrate per kg of live-weight gain. With this weight gain lambs are likely to be ready for slaughter at live weights of 36 kg at approximately three months of age. Cheapening the ration by feeding more good quality hay and less concentrate may drop the growth rate to 250 g/day, which will extend the feeding period for a month.

With early lambing the ewes are dry during the spring and summer and can be kept with little trouble at high stocking rates (18-20 ewes per ha) on grassland. Lambing in December and January will also avoid the busy period of spring cultivation.

Where the lambs are weaned and reared indoors there should be no problem with parasites, especially if the ewes are dosed before being housed in late autumn or early winter. Lambs reared outside on a catch crop or on grass, plus added hay or silage, will need dosing at regular intervals if the grass is not clean. Those lambs that have not been fattened by March can either be dosed or weaned onto clean pasture.

An analysis of costs for early lambing flocks by the Meat and Livestock Commission (MLC, 1980) showed that the top third flocks achieved a wider ratio between lamb sales and feed costs than average flocks. Costs are high in early lambing systems because of the level of concentrate fed to the lamb (55 kg to ewe, 33 kg to 1·37 lambs), and profitability depends on selling a high proportion of lambs between February and May. The top third sold more lambs during this time. However, the main determinants of the extra profitability were the extra lambs reared (11 per 100 ewes to the ram) and the higher stocking rate (an extra 2·2 ewes per ha). An extra 29 kg of N per ha of grassland was applied in the more profitable flocks, but because of the higher stocking rate, usage of nitrogen per head was the same (11 kg per ewe).

Lowland grassland

This heading covers all those flocks with sheep that rely on grassland and receive no subsidy. It has been subdivided into three: (a) intensive grassland, which is heavily fertilised and well stocked all the year round; all leys come into this category and some permanent pasture; (b) extensive grassland, which is less heavily fertilised and with a lower stocking rate. Grass under this heading would be predominantly permanent pasture; and (c) grass plus forage crops. Here a significant proportion of the flock's food comes from regularly grown forage crops such as swedes, rape or stubble turnips.

Intensive grassland

Sheep here are kept on the most productive grassland, often on land that will be used for cereals, and on land that may receive more than 200 kg N/ha/annum.

Type of ewe and ram

Many different breeds and crosses are kept on the lowlands but the objective is, or should be, the same, namely to sell as many lambs per unit area as possible, which means taking into account the ratio of ewe size and number of viable lambs born. Basically, only small ewes, such as Welsh Mountain, can afford to have singles and the larger the ewe the more lambs it must have to be nutritionally efficient (see Table 37.1). But there is a top limit; some crosses of

Table 37.1   Theoretical relationships between size and reproductive performance needed to obtain similar carcass output

Ewe size (kg)	Lambs sold per ewe to the ram	Stocking rate (ewes/ha)	Carcass wt. of lamb (kg)	Carcass output (kg/ha)
35	1·0	32	15	480
55	1·5	20	16	480
75	1·9	15	17	484
85	2·2	12	18	475

sheep available now, particularly half Finnish Landrace, will produce mean litter sizes of between 2·5 and 3·0 lambs; this implies a high proportion of litters of three or more. Although some ewes can rear three lambs naturally at pasture, not all can and very few will rear four without some bottle feeding by the shepherd. Thus, if the ewes produce a mean of 2·5, definite arrangements have to be made to rear a considerable proportion of the lambs artificially on cold milk substitute. And although this is a very satisfactory method of rearing lambs, it is not economic in this country because of the high price of milk substitute in relation to to the price of lamb meat. This means that the level of prolificacy to be aimed at for a grazing system, currently, is 2·2 which should result in between 180 and 190% lambs sold per ewe put to the ram. The indigenous cross breeds that can meet this reproductive target are the Greyface, Masham, Mule and Scottish Halfbred. The reproductive performance of the Welsh Halfbred is slightly lower but the ewe is smaller, enabling more to be kept per hectare. There are other crosses of a similar nature that use a prolific ram breed on a mountain ewe; the crossbreds are not always first crosses; quite often lambs from the Suffolk × Scottish Halfbred are retained for breeding purposes. Pure breeds that are found on the lowlands are the Clun, Devon Longwool and Romney Marsh. Finally, several hybrids have been produced during the last 20 years; their breeding has been based either on the Finnish Landrace for prolificacy or on the East Friesland for milk yield with an admixture of various British breeds. The use of a large ram (100 kg) has a proportionately much larger effect on the carcass weight of the crossbred lamb from a small ewe (35 kg) than from a large ewe (80 kg): litter size also has an effect.

In a survey conducted in 1972 by the Meat and Livestock Commission, only 19% of the lowland ewes were purebred (with the Romney and the Devon having the highest individual breed totals); the remaining 81% were crossbreds.

Of the ram breeds used in the lowlands as sires of prime lamb, the Suffolk is used most widely, with the Dorset Down next.

Type of pasture

Leys are normally sown for a specific purpose, such as early bite, a hay cut or maximum yield of dry matter under grazing, and therefore contain the pertinent grass or clover species. Of grass seed sown for agricultural purposes over 80% consists of ryegrasses, and of this 80%, approximately two-thirds is perennial ryegrass and one-third is Italian or hybrid ryegrasses. Other grasses used include Timothy, Cocksfoot and fescues. The amount of white clover sown (7%) has remained fairly constant but there has been a very marked decline in the use of red clover. The trend is to simpler mixtures.

On permanent pasture the proportion of ryegrass has increased during the last 20 years at the expense of *Agrostis* spp.

There is controversy as to whether leys are really that much more productive than permanent pasture. The productivity of permanent pasture will depend on the proportion of perennial ryegrass that it contains and the amount of fertiliser that it receives.

Reproduction

Most flocks, in which the aim is to fatten the majority of lambs from grass, are mated so that lambing coincides with the average date for the onset of rapid grass growth. This means that mating will take place considerably earlier (September) in the south-west than in the north-east (November). From the reproductive point of view, most lambs are produced by ewes in good body condition mated in the first half of the breeding season, but if a compact lambing is desired, without the aid of synthetic progestagens, then mating should be delayed until the third or fourth oestrus after the start of the flock breeding season; this allows the late breeders to start cycling.

Good management affects the number of lambs born and reared per ewe to the ram. The selling percentage of the average lowland flock (average for gross margin) was 137 and for the top third was 153.

Lamb growth rate, whilst being suckled, is influenced by a number of factors, one of the most important being litter size. Singles grow faster than twins, which grow faster than triplets, because they get more milk. Thus the more singles there are in the flock the faster will be the overall lamb growth rate but the less meat will be produced per ha. The more lambs that are produced and reared per ewe the higher the efficiency of conversion of food to meat, but the harder will be the management and the slower the growth rate of the lambs in the flock. The farmer has to decide, therefore, whether he wants a relatively easy grazing management system based on singles, which might still be intensive in relation to the stocking rate, or a potentially more productive system based on twins and triplets.

Grazing management

The aim of grazing management is to make full use of grass over a number of years and to send the desired number of lambs for sale when planned. The art of the grazier is to do outside with fields of herbage what the indoor feeder can do by controlling the amount and quality of food brought to the animals daily.

The simplest method of grazing ewes is to turn them into a field with their lambs at the beginning of the year and leave them there for the rest of the year – set-stocking. If the lambs are to be weaned, another field is required for them. The next stage up is paddock grazing where a number of separate fields are used during the year and the ewes and lambs are grazed rotationally. The frequency of moving will depend on the number and size of paddocks. A variation of this is to set up creep gates in the fences so that the lambs have access to a field before the ewes. This is rotational grazing with forward creep; a variant of this is sideways creep, but in sideways creep the ewes do not get access to the lamb paddocks and so cannot be employed as mowing machines, ensuring that clumps of grass get eaten down.

Intake of grass is regulated by the quantity available and its digestibility. Intake is depressed by a lack of grass because the amount of time spent grazing is finite and the shorter the grass the smaller the amount removed by each bite; intake is also reduced by low digestibility. Lambs tend to be fussier in their requirements than ewes and this is the advantage of using a forward creep; the ewes can be held back without the lambs being affected at the same time.

Stocking rate is the key to success in an intensive grazing system and the area for conservation can be used to buffer variations in annual grass production. A rotational system with a number of paddocks makes it possible for the grazier to vary the area conserved according to annual circumstances. If the grass is to be fully utilised then the area set aside for conservation needs to be brought back into the grazing rotation quickly (e.g. mid-June) so that the increased demand for grass by the lambs can be met by an increased available area, because at this stage of the year the rate of grass growth begins to decline.

The amount of grass produced is directly related to the amount of nitrogen applied and to the site (soil type and drainage). An intensive grass system should be able to produce 25–30 lambs per ha.

With ryegrass-white clover pastures the growth of the clover is much less predictable than with nitrogen-treated grass. The finan-

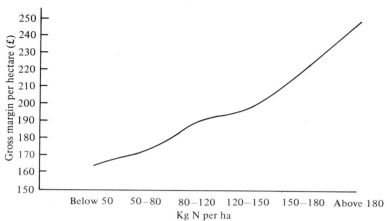

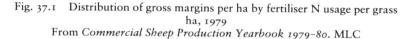

Fig. 37.1   Distribution of gross margins per ha by fertiliser N usage per grass
ha, 1979
From *Commercial Sheep Production Yearbook 1979–80*. MLC

cial advantage of using levels of nitrogen higher than the average for the intensive grazing system (138 kg N/ha) are shown in Fig. 37.1. The dry matter production response of grass in the cut situation to added nitrogen is linear up to about 180 kg N/ha, depending on soil type, but in the grazed situation the response is complicated by the amount of nitrogen deposited in the dung and urine, varying amounts of which can be recycled (approx. 8 kg N/ewe/annum), depending on weather conditions and soil type.

Lamb sales

Most of the lambs in this intensive lowland grass system are sold during the summer and autumn. With the emphasis on high stocking rates and numbers of lambs per ewe, fast lamb growth rate is not a feature. On average, about 50% are sold finished from grass and the other half are sold as store lambs or finished on catch-crops or root crops on the same farm.

Parasites

In a system such as this in which the majority of lambs are still present in July and August, there is a considerable risk of them being challenged by a heavy infestation of roundworms (*Ostertagia circumcincta, Haemonchus contortus, Trichostrongylus* spp.) after July, a risk that is increased by the high stocking rates and by the fact that the greater proportion of the lambs are twins and triplets, which will eat more grass in their lifetime than singles. As mentioned earlier, because intensive systems are more often carried out on arable land, a greater proportion of the grassland is temporary and therefore stands a better chance of being parasite-free or relatively uncontaminated. Lambs born after the end of March will be at risk from *Nematodirus* spp. in the spring, because between March and May is the period when a mass hatch of infective larvae occurs. However, *Nematodirus* infection is transmitted via the pasture from one crop of lambs to the next and the ewe plays a negligible role. Furthermore, lambs build up immunity faster to *Nematodirus* than they do to other important roundworm species. The resting of pastures for up to 30 days in rotational grazing is no better than set stocking in allowing the lambs to avoid infestation; but if clean grazing is to be found for lambs from July onwards two paddocks at least are required, one of which can then be conserved and grazed by the weaned lambs from July onwards. The problem with intensive systems which utilise a very high proportion of the grown grass efficiently is that the aftermath grazing is required as soon as it becomes available, which is likely to be in early June. A number of paddocks does help flexible management but the maintenance of clean and relatively clean pastures is a much more effective and cheaper policy of parasite control than the frequent use of anthelmintics to ewes and lambs. If the pastures become heavily infested with roundworms then it is better to rest the land from sheep for a year, either by grazing with cattle or by ploughing the land up and sowing something other than grass for a year.

Summary

The system is a high output one producing 1,000 kg of lamb live weight per ha (460 kg of carcass per ha) which is an efficient return from the land for an animal production system, considering that 80% of the grass grown goes to the ewe annually. In terms of milk production, assuming that a 70-kg ewe gives 3 kg of milk per day, then at 16 per ha, daily milk production per ha is 48 kg. (This may be compared with a dairy cow that gives 33 kg of milk per day and is stocked at 1·64 cows per ha: milk production is then 54 kg per ha per day.) The gross margin returns per ha from an intensive grassland system can be high and compare favourably with the return from barley.

Extensive grassland

The difference in emphasis between extensive and intensive is that annual stocking rates are lower in the former, less fertiliser is applied and more of the grass is likely to be either in a long ley or permanent pasture. Because stocking rates are lower, less attention is likely to be paid to grazing management and there may be greater risk of parasitic problems.

If the flocks on extensive grassland systems are not housed during the winter months, and if the reason for the land being in permanent pasture is poor drainage or rough land, then the extensive grassland system may well be difficult to manage in the winter because of bad poaching.

Type of ewe and ram

The reproductive performance of extensive lowland grassland flocks is very similar to that of the intensive lowland grassland flocks – an average of 140 lambs reared for every 100 ewes to the ram with a figure of 157 lambs by the top third flocks. Similar breeds of ewe are used and the crossing rams will either be the Down rams for fat and store lamb production (Suffolk, Dorset and Hampshire Down) or else the pure-bred rams for replacement stock.

Type of pasture and grazing management

Extensive flocks are kept on long leys and permanent pasture with overall ewe stocking rates of less than 7 per ha and grass stocking rates of about 7 per ha. This is less than half the number of ewes kept on the intensive systems, and the fertiliser applied is about 40% less. Although there has been an increase in perennial ryegrass in recent years, indigenous grasses still contribute about two-thirds of the cover in the average old sward (over 20 years) and about half in the swards aged 9–20 years old. The indigenous species most commonly found are *Agrostis* and *Poa* spp. and Yorkshire fog.

Because the extensive lowland sheep systems are found on the rougher parts of farms, management is likely to be a paddock grazing system, with few paddocks used in the summer, or set-stocking with less emphasis on fencing than in intensive systems. In the winter, if there is no adequate sheep housing, then the sheep may well have the run of the farm's grassland, with the flock being divided into small units, each kept on the same field from November to April. The problems with this system are poaching round feed troughs, the inability to provide a dry lying area for the young lambs and the reduction in grass yield in the spring due to over-grazing during the winter.

Lamb sales

A slightly higher proportion of the lambs is sold fat (61%) on extensive systems than on intensive (50%) but the number of lambs sold per ha is much less.

Parasites

Once permanent pasture or a long ley has been infected by parasites then the cheapest way to prevent parasites from becoming a problem to lambs is to graze the land with cattle in alternate years.

Grass and forage crops

In this system the grassland management is similar to that in the intensive or extensive system but more reliance is placed on forage crops for the autumn and winter period either for the ewes or for fattening lambs. About 15–20% of the flock's annual feed requirement is likely to come from forage, 11–20% from concentrates (depending on the stocking rate) and the remainder from grass.

Type of forage crop and place in the rotation

Forage crops for sheep are either annuals such as swedes or kale, occupying the field for twelve months, or catchcrops such as rape or stubble turnips, which are getting increasingly popular and can be fitted into a cereal rotation between the harvesting of winter barley in mid-July and the sowing of spring barley in March. Crops such as kale, fodder radish and mustard can be grazed from August onwards but it is in the months from October to February, when there is only slow grass growth or none at all, that forage crops are of most value. Dry matter yields of forage crops can be quite high (3–6 t DM/ha) and the material is of high digestibility (77–80 OMD). The problems of achieving a high intake are twofold. One is the high moisture content of the crop (dry matter 9–12%) and the other is the contamination by mud of the tops and roots. Often grazing takes place during wet conditions and the plants and animals get plastered with mud. The farmer has to compromise between maximum utilisation of the crop and animal performance. The relationship between utilisation and intake and live-weight gain will be discussed more fully in the section on store lambs. Some form of strip grazing is normally practised with ewes grazing forage and root crops in order to minimise crop spoilage. Alternatively, root crops can be harvested and fed indoors; this saves the flock from getting muddy but increases the work.

The brassicas and radish family contain S-methyl cysteine sulphoxide (SMCO) and thiocyanates which are known to cause temporary anaemia. They also contain glucosinolate compounds which cause goitre. Despite these potentially toxic compounds there is no evidence to suggest that the reproduction of ewes is impaired by these compounds. In fact recent evidence suggests that the mean litter size of ewes may be enhanced by 0·3 of a lamb per ewe for ewes mated on rape compared with a flock in similar body condition mated on perennial ryegrass. Although crops such as rape and stubble turnips can be held over for grazing in January and February there is a considerable risk that the crop may rot after frost damage. Kale is less susceptible.

Lowland arable flocks using by-products

The combination of sheep and cereals is common in Britain and the grassland side has already been described under intensive lowland flocks. However, there are many arable flocks where sheep have access at certain periods of the year to other crops, such as sugar beet tops, fresh vining pea haulms and a range of vegetable crops grown initially for human consumption, such as cabbages, Brussels sprouts and carrots. The latter become available for sheep either because the price drops, making the labour of gathering the crop uneconomical, or because too much was sown or the crop was spoiled by the weather and no longer fit for human consumption. If by-products are to form a regular part of the annual nutrition of the sheep flock, then a fairly large area has to be devoted to cash crops for the contribution to be reliable. Catchcrops and main crops such as swedes and fodder beet have not been included in this section, but the use of treated straw has.

On a farm mostly devoted to wheat and barley the temporary grassland is of short duration, two- or three-year leys being the most common. On lighter soils, for instance the chalk soils, the use of sheep for two or three years has often had a marked effect in boosting subsequent cereal yields.

The breeds of ewe and ram and the reproductive performance of the sheep are similar to those kept on intensive grassland. Lamb growth rate, stocking rate and grazing management will also be similar.

The types of by-product crop available are sugar beet tops (Oct.–Jan.), carrot tops and rejects (Sept.–Feb.), Brussels sprout stalks and leaves (Aug.–Dec.). Most of these by-products are available during the autumn and winter and are therefore suitable for maintaining dry ewes or for feeding them in early pregnancy. Alternatively they can be used for store lamb feeding. Table 37.2 shows the yield and feeding value of a selection of these by-products. With

Table 37.2   Yield and feeding value of by-products

Food	DM %	Metabolisable energy MJ/kg DM	Digestible crude protein g/kg DM	Digestible organic matter in dry matter DOMD %	DM yield t per ha
Carrot leaves	18	7·9	123	48	0·22
Sugar beet tops	16	9·9	88	62	3·16
Cabbage, drumhead	11	10·4	100	66	10·9
Vining pea haulms	21	8·7	95	51	1·9
Winter wheat straw	86	5·7	1	39	3·2
Spring barley straw	86	7·3	9	49	2·1

the exception of cabbage, which is unlikely to be grown specially for sheep regularly, the feeding value and dry matter yields of these by-products make them suitable for little other than maintenance. However, they may eke out the hay or silage allocated for the winter.

Untreated straw has a low digestibility, below 50, and a negligible crude protein content, but when treated with sodium hydroxide the digestibility can be improved by up to 10 units. In these circumstances it can be used as a substitute for up to 30% of the silage or hay ration.

One of the problems of grazing sheep across fields of sugar beet tops or carrot rejects is fencing, but the new lightweight, electrified fencing has made it much easier to fold sheep across unfenced fields at a relatively low cost. Grazing sheep on short-term leys or on cabbage or sugar beet fields should minimise the risk of parasites.

Upland flocks

Upland flocks are defined as those eligible for the basic rate of subsidy, and not the supplementary rate. In 1978 the basic rate for the upland sheep was only 18% less than the rate for hill sheep. The difference between upland and hill farms is that on upland farms the land is below 300 m and more of the area is fenced. Rough grazing on upland farms is less than 30% of the total farm area and the ratio of cattle to sheep is higher than on hill farms. In comparison with lowland farms there is poorer pasture production on upland farms because of higher altitudes, a shorter and colder grazing season, less frequent reseeding, less fertiliser and poorer drainage. The topography also prevents easy conservation so it is harder to integrate the hay or silage aftermath with the upland grazing system. All this leads to lower stocking rates than on the lowlands.

Pasture

No clear distinction is made between the various vegetation types on the uplands and on the hills. However, of the four most commonly recognised plant communities, (a) blanket bog (*Calluna, Eriophorum, Trichophorum*), (b) dwarf shrub heath (*Calluna*), (c) grass heath (*Nardus, Molinia, Deschampsia*) and (d) acid grassland (*Agrostis/festuca*), the uplands would contain a higher proportion of (d) and (c) than the hills. The commonest vascular plant species in the British uplands are *Festuca ovina* (67%), *Vaccinium myrtillus* (Bilberry) (66%) and *Deschampsia flexuosa* (61%). Common species (40–60%) are *Agrostis canina, Agrostis tenuis, Calluna vulgaris* (heather), *Galium saxatile* (heath bedstraw), *Juncus squarrosus* (heath rush), *Nardus strictus* (mat grass) and *Potentilla erecta* (tormentil). On upland limestone *Festuca ovina* is almost ubiquitous (93%) and *Thymus drucel* (wild thyme) is very common (70%). The objective with re-seeding is to introduce perennial ryegrass, cocksfoot, timothy and white clover; the last of these has been described as 'the cornerstone of hill land improvement'. Production of acid grassland can be enhanced by lime, phosphate and nitrogen, with white clover fixing nitrogen.

To extend the supply of forage in the uplands farmers are using catchcrops such as rape and stubble turnips. These have to establish quickly and mature early and the problem is often one of efficient utilisation in conditions of high rainfall and poor drainage, both of which lead to extremely muddy grazing conditions and crop soiling.

Flock performance

Just as on the lowlands, the types of sheep enterprise on the uplands can be divided into intensive and extensive flocks, with the extensive flocks being run on traditional lines using unimproved pastures and low levels of fertiliser and the intensive flocks being kept on improved pastures with higher fertiliser and stocking rates.

The pure breeds kept on the uplands are Scottish Blackface, Swaledale, Welsh Mountain, Beulah and Hardy Speckleface, Cluns, Kerry Hills and North Country Cheviots, and a similar range of crossbreeds to the lowlands. Performance of these ewes on the uplands is slightly inferior to their performance in the lowlands.

Stocking rate for the average upland flocks was 8·6 ewes per ha in 1978 compared with 10·6 for the lowlands. Not all this difference is caused by the use of less nitrogen, because at similar levels of nitrogen application upland stocking rates are lower. Just as in lowland flocks, the main factors explaining the success of the top recorded upland flocks were number of lambs reared and stocking rate.

It has been said that the payment of a subsidy on a per head basis rather than on a performance basis has led to overstocking of the uplands and reduced performance by the ewes and lambs. Against this is the fact that the ewe subsidy is only 10% of the gross output of upland flocks.

The pastures of the uplands can be improved with re-seeding, lime, phosphate and more nitrogen but at a cost and the comment has been made that because of the family nature of hill and upland farming, this may slow the pace of change, partly because of insufficient capital from family resources and partly because of a cautious attitude to borrowing. Added to this is the small size of many upland farms; for an economic return units ought to be at least 100 ha in size and able to stock 450 ewes and 30 beef cows. Upland farms are often used for the production of breeding stock. This is covered more fully in a later section.

Hill sheep

Hill farms have 95% or more of their land classified as rough grazing and depend mainly on the breeding ewe flock. The ratio of cattle to sheep is much lower than on upland farms. A supplementary rate is payable in addition to the basic rate on ewes of specified hardy breeds and crosses kept in flocks which are self-maintaining, made up of regular age groups and which comply with recognised practices of hill sheep farming in the district.

The pasture

The quality of rough grazing varies widely, with a high proportion dominated by *Calluna*, *Eriophorum* and *Trichophorum* and dwarf shrub heath with *Calluna*. There are less areas of grass heath (*Nardus*, *Molinia* and *Deschampsia*) and of acid grassland (*Agrostis/Fescue*). Annual yields can range from 1-4 t DM/ha and the amount utilised ranges from 0·1 to 0·6 t DM/ha.

This is in stark contrast to a well-fertilised lowland pasture on which sheep will eat 8–9 t DM/ha per annum from a total yield of 10–14 t DM/ha. There is a difference in the amount of dry matter utilised of ten to almost a hundredfold. The digestibility can vary from 35 to 70% and there is a marked seasonality of production, with three-quarters of the growth coming in May and June. After some years of grazing control improvement of the better areas has led to an increase of from 2,000 to 3,000 sheep grazing days per annum; on the *Nardus* and *Molinia* sites the increase has been from less than 2,000 to 2,500. Considerable selection has been carried out for winter-hardy grasses but the factors involved include low temperature, frost heaving, desiccation, disease and interaction with soil conditions such as compaction, water-holding capacity and nitrogen status. The effect of these can be modified by snow cover and the way in which the swards are managed. For instance, too much nitrogen in the autumn on an upland site may well reduce yield the following spring compared with a much lower rate of autumn nitrogen.

Animal performance

The two most common hill breeds of sheep in this country are the Scottish Blackface and the Welsh Mountain, to which may be added the Derbyshire Gritstone, the Herdwick, the Radnor and the Swaledale. The Welsh Mountain (35 kg) is a smaller breed than the Scottish Blackface (53 kg) but is also less prolific, having a mean litter size of 1·16 compared with 1·49 for the Scottish Blackface. In terms of number of lambs reared per 100 ewes to the ram the figure for the hills was 103 in 1978 compared with 129 for the uplands and 140 for the lowlands. Date of lambing is best geared to the start of grass growth, that is late April for the hills. Of the ewe lambs born, half will be needed as replacement animals and the remainder can be sold either to another hill farm, or if crossed with a ram of the Border Leicester or a similar breed, as a crossbred ewe to a kinder climate. The Scottish Blackface and the Welsh Mountain ewes are the maternal side of the popular Greyface and Welsh Half-bred crosses.

It is significant that in gross margin costings for hill flocks there is no atttempt to express the gross margin per ha but simply to leave the figure on a per ewe basis. Depending on the type of rough grazing the return per ha may well be lower than per ewe. Of the gross output the subsidy accounts for 18% of the total and the sale of lamb only 58%. Sale of other stock makes up another 18% and this covers mountain ewes that will have had four or five lamb crops and are then sold to an easier location. The wool to lamb sale ratio is higher for the hills.

Hill flocks and their lambs are dosed regularly against worms and also against liver fluke.

Problems

The main problem is the environment in the winter, in particular nutrition, and the problems are different for the main ewe flock, the ewe lambs that are to be retained and the lambs for fattening. Self-help feed blocks have been found to be useful during the winter, supplying much-needed protein and energy without the farmer having to resort to daily feeding. Intake is regulated by the proportion of salt in the block.

The ewe lambs need special care during their first winter if they are to develop properly. Two options are available to the hill farmer. Either he can pay for them to spend the winter on a lowland farm, giving the ewe lamb a better chance to grow and reducing his own stock numbers in the winter, or, if he has some suitable land, then he can either winter them on grass forage or else on a specially grown brassica crop. For the fattening lambs it may be necessary to erect a special indoor house where they can be fed cereals, hay or concentrates. This may be his only alternative to having to sell poorish lambs in the store market.

Flocks selling breeding stock

This system, unlike the preceding ones, is not based on an environment but on a product, or rather several products. It covers those farms that sell breeding stock, either ewes (at various ages from ewe lambs to broken-mouthed mountain ewes), or rams.

The sale of female breeding stock

This can be either the sale of purebred ewes or else, more importantly, of crossbreds, because this is an integral part of the stratification in this country.

Most crossbreds are sold at 18 months of age as gimmers (also known as shearlings or theaves). Ewe lambs can be bred from and it is easily argued that a selling percentage of 100 covers the cost of purchase in the first year. Provided they are fed properly, particularly during their first lactation, they will grow to their normal size and have just as many lambs in their lifetime as if mating had bee deferred another year. However, the number of ewe lambs actually lambing is erratic and the mortality of their lambs is higher. Purebred ewes are also sold but not in anything like the same quantities as crossbreds. Some farmers specialise in buying in ewe lambs, keeping them for a year unbred and then selling them at 18 months. The main source of mature purebred ewes is the hill and upland farms. After four or five lamb crops they are sold to the lowlands.

The sale of male breeding stock

The ram sales that hit the headlines are the top prices paid for pedigree stock, but the most numerous sales are of Down rams for the fattening of prime lambs for slaughter (the terminal fat lamb sire) and of these the majority sold are Suffolks, estimated to be mated to over one third of the national ewe flock, or over 4 m. ewes. The other nationally important ram breeds are the Scottish Blackface, the Welsh Mountain, the Border Leicester and the Blue-faced Leicester.

The plea from ram buyers is that there are no performance figures available and that although the ram may look well, this may be because he is a single, older than average or better fed. A limited amount of performance testing has been set up on ram lambs under a standardised environment to reduce the influence of differential rearing. The problem with performance testing of rams, as opposed to bulls, is that most lambs are sold for slaughter at 12–16 weeks of age and thus the predominant effect is that of the ewe's milking ability. However, the point has been made that in extensive sheep systems on poor land, wool is a relatively important product, and under these conditions selection of rams on yearling wool characters may be justified.

The average size of the ewe breeding flock is untypically small (for Suffolks it is 40, for Border Leicesters 24 and for Blue-faced Leicesters it is only 10).

Store lamb fattening

Store lamb fattening in the autumn may be based on catchcrops. These fit well into upland farms where the grass growing season is too short to fatten lambs and the catchcrop helps to extend the grazing season and to lessen dependence on conserved hay or silage.

The place of the catchcrop in the rotation is that it is normally sown following the harvesting of a winter cereal, usually barley, in mid-July. The catchcrop (fodder radish, rape or stubble turnip) is then sown as soon as weather permits, ready for grazing by mid-October. The crop can be grazed until January or later depending on the severity of the winter: it is then ploughed up and a spring cereal sown in its place. In this way the catchcrop is extra to the cereal rotation and provided it is used well can make a useful addition to farm profit. The crop is highly digestible (OMD 76–80%) and stays that way during the autumn, a period when the digestibility of grass is falling much lower. Catchcrops are succulent, with dry matter contents ranging from 10–14%. Yields range from 2–4 t DM/ha.

Other crops such as kale and swedes are often used for fattening store lambs in the autumn. These have higher yields of DM per ha (5–7 t), but of course the crop has to be sown in the spring. This means that the extra yield of the crop has to be set against the loss of winter barley – where land is suitable for cereal growing.

Lamb performance

The farmer has to decide whether he wants the lambs to grow or whether he wants the crop to be fully eaten by the lambs. The two aims are not compatible. With 50% utilisation of rape, lamb growth rates of 120 g/ha/day are easily achievable. This means that in 80 days a lamb of 30 kg will grow to 40 kg, a marketable weight. With 75% utilisation, lamb growth drops to 80 g a day. With higher utilisation the crop will provide only a maintenance ration and lambs will have to be finished later on hay or silage and concentrates, probably indoors. Grazing conditions can often be very muddy in the autumn, leading to considerable crop spoilage. A compromise between lamb performance and utilisation can be achieved by grazing the lambs ahead of either ewes or other stock. The lambs can then be moved on before intake falls and the other stock will ensure that crop wastage is minimal. Measured intake levels vary with amount of crop on offer, but under *ad lib* conditions 35-kg lambs will eat 0·9–1·1 kg of organic matter per head per day.

A great deal of success in making money out of store lamb fattening comes from careful buying. Healthy lambs that will thrive and put on weight and can be bought cheaply are the most profitable. The profit margin per lamb between 1971 and 1979 was about

10% of the purchase price. The farmers who made most money from store lambs paid less per kg liveweight for them at purchase, grew them slightly faster and for longer, and sold them at a heavier weight, but not at a higher price per kg liveweight.

Health problems

When lambs are changed from a grass diet to rape or stubble turnips there is a period of acclimatisation during which their liveweight stays constant or falls slightly. This check has been shown to be associated with a mild anaemia which has been linked with the occurrence of S-methycysteine sulphoxide (SMCO) and with the presence of thiocyanates. Additionally, thyroid glands have been shown to become enlarged in lambs grazing on catch-crops. However, the mild anaemia passes, performance improves and growth rates can be quite impressive for that age of lamb and that time of year.

Wool production

In a sense there is no special wool production system in Britain, not in the way there used to be, and not in the way there is in Australia with large numbers of wethers being kept for four or five years almost solely for their wool clip.

On lowland farms the amount of money received for wool is about 10% of the money for lamb sales; on hill farms the percentage is closer to 15%, but the value of the wool clip is only half that of the subsidy.

38

PIG PRODUCTION SYSTEMS

Pigs are kept exclusively for meat production, unlike cattle, poultry or sheep which can produce in addition to meat either milk, eggs or wool. The existence of one primary objective has not resulted in one type of production because pigs are among the most versatile of farm species. They can be housed in extremely intensive conditions without the distress caused by wet or fouled fur or feathers. Sows may be seen rooting outdoors in the bitter Canadian winter or rolling in mud wallows at the height of a Texan summer.

Wild pigs are distributed through many latitudes, from the temperate forests of Europe, which are the natural habitat of the wild boar *Sus scrofa*, to the semi-tropical and tropical climates of South-East Asia where the oriental species *Sus vittatus* and *Sus indicus* live. Feral pigs are found in many countries; they are regarded as serious pests in Australia where they have been known to round up, attack and eat lambs.

Historic and social background of pig domestication

Archaeological studies show that when early man developed a settled agricultural life, pigs were often associated with the group. Pigs are embedded in the culture of early China and ancient Egypt.

The pig fits in well with a peasant, rural economy. Indeed, in China today about a third of the world's pigs live with a quarter of the world's population. The pig's prodigious appetite allows it to develop generous fat reserves, which can support it and, after due process, the farmer, during periods of shortage or even famine. Also, since pig meat is easily cured with salt or smoke it provides a readily accessible meal at any time. The keeping of pigs was prohibited by the Law of Moses in the Old Testament of the Bible and in the Koran, which is why strict Jews and Muslims do not eat meat from pigs. Apart from 'faith', there were perhaps health reasons why pig meat was rejected by some. The pork tape worm, *Taenia solium*, is associated with poor hygiene and with the eating of only partly cooked 'measly' pork. Although to be burdened with a tape worm is unpleasant, far more serious is the complication when humans reinfect themselves directly with the egg stage. As in the pig, these eggs hatch to form the burrowing worm, which can encyst in any part of the body including the brain. Cysticercosis, as this form is called, may have been recognised very early as a disease occurring in connection with pigs. Fortunately, the disease is now almost unknown in developed societies where carcasses are regularly inspected and hygiene standards are high.

PIG MEAT

The market for pig meat is essentially in three parts: fresh pork, cured joints and processed meats. Some examples of these are listed in Table 38.1, and illustrated in Fig. 38.1.

Table 38.1   **Pork and bacon cuts**

	Anatomical region	Fresh Pork	Cured
1	Head	Head	Cheeks
2	Neck	Neck chops Butt	Butt
3	Upper fore-limb	Shoulder Blade bone	Collar 'Picnic' ham Fore slipper Shoulder bacon
4	Lower fore-limb	Hand Hand and spring	Small hock
5	Dorsal thorax	Rib chops Loin chops (+ kidney)	Back and rib rashers
6	Ventral thorax	Belly	Streaky rashers
7	Lumbar	Loin Loin chops Chump chops	Short back rashers
8	Abdominal Wall	Flank	Thin streaky rashers
9	*Longissimus dorsi*	Eye muscle (large muscle in chops)	Special 'loin' bacon
10	*Ileo-Psoas*	Fillets	—
11	Pelvic	Chump Chump chops	Top Back Gammon slipper Ham
12	Upper hind-limb	Gigot Leg	Corner Gammon Middle Gammon Ham
13	Lower hind-limb	Hock	Gammon hock Hock Ham
14	Feet	Trotters	—

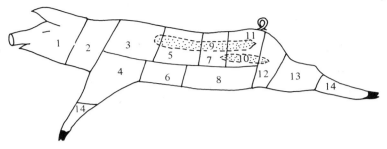

Fig. 38.1 Parts of carcass (numbered in Table 38.1)

There are several variants of nomenclature depending on the locality and many ways of butchering the carcass. Different retail products are reflected in specific markets for the live pig. The main classification is shown in Table 38.2. Historically, some of these markets were satisfied by different breeds of pigs, but the tendency over recent years has been for them to take the same basic type (Large White or Landrace or crossbreds between these two) at different weights.

Table 38.2 Live and dead weights of pigs for different markets

Live weight range (kg)	Dead weight* range (kg)	Market
10–20	(eaten whole)	Sucking pig**
58–68	40–50	Light pork
69–95	51–70	Pork, Cutters
80–102	59–77	Bacon
95–106	70–80	Heavy Cutters
106+	80+	Heavy Hogs
120+	88+	Cast Sows

* Weight of cold eviscerated carcass, includes head and abdominal fat; usually about 70–76% of live-weight after fasting overnight
** Very small but interesting connoisseur market

Carcass grading

Even within the same weight range pigs are not all of equal value per unit of carcass weight, because the demand is much greater for lean meat (muscle) than for fat (adipose tissue).

Essentially the purpose of grading is to distinguish carcasses containing more lean from those with less. In practice, direct measurement is too laborious except for reference and indirect measures are used. Because the pig requires subcutaneous fat to compensate for its deficiency of insulative hair, it tends to be spread relatively evenly over the body. Since subcutaneous fat is a relatively high proportion of the total fat, any standardised measurement of its thickness tends to predict with considerable accuracy the total carcass fat or its inverse, total lean. The most predictive measures tend to be those around the middle of the back overlying the *longissimus dorsi* or eye muscle, shown in Fig. 38.2.

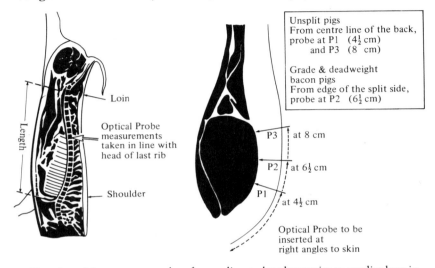

Fig. 38.2   Measurements taken for grading or by ultrasonics to predict lean in
live animals
Source: MLC

The single most accurate measure is P2 but this is marginally improved by including P1 and P3. Further small improvements are made if measures are taken at the mid-line over the shoulder and over the *gluteus medius* on the rump. In bacon pig production such measures are used to classify pigs into grades for differential payment, and are usually made by independent graders paid for by a levy on the slaughtered pigs.

The inclusion of length of carcass in grading is controversial. It has no relationship with leanness above that provided by backfat thickness.

Shortness in pigs can indeed be associated with fatness, but some of the leanest breeds, Pietrain, Hampshire and Duroc, are short compared with the White breeds, Landrace and Large White. It is unfortunate that to both the manufacturer and the farmer, grading tends to become an end in itself. Perhaps, in future, the UK will move towards the Canadian system in which all the measurements including live-weight are used in a prediction of lean content, for which the farmer is then paid.

The production of boars

Until recently, all male pigs except those reserved for breeding were castrated before three weeks of age. The main reason for this was the risk of boar taint in the meat. This musky odour is that of a chemical secreted initially by the testes (5α-androst-16-en-3-one) and it accumulates in the body fat. On cooking, it is released and can be detected by those with a good sense of smell. At present, there are conflicting views on the seriousness of the problem. Entire boars have a major advantage in that, compared with castrates, they are more efficient in converting feed to gain, and produce carcasses which are much leaner. Consumer surveys indicate that despite the odour, boar meat is usually preferred because of the overriding factor of leanness. It seems likely, with increasing public revulsion against the mutilation of animals, that boars slaughtered at 85 kg or less when the taint is slight, will become an increasing proportion of slaughtered pigs.

THE BIOLOGY OF REPRODUCTION IN THE SOW

The female pig is a mammal *par excellence*, having 12 to 16 mammae, and is capable of sustaining a lactation over two months during which 400–500 kg of milk may be produced. If piglets are weaned at an early age, that is, at less than three weeks, the sow can produce on average 2·5 litters and 25 offspring per year. Claims made for the Mei Shan pig of China suggest that the biological potential for the species may be as high as 50 piglets per sow annually.

Puberty

When pigs are kept undisturbed in established social groups, puberty normally occurs between 180 and 210 days of age when pigs weigh between 95 and 120 kg. Disturbance of the social order and some degree of stress, however, may bring much earlier puberty. The most powerful single influence occurs when gilts are moved into the presence of a mature boar. Boars emit through their salivary glands and in their urine musk-like pheromones which together with courtship activity appear to stimulate the gilt. Oestrus and ovulation can thereby be induced precociously and in some cases as early as 135 days or at about 65 kg live weight.

Oestrus and ovulation

After puberty, unserved sows regularly come into oestrus at intervals of about 21 days throughout the year, showing little seasonality. The reproductive tract of the sow is shown diagrammatically in Fig. 38.3. About 36 hours before ovulation the sow will show

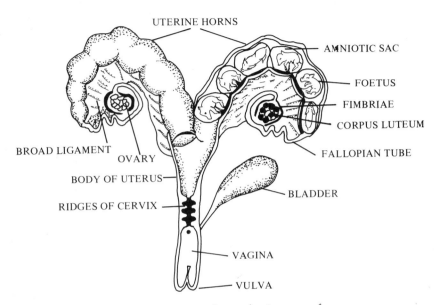

Fig. 38.3   Diagram of reproductive tract of sow

signs of oestrus, with reddening of the vulva, and will stand to the boar. Oestrus lasts about two and a half days. Some 8–12 follicles ripen on each ovary and in mid-oestrus these rupture and release the ova into the fimbriae of the Fallopian tube. As the eggs travel down the tube they will be fertilised if spermatozoa are present. Transport of sperm through the uterus to the Fallopian tube takes about 12 hours during which time the cells undergo a process of maturation. Sows are most fertile in relation to a single service at mid-oestrus and coincident with ovulation. Where possible it is advisable to have sows served at least twice with an interval of at least 8 hours.

Artificial insemination

AI is not used as routinely for pigs as for cattle partly because, unless the timing is perfect, poor rates of conception result. However, it is an invaluable means of introducing new genes into a herd without the attendant risk of disease which may be imported by buying a boar. The fresh, diluted semen can be sent by rail or post but it has a very limited life and should be used if possible within 24 hours. About 20 doses can be obtained from a single ejaculate and because semen can be obtained cheaply through AI centres from boars of the highest genetic merit it has been a major means of improving the national pig herd. The same general rules of timing apply as with natural service except that it is normally considered best to delay the first service until the sow is well established in heat and has ovulated since the life of the sperm in the female tract is reduced.

Pregnancy

Pregnancy or gestation is about 115 days in the sow, the majority farrowing between the 112th and 118th day. There is a tendency for sows bearing large litters to farrow early and those with small litters to be late.

Unfortunately, not all the eggs produce piglets at term and the reasons for the losses are not well understood. Some are due to failure in fertilisation since timing has been shown to be critical. Others are known to be lost in the stage between their arrival in the uterus and the beginning of implantation at about 13 days. Finally, some embryos or blastocysts are lost during the process of implantation, perhaps because of the existence of some mechanism to ensure regular spacing within the uterus.

The number of piglets born is often only about 75% of ovulated follicles. At present there appears to be no strategy of husbandry which alters this favourably and it is probable that the limitation to numbers born is set by the uterine response rather than by the ovary. There are some reports that overfeeding during early pregnancy can cause an increase in embryonic mortality, but the reasons for this are unclear.

Parturition

At birth, piglets weigh about 1·3 kg, with a normal range of 0·8–2·0 kg. This is a very variable figure since smaller piglets are typical of large litters (12 or more) whilst large piglets occur more frequently in small litters (eight or less). The smallest pigs tend not to be viable and it is rare for pigs of less than 800 g to survive. Because piglets are small in relation to the size of their dam (about 1%), obstetrical problems are much less common than in other large farm animals. Normally the sow lies on one side during parturition and in this position allows piglets already born to suck. It is important to avoid major disturbance of the sow during parturition since she may become very active and crush piglets by trampling or lying on them. Gilts, farrowing for the first time, may savage their piglets, apparently not realising what they are. Since up to 10% of newborn piglets may be crushed or savaged it is normal practice to confine sows in special crates which are designed to allow piglets to escape if she changes position (see Fig. 38.4). The crucial nature of management at farrowing has led some farmers to consider special maternity wings. These give cleanliness, and allow access for resuscitation of weak piglets and the reorganisation of litters into even groups.

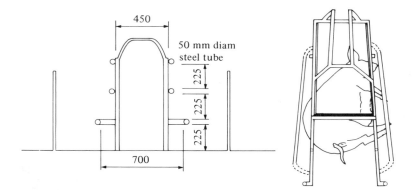

Fig. 38.4   Typical restraining crate for farrowing sow
(dimensions in mm)

Lactation

Until the 1970s weaning was normally at 6–8 weeks of age. With the increase in earlier weaning at four weeks or less, the lactational component of the sow's production has become of less importance. It is not negligible, however, and well-grown piglets at three weeks of age (7 kg live weight) are a reflection of good milk yield. During a heavy lactation, particularly over five or six weeks, sows may lose considerable body fat and this may lead to infertility.

Rebreeding the sow

During lactation the sow does not usually have a fertile oestrus. It is thought that the hormones associated with lactation suppress the pituitary hormones which control the oestrus cycle. However, in some herds there has been a measure of success in achieving con-current lactation and pregnancy by grouping sows and their litters at about three weeks after birth and introducing a mature boar. This is a similar theme to that of inducing puberty in gilts. It is not widely practised since it is difficult to know if sows have been effectively served and the piglets are often checked severely by the process of mixing.

If piglets are removed from the sow the milk supply stops. It is not necessary to do anything active to encourage 'drying off'. Normally sows come into fertile oestrus within 5–7 days of piglet removal. The frequently encountered problem of extended intervals between weaning and service is usually resolvable by encouraging the social stimulation mentioned previously and by ensuring that sows do not end their lactation in an emaciated state. As a last resort, it is possible to use hormone therapy to stimulate the cycle artificially, but this is a poor substitute for management.

THE YOUNG PIG
The neo-natal piglet

Piglets are born with no subcutaneous fat and almost no resistance to infection. Under normal farm conditions it is essential for them to receive colostrum from the sow, because it contains the important immunoglobulins, IgG and IgM. These are absorbed unchanged through the gut wall into the bloodstream and protect the piglet from a wide range of infections. After the first day the gut wall is no longer permeable to intact protein. Through the rest of lactation the sow continues to secrete immunoglobulin in the milk, particularly IgA. This is active in the gut protecting the piglet from gastro-enteric infections. For these reasons it has proved extremely difficult to wean piglets from the sow at ages earlier than two weeks without incurring severe losses due to infection.

Sows' milk is extremely rich in lipid containing between 7 and 8.5%. The newborn piglet uses the lipid with great efficiency and quickly acquires a covering of subcutaneous fat.

The period of highest risk to the piglet is that between birth and becoming established on a productive teat. During this time it may become severely chilled. The lower critical temperature of a newborn pig which has not had milk is about 34°C. Below this temperature it begins to lose body energy in increasing amounts. It is common practice to provide infra-red lamps or under-floor heating to give an area of high temperature in the farrowing pen. Unfortunately, this is only a partial solution for the piglet since its life depends not in lying under the lamp but in drinking at the udder. A high survival rate will result from ensuring that each piglet has a long drink at a teat before it seeks the comfort of the lamp. It is common practice to dose piglets reared indoors with iron either orally or by intramuscular injection (150–200 mg) to prevent anaemia.

Weaning the young pig

Weaning poses particular nutritional and environmental problems for the piglet. These tend to be increased the earlier it is done. In the wild, young piglets explore all objects with their snout and mouth. They will eat small quantities of many materials from a very early age, including feed, earth and the faeces of the dam. This has an important educational function not only at the cerebral level but also for the immune and digestive systems. The piglet must develop amylolytic enzymes for hydrolysing starch and powerful proteolytic enzymes for coping with a wide range of plant proteins. Just as importantly, it must distinguish between harmless food proteins and those of pathogenic bacteria which require a response by the immune system. These abilities are not acquired instantly and in a natural environment the process of weaning occurs over several weeks.

On the farm, weaning tends to be an event rather than a process. Piglets weaned at three weeks of age have only a tenuous residue of the passive immunity imparted by the mother's colostrum and only a partially-competent active immune system of their own.

Factors favouring successful weaning at three weeks or earlier are (a) exposure to the weaning diet (replaced daily) before weaning; (b) clean pens, preferably with perforated floors, to ensure dryness and minimal contact with faeces, or abundant straw; (c) a highly digestible diet (see later); (d) protection with an antibacterial agent if necessary; and (e) even warmth and a smooth pattern of ventilation.

This short list is difficult to achieve in practice, and many farmers are prepared to sacrifice the increased sow productivity and more efficient use of the lactation pens for the fewer problems which arise from weaning at 5–6 weeks of age. As usual, it is a question of the suitability of a system for a given set of resources.

NUTRITION

The formulation of diets can be a very precise science and many pig farmers are content to leave the complexities to the specialists of feedstuff compounders, but some insight into the principles and terminology of the subject is useful.

Digestive system

The pig is a simple-stomached animal and has no rumen, omasum or reticulum. The gut is similar to that of man except for the well-

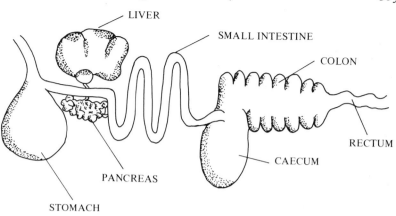

Fig. 38.5 Diagram of digestive system of pig

developed caecum replacing the small appendix of the human, and is shown diagrammatically in Fig. 38.5. Although the pig is usually described as omnivorous this does not mean that it can live on anything. The digestive system is adapted to the particular technique of foraging. Its highly specialised snout is used for unearthing roots and the subterranean storage organs of plants, which are particularly rich in starch and sugars. It also relishes earthworms, carrion and fruit. The mouth and tongue are poorly adapted to grazing and it eats grass and long forage only clumsily and slowly. These features, together with the absence of the rumen as a large receptacle, prevent the pig from eating the large quantities of roughage which so typify the ruminant. It is true, however, that the pig does have a limited capacity to ferment feed materials in the caecum, although this is not the primary organ for releasing dietary energy. The function of the caecum appears to be concerned mainly with fermenting incompletely digested starch, sugars, proteins, simple polysaccharides and hemicelluloses. The fermentation of the cellulose of plant cell walls is extremely limited and even dry sows offered nothing but high-quality fresh grass can barely meet their maintenance requirements.

Bulky feeds which contain starch or sugar in appreciable quantity can of course be used. Similarly, liquid skimmed milk or whey or the waste liquors from distilleries can be used if priced economically.

Energy from feed

The pig can digest only a proportion of the energy-yielding components of a normal diet. The indigestible material is excreted in the faeces together with bacterial cells and endogenous matter such as epithelial cells and mucus. The difference in dietary and faecal energy gives the digestible energy (DE), usually expressed as megajoules (MJ) per kilogram (kg) of diet. For example, pure starch contains 18·49 MJ DE/kg. DE is similar to the North American system of total digestible nutrients (TDN), 1 kg of TDN being equal to 18·24 MJ DE. Metabolisable energy (ME) is the DE less the urinary energy, although it can be given more precise definitions. For most pig diets

$$ME = 0·96 \times DE$$

Because little is achieved in precision by using ME rather than DE for pigs, the Working Party of the Agricultural Research Council on Nutrient Requirements of the Pig (ARC, 1981) chose to express dietary requirements in terms of DE. A typical value for a diet of 80% barley, 17% soya bean and 3% vitamins and minerals would be about 13 MJ DE/kg.

Protein in pig diets

By law the protein concentration in purchased pig diets must be declared. It is based on the multiplication of the N concentration by the factor 6·25. It is not, however, the concentration of protein *per se* which is important, but the amount of essential amino acids provided by the diet. There are ten amino acids which are essential for the growing pig and these are required in particular ratios which correspond closely with their proportions in the pig's own body. The amino acids in the proteins of cereals occur in quite different proportions.

The ARC (1981) gave estimates of the concentrations of amino acids required in an 'ideal' protein for the growing pig and these are compared with those in barley protein in Table 38.3.

Table 38.3 Comparison of concentration of essential amino acids in barley protein compared with those recommended for an 'ideal' protein by ARC (1981) shown in g/kg protein

Amino acid	Ideal	Barley	Amino acid	Ideal	Barley
Lysine	70	34	Isoleucine	38	43
Methionine + Cystine	35	33	Leucine	70	68
Threonine	42	35	Histidine	23	26
Tryptophan	10	12	Phenylalanine + Tyrosine	67	82
			Valine	49	54

Table 38.4 Examples of formulations of diets for different classes of pigs. Amounts expressed in kg per tonne (1000 kg) of diet unless otherwise stated

Live weight (kg)	Pregnant 100+	Lactating 100+	Early weaned 7–15	Follow-on 15–30	Growing 30–60	Finishing 60–110
Ingredients						
Protein-rich						
Ext. soyabean meal	60·0	120·0	–	180·0	150·0	150·0
White-fish meal	–	40·0	130·0	90·0	40·0	–
Meat and bone meal	30·0	–	–	–	40·0	–
Dried-skimmed milk	–	–	350·0	–	–	–
L-lysine hydrochloride	–	–	1·0	2·0	2·0	–
Energy-rich						
Ground barley	777·5	812·5	–	608	745·5	720·0
Weatings	100·0	–	–	–	–	100·0
Ground flaked maize	–	–	227	100	–	–
Breakfast oat flakes	–	–	227	–	–	–
Soyabean meal	–	–	60	–	–	–
Major minerals						
Di-calcium phosphate	15·0	15·0	–	7·5	10·0	15·0
Limestone flour	10·0	5·0	–	5·0	5·0	7·5
Salt	5·0	5·0	2·5	5·0	5·0	5·0
Trace element and vitamin premix	*	*	*	*	*	*
Analysis						
Crude protein (g/kg)	133	160	258	210	187	15·4
Digestible Energy (MJ DE/kg)	12·4	12·9	17·8	13·5	12·9	12·7
Total lysine (g/kg)	5·8	8·1	18·0	11·9	11·2	7·3

* Trace mineral and vitamin supplement should provide per tonne (1,000 kg) of diet the following amounts: vit. A, 8 m. IU; vit. D3, 1·5 m. IU; vit. E, 10,000 IU; vit. K (menaphthone) 2 g; Thiamine, 1·5 g; Riboflavine, 4 g; Nicotinic acid, 10 g; Pantothenic acid, 10 g; Pyridoxine, 2·5 g; Cobalamin (B12), 15 mg; Copper, 20 g; Zinc, 100 g; Magnesium, 30 g; Manganese, 20 g; Iron, 30 g; Cobalt, 1 g; Iodine, 1 g; and Selenium, 0·1 g.
For growing pigs if high levels of copper are permitted as a feed additive then the supplement may also include copper to give 175 g/tonne.

It can be seen that the greatest deficiency occurs with lysine which is only 48·5% of that considered to be ideal and in barley, as in all cereals, lysine is the most limiting amino acid. Merely adding synthetic L-lysine to it will improve the utilization of the protein by as much as 50%. The second most limiting amino acid in barley is threonine which is 83% of ideal.

Formulation of pig diets

The formulation of diets for pigs must take account of many factors. The most important of these are (a) nutrient requirements of class of stock; (b) concentration of DE; (c) ratio of protein to DE; (d) amino acids in protein (particularly lysine); (e) digestibility of protein and availability of lysine; (f) amount of fibre in consti-tuents; (g) palatability and acceptability; (h) presence of anti-nutri-tive factors e.g. trypsin inhibitors and lectins such as those in raw soyabean; and (i) cost.

The solution to the problem of choosing one of the many options for meeting the requirements at an acceptable cost is done com-mercially by the computing technique of least-cost ration program-ming. Examples of diets for each of six classes of pigs are given in Table 38.4. These meet the nutrient requirements and incorporate typical ingredients although they are not necessarily least-cost solutions.

Feeding growing pigs

Growing pigs even of the same weight vary greatly in their daily voluntary intake. It is affected not only by the diet but also by the breed and sex. When healthy pigs of high genetic merit are fed *ad lib* the intakes and growth rates shown in Table 38.5 may be obtained. The table also shows the ratio of feed: gain at each stage, which is an index of efficiency. It worsens during growth, partly because the tissues of the older pig contain less water and more fat

Table 38.5 The amounts of feed which can be eaten daily by pigs given a high quality diet *ad lib*, with the corresponding growth and the feed: gain ratio

Live weight kg	Daily feed MJ DE	Daily feed g	Daily gain g	Feed: Gain ratio
20	18	1,380	690	2·0
30	25	1,920	875	2·2
40	30	2,300	920	2·5
50	35	2,690	995	2·7
60	39	3,000	1,000	3·0
70	42	3,230	1,010	3·2
80	44	3,385	995	3·4
90	46	3,540	980	3·6

than those of younger animals, and also because the proportion eaten in excess of the maintenance requirement diminishes as the animal approaches maturity.

The grading of bacon pigs can be improved if the feed intake is restricted from about 50 kg live weight. The appropriate degree of restriction varies with the circumstances but a flat rate of about 2·7 kg/d (35 MJ DE) from this live weight would be a progressive restriction and would extend the growing period by about two weeks. Scales, which are about 75% of voluntary intake, are commonly used from 60 kg.

Rationing sows

Appropriate feeding scales for the pregnant sow are difficult to prescribe, because the environment and her past production affect her body condition. Between 25 and 36 MJ (1·9–2·8 kg) DE daily may be needed to sustain a gain of about 15 kg from service to service for the first four reproductive cycles. The exact amount should be judged by the condition of the sow. She should not become so emaciated that the ribs and spine are seen to protrude, nor so fat that the same bones cannot be distinctly felt with hand pressure.

During lactation sows may be fed to appetite or given a daily feed allowance of 4 kg plus 400 g for each piglet in the litter in excess of five.

Systems of feeding

Pigs of all ages may be offered feed in the form of meal, pellets or as slurry mixed with water. Whichever system is used, fresh water should always be available, since pigs easily die from salt poisoning if water is restricted too severely. Growing pigs are usually kept in groups of 10–40. Large groups tend to become uneven with the result that bullying and vices such as tail biting may occur. Pregnant sows can be kept in groups of up to 10 but it is best to have facilities for feeding them individually. In practice, many sows are housed in stalls or are tethered. This allows for easy management but greatly restricts the sow's freedom of movement.

Although the labour requirements of dry feeding may be low, wet feeding has a number of advantages. These include the potential to use bulky feeds such as cooked potatoes or swill, and liquid by-products such as skimmed milk or whey. Swill must be cooked properly under licenced conditions, since feeding uncooked waste has been responsible for the spread of foot-and-mouth disease and swine vesicular disease, which are both serious notifiable diseases requiring immediate slaughter of the pigs if confirmed.

Feed additives and growth promoters

Most feed additives are in four categories: (a) flavouring agents, (b) antibacterial agents, (c) hormones and (d) vaccines. Flavours are comparatively unimportant to the pig but may impress the farmer. Antibacterial agents are widely used both therapeutically to control enteric and respiratory disease and at lower levels to control the bacterial population of the gut. These currently include copper sulphate and antibiotics such as bacitracin, tylosin and virginiamycin, but this is not a comprehensive list. Certain antibiotics are not allowed for routine use because of their importance in human medicine. The most valid use for antibacterial agents is in early-weaning diets, but even then they are not essential. In general, they are a substitute for good hygiene. Although heavily advertised, they should be used with caution because of the dangers of organisms becoming resistant to several of them at once.

Analogues of the natural sex-hormones have been used in diets to improve leanness, but concern about the wide dissemination of such potent chemicals may greatly restrict their use in future.

Oral vaccination of piglets using killed bacteria is an attractive new development. There are ethical problems about stimulating natural immunity and there seems to be considerable scope for technological advance in this area. It has been used successfully to treat certain coliform scours in young pigs.

ENVIRONMENT AND HEALTH

Pigs are extremely robust animals, very resistant to stress and infection. There are, however, environmental circumstances and infections to which they succumb. Outdoors, pigs should always have the option of well-bedded, dry shelter. Indoors, pigs should be housed where possible within their thermoneutral range, that is below the temperature at which they become heat stressed and above the temperature where they must expend extra energy to maintain their body temperature. The latter is called the lower critical temperature and varies with feeding level, ventilation rate, group size, the weight of the pig and the insulation of the bed.

For pigs kept in groups and fed generously in good conditions, the approximate lower critical temperatures are at 20 kg live weight 23°C, at 50 kg 17°C, at 80 kg 14°C and at 100 kg 11°C. For pregnant sows fed close to maintenance and housed individually the lower critical temperature is about 20°C.

Two major respiratory diseases, enzootic pneumonia and rhinitis, can be greatly exacerbated under poor environmental conditions. However, these and other chronic diseases of pigs can be eliminated by a technique involving the removal of piglets at term by Caesarian section. The piglets are then reared away from contact with conventional pigs and used as foundation stock for 'minimal disease' herds. The highest degree of isolation and hygiene are required thereafter, but this approach has been used extensively by breeding companies to reduce the risk of the animals they sell spreading infections.

GENETIC IMPROVEMENT AND BREEDING

Scientific breeding has led to very rapid improvements in the rate and efficiency of growth in pigs. The M L C provide central facilities for measuring the individual growth and performance of potential breeding boars and also to test the various genetic 'packages' offered by commercial breeding companies. Traditional breeding practice was to select improved pigs *within* a specific breed. Such purebred pigs have been shown to lack hybrid vigour and to be slightly less fertile than crossbred pigs. Most modern breeding programmes are designed so that the dams of the slaughter generation are cross bred and there is now a tendency to breed crossbred boars as well.

Selection objectives which have been pursued successfully in breeding programmes have been lean-tissue feed conversion (food/ lean) and lean-tissue growth rate. The current state of the art is illustrated in Table 38.6, which shows mean figures over three years for one of the more successful breeding companies.

Table 38.6 Three-year means for pigs submitted by one company which scored consistently well on the Commercial Production Evaluation of the MLC

Slaughter Wt. (kg)	Feed:gain ratio	Live weight Gain (g/d)	Lean Gain (g/d)	Feed per kg lean (kg)
61	2·54	767	312	6·33
87	2·75	794	328	6·66
116	3·19	840	332	8·06

With the considerable improvement in efficiency of the growing pig, breeding companies are now turning their attention to improving the fertility of the sow and to improving the shape and distribution of lean in growing pigs.

In the future, the genetic resources of the many exotic breeds of pigs, such as the Mei Shan mentioned previously, may be introduced by specialised cross-breeding programmes.

PIG PRODUCTION AND SOCIETY

The consumption of pork increased by 50% in the UK during the decade 1970-80 and the total weekly consumption of pork, bacon and ham now exceeds that of either beef or poultry. This reflects the comparative cheapness and general acceptibility of pig meat. There are about 0·8 million sows in the UK and about 7 m. growing pigs. Increasingly these are accommodated on large specialist units and housed indoors. However large or efficient the unit, farmers cannot expect neighbours and water authorities to accept major pollution from effluent. One of the potential advantages of these large units could be that the capital cost of plant for the production of biogas from slurry may now be justified.

Recently, concern has been expressed that behind the closed doors of large units intensification is taken to unreasonable limits. It is extremely important that those who tend pigs should appreciate that care and thoughtfulness are not only morally reasonable, but that the pig's astonishing productivity can be realised only by meticulous attention to its physiological and social needs.

39

POULTRY PRODUCTION SYSTEMS

Introduction

Poultry husbandry has been through a remarkable period of development in recent years and is now the most intensive sector in livestock farming. Productivity at the same time has increased greatly. There is also a high degree of standardisation and uniformity of methods in poultry husbandry not only in Great Britain but world-wide. Nearly everywhere poultry bred and reared for meat are kept on 'built-up' litter in insulated, mechanically ventilated and artificially heated housing. Also, by far the majority of laying birds are reared and kept in multi-bird tiered cages usually under environmentally controlled conditions.

The reasons for these developments are not hard to find. Advances in genetics have enabled strains of birds to be produced which have an increasingly greater potential for egg production and/or growth at a very economical consumption of food. If such birds were kept under extensive conditions their potential could never be attained. Food costs, for example, would be very heavy. Cold weather would mean that a certain proportion of the feed would be used merely to keep the birds warm whilst a further proportion would be used for the birds' extra activity. There would also be considerable losses of food from wild life and vermin. Labour demands under extensive management are also much greater and more land is required. Control of disease may also be less effective when birds are kept outdoors. Contagion from other units is more likely and the administration of such vaccines and drugs as may be required is difficult to provide. It has been because of the very effective method of maintaining health indoors that there has been a pronounced growth of large poultry units, sometimes of mammoth size, which enable capital and labour cost per bird to be minimal.

This is where we find ourselves at present though poultry husbandry is an excitingly dynamic subject and changes are still taking place at a relatively rapid pace.

POULTRY BREEDS

Commercial egg layers are divided into two principal types: the light and heavy breeds. The light breeds, such as the White Leghorn, are small compact birds that weigh between 1·5 and 2 kg. They lay white-shelled eggs and being of modest size they are economical on food. The heavy breeds, predominantly based on the Rhode Island Red and Light Sussex, weigh about 2–2·5 kg. They lay the more popular brown-shelled egg but eat more food for their maintenance. On the other hand, the carcasses have more value for meat than those of light breeds, at the end of the laying season.

Pure breeds as such are rarely used as layers nowadays. The commercial producer of eggs purchases his stock from breeding companies that are specialists in the production of hybrids. These are genetically complex mixtures of strains and breeds of a light, heavy or light-heavy mixture, laying eggs with white, brown or tinted shells. The production of hybrid poultry is done by two fundamental procedures. First, closed families are bred by a process of close breeding and selecting carefully for a limited number of the most important characters. These genetically uniform families are then crossed in many combinations from which are chosen those that are of the highest quality to give the commercial end-product.

The same basic procedure is used for the production of meat chickens, using the White Rock and Cornish as the principal foundation breeds.

Geneticists who are concerned with the breeding of poultry have an extremely complex task. A good commercial egg layer must have the following qualities: a high yield of large eggs with shells and content of high quality throughout the laying period; feed requirement for maintenance and production should be low; and birds must have good liveability and disease resistance. Temperament should be quiet and broodiness should be minimal.

Chickens for meat (broilers) must have fast growth and good live weight, good food-conversion efficiency, good conformation, high liveability, disease resistance and flesh and skin colour as required by the market.

With poultry breeders the requirement is for the female to produce large numbers of fertile eggs with a good hatchability, whilst the males should have a good growth rate, excellent conformation and good food-conversion efficiency.

NUTRITION OF POULTRY

The digestive system of the chicken is very different from that of the mammal and in order to understand the principles behind their

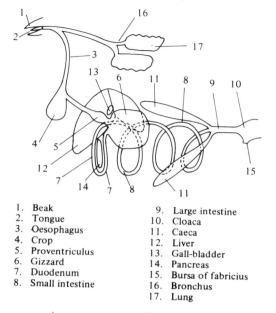

1. Beak	9. Large intestine
2. Tongue	10. Cloaca
3. Oesophagus	11. Caeca
4. Crop	12. Liver
5. Proventriculus	13. Gall-bladder
6. Gizzard	14. Pancreas
7. Duodenum	15. Bursa of fabricius
8. Small intestine	16. Bronchus
	17. Lung

Fig. 39.1   Diagrammatic representation of the digestive and respiratory systems
of the domestic fowl
From *Poultry Health and Management* by D. W. B. Sainsbury. Granada
Publishing. 1980

feeding it is helpful to study the way the bird digests its feed (Fig. 39.1).

The food is gathered by the beak which has at its edges tactile cells by which the bird decides whether it will accept the food or reject it – based on feel, taste and reflectivity. The food is swallowed whole passing down the oesophagus into the crop which is basically a container where the fibre is well softened and the food acidified. From the crop the food passes to the proventriculus which secretes hydrochloric acid and pepsin. This is the closest in form to the mammalian stomach. From here the food passes to the gizzard which is a strong muscular organ which grinds the contents to a paste-like mass with the aid of the insoluble grit present within it. The food is now prepared for the small intestine where the major digestion and absorption proceeds and then to the two caeca where most of the water absorption takes place. From here the material passes for evacuation to the cloaca, an organ which is also involved with the excretion of urine, the acceptance or delivery of semen and the passage of the egg to the outside.

Essentials of the feed

Poultry generally have no access to soil or herbage nor do they usually receive sunlight. Thus the food must be totally sufficient in all known nutrients and with a safe margin for any additional requirements such as the demands of very high productivity over and above the anticipated figure.

The first need is for the provision of sufficient energy and protein. The term now used for the assessment of energy is 'metabolisable energy' (ME), which is that part of the food which is truly available (or metabolisable) to the animal. Energy is usually measured in megajoules (MJ). Diets are classified as high energy (over 11·9 MJ ME/kg); medium energy (11·0–11·9 MJ ME/kg) or low energy (10·5–11·0 MJ ME/kg). In practice high-energy diets will often be used for small hybrid laying pullets, referred to previously, so that they can produce the very large number of eggs in relation to their size.

Though they may weigh only a little over 1·6 kg they may consume less than 110 g of food daily. Broilers also need high-energy diets as they must reach about 2 kg in seven weeks at a food conversion ratio of a little over two.

By contrast, for 'heavy' birds and for diets for rearing stock, high-energy diets would be wasteful and harmful, giving rise to over-fat birds. In particular the heavy broiler breeder must be checked from laying too early in its development and should receive a lower energy diet to assist towards this end.

In high-energy diets the principal cereals are maize and wheat but where a lower-energy feed is required the proportion of wheat by-product and barley and oats can be raised in relation to the wheat and maize to reduce the ME figure.

Where extremely rapid growth and weight gains are needed, as in broilers, the energy level can be raised by including in the ration up to about 5% of added fat.

Protein is required for the general body development in all growing birds and layers also need a high percentage since there is some 14% of protein in an egg. The young chick requires 19–23% of a high-quality protein in its diet for the first six weeks but thereafter the level can be reduced to 16–18% and in the case of broiler breeder stock, where fast growth is undesirable, to 14–16%.

It is important to know that birds will eat sufficient food to satisfy their *energy* requirements and this is one of the main factors limiting intake. Hence the consumption of a high-energy food will be lower than that of a low-energy one and the protein, vitamin and mineral content must be adjusted accordingly in the former type of ration.

The quality of the protein is also a vital factor. For chickens, the two most important amino acids, which form the constituent parts of the protein, are lysine and methionine and the proteins richest in these are those derived from animal origin. For this reason animal protein is almost always included in chick and broiler diets though it is not essential for laying birds on medium- or low-energy diets. In some diets synthetic amino acids are now being used as an economic method of adding these to improve the quality of the protein. The best-quality protein of all for poultry is white fishmeal, which may often be included in the rations of young chicks at levels of up to 10%, this amount being reduced as the birds get older. Meat meal products are also used. The best vegetable protein is soyabean meal, though it is relatively poor in methionine and should not be fed to chicks as the sole source of protein. A Table of nutritional standards for the principal poultry feeds is given in Table 39.1.

Vitamins

Housed poultry are totally dependent on the vitamins in their feed in the correct amount and any disturbance in their supply can cause most serious consequences, sometimes with striking suddenness. The principal vitamins required are as follows:

Vitamin A

This is the principal vitamin that ensures adequate growth and assists the birds' resistance to disease. It may be formed in the bird's body by synthesis from carotene present in green vegatable matter or yellow maize, or it may be fed direct by using vitamin A in synthetic form. Chick and breeder diets should contain approximately 12 m. international units (m.IU) per tonne of feed and laying diets 6 m.IU/tonne.

Vitamin D

Chickens can synthesise this vitamin from sunlight, but even under natural extensive or semi-intensive systems, the amount is totally unreliable and usually insufficient. Hence vitamin D is usually supplied in synthetic form in the ration. It must be given to poultry as vitamin D_3. Without sufficient of this vitamin birds cannot utilise efficiently the available calcium and phosphorus in the diet which are vital for bone formation and the production of good-quality egg shells. The requirement for vitamin D_3 in the diet is as follows: 3-4 m.IU/tonne of feed for chicks and 3 m.IU/tonne of feed for layers and breeders.

Vitamin E

A deficiency of this vitamin gives rise to diseases of the nervous system in chicks known as 'crazy chick disease'. It is also essential

Table 39.1 Nutritional standards for principal poultry feeds

	Crude protein %	ME, MJ/kg	Lysine g/kg	Methionine + cystine g/kg	Calcium %	Phosphorus %	Salt %	Zinc mg/kg	Sodium mg/kg	Manganese mg/kg	Vit. A m.I.U./ tonne	Vit. D m.I.U./ tonne	Choline mg/kg
Starting chicks	20	11·66	11	7·5	1·0	0·5	0·40	60	1500	100	12	3–4	1300
Early growers	15	11·25	8	6·0	1·0	0·4	0·40	60	1500	100	12	3	1300
Late growers	12	11·25	6	4·5	1·0	0·4	0·40	50	1500	100	8	3	1300
Light hybrid layers	16–19	11·66	8	4·6	3·6	0·5	0·40	50	1500	100	6	3	600
Medium hybrid layers	15–18	11·66	8	4·6	3·6	0·5	0·40	50	1500	100	6	3	600
Breeders	16	11·66	8	4·8	3·6	0·5	0·40	50	1500	100	10–12	3	1100
Broiler starters	23	12·88	12·5	9·2	1·2	0·5	0·40	50	1500	100	12	4	1300
Broiler finishers	19	12·92	10	7·3	1·0	0·5	0·40	50	1500	100	12	4	1300

to breeding stock for the good hatchability of their eggs. Vitamin E is present in cereals but possibly not in sufficient amount. In any case it is very easily destroyed by bad storage or rancid oil or fats in the food. The ration should contain up to 10 mg/kg of added vitamin E.

Vitamin K

This is an essential vitamin which is crucial for blood clotting and a deficiency leads to haemorrhages. It can be provided for the bird by green food, or lucerne meal or synthetic forms can be used. The requirements are for 2 mg/kg in the normal chicken but may be used in larger quantities to assist in the cure of diseases which cause haemorrhaging.

The B vitamins

These are well distributed in cereals but the vitamins are frequently added in synthetic form to ensure a known and sufficient source. The main functions of the B vitamins are to assist chicks in achieving their optimum growth. Vitamin B_1 is needed for the metabolism of carbohydrates whilst a deficiency of B_2 (riboflavin) causes a condition known as 'curled toe paralysis', which is not uncommon in chicks. Another B vitamin, Nicotinic Acid, is required especially for correct feathering and choline is associated with the correct metabolism of fat. Pantothenic acid deficiency leads to dermatitis. Vitamin B_{12} (cobalamin) is essential for the development of normal red blood cells and also for good hatchability. It is especially vital for young birds and breeders. Folic acid shortage can lead to anaemia and weaknesses of the legs.

Substantial quantities of B vitamins are found in the following foods: grains, seeds, fishmeal, liver meal, meat meal, soyabean meal, dried milk, dried yeast, sunflower meal, unheated offals. It should be noted, however, that B_{12} is found only in foods of animal origin, or it may be synthesised by bacteria from manure or deep litter, hence one of the assets of keeping birds on litter. Levels required by chicks are as follows, in mg per kg of food: B_1 1·0, B_2 5·0, Nicotinic Acid 28, Pantothenic Acid 10·0, Folic Acid 1·5, Choline 1300, B_{12} 0·02. Special mention should finally be made with reference to the B vitamin Biotin. In recent years there have been serious outbreaks of a disease known as 'Fatty Liver Kidney Syndrome' (FLKS). The inclusion of extra biotin in the ration protects birds from this condition. The total allowance of biotin should be up to 0·18 mg/kg of diet and because foods naturally do not contain such quantities, some extra biotin usually has to be added.

Minerals

Calcium and phosphorus

The main function of these two important minerals is for bone formation. They must not only be present in sufficient quantities but also in the correct proportions. For the young chick the ratio should be calcium to phosphorus 2:1 to 1:1 and not outside this range. There is some calcium in all foods but those of animal origin are generally much richer than those derived from vegetable sources. Phosphorus is also present in all foods but in cereals it may be in a form which is poorly absorbed so that supplements may be needed. It is also essential to have sufficient D_3 and manganese to assist absorption and utilisation of the calcium and phosphorus. The young chick needs a minimum of 1% of the diet as calcium and 0·5% as available phosphorus, whilst the laying bird requires 3·5% of calcium since this is the main constituent of the shell of the egg.

Other essential minerals are manganese, which forms an essential link in the calcium metabolism of the bird, iron, cobalt and copper, all of which are essential elements for the formation of haemoglobin, iodine, which is vital for good hatchability of the fertile egg and salt which is needed at levels up to 0·5% of the ration and is required for protein digestion and normal growth.

Water

Poultry have very considerable demands for water and in most systems of poultry production water should be provided *ad lib* and with easy availability. Production will be severely curtailed if there is any interruption of supplies and in hot weather denial of water will accelerate the danger of death. This must be borne in mind, particularly where there is a deliberate restriction of water intake. There is, for example, a relatively new system to control the water consumption of growing breeders on restricted diets in order to prevent excessive intake of water in the absence of feed. Too great a restriction may have disastrous consequences in very hot weather.

Practical feeding of poultry

Balanced rations may be fed to poultry in several ways – an all-dry mash, a mixture of dry mash and grain, wet mash, crumbs or pellets or various combinations of these. All-dry mash feeding is especially useful to the farmer who makes up his own rations and because it takes longer to consume it can help to prevent vices such as feather pecking and cannibalism. Wet mash feed is little used but is very palatable and favoured by the birds. It involves a minimum of wastage.

A mixture of grain balanced with mash is very suitable for birds on deep litter or built-up straw because the retrieval of the grains scattered over the litter exercises the birds, helps to prevent vices and boredom and tends to create better conditions in the litter.

Pellets and crumbs are rather favoured in modern poultry feeding systems as there is little waste and since every crumb and pellet is a completely balanced diet in itself there is no chance of the bird practising selective feeding. Nevertheless, because of the advantages of dry meal feeding just referred to, it is likely to be the foodstuff of choice for birds in cages.

Construction of a poultry ration

The complete compounded poultry ration is more complex than almost any other diet used on the farm and relatively few farmers mix their own diets because of the considerable problems involved in doing so.

In the formulation of a compound feed various procedures must be undertaken. First, one must know the nutrients required by the type of bird to be provided for. Then a diet can be built up from a knowledge of the composition and nutritive value of the commonly available foodstuffs and the range of normal inclusion rates of these feedstuffs in practical poultry rations. In the case of the compounder, all this information is put on to a computer and the final choices are rapidly obtained.

It should be emphasised that there are other factors besides the main nutrients alone which must be considered. The content of indigestible organic material is important. A bird's intestine must have a fibrous content in the diet to maintain its natural functioning. In young chicks this proportion is about 13% rising to 15% in the adult. There must also be a correct ratio between the energy and protein which varies depending on the class or age of bird. For example, with laying birds the ratio between the metabolisable energy (k J/kg) and the crude protein of the diet should be 782:1. Growers require a ratio of approximately 736:1, chicks 598:1 at first and then increasing to 736:1 at about eight weeks.

In addition it is necessary to have up-to-date costs of all the likely constituents so that a completely economical ration can be made. In the end a diet has to be formulated that is of minimum, or at least, favourable cost, satisfies all the standards given, does not go outside the minimum and maximum inclusion rates decided for any of the ingredients and represents reasonable uniformity from batch to batch.

Some typical feeding routines are as follows:

Broilers

These are always fed *ad lib* from day-old to finishing. The broiler 'starter crumb' usually contains 22–3% of crude protein and may be fed for the first three weeks. Then a grower 'pellet' is used containing about 20% protein, followed by 'finisher' pellets of 18% protein.

Replacement stock for egg production

These are normally fed *ad lib* from hatching to eight weeks on baby chick crumbs (18–20% protein) and then on early growers pellets (14–16% protein) to 12–14 weeks, followed by later grower pellets reduced to 12% to the point of lay at 18 weeks. From two weeks onwards mash may be used instead of crumbs or pellets. Grain balancer rations are also appropriate. It is likely that restricted diets will become increasingly common practice in the future as a means of economising on feed.

Commercial egg layers

These are usually fed *ad lib* with the exception of any addition of grain. Protein percentages vary from about 15% for birds on pasture in summer, to 18% for high-energy intensive rations. Whilst they may be fed with any form of ration, mash is nevertheless favoured for birds housed in cages.

Breeding stock

The parent stock of commercial egg layers are fed in the same way as other intensively kept laying birds. Extra quantities of vitamins and minerals are given to ensure good fertility and hatchability. Rations for high egg-producing strains should contain up to 19% of the best-quality protein. The breeding stock for heavy broiler birds are prone to become too fat and should receive 16–18% of protein in the ration. Various techniques have been developed to restrict the food intake in such birds and the poultryman must follow the precise details as laid down by the breeding company, especially as recent considerable advances in the productivity of breeders have made extra attention to feeding critically important and the individual needs of each strain must be met.

Feeders

Both hand-operated and automatic feeders are available. Space requirements are as follows:

Trough feeders:
1-4   weeks 25 mm per bird
4-10   ,,   50-75 mm per bird
10-20   ,,   75-100 mm per bird.

The requirements for tubular feeders of 300-400 mm diameter are:

1-4   weeks one per 35 birds
4-10   ,,   one per 25 birds
10-20   ,,   one per 20 birds.

Drinkers

Poultry drinkers must be automatic, clean, free from excessive splashing and not liable to flood. Recommended space allowances are 12 mm per bird up to 12 weeks of age and 25 mm thereafter. The main types of drinkers are circular in shape and are suspended by cords or chains from the roof so they may be adjusted in height according to the size and age of the birds.

THE ENVIRONMENTAL REQUIREMENTS OF POULTRY

Temperature

The day-old chick requires an ambient temperature of 35°C which may be reduced by 3°C a week. These relatively high temperatures are provided by artificial sources of heat, usually by radiant gas brooders (heaters) which give a humid warmth with easy adjustability, which is essential in good rearing.

After the brooding period of about six weeks, when artificial heating is no longer required, the temperature in the poultry house will be regulated only by good thermal insulation of the building and controlled ventilation, either artificial with fans or by natural draught. From six weeks, when the temperature will be ideally within the range of 18-21°C, birds will thrive most economically if they continue at an ambient temperature fairly close to this range, though it is acceptable in rearing birds for laying and breeding to allow the temperature to be reduced to 13°C to assist in the provision of good ventilation and litter conditions.

Layers in cages are best housed at a temperature of at least 21°C

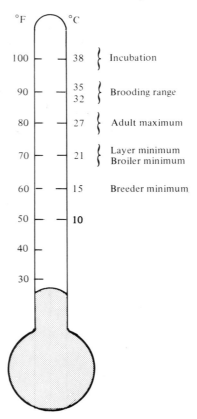

Fig. 39.2   Optimum ambient temperatures for chickens of different ages From *Poultry Health and Management* by D. W. B. Sainsbury. Granada Publishing. 1980

if they are to make the most sparing use of their feed, though lower temperatures are permissible with birds kept on deep litter or straw.

Humidity

It is of advantage to maintain the relative humidity within a range of 30–80%. Too high a humidity gives rise to condensation on the surfaces of the building, damp litter, cooling of the birds themselves and an increased risk of disease. A very low humidity causes bad feathering, drying of the mucous membranes and thereby the easier invasion of the body by harmful disease-causing micro-organisms.

Lighting

The development of the reproductive (egg-laying) system of the chicken is stimulated by increasing amounts of light, as naturally occurs in springtime but is depressed when this is reduced in the autumn. The modern layer, under the stimulus of spring-like conditions, tends to lay before sufficient bodily development has taken place to support egg production properly and it will not be able to lay either the number of eggs or the larger sizes of which it would otherwise be capable. An autumn-like pattern or even a constant but short day-length will allow the body to develop properly before the bird starts laying. Thereafter, to stimulate maximum production, the procedure is to give a weekly increase of light duration of about 20 minutes up to a maximum of 16–18 hours. Artificial lighting is essential to achieve this at all seasons, though by rearing chicks in the autumn the natural advantages of seasonal changes can be made use of.

For example, a suitable programme for layers is:

0–2 weeks   18 hours light, 6 hours darkness.
2–18 „   6 hours light, 18 hours darkness.
19–22 „   Increase light by 45 minutes per week to give a good stimulus to the first part of production.
23 weeks onwards   Increase light by 20 minutes per week until a maximum of 18 hours light per day is achieved.

For broilers the usual pattern is to have 23 hours lighting and 1 hour darkness in each 24 hours, the short period of darkness being necessary to 'train' birds to darkness. If this is not done and the light is suddenly withdrawn for some reason, a 'pile-up' is a likely consequence, the birds tending to crowd into corners and suffocate.

The lighting procedure

For proper artificial control exclusion of all natural light must be complete; for this reason most poultry houses are now windowless and all the ventilators are baffled against natural light entry (Fig. 39.3).

Light intensity in the house should be uniform and at least 10–16 lux for layers. On every lighting curcuit there should be a dimmer so the light intensity can be varied. For example, with broilers the chicks are started at a bright intensity of about 20 lux but later it is much reduced to as little as 0·2 lux since this will reduce hyper-activity, reduce the likelihood of vices and will encourage the most economical growth.

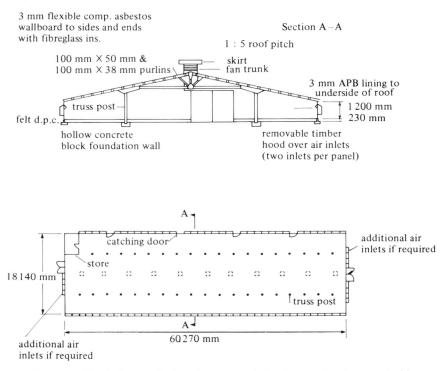

3 mm flexible comp. asbestos
wallboard to sides and ends   Section A–A
with fibreglass ins.

                                                1 : 5 roof pitch

100 mm × 50 mm &   skirt
100 mm × 38 mm purlins   fan trunk

                                                3 mm APB lining to
                                                underside of roof

        truss post   1 200 mm
felt d.p.c.   230 mm

hollow concrete   removable timber
block foundation wall   hood over air inlets
                                         (two inlets per panel)

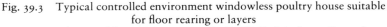

                            A

        catching door   additional air
                                               inlets if required
        store

18 140 mm

                                    truss post

                            A
                     60 270 mm

additional air
inlets if required

Fig. 39.3   Typical controlled environment windowless poultry house suitable
for floor rearing or layers
From *Poultry Health and Management* by D. W. B. Sainsbury. Granada
Publishing. 1980

SYSTEMS OF MANAGEMENT

Floor housing: single stage

Chicks may be reared on litter on the floor from day-old until they
are ready to enter the laying quarters. There is no stress caused by
changes in housing and resistance to parasitic diseases, such as
coccidiosis, tends to build up gradually so the risk of disease from
this type of source is reduced. Birds may be reared for any subsequent laying method in this way and it is also a popular arrangement to take breeding stock and some commercial egg layers from
day-old through to the end of the laying period in the same quarters.
With these systems of rearing, a floor area of 0·2 m² per bird

should be allowed for the lighter breeds and up to 0·24 m² for the heavier breeds, the maximum figure being required for broiler breeding stock. The floor area per bird can be reduced by about one-third by using housing of partly deep-litter and partly slats or wire. With this arrangement a useful proportion of the droppings fall into a pit under the slats or wire and the deep-litter remains in a much better condition. Also, by placing the feed and water points largely on the slats the birds are not only encouraged to make more use of this area but water spillage goes safely into the droppings pit.

Where there is a combination of litter and slats, the young chicks can be brooded on the slats which are covered at first with litter. After a few days the chicks are given access to the littered area. More often they are brooded on the litter and later, at a few weeks of age, they are trained to use the slats by the provision of a sloping run-up and the addition of water and food troughs to the raised area. Rearing on slats and wire alone can be used but is uncommon now as it presents husbandry difficulties, such as feather pecking, cannibalism and hyper-excitability.

Cage rearing

Single-stage rearing is also carried out very successfully in special pullet-rearing battery cages in which the birds are taken right through from day-old to placement in the laying cages. The usual system is to start by putting all the chicks in the top tiers, where the birds can more easily be kept at the correct high temperature and in which management and observation are more easily done. The birds can then be spread through all the tiers within a few weeks.

Birds that are reared in cages should be housed subsequently in battery cages and not on floor litter systems, since they are likely to have a poor resistance to parasitic infections having probably had no challenge at any time and hence no chance to build up resistance.

Cages are produced that are so designed that chicks can be brooded in them and yet they may also serve as laying cages. They have, as in the floor systems that parallel this, advantages in terms of labour economy. In addition they tend to ensure that the chicks are in small groups and also parasitic diseases are unlikely and the food and water are very accessible. Great care must be taken to choose good equipment that has all food and water points easily adjustable. The floors may be 20 mm wire netting or 25 m × 12·5 mm, 16-gauge welded wire mesh. When the chicks are placed in these cages a temporary flat floor is used, later to be replaced by the sloping floor required for layers.

Multi-stage systems

There are a number of systems that brood chicks in a special house then move them on to separate accommodation when the period of artificial heating – about six weeks of age – is over. Many of these alternative systems are traditional and gradually disappearing but may still be in use whilst the equipment is available. For example, the chicks may be reared in a tiered brooder for the first period of four weeks or so. This consists of two or three warmed brooder compartments, largely enclosed and placed one on top of another with wire sided and roofed runs in front. The base is also of wire with droppings trays underneath. A typical size of tier brooder is 3 m × 1 m, taking about 50 chicks to four weeks of age.

From tier-brooders the birds can go on to any other of the systems which are dealt with subsequently. However, it is always best to reduce the change from one system to another to as small a difference in its nature as possible and if tier-brooders are used a very suitable arrangement is to place the chicks straight into laying cages which are specially designed to cope with the feeding and modified flooring required for younger birds.

Outdoor rearing

A system once very common is to rear chicks from hatching to about four weeks either in tier-brooders or on the floor below an ordinary radiant brooder. They may then be transferred to a hay-box brooder. This consists of a covered compartment 1 m × 1 m with a run approximately twice as long in front. No artificial heating is required because the covered section is packed round with straw or hay to conserve the chicks' heat and prevent draughts. The latter part is of solid wood sides and roof but the run has wire mesh sides and roof. The units may either be moved over the pasture or used as fixed units mounted on straw bales.

The system is healthy and can produce hardy, vigorous stock at an economic cost as no heat is used, but it is unfortunately labour intensive. After the hay-box stage birds are normally reared in 'range shelters', or indeed they may be brought to range shelters from other systems of management, especially if they have been reared on one of the floor systems. Range shelters are simple, easily made and light structures, consisting of an ark approximately 3 m × 2 m with a height of 1·7 m at the ridge and 0·55 m at the eaves. A unit of this size will take 60 growers from eight weeks to 18 weeks of age. The roof can be built of metal, plywood, hardboard or timber and felt or even felt on wire for economy. The sides are of wire but a large overhang on the roof protects the stock from wind and rain.

The management of broiler (meat) chickens

Systems used for rearing broilers are probably more standardised than any other arrangement. Almost invariably the birds are reared from day-old to about 50 days in a controlled environment house, which is fully insulated and with mechanical ventilation. The birds will be kept on built-up litter of wood shavings or straw or a mixture of the two types. The litter must be kept extremely soft and friable or the flesh may be bruised or damaged. A typical broiler house has about 20,000 birds and is some 15 m wide. The density of stocking gives a maximum live weight at the end of the growing period of about 34 kg/m². The lightest weight of bird is marketed at about 1·4 kg live weight but the most common weight now sought by the market is a 1·8–2·0 kg bird. Food and water are provided *ad lib* and light, after the first two weeks, is kept dim in order to prevent over-activity and vices developing.

LAYING SYSTEMS

Free range

Only a very few layers are now kept on free range. Nevertheless certain purchasers of eggs are still willing to pay a premium price for the genuine free-range egg so it can still be profitable under such circumstances to keep birds in this way. With free-range systems housing is simple and cheap but any real environmental control is impossible. The birds may be housed in movable or fixed houses of various types, including those with slatted floors.

Free-range systems may be healthy but it is entirely wrong to believe they are inherently healthier than indoor systems. They have serious disadvantages in that the birds and the stockmen are at the mercy of the elements and winter production is almost certain to be particularly poor. It is thus difficult to see them having any real future since food consumption is about twice that when birds are kept under intensive systems.

Semi-intensive system

Another of the traditional systems is the semi-intensive. This consists of a fixed hut placed in one or two or more outside runs to be used in rotation. A typical unit consists of 50 birds kept in a house 3 m × 3 m with two runs each 10 m × 10 m, the birds alternating between them at six-monthly intervals. However, the cost of such a system is high both in capital and labour while the climatic control over the birds is poor. Even when alternating between two runs a certain build-up of disease may occur. Eggs from this system are often dirty. In many respects it is the worst system of all but it can be used by the domestic poultry keeper where the normal economics of the market place do not operate.

FULLY INTENSIVE HOUSING SYSTEMS

The deep-litter system

One of the first really successful methods of housing layers intensively was the built-up or deep-litter system. After many years of use it remains a satisfactory system but of high capital cost. Because of this, commercial egg layers are rarely kept on deep-litter but breeders and broilers are. The litter is either wood shavings or chopped straw to a depth of about 300 mm.

In the first designs of deep-litter houses, movable perches, drinkers and feeders were used so that, as far as possible, droppings and water splashings were distributed as evenly as possible by moving these around the house. However, under these systems birds require about 0·27 m²–0·36 m² of floor space per bird if the litter is to work properly. It is possibly a cheaper arrangement to have a slatted area over a droppings pit with perches and food and water troughs situated here as well.

In this way the litter receives proportionately less droppings so the density of the stocking can be increased. If the nest boxes are suitably disposed with feeding and egg collection by hand, the two operations can be combined in a single movement round the house. Automatic feeding is more usual and there is now at least one system for satisfactory automatic egg collection, so the 'chores' can be quite well reduced.

With an area of deep litter combined with a droppings pit of roughly equal proportions, 0·18 m² per bird is sufficient for 'heavies' and 0·14 m² is enough for the 'lights'. More recently, breeding stock have been kept on deep litter without a droppings pit but sometimes with a source of space heating to keep the litter in good condition. However, even without a space heater, provided the insulation is very good and the ventilation of the highest standard, it is possible to manage without a droppings pit. Its absence cheapens costs and makes management simpler and cleaning out a much easier process.

The covered strawyard

The covered strawyard is a much cheaper version of the deep-litter house, using a thick bedding of about 400 mm of straw as the litter and a different approach to environmental control. The construction of the building is simple. The only essential is a shed or covered yard, uninsulated and with free ventilation without draughts on the birds. Moveable perches are desirable as they spread the droppings evenly and take the birds off the floor at night giving a better environment. The usual stocking density is approximately 0·3 m² per bird.

The slatted floor system

An arrangement was developed in the UK some 15 years ago in which birds were kept on fully slatted or wire covered floors. It is the most concentrated method of housing layers on the floor at a stocking rate of 0·09 m² per bird. Labour requirements are minimal and housing costs can be competitive but the system has not been very successful. Problems of eggs laid on the floor arose and vices tended to be serious. Nevertheless in some parts of the world the system has been successfully adapted and used.

Laying cages

The system of keeping birds in cages, often three or four tiers high, exceeds all other housing systems in popularity. This is due to several factors. It has been due partially to the breeding of smaller birds which can be kept housed much more densely than hitherto and also due to the fact that whereas one bird per laying cage was the normal practice in the original forms of the cage, it is possible to keep 2–25 birds per cage at a greatly reduced cost compared with all other systems, apart from the strawyard. The most popular cage systems house 3–5 birds per cage.

With a multi-bird cage system the total floor area per bird can be as little as 0·06 m². Cages vary in width as follows: 225–300 mm for a single light hybrid; 300–25 mm for a single heavy hybrid or two light hybrids; 350–75 mm for two heavier hybrids or three light hybrids, and 525 mm for five light hybrids or four heavy hybrids. Cages are usually 450 mm deep and 450 mm high at the front, sloping to 350 mm high at the back.

The various cage systems

There are several different types of cage that can be used (Fig. 39.4). The three major groups are as follows: stacked cages, one on top of another, are three or four tiers high and are mostly cleaned mechanically by a scraper, a plastic belt or occasionally by using disposable paper belts. Some are still hand cleaned. Occasionally

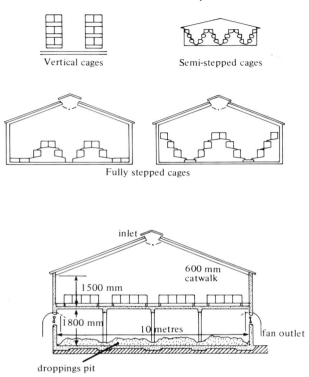

Vertical cages

Semi-stepped cages

Fully stepped cages

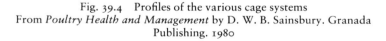

Flat deck cages over deep pit for droppings

Fig. 39.4   Profiles of the various cage systems
From *Poultry Health and Management* by D. W. B. Sainsbury. Granada
Publishing. 1980

more than four tiers are used, even up to six or seven, and an intermediate catwalk is required to service the top tiers.

The next group of cages is the 'Californian' or 'stair-step', often called the 'deep-pit' system because of the way it is installed (Fig. 39.5). These cages are staggered so that the droppings go into a large pit under the cages and can build up right through the laying period when they are cleaned out by tractor and fore-loader. They also may be two or more tiers high, three being usual and four now becoming popular. A recent form of 'stair-step' is the 'semi' form in which the upper cages are set partly over the lower ones, thus economising on space.

Finally, there is the 'flat deck' system which is of single-tier cages, usually in 2 m wide blocks, consisting of four cages with a double

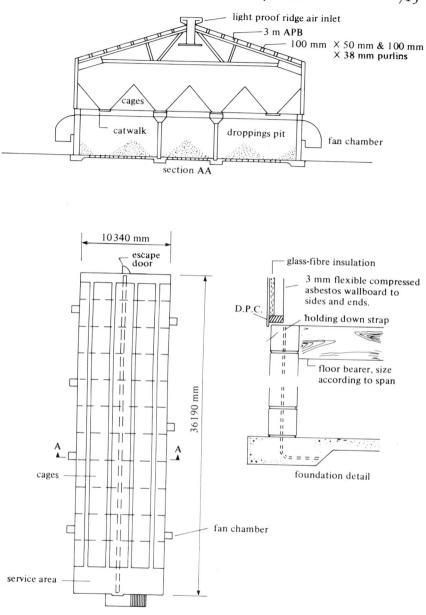

Fig. 39.5 'Deep-pit' house for layers
From *Poultry Health and Management* by D. W. B. Sainsbury. Granada
Publishing. 1980

line of automatic feeders and belt egg collectors running between the inner and outer cages. This is the most expensive system but the labour cost can be very low and the mechanical installation is simple. Cleaning may be done either by a scraper over a shallow pit or the droppings can be in a 'deep-pit'.

With tiered or stacked cages, adequate passageways between the rows are essential for ease of management and to promote good air movement. A 1 m passageway is ideal with 600 mm minimum. Feeding in cage systems is usually by continuous troughs filled by travelling hoppers which are moved either by hand or mechanically to fill the tiers simultaneously. The feed hoppers themselves may be filled automatically from an overhead auger operating from a bulk food bin.

Drinkers

There are three arrangements to give automatic and *ad lib* supplies of water: a continuous trough, cups or nipples. The continuous trough is served by a drip feed or tippler tank at the end. The most popular system, however, is the 'nipple' drinker, which is a valve, fitted into plastic piping, which the bird can depress to produce a modest flow of water. It is simple, clean and cheap. One nipple is required for four birds. The cup drinker is a small cup, up to 50 mm in diameter, and filled by a valve which the bird operates automatically when drinking. It is quite a good compromise between the two systems, being clean, hygienic and providing a 'dip' for the beak. There is some evidence that birds 'do' better if they have water visibly in front of them.

WELFARE

It is considered by some people that poultry kept in normal multi-bird cages are provided with so little space that their maintenance under these conditions constitutes cruelty. Alternative systems have been suggested for this reason. Some of those mentioned already are humane, viz. the deep litter and covered strawyard systems, but tend to be more expensive in either capital and/or running costs. In order to overcome some of the criticisms of the normal cage, a so-called 'Get-Away' cage has been designed to give a larger and more interesting environment for the bird. This cage has three areas – a horizontal wire floor, a nest box for laying, and separate perch areas. The attempt is to provide an environment which is more akin to the 'natural' state insofar as there are perches for roosting, a separate area for laying, and other facilities for feeding and drinking. The full evaluation of this system has yet to be made, and likewise the 'Aviary' system, another attempt to provide a satisfactory alternative to the battery cage. This is in effect a deep-litter house with additional raised areas within the house for the birds to feed, roost and exercise in. This means that a much increased number of birds can be housed for the floor area and the house can also be kept warmer.

40

OTHER ANIMAL
PRODUCTION SYSTEMS

The production of animal products from British agriculture has been based traditionally on cattle, sheep, pigs and poultry. These are but a small selection of the possible species and there are many others which can play a useful part in the provision of milk, meat, eggs and a number of other commodities. The widening of horizons through improved communications and travel opportunities has encouraged both farmers and consumers to explore the variety of the animal kingdom more fully. Some of the species have been found to make better use of resources such as upland grassland and waste materials than the traditional species. Not all the species have yet been developed to the point where their full agricultural potential can be exploited but many are already fulfilling a useful if limited role in British agriculture.

DEER FARMING

The consumption of venison in the UK is extremely low despite the fact that it is a comparatively high-protein meat with low fat content. The potential sales of venison are usually considered to be at the luxury end of the market with heavy emphasis on exports and farm gate outlets. The majority of the venison sales still come from hunted animals which are killed during the culling of herds kept on large estates. However, the farming of deer has developed in recent years and the British Deer Farming Association now has approaching 100 farmer members. At the same time the Highlands and Islands Development Board is investing capital in projects to test the financial viability of deer farming in the west of Scotland.

Deer are well adapted to ranging over large areas of the uplands in search of food supplies. Grass forms a major part of their diet but they also utilise heather, particularly in winter. They are also well adapted to severe winters when they may mobilize body protein as a source of energy. The young calves are produced in the favourable weather of June and they grow as rapidly as most other animals (Table 40.1).

Table 40.1   Time taken by young animals to
double their birthweight (*days*)

Lambs	15
Deer Calves	20
Goat Kids	22
Calves	47
Foals	60

In the wild, the ratio of adult male stags to female hinds is almost equal but in farm-based breeding populations the number of stags is reduced to about one for every 10 hinds. The meat animals are usually Red Deer and are slaughtered at about 2½ years of age when the carcass weight of each animal would be between 45 and 50 kg.

One of the major costs of setting up a deer farm is in fencing to curb the animals' wide-ranging habit. Fences of 1·8 m high are generally required for Red Deer. However, besides venison the animals also yield skins and antlers for sale and there is usually a good market in the sale of breeding stock. Some analyses indicate that there are some Highland situations in which it would be more economic to farm deer than sheep and this is probably also the case in the Lowlands. Much of the farm-produced venison is currently sold to consumers at the farm gate although there are market outlets to hotels and restaurants as well as in exports to the rest of Europe, particularly West Germany. New Zealand is one of the few other countries where deer are farmed on any scale; there the national herd has reached 100,000 whilst that in the UK is probably around 10,000.

RABBITS

Rabbits have traditionally been hunted on farmland and shot to prevent damage to crops. The animals that were shot were cooked and eaten, but rabbit meat came to be associated with the diets of the rural poor. Rabbit meat enjoyed some revival during wartime but the advent of the disease myxomatosis in the late 1950s encouraged consumer resistance to its consumption. In the UK no more than 100 g per person per year is eaten and this level has changed little in the last twenty years.

There have, however, been changes in rabbit meat production; many rabbits are now housed in cages in draught-proof buildings and feed can be supplied as a pelleted ration containing some (up to 30%) dried grass. Such rations can be formulated to supply all the nutritional needs of the young growing rabbit if they provide 15% crude protein, 10.5 MJ metabolisable energy per kg and between 5% and 9% fibre.

New Zealand White and Californian are the main breeds of rabbit employed, although there are some commercial hybrids which are said to produce larger numbers of offspring as well as having improved body conformation for meat production. Where a 14-hour daylength is maintained, rabbits can produce young all the year round and, whilst it is theoretically possible for them to have 11 litters of eight to 10 young, the national average for commercial rabbit units is probably under 40 young per doe per year.

The young rabbits take about 10 weeks to reach a weight of 2 kg, when they are usually delivered to a packing station for slaughter, processing and packing. Carcasses may be sold fresh but many are frozen and stored prior to sale. Much of the output from UK rabbit farms is exported to continental Europe where the market for rabbit meat is much stronger and there is far less consumer resistance. As well as meat, rabbit farmers also produce rabbits for medical research purposes and rabbit skins for the fur trade. These skins can create a storage problem when the fashion trade has a low demand for fur and they rarely contribute any significant proportion to the rabbit farmer's income.

OTHER FUR ANIMALS

Other animals which are farmed for fur production in the UK include foxes, mink and polecats. Some 70 farms are involved in the production of about 256,000 skins, the majority of which are sold on the London markets for export. Such farming requires considerable financial investment and skill. The animals depend on a diet of poultry, meat and fish offals, which have to be carefully prepared and combined with pre-cooked cereal. Fur can also be produced from chinchilla, which are herbivorous rodents, that are reared on pelleted diets and hay. The Ranch Chinchilla originated in America but has been bred in this country for over 20 years, and is a possible means of providing a small income from pelts. The standard colour is grey with white underparts, but can vary from dark to light, with other mutation colours becoming available: size is that of a large guinea-pig or small rabbit. Polygamous mating is practised for commercial purposes but the reproductive cycle is slow, gestation taking 111 days. The average chinchilla group produces two live young per female per year, this rate of production keeping the fur in the luxury class. The live chinchilla can release its fur in tufts when startled or frightened and therefore must be very carefully handled. After dressing, the pelt is quite firm and strong and tufts will not slip free from the leather. Pelts are often marketed through specialised dealers in Denmark.

HORSES

Whilst horses are single-stomached and do not ruminate, they consume much the same materials as cattle and differ little in their capacity to digest them. Traditionally, in this country, horses are kept at pasture and supplemented with cereals or hay according to the work they do or the stage of their lifecycle. Whilst in many European countries, particularly France and Italy, horsemeat represents a conventional source of food it seems unlikely that consumer resistance in Britain will recede to any extent. It is perhaps more likely that some farmers will return to the practice of keeping working horses in order to eke out expensive tractor fuel. Such a development might mean that less spare grassland would be available for renting to leisure-horse owners.

FISH

Fish have the advantage over other food animal species that they are cold-blooded so they do not have to expend feed energy in keeping warm. This means that their conversion of feed supplies into flesh can be very efficient and this is further improved by the buoyancy of water which reduces the energy required for movement and support. The disadvantage of fish is that they need a diet which is high in protein and, in the farm context, they need a relatively constant aquatic environment which tends to make the capital costs of setting up a fish farm relatively high.

The culture of fish now contributes some 5 m. tonnes (6%) of fish to the annual world harvest. Freshwater species of carp have been cultured since ancient times and in many parts of the world, notably the Far East, they still comprise a large proportion of the farmed fish. In Europe, trout and salmon are more commonly farmed and the rainbow trout, *Salmo gairdneri*, has become one of the most popular for farming purposes. Open ponds are most commonly used, although a variety of shapes and lining materials can be employed. A good reliable supply of neutral or slightly alkaline water is also an essential prerequisite. Both rainbow trout and brown trout, *Salmo trutta*, can be adapted to marine water in which they can be kept until they are sexually mature. At this point, salt water becomes damaging so that they have to be removed and sold or moved to freshwater sites. Marine fish farming has the advantage that the sea is a more uniform environment than fresh water and the initial capital cost of establishing a fishery for a given throughput of fish is much less than for freshwater farms. Most marine fish farming is carried out in flexible mesh cages suspended from a floating collar although there are some rigid-walled cages. In both freshwater and marine situations the majority of the fish are fed on specially formulated pelleted diets. The location of fish culture in the sea makes it possible to produce a range of shellfish which often derive some of their feed supply from filtering out organic matter from sewage outflows. Oysters, mussels, shrimps and prawns are the chief possibilities and in some situations their rate of productivity is improved by seawater which has been warmed by the outflow from factories and power stations.

INVERTEBRATES

Being cold-blooded, invertebrates have many of the same advantages as fish in terms of the efficiency of food conversion. However, few of them have been farmed to any extent in Europe and fewer still are used as food or feed resources. Anglers depend on a steady supply of worms and maggots as bait and these can be reared relatively easily on waste organic material such as rabbit faeces. It is possible to breed some species of earthworm, notably *Eisenia foetida*, on waste organic materials and in some parts of the world the worms produced have been processed into a protein supplement for animal feeds.

Snails, particularly the Roman snail, *Helix pomatia*, have been cultivated more than most other invertebrates. In France, they were traditionally collected and kept in small field enclosures where they could graze and be fed additional materials. At the end of the grazing season, they were harvested for the restaurant trade. In Britain, there has been no tradition of snail-eating and culture techniques exist only in the hands of a few entrepreneurs. The snails which are available in restaurants are almost all imported but in principle there is no reason why they could not be home produced.

The commercial production of honey from the many varieties of the honey bee *Apis mellifera* represents the best use that has been made so far of invertebrate species. The average harvest of honey from the static hives used in the UK is 15–20 kg per hive per year (Crane, 1975) but very much more than this has been recorded from Australian hives. The proportion of the energy available in the nectar of flowers which ends up as honey for human consumption is very small but it is difficult to imagine it being harvested in any other way. The other main advantage of bees is that the hives can be strategically sited to enable the bees to pollinate cultivated crops.

GOATS

Like cattle, goats are dual-purpose animals from which both milk and meat can be obtained. They serve a useful purpose amongst the world's many small farmers because they are prepared to eat materials which are not readily consumed by cattle and because their milk and meat yield is lower and therefore more easily coped with on a family scale – particularly in warmer climates. In Britain, many smallholders keep one or two milking goats and the animal is almost never kept for meat. The total value of the UK goat herd is difficult to estimate because many are on holdings which are too small to be recorded in the agricultural census. This census shows that there were 9,900 goats on recorded holdings in 1975 (MAFF, 1976). In 1978, the British Goat Society had 7,938 registered goats and there are probably three or four unregistered adult goats for each one that is registered.

Whilst the current consumption of goatmeat and milk in the UK is so small that it is not identified in the National Food Survey, it is not clear whether this is primarily the result of lack of supply or due to consumer resistance. The growth in the number of UK citizens who originate from countries where goatmeat and goat's milk and its products are more commonly consumed could stimulate internal demand. Equally, the entry of Spain and Greece, where goatmeat is eaten (Table 40.2), into the EEC may encourage the use of this meat and possibly generate export possibilities, if not for meat at least for breeding stock.

Table 40.2   Amount of goat meat consumed in various countries (kg/head/year) 1964/5–1966/7 average

Greece	3·0
Italy	0·1
Spain	0·4
Gambia	1·6
Indonesia	0·1
Morocco	2·1
Kenya	2·5
India	0·6

Source: F.A.O. (1971)

The meat also has the advantage that it is relatively low in fat. Goat's milk has useful anti-allergic properties, resulting from the absence of gamma-globulins, which give it a steady market outlet to people who suffer from eczema, asthma and hay fever.

Goats have a considerable ability to digest fibre and are often considered to be more useful as browsers than as grazers. In temperate environments, there has been relatively little investigation of the differences in digestive efficiency between the goat and other ruminant animals. Summaries of data collected in tropical environments suggest that, as roughage feed quality declines, the digestibility of fibre by the goat improves and efficiency of utilisation of such fibre is better in goats than in sheep. More recently, it has been suggested that, because of their lack of wool and the high content of lean tissue in the carcass, goats have a much higher daily requirement for metabolisable energy for maintenance than sheep. A review of the limited literature reporting experiments on comparative digestibility of sheep and goats indicates no evidence that goats and sheep in non-tropical environments differ in their ability to digest herbage materials. A direct comparison of the potential output of goatmeat from grassland, against that from other meat-producing species, appears to be overdue.

As far as dairy goats are concerned, whilst they are known to have a relatively high feed intake compared with other ruminants, research in France suggests that their milk production capacity exceeds that of the cow. Most of the milk produced by goat owners in the UK is on a small scale and therefore often consumed on the holding. Many owners also market some milk locally either fresh or frozen in polythene packs; others make yoghurt or cheeses for sale through 'health' or 'whole' food shops. The commonest milk-producing breed on UK holdings is the British Saanen.

POULTRY

The production of eggs and poultry meat in Britain is principally from hybrid domestic hens but around 1 m. ducks, 6 m. turkeys and 100,000 geese are also kept. Small numbers of other species are also farmed, notably quail and game birds such as partridge, grouse and pheasant which are reared for sporting purposes.

Turkeys are traditionally reared to supply the seasonal trade at Christmas and to a lesser extent at Easter. The advent of domestic freezers has modified these peak demands whilst diversification into processed turkey products has generated new markets so that turkeys are now marketed throughout the year. The continuous production derives mainly from battery-reared birds whilst many farmers batch-rear turkeys on crumbs and/or pelleted feed in covered yards or sheds for the seasonal market. The annual output of turkeys from both rearing methods using hybrids is more than 15 m. birds.

Much smaller numbers of ducks and geese are produced commercially although both are popular for use in large gardens and on smallholdings. Aylesbury and Pekin are the best breeds for table duckling and Embden and Toulouse geese suit most conditions. They are usually kept in simple free-range conditions, possibly with sheds for shelter, and both ducks and geese suffer from the disadvantage of being difficult to pluck. Duck eggs are also produced for human consumption but the quantities (around 3 m. dozen per year) are very small compared with the output of hen's eggs (around 1,100 m. dozen).

The smallest British game bird, the Common or Japanese Quail (*Coturnix coturnix*), is now rarely found in the wild but in recent years it has been reared indoors in cages to supply a specialist market for oven-ready birds and quail eggs. Cocks and hens can be kept in the same cage and the hens may lay up to 300 eggs per year but fertility declines rapidly over a long laying period. The birds can also be kept in pens on shavings instead of in cages. The eggs are set in incubators after which they take 17 days to hatch. The young quail are no bigger than bumble bees and need a high protein (28–30% crude protein) starter ration as well as an environmental temperature of 32°C. At about 10 days a finisher ration of 19–20% crude protein content is introduced and the temperature gradually reduced to ambient by the time the chicks are five weeks of age. The young quail can begin to lay eggs at six weeks but they are not fully grown until they are 12 weeks of age. The meat birds are killed when they weigh 200–225 g. The market for quail is less seasonal than for most game and, during most of the year, oven-ready birds are available in supermarkets. There are only about six quail producers dealing in significant numbers of birds or eggs. Much of the produce is sold to restaurants via London wholesalers.

GAME

The production of food from game animals, principally birds, has become an important component of income on many farms and estates where shooting rights are sold or rented for a season. The supply of birds for shooting is maintained by rearing birds on chick crumbs in captivity and releasing them. Pheasant and partridge adapt fairly readily to farmland but grouse do not and it is almost impossible to restock with them. The management of farmland habitats to provide shelter and food for game birds is another way of maintaining continuity of supply; many shelter belts of trees and hedges as well as food plants such as artichokes are planted for this purpose.

PART VI

THE ROLE OF AGRICULTURE

41

THE CHANGING SHAPE OF AGRICULTURE

WORLD CONTEXT OF UK AGRICULTURE

From 1952 to 1978 total agricultural production increased by the order of 100% in both the developed and developing world. However, the population increase was small in the developed world and much larger in the developing world. The result was that per caput food production increased in the developed countries but not by as much as the total agricultural production. In the developing countries per caput food production hardly increased at all. The prospective annual growth rates of population and gross domestic product (GDP) for the different parts of the world indicate that they will increase more slowly in the developed than in the developing countries. In a thorough analysis of the possibilities for expanding agricultural production in the developing world FAO came to the conclusion that over the next 20 years self-sufficiency ratios for nearly all commodities will fall. The decline in self-sufficiency ratios is likely to be particularly severe for cereals and livestock products.

A significant feature of agriculture in the developed world over recent decades has been the decline in the number of countries with surplus cereals for export. North America (especially the USA) is now the one area of the world with substantial quantities of cereals for export upon which the rest of the world relies. Perhaps it should be noted that the EEC is now an exporter of cereals and has the technical and resource potential to grow and export substantial quantities of cereals. The world balance of food consumption and production is very precarious (Table 41.1). The possibility of a world shortfall of cereals becomes progressively more likely as the century progresses.

At the time of the world food crisis in 1973/4 the world tried to establish a strategic reserve of cereals to meet future emergencies. Though a cereal reserve is still a desirable goal there is little enthusiasm to do much about it among those countries with the cereals and the resources to achieve it. It looks as though the world will teeter from one cereal crisis to the next with the developing countries suffering intermittent starvation and the developed world paying an even higher price for cereals. Even if one assumes that

Table 41.1 World cereals: production, demand and balances (Million tonnes)

	1975	1990	2000
Developing countries			
Production	366	589	788
Demand	400	642	877
Imports	52	86	152
Exports	18	34	62
Balance	−34	−53	−89
Developed countries production surplus			
Average expected		189	202
Maximum		210	292
Minimum		76	29

Source: Agriculture: Toward 2000 (1979) FAO: Rome

technically the world can make cereal production and consumption balance there still remains the serious question of who will pay? There is no indication of how countries in the developing world will afford to pay for the cereal imports they will need. Those countries with oil and mineral resources will have the necessary foreign exchange to buy cereals but many developing countries are among the poorest in the world and already in serious economic difficulties. The countries likely to have cereals for sale are high-production-cost countries. Though they may be able to afford to give away large quantities of cereals they may not be willing to use such development aid as they provide in this way. Food aid on a continuing basis is no longer considered to be a help to a developing country. It is clearly helpful in an emergency but over long periods food gifts may inhibit the development of food production in a developing country. Therefore aid is likely to be used to stimulate developing countries to increase their own food production rather than to import more food.

In developed countries a large proportion (up to 50%) of the cereals produced are fed to livestock for conversion to protein-rich foods, mainly milk, meat and eggs. It has been suggested that since most people in those countries eat far more protein than they need (though not necessarily than they desire) the problem of world cereal shortages could be solved by reducing the level of animal production. The cereals not fed to livestock could be transferred to developing countries, though the question of who pays remains unanswered. A reduction in livestock production, if accompanied by a decline in the demand for cereals might also lower the world market price for cereals so making it easier for developing countries to purchase supplies.

An alternative way of decreasing the demand for cereals may be to reduce the quantities of cereals fed to livestock but maintain the level of animal production by feeding more forages and crop by-products. Such changes, though technically feasible, will not occur as a result of exhortation but as a result of changes in financial stimuli and legislation applied to farmers.

RESOURCES FOR UK AGRICULTURE

Land

It is anticipated that the transfer of farmland to housing, roads, industry and other urban uses will continue at a rate of 15,000–25,000 ha per year. The transfer of farmland to forestry may be in the range of 25,000–50,000 ha per year. Much of the land lost to agriculture will be of below-average quality. This loss can be made good if the output on the remaining farmland increases by 1.5% per year. This should be possible for it is much lower than the historic rate of increase in output which has been nearly 3% per year. This may seem a satisfactory situation but there are three reasons for caution. First, as a consequence of higher energy prices there may be an incentive to practise more land-extensive forms of agriculture. Second, a reduction in the farm land area reduces the level of food and material self-sufficiency which can be achieved. A high level of food self-sufficiency has been a high priority in the UK for many years but membership of the EEC may make this a less desirable objective. Third, forestry is likely to become an increasingly attractive enterprise taking a substantial area of poor quality land and putting pressure on the best quality land for food production.

The UK is a very densely populated part of the world with no hope of becoming self-sufficient in both food and timber needs. The pressure to use land for a wide range of uses, including farming, forestry, buildings, transport, water supplies, conservation, tourism and recreation and mineral extraction means that ways of integrating these uses must be developed. The UK cannot afford to devote much land to single purposes so that the management of farms will embrace to a greater extent than at present a range of activities other than food production.

Energy

The effect of a rise in energy cost relative to other resources is likely to have one of three effects. These are a rise in the end product price to take account of the higher energy cost, attempts to conserve on energy use whilst trying to maintain levels of output, and the substitution of fossil energy by other cheaper resources such as land and labour. There is strong consumer resistance to food price rises and consumption of some of the more expensive products may fall. The opportunity to replace energy by land in the UK is, as shown earlier, very limited but there may be considerable scope for replacing fossil energy by labour if the tasks are not boring, dirty or do not involve unsocial hours. The greatest savings in energy use are likely to come from changes in cultivation and fertiliser practices and from an accumulation of small savings in every aspect of farming. The drying of crops and the heating of glasshouses are two activities which are likely to be severely modified or curtailed. The use of all sorts of farm by-products, such as straw, and vegetable and animal wastes will be used for local energy generation for use on the farm or in nearby communities. Farms may become the source of crops for fuel production, and for substrates for other industrial processes. High-energy crops such as cereals, sugar beet and potatoes seem superficially attractive for this purpose but trees will produce more biomass per ha per year. Short-rotation tree coppicing may be developed on lowland and upland for this purpose.

Fertilisers

The UK is self-sufficient in supplies of potash fertiliser but all phosphates have to be imported. There is some concern that supplies of phosphate might be interrupted, albeit briefly, for political or economic reasons, but soil phosphorus levels in the UK are generally satisfactory as a result of steady phosphorus applications over the past 100 years. The UK is also self-sufficient for nitrogen fertilisers but the main concern is their cost in the future, because of the large amount of fossil energy at present used in their manufacture. Less energy-intensive methods of production will be sought as well as means of reducing the levels of nitrogen application to crops.

Labour

It is very difficult to forecast to what extent the low labour requirement in agriculture will continue in the UK. For some enterprises such as cereal production the amount of labour needed may decline further, but for vegetables, fruit and livestock greater labour use to substitute for fossil energy and to provide essential husbandry for high performance may be expected. For all enterprises the level of skilled labour needed will increase.

The specialist skilled nature of many farm practices, such as pest control, and the uneven level of labour requirement throughout the year has offered an opportunity for the introduction of contract services and labour. The demand for such services tends to be seasonal and unless a contractor can provide a sequence of services throughout the year the problem of how to provide employment for a stable size of work force is merely passed from farmers to contractors.

Capital and business structure

Much of the capital in agriculture is privately owned either in land or in machinery, buildings and stock. The levels of earnings on capital in land have been traditionally low (1-2%) and have been accepted largely because of the compensation of capital appreciation on land values. The earnings on working capital have been higher (8-10%). The main threat to the continuation of private capital in agriculture is seen to be the introduction of more extensive forms of capital taxation and the possibility of extending rating to farming. However, it seems likely that even if additional tax burdens of these sorts are applied, farming will remain largely in private hands and financed largely by private capital in small businesses.

For the forseeable future the main unit in farming is likely to be the family farm of 100-300 ha with few, if any permanent employees. Owner occupancy and tenancy from traditional landlords are likely to remain the dominant features though institutional ownership may well increase by about $\frac{1}{2}$% of the land area per year. Part-time farming is also likely to increase as a result of a densely populated country leading to close proximity of rural and urban activities, of the security and stability of income which arises from earnings being derived from several sources and of the now obvious benefits in social and environmental terms of small-scale farm businesses.

CROP PRODUCTION

Market requirements

The requirement for cereals for baking, brewing and for animal feeding has remained firm with substantial incentive to produce more hard wheat suitable for bread making and to replace wheat imported largely from North America. There has also been an aim of replacing imported maize with home-grown wheat and barley. This situation is unlikely to change, if the total world demand for cereals mentioned earlier is taken into account. Sugar beet has been an important crop in the EEC and as a replacement for cane sugar from the tropics. Reliance on sugar beet for supplies of sugar is likely to remain though the consumption of sugar may fall and some could be replaced by sweeteners, such as fructose syrup

derived from cereals. The UK is self-sufficient in main-crop potatoes and it seems likely that supplies of earlies will come mainly
from Mediterranean and tropical countries. Vegetables are likely
to increase in demand but the opportunities for meeting other than
the main season winter and summer requirements will be limited
by weather. Glasshouse production in the UK, because of the high
fossil fuel requirements, does not appear competitive with production of fresh fruit and vegetables in Mediterranean and tropical
countries followed by air transport. The production of peas, carrots, onions and sprouts as well as of currants has been mechanised
so that large areas can be grown and much labour is no longer
needed. There is a limited range of soft and top fruit which can be
grown in the UK and importation of many citrus and other fruits
is likely to continue. The main crop in which the UK is deficient
and for which demand will remain high is a high-oil/high-protein
crop. The oil is needed for cooking and margarine production and
the high protein residue for feeding to animals in lieu of soyabeans.
Oil-seed rape is being grown to meet this need but other crops need
to be developed.

The advent of more spare time for the general public has led to
a greater production of fruit and vegetables in gardens and to the
development of pick-your-own systems. The problems of gluts
from garden production are overcome in the home by the deep-
freezer.

Grass is the predominant crop in the UK and so far has been
much neglected and underexploited. Its use for animal production
is likely to increase in future, as will the use of other forage crops
such as maize but especially leguminous crops such as lucerne and
the clovers.

Mention must be made of the possibility of the use of crops as a
source of fuel in future. Whilst crop by-products such as straw are
being used in this way the purposeful growing of cereals, potatoes
or sugar beet for fuel production still looks financially unattractive.
The most likely crop for this purpose will be trees grown as coppice
or even in more mature stands of coniferous and other species.

Production methods

The need to constrain energy use in all forms, to carry out the
minimum of cultivation and harvesting operations and to be prudent with the use of chemicals, both as fertilisers and as pest and
weed destroyers will be the main determinants of crop production
in the next two decades. The higher relative cost of energy as fuel
and in the production of machinery, fertilisers, especially nitrogen,
and chemicals will stimulate production practices which minimise
their use even to the extent of reducing yields per unit area.

Amongst practices likely to be adopted in these circumstances are the reduction and combination of cultivation operations such as harrowing, drilling, fertilising and rolling; more use of 'no-ploughing', direct drilling and precision fertiliser application; a closer matching of the size of machinery to the power required to do the work; a much closer regular watch on the growing crop to be able to assess the precise need for chemical pest controls; the replacement of chemicals for pest control by the reintroduction of crop rotations, by biological control techniques and by growing sequences of varieties with different disease susceptibilities. The high cost of developing and testing for safety new chemicals for pest control and the increasingly stringent constraints on where, when and how they may be used will be a strong incentive for the adoption of alternative means of reducing crop and crop storage losses.

Fertilisers are essential for high yields, and whilst the costs of phosphorus and potassium fertilisers are not a large element of crop production cost nitrogen fertilisers are. In future nitrogen fertilisers will be applied with greater precision to minimise nitrogen leaking into water courses and nitrogen losses to the atmosphere and to give maximum effect on crop growth. The more appropriate precision application of animal faeces on land, the more extensive growing of leguminous crops and the ploughing in of crop residues will all help to reduce the amount of nitrogen and other fertilisers needed.

Grass and forage crop production and conservation can be improved, perhaps doubled, on farms merely by the application of technology already developed. Undoubtedly the substitution of cereals by forage can reduce the cost of ruminant animal production. It seems likely that mixed swards involving legumes, medium- rather than short-term leys and permanent pasture will make an increasing contribution to animal feed supplies. The importance of grassland in the eastern drier areas and in the more marginal hill areas is likely to decline and to increase in areas less suited to arable farming, in the west, the midlands and the better uplands. Forage utilisation and conservation methods which use a lot of fossil energy, such as drying, are likely to be replaced by greater emphasis on direct grazing, silage and on improved hay-making techniques which make better use of wind and sun.

The potential for crop production

The rise in yield per unit area of most crops has been of the order of 1%–3% per year for the last 30 years. There is some evidence that the increase has been slowing recently and several reasons have been advanced for this. First there is an upper limit to the yield of any crop and the closer the limit is approached the more difficult it is to find high-yielding crop varieties, to improve the growing conditions and to optimise harvesting. Second, the period 1950–70 is now seen as a period of exceptionally stable and favourable climate for crop growth in the major cereal-growing areas of the world. The periods before and after seem to have been less stable and it is more difficult to obtain consistently high yields in a climate which fluctuates markedly even if the averages for temperature, rainfall, sunshine and so on stay the same. Third, as the relative cost of the major inputs to crop production change, there may be profit in using more land to substitute for other resources. The pressure for increased yield per unit area may decline.

In spite of these pressures, it is worth noting that crop yields on test plots on experimental stations are frequently three times those of the average farm, whilst the best farms have yields which are twice the average. Clearly there is considerable scope for increasing yields with existing technology even if local soil and weather conditions continue to cause substantial variation between farms. The development of new crop varieties, of more accurate timing of application of seed, fertiliser and water to the soil, and of maintaining plant health will certainly enable crop yields to increased by at least 1% per year for the next 20 years. The introduction of new crops such as legumes e.g. high-oil/high-protein lupins and hybrids e.g. triticale (a cross of wheat and rye) may lead to new levels of crop output. However, in terms of total biomass produced per unit area the highest yields are always obtained from perennial crops such as grass and trees.

ANIMAL PRODUCTION
Market requirements

The consumption of most animal products in the UK, apart from cheese and chicken, is stagnant or declining and the outlook is for a continuation of these trends. The opportunity to export animal products apart from cheese and lamb is also limited. The final consumer has a number of considerations which determine the purchasing pattern of animal products.

For reasons associated with a more sedentary life-style and a greater recognition of the relationship between diet and health there is a desire for a lower intake of animal fats in meat, butter and possibly milk. Most homes now have a refrigerator and the ownership of deep freezers is increasing rapidly. Many families do not wish to spend so much time preparing and cooking in the house and wish to eat out. Together these facts lead to a great desire for convenience, prepackaging, long storage, bulk buying and infrequent purchasing. A greater involvement of supermarkets and the food manufacturing industry in animal processing has already shown that their product requirements from farmers are not necessarily the same as the requirements of the traditional wholesaler-retailer-consumer chain. For instance, the shape and composition of beef animals for mincing into beefburgers and prepared steak puddings can be very different from those required for the provision of roasting joints and steaks. Similarly, the supermarket sales of milk of various compositional qualities and flavours modifies the incentives offered to farmers to produce milk of a rather limited range of composition. In a saturated market for traditional goods it is often possible to sell new products on the basis that consumers enjoy variety. This is the case for animal products. The expansion of sales of rabbit, turkey, venison and trout as comparatively reasonably priced meats in competition with beef, pork and lamb may well lead to an increase in total meat sales. However, it is also likely that the partial replacement of meat and milk consumption by vegetable products simulating animal products will increase beyond the present level of 5–10% of the market. Indeed it is argued that perhaps 30% of the present animal product consumption will be replaced by simulation vegetable products by the end of the century.

Production methods

Apart from alterations in product needs arising in the market, there are significant changes in the balance of availability and cost of resources which can be expected to modify production systems. Foremost is the rise in energy cost which not only affects fuel and electricity but also the other inputs like buildings and machinery which need large energy inputs in their manufacture. At the same time the consequences of automation in industry and agriculture, which is often a means of reducing energy and labour, have given rise to a substantial surplus of labour in the economy. Whilst superficially such a situation might portend a substantial reversal of the trend for reduced labour use in animal production, this may not happen. Not only is the work financially unattractive, farm labour being amongst the most lowly paid group in society, but the type of work with long unsocial hours often in dirty and wet conditions does not appeal these days. It seems more likely that the introduction of automation and micro-chip technology, by mechanising the dirty repetitive work will reduce the need for manual labour but at the same time allow the staff to concentrate on the husbandry aspects of stock keeping. Though there may be no increase in the number of workers, it is clear that their ability to cope with more complex situations, the need to be more highly trained and to have a better appreciation of the principles of livestock production will increase.

Apart from energy and labour, land either for grazing or to produce feed for grazing systems is the other major input to animal production. Development of intensive systems has meant a reduction in the proportion of feed eaten as grazing by animals and an increase in conserved forage, grains and legumes fed in zero-grazed units. This has happened because it has proved profitable to divorce the animals from the land on which their food is grown, in many instances by thousands of miles. There are strong indications that the very large and the intensive units are no longer so profitable. A return to more extensive systems, to closer proximity between the animal and its feed supply to save transport, and to more modest numbers of animals per unit is likely to occur in the coming years.

The intensification of animal production was associated, particularly with dairy cattle, with a migration eastward from the traditional wetter western grazing areas to the drier eastern counties. The pressures on land use for timber in the hills and uplands and for food and fuel as well as feed crops in the arable areas is likely to lead to the concentration of more extensive animal production in the western and upland grazing areas which are not suitable for arable farming. For purely technical reasons there seems little advantage in maintaining very low output sheep and cattle systems on very marginal land suitable for trees, when more productive grazing land is being underused in the lowlands.

The pressure on farmers from the general public to avoid the polluting effects of livestock, as well as the changing balance of cost of land, energy and minerals will lead to substantial changes in the design of animal buildings and in the treatment of animal faeces. So far as possible buildings will be of simple multipurpose designs incorporating the minimum of ventilation and insulation. Production systems will be planned either to return faeces directly to the land at times of year when they will benefit crops and minimise air and water pollution, or to use faeces for the production of methane for heating, lighting or power. The residue from faeces used for energy generation can be used on crops as a fertiliser.

Public concern about the welfare of farm animals which has already led to codes of practice for stock systems will certainly bring about further changes in husbandry. For instance, the systems of veal calf production, of battery egg production and of sow stalls and tethering will have to be altered and perhaps abolished. The acceptable standards of animal husbandry will be based on ethical judgements by the whole population. Whilst the views of farmers as to what is acceptable, of scientists as to what may give rise to animal stress and suffering and of economists as to what may happen to animal product prices and demand may be taken into account, these will not be over-riding considerations. The public and their elected representatives will decide and more stringent controls on animal welfare will be applied.

The potential for animal production

There is plenty of evidence that the yield per animal and per unit area from animal production can be increased at the present rate of 1-2% per year. There seems little possibility of changing the efficiency of separate metabolic processes in animals but it has been clearly demonstrated that it is possible to alter the balance of components produced such as between bone, muscle and fat in meat animals or between fat, protein and lactose in milk. There are also possibilities of achieving the same output using less or different combinations of inputs. In particular it is estimated that it would be possible to double the livestock output from the present area of grassland in the UK if known technology was used appropriately. The possibilities of breeding more uniform animals of outstanding performance and of reducing the effects of subclinical disease could have the effect of doubling the rate of improvement of stock performance. The use of animal species not recently given much attention for food production in the UK should be explored. For example these include species of deer, geese, certain snails and grass carp.

RESEARCH POTENTIAL FOR AGRICULTURE

Five subjects at various stages of investigation should be mentioned as having the greatest potential to cause change in the structure and performance of agriculture. These are research on nitrogen fixation by plants, photosynthesis, biological disease control, genetic engineering and micro-electronics. The possibility that the property of nitrogen fixation from the air, by bacteria in the nodules on leguminous plants, might be transferable to other plants has been the subject of intense research for some time. It is now clear that a number of bacteria, including some on the roots of maize plants, are capable of fixing nitrogen from the atmosphere. This knowledge may lead to the development of a range of crops with nitrogen fixing properties or to the development of bacterial nitrogen fixation in continuous industrial processes. The energetic efficiency of the latter is not yet clear and may be so low as to make such an option uncompetitive with other chemical means of producing nitrogen fertiliser.

Differences in the photosynthetic efficiency of plants have been observed, especially between C_3 and C_4 plants, which have slightly different chemical pathways for converting solar energy into plant tissue. Research may lead to the development of more crops with the photosynthetic properties of more efficient C_4 plants or even to the establishment of artificial 'plants' with photosynthesis properties derived from isolated chloroplasts. Such artificial plants may eventually lead to the means of producing hydrogen and oxygen for fuels as well as carbohydrate for food and materials from solar energy.

Examples of using viruses, bacteria and other natural predators to control the pests and weeds in farm crops and animals are increasing. Research in this field may enable a wide range of biological control methods to replace a good many of the chemical control methods now in use. However, though it may be possible to find natural predators, the difficulties of ensuring that the organism used to control the primary pest does not itself become a pest should not be under-rated.

Genetic engineering is a term used to include several potentially important areas of development, such as the possibility of replicating outstanding individual plants or animals so that whole fields or enterprises may be populated by uniformly high-yielding, disease-resistant varieties. Also it may be possible to transfer chromosome material by genetic manipulation from one individual to another or from one species to another. The breeders ability to change the production potential of crops and animals will be enhanced. Genetic engineering also includes the manipulation of viruses and bacteria for use in industrial processes producing a wide range of substances which may be used to improve crop and animal production or even to replace it. There is the possibility of producing starches, fats and proteins and other substrates for use in the food and materials industry directly from bacteria, fungi and algae. Whatever organism is used it will require energy to carry out the process. Increasingly this will be solar rather than fossil energy.

Micro-electronics are already changing our lives in many ways. In agriculture two main effects, one direct and one indirect, can be seen. The direct effect is on farming practice where micro-electronics are allowing the automation of manual procedures to give more precision, better information and a lower use of inputs for the same output. In particular, electronics are leading to a reduction in the labour needed and to an increase in the mental ability of the remaining labour. The indirect effect is on the way the general public live. Micro-electronics are changing the style of shopping, the extent of 'leisure' time and the eating habits of the population. These in turn will affect the products required from farming. For example, people with more leisure may enjoy more gardening as a hobby. A combination of more gardening and the deep-freezer means more domestic food production and a lower need for vegetables to be produced on farms.

Whilst research in the past has had the main effect of increasing yields of traditional crops and animals, it may in future have more striking effects on the products required from farming and on the structure of farming methods. Far from increasing the conflict between farming and other land use interests it may reduce it by allowing our food needs to be produced from a much smaller area, so leaving more space for other activities and requirements.

42

AGRICULTURE, SOCIETY
AND THE CITIZEN

In many countries, agriculture is of direct concern to the vast majority of the people (see Table 42.1) and, even for the world as a whole, nearly half the population is involved in it. By contrast, as has been shown in Chapter 9, relatively few people are directly engaged in UK agriculture, although considerably more are directly engaged in the processing, packaging and distribution of agricultural products. Even in Europe, the UK is at one end of the range for the proportion of the population directly involved (Table 42.2). Since this situation has arisen historically as a result of relatively

Table 42.1 **Number of people directly concerned with agriculture worldwide (1979)**

Country			Number of people '000	
		Total	Agricultural	%
Bangladesh	(1)	86062	72398	84·1
	(2)	29392	24725	84·1
Tanzania	(1)	17382	14179	81·6
	(2)	7132	5818	81·6
Uganda	(1)	12796	10421	81·4
	(2)	5277	4297	81·4
Ethiopia	(1)	31773	25320	79·7
	(2)	13153	10482	79·7
India	(1)	678255	434047	64·0
	(2)	261087	167046	64·0
China	(1)	945018	572748	60·6
	(2)	440220	266804	60·6
World	(1)	4335310	2024410	46·7
	(2)	1788240	820317	45·9

[1] Agricultural population is defined as all persons depending for their livelihood on agriculture. This comprises all persons actively engaged in agriculture and their non-working dependants.

[2] Economically-active population is defined as all persons engaged or seeking employment in an economic activity, whether as employers, own-account workers, salaried employees or unpaid workers assisting in the operation of a family farm or business.

Economically-active population in agriculture includes all economically-active persons engaged principally in agriculture, forestry, hunting or fishing.

Source: FAO (1980)

Table 42.2 Number of people directly concerned with agriculture in some European countries in 1979

Country	Number of people '000		
	Total	Agricultural	%
Ireland	(1) 3271	702	21·5
	(2) 1247	268	21·5
Spain	(1) 36356	6470	17·8
	(2) 12960	2352	18·2
France	(1) 53509	4880	9·1
	(2) 22906	2089	9·1
Denmark	(1) 5099	377	7·4
	(2) 2438	180	7·4
Germany FR	(1) 61200	2650	4·3
	(2) 28643	1240	4·3
Belgium-Luxemburg	(1) 10210	333	3·3
	(2) 3957	129	3·3
UK	(1) 56076	1173	2·1
	(2) 25919	542	2·1

Notes 1 and 2 and Source as for Table 42.1

cheap and abundant oil and policies that have encouraged capital investment, it may now be asked whether the past trend will continue, stop or even reverse somewhat.

In any event, there are other reasons why agriculture may be regarded as of vital importance to the majority of the population. First, we are all consumers of agricultural products. Secondly, a considerable number of people are processors or preparers of food at a household level. Thirdly, we are all concerned with the contribution that agriculture makes to the wealth of the nation. Fourthly, many people live, work or play in an environment affected by agriculture and its activities. Finally, we all tend to be affected by the operations of any major landowner.

THE CONTRIBUTION OF AGRICULTURE TO THE CONSUMER

The importance of UK agriculture to the UK consumer can be judged only from *average* values and quantities (Table 42.3) and the variation may be considerable. Nevertheless, the proportion of the food consumed that comes from UK agriculture is of the order of 55%.

Food quality also matters, however, and there are strong, and apparently increasing, pressures concerned with the consumer's view of the quality and freshness of the food on offer. This has to do with processing, as with wholemeal versus white bread, and

Table 42.3   Importance of U K agriculture to the consumer

Product	% Self-sufficiency 1977
Beef and Veal	73
Pigmeat	65
Mutton, Lamb and Goatmeat	57
Poultry Meat	106
Eggs	100
Whole Milk	100
Butter	32
Cheese	68
Skimmed Milk Powder	202
Condensed Milk	127
Wheat	61
Barley	121
Oats and Mixed Summer Cereals	99
Rye and Meslin	65
Maize	–
Vegetable Fats and Oils	6
Vegetables	72
Total Citrus Fruit	–
Other Fresh Fruit	26
Potatoes	95
Sugar	40
Wine	–

Source: MAFF (1980a)

brown versus white sugar, with the choice of products, such as brown versus white eggs, and the methods of production, involving attitudes to herbicides, pesticides, hormones and drugs of all kinds. Many of these aids to farming are thought of as 'unnatural' in senses that are not affected by the argument that agriculture itself is not natural. There are worries about chemical residues in the product and in the environment and there are worries about the flavour, keeping quality and nutritional value of products produced with such 'artificial' aids. Most of these worries relate to possible effects on human health and it would be foolish to ignore such evidence as exists that some products of modern agriculture may have adverse consequences to health. This argument is often extended to the point that 'artificial' (or 'chemical') fertilisers should not be used and that 'organic' farming should be pursued as an alternative.

In a country where food is plentiful, therefore, it would be a great mistake to consider food supply only in quantitative terms.

Table 42.4   Value of Pet Feeds in 1977

	£m.
Canned feed for dogs	144·3
Canned feed for cats	120·5
Biscuits and Meal	41·3
Other	11·8
Total	317·9

This is approximately the same amount as is spent on tea or detergents

Source: Vet. Rec. (1978)

But food is not the only product of agriculture and significant contributions are made to households in terms of feed for pets (Table 42.4) and ornamental plants (Table 42.5) and to industry in terms of raw materials for clothing and furnishings (wool production alone was valued at £35 m in 1979).

Table 42.5   Value of food products from agriculture and horticulture compared with the value of ornamental plants (1978)

Products	Value (£ m.)
Farm Crops	1,512
Horticultural fruits and veg.	615
Livestock	2,609
Livestock products (milk and eggs)	2,021
Total food products	6,838
Flowers, bulbs and nursery stock etc.	129 1·9%

Sources: Annual Rev. of Agriculture (1980) HMSO
MAFF (1980b)

In the future, particularly as a result of increasing energy prices, the relative importance of the contribution made by agriculture to raw materials and directly to fuel supplies might be expected to increase.

PEOPLE AS PROCESSORS

The proportion of food that enters the household in a raw state is considerable but difficult to calculate (Table 42.6). If the compari-

Table 42.6   Food that enters the household in a raw state

Food	g fresh weight per person per week
Eggs	227
Beef and veal	235
Mutton and lamb	111
Pork	95
Poultry	142
Offals	32
Fish, fresh	57
Potatoes	1,251
Fresh green vegetables	382
Tomatoes	104
Other fresh vegetables	284
Apples	199
Oranges	83
Other citrus fruit	52
Other fresh fruit	181
Nuts	14
Total raw food	3,449
Total weekly consumption	9,740
% Raw food	35

Source: CSO (1980)

sons have to be made in monetary terms, it automatically intro-
duces a bias in favour of processed foods. If, on the other hand,
comparisons are made on a weight basis, wet foods such as milk
and potatoes have disproportionately large effects.

In addition, all other purveyors of cooked meals also take a
proportion of raw foods and thus engage in processing and preparation, involving people, time and fuel.

Incidentally, for much of the year, in a well-insulated house it is
possible to argue that very little of the heat generated in cooking
need be lost. Its contribution to space heating may be regarded
as representing a sizeable reduction in the total cost of heating
or, alternatively, can be regarded as making the cost of cooking
negligible.

GROSS DOMESTIC PRODUCT (GDP)

Agriculture is reckoned to contribute about 2·3% of the GDP: the
food industry has to be considered separately, since it uses large
quantities of imported raw materials.

It is thus one of the biggest producers of our national wealth and
noticeably so in relation to the number of people and amount of
capital involved.

The contribution of agriculture also concerns the nation in terms of its effect on degrees of self-sufficiency, especially for food. Economically, there is often a case for growing food where it is best, or most cheaply grown, even if this is in another country. The EEC accepted this argument for location of production within the community, but it does cut across some national interests or the interests of powerful or vocal groups within nations.

Notions of security, against war, disaster, monopolies and political pressures, also have a bearing on the degree of self-sufficiency aimed at. However, it is important to distinguish between the degree of self-sufficiency that it is desirable to be capable of and that which it is desirable to achieve at any one time.

AGRICULTURE AND THE ENVIRONMENT

Agriculture influences the environment in four main ways: visually; by producing odours; by making noise; and by effects on our total way of life.

Visual effects

These are the most widespread and can be observed by all who travel through the countryside. The major features that strike the observer tend to relate to changes more than to the actual state of farmland. Informed observers notice whether land is well farmed or not but more people notice the removal of hedges, the absence of livestock formerly seen, the arrival of new buildings or tower silos and changes in crops (or even weeds, such as charlock) that affect the general appearance of fields. Recent studies suggest that, between 1950 and 1980, the rate of hedge removal has averaged about 8,000 km/yr and that over Britain as a whole, nearly 20% of the hedges have disappeared.

There may be a need to distinguish between opposition to change as such and to changes that are in some way deleterious to the land as a visual amenity.

Odours

As with all these features, agriculture contributes to our advantage as well as to our disadvantage but it has to be admitted that the noticeable bad odours (attending the application of pig slurry, for example) tend to have a more powerful (if local) influence than the good ones, such as new-mown hay.

Noise

The same is generally true of noise. Pleasant noises (e.g. cowbells, lowing of cattle) are so *because* they are not especially penetrating.

They merge into the background, rather like birdsong. On the other hand, machinery noise can be very loud and unpleasant and, in some cases, such as grass-driers, may affect quite a large area. The fact is that if a cowbell was as loud as a grass-drier, it, too, would be the subject of complaint.

Other effects

A major example of less obvious effects on the total environment is the concern that people have in relation to animal welfare. This is a problem that gives rise to much emotion and a dangerous and unhelpful polarisation of views. On the one hand, there are people who believe that some farm practices are cruel but are pursued for reasons of profit. On the other, there are those who think that these views are exaggerated, that there is no real cruelty (and that adequate legislation exists to prevent or control it) and that the farm practices complained of are essential in order to keep the cost of food down.

This is not so much a sterile debate but hardly a debate at all, since those involved tend not to talk to each other, even less to listen to each other. The fact is, of course, that we can all imagine practices and methods of animal production (by no means all of them modern – e.g. castration) that we would not tolerate in a civilised society because we judge them to be probably cruel but certainly unacceptable. We would all draw the line somewhere but we would differ in exactly where. The need therefore is to arrive at lines that are generally acceptable. It would be appalling if nobody cared but we cannot expect easy agreement in such matters.

The point to be made here, however, is that agricultural activities can affect the quality of life in such ways as this, in addition to effects on the physical environment.

LAND OWNERSHIP

The demand for land, to which agriculturists contribute, influences its price and may, in this way, have an effect on house prices and rents and thus on a very large section of the community.

In the conditions of developing countries, very marked effects can be seen on land prices and rents as a direct reflection of technological improvements, and sometimes as a result of increased potential, without waiting to see if this is necessarily or always achieved.

Such increases can adversely affect small farmers who lack resources, skills, inputs, capital or credit.

In this kind of way, agricultural practices can influence farm size, number of owners, tenants and employees, ease of entry into farming or farm management, social structures such as villages, and thus be a major social force.

AGRICULTURAL POLICY

It is the role of agricultural policy to reconcile all these legitimate interests, within a framework of legislation, regulations, technical advice and financial subsidies that are designed primarily to encourage a strong agricultural industry producing what is required.

Many of the interests often conflict and some are not well represented at any one time. In all this a balance has to be struck that allows change to occur. Much change is desirable, and not only for economic reasons, but it has to be recognised that the U K agricultural industry is in competition with that of other countries, especially within the E E C.

Change is thus inevitable but particular changes are not. Just because something is technically possible, it does not mean that it is right or desirable. Judgement has to be exercised and this can take into account all the current values of society and of individual producers.

Currently, a very good example is the technical possibility of mini-computers on farms. Changes in this direction seem likely but their desirability has to be judged by farmers individually. This is quite difficult because typically such developments are themselves subject to rapid change, in the size, complexity, cost and availability of hardware and, in this particular example, separate developments in supporting software.

It may be expected that the rate of *possible* change will continue to accelerate and this will further increase the urgent need for the implications of change – including changes in policy – to be carefully and continually examined. It is very important to try and forsee the consequences of change before it is instituted or encouraged, and it is extremely difficult to do for an industry as complex as agriculture.

It is a kind of research that, in relation to its importance, has been somewhat neglected in the past, although the Centre for Agricultural Strategy (C A S, 1976, 1978, 1979, 1980a and b, 1981) has been active in this field and University Departments of Agricultural Economics have regularly published assessments and studies of economic developments.

Assessments can work only on information, however, and, in some cases, the relevant research has not been carried out. This has been notably so in relation to animal welfare where the House of Commons Select Committee on Agriculture (see House of Commons Agriculture Committee, 1981) found that the necessary research has not been done on alternative production systems for it to form a judgement about what was feasible.

It is, of course, a major function of research to investigate possibilities, so that policy makers and producers can choose amongst well-documented alternatives.

This depends upon research embracing all the objectives of concern to the producer and the policy maker (who has to represent other interests as well). Such objectives are not only technical, or even economic, but include many other considerations, at least as major constraints.

Non-agriculturists may thus influence the future shape and structure of the agricultural industry but, in general, they could not be described as well-informed about the subject.

This ought to be a major concern of those involved in agricultural education.

AGRICULTURAL EDUCATION AND SOCIETY

The Institutions currently engaged in teaching Agriculture are given in Appendix Table 42.1 (pp. 804–5), with an indication of the level of education involved. The UK is well served in this regard and many of these institutions play an additional role in educating the non-agricultural community about agriculture.

Many would do more if the resources were available.

Farmers, too, are engaged in this process: many of them open their farms to visitors, especially parties of school children, and, in this way, greatly increase the town-dweller's understanding of farming. The extension service (ADAS) also has an increasingly important role in this area.

But the need is enormous. Most people in the UK have very little to do with agriculture and are ignorant of what it is about, how it operates, how important it is and, indeed, why it should matter to them.

The same could be said, of course, about many other industries and we cannot all know much about all other walks of life, however important these activities may be.

Agriculture is rather different, however, because it can genuinely be said to affect everyone and in several different ways.

The task of increasing understanding of agriculture on this scale is so large that it will not be achieved in any one way. Many channels of communication, and perhaps especially television, will be needed and all will depend upon the ability and willingness of agriculturists to communicate with the community at large.

REFERENCES

Ch. 2 Bronowski, J. L. (1973). *The Ascent of Man*, 77. British Broadcasting Corporation, London.
FAO (1980). *Production Year Book*. Vol. 33. Food and Agriculture Organisation of the United Nations, Rome.
Nove, A. (1977). *World Development*. Vol. 5, 417–24.
Rado, E., and Sinha, R. (1977). *World Development*. Vol. 5, 488.

Ch. 3 Buringh, P. (1977). *World Development*. Vol. 5, 477–85.
FAO (1980). *Production Year Book*. Vol. 33. Food and Agricultural Organisation of the United Nations, Rome.
Mauldin, W. P. (1977). *World Development*. Vol. 5, 395–405.
Preston T. R. (1980). Public Lecture, University of Newcastle-upon-Tyne.
US Report (1967). 'The World Food Problem', a report of the President's Science Advisory Committee, Vol. 11. The White House, Washington.

Ch. 4 Northfield (1979). 'Report of the Committee of Enquiry into the Acquisition and Occupancy of Agricultural Land'. HMSO Comnd 7599.
Nove, A. (1977). *World Development*. Vol. 5, 417–24.

Ch. 5 'Annual Review of Agriculture' (1981). HMSO Cmnd 8132.
Edgar, C. D. (1979). *J. Royal Agric. Soc. England*.
Ernle, Lord (1972). *English Farming, Past and Present*. Heinemann, London.
Hall, A. D. (1913). *A Pilgrimage of British Farming*. John Murray, London.
Murray, Sir Keith (1955). *Agriculture, Official History of the Second World War*. HMSO and Longmans Green & Co., London.
Northfield, Lord (1979). 'Report of the Committee of Enquiry into the Acquisition and Occupancy of Agricultural Land.' HMSO Cmnd 7599.
Pawson, H. C. (1957). *Robert Bakewell*. Crosby Lockwood, London.

Trevelyan, G. M. (1942). *English Social History*. Longmans Green & Co., London.

Winifrith, Sir John (1962). *The Ministry of Agriculture, Fisheries and Food*. George Allen & Unwin, London.

Ch. 6   Agriculture EDC (1977). 'Agriculture into the 1980s – Land Use.' A report by the Agriculture EDC of the National Economic Development Office, NEDO, London.

Burnham, C. P. (1970). 'The regional pattern of soil formation in Great Britain'. *Scottish Geographical Magazine*, 86, 25–34.

CAS (1980). 'Strategy for the UK forest industry.' CAS Report 6. Centre for Agricultural Strategy, Reading.

Coppock, J. T. (1971). *An agricultural geography of Great Britain*. Bell, London.

Curtis, L. F., Courtney, F. M. and Trudgill, S. T. (1976). *Soils in the British Isles*. Longman, London.

Jollans, J. L. (ed.) (1981). 'Grassland in the British economy.' CAS Paper 10. Centre for Agricultural Strategy, Reading.

Lund, P. J. and Slater, J. M. (1979). 'Agricultural land: its ownership, price and rent'. *Economic Trends*. No. 314, 97–110, December 1979.

Mackney, D. (1974). 'Land use capability classification in the United Kingdom'. *Land capability classification*. MAFF Technical Bulletin 30, HMSO, London.

MAFF (1975). 'Agricultural statistics, United Kingdom, 1974.' HMSO, London.

MAFF (1976). 'Agricultural land classification of England and Wales. The definition and identification of sub-grades within Grade 3.' ADAS Technical Report 11/1. MAFF, Pinner.

MAFF (1978). 'Agricultural statistics, United Kingdom, 1975.' HMSO, London.

MAFF (1979). 'Report of the committee of inquiry into the acquisition and occupancy of agricultural land.' Cmnd 7599. HMSO, London.

MAFF (1981). 'Annual review of agriculture, 1981.' Cmnd 8132. HMSO, London.

Royal Commission on Common Lands, 1955–58 (1958). Report. HMSO, London.

Ch. 7   Blanc, M. L. (ed.) (1973). 'Protection against frost damage'. Technical Note 51, World Meteorological Organisation, Geneva.

Davies, J. W. (1975). 'Mulching effects on plant climate and yield.' Technical Note 136, World Meteorological Organisation, Geneva.

Smith, L. P. (1976). 'The agricultural climate of England and Wales'. M A F F Technical Bulletin 35. H M S O, London.

Smith, L. P. and Trafford, B. D. (1976). 'Climate and drainage'. M A F F Technical Bulletin 34. H M S O, London.

Ch. 8   Bruce, R. (1979). 'Prospects for Cows and Corn'. Paper to Midland Bank/I C I Conference, Centre for European Agricultural Studies, Wye College.

Harrison, A. (1975). 'Farmers and Farm Businesses in England'. Misc. Study No. 62, 1975, University of Reading, Department of Agricultural Economics & Management.

Hill, B. and Kempson, R. E. (1977). 'Farm Buildings' Capital in England and Wales'. Wye College.

M A F F (1975). 'Farm Management Survey: Farm Liabilities and Assets, 1973-74'. H M S O, London.

M A F F (1979). 'Farm Incomes in England and Wales, 1977-78'. H M S O, London.

M A F F (1980). 'Farm Incomes in England 1978-79'. No. 32. H M S O, London.

M A F F (1981). 'Farm Incomes in England 1979-80'. No. 33. H M S O, London.

Nix, J. (1977) (1978) (1979). 'Farm Management Pocketbook'. 9th and 10th editions. Wye College.

de Paula, F. C. (1976). 'Farm finance and fiscal policy'. A M C, London.

Ch. 9   Agricultural Training Board (1980). 'Annual Report 1979/80'.

Agriculture E D C (1972). 'Agricultural Manpower in England and Wales'. National Economic Development Office.

Agriculture E D C (1977). 'Agriculture into the 1980s: Manpower'. National Economic Development Office.

Britton, D. K., Burrell, A. M., Hill, B. and Ray, D. (1980). 'Statistical Handbook of U K Agriculture'. Wye College.

Harrison, A. (1975). 'Farmers and Farm Businesses in England'. University of Reading, Department of Agricultural Economics & Management, Misc. Study No. 62.

Jones, G. E. and Peberdy, M. A. (1979). 'Into Work: A Study of School-leavers entering Agriculture'. University of Reading, Agricultural Extension & Rural Development Centre.

M A F F (1976). 'E E C Survey on the Structure of Agricultural Holdings, 1975: England and Wales'. Government Statistical Service.

M A F F (1980). 'Agricultural Labour in England and Wales. Earnings, Hours and Numbers of Persons' (including the Report of the Wages and Employment Enquiry, 1979).

Sparrow, T. D. (1972). 'The use of agricultural manpower'. *Journal of Agricultural Labour Science* 1 (2), 3-10.

Ch. 10   'Annual abstract of statistics'. H M S O, London.

'Annual review of agriculture'. H M S O, London.

Blaxter, K. L. (1975). 'The energetics of British agriculture'. *J. Sci. Food Agric.* 26, 1055–64.

'Digest of United Kingdom energy statistics' (published annually). H M S O, London.

'Fertiliser statistics 1979'. Fertiliser Manufacturers' Association, London.

Green, M. B. (1978). *Eating oil.* Westview Press, Boulder, Colorado.

Leach, G. (1976). *Energy and Food Production.* I P C Science and Technology Press, Guildford, Surrey.

Lewis, D. A. and Tatchell, J. A. (1979). 'Energy in U K agriculture'. *J. Sci. Food Agric.* 30, 449–57.

'Output and utilisation of farm produce in the United Kingdom' (various years). H M S O, London.

White, D. J. (1979). 'Efficient use of energy in agriculture and horticulture'. *Agric. Engr.* 34, 67–73.

Ch. 11   Church, B. M. (1981). 'Use of fertilisers in England and Wales, 1980'. Report of Rothamsted Experimental Station for 1980, Part 2, 115–22.

Johnston, A. E. and Whinham, W. N. (1980). 'The use of lime on agricultural soils'. *Proceedings of the Fertiliser Society.* No. 189.

M A F F (1978). 'Guide to current fertiliser regulations'. Leaflet AF 53.

Royal Commission on Environmental Pollution (1979). 7th Report. Agriculture and Pollution, Command 7644. H M S O, London.

Shaw, K. (1979). 'Fertiliser recommendations'. M A F F G F1. H M S O, London.

Ch. 12   Brouk, B. (1975). *Plants Consumed by Man.* Academic Press, London.

C A S (1980). 'Strategy for the U K Forest Industry'. C A S Report 6. C A S, University of Reading.

Hubbard, C. E. (1968). *Grasses.* 2nd edn. Penguin Books, Harmondsworth.

Jewiss, O. R. (1981). 'Shoot development and number', Chap. 5 in *Sward Measurement Handbook*, eds. J. Hodgson *et al.* B.G.S., 93–114.

Spedding, C. R. W. and Diekmahns, E. C. (1972). *Grasses and Legumes in British Agriculture.* C A B, Farnham Royal.

Weaver, J. E., Jean, F. C. and Crist, J. W. (1922). *Development and Activities of Roots of Crop Plants.* Carnegie Institute, Washington.

Westwood, M. N. (1978). *Temperate-zone Pomology.* W. H. Freeman & Co., San Francisco.

Ch. 13 Allen, D. and Kilkenny, J. B. (1980). *Planned beef production.* Granada, St Albans.

Federation of UK MMBs (1980). *UK Dairy Facts and Figures 1980.*

Holmes, W. (1970). 'Animals for food'. *Proc. Nutr. Soc.* 29, 237.

Jenness, R. and Sloan, R. E. (1970). 'The composition of milks of various species: a review'. Review Article No. 158. *Dairy Sci. Abstr.* 32 (10), 599–612.

MAFF (1973). 'Commercial Rabbit Production'. Bulletin 50. HMSO, London.

Maynard, L. A. and Loosli, J. K. (1965). *Animal Nutrition.* 6th edn. McGraw Hill, New York.

Meat and Livestock Commission (1976). 'British beef cattle'. MLC, Bletchley.

Meat and Livestock Commission (1979). 'Data summaries on beef production and breeding'. MLC, Bletchley.

Meat and Livestock Commission (1981a). 'UK meat and livestock statistics 1981'.

Meat and Livestock Commission (1981b). 'Commercial beef production yearbook 1980–81'. MLC, Bletchley.

Milk Marketing Board (1980). 'Report of the breeding and production organisation 1979/80'. No. 30. MMB, Thames Ditton.

Robinson, M. A. and Wiggins, J. F. (1971). *Animal Types 2 Vertebrates.* Hutchinson, London.

Royal Smithfield Club (1964). 'Major beef project – A comparison of the growth of different types of cattle for beef production'. Royal Smithfield Club, London.

Spedding, C. R. W. (1971). *Grassland Ecology.* OUP, Oxford.

Spedding, C. R. W., Walsingham, J. M. and Hoxey, A. M. (1981). *Biological Efficiency in Agriculture.* Academic Press, London.

Ch. 14 JCO (1976). 'Protein Feeds for Farm Livestock in the UK'. Joint Consultative Organisation for Research and Development in Agriculture and Food. Report No. 2. A. R. C., London.

Thomas, C. and Wilkinson, J. M. (1977). 'Maize Silage for Beef Production'. Grassland Research Institute Information Leaflet No. 7.

Ch. 15 Bunyan, P. J. and Stanley, P. I. (1980). 'Assessment of the environmental impact of new pesticides for regulation purposes.' Proc. Brit. Crop Prot. Conf. 1979 Pests and Diseases, Vol. 3, 881–91.

Cook, R. J., Jenkins, J. E. E. and King, J. E. (1981). 'The deployment of fungicides in cereals'. *Strategies for the control of cereal disease* (ed.) Jenkyn, J. F. and Plumb, R. T. pp 9–99. Blackwell Scientific Publications, Oxford.

Cramer, H. H. (1967). 'Plant protection and world crop production.' *Pflanzenschutz-Nachrichten "Bayer"* 20 (I).

Gough, H. C. (1977). 'Pesticides on crops – some benefits and problems'. *Ecological effects of pesticides* (ed.) Perring, F. H. and Mellanby, K. Linnean Society Symposium Series No. 5, 7–26.

MAFF (1983). 'Approved products for farmers and growers'. HMSO, London.

Ch. 18   Barker, J. W. (1981). *Agricultural marketing*. OUP, Oxford.

EIU (1981). 'Meat'. Retail Business No. 283. Economist Intelligence Unit.

MLC (1980). 'Livestock marketing methods'. Meat and Livestock Commission, Bletchley.

MMB (1980). *Dairy Facts and Figures*. Milk Marketing Board, Thames Ditton.

Palmer, C. M. (1975). 'Distributive margins for meat in Great Britain'. University of Exeter, Agricultural Economics Unit Report No. 194.

Tanburn, J. J. (1981). 'Food distribution: its impact on marketing in the '80s'. CCAHC, London.

Watts, B. G. A. (1982). 'Structural adjustment in the UK food manufacturing industry over the past twenty years'. Paper given at OECD Symposium 'The Adjustment and Challenges Facing the Food Industries in the 1980s', 11–14 January 1982, Paris.

Williams, R. E. (1980). 'Milk marketing in a European framework'. *J. A. E.* Vol. XXXI No. 3.

Ch. 19   Biscoe, P. V. & Gallagher, J. N. (1978). 'A physiological analysis of cereal yield'. *Agric. Progr.* 53, 34–50.

Bleasdale, J. K. A. (1973). *Plant physiology in relation to horticulture*. MacMillan Press Ltd.

Bullen, E. R. (1977). 'How much cultivation?' *Phil. Trans. R. Soc. Lond.* B. 281, 153–61.

Dyson, P. W. & Watson D. J. (1971). 'An analysis of the effects of nutrient supply on the growth of potato crops'. *Ann. appl. Biol.* 69, 47–63.

Evans, L. T. (1975). 'The physiological basis of crop yield'. *Crop Physiology* (ed.) Evans, L. T. CUP, Cambridge.

Harris, P. M. (1972). 'The effect of plant population and irrigation on sugar beet'. *J. Agric. Sci., Camb.*, 78, 289-302.

Heath, S. B. & Roberts, E. H. (1981). 'The determination of potential crop productivity'. 17-49. *Vegetable Productivity* (ed.) Spedding, C. R. W. Macmillan, London.

Hill, T. A. (1977). 'The biology of weeds'. Institute of Biology's Studies in Biology. No. 79. Edward Arnold, London.

Jaggard, K. W. & Scott, R. K. (1978). 'How the crop grows from seed to sugar'. *British Sugar Beet Review* 46 (4), 19-22.

Luckman, W. H. & Metcalf, R. L. (1975). 'The pest management concept'. *Introduction to Insect Pest Management* (eds.) Metcalf, R. C. & Luckman, W. H. John Wiley & Sons.

Monteith, J. L. (1977). 'Climate and the efficiency of crop production in Britain'. *Phil. Trans. R. Soc. Lond. B.* 281, 277-94.

Penman, H. L. (1949). 'The dependence of transpiration on weather and soil conditions'. *J. Soil. Sci.* 1, 74-89.

Penman, H. L. (1971). 'Irrigation at Woburn – VII'. Rothamsted Exp. Stn. Rep. for 1970, part 2, 147-70.

Perry, D. A. (1970). 'Seed vigour and field establishment'. *Hortic. Abst.* 42 (2), 334-42.

Russell, E. W. (1973). *Soil Conditions and Plant Growth.* 10th edn. Longman, London.

Samways, J. (1981). 'Biological Control of Pests and Weeds'. Institute of Biology's Studies in Biology no. 132. Edward Arnold, London.

Scott, R. K. & Allen, E. J. (1978). 'Crop physiological aspects of importance to maximum yields – potatoes and sugar beet'. *Maximising Yields of Crops*, 25-30. Proceedings of symposium organised jointly, by ADAS and ARC, 17-19 January 1978. HMSO, London.

Sibma, L. (1977). 'Maximization of arable crop yields in The Netherlands'. *Neth. J. agric. Sci.*, 25, 278-87.

Silvey, V. (1978). 'The contribution of new varieties to increasing cereal yield in England and Wales'. *J. natn. Inst. agric. Bot.*, 14, 367-84.

Wit, C. T. de, Laar, H. H. van & Keulen, H. van (1979). 'Physiological potential of crop production', pp. 47-82. *Plant Breeding Perspectives* (eds.) Sneep, J. & Hendriksen, A. J. T. (coed) Holbek, O. Centre for Agricultural Publishing & Documentation, Wageningen.

Zaag, O. E. van der (1972). 'Dutch techniques of growing seed potatoes', 188-205. *Viruses of potatoes and seed potato production* (ed.) Bokx J. A. de. Centre for Agricultural Publication and Documentation, Wageningen.

Ch. 21   Bircham, J. S. and Hodgson, J. (1981). 'The dynamics of herbage growth and senescence in a mixed-species temperate sward continuously grazed by sheep'. Proceedings of the 14th International Grassland Congress, Lexington, 1981.

Jones, R. J. and Sandland, R. L. (1974). 'The relation between animal gain and stocking rate. Derivation of the relation from the results of grazing trials.' *J. agric. Sci. Camb.*, 83, 335–42.

Ch. 25   Snoad, B. (1980). 'The origin, performance and breeding of leafless peas'. *ADAS Quarterly Review*, 37, 69–86.

Ch. 26   ADAS (1980). 'Chemical weed control in oilseed rape'. Booklet 2068, MAFF, London.

Bunting, E. S. (1974). 'New arable crops – retrospect and prospects'. *J. Royal Agric. Soc. England*, 135, 107–19.

MAFF (1979). 'Fertilizer recommendations for agricultural and horticultural crops'. 2nd edn. HMSO, London.

National Institute of Agricultural Botany (1981). 'Varieties of oilseed rape'. Farmers Leaflet No. 9, NIAB.

Ch. 29   Green, J. O., Corrall, A. J. and Terry, R. A. (1971). 'Grass species and varieties'. Grassland Research Institute Tech. Rep. No. 8. GRI, Hurley.

Ch. 34   Dennington, V. N. and Chadwick, M. J. (1979). 'An assessment of the potential of derelict and industrial wasteland for the growth of energy crops'. Report to the Department of Energy.

Lawson, G. J., Callaghan, T. V. and Scott, R. (1980). 'Natural vegetation as a renewable energy resource in the UK'. Report to the Department of Energy.

Lucas, N. G. (1978). 'Whole crop harvesting'. *Power Farming*, August 1978.

MAFF (1980a). Report on 'Straw as a fuel' Conference, Oxford, February 1980.

MAFF (1980b). 'Agricultural and horticultural returns – First results of the June 1979 census'.

White, D. J. (1980). The potential uptake of straw as fuel in agriculture and horticulture. In Report on MAFF 'Straw as a fuel' conference. Oxford, Feb. 1980.

Ch. 36   Baker, R. D., Kilkenny, J. B., Spedding, A. W. and Tayler, J. C. (1967). *Beef production: an intensive grassland system using autumn-born calves*. MLC, Bletchley.

Broadbent, P. J. (1971). 'The nutritive requirements of the suckler cow and the choice of calving date in relation to feed supplies'. Paper No. 3. Suckler herd management course. N. Scot. Coll. Jan., 1971

Hardy, R. (1981). 'A system of indoor finishing of young bulls on grass silage supplemented with barley'. *Anim. Prod.* 32: 359 (Abstr.).

Kilkenny, J. B. and Dench, J. A. L. (1981. 'The role of grassland in beef production'. In: Jollans (Ed.) *Grassland in the British economy.* CAS paper 21. Reading: Centre for Agricultural Strategy.

MLC (1979). *Data summaries on beef production and breeding.* MLC, Bletchley.

MLC (1981). *Commercial beef production yearbook, 1980–1.*

Wilkinson, J. M. and Tayler, J. C. (1973). *Beef Production from Grassland.* Butterworth, London.

Ch. 37   Hopkins, A. (1979). 'The botanical composition of grasslands in England and Wales: An appraisal of the role of species and varieties'. *J. Royal Agric. Soc. England,* 140, 140–50.

MAFF (1980). 'Grazing plans for the control of stomach and intestinal worms in sheep and in cattle'. Booklet 2154.

MLC (1979). 'Data summaries on upland and lowland sheep production'.

MLC (1980). *Commercial sheep production yearbook, 1979–80.*

Wilde, R. M., Young, N. E. and Newton, J. E. (1980). 'Grass-lambs. An intensive system of lamb production from temporary grassland'. Farmer's Booklet No. 2. GRI, Hurley.

Ch. 38   ARC (1981). *The nutrient requirements of pigs.* Commonwealth Agricultural Bureaux, Farnham Royal.

Ch. 39   Crane, E. (1975). *Honey.* Heinemann, London.

Devendra, C. (1978). 'The digestive efficiency of goats'. *World Review of Animal Production* 14 (1), 9–22.

FAO (1971). 'Food balance sheets, 1964–66'. Food and Agriculture Organisation, Rome.

MAFF (1976). 'Agricultural returns – England and Wales, regions and counties – final results of the June 1975 census'. MAFF, Guildford.

Mohammed, H. H. and Owen, E. (1980). 'Comparison of the maintenance energy requirement of sheep and goats'. Paper No. 69, Winter Meeting, BSAP, Harrogate.

Ndosa, J. E. M. (1980). 'A comparative study of roughage utilisation by sheep and goats'. M.Phil. Thesis, University of Reading.

Olsson, N. O. (1969). 'The nutrition of the horse'. Cuthbertson, D. (ed.) *Assessment of and factors affecting requirements of farm livestock*. Part 2 of *Nutrition of animals of agricultural importance*. Pergamon Press, Oxford.

Sabine, J. R. (1978). 'The nutritive value of earthworm meal'. *Utilization of soil organisms in sludge management*. 122–30. National Technical Information Services, Springfield, Illinois.

Ch. 40   Crane, E. (1975). *Honey*, Heinemann, London.

FAO (1971). *Food Balance Sheets*, 1964–66 Rome: Food and Agriculture Organisation.

MAFF (1976). *Agricultural returns – England and Wales, regions and counties – final results of the June 1975 Census*. *Guildford*: Ministry of Agriculture, Fisheries and Food.

Ch. 42   CAS (1976). *Land for agriculture*. CAS Report 1. Centre for Agricultural Strategy, Reading.

CAS (1978). *Strategy for the UK dairy industry*. CAS Report 4. Centre for Agricultural Strategy, Reading.

CAS (1979). *National food policy in the UK*. CAS Report 5. Centre for Agricultural Strategy, Reading.

CAS (1980a). *Strategy for the UK forest industry*. CAS Report 6. Centre for Agricultural Strategy, Reading.

CAS (1980b). *The efficiency of British agriculture*. CAS Report 7. Centre for Agricultural Strategy, Reading.

CAS (1981). *Grassland in the British economy*. Proc. Symp. organised by the Dept. of Agriculture & Horticulture, Dept. of Agricultural Economics & Management, University Reading, Grassland Research Institute and CAS. 15–17 September 1980.

CSO (1980). *Annual abstract of statistics*. Government Statistical Service. HMSO.

FAO (1980). *Production Yearbook 1979*. Vol. 33.

House of Commons Agriculture Committee (1981). Animal welfare in poultry, pig and veal calf production. HMSO, London.

MAFF (1980a). 'EEC agricultural and food statistics 1975–1978'.

MAFF (1980b). 'Output and utilization of farm produce in the United Kingdom 1973 to 1979'. Government Statistical Service, HMSO, London.

Veterinary Record (1978). 'Economic aspects of pet food'. *Vet. Rec.*, 102 (25), 539–40.

FURTHER READING

Ch. 1–5   Allaby, M. (1977). *World Food Resources*. Applied Science Publishers, London.

Atlas of Earth Resources. Population, Food, Energy, Minerals and Development: a source book of information and options. Mitchell Beazley, London (1979).

Børgstrom, G. (1973). *World Food Resources*. Intertext Books.

CAB (1980). *Perspectives in World Agriculture*. CAB, Farnham Royal.

Cox, G. W. and Atkins, M. D. (1979). *Agricultural Ecology*. W. H. Freeman & Co., San Francisco.

Duckham, A. N. and Masefield, G. B. (1970). *Farming Systems of the World*. Chatto & Windus, London.

Duckham, A. N., Jones, J. G. W. and Roberts, E. H. (1976). *Food Production and Consumption*. North Holland and American Elsevier.

Edwards, A. M. and Rogers, A. (1974). *Agricultural Resources*. Faber & Faber, London.

Grigg, D. B. (1974). *The Agricultural Systems of the World*. CUP, Cambridge.

Ruthenberg, H. (1977). *Farming Systems in the Tropics*. 2nd edn. OUP, Oxford.

Webster, C. C. and Wilson, P. N. (1966). *Agriculture in the Tropics*. 2nd edn. Longman, London.

Ch. 6   Advisory Council for Agriculture and Horticulture in England and Wales (1978). *Agriculture and the Countryside*. MAFF, Pinner.

Buckman, H. O. and Brady, N. C. (1974). *The Nature and Properties of Soil*. Macmillan, London.

CAS (1976). *Land for Agriculture*. CAS Report 1. Centre for Agricultural Strategy, Reading.

Edwards, A. M. and Rogers, A. (1974). *Agricultural Resources*. Faber & Faber, London.

Selman, P. (1978). *The application of soil survey and land capability classification in planning*. Planning Monograph No. 1. Planning Dept., Glasgow School of Art, Glasgow.

Tranter, R. B. (ed.) (1978). *The Future of Upland Britain*. CAS Paper 2. Centre for Agricultural Strategy, Reading.

Tranter, R. B. (ed.) (1981). *Small Farming and the Nation*. CAS Paper 9. Centre for Agricultural Strategy, Reading.

Whitby, M. C. and Willis, K. G. (1978). *Rural Resource Development: An Economic Approach*. Methuen, London.

Ch. 7   Smith, L. P. (1975). *Methods in Agricultural Meteorology*, Elsevier, Amsterdam.

Smith, L. P. (1976). *The Agricultural Climate of England and Wales*. MAFF Technical Bulletin 35. HMSO, London.

Mount, L. E. (1979). *Adaptation to Thermal Environment*. Edward Arnold, London.

Caborn, J. M. (1965). *Shelterbelts and Windbreaks*. Faber & Faber, London.

Ch. 8   Harrison, A. (1975). *Farmers and Farm Businesses in England*. Misc. Study No. 62. Department of Agricultural Economics & Management, University of Reading.

Centre for Agricultural Strategy (1978). *Capital for agriculture*. CAS Report 3. Centre for Agricultural Strategy, Reading.

Harrison, A. et al (1981). *Factors influencing ownership, tenancy, mobility and use of farmland in the United Kingdom*. Commission of the European Communities. Information on Agriculture No. 74.

Ch. 9   *The Drift from the Land*

Bessell, J. E. (1972). *The Younger Worker in Agriculture: Projections to 1980*. National Economic Development Office.

Black, M. (1968). 'Agricultural labour in an expanding economy'. *Journal of Agricultural Economics*, 19. 59–76.

Cowie, W. J. G. and Giles, A. K. (1957). *An Inquiry into Reasons for 'The Drift from the Land'*. University of Bristol Department of Economics (Agricultural Economics), Selected Papers in Agricultural Economics, 5 No. 3.

Gasson, R. (1974). *Mobility of Farm Workers*. University of Cambridge Department of Land Economy, Occasional Paper No. 2.

McIntosh, E. (1972). 'A survey of workers leaving Scottish farms'. *Scottish Agricultural Economics*, 22. 147–52.

Mackel, C. J. (1975). 'A survey of the agricultural labour market'. *Journal of Agricultural Economics*, 26. 367–81.

Newby, H. (1977). *The Deferential Worker: A Study of Farm Workers in East Anglia*. Allen Lane, London.

Wagstaff, H. R. (1971). 'Recruitment and losses of farm workers'. *Scottish Agricultural Economics*, 21. 7–16.

The *Agricultural Tied Cottage*

Fletcher, R. *(1969)*. 'The agricultural housing problem'. *Social and Economic Administration*, 3. 155–166.

Gasson, R. (1975). *Provision of Tied Cottages*. University of Cambridge Department of Land Economy, Occasional Paper No. 4.

Giles, A. K. and Cowie, W. J. G. (1960). 'Some social and economic aspects of agricultural workers' accommodation'. *Journal of Agricultural Economics*, 14. 147–69.

Irving, B. and Hilgendorf, L. (1975). *Tied Cottages in British Agriculture*. Tavistock Institute of Human Relations, London.

Jones, A. (1975). *Rural Housing: The Agricultural Tied Cottage*. Social Administration Research Trust, Occasional Papers on Social Administration No. 56.

Newby, H. (1977). op. cit.

Earnings of Agricultural Workers

Brown, M. and Winyard, S. (1975). *Low Pay on the Farm*. Low Pay Unit.

Newby, H. (1977). op. cit.

Newby, H. Bell, C., Rose, D. and Saunders, P. (1978). *Property, Paternalism and Power: Class and Control in Rural England*. Hutchinson, London.

Wage Bargaining

Groves, R. (1949). *Sharpen the Sickle! The History of the Farm Workers' Union*. Porcupine Press.

Mills, F. D.. (1964). 'The National Union of Agricultural Workers'. *Journal of Agricultural Economics*, 16. 230–58.

Newby, H. (1977). op. cit.

Ch. 10   Fluck, R. C. and Baird, C. D. (1980). *Agricultural Energetics*. AVI Publ. Co., Connecticut.

IFIAS (1974). *Energy Analysis*. Workshop Report No. 6, International Federation of Institutes for Advanced Study, Stockholm, Sweden.

Stout, B. A., Myers, C. A., Hurand, A. and Faidley, L. W. (1979). *Energy for World Agriculture*. FAO, Rome.

Ch. 11   Cooke, G. W. (1967). *The Control of Soil Fertility*. Granada Publishing Ltd,, St Albans.

Cooke, G. W. (1971). 'Fertilisers and Society'. *Proc. Fertiliser Soc.*, No. 121.

Cooke, G. W. (1975). *Fertilizing for Maximum Yield*. 3rd edn. Granada Publishing Ltd, St Albans.

Cooke, G. W. (1980). 'Changes in Fertiliser Use in the UK from 1950 to 1980.' *Proc. Fertiliser Soc.*, No. 190.

MAFF, (1976). *Organic Manures*. Bulletin 210, London, HMSO.

MAFF (1979). *Farm Waste Management: Profitable utilisation of livestock manures*. Booklet 2081.

Russell, E. W. (1973). *Soil conditions and plant growth*. 10th edn. Longman, London.

Ch. 12   Janick, J., Schery, R. W., Woods, F. W. and Ruttan, V. W. (1969). *Plant Science*, W. H. Freeman, San Francisco.

Kirby, R. H. (1963). *Vegetable Fibres: Botany, Cultivation and Utilization*. World Crops Series. Leonard Hill, New York.

Martin, J. H., Leonard, W. H. and Stamp, D. L. (1976). *Principles of Field Crop Production*. 3rd edn. Macmillan, New York and London.

Schery, R. W. (1972). *Plants for Man*. 3rd edn. Prentice-Hall, New Jersey.

Ch. 13   Bowman, J. C. (1977). *Animals for Man*. Studies in Biology No. 78, Edward Arnold, London.

Cole, H. H. and Ronning, Magnar (eds.) (1974). *Animal Agriculture*, W. H. Freeman, San Francisco.

Spedding, C. R. W. (1975). *The Biology of Agricultural Systems*, Academic Press, London and New York.

Ch. 14   Broster, W. H. and Swan, H. (1979). *Feeding Strategy for the High Yielding Dairy Cow*. Granada, St Albans.

Felwell, R. and Fox. S. (1978). *Practical Poultry Feeding*. Faber and Faber, London.

Whittemore, C. T. and Elsley, F. W. H. (1979). *Practical Pig Nutrition*. 2nd edn. Ipswich Farming Press.

Wilson, P. N. and Brigstocke, T. D. A. (1981). *Improved Feeding of Cattle and Sheep*. Granada, St Albans.

Ch. 15   Fryer, J. D. and Makepeace, R. J. (eds.), *The Weed Control Handbook*: Vol. II, Recommendations, 8th edn. (1978). Blackwells, Oxford.

Green, M. B., Hartley, G. S. and West, T. F. (1977). *Chemicals for Crop Protection and Pest Control*. Pergamon Press, Oxford.

Matthews, G. A. (1979). *Pesticide Application Methods*. Longman, London.

Roberts, H. (ed.), *Weed Control Handbook*: Vol. I. Principles, 7th edn (1982). Blackwells, Oxford.

Scopes, N. (ed.), *Pest and Disease Control Handbook* (1979). Blackwells, Oxford.

Worthing, C. R. (ed.), *The Pesticide Manual*, 6th edn. (repr. 1980).

Ch. 16 Butterworth, B. and Nix, J. (1982). *Farm Mechanisation for Profit*. Granada, St Albans.

Cox, S. W. R. (1982). *Microelectronics in Agriculture and Horticulture*. Granada, St Albans.

Culpin, C. (1981). *Farm Machinery*, 10th edn. Granada, St Albans.

Shippen, J. M., Ellin, C. R. and Clover, C. H. (1980). *Basic Farm Machinery*, 3rd edn. Pergamon Press, Oxford.

Ch. 17 Noton, H. N. (1983). *Farm Buildings*, College of Estate Management, Reading.

Weller, J. B. (1972). *Farm Buildings*. Vols. 1 and 2. Crosby Lockwood Staples, St Albans.

Ch. 18 Barker, J. W. (1980). *Agricultural Marketing*. OUP, Oxford.

Burns, J. A., McInerney, J. P. and Swinbank, A. (1983). *The Food Industry: Economics and Policy*. Heinemann Technical Books in association with Commonwealth Agricultural Bureaux, Slough.

Tanburn, J. J. (1981). *Food Distribution: Its Impact on Marketing in the 1980s*, CCAHC, London.

Ch. 19 Bleasdale, J. K. A. (1973). *Plant Physiology in relation to Horticulture*. Macmillan Press Ltd.

Davies, D. B., Eagle, D. J. and Finney, J. B. (1977). *Soil Management*. 3rd edn. Farming Press Ltd.

Hill, T. A. (1977). *The Biology of Weeds*. Studies in Biology No. 79. Institute of Biology. Edward Arnold, London.

Hurd, R. G., Biscoe, P. V. and Dennis, C. (1980). *Opportunities for increasing Crop Yields*. Pitman. Advanced Publishing Program, Boston, London, Melbourne.

Lawrence, W. J. C. (1968). *Plant Breeding*. Studies in Biology No. 12. Institute of Biology. Edward Arnold, London.

Luckwill, L. C. (1981). *Growth Regulators in Crop Production*. Studies in Biology No. 129. Institute of Biology. Edward Arnold, London.

Nash, M. J. (198). *Crop Conservation and Storage*. Pergamon Press, Oxford.

Russell, R. S. (1977). *Plant Root Systems*. McGraw-Hill, Maidenhead.

Samways, M. J. (1981). *Biological Control of Pests and Weeds*. Studies in Biology No. 132. Institute of Biology. Edward Arnold, London.

Wheeler, B. E. J. (1976). *Diseases in Crops*. Studies in Biology No. 64. Institute of Biology. Edward Arnold, London.

Withers, B. and Vipond, S. (1974). *Irrigation Design and Practice*. Batsford, London.

Ch. 20   Dalton, D. C. (1980). *An Introduction to Practical Animal Breeding*. Granada, St Albans.

McDonald, P., Edwards, R. A. and Greenhalgh, J. F. D. (1981). *Animal Nutrition*. 3rd edn. Oliver & Boyd, Edinburgh.

Parker, W. H. (1980). *Health and Disease in Farm Animals*. 3rd edn. Pergamon Press, Oxford.

Swenson, M. J. (1977). *Dukes' Physiology of the Domestic Animal*. 9th edn. Cornell University Press, Ithaca, NY.

Ch. 21   Langer, R. H. M. (1972). *How Grasses Grow*. Institute of Biology. Studies in Biology No. 34. Edward Arnold, London.

Morley, F. H. W. (1981) (ed.). *Grazing Animals*. World Animal Science. Book 1, Elsevier, Amsterdam.

Spedding, C. R. W. (1971). *Grassland Ecology*. Clarendon Press, Oxford.

Spedding, C. R. W. (1975). *The Biology of Agricultural Systems*. Academic Press, London.

Ch. 22   Barnard, C. S. and Nix, J. S. (1973). *Farm Planning and Control*. CUP, Cambridge.

Giles, A. K. and Stansfield, M. J. (1980). *The Farmer as Manager*. Allen & Unwin, London.

Norman, L. and Coote, R. B. (1971). *The Farm Business*, Longman, London.

Sturrock, F. G. (1971). *Farm Accounting and Management*. Pitman, London.

Ch. 23   Barker, J. W. (1980). *Agricultural Marketing*. OUP, Oxford.

Kohls, R. L. and Ohl, J. N. (1980). *The Marketing of Agricultural Products*. 5th edn. Macmillan, London.

Ch. 24   Briggs, D. E. (1978). *Barley*. Chapman & Hall, London.

Milthorpe, F. L. and Ivins, J. D. (1966). *The Growth of Cereals and Grasses*. Butterworth, London.

Ch. 25   Hebblethwaite, P. D. and Davies, G. M. (1973). *The Production, Marketing and Utilisation of the Field Bean*. RHM Agriculture.

Snoad, B. (1980). 'The Origin, Performance and Breeding of Leafless Peas.' *ADAS Quarterly Review*, 37. 69–86.

Ch. 26   Green, C. G. (1981). *Oilseed Rape Book*. Cambridge Agricultural Publishing.

Holmes, M. R. J. (1980). *Nutrition of the Oilseed Rape Crop.* Applied Science, London.

Ch. 27   Draycott, A. P. (1972). *Sugar Beet Nutrition.* Applied Science, London.

Harris, P. M. (ed.) (1978). *The Potato Crop.* Chapman & Hall, London.

Ch. 28   MAFF/ADAS (1973). Short Term Leaflet 93. *Maize Production for Silage and Grain.*

Bastiman, B. and Slade, C. R. F. (1978). 'The Utilisation of Forage Crops – A Review of Recent Work Undertaken by ADAS'. *ADAS Quarterly Review*, 28. 1–12.

Bunting, E. S. *et al.* (1978). *Forage Maize: Production and Utilisation.* ARC, London.

Corrall A. J. *et al.* (1977). GRI Technical Report No. 22. *Whole Crop Forages.*

Ch. 29   Green, J. O., Corrall, A. J., Terry, R. A. (1971). *Grass Species and Varieties.* Technical Report No. 8. Grassland Research Institute.

Hubbard, C. E. (1954). *Grasses.* Penguin Books, Harmondsworth.

Ch. 30.   Bleasdale, J. K. A. and Salter, P. J. (eds.) (1979). *Know and Grow Vegetables.* OUP, Oxford.

Bleasdale, J. K. A. and Salter, P. J. (eds.) (1982). *Know and Grow Vegetables 2.* OUP, Oxford.

Robertson, J. (1974). *Mechanising Vegetable Production.* Farmers Press, Ipswich.

Ch. 31   Burgess, A. H. (1964). *Hops.* Interscience Publishers, New York.

Steer, P. (1980). *Quick Guide to Soft Fruit Costings.* MAFF, Cambridge.

Steer, P. (1980). *Quick Guide to Top Fruit Costings.* MAFF, Cambridge.

Westwood, M. N. (1978). *Temperate-Zone Pomology.* W. H. Freeman, San Francisco.

Ch. 32   MAFF (1967). *Horticulture in Britain, Part I. Vegetables.* HMSO, London.

MAFF (1970). *Horticulture in Britain, Part II. Fruit and Flowers.* HMSO, London.

Ch. 33   Edmunds, J. (1980). *Container Plant Manual*. Grower Books, London.
Lamb, J. D., Kelly, J. C., Bowbrick, P. (1975). *The Nursery Stock Manual*. Grower Books, London.

Ch. 34   Carruthers, S. P. (1981). The Productivity of Catch Crops Grown for Fuel. *Energy from Biomass*, 97-102. Eds. Palz, W., Charter, P. & Hall, D. O. Applied Science, London.
Slesser, M. and Lewis, C. (1979). *Biological Energy Resources*. E. & F. N. Spon, London.

Ch. 35   Castle, M. E. and Watkins, P. (1979). *Modern Milk Production*. Faber & Faber, London.
Kilkenny J. B. and Herbert, W. A. (1976). *Rearing Replacements for Beef and Dairy Herds*. Milk Marketing Board, Thames Ditton.
Parker, W. H. (1976). *Health and Disease in Farm Animals*. 2nd edn. Pergamon Press, Oxford.
Russell, K. (1980). *The Principles of Dairy Farming*. 8th edn. Farming Press, Ipswich.
Thiel, C. C. & Dodd, R. H. (1979). *Machine Milking*. Technical Bulletin 1. NIRD and Hannah Research Institute.

Ch. 36   Allen, D. and Kilkenny, J. B. (1980). *Planned Beef Production*. Granada, St Albans.
Bowman, J. C. and Susmel, P. (eds.) (1979). *The Future of Beef Production in the European Community*. Martinus Nijhoff, Netherlands. EUR 6363 E.N.
Jollans, J. L. (ed.) (1981). *Grassland in the British Economy*. C A S Paper 2, Centre for Agricultural Strategy, University of Reading.
Meat and Livestock Commission. (1981). *Commercial Beef Production Yearbook*, 1980-1. Meat and Livestock Commission, Bletchley.
Preston, T. R. and Willis, M. B. (1974). *Intensive Beef Production* (2nd rev. edn.). Pergamon Press, Oxford.

Ch. 37   Fell, H. B. (1979). *Intensive Sheep Management*. Farming Press, Ipswich.
Owen, J. B. (1976). *Sheep Production*. Baillière Tindall, London.
Spedding, C. R. W. (1970). *Sheep Production and Grazing Management*. 2nd edn. Baillière Tindall, London.
Speedy, A. W. (1980). *Sheep Production: Science into Practice*. Longman, London.

Ch. 38   English, P. R., Smith, W. J. and MacLean, A. (1977). *The Sow: Improving her Productivity.* Farming Press, Ipswich.

Goodwin, D. H. (1973). *Pig Management and Production.* Hutchinson, London.

MAFF (1977). *Pig Husbandry and Management.* Bulletin 193. HMSO, London.

Thornton, K. (1978). *Practical Pig Production.* Farming Press, Ipswich.

Whittemore, C. T. and Elsley, F. W. H. (1977). *Practical Pig Nutrition.* Farming Press, Ipswich.

Ch. 39   Feltwell, R. and Fox, S. (1978). *Practical Poultry Feeding.* Faber & Faber, London.

MAFF (1973–8). *Poultry Bulletins on Incubation, Poultry Nutrition, Poultry Management.*

Nesheim, M. C., Austic, R. E. and Card. L. E. (1979). *Poultry Production.* 12th edn. Lea and Febiger, Philadelphia.

Sainsbury, D. W. B. (1980). *Poultry Health and Management.* Granada, St Albans.

Ch. 40   Banks, S. (1979). *The Complete Handbook of Poultry Keeping.* Ward Lock, London.

Blaxter, K. L. et al (1974). *Farming the Red Deer.* First report of an investigation by the Rowett Research Institute and the Hill Farming Research Organisation. HMSO, Edinburgh.

Gall, C. (ed.) (1981). *Goat Production.* Academic Press, London.

Hickling, C. F. (1971). *Fish Culture.* 2nd edn. Faber & Faber, London.

Lang, J. (1981). 'The Nutrition of the Commercial Rabbit'. *Nutrition Abstracts and Reviews.* Series B. *51* (4) 1977, 302. Commonwealth Bureau of Nutrition.

Parkins, R. J. and others (1979). *Commercial Rabbit Production.* Bulletin No. 50. MAFF 10th edn. HMSO, London.

Sandford, J. C. (1979). *The Domestic Rabbit.* 3rd edn. Granada, St Albans.

Stevenson, J. P. (1980). *Trout Farming Manual.* Fishing News Books, Farnham.

Walsingham, J. M. (1981). 'The Use of Invertebrates in Recycling'. Spedding, C. R. W. (ed.). *Vegetable Productivity.* Institute of Biology Symposium No. 25, Macmillan, London.

Ch. 41   FAO (1979): *Agriculture: Toward 2000.* FAO C79/24, Rome.

Ch. 42   Brown, L. R. and Finsterbusch, G. N. (1972). *Man and His Environment: Food.* Harper & Row, New York and London.

George, S. (1977). *How the Other Half Dies*. Penguin Books, Harmondsworth.

Mellanby, K. (1975). *Can Britain Feed Itself?* Merlin Press, London.

Spedding, C. R. W. (1979). *An Introduction to Agricultural Systems*. Applied Science, London.

APPENDIX TABLES

Appendix Table 6.1 The new soil classification for England and Wales

Major Group	Group
Lithomorphic (A/C) soils Normally well-drained soils with distinct, humose or organic topsoil and bedrock or little altered unconsolidated material at 30 cm or less	*Rankers* With non-calcareous topsoil over bedrock (including massive limestone) or non-calcareous unconsolidated material (excluding sand) *Sand-Rankers* In non-calcareous, sandy material *Ranker-like alluvial soils* In non-calcareous recent alluvium (usually coarse-textured) *Rendzinas* Over extremely calcareous non-alluvial material, fragmentary limestone or chalk *Pararendzinas* Over moderately calcareous non-alluvial (excluding sand) material *Sand pararendzinas* In calcareous sandy material *Rendzina-like alluvial soils* In calcareous recent alluvium
Brown soils Well drained to imperfectly drained soils (excluding Pelosols) with an altered sub-surface (B) horizon, usually brownish, that has soil structure rather than rock structure and extends below 30 cm depth	*Brown calcareous earths* Non-alluvial, loamy or clayey, with friable moderatly calcareous sub-surface horizon *Brown calcareous sands* Non-alluvial, sandy with moderately calcareous sub-surface horizon *Brown calcareous alluvial soils* In calcareous recent alluvium *Brown earths (sensu stricto)* Non-alluvial, non-calcareous loamy, with brown or reddish friable sub-surface horizon

Major Group	Group
	Brown sands Non-alluvial, sandy or sandy gravelly
	Brown alluvial soils Non-calcareous in recent alluvium
	Argillic brown earths Loamy or loamy over clayey, with sub-surface horizon of clay accumulation, normally brown or reddish
	Paleo-argillic brown earths Loamy or clayey, with strong brown to red sub-surface horizon of clay accumulation attributable to pedogenic alteration before the last glacial period
Podzolic soils Well drained to poorly drained soils with black, dark brown or ochreous sub-surface (B) horizon in which aluminium and/or iron have accumulated in amorphous forms associated with organic matter. An overlying bleached horizon, a peaty topsoil, or both, may or may not be present	*Brown podzolic soils* Loamy or sandy, normally well drained, with a dark brown or ochreous friable sub-surface horizon and no overlying bleached horizon or peaty topsoil
	Gley-podzols With dark brown or black sub-surface horizon over a grey or mottled (gleyed) horizon affected by fluctuating ground-water or impeded drainage. A bleached horizon, a peaty topsoil or both may be present
	Podzols (sensu stricto) Sandy or coarse loamy, normally well drained, with a bleached horizon and/or dark brown or black sub-surface horizon enriched in humus and no immediately underlying grey or mottled (gleyed) horizon or peaty topsoil
	Stagnopodzols With peaty topsoil, periodically wet (gleyed) bleached horizon, or both, over a thin ironpan and/or a brown or ochreous relatively friable sub-surface horizon
Pelosols Slowly permeable non-alluvial clayey soils that crack deeply in dry seasons with brown, greyish or reddish blocky or prismatic sub-surface horizon, usually slightly mottled	*Calcareous pelosols* With calcareous sub-surface horizon
	Argillic pelosols With sub-surface horizon of clay accumulation, normally non-calcareous
	Non-calcareous pelosols Without argillic horizon

Appendix Table 6.1 *Continued*

Major Group	Group
Gley soils	
With distinct, humose or peaty topsoil and grey or grey-and-brown mottled (gleyed) sub-surface horizon altered by reduction, or reduction and segregation, of iron caused by periodic or permanent saturation by water in the presence of organic matter. Horizons characteristic of podzolic soils are absent	1. Gley soils without a humose or peaty topsoil, seasonally wet in the absence of effective artificial drainage
	Alluvial Gley soils In loamy or clayey recent alluvium affected by fluctuating ground-water
	Sandy gley soils Sandy, permeable, affected by fluctuating ground-water
	Cambic gley soils Loamy or clayey, non-alluvial, with a relatively permeable substratum affected by fluctuating ground-water
	Argillic Gley soils Loamy or loamy over clayey, with a sub-surface horizon of clay accumulation and a relatively permeable substratum affected by fluctuating ground-water
	Stagnogley soils Non-calcareous, non-alluvial, with loamy or clayey, relatively impermeable sub-surface horizon or substratum that impedes drainage.
	2. Gley soils with a humose or peaty topsoil normally wet for most of the year in the absence of effective artificial drainage
	Humic-alluvial gley soils In loamy or clayey recent alluvium
	Humic-sandy gley soils Sandy, permeable, affected by high ground-water
	Humic gley soils (sensu stricto) Loamy or clayey, non-alluvial, affected by high ground-water
	Stagnohumic gley soils Non-calcareous, with loamy or clayey, relatively impermeable sub-surface horizon or substratum that impedes drainage
Man-made soils With thick man-made topsoil or disturbed soil (including material recognisably derived from pedogenic horizons) more than 40 cm thick	*Man-made humic soils* With thick man-made topsoil
	Disturbed soils Without thick man-made topsoil

Appendix Table 6.1 Continued

Major Group	Group
Peat soils With a dominantly organic layer at least 40 cm thick formed under wet conditions and starting at the surface or within 30 cm depth	*Raw peat soils* Permanently waterlogged and/or contain more than 15 per cent recognisable plant remains within the upper 20 cm *Earthy peat soils* With relatively firm (drained) topsoil, normally black, containing few recognisable plant remains
Raw soils With no distinct pedogenic horizons other than a superficial organo-mineral or organic layer less than 7·5 cm thick or a buried horizon below 30 cm depth	*Raw sands* Raw sandy soils, chiefly dune sand *Raw skeletal* Very stony and/or very shallow raw soils, including screes, mountain-top detritus and shingle beaches.

Avery, B. W. (1973). Soil classification in the Soil Survey of England and Wales. *Journal of Soil Science*, 24, 324–8.

Appendix Table 6.2 **Agricultural land classification in Britain**

(1) *The MAFF Agricultural Land Classification of England and Wales*
The explanatory note to the Classification published by MAFF states:
'The grading is on the basis of the physical quality alone. Other less permanent factors such as the standard and adequacy of fixed equipment, the level of management, farm structure and accessibility have not been taken into account. It follows that the grades give no indication of the relative values of farms located on them, either as a source of income or capital, since these values will usually depend largely on the shorter term factors mentioned above'.

The grades can be described briefly in the following way:
Grade 1 Land with very minor or no physical limitations to agricultural use. Capable of growing most crops including the more exacting horticultural crops.

Grade 2 Land with some minor limitations which exclude it from Grade 1. Capable of growing a wide range of crops.

Grade 3 Land with moderate limitations due to the soil, relief or climate, or some combination of these factors which restrict the choice of crops, timing of cultivation or level of yield. Grass and cereals are usually the principal crops.

Grade 4 Land with severe limitations due to adverse soil, relief or climate, or a combination of these. Generally only suitable for low output enterprises.

Grade 5 Land generally under grass or rough grazing, except for pioneer forage crops and including land unfit for vegetation, such as bare rock outcrop.

MAFF (1974). *Agricultural land classification of England and Wales*. Pinner: MAFF.

Appendix Table 6.2 *Continued*

(2)   *The Land Use Capability Classification of the Soil Survey of England and Wales*

Capability class and degree of limitation to agriculture	Principal agricultural crops grown	Comment
1 Very minor.	All usual British crops including horticultural produce.	High yields obtainable. Ground suitable for most enterprises.
2 Minor.	All usual British crops. Increased risk of failure jeopardises many horticultural crops.	High yields obtainable with good management. Suitable for most enterprises.
3 Moderate.	Cereals (all types), grass, forage crops (turnips, kale, etc.) potatoes.	Yields equal to those of Class 1 may be obtained from selected crops under good management. Ground suitable for most enterprises.
4 Moderately severe.	Grass dominant, restricted cereals (mainly oats, some barley).	High yields of a very restricted range of crops possible under good management. Suitable for forestry and recreation.
5 Severe.	Improved grass is the only crop apart from the occasional break crop.	High yield of grass products possible, but risk of failure high. Suitable for forestry and recreation.
6 Very severe.	No cropping. Not improvable, except by aerial spray.	Extensive stock ranching (sheep, deer, cattle). Care needed in grazing management to prevent sward deterioration. Forestry in parts; some recreational pursuits.
7 Extremely severe.	No cropping.	Some grazing by hardier stock (sheep, deer) but season restricted to less than five months. No forestry possible; some recreational pursuits.

Bibby, J. S. & Mackney, D. (1969). *Land Use Capability Classification.* Technical Monograph No. 1. Rothamsted and Craigiebuckler: The Soil Survey.

Mackney, D. (1974). Land use capability classification in the United Kingdom. In: *Land capability classification.* MAFF Technical Bulletin 30. London: HMSO.

Appendix Table 9.1 Composition of the farm labour force by region in 1978

| | Farm family members | | Non-family workers | | |
	Farmers	Others	Regular	Seasonal	Total Numbers
	%	%	%	%	'ooos
England	38·2	17·1	30·2	14·5	530·9
East	28·8	12·2	42·1	16·9	91·9
South east	27·4	11·9	43·2	17·5	86·1
East Midland	35·8	15·1	35·3	13·8	63·3
West Midland	40·0	18·0	26·0	16·0	74·5
South west	42·5	19·3	26·1	12·1	101·9
North	45·0	20·2	24·2	10·6	50·1
Yorks/Lancs	40·9	18·3	26·5	14·3	63·2
Wales	50·6	24·9	11·0	13·5	67·1
Scotland	32·3	24·0	36·6	7·1	73·2
North west	48·8	30·0	16·4	4·8	12·8
North east	35·6	22·8	36·2	5·4	14·0
South east	22·1	16·4	52·9	8·6	21·6
South west	30·8	28·3	33·0	7·9	24·8
Northern Ireland	59·9	25·1	7·3	8·1	65·9
United Kingdom	39·3	18·6	28·8	13·3	737·0

(*Sources:* Ministry of Agriculture, Fisheries and Food,
Final results of June 1978 Census
Department of Agriculture and Fisheries for Scotland,
Agricultural Statistics 1978 Scotland
Department of Agriculture Northern Ireland
Statistical Review of Northern Ireland Agriculture 1978.)

Appendix Table 9.2 Average hourly earnings of agricultural workers compared with those of other manual workers in the United Kingdom

	Agriculture[1] p per hour	Other industries[2] p per hour	Agriculture as per cent of other industries
1961	21·5	32·4	66
1963	24·0	35·2	68
1965	28·0	41·7	67
1967	31·0	46·2	67
1969	36·1	53·4	68
1971	45·1	69·2	65
1973	61·7	89·7	69
1975	98·2	136·7	72
1976	112·8	152·2	74
1977	122·6	164·9	74
1978	139·4	188·9	74

(*Source:* Ministry of Agriculture, Fisheries and Food Wages and Employment Enquiries)

[1] Figures for farm workers' hourly earnings relate to hired full-time men and are based on April/March years, October being the mid-point.

[2] Industrial earnings are for full-time adult male workers for a single week in October. Agriculture, coal mining, British Rail and certain other industries are excluded.

Appendix Table 13.1 Live weight (kg), skeletal size (height at shoulders, cm) and fatness (depth of back-fat, mm) of pure breeds of beef cattle in the UK, recorded on commercial farms (bulls and heifers) and in performance tests (bulls).

Breed	Birth weights		Weight at 400 days age		Performance tests (bulls under standard feeding)					
					400 days age		One year of age			
					Live weight		Height		Fatness	
	Bulls	Heifers	Bulls	Heifers	Mean	Range	Mean	Range	Mean	Range
Highland	—	—	268	175	—	—	—	—	—	—
Galloway	27	26	339	325	—	—	—	—	—	—
Aberdeen Angus	29	27	385	273	470	385–539	110	102–118	6·5	3·0–11·5
Beef Shorthorn	31	28	400	269	501	441–562	123	116–130	6·8	3·2– 9·0
Luing	—	—	425	318	—	—	—	—	—	—
Hereford	36	33	434	297	495	399–590	112	101–123	6·9	1·9–14·0
Sussex	36	34	438	298	528	439–610	116	107–123	5·1	2·9– 8·0
Welsh Black	36	34	443	287	524	467–626	119	114–128	4·5	2·4– 8·4
Limousin*	39	37	441	354	537	473–606	123	118–129	1·9	1·4– 2·7
Devon	37	35	471	278	531	450–657	116	108–123	7·2	3·1–12·9
Lincoln Red	37	35	502	333	534	476–613	122	115–127	8·3	5·3–12·9
South Devon	43	40	525	335	600	494–728	127	118–136	3·6	1·8– 7·6
Simmental	42	39	528	404	612	530–677	128	116–137	3·5	1·9– 5·8
Charolais	45	43	565	420	622	508–712	124	116–130	3·1	1·2– 6·1

Source: MLC (1976) British Beef Cattle MLC, Milton Keynes (Performance Test reports for 1974/75)

* MLC (1980) Bull Performance Test Report 1978/79 12th Annual Summary

Appendix Table 15.1 Pesticides

This table includes brief details of the more important and widely used pesticides listed alphabetically under their common name. In the second column the following abbreviations are used. A acaricide; F fungicide; G growth regulator; H herbicide; I insecticide; L livestock use; M molluscicide; N nematicide or nematostat; R repellent; S sterilant (usually for soil) V vertebrate control—mammals or birds.

If there is no entry under 'chemical group' there are few or no related compounds or the name is self-explanatory and the compound is usually listed under 'miscellaneous' in reference books.

The toxicity rating is arbitrarily based on the acute oral toxicity of the ai to the rat, and is only a very rough guide to the potential hazard of the formulated product. Up to an LD 50 of 100 mg/kg is regarded as high, between 100 and 1000 mg/kg as moderate, and above 1000 mg/kg as low. These are fairly conservative ratings and when the figure selected is borderline or variable, some discretion has been exercised according to experience.

The main uses in the last column are largely based on the Approved Products booklet (MAFF 1983) but a few major uses in other countries are sometimes included.

Common name	Use	Chemical group	Mode of action	Toxicity	Main uses and general comments
aldicarb	I,N	carbamate	systemic	high	control of nematodes, aphids and other pests on potatoes, beet, onions and ornamentals
aldrin	I	OC	contact	high	control of soil pests especially wireworms in potatoes; use restricted due to environmental persistence.
2-aminobutane	F		fumigant	low	fumigation of seed and ware potatoes in store for control of skin spot and gangrene; control of fungus diseases on citrus fruits in store.
aminotriazole	H	heterocylic nitro compound	translocated	low	control of couch and many grass and broad-leaved weeds before planting or sowing and around established top fruit; often used in mixtures.
amitraz	A,I,L		contact	moderate	control of red spider mites on apple and pear and pear sucker; and of ticks on sheep and cattle.
ammonium sulphamate	H		translocated and soil acting	low	control of weeds, coppice and standing timber on non-cropped land, on land to be planted and as a local application on rough grazing and forestry.

Appendix Table 15.1 Continued

Common name	Use	Chemical group	Mode of action	Toxicity	Main uses and general comments
asulam	H	carbamate	translocated and soil acting	low	control of docks in established grassland, orchards and non-cropped land; control of bracken in hill and grazing land and forestry; used in mixtures and as an aquatic herbicide.
atrazine	H	triazine	soil acting	low	control of many germinating weeds pre- and post-emergence in field beans and maize; also in established tree, bush and cane fruits and ornamental trees and shrubs; often in mixtures.
azinphos-methyl	I,A	OP	contact	high	wide-spectrum insecticide and acaricide for fruit and vegetables; in UK only available in mixtures with demeton-S methyl sulphone; fairly persistent.
aziprotryne	H	triazine	soil acting	low	control of wide range of annual weeds post-emergence in brassicas and onions and pre-emergence in peas.
barban	H	carbamate	translocated	moderate	control of wild oats and blackgrass post-emergence in wheat and barley (some cultivars sensitive), beans, peas and beet; also in mixtures.
benazolin	H	heterocyclic compound	translocated	low	used mainly in mixtures to control many broad-leaved weeds especially black bindweed, and cleavers post-emergence in undersown cereals, leys and permanent grass.
bendiocarb	I	carbamate	contact and stomach	high	control of cockroaches and other nuisance pests in buildings; seed treatment on beet.
benodanil	F	benzene derivative	systemic	low	control of rusts on cereals and other crops.
benomyl	F	benzimidazole (carbamate)	systemic	low	control of many fungus diseases in most crops
benzoyl prop-ethyl	H	amide	translocated	low	control of wild oats post-emergence in wheat, field beans, oil-seed rape and mustard.

binapacryl	F,A	nitrophenol derivative	contact	moderate	control of mildew and red spider mite on apples
bromacil	H	diazine	contact and soil acting	low	total weed control on non-cropped areas and selective control in soft fruit; also in mixtures.
bromophos	I,L	OP	contact and stomach	low	broad-spectrum insecticide used mainly as a sheep dip in the UK.
bromoxynil	H	benzonitrile	contact	moderate	used mainly in mixtures for control of broad-leaved weeds post-emergence in cereals mainly.
bupirimate	F	pyrimidine derivative	systemic	low	control of powdery mildew on apple, blackcurrant and glasshouse roses.
captafol	F	trichloromethyl mercapto compound	protective	low	control of potato blight and *Rhynchosporium* on barley.
captan	F	trichloromethyl mercapto compound	protective	low	control of apple and pear scab, black spot of roses, and as a seed treatment with gamma-HCH.
carbaryl	I,G,L	carbamate	contact	moderate	caterpillars, capsid bugs etc on apple, and pea moth on peas; killing earthworms in turf; fruit thinning of apples; control of pests in small animal houses including poultry.
carbendazim	F	benzimidazole (carbamate)	systemic	low	for many diseases including eyespot on wheat and barley, Botrytis on fruit and vegetables, and powdery mildew on fruit crops.
carbetamide	H	carbamate	foliar and soil acting	low	post-emergence control in winter of wild oats, blackgrass and other weeds in legumes, brassicas and other crops.
carbofuran	I,N	carbamate	systemic	high	control of many soil insects.
carbophenothion	I,L,A	OP	seed treatment	high	seed treatment against wheat bulb fly; as an acaricide on citrus; sheep dip.
carboxin	F	oxathiin derivative	systemic	low	mainly used as a seed treatment with organo-mercurials and sometimes gamma-HCH.

Appendix Table 15.1 Continued

Common name	Use	Chemical group	Mode of action	Toxicity	Main uses and general comments
chlorbromuron	H	urea	contact and soil acting	low	controls a wide range of annual weeds pre- and post-emergence in carrot, parsnip and potato.
chlorfenprop-methyl	H	halo-alkanoic acid	translocated	low	control of wild oats in cereals.
chlorfenvinphos	I,L	OP	contact	high	control of a wide range of soil pests especially root flies, used in form of granules or liquid and as seed treatment for wheat bulb fly.
chloridazon (pyrazon)	H	diazine	contact and soil acting	low	control of many germinating weeds pre-sowing or in mixtures pre-emergence in beet etc.
chlormequat	G			moderate	for reducing straw length in cereals.
2 chloroethyl phosphonic acid (see 'ethephon')					
chloroxuron	H	urea	soil acting	low	control of many germinating weeds in strawberries and chrysanthemums.
chlorpropham	H,G	carbamate	soil acting and fumigant	low	control of many germinating weeds in some bulb crops, onions, carrots and lettuce; also used in mixtures and for the control of sprouting in stored ware potatoes.
chlorpyrifos	I,L	OP	contact, stomach poison and vapour	high	broad spectrum insecticide for control of many fruit and cereal pests, root flies of vegetables; sheep dip.
chlortoluron	H	urea	translocated, soil acting	low	control of blackgrass and other weeds in winter wheat and barley pre- or post-emergence, and wild oats pre-emergence.
copper compounds	F,M,L		protective and contact	moderate	broad-spectrum fungicide in many forms and mixtures (e.g. Bordeaux) though largely replaced by synthetics now; limited use as a molluscicide but still important as a feed additive for pigs.

Name	Type	Chemical group	Action	Toxicity	Notes
cyanazine	H	triazine	foliar and soil acting	high	control of annual broad-leaved weeds and annual meadow grass pre-emergence in peas and maize, or post-emergence in mixtures on peas.
cyanides	I,V		fumigant and stomach	high	used in various forms to produce hydrogen cyanide; main use is control of rabbits by injecting burrows but limited use as grain and glasshouse fumigant.
cycloate	H	carbamate	soil acting	low	control of wild oats and other annual weeds when used with lenacil pre-drilling of beet and related crops.
cyhexatin	A	organo-tin compound	contact	moderate	control of red spider mites on fruit and glasshouse crops
cypermethrin	I	synthetic pyrethroid	contact	low	broad-spectrum insecticide for many pests on fruit and vegetables.
2,4-D	H	phenoxyacetic acid	translocated	low	effective against many common annual weeds and still one of the most widely used herbicides and the basis of many mixtures.
dalapon	H	halo-alkanoic acid	translocated	low	used alone or in mixtures for control of perennial and annual grasses, pre-sowing, pre- and post-emergence in various crops, uncropped land and around tree, bush and cane fruits and forest trees.
daminozide	G			low	plant growth retardant used to control vegetative growth of fruit trees, and reduce stem length of ornamentals.
dazomet	S	diazine	soil fumigant	moderate	soil sterilant for checking weeds, and controlling soil fungi and nematodes under glass.
2,4-DB	H	phenoxybutyric acid	translocated	low	control of many broad-leaved annual and perennial weeds alone or in mixtures; can be used on leguminous crops.
DDT	I	OC	contact, stomach and residual	moderate	broad-spectrum insecticide but owing to environmental and resistance problems, its use is generally restricted.
deltamethrin	I	synthetic pyrethroid	contact	high	broad-spectrum insecticide for many pests on fruit, vegetables and glasshouse crops.

Appendix Table 15.1 *Continued*

Common name	Use	Chemical group	Mode of action	Toxicity	Main uses and general comments
demephion	I,A	OP	systemic	high	control of aphids, red spider mites etc on various crops.
demeton-S-methyl	I,A	OP	systemic	high	control of aphids, red spider mites and other pests on most crops; most widely-used aphicide. In plants metabolised to the sulphone with similar properties (see azinphos-methyl)
derris	I	plant product containing rotenone etc	contact and stomach poison	low	broad-spectrum insecticide of limited efficacy and persistence; now mainly used in gardens or integrated control.
desmetryne	H	triazine	translocated contact and soil acting	low	control of annual broad-leaved weeds post-emergence in brassicas.
di-allate	H	thiocarbamate	soil acting	moderate	control of blackgrass and wild oats pre-sowing in brassicas and beet.
diazinon	I,A,L	OP	contact	moderate	broad-spectrum insecticide used mainly for the control of root flies in vegetables.
dicamba	H	benzoic acid	translocated	low	used mainly in mixtures to control many broad-leaved weeds post-emergence in cereals and grassland.
dichlobenil	H	benzonitrile	soil acting	low	total weed control in land not intended for cropping; selective control of many weeds in fruit, ornamental trees, shrubs and forestry.
dichlofluanid	F	dichloromethyl mercapto compound	protective	moderate	control of Botrytis and other diseases on soft fruit and glasshouse crops.
dichloropropene	N,S	OC	soil fumigant	moderate	control of nematodes in the field or under glass either alone or in mixture (D-D) with dichloropropane; interval must elapse between treatment and sowing or planting.
dichlorprop	H	phenoxypropionic acid	translocated	moderate	control of many broad-leaved weeds either alone or in mixtures.

dichlorvos	I,L	OP	fumigant (also contact & stomach)	high	broad-spectrum insecticide of short persistence for use on many crops; widely used in various forms as a fly killer in domestic and industrial premises and animal houses; safely used in this way.
diclofop-methyl	H	4,phenoxyphenoxy propionic acid	foliar-acting	low	control of wild oats post-emergence on a wide range of crops.
dicofol	A	OC	contact	moderate	control of red spider mites on fruit and under glass.
dieldrin	I	OC	contact and stomach	high	control of soil pests but use restricted owing to environmental persistence; formerly used as a sheep dip.
difenacoum	V	coumarin	anticoagulant	high	used in baits against warfarin-resistant rodents.
difenzoquat	H	heterocyclic compound	translocated	moderate	control of wild oats post-emergence in maize and cereals.
diflubenzuron	I		interferes with chitin formation	low	control of caterpillars and some other pests of top fruit.
dikegulac	G		systemic	low	reduces apical dominance thus increasing side branching and flower bud formation on ornamentals.
dimethoate	I,A,L	OP	systemic	moderate	controls aphids and red spider mites on a wide range of crops; also sawflies, some diptera and thrips; sheep dip.
dinocap	F,A	dinitrophenol derivative	protective fungicide and contact acaricide	low	control of powdery mildews on apple, soft fruits, hops, etc; suppresses red spider mites.
dinoseb	H	phenol	contact and soil acting	high	control of broad-leaved annual weeds post-emergence in legumes and as salts in cereals and field beans; in oil used for desiccation of seed crops and potato haulm.
diquat	H	bipyridilium	contact	moderate	desiccation of seed crops and potato haulm; aquatic herbicide; becomes ineffective after contact with soil.
disulfoton	I	OP	systemic	high	used as granules to control aphids in field beans, brassicas and other crops; also carrot fly.

Common name	Use	Chemical group	Mode of action	Toxicity	Main uses and general comments
ditalimfos	F	OP	protective	low	control of powdery mildews on spring barley and apples, some control of apple scab.
dithianon	F	quinone	protective	moderate	control of apple and pear scab.
diuron	H	urea	soil acting	low	total weed control on land not intended for cropping or selective control on top and soft fruit and ornamental trees and shrubs; often in mixtures.
dodine	F	surfactant	protective	low	control of apple and pear scab.
endosulfan	I,A	OC	contact and stomach	high	use limited because of environmental persistence but very effective against eriophyid mites on soft fruit especially blackcurrants and strawberries; other pests also controlled; very toxic to fish but less toxic to bees.
ethirimol	F	pyrimidine derivative	systemic	low	control of mildews on barley and other crops.
ethofumesate	H	heterocyclic compound	soil acting and translocated	low	control of weed grasses and broad-leaved weeds in grass crops and when mixed with other herbicides controls germinating weeds pre- and post-emergence in beet etc.
"ethephon" (no BSI common name)	G	(2 chloroethyl phosphonic acid)	systemic	low	accelerates pre-harvest ripening of fruit and vegetables.
fenitrothion	I	OP	contact	moderate	controls a wide range of insects on various crops especially fruit; very effective against rice borers; used for control of grain pests by admixture with grain.
fenoprop	H	phenoxypropionic acid	translocated	moderate	control of woody and other weeds in grassland and forestry.
fenpropimorph	F	oxazine derivative	systemic	low	control of mildews, rusts and Rhynchosporium on cereals.

fentin	F	organotin	contact	moderate	as acetate or hydroxide alone or mixed with maneb for control of potato blight; anti-feedant effect on some insects
flamprop-isopropyl flamprop-methyl	H	amide	translocated	low	both used for control of wild oats post-emergence in wheat and the isopropyl form for barley
fonofos	I	OP	contact	high	control of soil insects especially cabbage root fly
formothion	I,A	OP	contact	moderate	control of aphids and other insects and mites on various mainly horticultural crops
gamma-HCH (lindane) (formerly gamma BHC)	I	OC	contact and fumigant	moderate	less persistent and hence more widely used than other OCs; used as dusts, sprays, smokes and as seed treatment for many crop and stored product insect and mite pests; can cause tainting of some crops and phytotoxic to cucurbits
glyphosate	H	glycine derivative	translocated	low	control of couch and other annual and perennial-weeds, pre-drilling on all outdoor crops, top fruit in dormant season; to kill grass before reseeding (very effective against bracken); forest use and aquatic herbicide
4-(indol-3-yl) butyric acid	G			low	to encourage rooting of cuttings
ioxynil	H	benzonitrile	contact	moderate	general weed control in newly-sown turf, onions and leeks; in mixtures to control a wide variety of weeds in various crops
lenacil	H	diazine	soil acting	low	control of many annual weeds in beet and related crops both alone and in mixtures; post-planting in bulbs and soft fruit
linuron	H	urea	contact and soil acting	low	control of wide range of weeds pre-emergence in spring cereals and potatoes; pre- or post-emergence in carrots, celery etc; often used in mixtures.
malathion	I,A,L	OP	contact	low	broad spectrum insecticide for use on most crops, in grain stores and admixed with grain; control of flies in animal houses
maleic hydrazide	G	diazine	translocated	low	suppression of grass growth, shoot and sucker growth on some trees and of onion sprouts in store

Appendix Table 15.1 Continued

Common name	Use	Chemical group	Mode of action	Toxicity	Main uses and general comments
mancozeb (zinc and maneb complex)	F	dithiocarbamate	protective	low	control of potato blight, blackcurrant leaf spot and apple scab and other foliage diseases
maneb	F	dithiocarbamate	protective	low	as for mancozeb plus blight and leaf mould on tomato, black spot of roses and tulip fire
MCPA	H	phenoxyacetic acid	translocated	low	effective against many common annual weeds and still one of most widely used herbicides: basis of many mixtures
MCPB	H	phenoxybutyric acid	translocated	low	control of many broad-leaved annual and perennial weeds; especially useful on legumes
mecoprop	H	phenoxypropionic acid	translocated	moderate	control of many broad-leaved weeds post-emergence in cereals, other crops and top fruit; often in mixtures
mercury compounds	F		contact	high	because of environmental risks only some special uses are retained e.g. mercuric oxide for apple canker, mercurous chloride for onion white rot, club root of brassicas and certain turf diseases; organomercury compounds are still the main basis of cereal seed treatments though replacements are being sought
metaldehyde	M		contact and stomach	moderate	used mainly in a bait to control slugs and snails
methabenzthiazuron	H	urea	translocated and soil acting	low	control of blackgrass and some annual weeds pre-emergence in winter barley and oats, pre- or early post-emergence in winter wheat
metham-sodium	S		soil fumigant	moderate	control of various soil diseases and nematodes mainly under glass; several weeks interval needed before planting
methidathion	I,A	OP	contact	high	controls a wide range of insects but very effective against scale insects; main use in UK for aphids on hops
methiocarb	M,I	carbamate	contact and stomach	high	used as bait in pellet form for slug and snail control in field crops; also strawberry seed beetle

methomyl	I,N	carbamate	contact and systemic via root	high	main use in UK as aphicide on hops, nematicidal value
methyl bromide	S		soil and space fumigant	high	general soil sterilant mainly used against soil fungi and nematodes under glass; also for space and stored product fumigation; up to 2% chloropicrin may be added as a warning gas
metoxuron	H	urea	translocated and soil acting	low	control of blackgrass and other weeds post-emergence in winter barley, wheat and carrots; also for potato haulm destruction; often in mixtures
metribuzin	H	triazine	translocated and soil acting	low	for wide range of crops including soyabean; in UK for many annual broad-leaved and grass weeds pre- and post-emergence on potato
mevinphos	I,A	OP	systemic and contact	high	control of aphids on many crops where rapid kill needed close to harvest, as it has short persistence
monolinuron	H	urea	contact and soil acting	low	control of wide range of annual weeds pre-emergence in french beans and potato; mixed with paraquat suppresses perennial grasses
nicotine	I	pyridine derivative	contact and fumigant	high	formerly used to control aphids and other pests mainly on horticultural crops; superseded by OPs
nitrothal-isopropyl	F		contact	low	available mainly as a mixture with sulphur or other fungicides to control powdery mildew and scab on apple
oxamyl	N,I	carbamate	contact and systemic with some downward translocation	high	control of nematodes in potatoes, sugar beet and onions; some control of aphids and other insects
oxydemeton-methyl	I,A	OP	systemic and contact	high	control of aphids, red spider mites and some other pests on most agricultural and horticultural crops
paraquat	H	bipyridilium	contact	high	controls a wide range of annual broad-leaved weeds and grasses and kills the tops of perennial weeds in non-cropped land, established orchards and dormant soft fruit; for stubble treatment and grassland before direct drilling; often used in mixtures; inactivated by soil

Common name	Use	Chemical group	Mode of action	Toxicity	Main uses and general comments
permethrin	I	synthetic pyrethroid	contact	low	control of caterpillars and other pests on top fruit, brassicas and peas
phorate	I	OP	systemic	high	control of many pests especially aphids, diptera, caterpillars and wireworms on a wide range of crops; mainly used as granules
phosalone	I,A,L	OP	contact	moderate	control of aphids, caterpillars and red spider mites on apples; useful on brassica seed crops near flowering as it is less harmful to bees than alternatives; sheep dip
phosphamidon	I	OP	systemic	high	main use for sap-feeding insects including sugarcane and rice stem-borers; as aphicide on peas and beans in UK
pirimicarb	I	carbamate	contact, fumigant and some systemic action	moderate	selective aphicide, fast-acting; for control of aphids on many field and horticultural crops; short persistence
pirimiphos-methyl	I,A	OP	contact and fumigant	low	broad spectrum of activity against field and stored-product pests; can be admixed with grain
propachlor	H	amide	soil acting	low	control of germinating annual weeds in vegetables and some other crops
propham	H	carbamate	soil acting	low	control of germinating weeds especially wild oats, pre-sowing in peas and sugar beet; in mixtures with other crops
propiconazole	F	triazole derivative	partial systemic	low	controls most foliar, stem and ear diseases of wheat and barley
propineb	F	dithiocarbamate	protective	low	control of potato blight, downy mildew on hops and apple scab
propoxur	I,L	carbamate	contact	high	quick knock-down effect; main use in UK is control of aphids on hops; also to control flies in animal houses
propyzamide	H	amide	soil acting	low	selective control of many annual and some perennial weeds in fruit, forestry and some other crops

pyrazon (see chloridazon)

pyrazophos	F	OP	systemic	moderate	control of powdery mildews especially on apple and hops
pyrethrins	I		contact	low	quick knock-down effect; mainly used to control flies in domestic and industrial situations; often used with synergists or mixed with other insecticides; short persistence
simazine	H	triazine	soil acting	low	control of many germinating weeds pre- or post-emergence in field beans and maize; in established fruit, ornamental trees and shrubs, and on land not intended for cropping
sulphur	F,A		contact, fumigant and protective	low	control of powdery mildew on fruit except on sulphur-shy cultivars; also in mixtures
2,4,5-T	H	phenoxyacetic acid	translocated	low	effective against woody plants and to prevent establishment of scrub in grassland; subject of great controversy; in UK, TCDD content limited and after detailed investigation cleared for use in forestry and grassland
2,3,6-TBA	H	benzoic acid	translocated	low	control, usually in mixtures, of many broad-leaved annual and perennial weeds, especially deep-rooted ones
TCA	H	trichloracetic acid	soil acting	low	control of couch, wild oats and volunteer cereals pre-sowing or pre-planting in various crops
tetrachlorvinphos	I,L	OP	contact and residual	low	selective insecticide controlling caterpillars and dipterous pests of fruit, rice, cotton and maize; useful against flies in animal houses
tetradifon	A	OC	contact, ovicidal	low	controls all stages of red spider mites on fruit and under glass
thiabendazole	F,L	benzimidazole (carbamate)	systemic	low	controls many fungi but main use in UK is against storage rots of potatoes; also anthelmintic
thiofanox	I	carbamate type	systemic	high	control of aphids on potatoes and sugar beet

Appendix Table 15.1 *Continued*

Common name	Use	Chemical group	Mode of action	Toxicity	Main uses and general comments
thiophanate-methyl	F,L	carbamate	systemic	low	control of eyespot on winter wheat and barley; *Rhynchosporium* on barley; mildew, scab and storage rots of apple; also anthelmintic
thiram	F	dithiocarbamate type	protective	moderate	control of Botrytis on various horticultural crops; cane spot of raspberry, rusts on ornamentals; seed treatment of beans, peas and maize
triademefon	F	triazole derivative	systemic	moderate	control of powdery mildew, *Rhynchosporium*, and brown rust of barley, yellow rust of wheat, mildew of oats
tri-allate	H	thiocarbamate	soil acting	low	control of blackgrass and wild oats pre-emergence in barley, wheat, sugar beat, beans and peas; control of wild oats post-emergence in barley, wheat and winter beans
triazophos	I,A	OP	contact	high	broad spectrum insecticide and acaricide controlling a wide range of pests on many crops; good control of seed weevil on brassica seed crops but use limited by toxicity to bees
trichlorphon	I,L	OP	contact and stomach	moderate	controls caterpillars on many crops, cabbage root fly in sprout buttons; warble fly on cattle

tridemorph	F	oxazine derivative	systemic	low	control of mildew on cereals and in mixtures with other fungicides for yellow rust and *Rhynchosporium*
trietazine	H	triazine	soil acting	low	used in mixtures with linuron to control annual broad-leaved weeds, pre-emergence in potato; with simazine for annual grasses and broad-leaved weeds pre-emergence in peas, field beans and strawberry
trifluralin	H	dinitroaniline	soil acting	low	control of many annual germinating weeds when incorporated into soil prior to sowing or planting most vegetable and some soft fruit crops; also in mixtures
triforine	F	piperazine derivative	systemic	low	control of mildew on spring barley as seed treatment and spray; for various diseases of apple; one of few systemics with both upward and downward movement in plants
vamidothion	I,A	OP	systemic	high	control of aphids, red spider mites and some other pests on fruit, particularly effective against woolly aphid
vinclozolin	F	contact	low	selective fungicide effective against Botrytis on strawberry and some other diseases of fruit, vines and ornamentals
warfarin	V	coumarin type	anti-coagulant	high	rodenticide; rats and mice are resistant in some areas
zineb	F	dithiocarbamate	protective	low	control of potato blight, downy mildew and Botrytis on various crops, rusts on carnations and chrysanthemums

Appendix Table 22.1   **An example of the arithmetic framework required for the application of budgetary control to milk sales**

Planned sales:
    No. of cows × litres of milk per cow × pence per litre = £
less Actual Sales:
    No. of cows × litres of milk per cow × pence per litre = £

Discrepancy:	±	No. of cows		litres of milk per cow		pence per litre	
Value of discrepancy:		£	+	£	+	£	= £

Appendix Table 22.2   **An example of the framework required for the calculation of cereal and a grazing livestock gross margin (G.M.)**

Cereal (G. M.) per hectare		Dairy Cows G. M. per hectare	
Gross Output (G. O.)	£	Gross Output	£
tonnes at £ per tonne =		litres per cow at p. per litre =	
Variable costs (V. C.)		plus value of calf	
kg. at p. per kg. =			
Seed		sub total	
Fertiliser		less annual value of cow	
Sprays		depreciation	
Total V. C.			
Gross Margin (G.O. less V.C.)			
		Variable costs per cow	
		Concentrates	
		Forage crop V.C.	
		Sundries	
		Total V.C.	
		Gross Margin per cow	
		(G.O. less V.C.)	
		Hectares per cow	
		Gross Margin per hectare (G.M. per cow ÷ hectares per cow)	

Appendix Table 22.3   An example of layout for a complete farm budget

Inputs	£	Outputs	£
Purchased feed: (productive units × quantity per unit × cost per unit)		Crops (productive units × yield per unit × market price)	
Purchased seed: as for feed		Milk as for crops	
Fertilizer: as for feed		Cattle) Productive Units ×) yield per unit ×	
Rent and rates: actual level, allowing for possible increase		Sheep) market price (or,) if more aproppriate, Pigs) value increase).	
Power and Machinery: previous figures appropriately adjusted, or separate estimates for each item involved.		) Allow for replacement Poultry) stock by specifying) sales (e.g. culls and) surplus young stock)) and purchases, if	
Labour: calculate person by person, allowing for increases, plus casual, etc.		) replacements not home) reared.	
Other costs: previous figures, appropriately adjusted, or separate estimates.		Sundries itemise and value	
TOTAL	£	TOTAL	£

Appendix Table 22.4   An example layout for a partial budget

Gain from change		Loss from change	
What fresh revenue will be added?	£	What fresh costs will be incurred	£
What previously incurred costs will be avoided?	£	What previous revenues will be lost?	£
Total gain	£	Total loss	£

Appendix Table 22.5   **An example layout for a cash flow budget**

ITEM	◄——— T I M E P E R I O D S ———►			
Receipts itemised ——►				
Sub total	say 30,000	say 20,000		
Expenses itemised ——►				
Sub-total	say 20,000	say 25,000		
Balance for each time period	+ 10,000	– 5,000		
Cumulative Balance	+ 10,000*	+ 5,000	etc.	etc.

*assuming no opening positive or negative balance.

Appendix Table 24.1   **Description of the Zadoks Scale**

0	GERMINATION
00	Dry seed
01	Start of water absorption (imbibition)
02	–
03	Water absorption complete, seed swollen
04	–
05	Root (radicle) emerged from seed
06	–
07	Shoot (coleoptile) emerged from seed
08	–
09	Leaf just at coleoptile tip
1	SEEDLING GROWTH
10	First leaf through coleoptile
11	First leaf unfolded
12	2 leaves unfolded
13	3 leaves unfolded
14	4 leaves unfolded
15	5 leaves unfolded
16	6 leaves unfolded
17	7 leaves unfolded
18	8 leaves unfolded
19	9 or more leaves unfolded

Appendix Table 24.1 *Continued*

2   TILLERING

20 Main shoot only
21 Main shoot and 1 tiller
22 Main shoot and 2 tillers
23 Main shoot and 3 tillers
24 Main shoot and 4 tillers
25 Main shoot and 5 tillers
26 Main shoot and 6 tillers
27 Main shoot and 7 tillers
28 Main shoot and 8 tillers
29 Main shoot and 9 or more tillers

3   STEM ELONGATION

30 Pseudostem erection
31 1st node detectable
32 2nd node detectable
33 3rd node detectable
34 4th node detectable
35 5th node detectable
36 6th node detectable
37 Flag leaf just visible
38 –
39 Flag leaf ligule/collar just visible

4   BOOTING

40 –
41 Flag leaf sheath extending
42 –
43 Boots just visibly swollen
44 –
45 Boots swollen
46 –
47 Flag leaf sheath opening
48 –
49 First awns visible (where appropriate)

5   EAR EMERGENCE

50
51 } First spikelet of ear just visible

52
53 } $\frac{1}{4}$ of ear emerged

54
55 } $\frac{1}{2}$ of ear emerged

56
57 } $\frac{3}{4}$ of ear emerged

58
59 } Emergence of ear complete

6   FLOWERING (Anthesis)

60
61 } Beginning of flowering (not easily detected in barley)

62 –
63 –
64
65 } Flowering half-way

66 –
67 –

Appendix Table 24.1   *Continued*

68 } Flowering complete
69 }

7	MILK DEVELOPMENT
70	–
71	Grain (Caryopsis) water ripe
72	–
73	Early milk
74	–
75	Medium milk
76	–
77	Late milk
78	–
79	–

8	DOUGH DEVELOPMENT
80	–
81	–
82	–
83	Early dough
84	–
85	Soft dough (finger-nail impression not held)
86	–
87	Hard dough (finger-nail impresion held, ear loosing green colour)
88	–
89	–

9	RIPENING
90	–
91	Grain hard (difficult to divide by thumb-nail)
92	Grain hard (can no longer be dented by thumb-nail)
93	Grain loosening in day time
94	Over-ripe straw dead and collapsing
95	Seed dormant
96	Viable seed giving 50% germination
97	Seed not dormant
98	Secondary dormancy induced
99	Secondary dormancy lost

Appendix Table 26.1 Recommended list of oil-seed rape varieties, 1981

Winter varieties

Based on data from trials 1978–1980.
Differences in yield of less than 5%, and differences in oil content of less than 0·4%, should be treated with reserve.

Varieties classified for General use (G) Provisional Recommendation (P)	Relative Seed Yield at 9% Moisture Content	% Oil Content at 9% Seed Moisture Content	Shortness of Stem	Resistance to Lodging	Earliness of Flowering	Earliness of Ripening	Resistance to Stem Canker	Resistance to Light Leaf Spot	Resistance to Downy Mildew
Recommended:									
G Jet Neuf	100	39·2	7	7	4	5	8	5	6
G Rafal	101	39·5	8	8	7	7	7	7	4
Provisionally Recommended:									
PG Elvira	98	40·0	5	5	5	5	5	8	7
PG Norli (Only 2 years trials)	102	39·6	5	6	3	4	(6)	(8)	(8)
Mean	100 = 3·14 t/ha								

Disease scores in brackets are based on limited data.
A high figure indicates that the variety shows the character to a high degree.

Appendix Table 26.1 Continued

Spring varieties

Based on data from trials 1978–1980.
Differences in yield of less than 5%, and differences in oil content of less than 0.8%, should be treated with reserve.

Variety	Relative Seed Yield at 9% Moisture Content	% Oil Content at 9% Seed Moisture Content	Mean Glucosinolate Content*	Shortness of Stem	Resistance to Lodging	Earliness of Flowering	Earliness of Ripening
Brutor	106	38.3	106	7	7	6	4
Christa	101	39.4	102	5	5	3	3
Cresor	105	38.0	101	7	6	4	4
Duplo	94	40.2	16	5	6	4	4
Fido	104	40.2	32	7	6	3	6
Gulliver	104	38.9	122	7	6	7	6
Line	91	40.5	6	5	6	3	3
Loras	99	41.1	13	6	6	5	6
Mary	92	39.1	18	6	6	7	6
Olga	96	38.0	107	6	4	7	7
Willi	108	40.0	68	5	4	4	5
Mean	100 = 2.08 t/ha						

* μ moles per g de-fatted meal (dry matter basis).
A high figure indicates that the variety shows the character to a high degree.

National Institute of Agricultural Botany (1981) Varieties of Oil-seed Rape, 1981 Farmers Leaflet' No 9, NIAB.

Appendix Table 26.2 Recommended fertiliser rates as N, P_2O_5 and K_2O for oil-seed rape

N, P or K Index	N								P_2O_5	K_2O
	Winter oil-seed rape				Spring oil-seed rape				Both crops	
	Sandy soils and shallow soils on chalk		Other soils		Sandy soils and shallow soils on chalk		Other soils		All soils	
	Seed bed	Top dressing	Seed bed	Top dressing	Seed bed	Top dressing	Seed bed	Top dressing	Seed bed	Top dressing
	kg/ha	kg/ha	kg/ha	kg/ha	kg/ha	kg/ha	kg/ha	kg/ha	kg/ha	kg/ha
0	60	225	60	200	50	150	150	Nil	75	75
1	50	200	50	175	50	125	125	Nil	40	40
2	40	175	40	150	50	100	100	Nil	40	40
3	—	—	—	—	—	—	—	—	40	Nil
Over 3	—	—	—	—	—	—	—	—	Nil	Nil

Ministry of Agriculture, Fisheries & Food (1979) Fertiliser recommendations for Agricultural and Horticultural crops. 2nd Ed., HMSO, London

Appendix Table 26.3 Suitability of herbicides for winter and spring oil-seed rape

| Herbicide | | Oil-seed rape crop | |
Common name	Method of use*	Autumn sown	Spring sown
alachlor	Pre.em	✓	✓
aminotriazole	Pre.s-S	No	✓
barban 1·25%	Post.em	✓	✓
benazolin + 3, 6-di-chloropicolinic acid	Post.em	✓	No
benzoylprop-ethyl	Post.em	✓	✓
butam	Pre.em	✓	No
carbetamide	Post.em	✓	No
carbetamide + dimefuron	Post.em	✓	No
dalapon sodium	Pre.s-S or Post.em	✓	✓
triallate	Pre.s-S	✓	✓
diclofop-methyl	Post.em	✓	✓
glyphosate	Pre.s-F	✓	✓
paraquat	Pre.s-F	✓	✓
propachlor	Pre.em or Post.em	✓	✓
propyzamide	Post.em	✓	No
TCA	Pre.s-S or Pre.em	✓	✓
trifluralin	Pre.s-S	✓	✓

* Pre.s-S: Pre-sowing, soil acting
  Pre.s-F: Pre-sowing, foliar acting
  Pre.em: Pre-emergence of crop
  Post.em: Post-emergence of crop

Agricultural Development and Advisory Service (1980) Chemical Weed Control in Oil-seed Rape. Booklet 2068 M A F F., London.

Appendix Table 31.1 Recommended rates of fertilisers per annum (kg/ha)

ORCHARD FRUITS

N — Summer rainfall

	<350 mm (b) C	<350 mm GHS	<350 mm G	>350 mm C	>350 mm GHS	>350 mm G
Apples (dessert)	40	60	120	30	40	90
Apples (culinary)	100	120	180	60	90	120
Cherries		120	150		90	120
Pears	90	120	150	50	70	100
Plums						

P, K, Mg (Orchard fruits)

Index (a)	P₂O₅ kg/ha	Index (a)	K₂O kg/ha	Index (a)	Mg kg/ha
0	80	0	220	0	60
1	40	1	150	1	40
2	20	2	80	2	30
>2	Nil	>2	Nil	>2	Nil

SOFT FRUITS

N — Summer rainfall

	<350 mm	>350 mm
Blackcurrants	140	70
Raspberries (c)	100	50
Strawberries	Nil (d)	Nil
Grapes	40	20

P and K

Blackcurrants

Index (a)	P₂O₅ kg/ha	Index (a)	K₂O kg/ha
0	110	0	250
1	70	1	180
2	40	2	120
3	40	3	60
<3	Nil	<3	Nil

Strawberries

Index (a)	K₂O kg/ha
0	220
1	150
2	80
3	Nil
<3	Nil

Appendix Table 31.1 Continued

	N Summer rainfall		P Index(a)kg/ha		K Index (a)kg/ha		Mg Index(a)kg/ha	
Hops	0	225	0	300	0	450	0	60
	1	200	1	250	1	375	1	30
	2	150	2	200	2	300	2	Nil
			3	150	3	225	>2	Nil
			4	100	4	150		
			5	50	5	75		
			6	50	6	Nil		
			7	Nil	7	Nil		

(a) Based on soil test
(b) C—cultivated
GHS—grass/herbicide strip
G—grass
(c) After establishment, nitrogen rates are reduced 25 kg/ha
(d) 40 kg N/ha are required where beds are kept down for more than 2 years
Adapted from MAFF (1979) 'Fertiliser Recommendations for Agricultural and Horticultural Crops'. HMSO, London

Appendix Table 31.2 Irrigation guide for fruits and hops

a) Time of application

Strawberries	May to June and August to September
Plums, cherries and hops	May to July
Apples, blackcurrants and pears	May to August
Raspberries	June to July

b) Recommended rates

		Irrigation plan for soil types		
		A	B	C
Rooting depth	Crop	mm available water per 30 cm		
		less than 40 mm	40 to 65 mm	65 mm or more
less than 45 cm or very gravelly soil of any depth	All fruit crops and hops	25 mm at 25 mm SMD*	25 mm at 25 mm SMD	50 mm at 50 mm SMD
45 cm—60 cm	All fruit crops and hops	25 mm at 25 mm SMD	50 mm at 50 mm SMD	50 mm at 75 mm SMD
60 cm—more than 120 cm	All fruit crops and hops except mature fruit trees	50 mm at 50 mm SMD	50 mm at 75 mm SMD	50 mm at 75 mm SMD
60 cm—more than 120 cm	Mature fruit trees	50 mm at 75 mm SMD	50 mm at 75 mm SMD	75 mm at 125 mm SMD
more than 120 cm	Mature fruit trees	50 mm at 75 mm SMD	75 mm at 125 mm SMD	No irrigation

Adapted from MAFF (1974) 'Irrigation'. Bull. 138. HMSO. London.
* Soil moisture deficit.

Appendix Table 42.1   Institutions teaching agriculture or closely related subjects

England and Wales Level	Institution
Postgrad Diploma	University of Cambridge
Degree	University of Exeter
,,	University of Leeds
,,	University of London
,,	University of Manchester
,,	University of Newcastle upon Tyne
,,	University of Nottingham
,,	University of Reading
,,	National College of Agricultural Engineering
,,	University College of Wales Aberystwyth
,,	University College of North Wales Bangor
Higher National Diploma	Harper Adams Agricultural College
,,	Lancashire College of Agriculture
,,	Royal Agricultural College
,,	Seale-Hayne Agricultural College
,,	Shuttleworth Agricultural College
,,	Welsh Agricultural College
,,	Writtle Agricultural College

In addition most counties have Colleges of Agriculture.

More detailed information is available in a booklet published by the Department of Education & Science, Elizabeth house, York Road, London SE1 7PH called "Agriculture, Horticulture and Forestry Courses. List of full-time and sandwich courses in England and Wales".

Scotland Level	Institution
Degree	University of Aberdeen
,,	University of Edinburgh
,,	University of Glasgow
Higher Diploma	West of Scotland Agricultural College
,,	North of Scotland College of Agriculture
,,	East of Scotland College of Agriculture
Diploma	West of Scotland Agricultural College
,,	North of Scotland College of Agriculture
,,	East of Scotland College of Agriculture
,,	Oatridge Agricultural College

There are many colleges giving other courses in agriculture.

More detailed information is available in a booklet called "Opportunities in Agriculture 1981" published by The Scottish Agricultural Colleges Publ. No. 79, available from Department of Agriculture and Fisheries for Scotland, Chesser House, Gorgie Road, Edinburgh EH11 3AW.

Appendix Table 42.1 *Continued*

Northern Ireland Level	Institution
Degree	Queens University of Belfast
Diploma	Loughry College of Agriculture and Food Technology
,,	Greenmount Agricultural and Horticultural College

Further information can be obtained from Department of Education, Rathgael House, Balloo Road, Bangor, Co. Down BT19 2 PR.

INDEX